Practical Data Analytics for Innovation in Medicine

Building Real Predictive and Prescriptive Models in Personalized Healthcare and Medical Research Using AI, ML, and Related Technologies

ELSEVIER *science & technology books*

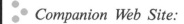

Companion Web Site:

https://www.elsevier.com/books-and-journals/book-companion/9780323952743

Practical Data Analytics for Innovation in Medicine

Gary D. Miner, Linda A. Miner, Scott Burk, Mitchell Goldstein, Robert Nisbet, Nephi Walton, Thomas Hill

Resources available:

- This book's COMPANION WEB PAGE contains the following items:

  1. COMPANION WEB PAGE - Selected Chapters from 1st Edition:
  Chapters written for the first edition published in 2014-2015 that are NOT included include in this 2nd Edition, but are still important to understanding the overall story of the background to use of 'Predictive Analytics' and 'Digital processes' in medicine, are included for your reading.

  2. COMPANION WEB PAGE - Tutorials from the 1st Edition:
  All of the Tutorials and Case Studies that were published on the paper pages of the 1st edition of this book (published in 2014-2015) are NOT included in the paper pages of this 2nd edition, but are available in their entirety on this book's COMPANION WEB PAGE, including data sets, where applicable.

  3. COMPANION WEB PAGE - Software use instructions:
  Information providing the Internet on-line URL – LINKS for obtaining downloads or evaluation copies (when available) of 'Predictive Analytic' software that is used / illustrated in the TUTORIALS and CASE STUDIES is provided here, to assist the reader in using this software, where interested.

  4. COMPANION WEB PAGE - Miscellaneous resources:
  Various background resource material, either in the form of URL LINKS and/or PDF documents may be provide here as deemed appropriate by the authors of this book; this may include additional items added to this page in future years. Currently APPENDIX A items are included here.

ELSEVIER

ACADEMIC PRESS

Practical Data Analytics for Innovation in Medicine

Building Real Predictive and Prescriptive Models in Personalized Healthcare and Medical Research Using AI, ML, and Related Technologies

Second Edition

Gary D. Miner, PhD; Linda A. Miner, PhD
Scott Burk, PhD; Mitchell Goldstein, MD
Robert Nisbet, PhD; Nephi Walton, MD
Thomas Hill, PhD

Linda A. Miner, *Originating Author & Editor*
Robert Nisbet, *Integrating Author & Editor*
Gary D. Miner, *Coordinating Author & Editor*

Guest—Authors:

Anna J. C. Russell-Toner, MComm(Statistics), CEO; *The Boss Lady*, The Data-Shack; Head-Office: York, UK; also offices in South Africa & Hong Kong; http://www.data-shack.co.uk

Billie Corkerin, CPOT; *I-CARE HOME Trainer*; OMEG, Jenks, OK, https://www.omeg2020.com/

John B. C. Tan, PhD; *Data Scientist*, Department of Pediatrics, LOMA LINDA UNIVERSITY I School of Medicine, Loma Linda, CA, https://medicine.llu.edu/

Fu-Sheng Chou, MD, PhD; *Neonatologist and Assistant Professor*, Department of Pediatrics LOMA LINDA UNIVERSITY I School of Medicine, Loma Linda, CA, https://medicine.llu.edu/

Rebekah Leigh; *Ms2 Medical Student*, MD Candidate, Class of 2024, LOMA LINDA UNIVERSITY I School of Medicine, Loma Linda, CA, https://medicine.llu.edu/

Harsha K. Chandnani MD, MBA, MPH; *Pediatric Intensivist*, Department of Pediatrics LOMA LINDA UNIVERSITY I School of Medicine I Children's Hospital, Loma Linda, CA, https://medicine.llu.edu/, https://lluch.org/

ELSEVIER

ACADEMIC PRESS
An imprint of Elsevier

Academic Press is an imprint of Elsevier
125 London Wall, London EC2Y 5AS, United Kingdom
525 B Street, Suite 1650, San Diego, CA 92101, United States
50 Hampshire Street, 5th Floor, Cambridge, MA 02139, United States
The Boulevard, Langford Lane, Kidlington, Oxford OX5 1GB, United Kingdom

Notices

Knowledge and best practice in this field are constantly changing. As new research and experience broaden our understanding, changes in research methods, professional practices, or medical treatment may become necessary.

Practitioners and researchers must always rely on their own experience and knowledge in evaluating and using any information, methods, compounds, or experiments described herein. In using such information or methods they should be mindful of their own safety and the safety of others, including parties for whom they have a professional responsibility.

To the fullest extent of the law, neither the Publisher nor the authors, contributors, or editors, assume any liability for any injury and/or damage to persons or property as a matter of products liability, negligence or otherwise, or from any use or operation of any methods, products, instructions, or ideas contained in the material herein.

ISBN: 978-0-323-95274-3

For Information on all Academic Press publications
visit our website at https://www.elsevier.com/books-and-journals

Publisher: Stacy Masucci
Acquisitions Editor: Rafael E. Teixeira
Editorial Project Manager: Sam Young
Production Project Manager: Swapna Srinivasan
Cover Designer: Vicky Pearson

Typeset by MPS Limited, Chennai, India

Dedication

A DUAL DEDICATION — to the PAST and to the FUTURE: **This book is dedicated to two important people: Joseph M. Hilbe, J.D., Ph.D., a co-author of our 1st Edition that passed away too-soon a few years ago; and Rafael Teixeira, Senior Acquisitions Editor for Elsevier responsible for Medical Informatics and Biostatistics among other health areas.**

Joe Hilbe became "our friend" and "mentor" for all the co-authors of the first edition, guiding the tone and structure of the book. Joe would attend the same "Statistical conferences" as co-authors Gary and Linda Miner, and oftentimes Thomas Hill, so we had frequent in-person contact with Joe during the writing of the first edition, in addition to On-Line Meetings and digital communications. We greatly missed Joe's guidance during the writing of this 2nd edition.

Rafael Teixeira contacted the authors over two years ago (March of 2020, as I recall) asking if we'd consider writing a 2nd Edition. THIS SURPRISED US … as the 1st edition was NOT selling the thousands of copies as did our 2009 and 2012 books on "Data Mining" and "Text Mining," perplexing the authors, as we had thought that the release of the first edition in 2015 was at the "crest of a wave" of interest in using predictive analytics and other digital means of "re-engineering health care delivery." It appeared we were wrong. BUT Rafael had noticed that chapters and passages of our 1st edition were being accessed numerous times via SCIENCE DIRECT, and even made an Excel sheet providing the "number of hits" for each chapter and tutorial. Rafael said it was "time for a 2nd edition"!!! … and as we spent the past two years writing this 2nd edition, we discovered his prediction was "right on," as now in 2022 and forthcoming years it appears that medicine is finally ready to put predictive analytics and all sort of digital means to uses to provide better and more accurate diagnoses and treatment plans for patients that receive the care from these medical providers.

So, THANKS to JOE and RAFAEL for their wisdom and abilities to act on their insights to help the world be a better place.

Contents

12. Using data science algorithms in predicting ICU patient urine output in response to diuretics to aid clinicians and healthcare workers in clinical decision-making   257

Anna J.C. Russell-Toner

13. Prediction tool development: creation and adoption of robust predictive model metrics at the bedside for greatly benefiting the patient, like preterm infants at risk of bronchopulmonary dysplasia, using Shiny-R   325

John B.C. Tan, Rebekah M. Leigh and Fu-Sheng Chou

About the authors

Linda A. Miner, PhD, earned her bachelor's and master's degrees at University of Kansas, her doctorate at the University of Minnesota, and completed postdoctoral studies in psychiatric epidemiology at the University of Iowa. She spent most of her career as an educator, in teacher education, statistics, and research design. She spent nearly 2 years as a site coordinator for a major (Coxnex) drug trial. For 23 years, she was a program director at Southern Nazarene University, Tulsa. She directed six programs: two undergraduate programs in business (Organizational Leadership, & Business Administration), and one in psychology (Family Studies and Gerontology). Her graduate program direction included the Master of Science in Management (MSM), the Master of Business Administration (MBA), and the Master of Business Administration in Health Care (MBAHC). She has authored or coauthored numerous articles and books, including with Gary and others; the first book concerning the genetics of Alzheimer's, *Alzheimer's Disease: Molecular Genetics, Clinical Perspectives and Promising New Research*. L.A. Miner authored some of the tutorials in the first two predictive analytic books published in 2009 and 2012 by Elsevier. For 10 years, she served as a Community Faculty for Research and Data Analysis at IHI Family Practice Medical Residency Program in Tulsa, OK. She taught predictive analytics online, including "healthcare predictive analytics," for the University of California-Irvine. At present, Dr. Miner is professor emeritus, professional and graduate studies, Southern Nazarene University and serves on the editorial board of The Journal of Geriatric Psychiatry and Neurology.

Scott Burk, PhD, is founder of It's All Analytics (itsallanalytics.com), where he advises companies on creating the optimal data, AI, and analytics architecture to maximize their objectives. He stays active by writing and teaching as well. He is the author of four books on AI, data science, and analytics including the It's All Analytics Series, the Executive Guide for AI and Analytics, and Practical Data Analytics for Innovation in Medicine. He currently teaches in the MS of Data Science program at CUNY and has taught at Baylor and Texas A&M. He has held executive and VP roles as well as hands on positions in startups to Fortune 50 enterprises.

His experience is primarily in solving difficult AI, statistical, and analytical problems at companies, such as Texas Instruments, Dell, Paypal, EBay, Overstock.com, healthcare companies, and many others. Scott has a Bachelor's in biology and chemistry, Master's degrees in finance, statistics and data mining, and a PhD in statistics. Data has been the thread that has tied his professional experience together.

Scott resides in Central Texas.

He is the senior author of the *It's All Analytics* series of books; three of the four books in the series have been released in June, 2022 (https://www.routledge.com/The-Executives-Guide-to-AI-and-Analytics-The-Foundations-of-Execution/Burk-Miner/p/book/9781032007946#) and July 2020 (https://www.routledge.com/Its-All-Analytics-The-Foundations-of-AI-Big-Data-and-Data-Science-Landscape/Burk-Miner/p/book/9780367359683?source = igodigital#) and September 2021 (https://www.routledge.com/It-s-All-An-alytics—-Part-II-Designing-an-Integrated/Burk-Sweenor-Miner/p/book/9780367359713). His author page with more details is found at: https://www.amazon.com/Scott-Burk/e/B08FG8677J/ref = aufs_dp_mata_dsk.

Mitchell Goldstein, MD, attended the University of Miami's Honors Program in Medical Education and completed his pediatric residency at the University of California, Los Angeles, and Neonatal–Perinatal Medicine fellowship at the University of California, Irvine. Dr. Goldstein is board-certified in Pediatrics and Neonatal–Perinatal Medicine. He is a Professor of Pediatrics at Loma Linda University Children's Hospital.

Dr. Goldstein is the editor-in-chief of Neonatology Today (NeonatologyToday.net), the chairman of Physicians Against Drugs Shortages (PADS) (http://www.physiciansagainstdrugshortages.com/), and medical director of the National Coalition for Infant Health (NCfIH) (http://www.infanthealth.org). He is also the national chair of the American Academy of Pediatrics Multidisciplinary Action Group 2 Section Forum Management Committee and past president (aap.org) of the National Perinatal Association (NPA) (http://www.nationalperinatal.org).

Dr. Goldstein's most important trait is his tireless devotion to patient access issues, and he has been a vocal advocate for "right" sizing technology. When his hospital and many others nationwide could not purchase medical devices vital to patient care, Dr. Goldstein wrote senators and members of Congress. He testified in front of a United States Senate Sub-Committee of the Judiciary regarding the practices of the Group Purchasing Organizations and the restriction of access to vital medical devices and pharmaceuticals. Dr. Goldstein's vigilance in modifying these practices and improving patient access to medical supplies and pharmaceuticals made the front page of the *New York Times* and was featured on 60 minutes.

Dr. Goldstein advocates improving the physician–industry relationship as crucial in providing exceptional care. He supports transparency, dialog, and better engagement with industry to ensure that the most advanced pharmaceuticals and medical devices are developed and available for the most at-risk pediatric patients.

Thomas Hill, PhD, is a senior director for Advanced Analytics and Data Science in the TIBCO Analytics group. He previously held positions as an executive director for analytics at Statistica, within Quest's and at Dell's Information Management Group. He is a cofounder of StatSoft, Inc., the creator of the Statistica line of products. At StatSoft he was responsible for building out Statistica into a leading analytics platform adopted by organizations worldwide, serving as Senior Vice President for Analytic Solutions for over 20 years until the acquisition by Dell in 2014.

Dr. Hill received his Vordiplom in psychology from Kiel University in Germany, earned an MS in industrial psychology and a PhD in psychology from the University of Kansas. He was on the faculty of the University of Tulsa from 1984 to 2009, where he conducted research in cognitive science and taught data analysis and data mining courses. He has received numerous academic grants and awards from the National Science Foundation, the National Institute of Health, the Center for Innovation Management, the Electric Power Research Institute, and other institutions.

Over the past 20 years, Dr. Hill has guided the development of advanced analytic software and solutions; his teams have also completed diverse consulting projects with companies from practically all industries in the United States and internationally on identifying and refining effective data mining and predictive modeling/analytics solutions for diverse applications.

Dr. Hill has published widely on innovative applications for data science and predictive analytics, and authored over 15 patents on innovative analytic technologies and applications. He is a coauthor of *Statistics: Methods and Applications* (2005), the *Electronic Statistics Textbook* (a popular online resource on statistics and data mining), a coauthor of *Practical Text Mining and Statistical Analysis for Non-Structured Text Data Applications* (2012), *Practical Predictive Analytics and Decisioning Systems for Medicine* (2014), *Practical Data Analytics for Innovation in Medicine* (2022, in press), and *Ten Things to Know About ModelOps: Successful Strategies* (2022, in press); he is also a contributing author to the popular *Handbook of Statistical Analysis and Data Mining Applications* (2009).

Robert Nisbet, PhD, was trained initially in ecology and ecosystems analysis. He has over 30 years' experience in complex systems analysis and modeling, most recently as a researcher in forest growth modeling (University of California, Santa Barbara). In business, he pioneered the design and development of configurable data mining applications for retail sales forecasting, and Churn, Propensity-to-buy, and Customer Acquisition in Telecommunications, Insurance, Banking, and Credit industries. In addition to data mining, he has expertise in data warehousing technology for Extract, Transform, and Load (ETL) operations, Business Intelligence reporting, and data quality analyses. He is lead author of the *Handbook of Statistical Analysis & Data Mining Applications* (Elsevier Academic Press, 2009, 2017), and a coauthor of *Practical Text Mining* (Elsevier Academic Press, 2012). He served for 9 years as an Instructor in the Data Science Certificate Program at the University of California, Irvine, teaching courses in Effective Data Preparation, and Applications of Predictive Analytics. Currently, he is modeling the presence of precancerous cancer polyps with clinical data at the UC-Irvine Medical Center.

Dr. Nephi Walton, MS, MD, FACMG, FAMIA, earned his MD from the University of Utah School of Medicine and a Master's degree in biomedical informatics from the University of Utah Department of Biomedical Informatics where he was a National Library of Medicine fellow. His Master's work was focused on data mining and predictive analytics of viral epidemics and their impact on hospitals. He was the winner of the 2009 AMIA Data Mining Competition and has published papers and coauthored books on data mining and predictive analytics. He completed a combined residency in Pediatrics and Genetics at Washington University in St Louis. He is boarded in both clinical genetics and clinical informatics. He led several research initiatives in genomics and informatics at Geisinger prior to joining Intermountain Precision Genomics in March of 2020. At Geisinger he successfully completed one of the first pilot integration of genomics data into the EPIC electronic health record system for both pharmacogenomics and CDC tier one genetic conditions. He currently serves as the associate medical director of Intermountain Precision Genomics where he leads the HerediGene Genomic Sequencing Return of Results program. He also serves as the associate medical director of Intermountain's sequencing laboratory. He is the chair of Genomics and Translational Bioinformatics for the American Medical Informatics Association and has presented at several meetings on translating the use of genomics into general medical practice and the use of artificial intelligence in genomics, two areas he is actively pursuing at Intermountain Healthcare.

Dr. Gary D. Miner received a BS from Hamline University, St. Paul, MN, with biology, chemistry, and education majors; an MS in zoology and population genetics from the University of Wyoming; and a PhD in biochemical genetics from the University of Kansas as the recipient of a NASA predoctoral fellowship. He pursued additional National Institutes of Health postdoctoral studies at the University of Minnesota and University of Iowa, eventually becoming immersed in the study of affective disorders and Alzheimer's disease. In 1985, he and his wife, Dr. Linda Winters-Miner, founded the Familial Alzheimer's Disease Research Foundation, which became a leading force in organizing both local and international scientific meetings, bringing together all the leaders in the field of genetics of Alzheimer's from several countries, resulting in the first major book on the genetics of Alzheimer's disease. In the mid-1990s, Dr. Miner turned his data analysis interests to the business world, joining the team at StatSoft and deciding to specialize in data mining. He started developing what eventually

became the *Handbook of Statistical Analysis and Data Mining Applications* (coauthored with Drs. Robert A. Nisbet and John Elder), which received the 2009 American Publishers Award for Professional and Scholarly Excellence (PROSE). Their follow-up collaboration, *Practical Text Mining and Statistical Analysis for Non-structured Text Data Applications*, also received a PROSE award in February of 2013. Overall, Dr. Miner's career has focused on medicine and health issues, so serving as the "project director" for *Practical Predictive Analytics of Medicine* fit his knowledge and skills perfectly. Gary also serves as VP and scientific director of Healthcare Predictive Analytics Corp; as merit reviewer for Patient Centered Outcomes Research Institute that awards grants for predictive analytics research into the comparative effectiveness and heterogeneous treatment effects of medical interventions including drugs among different genetic groups of patients. Dr. Miner has taught online classes in "Introduction to Predictive Analytics," "Text Analytics," and "Risk Analytics" for the University of California-Irvine, and other classes in medical predictive analytics for the University of California-San Diego. Since retiring in 2016 from Dell Software as Senior Analyst-Healthcare Applications Specialist for Dell's Information Management Group. Dr. Miner has continued to write books, including the *It's All Analytics* series (2020; 2021; 2022; and the fourth is expected 2023 or 2024), a 2nd edition of the popular 2009 *Handbook of Statistical Analysis & Data Mining Applications* (2019), and completing in 2022 this 2nd edition of the 2015 "*Big Medical Book*" (i.e., it weighs about 8 pounds and makes a "good doorstop"!), which is expected to be released in September of 2022 (https://www.elsevier.com/books/practical-data-analytics-for-innovation-in-medicine/miner/978-0-323−95274-3). Additionally a 3rd edition of the 2009 "*Data Mining Handbook*" is currently being written for Elsevier Academic Press with an expected release date sometime in 2023 or 2024.

His author page with more details is found at: https://www.amazon.com/Gary-D-Miner/e/B003CRRAI8/ref = aufs_dp_mata_dsk.

Foreword for the 2nd edition—
John Halamka

Since January 2020, I have served as President of Mayo Clinic Platform, leading a portfolio of platform businesses focused on transforming healthcare by leveraging artificial intelligence, connected healthcare devices, and a network of trusted partners.

Why would a healthcare system establish a platform as part of its 10-year Cure, Connect, and Transform plan to create a Bold Forward strategy?

The answer is that the lifetime data of past patients can inform future care plans, new discoveries, and wellness strategies.

We have new sources of data from the devices that we wear, devices that we carry, and devices in our homes. Integrating clinical data, financial data, imaging data, telemetry, and genomics requires new techniques. As I said in the foreword to the 1st Edition, our challenge is no longer storage, computing power, or interoperability, it's turning our petabytes of data into information, knowledge, and wisdom.

The challenges are many—how to protect privacy, how to find signal in noise, and how to curate the data so that models are based on the best training set possible.

Just as the 1st Edition was a roadmap to supporting care with predictive analytics, the 2nd Edition goes further to provide practical examples, case studies, and methodologies supporting the entire model development life cycle.

I think of "Machine Learning Ops" as the process of "gathering" multimodal data, "discovering" models from integrated/curated longitudinal data, "validating" those models from heterogeneous data that represent the target population where models will be used, and "delivering" the output of models in workflow.

The 2nd Edition has been updated to reflect the latest digital health demands from stakeholders in the post-COVID new normal. Technology, policy, and culture changes from the COVID era have encouraged patients to be data contributors, moved care from physical to virtual, and motivated us to deploy machine learning models so that everyone in a shrinking workforce practices at the top of their license.

I believe you will find this book to be an invaluable resource as healthcare organizations become more data-focused in preparation for the market demands of 2030.

As Warren Buffet has said, the best way to thrive in the future is to be exceptionally good at something. Competence in data science is something to be exceptionally good at.

Dr HALAMKA JOB POSITIONS - EXPERIENCE:

Since January 2020: President of Mayo Clinic Platform, Rochester, Minnesota, Scottsdale, Arizona, and Jacksonville, Florida, and Mayo Clinic Global Outreach

Formerly: Professor of Medicine and Chief Information Officer at the Beth Israel Deaconess Medical Center, Boston, Massachusetts; Chief Information Officer and Dean for Technology at Harvard Medical School; Chairman of the New England Health Electronic Data Interchange Network (NEHEN); CEO of MA-SHARE (the Regional Health Information Organization); Chair of the US Healthcare Information Technology Standards Panel (HITSP); and member of the Board of the Open Source Electronic Health Record Agent.

Author of *The Best of CP/M Software* (Longman, 1984); *Real World Unix* (Longman, 1984); and *Espionage in the Silicon Valley* (Longman, 1985), and of the blog Geekdoctor: Life as a Healthcare CIO (http://geekdoctor.blogspot.com/).

Author of more recent books and blogs:

BOOKS: *Geek Doctor: Life as Healthcare CIO* (Rutledge, 2014); *Realizing the Promise of Precision Medicine—The Role of Patient Data, Mobile Technology, and Consumer Engagement* (Elsevier, 2017); *The Transformative Power of Mobile Medicine: Leveraging Innovation, Seizing Opportunities and Overcoming Obstacles of mHealth* (Elsevier—Academic Press, 2019); *Reinventing Clinical Decision Support—Data Analytics, Artificial Intelligence, and Diagnostic Reasoning* (Routledge—Taylor & Francis Group, 2020); *The Digital Reconstruction of Healthcare: Transitioning from Brick and Mortar to Virtual Care* (Routledge—Taylor & Francis Group, 2021).

BLOGS: https://www.healthcareitnews.com/blog/advice-new-national-coordinator (2014); https://www.mobihealthnews.com/news/north-america/john-halamka-heading-mayo-clinic-google-snipes-facebook-health-project-manager (December, 2019); https://www.opennotes.org/partner/john-d-halamka-md-Ms/ (January, 2020); https://www.mayoclinicplatform.org/blog/ (November, 2021 forward. . .).

https://www.blogger.com/profile/04550236129132159307.

LinkedIn: https://www.linkedin.com/in/johnhalamka/.

TWITTER: https://twitter.com/jhalamka?lang = en.

John Halamka, MD, MS

Foreword for the 1st edition by Thomas H. Davenport

I am pleased that you have purchased or picked up this book, because it covers one of the most important topics in healthcare and human society overall. What is more important than using data and scientific evidence to improve healthcare? As the many authors demonstrate throughout this volume, their topic has implications for not only our personal health, but also our finances, our happiness, and our legacy as a country and society.

Still, I suspect as you gaze upon this hefty tome you might find it a little overwhelming. Even if you're an expert in the field of healthcare informatics and analytics, there is likely to be much in this book that you don't know about. You may even feel that the term "practical predictive analytics" is something of an oxymoron.

However, dear reader, do not be dissuaded from diving into this book because of these concerns. First, the authors of most chapters are careful to define their terminology and stay away from highly technical expositions. An educated layperson can learn much from the book, and experts will have no problem with the great majority of chapters. The tone is relatively light in most chapters (at least, given the very serious subject), and there are even a few cartoons!

Secondly, the topics and terms are highly related. The reader comes to understand the relationships between and among various types of healthcare analytics, informatics, and tools and approaches for medical decision-making. Upon reading several chapters I realized, for example, that "predictive analytics" and "personalized medicine" are two sides of the same coin. These are two terms that I'd been familiar with for a while, but I never thought about their close relationship. Predictive analytics is simply the way we achieve personalized medicine; we are predicting how a particular patient will respond to a medicine or treatment, given that person's genotype and phenotype. It follows that the personalization of medicine is a probabilistic activity just like any other form of predictive analytics.

Thirdly, the book's chapters are replete with examples, and two entire parts—the second with tutorials, and the third with case studies—contain rich detail on how these ideas can be implemented in particular settings. I found the case studies particularly useful because they include factors that go well beyond the theory of predictive analysis in that particular domain. Many address issues such as culture, resistance by doctors and other practitioners, regulatory constraints, and other aspects of the institutional context in which predictive analytics are applied to healthcare.

Finally, you may find the volume of content in the book to be rather intimidating, as did I. However, I realized that the notions of personalization and prediction have relevance in navigating this book as well. Not all the content and chapters will be of equal interest to all readers. In a perfect world, we'd have a predictive analytics-based "configurator" that would tailor the content to each individual reader. Short of that, however, the Contents list can serve as your content personalization engine. I encourage you to use it aggressively. In fact, I confess not to have perused all of the chapters herein, but I have checked out most of them. I can safely assure that they represent a high level of content quality on this most important of topics. Healthcare professionals, quantitative analysts, and anyone interested in the future of medicine will be consulting it for many years to come.

Thomas H. Davenport

Babson College, Research, International Institute for Analytics, Wellesley, MA, United States
MIT Center for Digital Business, Boston, MA, United States

Foreword for the 1st edition by James Taylor

Those of us working in analytics and decision management have long regarded medicine as a prime target area. Like many industries, medicine is seeing an explosion in available data as medical records are digitized, medical equipment monitors patients in real time, and medical research is provided in machine-readable formats. If these data can be effectively applied to make better decisions, then lives as well as money can be saved. Applying predictive analytics and decision management to diagnosis, treatment selection, cost management, fraud detection, and more will result in consistent, effective, cost-efficient medical outcomes. These decisions will be based not only on medical research and the skills of medical professionals, but also on the analysis of an extensive and rapidly growing body of data. A book on predictive analytics and decisioning systems in medicine, then, is both timely and necessary.

This book provides an introduction to both the healthcare industry and the predictive analytic process. It discusses the ongoing digitization of electronic medical records and how to use these systems, as well as the regulatory and research framework within which medicine operates today. A wide-ranging collection of examples shows the potential for predictive analytics and decisioning, while dozens of tutorial examples walk through real-world uses of predictive analytics in medicine. Critically, throughout the book, the authors keep the reader focused on the need to use data to improve decision-making, not just to make predictions.

There is a deep, across-the-board need for predictive analytics and more advanced decisioning systems in healthcare. This book offers a comprehensive look at the challenges, opportunities, techniques, and technologies.

James Taylor
CEO Decision Management Solutions, Palo Alto, CA, United States;
https://www.linkedin.com/in/jamestaylor/

Author of Decision Management Systems: A Practical Guide to Using Business Rules and Predictive Analytics *(IBM Press, 2011).*

Foreword for the 1st edition by John Halamka

At Beth Israel Deaconess Medical Center, I oversee 3 petabytes of healthcare data. Big Data is no longer "big," since I can easily store, protect, and curate multiple petabytes growing at 25% per year. My challenge is turning data into information, knowledge, and wisdom.

Providing analytics and decision support to clinicians depends on good data, the appropriate business intelligence tools, and actionable visualizations. This book is a primer and roadmap for the entire process of supporting care with predictive analytics.

It includes an important discussion of healthcare data capture, highlighting variations in data quality and the usefulness of data gathered by different people in different workflows. Case studies illustrate how data can change care, getting the right care to the right patients at the right time.

The book also captures major trends, such as mobile technologies to gather patient-generated data and provide alerts/reminders to providers. It includes emerging technologies such as the transformation of unstructured data to structured data using natural language processing and inference such as IBM's Watson.

Healthcare IT industry pundits believe the next 5 years will belong to social networking, mobile, analytics, and cloud technologies. This book provides practical advice that helps navigate many of these trends as we all work to implement healthcare reform, supported by predictive analytics.

John Halamka, MD, MS

*Professor of Medicine and Chief Information Officer at the Beth Israel Deaconess Medical Center,
Chief Information Officer and Dean for Technology at Harvard Medical School,
Chairman of the New England Health Electronic Data Interchange Network (NEHEN),
CEO of MA-SHARE (the Regional Health Information Organization),
Chair of the US Healthcare Information Technology Standards Panel (HITSP),
and member of the Board of the Open Source Electronic Health Record Agent.*

Author of The Best of CP/M Software *(Longman, 1984)*; Real World Unix *(Longman, 1984)*;
and Espionage in the Silicon Valley *(Longman, 1985)*, and of the blog
Geekdoctor: Life as a Healthcare CIO (http://geekdoctor.blogspot.com/).

Preface and overview for the 2nd edition

Joseph Hilbe, the renowned statistician among the coauthors of the 1st Edition, passed away in April of 2017, way too early in his career, leaving a void among our group of coauthors. Scott Burk, a PhD in statistics and Master's in predictive analytics, was chosen to replace Joe (as the "master Statistician") for this 2nd Edition.

The 1st Edition contained a lot of health policy and heath accrediting agencies among its early chapters. This, along with the predictive analytic main thrust of the 1st Edition, made for a very "big book"—the hardcover actually weighs about 8 pounds, and does make a "good doorstop," but not something that a person wants to carry around with them for reading in their spare moments. (Though one of the UCI students carried it around with him when talking to Washington legislators about the urgency of incorporating PA into support for research in medicine. We considered his efforts laudable.) However, it became imperative that if we produced a 2nd Edition it should be a "smaller sized book," one that could easily be read and studied without causing arthritis flareups.

Thus, for this 2nd Edition, and since the 1st Edition would no longer be sold by Elsevier when the 2nd was released, we took most of the "policy/accrediting agencies" chapters out of the 2nd Edition paper copy, but are putting them on a companion web page, so that the very interested reader can have them available, as per the reader's specific interests. The 1st Edition came out in 2015 when the authors thought/"hoped" that the medical industry was ready to put into play predictive analytics, but unfortunately this 1st Edition did not hit the crest of the wave of predictive analytic interests. However, now in 2023 and forward, the medical industry is poised and ready to rapidly adopt predictive analytics which is the backbone to the *digital age of medicine*. Thus, the approximately 30 Tutorials—Case Studies from the 1st Edition are most useful (although they took up numerous paper pages), and need to be available to 2nd Edition readers, and thus they can be found on the companion web page as well.

The 2nd Edition has a *new PART 3* of seven chapters that come under the category of *practical application examples*. These examples are like cases studies and give rise to *N-of-1 studies* that instrumentally involve patients in determining their individual best practice. The examples cover factual issues in healthcare delivery that need *innovative solutions*—solutions that need *digital* processing in the right manner to develop *new gold standards of care*. These examples include a glaucoma patient taking their own IOP (eye pressure measurement) so that a truly accurate diagnosis and ongoing treatment decisions could be reached. The next logical step is the development of cost-effective, self-measuring eye devices and a change in provider attitudes that allow the patients to be *coproducers of data and codecision-makers* in diagnosis and treatment. Another example shows how a patient that would normally be in the ICU for monitoring urine output following various diuretic drugs can now do this at home. The logical next step is the development of a *home device* that makes this happen cost-effectively and provides the provider with *real-time data—information* to make treatment decisions at the right time for each patient. There are several other examples involving cancers (colon, pancreatic, lung), and neonatal data collection (again "averages and other statistical measure or stats" are needed instead of periodic—point EHR data to make accurate and life-saving decisions), that illustrate *how data are collected* and *how both the provider and the patients use data to makes decisions* will change.

The 2nd Edition ends with Part 4 which includes several advanced chapters on *how to gain insight from data, and how to manage models* in a big healthcare industry setting, *how variability in data* may be more important in DX and treatment decisions instead of the *point data* which are what are now generally available in the EMRs. There is also a chapter on *analytics architectures* needed in the 21st century; and a discussion of *Predictive versus Prescriptive versus Causal models*. And finally in the last chapter, Chapter 26, the authors speculate (but with data and information that points the way) on the future—what 21st century healthcare and wellness will look like in this new *digital age*—which we predict will see a mushrooming of *digital wearable devices* and other *data gathering devices* in the next few years. In fact, we predict this will start happening at an ever-increasing rate during the year 2022 as we finish writing and editing this 2nd Edition of this book.

The year 2022 has been proclaimed the year of the real beginning of *patient centered medical care **delivery***, by many forecasters in their blogs posted during the last few weeks of December 2021. Even PCORI in their Feb 2, 2022, webinar to introduce what they would be doing in 2022 and the future presented the following illustration at the beginning of their presentation:

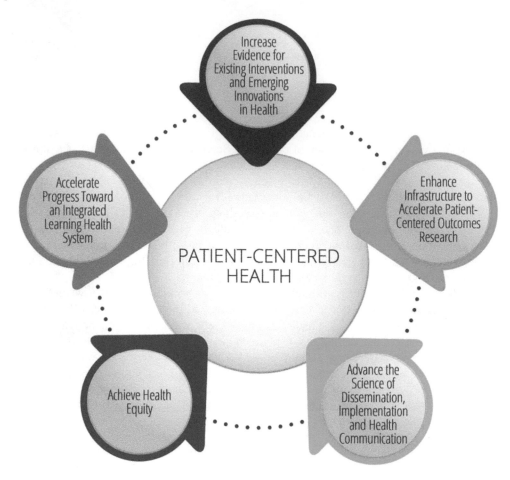

Figure Introduction. 1. Even PCORI in early 2022 is putting "patient-centered health" at the center of their goals, as a healthcare delivery research funding organization. This figure is presented again, in the last chapter of this book, Chapter 26, as a reminder of what 2022 and future years are all about in healthcare delivery (Adapted from, and credit given to: Cook, N.L.; Feb 2, 2022; in "Opening Remarks" at: PCORI-2022-and-Beyond-Webinar-Presentation-Slides-020222; https://www.pcori.org/sites/default/files/PCORI-2022-and-Beyond-Webinar-Presentation-Slides-020222.pdf).

As was stated in the preface to the 1st Edition of this book: One of the biggest challenges to evidence-based medicine (EBM) implementation is the set of limitations associated with the "gold standard" of randomized control trials, including high cost, length of time, often the significant bias, the enormous problem of the lack of access to unpublished (possibly suppressed) data, and the problems of using results that are not applicable very often to the individual patient under treatment. Many practicing physicians realize how little good evidence we have for the diagnosis and treatment of patients. If patients knew how much their physicians don't know, they might not want to see them. We need desperately to know the right test, and the right treatment, for the right patient at the right time. The goal of this book is to explain the enormous power of current technology in the search for evidence, which avoids those problems, and can provide timely, relevant, and practical information to use in the clinic and at the bedside.

So, your authors for this book are quite certain now in their prediction that the years 2022, 2023, 2024, and succeeding years up to 2030 will finally see healthy adoptions of predictive analytics, including AI (Artificial Intelligence) and ML (Machine Learning) data analytics in the field of medicine and in the delivery of healthcare, making it more accurate, safe, reliable, and tailored to the individual. This will start with a mushrooming of patient data-gathering digital devices beginning in the year 2022; this 2nd Edition is scheduled to be released in the fall of 2022, so as you readers begin studying this book in late 2022 and 2023, you should be able to see if our predictions were correct.

Preface to the 1st edition

Everyone is fascinated by predictions; everybody wants to know the future. Some people go to psychics, some go to mystics, others go to philosophers, to soothsayers, to television commentators, to a Nostradamus quatrain, or even to the ancient Incas. Recently, a new source of predictions has risen in our perceptions—*predictive analytics*.

What is different about predictive analytics, compared with the many other claims of future knowledge? Predictive analytics is based on relatively objective analyses of past data, following the principle that the best guide to the future is the past. All other predictions of the future are based on highly subjective (maybe even fraudulent) systems that their practitioners claim are predictive of (or at least relevant to) the future.

Why is this distinction so important?

First, it is impossible to know the future with certainty. Even those future events that we believe are absolutely certain can be thwarted from some previously unknown influence. We can't know everything in the present, and we certainly can't know anything for certain in the future.

Secondly, predictive analytics can expose new events that might happen in the future. The most intelligent approach to preparing for future events is to study past events, and be prepared for them to happen again. But, what if new events happen in the future? No other system can predict new events, with any degree of credibility.

Thirdly, predictive analytics allows different people to look at the same information, predict new events in the future, and compare their findings. These predictions can be analyzed according to their assumptions, methods, and outcome probabilities to decide which outcome is the best prediction of the future, even if we can't know for certain.

Modern medicine: an exercise in prediction and preparation

Modern medical treatment is composed of two general activities: (1) diagnosis, and (2) treatment. Both of these activities involve predictions. In the past, some medical practitioners based their predictions on naturopathic and nutrition philosophies. More modern practitioners base their predictions on results from the pharmaceutical industry's clinical trials. Some practitioners may trust the government to digest all of the information and clinical trial outcomes, and mandate the acceptable course of action. Other practitioners, such as medical researchers, just want to do it themselves, creating their own knowledge.

This book is based on the premise that the future becomes evident (as far as we can know it) by careful analysis of patterns in past data. This past evidence can provide trajectories of change in medical conditions and outcomes. For example, a tree that bends toward the sun at its apogee in the sky during the active growing season will continue to grow in that direction; that is the way living things in the world work. We can mark the starting point of growth, measure the angle and rate of tree growth, and (barring some catastrophic event) we can calculate where the top of the tree will be in 20 years, and how much wood it will provide if harvested.

Topol (2012) maintained that the starting point for human responses is the individual genome; from that point, our medical lives can be predicted. In this book, we will show how genomic information can be analyzed to predict a bewildering variety of medical outcomes.

Regardless of the starting point, we must follow some methodologies for analysis of past evidence in order to make predictions of any future medical outcome—but these methodologies must all be compatible with each other. A recent issue of the AARP Bulletin posed this rather startling observation:

"If home building were like healthcare, carpenters, electricians, and plumbers each would work with different blueprints, each with very little coordination" (AARP, 2012).

The resulting building constructed with such a disorganized methodology would not be very useful, nor would it persist for very long. But our healthcare system is like that! Many patients find their healthcare systems rather opaque, disorganized, and hard to navigate in order to produce acceptable outcomes.

This situation in US healthcare must change.

Wasted costs in US healthcare systems

Our US healthcare system is not only disorganized and inefficient; it is also very costly, and a significant amount of the cost is wasted (AARP, 2012). Some of these wasted costs include:

- $210 billion Unnecessary services
- $190 billion Insurance overheads and bureaucratic costs
- $130 billion Preventable errors/mistakes
- $105 billion Excessive prices
- $75 billion Fraud
- $55 billion Missed prevention opportunities

This book provides many useful methodologies that can be orchestrated to work together in future medical decision systems for building consistent medical outcomes without major cost wastage. This book also provides many tools for use by medical practitioners to make accurate predictions. The focus of these predictions is on anything related to medical or healthcare. You might be a practicing physician seeking correct and accurate medical outcomes, or you might be a hospital staff member who must demonstrate safety to the accreditation organizations, or you might be an analyst in a healthcare insurance organization who wants to predict the frequency and severity of loss risk.

For all of those activities (and many more), *Practical Predictive Analytics and Decisioning Systems for Medicine* provides the data and the step-by-step tutorials that will put power into the hands of practitioners.

References

AARP, 2012. Caring for the caregiver. AARP Bull. 53, 9.

Topol, E., 2012. The Creative Destruction of Medicine: How the Digital Revolution Will Create Better Health Care. Basic Books, New York.

Acknowledgment

The seven coauthors express gratitude to many for the tremendous efforts toward the completion of this book.

The guest authors of chapters, case studies, and tutorials added their unique expertise through their efforts; for this we owe a debt of gratitude.

The editorial expertise of the Elsevier and Academic Press staff is greatly appreciated: thanks to Rafael Teixeira, Senior Acquisitions Editor in the São Paulo, Brazil Elsevier office for studying our 1st Edition and realizing the potential and great need for a 2nd Edition of this book; his request to us authors in early 2020 to write a 2nd Edition was a "Big Surprise" to us, but we welcomed and embraced it! Rafael and authors Gary Miner, Bob Nisbet, and Linda Miner had many email communications during 2020 and early 2021 before we got all the details arranged. However, authors Linda and Bob had already written the first drafts of their new and revised chapters during early 2020, and the rest of us joined in during 2021, finishing the first drafts by December 31, 2021 and then polishing them during the first 3 months of 2022 as Bob Nisbet wrote the Preambles and Postscripts to each chapter, drawing all the contents of the book together as a whole complete easy-to-follow story. Rafael followed this all the way; without Rafael, there would have been no 2nd edition: Thanks Rafael!

Then, with the full chapter manuscripts submitted, Samuel Young, Editorial Project Manager in the San Diego, California Elsevier—Academic Press office took over day-to-day book production activities. We thank Sam for his attention to detail, staying up at late hours to get all the manuscripts into the Elsevier EMSS system, and gracefully putting up with authors—project manager Gary Miner for all his "way-word methods" in getting the final drafts to the San Diego office: Thanks Sam!

Then in April 2022, Swapna Srinivasan, Project Manager for this 2nd Edition book production, took over on a day-to-day basis, working between author Gary Miner and the copyeditors and typesetters to get all aspects of the book in proper order. Swapna is with the Chennai, India Elsevier office in the Health Content Management division. Without Swapna's continual diplomatic prodding of Gary to get the 2nd Edition Foreword and final Front Matter into their offices, the book could have easily gotten off-schedule: Thanks Swapna!

Finally, the authors acknowledge and thank their families, who sacrificed time with their loved ones and offered their support throughout the nearly two-and-a-half years during which the writing of this 2nd Edition book was "in process."

Guest Chapter Author's Listing

Guest — Authors

Anna J.C. Russell-Toner, MComm(Statistics), CEO; *The Boss Lady*, The Data-Shack; Head-Office: York, UK; also offices in South Africa & Hong Kong; http://www.data-shack.co.uk

Billie Corkerin, C.P.O.T.; *I-CARE HOME Trainer*; OMEG, Jenks, OK, https://www.omeg2020.com/

John B. C. Tan, PhD; *Data Scientist*, Department of Pediatrics LOMA LINDA UNIVERSITY | School of Medicine, Loma Linda, CA, https://medicine.llu.edu/

Fu-Sheng Chou, MD, PhD; *Neonatologist and Assistant Professor*, Department of Pediatrics LOMA LINDA UNIVERSITY | School of Medicine, Loma Linda, CA, https://medicine.llu.edu/

Rebekah Leigh; *Ms2 Medical Student*, MD Candidate, Class of 2024 LOMA LINDA UNIVERSITY | School of Medicine, Loma Linda, CA, https://medicine.llu.edu/

Harsha K. Chandnani MD, MBA, MPH; *Pediatric Intensivist*, Department of Pediatrics LOMA LINDA UNIVERSITY | School of Medicine | Children's Hospital, Loma Linda, CA, https://medicine.llu.edu/, https://lluch.org/

Mahmood H. Khichi, MD, FAAP, MBA; *Critical Care Provider*, Texas Children's Hospital, TX, United States

Cynthia H. Tinsley, MD, FAAP, FSCCM; LOMA LINDA UNIVERSITY Children's Hospital, *Pediatric Critical Care Medicine Division Chief*, CA, United States

Endorsements and reviewer Blurbs—from the 1st edition

With healthcare effectiveness and economics facing growing challenges, there is a rapidly emerging movement to fortify medical treatment and administration by tapping the predictive power of big data. And it works! Predictive analytics bolsters patient care, reduces cost, and delivers greater efficiencies across a wide range of operational functions. In-depth and eye-opening, this seminal tome serves both the healthcare professional and the analyst: If you are a healthcare provider, researcher, or administrator, this handbook will motivate and guide your data-crunching; if you are an analytics expert, this industry overview will illuminate the pertinent background you need from the complex and dynamic healthcare industry. To get a grip on the predictive healthcare revolution, one must begin with this book's comprehensive 26 chapters and 33 hands-on tutorials. (February, 2014)

Eric Siegel, PhD, founder of Predictive Analytics World and author of Predictive Analytics: The Power to Predict Who Will Click, Buy, Lie, or Die.

Review by Dr. Lonny Reisman

Medical records and practice have changed little in the past two decades, from when I was a practicing cardiologist in the 1990s using paper charts, to today with electronic medical records that are little more than digital file cabinets. The science and technology behind medicine, however, have changed dramatically and this new book shows how the gap between practice and science is being bridged—through predictive analytics and decisioning systems.

The book gives clear evidence of the need for predictive analytics, citing the number of preventable deaths from nosocomial infection, increased morbidity from misdiagnosis, and other tragedies in our healthcare system. My personal experience gives a similar perspective. As a practicing cardiologist in the 1990s, one of my frustrations was the inability to access important information about my patients from other parts of the health system. My "Ah-Ha" moment came during a meeting, when colleagues were discussing the case of a 42-year-old male who had a stroke, was stabilized, and released from the hospital. While others thought the shortened hospital stay was a success, my focus was on why we didn't prevent the stroke in the first place? My work had shown the data and evidence existed to prevent the problem before it happened, but not where it was needed and not in a useful format. I realized that aggregating that data, and combining it with scientific evidence, would allow us to predict health risks and identify opportunities to improve care and prevent serious illness. That revelation, 15 years ago, led to our patented clinical decision support tool—a breakthrough in combining claims data with clinical science to enhance the expertise of the physician and patient experience at the point of care. At the time our software application just seemed like the right thing to do for the right reasons. We only learned later it was a pioneering effort in linking clinical science with patient data to improve care—moving from reactive medicine to proactive prevention and care.

Dr. Linda A. Miner and her coauthors chronicle this and other achievements in medicine, even going back over 2000 years to put our current efforts in context. This groundbreaking book illustrates our first steps at predicting the future, with traditional statistical methods that attempt to stratify populations, but focus more on financial costs than clinical care with limited predictability (with 10%—20% considered a "high level" of predictability). The authors then meticulously walk through the current struggle to digitize health records and glean useful information from the data that is slowly beginning to accumulate. All the "usual suspects" are implicated in this review, including the Affordable Care Act, trade associations, accrediting bodies, standards organizations, health technology vendors, and hospitals and physicians. But the authors go far beyond our current activities, and forge a bridge to the future by showing how new technologies such as predictive analytics are being used to personalize medicine, enable consumer-driven care, and truly predict the future course of disease and treatment for the individual.

This book is being published at a critical inflection point in healthcare. As the CEO of a major health insurance company said, we are seeing the "end of [health] insurance companies," and "health information technology is helping put health insurance out of business" (Bertolini, 2012). Not only are health insurance companies coming under increasing pressure, but the entire practice of medicine is entering a new era, with the "creative destruction of medicine" (Topol, 2012). These tectonic changes require new thinking and new approaches, which are thoroughly described in this book.

Our previous work in financing and delivering healthcare has focused on populations, and predicting risk and care for large groups of individuals that in reality were genetically diverse. But this approach has inherent limitations as it shows large trends rather than individual traits, and says nothing about how to reach out, engage the individual person, and improve their care. In fact, many believe you can only predict risk for populations—and this may be true if you only have the tools of traditional statistics. The authors show how new predictive analytics tools and methods—many of which are revolutionizing other industries such as banking, consumer retail, and homeland security—can be used to predict risk and response of the individual—allowing us to tailor medicine not just to an individual's phenotype and genotype, but also to personal preferences—truly a "market of one."

These new technologies are becoming a necessity as we move from medical practice relying solely on individual physician experience to a new age of informatics enabled by the enormous volume, velocity, and variety of data—and the ability to generate real value from all this technology. As the authors point out, for thousands of years the practice of medicine has relied on "the most powerful nonlinear processing engine in the universe, the human brain." That is what we used in the 1990s. With scientific discovery exceeding the capacity of the individual physician to compile all the research discoveries in the world and understand every nuance of the individual patient, it is imperative we use new data analytics technologies to assist. Dr. Linda A. Miner and her group of authors give us both the vision and the tools to do just that.

Bertolini quotes:

http://www.forbes.com/sites/sallypipes/2012/03/19/the-end-of-private-health-insurance-in-america/

http://healthpopuli.com/2012/10/09/from-fragmentation-and-sensors-to-health-care-in-your-pocket-health-2-0-day-1/ (March, 2014)

Dr. Lonny Reisman, MD

Lonny Reisman, MD, FACC, founder and CEO, HealthReveal (2015−21)—http://www.healthreveal.com—https://www.hmpgloballearningnetwork.com/site/jic/news/acc-healthreveal-collaborate-speed-adoption-clinical-guidelines-and-optimize-patient-outcomes—acquired by Accolade on September 20, 2021: https://www.accolade.com/

PREVIOUSLY: Executive Vice President and Aetna's Chief Medical Officer (2008−2014), Founder and CEO of ActiveHealth Management (an Aetna subsidiary) (1998−current).

Early Bio: From 1991 to 1998, Dr. Reisman was a principal in the Managed Care Group of William M. Mercer where he led numerous consulting engagements with Fortune 500 corporations, healthcare providers, suppliers, and payers that focused on managing the demand for healthcare resources. Dr. Reisman was an attending physician at New York Hospital and St. Luke's−Roosevelt Hospital Center between 1987 and 1999, and was a cardiology fellow at the University of Chicago from 1985 to 1987. He received his undergraduate degree from the State University of New York at Stony Brook and his medical degree from Tel Aviv University.

References

Bertolini, M. quoted by: Potter, W. March 5, 2012. ANALYSIS: The end of health insurance as we know it?. In: The Center for Public Integrity - https://publicintegrity.org/health/analysis-the-end-of-health-insurance-as-we-know-it/. Based on a You-Tube Video he made with HIMSS at Major Health Conference in Las Vegas on February 22, 2012: https://www.modernhealthcare.com/article/20120222/VIDEO/302229933/video-news-live-himss-2012-interview-with-mark-bertolini-chairman-president-and-ceo-of-aetna-6-29; Video News: Live@HIMSS 2012 Interview with Mark Bertolini, chairman, president and CEO of Aetna (6:29).

Topol, E. 2012; The Creative Destruction of Medicine How the Digital Revolution Will Create Better Health Care; New York: Basic Books.

Instructions for using software for the tutorials—how to download from web pages—for the 2nd edition

The URL for this book's companion web page is: http://booksite.elsevier.com/9780323952743/

The "Tutorials and Case Studies" (only on the companion web site for this 2nd Edition book) illustrate the use of predictive analytic software from various sources. You, the reader, can oftentimes obtain "free evaluations" of software and/or low-priced academic copies of software to use in working through these tutorials, if you wish to study and replicate them. Some of the tutorials use TIBCO-STATISTICA, others use SAS-Enterprise Miner or R-Rattle, and others use KNIME. This software is not provided on this book's companion web page, but you can use a search engine to find the SAS, KNIME, and R websites and download what is available or find out how to get an evaluation copy (if available). In fact, in the tutorial using R-Rattle, instructions are provided on how to do this (realizing that the specific URL may have changed by the time you are reading this, and you will have to search further). KNIME and R are "freeware" or have free versions available to download from their web pages.

Below are links to possible sources for such software (but remember that software companies frequently change the form and conditions for "trial software," so the links may change after this book publishes, and you may have to "click around" to find the current online links):

1. TIBCO-STATISTICA:
   - Academic copy: https://edelivery.tibco.com/storefront/eval/tibco-statistica-ultimate-academic/prod11780.html
   - Commercial Evaluation copy: https://www.tibco.com/resources/product-demonstration-request/tibco-data-science-statistica-demonstration?_bm = b&_bn = g&_bt = 391695050093
2. SAS-Enterprise Miner:
   - https://www.sas.com/en_us/home.html
   - https://www.sas.com/en_us/trials.html
   - ACADEMICS: https://www.sas.com/en_us/software/on-demand-for-academics.html
3. R (and R-Rattle):
   - HOME R SOFTWARE PAGE: https://www.r-project.org/
   - R-RATTLE: https://cran.r-project.org/web/packages/rattle/index.html;
   - R-RATTLE QUICK START GUIDE: https://cran.r-project.org/web/packages/rattle/vignettes/rattle.pdf
4. KNIME:
   - https://www.knime.com/downloads
   - https://www.knime.com/downloads/download-knime

Prologue to Part I

This book comprises three basic areas:

1. Context and opportunities
2. Practice
3. Theory

Part I is primarily concerned with the first area.

Chapters 1–10 (basic area number one) acquaint the reader with the many facets of healthcare and explain the various aspects and organizations that, directly or indirectly, affect how medical and healthcare data are captured.

These chapters offer the perspectives of history, including where we are at present and where we might be heading. The reader can become grounded in the context of medical research by examining the history, the organizations that arose, the arising needs from our cultural milieu, the attempts at answers to those needs, and where those answers may have fallen short.

Predictive analytics requires data, and good data. It is the task of the data analyst to secure those data just as though they were gold. However, predictive analytics is not enough. Predictive analytic methods seek to anticipate good outcomes, and good outcomes for individuals. Predictive models are only as good as the data that are processed by the models. The chapters in Part I demonstrate the flow of data through history, including the laws that were meant to inhibit and those that were meant to increase the flow. We as researchers must know our context in order to apply our predictive analytic art to the canvas of individual outcomes.

In summary, data ("good data") are essential to the precision of predictive analytics. Good data are necessary for unleashing effective models from which excellent decisions can be made.

You, the reader, may wonder why we did not start this book with discussions of predictive analytic algorithms, and predictive analytic models and decisioning. But good decisioning can only come after good modeling, and both of these cannot be obtained if one does not have good data as a starting point.

Part I

Historical perspective and the issues of concern for health care delivery in the 21st century

Chapter 1

What we want to accomplish with this second edition of our first "Big Green Book"

Linda A. Miner

Chapter outline

Prelude

There are many books available on the subject of medical data analysis, therefore, you might wonder why so many authors spent so much time writing this one. It is not because the authors need books to bolster their academic positions (none of the authors are academics). All of the authors are professionals active either in the practice of medicine or the analysis of medical data. Four specific reasons for writing this book are presented in the Conclusions section. The overarching reason for writing this book discussed in this chapter is the need to serve the great need for analytical information to support patient-directed medicine by showing how to do it, and ways to do it right!

Purpose/summary

Dear Reader, you may find our original reasons for writing this book in the first edition, still available for sale (and you may find selected content from the first edition located online in the companion webpages, included with your purchase of the second edition). Here, you will find a brief outline of the original text, along with the intent of this second edition.

The first edition included nearly encyclopedic information and the latest (at the time) research concerning each topic of the book. In the second edition we are focused more on *how to do* predictive analytics (PA), machine learning, and artificial intelligence—more of the practical aspects rather than the rationale, although briefer rationales are still given.

Now there is a use for the encyclopedic versions of the first edition as it fits well into PAs projects. The provided backgrounds both in the first edition and in this edition for topics of interest can give one a start on several ideas for research. Developing research ideas is particularly useful for students, especially for medical residents who are required to carry out a project during their residencies.

I have kept the updated long version of Chapter 9 (now Chapter 7) in the second edition because I believe patient-directed medicine is coming into its own. Patients might want to conduct studies on themselves, as explained as N of 1 studies, in Chapter 7, and several other chapters, including the last chapter, Chapter 26. N of 1 studies can be rigorously conducted. For example, Halamka (2022) explained the use of crossover designs in N of 1 studies, allowing physicians to pinpoint patient treatment outliers in their practices. Patient responsibility is ripe for PA research and the tutorial example outlines from some of the first edition tutorials may stimulate creative ideas for research. In addition, in Chapter 7, there are boxes highlighted after most sections suggesting areas for traditional and PA research. Those suggestions could prove to be useful particularly for resident physician projects. Others may want to follow up on the idea of writing programs for wearable devices that would allow patients to analyze their own data at home. Such an activity might prove to be not only beneficial to patients but also lucrative for the designer.

The original reasons for writing the book were not only how to conduct PA research, but why it was important to do so. These reasons included flawed research needing help, helping to meet accreditation standards using six sigma and lean management, meeting changes in technological organizations, and meeting future demands (such as data analysis programs for wearable devices), all of which could benefit from PAs. (See Fig. 1.1.)

First reasons for our writing this book

Examined were:

1. flaws in publications, retractions, misconceptions in the research process in medicine
2. common flaws in medical research design
3. the necessity of quality data for quality research
4. understanding biases and research gaps
5. the necessity of research for achieving various accreditations
6. the horrors of inoperability
7. standards and commissions related to the original Affordable Care Act
8. organizations related to quality and safety
9. six sigma and lean tools
10. biomedical informatics
11. organizational and payment changes
12. the need of granularity of one—predicting to the individual patient
13. emphases on personalized medicine and patient-driven self-care
14. the future of medical research—genomics, nanotechnology, and wearable health devices

Highlighted new material

Throughout this chapter, there are some revised ideas from the original Chapter 2 that might be of interest here related to the pandemic of COVID-19 that we have all experienced in 2020 and beyond—at least until the writing of this book.

The first idea concerns examples of retractions (explored in the first original reason for the book, which was to help overcome flawed research):

FIGURE 1.1 Original reasons for writing the book.

One retraction that made the news was the COVID-19 hydroxychloroquine (HCL) research retracted by the New England Journal of Medicine, June 4, 2020, because they were unable to obtain the data from the study. One can read about this retraction at this website: https://www.nejm.org/doi/full/10.1056/NEJMc2021225 (Mehra et al., 2020a,b). Another study published in the Lancet, suggesting the medication was dangerous, was also retracted on June 5, 2020, and can be found at this website: https://www.thelancet.com/journals/lancet/article/PIIS0140-6736(20)31324-6/fulltext (Mehra et al., 2020a,b). It seemed that it was hard for researchers to conduct valid research due to the political nature of the subject in both directions, which in one way and another undoubtedly subtly biases any research. Politics seemed to affect nearly everything related to the diagnosis, spread, and treatment of the disease in 2020 and lingered into 2021. We should learn something here it seems to me. For example, it became apparent that the disease was in two parts—early illness and late, each demanding different treatment. Unfortunate politics, not understanding the nature of the disease, and conflicting information confused and, at times, alarmed patients. Retractions served to confuse further. At this writing, there is much conflict in knowing what to do. Conspiracies ran rampant further confusing people who were unacquainted with how to evaluate research.

Honest authors admit to flaws in their work and one such author was Frances Arnold, who shared in a 2018 Nobel Chemistry Prize for work in the journal, Science, on "enzymatic synthesis of beta-lactams." She was unable to replicate her findings and being a true and honest researcher, retracted her paper herself. Please see this website: https://retractionwatch.com/2020/01/02/nobel-winner-retracts-paper-from-science/#more-118672 (Marcus and Oransky, 2020).

Discovering and reporting one's own flaws is much better than having someone else find and report them (first page of original Chapter 2 revision online). Worse is knowing that the flaws exist but not owning up to them because one wants to publish or for some other personal gain if those flaws are never revealed. Self-reporting of discovered flaws is an important standard that all worthy research needs.

Retractions of medical articles seem to have reduced since our first edition but that might not mean that the research has improved. It may be that flawed research is not recognized as much as it was in the past, which could also mean the research situation is getting worse. There is tremendous pressure in many academies to publish (or perish, of course), and the more papers in prestigious journals the person has published, the more the person makes it up the academic ladder—and the more money is awarded to the person in terms of grants. There is one other aspect of today's research that one encounters as mentioned above—being politically correct for whatever "side" seems to be in power, or even greed itself, can influence. During the pandemic, we heard over and over, "Follow the science!" However, following the "science" may, in fact, mean make sure the findings agree with the current popular views of the world. If one understands science and what constitutes good research, then following the research makes a great deal of sense, but "good" science might be influenced by one's bias, or bias of the predominant world view. One needs to read studies that are published and understand how to evaluate each. In our epidemiology courses for our postdocs, Gary and I would join in weekly sessions reviewing recent research and finding flaws. It might sound a bit negative, but it was a great exercise in evaluating.

With the above in mind, we yet need to look at the structure of research designs, both traditional and designs more in keeping with the newer predictive, AI, machine learning models. The following are quite popular in the literature, though PAs is going strong.

Oftentimes, in studies, we start by examining descriptive statistics—finding means, looking at graphs of the raw data and so on. The chart in Fig. 1.3 by Dr. Bob Nisbet is one such example. Note that he used a workspace from KNIME (Download). In Chapter 16, Bob expertly exposes the methodology of displaying data using that excellent and free program. Excellent examination of data and using descriptive data often leads one to excellent variables for predictive studies.

Descriptive statistics, data organization, and example

Often today, one uses a program to quickly transform data into descriptive statistics. The raw data for a data set need to be organized in rows, with each row representing one case. Fig. 1.2 shows how the data are organized in rows (from Chandnani et al., 2018a,b). In preparation for predictive analyses, often an outcome variable (dependent variable) is selected and is measured. In PAs, the dependent variable is predicted by the independent variables. Outcome variables might be in interval data (such as scores on an instrument, numbers of cases for each date as the example below shows from), nominal (such as deceased vs survived), or ordinal (ranked data). Generally, the other variables will be potential predictors and are called the independent variables.

Various programs can be used to analyze data. Fig. 1.3 shows how Dr. Bob Nisbet analyzed some COVID data that he found on the Internet, using the freeware program, KNIME (Download) (https://www.knime.com/downloads).

Fig. 1.4 shows one of Dr. Nisbet's line chart outputs.

Now looking at types of studies that help control for bias, the following explanations are provided.

Example of Outcome Variable

11 outcome
Survived
Survived
Died
Died
Survived
Survived
Died
Died
Survived
Survived
Died
Died
Survived
Died
Survived
Survived

Examples of Predictor Variables

68 low sodium (serum) (mmol/l)	69 low sodium (whole) (mmol/l)
150	147
147	144
140	137
138	135
138	135
148	145
139	136
139	136
134	131
140	137
138	135
133	130
135	132
129	126

FIGURE 1.2 How data are arranged.

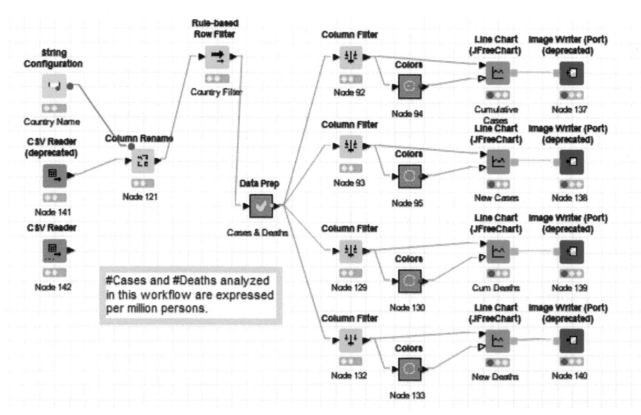

FIGURE 1.3 KNIME workspace for descriptive data by Dr. Bob Nisbet, examining new cases per day in the United States. *Data come from (https:// github.com/nytimes/covid-19-data).*

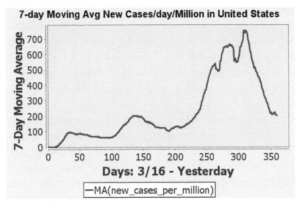

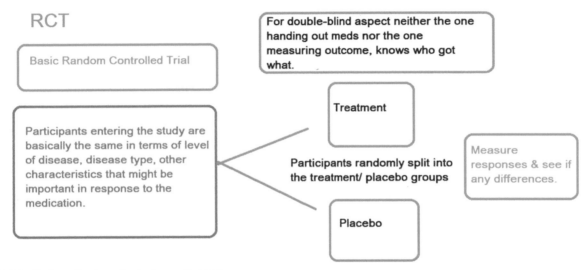

FIGURE 1.4   Dr. Nisbet's line chart outputs, 3—16—20 (**GitHub, Times**). *Data from Massachusetts https://github.com/nytimes/covid-19-data; GitHub (Times).*

FIGURE 1.5   Outline of basic randomized controlled trial.

Randomized controlled trials (RTC) are the gold standard for determining possible differences between treatments in groups of subjects, and PA for making predictions to the individual, rather than demonstrating differences between treatment versus control groups. If a study is truly controlled and the subjects randomized to treatment, and the measurements are accurate, and the subjects represent subjects in the population of the study, then there should not be room for political persuasion.

Randomized controlled trials

From observation to randomized controlled trials—eliminating bias

To follow the science, we need to know what studies are more desirable than others in terms of being able to trust the conclusions. For testing overall efficacy of a treatment or drug, one spans the gamut from observational studies, correlational studies, comparative studies, and controlled studies. Considered the best is the *randomized double-blind, controlled study* (RCT). (Please see Fig. 1.5 for a basic outline or comparing a medication to placebo.)

We often start by noticing or observing phenomenon after the fact, *retrospectively*, which we then wonder if there is something real (or causative) going on. Correlational studies and cluster analyses are used quite often to tease out relationships that might prove causative. For retrospective studies, we look at patterns after the fact. We trace back, for example, if one had a disease or did not and then look back to see what medications they were taking. Or perhaps we look at patterns of patients before either death or survival. Those patterns could be causative or could be simply happenstance.

As another example, during the pandemic people on certain medications seemed to have fewer symptoms than people not on those medications. Such observations could lead to a more structured and more "valid" type of *prospective* study in which participants, equivalent in important characteristics, are randomly sorted into groups either getting the certain medications or not getting them and then observing what happens to them. We simply cannot trust a study that does not randomize or that gives the medicine to everyone. Such a study could serve as generating an idea of what might be happening but needs an RCT of adequate size to even begin to understand the truth of the matter.

Once the participants are randomly separated into groups, all groups are given medications that look the same (the placebo looks just like the active medication, for example), and then data on symptoms are collected after the groups have been on the "medications" for the same amount of time. To make the study a double-blind study, neither the participants nor those handing out the medications, or those gathering the observations, know who received which medication. Double-blind studies help to eliminate possible biases (even unconscious desires for the medicine to work). The double-blind, random assignment (controlled) study is considered the gold standard for comparative studies. Achieving the permissions, the cases, the funding for such is not an easy task and is often a reason that the "science" sometimes must rely on prepublications, correlational studies, and retrospective studies, such as *meta*-analyses, all of which could contain bias. One other detestable thing that can happen is if a person (or persons) responsible for the study fabricates data or in some way skews the outcome. In other words, is dishonest. This can happen and is the reason why studies really need to be replicated by other researchers. Unfortunately, replications are not often published unless they can demonstratively refute the first paper.

One does not have to be dishonest to make mistakes that are common. One such mistake is the family-wise error, described in Chapter 7, in which tables of comparisons are made from data and then the researcher picks up on the ones that seem "significant" or with a p value less than .05 and deems the other 10 or 20 contrasts not significant. This is not a dishonest practice but is one of simply not knowing and is seen often in published literature. Multiple significance tests are something to watch for.

Basic predictive analytics and example

PAs, which some call prescriptive analytics, provides a model for predicting accurately to the individual. Basically, the terms are the same thing. It is not enough, in medicine, to know, for example, that one medication works better than another in some population. People do not simply fall to the mean of a group in terms of their responses to medications and/or treatments. Rather, we need to be predicting what works best for *each person*. The medical community is beginning to wake up to PA, but it has been slow coming. When this book first came out in 2015, most in the medical community were not even looking for PAs. Now it is becoming much more done and recognized.

PA can use both retrospective and prospective data. PA involves machine learning algorithms and various uses of artificial intelligence for developing a prediction algorithm. The goal is to predict some outcome for individuals rather than for a group. The outline below shows the basic steps for PA.

Again, the data are gathered in rows, with each row a different individual. The data comprise an outcome variable and as many predictor variables as might be important (in the literature of case control studies), and that might have been informally observed and thought to be important or which seem to be important because of known physiological processes known about the condition (See Fig. 1.5). Also included might be scores from known instruments which have attempted prediction in the past, to see if those attempts might be streamlined or increased in accuracy. Each variable should be measured for each individual case or person. Fig. 1.6 shows the outline of the basic steps for PAs. Perhaps semantics, but it seems to me that prescriptions follow the predictive algorithm from the PA process, hence I would not call it prescriptive analytics unless the prescriptive analytics have used the same rigorous procedures as PAs.

Like RCT, one divides at random the data set into three subsets—training, testing, and validation groups, using some percentages such as 50% of the cases for the training group, 25% for the testing group, and the additional 25% for the validation or holdout group. The training group is first group one uses. I like using Statistica because one can use the "feature selection and root cause analysis" program on the training data to give an idea of which predictors will be the best. Next, using only the predictors that seem to be predicting, the routine "data mining recipes" is employed to try many algorithms at the same time, using v-fold methodology (so that the errors resulting from multiple, family-wise analyses are reduced). The output of the data mining recipes allows one to see which of the predictive models likely would be the most accurate model to use, say boosted trees.

Basic Predictive Analytics Procedure

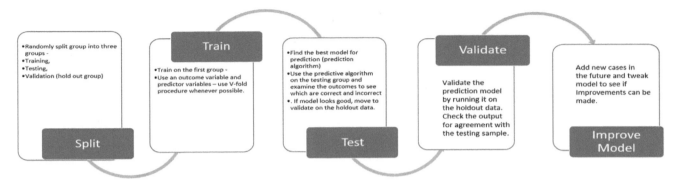

FIGURE 1.6   Basic predictive analytics. *Procedure by Linda A. Miner.*

Example

The example used below is from a study of survival versus death of subjects entering the hospital with DIC symptoms.

Using the training (called the analysis sample in the matrix below) sample and testing sample together (but labeled so the machine knows which is which), from the example above, one could do interactive boosted trees, again using v-fold methodology, on the training data, and then test the model on the testing data (heretofore, not used). One may examine the classification matrices to note the hits and misses. It is good if there is basic agreement between the two. Tables 1.1 and 1.2 show two such matrices. The outline of this study is in the examples section from (Chandnani et al., 2018a,b).

Lastly, one asks the machine to produce the PMML (or other language) of the predictive algorithm thus far, to deploy on the holdout sample (validation group). If the predictions hold true, then one inputs the independent variables from an unknown case to arrive at the prediction of whether the individual is likely to die or not. Such a prediction can be quite useful in deciding what to do in the attempts to save the person.

Research standards common to both traditional and predictive analytics

For both types of research, traditional, predictive (AI and learning models), data sets should contain full data points. This means that the samples being researched should truly represent the populations of interest (when conducting inferential statistics) and that the data for each subject should be complete across the variables (when conducting any type of research, but particularly PA). In our real world of research, these two standards are often difficult to achieve. (Please see the new DIC example mentioned in Chapter 9 and in the tutorials of this individual.) We are not trying to generalize to a greater population, while attempting to accurately predict to the individual. Regardless, if patients in one hospital differ greatly from patients in another hospital, prediction models that work for patients in the first hospital might not predict to individuals in the second. One would need to test and modify the model in much the same way as initially developing the model. If the model does, indeed, predict as accurately within the second hospital, then the model is validated, and everyone is happy (within those hospitals at any rate). Even better is if the model moves accurately across many hospitals, making the predictive algorithm ever more valuable. Lack of data is a serious deterrent to developing accurate models and withholding data to preserve one's turf can happen (likely an understatement). An excellent way of obtaining data is to work with colleagues within the same system, particularly if the system is large, including many hospitals.

Pandemic as related to research standards and accurate data

When the pandemic hit, very little was known about the novel COVID-19 virus. It was difficult to obtain accurate information. Unfortunately, we learned the lesson of poor data when the case—fatality ratio was initially inaccurate and led to perhaps, unnecessary fear, and panic. If the ratio were truly 3%, which was about the average reported in the beginning of the pandemic, and the death rate for common flu is typically 1%, then Coronavirus-19 would have been 30 times more deadly than the flu. Assuming COVID-19 was also highly infectious, left without social distancing, the

TABLE 1.1 Classification matrix for the training sample.

		Classification matrix (Training and Testing Data from original imputed data 10-28) - Analyis Sample Response: outcome Analysis sample;Number of trees: 200				
	Observed	Predicted Survived	Predicted Died	Row Total		
Number	Survived	1881	310	2191		
Column Percentage		92.52%	34.83%			
Row Percentage		85.85%	14.15%			
Total Percentage		64.35%	10.61%	74.96%		
Number	Died	152	580	732		
Column Percentage		7.48%	65.17%			
Row Percentage		20.77%	79.23%			
Total Percentage		5.20%	19.84%	25.04%		
Count	All Groups	2033	890	2923		
Total Percent		69.55%	30.45%			

TABLE 1.2 Classification matrix for the testing sample.

		Classification matrix (Training and Testing Data from original imputed data 10-28) (Test sample) Response: outcome Test set sample;Number of trees: 200				
	Observed	Predicted Survived	Predicted Died	Row Total		
Number	Survived	3722	645	4367		
Column Percentage		92.40%	36.67%			
Row Percentage		85.23%	14.77%			
Total Percentage		64.32%	11.15%	75.46%		
Number	Died	306	1114	1420		
Column Percentage		7.60%	63.33%			
Row Percentage		21.55%	78.45%			
Total Percentage		5.29%	19.25%	24.54%		
Count	All Groups	4028	1759	5787		
Total Percent		69.60%	30.40%			

result would have been an astronomically high number of deaths in the country/world. With that inaccurate death rate, shutting down the country would have been a good decision no doubt. As time went on, however, it appeared that many people had been infected but did not realize it because for them there were no symptoms. If all those people had been known by the epidemiologists, and reported, the denominator (total number of ill) would have then been much larger, and the virus might not have been any more deadly than the normal flu (deaths attributed to COVID/number of those ill). Unfortunately, attributed deaths were not accurately determined, but by the end of the pandemic, basically in 2021, the number was more accurately determined. Herd immunity was eventually on its way to being established, thanks to vaccinations. In assuming the worst, unfortunately, the country was essentially locked down (some states worse than others) and as a result many thousands of small businesses were ruined, jobs were lost, personal depression increased, children were denied in-person schooling, vital medical appointments were put off, and many deaths, including suicides, child abuse, and so on, may have resulted from the lockdown itself. So inaccurate data has consequences. If we learned anything, it seems that obtaining accurate data is paramount.

What we learned over 2020. The virus appeared to be two disease states—early stage and later stage. Treatment for the period differed over the course of the disease. Hindsight is always better it seems. Perhaps we have learned from our pandemic experience lessons to apply in our future. PAs would have been useful in identifying those most likely in need of what services.

Especially for the second edition

What we would still like for this book (first and second editions) to accomplish are valid from the first edition but are extended in the second:

1. We provide a basic outline of medical research methods and PAs in Chapter 1.
2. We review where we have come from and the development of statistical analysis in Chapter 2.
3. We update the understandings of medical bioinformatics in Chapter 3.
4. And then extend data processing models in medical bioinformatics by standardizing them in Chapter 4.
5. Once we can access the data! (Chapter 5).
6. We stimulate research and analysis methods that will help people get well, as in Chapters 6 and 7, and in those chapters we introduce and emphasize the concept of N of 1 studies.
7. We want to provide tools for evaluating innovations in genomics and technology, as in Chapters 3 and 10.
8. We stimulate imagination from newer ideas such as in Chapter 9 in which the process of combining publicly available data sets, or other disparate data sets from organizations, or various "omics," such as metabolomics, allows one to generate deeper understandings of complex levels.
9. We summarize the regulatory environment for research and focus on developing standards of care across the board as in Chapter 8.

    Further in all the chapters:

1. We hope to help physicians and medical researchers learn to apply PAs.
2. We hope to promote PAs because these approaches are so inherently interesting. Searching for patterns and making predictions are fascinating in and of themselves.
3. We hope to promote clustering and other methods of exploratory research, which help to determine decision rules for best practices as applied to populations versus individuals.

Chapter conclusion

In conclusion, the reasons for writing this book can be classified according to:

1. Finding solutions, and making well researched decisions
2. Promoting exploratory research for the future
3. Promoting predictive research
4. Relating medical research to governmental regulations and accreditation guidelines.

    This book will demonstrate *practical solutions drawn from predictive analytics methodologies*, which allow practitioners at all levels to:

1. Make timely decisions concerning patient-centered medical delivery.
2. Build accurate models to optimize processes in health care administration, health care delivery, and as patients wind their ways though the unknown as they endeavor to take care of themselves in the first place.

    To accomplish those lofty goals, we have focused this second edition more on the *how* and less on the surrounding literature milieu, more on *practical applications*, research ideas, and outlines for analysis using various platforms. Then, we hope, that you, the reader, will find resources for making your own applications and accomplish the goals of this book for yourself.

Postscript

Most scientific knowledge is gained by studying examples. These examples are couched primarily in the form of hypothesis generation and testing studies. This book will help you to formulate hypotheses properly and choose the right techniques and platforms to analyze them. The old adage of "garbage in...garbage out" certainly applies to medical analytics studies focused on the wrong data analyzed in the wrong ways. This book provides many examples of how wrong data and wrong analytics methods can prevent generating meaningful results, and some examples of appropriate methods employed in a number of case studies in Chapters 11−17. We hope that this book will guide you to select the right data and the right analytical methods to generate results that will be useful to guide the progress of personalized medicine during the rest of the 21st century.

References

Chandnani, H., Goldstein, M., Khichi, M., Miner, L., Tinsley, C., 2018a. Prediction model for mortality in patients with disseminated intravascular coagulopathy based on pediatric ICU admission: three factors identified. In: Proceedings of the Ninth World Congress on Pediatric Intensive and Critical Care, Singapore. PICC8-0680. Oral Presentation.

Chandnani, H., Goldstein, M., Tinsley, C., Khichi, M., Miner, L., 2018b. Developing a prediction model for mortality in patients with disseminated intravascular coagulopathy based on pediatric ICU admission. In: Pediatric Academic Societies Annual Meeting, Toronto, Canada. PAS2018: Poster Presentation.

GitHub, N.Y., Times/Covid-19-data. <https://github.com/nytimes/covid-19-data>.

Halamka, J., February 8, 2022. N of 1 studies can make patient care more personalized. Mayo Clinic Platform, N of 1 studies can make patient care more personalized—Mayo Clinic Platform.

KNIME Download <https://www.knime.com/downloads>.

Marcus, A., Oransky, I, January 2, 2020. Nobel winner retracts paper from Science. Retraction Watch <https://retractionwatch.com/2020/01/02/nobel-winner-retracts-paper-from-science/#more-118672>.

Mehra, M.R., Desai, S.S, Kuy, S., Henrey, T.D., Patel, A.N. June 4, 2020a. Retraction: Cardiovascular disease, drug therapy, and mortality in covid-19. N. Engl. J. Med. DOI: 10.1056/NEJMoa2007621. <https://www.nejm.org/doi/full/10.1056/NEJMc2021225>.

Mehra, M.R., Desai, S.S, Ruschitzka, F., Patel, A.N. May 22, 2020b. Hydroxychloroquine or chloroquine with or without a macrolide for treatment of COVID-19: a multinational registry analysis. Lancet. 10.1016/S0140-6736(20)31180-6 <https://www.thelancet.com/journals/lancet/article/PIIS0140-6736(20)31324-6/fulltext>.

Chapter 2

History of predictive analytics in medicine and healthcare

Robert Nisbet

Chapter outline

Prelude

The underlying purpose of predictive analytics in medicine is to predict and direct decision-making in diagnosis and treatment. The central element in this decision-making process is medical information. The original medium of this information was the mind of the physician. But quite early in history, some of this medical head knowledge was committed to writing. The purpose of this chapter is to trace the history of written forms of information storage, to serve as a foundation for the present conversion of it to digital forms.

Outline

1. Five examples in history of medicine, in which kings and emperors mandated the collection of medical information in written form to serve diagnosis and treatment of common people:
   a. Many Pharaohs of Egypt: written in hieroglyphics on internal pyramid walls;
   b. King Adad-apla-iddina of the Middle Babylonian period (1068–1047 BCE) decreed that existing medical records be collected to form a corpus of 40 tablets, referred to as the *Diagnostic Handbook*;

Practical Data Analytics for Innovation in Medicine. DOI: https://doi.org/10.1016/B978-0-323-95274-3.00011-7

    c. King Solon of Greece: Commissioned Hippocrates to create the Hippocratic Corpus;

    d. Roman Emperor Commodus: Commissioned the physician Galen to expand and preserve the Hippocratic Corpus;

    e. King Chosroes of Persia supported the translation of the Hippocratic Corpus into Arabic, where it stayed for almost 1000 years.

2. The Arabian works were translated into Latin during the Reformation, beginning with Erasmus in 1525, and remained as the standard in medical treatment until the rise of modern medicine at the beginning of the 20th century.

3. Medical Case books as a basis for case-related medical records.

4. Mandate by Obama and the Affordable Care Act (ACA) of 2009 to convert written medical records (on physician's bookshelves) to digital format by 2014.

Introduction

The goal of this chapter is to provide a foundation in the history of medical practice for the development of decision tools that guide the determination of correct diagnosis and treatment of human ailments. In general terms, these tools can be classified as:

1. *Bodies of knowledge* that govern the nature of medical diagnosis and treatment; and
2. *Analytical Approaches and Decision systems* that integrate diverse knowledge elements and direct the formation of accepted medical practices

    These two tool types are related, in that the decision systems are based on existing bodies of knowledge available to the physician. In Part I, we will discuss the various types of bodies of knowledge in the history of medicine, and see how they affected the quality and appropriateness of medical treatment. In Part II, we will discuss various ways that information resident in the bodies of knowledge has been combined, stored, analyzed, and used to express predictions of diagnosis and treatment. Prior to the invention of the computer, analysis and decision-making processes were performed entirely in the most powerful nonlinear processing engine in the universe, the human brain—the brain of the physician. In Part II, we will discuss how computers enable us to build and process complex mathematical predictive algorithms, and deploy them in medical decision systems. We will focus on computers and medical databases, and the development of best-practices documents among specialties in Personalized Medicine.

    The reason for considering the development of bodies of written medical information in this digital age is to highlight the various types of bodies of medical knowledge, show how they have been used in the past, and to provide guidance for using future medical information in digital form. Bodies of written knowledge are of two types: (1) subject-related documents; and (2) case-related documents. Both of these sources of medical knowledge serve as major components of the knowledge infrastructure of modern medical practice. Proper discussion of medical decision systems today (and in the future) must include an understanding of all of the components of the system of medical practice on which they are built. In addition, discussion of the history of the development of the current medical practice system can yield valuable insights to help us meet the technical and political challenges of today. One such insight provided by the study of medical history is the relationship between past royal decrees and the collection of medical documents together into comprehensive bodies of information for the sake of diagnosis and treatment of all people in a society. This insight is particularly relevant to the challenge by current legislation that requires the digitizing of physician medical case records to form an electronic health record (EHR).

    The availability of EHRs and digital medical knowledge databases will provide the foundation required by analytical medical decision and management systems of today and in the future. The overall goal of this book (described more fully in the Introduction to this book) is to show how to use advanced analytics to build and use a medical decision system that draws upon all available digital sources of information to increase the effectiveness of medical diagnosis and treatment.

Part I. Development of bodies of medical knowledge

The process of determining the appropriate treatment is conditioned by the training of the physician in medical school plus his evaluation of the aspects of the case in relation to:

1. personal experience;
2. personal judgment; and
3. written and/or digital records of medical and healthcare information from previous cases

Personal experience and personal judgment have always been important factors affecting the success of medical diagnosis and treatment; they must be kept central to the practice of medicine. The third aspect, development and use of records of medical and healthcare information, forms the foundation of responsible medical practice to all members of a society. The various methods and tools used to integrate these three aspects of medical practice characterize the entire history of medical practice.

The recognition of the need to collect and maintain written medical records extends back to the dawn of recorded history. We will confine our discussion to cultures in the West and Middle-East. The practice of Medicine in Far Eastern cultures developed almost independently from that in the West and Middle East, until medical practices and technology were imported from the West in the 20th century.

From the very beginning, medical diagnosis and treatment have been essentially prediction problems. Individual treatment specialists in primitive cultures (e.g., medicine men/women) learned to perform their responsibilities from experience and word-of-mouth. There was very little cooperation among them, and sometimes there was animosity. The results of this medical practice environment generated a wide variation in the nature, quality, and availability of techniques in medical practice, and difficulty in the training of new specialists. Five examples are given to show how rulers in ancient cultures have managed these problems with medical care in the past. These examples provide insights into how societal forces generated royal responses to the problems in medical care, how the responses were expressed in specific cultures, and how they can guide us in our response to the challenges in medicine and healthcare that face in our society today. Our approach to this daunting task will include discussion of the following topics:

1. earliest medical records in Ancient Cultures;
2. classification of Medical Practices in Ancient and Modern Cultures;
3. medical practice documents in major ancient world cultures;
4. summary of Royal decrees of medical documentation in Ancient Cultures;
5. effect of the Middle Ages on medical documentation; and
6. rebirth of interest in medical documentation during the Renaissance.

Earliest medical records in ancient cultures

The oldest official medical documents

There is some controversy about the identification of the earliest manuscript of medical treatments. Many scholars point to the Code of Hammurabi (~1700 BCE), containing accounts of various surgery procedures, recommended fees for service, and penalties for malpractice (Sigerist, 1951). But there are earlier Sumerian cuneiform texts that date to the reign of King Ur III (c. 2100 BCE) in the early Babylonian Empire, which contain instructions for medical treatments, without the diagnostic information for use by experts. There are earlier records of spells and incantations among the Pyramid Texts, caused to be written by King (Pharaoh) Unas on the walls of his pyramid, who was the last king of the 5th dynasty in Egypt (~2300 BCE). The text includes many religious spells and incantations aimed at assuring the well-being of the Pharaoh in the afterlife. Some evidence exists (if it is valid) that a very early Pyramid text (the Hearst Medical Papyrus dated about 2500–2000 BCE) contained some references to medical treatment (Hinrichs, 1905). This text mentions the Egyptian god Thoth, who gave the Egyptian physicians healing arts. Herodotus (484–425 BCE) collected of many documents of Egyptian medical practice, which he called "Hermetic books," because he identified the Egyptian god Thoth with the Greek god Hermes (Dawson, 2010). Clemens Alexandrinus (CE 215) claimed that he collected 42 such Hermetic books on anatomy, illnesses, eye diseases, gynecology, drugs, and surgical instruments. If any of these Hermetic books existed, none survive (Sigerist, 1951).

Classification of medical practice among ancient and modern cultures

Development of medical practice appears to follow a similar pattern in all cultures studied (Sigerist, 1951). The earliest phase of medical practice in a given culture presumes supernatural causes and prescribes spiritual treatments for ailments and diseases (Supernaturalistic Medicine). The second phase abandons (at least partially) the Supernatural Medicine approach, and accepts the Naturalistic Medicine approach, in which causes for ailments and diseases are sought among causes that can be studied in the natural world. The final phase is Scientific Medicine, based on

Supernaturalistic Medicine ➡ Naturalistic Medicine ➡ Scientific Medicine

FIGURE 2.1 The normal course of development of medical treatments in many cultures.

accumulated medical informational documents, and results of scientific tests. Most cultures follow the pattern shown in Fig. 2.1.

Even though this development trend shown in Fig. 2.1 describes the long-term trend in a culture, it is often the case that these approaches may be practiced at the same time by different segments of the culture (Sigerist, 1951). This was the case for Supernaturalistic Medicine and Naturalistic Medicine in Egypt and Mesopotamia. We find elements of Naturalistic Medicine along with Scientific Medicine in practice in many cultures today, even in the United States in the form of homeopathic medicine and conventional medicine.

Medical practice documents in major world cultures of Europe and the Middle East

Medicine of some sort has been practiced in every ancient culture. In this chapter, we will confine the discussion to the developments of medical practice documents in Europe and the Middle East:

1. Egypt
2. Mesopotamia
3. Greece
4. Rome
5. Arabia

Some naturalistic medical texts were written in China c. CE 600 (Unschuld, 1985), and in India even earlier than that (Sigerist, 1951), but these texts don't appear to represent ideas and relationships to predictive analytics that are not reflected in documents from countries of Europe and the Middle East

Egypt

Many pyramids in Egypt contain hieroglyphic scripts with medical content, the earliest of which appears on tombs of the 5th Dynasty, in about 2300 BCE (Faulkner, 1969). These scripts contain many incantations and spells designed to help the Pharaoh in his afterlife. Concepts of detailed examination, diagnosis, and prognosis (medical treatment) may have arisen in very early Egypt. The Edwin Smith Surgical Papyrus (c. 1700 BCE; see Fig. 2.2) contains very detailed descriptions of 48 medical cases involving head and body wounds, sprains, and tumors. Serving as the founder of Egyptology, Breasted (1967) claimed that the Edwin Smith document is composed of a copy of an earlier document dating back to 3000–2500 BCE in the form of 69 explanatory notes on the original document. If this document existed, it would be the earliest medical record of surgical information.

The Ebers Papyrus of 1550 BCE. Is the largest among the most ancient Egyptian medical documents known (Bryan, 1930). It contains over 700 spells and incantations for turning away evil disease-causing demons. The Hearst Medical Papyrus is thought to date back to 2000 BCE or beyond (Reisner, 1905), but some controversy exists about the date.

According to Sigerist (1951), other known Egyptian medical papyri included:

1. Kahum Papyrus (1900 BCE)—On Gynacology and pregnancy
2. London Papryus (1350 BCE)—On recipes and incantations
3. Berlin Papyrus (1350 BCE)—On medical tests, pregnancy, and fertility
4. Chester Beaty Papyrus (1250 BCE)—Drug recipes and diseases

Herodotus identified the Egyptian god Thoth (mentioned in the Hearst Papyrus) with the Greek god Hermes, the god of healing. We know of many of these works through the writings of Herodotus and Clement of Alexandria. They referred to them as the Corpus Hermiticum, or the Hermetic books. Six of the 42 Hermetic books were on medical subjects of physiology, male and female diseases, anatomy, drugs, and instruments (Zeller, 1886).

Thus it appears that naturalistic and supernaturalistic medicine were practiced side by side in ancient Egypt, as in Mesopotamia and Greece (see below).

FIGURE 2.2 Edwin Smith papyrus. *http://designblog.nzeldes.com/2011/03/hats-off-to-ancient-egyptian-medicine/.*

Mesopotamia

The earliest medical text in Mesopotamia (associated with King Ur III—c. 2100 BCE) contained instructions for treatments of patients without diagnostic information, intended for use by experts only (Oppenheim, 1962). It is unclear whether the Babylonians were the first to document the concepts of detailed examination, diagnosis, and prognosis (medical treatment), or that they just imported them from the Egyptians. In either case, enough medical knowledge was accumulated by the Middle Babylonian period (1532–1000 BCE) to support separate texts for diagnosis and treatment.

During his reign at the end of this period, King Adad-apla-iddina (1068–1047 BCE) decreed that existing medical records be collected to form a corpus of 40 tablets, referred to as the *Diagnostic Handbook*. Some of these tablets in the handbook listed treatments based on the number of days the patient had been sick (Tablet 15 and 16) and the pulse rate (Tablet 21). This and other similar numerical information used to compose diagnoses may represent the first bioinformatics "models" in medicine. We can see this same integration of medicine and numbers in the *Sakikku*, a large medical treatise of about 40 tablets divided into five parts (Neugebaur, 1957). Part II enumerates symptoms of diseases according to color, temperature, and movements of body parts. Part III describes treatments that were prescribed as a function of numerical information from observations described in Part II, combined with specific information about the disease course, phase, amelioration/aggregation relative to the time of day.

Sumerians used a symbol for "not" in the place of zero, as a placeholder in their sexagesimal number system based on 60, rather than 10 (Kaplan, 2000). For example, the number 2013 was represented by a string of symbols for 2, followed by the symbols for "not," 1 and 3. They did not understand the concept of zero (that came later from the Moors in 11th century Europe), but they could do geometry with this system. There is evidence to suggest that they understood the geometry of the Pythagorean Theorem 1000 years before Pythagoras (Oppenheim, 1962; Kaplan, 2000). Evidence of this rather sophisticated number system combined with their detailed medical knowledge suggests that diagnoses based on the *sakikku* were quantitative. To that extent, one of the roots of predictive analytics in medicine extends back to ancient Babylonia.

A vast library of over 30,000 clay tablets were found in Nineveh, collected into a library by Asherbanipal, successor to King Sennacherib (Polastron, 2007). Over 700 of these tables contained medical information. The tablets were distributed in many rooms according to a classification methodology. This library was the not the first of its kind, but it is the first known library having most of the attributes of a modern library (Johnson, 1970). This collection was a great interest of Asherbanipal; he was not above conquest to obtain new additions to it. To this extent, we can view that his collection arose from a royal mandate, and it serves as a second example of the collection of medical texts by royal decree.

The Babylonian medical tradition continued in practice until about 200 BCE, after which it dropped out of awareness of medical practitioners. This decline of the Babylonian medical tradition in Asia Minor was contemporary with the rise of the Hippocratic Corpus in ancient Greece. Maybe these two events were related.

Greece

Medicine in Preclassical Greece

The cultural mindset and worldview in place can greatly affect the course of development of concepts and practices in a culture, sometimes in multiple directions. We can see two good examples of this human response phenomenon in the development of medical practice in Greek history (and another one in Rome—see below). Beginning in about 600 BCE, medicine became organized around two centers of the same supernaturalistic worldview: (1) the temple cult worship of the god-man, Aesculapius (a man who was "elevated" to the status of a god); and (2) the philosophy of the man Pythagoras.

1. Asclepius was a demi-god in Greek mythology, son of the god Apollo and a mortal mother, Coronis. He became the god of medicine, and his followers held that aliments and diseases could be healed by prayers and sacrifices, particularly in the temples erected for that purpose. Homer included a man named Aslepias in his story of the Iliad, as a physician to wounded men at the battle of Troy. But in later years, he became "elevated" to the status of a god.
2. Pythagoras strived to create a balance among opposing forces acting on people. He quantified these forces with numbers, and analyzed these numbers (in terms of arithmetic) to guide medical treatments in terms of numbers. One of his treatments was to prescribe harmonic frequencies and music to treat human conditions. He invented instruments and with the use of sound and vibration he was able to bring an individual's attention to the awareness of their Divine Nature in order to facilitate the healing process. Pythagorus was a "Renaissance man," in that his ideas ruled Philosophy, Mathematics, Music, as well as Medicine throughout the "Golden Age" of Greece (~550–300 BCE).

Both of these forms of supernaturalistic medicine were practiced contemporaneously in Greece beginning about 600 BCE and extending to about 450 BCE, when a new approach was introduced by Empedocles via his philosophy (Sigerist, 1951). The previous belief system based on the philosophy of Parmenides held that the senses are not reliable, and that medical diagnosis and treatment should be found in supernaturalistic practices. Empedocles disagreed by stating that our senses are indeed a reliable guide to truth. He introduced the concept that all things were composed of four elements, Earth, Air, Fire, and Water. This belief was the first example of "atomistic" thinking in the Greek culture. This burgeoning culture permitted such philosophical disagreement, and the stage was set for Hippocrates (460–370 BCE), who formulated the principle of the "true mixture of elements." These elements (he called "humors") were black bile, yellow bile, phlegm, and blood. They were different from the four elements of Empedocles, but they followed the same philosophy of atomism (Empedocles) and balance (Pythagorus). Hippocrates believed that ailments and disease were caused by the wrong balance of the four humors, therefore, he sought medical treatments that restored the balance of humors to form the "true" mixture for human health. Naturalistic medicine had arrived in ancient Greece.

The result of this philosophical disagreement (and the medical practices that it induced) relegated supernaturalistic medical practice to the Asclepian temple, and pushed Hippocratic practices to outside the temple (Sigerist, 1951). Thus in Classical Greece, supernaturalistic medicine was practiced almost side by side with naturalistic medicine, following two very different worldviews. We see reflections of this same tension today between Homeopathic (or Wholistic) medicine practiced in a worldview dominated by modern scientific medicine, which developed from the naturalistic medicine on which Homeopathic medicine is based. This similarity between Ancient Greece and modern America of simultaneous practices arising from different worldviews supports Solomon's claim that "…there is nothing new under the sun" (Eccl. 1:9).

At this point, you may be wondering how all of this relates to the history of Predictive Analytics. The reason is related to the biggest problem with early Greek medicine, even in its naturalistic state—its fragmentation, the lack of standardization, and the "cure" of both. Many early Greek medical texts included much biological and medical information and described treatments for ailments and diseases, but they were scattered throughout the country in many different locations under the control of different physicians. Similarly, most medical treatment information today is scattered in the offices of individual physicians. The solution to the problem in ancient Greece (see below) followed that in Mesopotamia, and it preshadows the solution to our modern problem mandated by the ACA in 2009. This solution involved the collection and standardization of medical information, which became the basis for responsible medical practice in Classical Greece; it may become so soon in America in digital form. If it is true that those who ignore the mistakes of the past are bound to repeat them, then it is logical to expect that some solutions to present problems can be found in history. This is true of the history of Mesopotamia, and Greece. It came about in Ancient Greece in this manner.

There was a serious problem in Athens, one of the two dominant city-states in the 5th century BCE (along with Sparta). The current form of democracy generated serious inequities among the people, resulting in an environment of strong social unrest. They elected Solon as King, a statesman and poet, to restructure the early democracy of Athens to reduce these inequities, and bring a measure of civil peace to the city-states. At that time, repressive laws forced the poor into slavery to the rich. Solon designed an economic program was called the *seisachteia* or the "*shaking off of burdens*" to release the lower classes from the burden of debt to those in the wealthy classes (Hammond, 1961). By canceling and reducing debts and abolishing a system of mortgage which had turned many poor landowners into virtual slaves, Solon significantly reduced the huge social and economic gap between the rich and the poor, the source of the social unrest. His concern to elevate the miserable state of the poor led Solon to decree that all medical texts be gathered together into one corpus, making them available to everyone, including the poor. This job was commissioned to the Physician Hippocrates (Fig. 2.3).

Hippocrates and classical Greece

Hippocrates organized existing medical texts (and wrote some of them) in an attempt to integrate the previous philosophical concepts of Empedocles—the four elements, Philistion—the body is separate from the soul, and must be treated as such, and Diogenes—the soul, or the *pneuma*, was the vehicle of life (Wellmann, 1901). This group of documents included detailed discussions of brain, lungs, heart, liver, and blood, together with recommended treatments that became known as the *Hippocratic Corpus* (about 400 BCE). The corpus was composed of about 60 documents, and it represents the first widely distributed and integrated repository of medical information in the world that contained guides to diagnoses and treatments of ailments and diseases. A previous example of this sort of repository in the form of the Babylonian *Sakikku* documents were written on clay tablets and stored in a royal library, which was relatively inaccessible to common people.

Specific volumes of the Hippocratic corpus were devoted to the four humors: black bile, yellow bile, phlegm, and blood. In this regard, he followed the general approach of Empedocles, explaining medical illnesses as an imbalance between the four basic elements, but the elements in medicine were the four humors. Other volumes covered information and treatment for fractures, head wounds, gynecology, epidemics, obstetrics, ophthalmology, the heart, veins, and bones. Hippocrates used tools also. He mentions using a rectal speculum to observe a rectal fistula (Volume iii, pg. 331), shown in Fig. 2.4.

FIGURE 2.3 Roman coin of the 1st century CE from the island of Cos (birthplace of Hippocrates) showing his bust as of about 377 BCE. *British Museum, Coins and Medals catalogue number: GC18p216.216. (See why this Roman coin is important in the discussion of Roman medicine below.)*

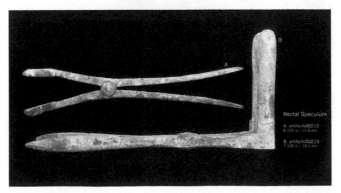

FIGURE 2.4 Rectal speculua of the type used by Hippocrates. *http://www.hsl.virginia.edu/historical/artifacts/roman_surgical/.*

Particularly relevant to the history of predictive analytics is that this repository of knowledge and practices arose largely due to concern for the poor in Greece (as it did in Mesopotamia). We see a reflection of this concern today in the current controversy over access to health information and healthcare in America. Evidently, our concern today is not purely a modern phenomenon—its roots extend back to ancient Greece and even to the cradle of civilization in Mesopotamia.

Ancient Rome

By winning the Second Punic War in 201 BCE with the victory over Hannibal of Carthage, Rome was transformed from a relatively loose confederacy into a permanent, expansionary war machine (Muhlberger, 1998). http://www.nipis-singu.ca/department/history/muhlberger/2055/l33anc.htm.

One result of this unification was a push to declare war on Macedon. Why do that? Rome was very busy with con-solidation of its new territory in Gaul (Spain and France), and rebuilding Italy after devastation of the war with Hannibal. Two reasons were offered by Eckstein (1987): (1) Rome was just coming to the aid of one of its allies being attacked by King Philip V of Macedon (even though it appears that Rome's emissaries engineered it); and (2) Rome had become a conscious imperialistic power in the Mediterranean world, and they lusted for conquest of Greece. Muhlberger (1998) claims that these two theories tend to divide those who like the Romans from those who don't.

For whatever reason, Greece was conquered in a short space of about 60 years, and the existing social structure of Greece crumbled, and became largely Roman. Along with this transformation, much of Greek medical literature and art was pushed aside and ignored. This purge included the Hippocratic Corpus, because early Romans believed that divina-tion was the way to cure illnesses. In other words, they were stuck on supernaturalistic medicine. Then, along came Claudius Galen (Fig. 2.5).

Library of Congress

FIGURE 2.5 Bust of Galen. *US Library of Congress.*

Galen

Claudius Galen (CE 131–201) was a Greek physician from Pergamum, who went to Rome, then studied at the famous medical school in Alexandria, Egypt. He was trained in the medicine of Hippocrates, and he revived interest in the Hippocratic Corpus (Sigerist, 1951). He stressed clinical observation of patients, during which he examined patients very closely to make his diagnosis and prescribe what he thought was the appropriate treatment. In his practice, he accepted the Hippocratic view that disease was the result of an imbalance between the four humors, and developed treatments aimed at restoring the "normal" balance among the four humors. Galen adopted many of the medical instruments described by Hippocrates. Fig. 2.6 shows some bone levers that may have been used to lever fractured bone into place.

Jealousy by his rivals, and prejudice against Greeks among the Romans, caused him to flee Rome in CE 166, but he returned at the request of the Emperor Commodus. Because he served as the personal physician to Commodus and later to Septimus Severus, he could work freely. The tide of imperial opinion turned so strongly back to Hippocrates and his work, that a coin was minted with his portrait (Fig. 2.3); on the reverse side is shown the serpent entwined staff of Asclepius, which many centuries later was incorporated into the symbol of American Medical Association and in the logo of the World Health Organization. Mintage of this coin demonstrates the change in attitude of the Romans toward Greek Medicine, fueled by his success under the royal aegis. We can view this revival of interest in the Hippocratic Corpus, as the fourth example of a royal mandate for the collection and dissemination of medical documents for the benefit of the people.

Galen continued to add to the Hippocratic Corpus by writing many books and treatises himself. He was affected greatly by prominent Greek philosophers, including Aristotle and Plato. He took aspects from each Greek school of thought and combined them with his original thinking. In this manner, Galen viewed medical practice as interdisciplinary. This attitude is seen clearly delineated in his book, "The Best Physician is also a Philosopher." In this regard, Galen presaged the approach of modern scientists and thinkers, in which important ideas of many schools of thought are combined to help explain complex systems in the world (see Nisbet et al., 2009). The books that he collected and wrote were still being used in the Middle Ages and for many medical students, they were the primary source of information on medicine, particularly in the Arab world (Swain, 1996).

The Roman poet Horace quipped that *"Graecia capta ferum victorem cepit et artis intulit agresti Latio"* ("Greece, the captive, took captive her savage conqueror and brought her arts into rustic Rome"—Horace, c. 14 BCE, Epistles 2.1.156–157). What did Horace mean by that statement? The most common interpretation is that even the Romans realized that while Rome vanquished Greece militarily, Greece captivated the great interest of "rustic" Romans in the arts and sciences, and in the process "civilized" them. It might be said that Greece was the cradle of Western civilization, and that Rome was simply the vehicle that brought it into all of Europe over the roads built by the Romans. One of the most influential men in that process was Galen the Physician. It may not be an overstatement to say that Galen bought Rome (and hence all of Europe) out of supernaturalism and into naturalistic medicine almost single-handedly following the work of Hippocrates, and the direct effect of his work on the practice of medicine persisted until the 20th century CE. The spirit of Hippocrates lives on in the Hippocratic Oath taken by every physician before entering into practice in America.

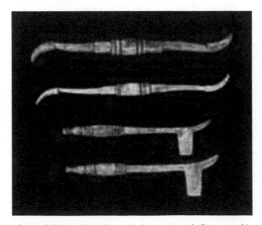

FIGURE 2.6 Various bone levers in use at the time of Galen. *http://www.hsl.virginia.edu/historical/artifacts/roman_surgical/.*

Arabia

The development of Arabian medicine is closely related to the history of Islam. In CE 622, Mohammed united the warring tribes of Arabia through a common religious and social system (Shanks and Al-Kalai, 1984). Medicine of the early Islamic and Umayyad period (CE 661–750) was largely supernaturalistic, which included three principle treatments: (1) administration of honey; (2) bloodletting through collection in a cup; and (3) cautery (sealing blood vessels with fire).

A rather curious sequence of events happened in the Eastern Orthodox Christian church, which brought Greek and Roman medicine into the Arab world, and maintained it until the 16th century.

1. As Bishop of Constantinople, Nestorius denied that Mary was the Mother of God, and was excommunicated at the Council at Ephesus in 431. Nestorius died shortly thereafter. But his followers fled east and founded the Nestorian Church, several medical schools, and a Nestorian center at Nisibis in Arabia.
2. The Persians at Nisibis warmly embraced the schools and the Nestorian center, and King Chosroes founded the university at Jundi Shapur, which combined Indian philosophy with the Greek medicine brought by the Nestorians in their school. This action can be considered as another example of collection of medical documents by royal decree, but the purpose was for education, not the provision of medical care to the poor.
3. With the defeat of Heraclitus, Emperor of the Eastern Roman Empire, the Arabs expanded their empire under the Umayyad Caliphate. This became an age of reawakening of the Greek arts and medicine, and ushered in a second golden age of Greek culture.

Even though (maybe because) Nestorius was branded as a heretic by the Eastern Orthodox Church, Nestorians were accepted in the Arab world as a Christian sect. Several other Christian men and families were also instrumental in the vending of Greek medical arts to the Arab world. Eight generations of the Christian Syriac-speaking Bakhtishu' family served as court physicians to the caliphs in Baghdad from about CE 770 to about CE 1050 (Savage-Smith, 1994). These men were instrumental in translating Greek medical texts into Arabic. During this period, the Baghdad House of Wisdom (*Bayt al-Hikmah*) was founded to encourage the collection (in the form of a library) and translation of Greek works into Arabic. The most prodigious scholar in this library was Hunayn ibn Ishaq al-'Ibadi, another Syriac-speaking Christian, who translated almost all of the Greek medical works, half of Aristotle's works, and even the Jewish Septuagint. Arabs of the Golden Age of Arab culture were not at all adverse to Jewish and Christian literature existing in their midst, and they appear to have even promoted translations of it. Ishaq also included 95 Syriac and 34 Arabic versions of Galen's works. Fig. 2.7 shows two pages of Ishaq's Arabic translation of Galen's introductory treatise on the skeletal system. Note that this book and other books up to the invention of the printing press were handwritten. Today, it is difficult for us to imagine the monumental impact of the printing press on the dissemination of knowledge throughout the world.

FIGURE 2.7   A very rare copy of Hunayn ibn Ishaq's Arabic translation of Galen's introductory treatise on the skeletal system, On *Bones for Beginners*, cited in Latin as *De ossibus ad tirones* at the beginning of the treatise. *NLM Ms P26, opening at fols. 62b-63a Courtesy of the National Library of Medicine.*

The importance of Arabian medicine lies not in its originality (which it isn't), but in that it was the vehicle that faithfully preserved the knowledge and arts of the Greeks by translating them into Arabic through Syriac versions. A major consequence of these translations is the preservation of the content of the Greek manuscripts. During the Renaissance, European scholars had no access to the original Greek texts, and so they translated the Arabic sources into Latin, the European scholarly language of the day. The availability of the Hippocratic Corpus and other Greek medical documents in the Latin language made them readily available to western scholars. It is interesting to note that this process was facilitated in the Arab world largely by Christians, who vended the preserved Greek medical and scientific works to Christian Europe, thus paving the way for the development of modern science and medicine.

Summary of royal medical documentation in ancient cultures

Because of the tension among individual medical treatment specialists in ancient cultures, and the lack of uniformity among them, many ancient rulers sought to codify, standardize, and disseminate medical information in the form of written records to aid in medical diagnosis and treatment for all people in their societies. Many of these official documents exist today, and together they form the context in which modern medical records developed. These standardized documents were mandated by past rulers, because they would not have been developed apart from official mandates (cf. the previous discussion of Kings Adad-apla-iddina and Asherbanipal in Mesopotamia, King Solon in early Greece, Commodus Caesar in Rome, and Emperor Chosroes of Persia).

From this viewpoint, we can understand some of the force behind the laws of modern countries that enable the collection and standardization of medical records in one place (e.g., the National Health Service of the United Kingdom in 1948; and in the United States, the Medicare Act or 1965 and the ACA of 2009). In addition, this historical context might lead us to believe that further centralization and standardization of medical and healthcare information in the United States is inevitable. The shadow of this inevitability lays particularly heavy upon many physicians in the United States today, who must convert their written medical records operations to digital format during the next several years (a provision of the ACA), or lose increasingly large proportions of their Medicare reimbursements. Some medical specialties (e.g., Dermatology) require handwritten figures of the body to be included in physician's notes, documenting locations of skin problems. While these diagrams do not currently exist in electronic clinical data entry programs, they could be created by graphics routines that permit annotation of locations of past treatments. But, these graphics routines may be slow in the development of software programs to implement digital record-keeping.

Due to the cost of digitizing and the initial lack of required software features available (i.e., graphics), some physicians will choose to delay document conversion and incur the cost of reduced Medicare reimbursements. Others who can do it will "bite-the-bullet" and do it as expeditiously as possible. But all practicing physicians will do it, eventually, as costs decline and the richness of software features increases. This conclusion is formed simply by the recognition that the realities of human nature and our need for medical care will interplay with the political and pragmatic realities in today's society in ways similar to those that occurred in other societies in the past. Technology that controls how it is done changes over time; human nature and basic human needs do not.

One of the themes of this book is the need to maintain current levels of personalized medicine AND provide EHRs for use in diagnosis and treatment by all physicians. Only in this way, can we leverage the power of predictive analytics to guide diagnosis and treatment across the entire landscape of medical practice in the United States. Certainly, it will be traumatic to convert all physician written records to digital format, but it can be done, and indeed must be done. That is the modern expression of the pattern we can see in history of rulers mandating the collection and standardization of medical records at many times in the past.

Despite significant differences between the Democratic and Republican agendas on domestic affairs in US politics, it is clear that it is Democratic political force that is driving this conversion. It appears that the administrations of Lyndon B. Johnson (The Medicare Act) and Barack Obama (the ACA) are just following in the footsteps of many ancient rulers to standardize bodies of medical information and medical care and make benefits based on them available to all citizens.

Effects of the middle ages on medical documentation

Between the fall of Rome (in CE 476, according to Gibbon, 1906) and the Renaissance (beginning in the 14th century), the development of medical knowledge came to a halt, and it was kept alive primarily in Arabia. In Europe, much of the knowledge of the Greeks and Romans was lost. The reason appears to be that Europeans (and particularly the early Britons) despised Rome, and destroyed or covered up anything pertaining to Rome (even Greek culture and knowledge promoted by Rome after Commodus). This disparagement and destruction of all things Roman after the fall of Rome paralleled the destruction by early Romans of all things of Greece after its conquest. Only in the Arab world were Greek and Roman medical documents preserved (see above). This situation prevailed for almost 1000 years, up to the Renaissance.

Rebirth of Interest in medical documentation during the renaissance

The Renaissance was marked by two pivotal events that served to begin the breakdown of the stasis and the mindset controlled by the thinking of the Middle Ages: (1) the printing press; and (2) the Reformation (Eisenstein, 1991).

The printing press

The use of movable type was invented in China in 1048 (Taylor, 1986). But, the concept of movable type did not surface in Europe until Gutenberg combined that concept with the screw press (Fig. 2.8). This kind of press was used previously in wine and olive presses, but Gutenberg was the first to adapt that technology to printing.

Thomas Carlyle (1836) quipped in his novel, *Sartor Resartus*, that

He who first shortened the labor of copyists by device of movable types was disbanding hired armies, and cashiering most kings and senates, and creating a whole new democratic world: he had invented the art of printing.

The effect of growing prosperity in the 15th century promoted the rise of literacy. Combined with the spread of Renaissance thinking, increased literacy moved many people to want to read. The invention of the printing press by Gutenberg in 1440 permitted the dissemination of new ideas quickly and accurately in written form, and permitted this expanding body of readers the opportunity to learn and adopt new ideas of the Renaissance.

The Protestant Reformation

The breakdown of the absolute control of religious and philosophical thinking in Europe by Luther and Calvin led to the introduction of new ideas into society. Luther and Calvin returned the church to the primacy of the Bible. All doctrines of the Catholic Church were rejected unless sanctioned directly by the Bible, and the purity of the Bible was based on Arab translations of the Greek texts, retranslated back into Greek. This awakening of interest in the Greek language prompted a parallel response in medical science (Porter, 1999).

The combination of the printing press, the consequent rapid spread of Renaissance thinking, and new ideas fueled by the Reformation generated a social climate that encouraged an increasing body of scientific and medical inquiry.

FIGURE 2.8   Johannes Gutenberg and his printing press. *Michael Halbert, Inkart.com.*

Erasmus (1466–1536) drew upon both of these pivotal events to lead Europe back to the Greek medical works of Hippocrates and Galen.

Erasmus

Erasmus was a former monk, who left his monastery to lead European scholarship for more than three decades, and established Greek as the standard for literary and theological studies (Porter, 1999). He translated the Arabian versions to produce the first modern Greek edition of the Hippocratic corpus of books in 1525. The Reformation, dawning at the same time, spurred Erasmus to throw off the constraints of the church that taught men to avoid invading the "sanctity" of the human body in the study of anatomy. The huge success of the reintroduction of Galen's medical works fueled Renaissance natural philosophers to become more inquisitive about human bodies. Thus the stage was set by the availability of the printing press, and the loosened ties to the church, for Erasmus to turn inquisitive minds toward a renewed interest in human anatomy.

Human anatomy

The recovered Greek texts supported the idea that ancient medicine was the right approach, and that scholars were the rightful guardians and interpreters of it (Porter, 1999).

Andreas Vesalius (1514–1564)

Vesalius was a Flemish anatomist who dissected the bodies of executed prisoners (human vivisection was still taboo at that time). He wrote many books on anatomy, including "De humani corporis fabrica" ("On the fabric of the human body"; see Fig. 2.9). Despite his respect for Ancient Greek medicine, his detailed drawings helped to correct some misconceptions of Galen and the Greek physicians, who dissected only animal bodies (Porter, 1999).

FIGURE 2.9 From Andreas Vesalius *De humani corporis fabrica* (On the fabric of the human body). Basel, Switzerland, 1555 (2nd edition). *Image of the document in the Huntington Library Collections, San Marino, CA.*

Thanks to the printing press, and published information about discoveries made by people such as Vesalius, it was very difficult for the established Catholic Church to stop the spread of this "heretical" information, having been severely weakened by the Reformation. The *Fabrica* laid the foundation that guided studies of human anatomy based on direct observation, rather than analogies from animal dissection. This new approach, the practice of "anatomizing" became a common concept in medicine for several centuries (Porter, 1999).

William Harvey (1578–1657)

William Harvey was an English physician known for his complete description of the systematic circulation of blood by the heart. Blood circulation was known in the Arab world by the work of Ibn-al-Nafis (1213–1288) of Damascus, but his description was short and incomplete compared to that of Harvey (Porter, 1999).

Medical documentation after the enlightenment

After the Renaissance and the Reformation spurred new thinking by scholars about the world around us, many medical documents were written on almost every conceivable aspect of medical science. This body of documents became source of information and the basis for training of physicians in medical schools. But, the sheer volume of documents produced in the 18th and 19th century was difficult to absorb and incorporate into medical practice. The variety of cases presented to the physician defied rigid characterization in publication in medical documents. These two challenges to physicians led to the development of medical case documentation standards, beginning in the 18th century, and to the development of medical databases in the 20th century. A short history of case medical documentation is presented below. Discussion of medical databases will be presented in Part II.

Medical case documentation

Both Hippocrates and Galen recorded information on specific medical cases, but they were used only as examples of general principles and practices. These accounts of specific cases were recorded primarily for the purpose of strengthening the reputation and acceptance of specific physicians, rather than objectively reported for the sake of the case (Lloyd, 2006). Strangely enough, the earliest record of medical case documentation in the modern age comes from Astrology, in the writings of Simon Forman's casebooks (Cook, 2001). These writings were intended to record the 10,079 consultations between March 16, 1596 and September 5, 1603. These casebooks contained detailed information about clients, and the advice given to them based on astrological relationships. At least 90% of the questions were about health and disease.

The early casebooks functioned as an important step in the development of modern medical records for individuals. Unfortunately, the individual case histories became resident on the physician's bookshelves, and were seldom gathered together for use by others. The modern development of the EHR concept has one "foot" anchored in early casebooks, and the other one anchored in the physician's record library.

The development of the National Library of Medicine

As early as 1818, the Surgeon General of the Army began to collect books for use in his work (Miles, 1992). This collection grew and became known as the Library of the Surgeon General's Office. The library continued to expand until it contain over 17,000 volumes, when in 1872, Surgeon General of the Army, Billings opened it to the public as the National Medical Library, and became its Director in 1883 (Miles, 1992). The library was awarded a building of its own on the Capitol Mall in 1886. The original goal of the library under the Surgeon General of the Army was to collect every medical document formally published in America in the form of books and pamphlets; that goal has continued through to the modern form of the library as the National Library of Medicine.

After WWII the library modernized, particularly under the directorship of Frank Bradway Rogers, who was trained in library science. Library organization became more hierarchical and academic in its focus on classification and cataloging. In 1975, Rogers was awarded the honorary degree of Doctor of Science by the University of Toledo (Miles, 1992). Beginning in 1950, the library forged strong relationships with the American Medical Association (AMA) and received over 75,000 journal copies and publications from the AMA. This huge influx of documents exacerbated the problem of cataloging and indexing to produce the yearly *Index-Catalogue*. A simple listing of all publications (the *Current List of Medical Literature)* was to take its place. But, various supplements were published until 1961, when the

FIGURE 2.10 Slips of paper bearing citations for the Current List of Medical Literature being shingled by Hertha E. Bishop. *From Miles, W.D., 1992. A History of the National Library of Medicine: The Nation's Treasury of Medical Knowledge. U. S. Government Printing Office.*

final supplement to the *Index-Catalogue* contained 579,566 author titles, 538,509 book titles, and 2,566,066 article titles.

The *Current List* was produced by the "shingle" method developed in the Department of Agriculture. Indexers scanned articles, and transferred author and subject names to forms, which were passed to typists, who annotated each article with author names and subjects on separate slips of paper. These slips of paper were alphabetized and pasted onto a larger paper in an overlapping fashion as shingles on a house to form author and subject pages (Fig. 2.10). The pages were accumulated for a month, and composed into the *Current Index*.

The challenge was to find a way to reduce the massive work necessary to shingle each publication of the *Current List*. This approach was abandoned in favor of a new punch card method that could be sorted mechanically by machines. Use of punch cards using the new Holorith code led directly to the development of computers, as described below.

Part II. Analytical decision systems in medicine and healthcare

In the latter half of the 20th century, two themes developed in medicine that have yet to be integrated:

1. Computers and Medical Databases
2. Development of best-practices documents among specialties in Personalized Medicine

Computers and medical databases

Prior to the advent of the computer, storage of information was done in print format. Computers permitted the electronic expression of information, permitting storage of data over a period of time. This storage was accomplished initially by magnetic cores or drums. With the invention of the Winchester 30–30 hard drive in 1956, storage volumes and storage times increased significantly; they have increased exponentially ever since.

One of the earliest movements away from storage of information in text form was the invention of the punch card, in which textual information (including numbers) was coded into a series of holes in a paper card. The code was designed by Holorith in 1889 (Randell, 1982). The Director of the Army Surgeon General's Library, John S. Billings, recommended that the US Census Bureau should store the 1890 Census information with punch cards, using the Holorith coded, giving birth to the Holorith Company (Collen, 2012).

Thomas Watson took control of the Holorith Company in 1924, and renamed it International Business Machines (IBM), to focus on *Informatics* (coined by IBM), Information Science, and Data Communication Augarten (1984). The stage was set for the computer revolution, notwithstanding Thomas Watson Sr.'s reputed reluctance to go there. The ensuing development in computers at IBM was fueled by the efforts of his son, Thomas Watson, Jr.

The development of the Electronic Numeric Integrator and Calculator (ENIAC) was developed during World War II to help track ordinance trajectories (Rosen, 1969). Shortly after the war, this technology was adapted to peacetime uses in the form of UNIVAC to store the 1950 Census data. By this time, Thomas Watson, Jr. had become President of IBM, and it didn't take long for him to get into the act by developing the IBM 701 (1952) and the IBM 704 (1954) to support the Korean War effort (Blum, 1983). This effort led to the development of the Formula Translation language (FORTRAN), which became the primary computer programming language for scientific purposes until the 1990s.

Medical science and computer storage and retrieval capabilities came together with the development of the Massachusetts Utility Multi-Programming System (MUMPS) at the Massachusetts General Hospital in 1967. MUMPS become the most common programming language for medical computer applications (Barnett et al., 1981).

Early medical databases

One of the first medical databases was written in MUMPS at the Beth Israel Hospital in Boston (Bleich, 1969). This system used the hospital clinical database as a knowledge base, input data about a patient's acid–base balance, and recommended appropriate treatments. This system was particularly noteworthy, because it was built for the expressed purpose of determining appropriate treatments. By the 1980s medical knowledge bases were relatively common in specific institutions. One of particular mention was Infernet at the University of California, San Francisco. Infernet was composed of a cardiac medical knowledge base and a Bayesian algorithm to infer the proper treatment of the cardiac patient. This system was a harbinger of things to come in the 21st century.

Medical literature databases

Printed documents were also collected by every medical school to form libraries. These libraries were linked together in 1965 by the National Network of Libraries of Medicine (NN/LM), coordinated by the National Library of Medicine. The NN/LM provides access to biomedical information not available in NLM. This combined library system provides interlibrary loans, and currently provides online access to many digitized medical documents. NLM now partners with over 20 nations to serve consumers and health professionals around the world.

MEDLINE (Medical Literature Analysis and Retrieval System) is a bibliographic database of biomedical and life sciences, which includes citations for articles from scientific journals of medicine, healthcare, biochemistry, biology, and microbiological evolution, accessible through the PubMed search interface from the NLM (http://www.ncbi.nlm.nih.gov/pubmed/). PubMed includes lists of books as well as the journal articles of MEDLINE. The current push by Google to digitize many of the world's works of literature (including previous important medical books) may be a harbinger of the future, when most printed documents may be available online. When this happens, the NN/LM with MEDLINE and PubMed will comprise the largest online medical database in the world.

National Library of Medicine list of online medical databases

The NLM provides access to a large list of medical databases (https://www.nlm.nih.gov/services/databases_abc.html).

Some of these databases are accessible online, but many of them are accessible only on-site at the NLM facility at 8600 Rockville Pike, Bethesda, Ms 20894.

Other medical research databases

1. Scopus—an abstract and citation database hosted by Elsevier Publishers including hundreds of medical journal articles.
2. EBSCO—a library service company that vends a number of special-purpose medical databases.
3. University libraries, like the University of Illinois Chicago, can provide access to an extensive list of medical databases (https://researchguides.uic.edu/publichealth).
4. EMBASE—a proprietary research database that also includes in its corpus PubMed. It can also be accessed by other database providers such as OVID.
5. The Cochrane Library—a proprietary research database hosted by J. Wiley & Sons.
6. PubMedCentral (PMC)—the free open-source access branch of PubMed.
7. UpToDate—provides detailed clinical reviews on medical topics.
8. MedScape Online Medical Databases—free access provided by the National Center for Biotechnology Information (NCBI).
9. CDC Stephen B. Thacker Library—medical article digests, and COVID-19 research articles.

Bills of Mortality in London, United Kingdom

Boice (2020) cites that the Bills of Mortality were published by the Worshipful Company of Parish Clerks in London. The weekly Bill of Mortality was a single sheet of paper that listed the mortality figures for each of the 130 parishes of London on one side, and listed the various causes of death on the other side.

The Bills of Mortality were read by many people even outside of London. Daniel Defoe referred to them in his novel *A Journal of the Plague Year* (Defoe, 1722). Defoe cites the deaths of two men in the Long Acre district of London, the names of which city physicians gave to the parish clerk in December of 1664 to be published in the "weekly bill of morality." These two deaths were but the harbinger of 97,000 deaths in London during the Great Plague year of 1665.

The Bills of Mortality provided a stable inventory of terms that would have minimized reporting error during the compilation and printing of the Bills. There are references to the "forms" used to record the mortality data. Therefore, it appears that the parish clerks were using preprinted forms to speed the process of data collection at least by the end of the 17th century (Boyce, 2020). These preprinted forms may be the earliest evidence we have of the use of standard medical forms for recording and storing of medical information. See Chapter 4 for more information on the Bills of Mortality.

Best practice guidelines

This proliferation of medical research documents in printed and electronic form has made it rather difficult for a physician to keep up with developments in a given medical specialty. Many medical specialty groups have sought to solve this problem by developing and issuing evidence-based best-practice guidelines. These weighty documents suggest that care can be provided to patients based on rigorous science and medical experts whose judgment is not motivated in any way by self-interest (Brawley and Goldberg, 2011). Worldwide, over 300 organizations have issued over 2300 guidelines in their respective specialties. Grilli et al. (1978) reported that in the processes used to prepare 431 guideline documents, 67% did not describe the participants (patients and physicians), and 88% of the process documents did not provide adequate literature citations to support the guideline recommendations. Only 5% of the documents met these criteria.

Many of the guidelines are written by the US Preventive Services Task Force (USPSTF), an independent panel that advises the US government. Brawley and Goldberg (2011) cites a recent USPSTF recommendations on breast cancer screening, which specifically excluded radiologists from the group who wrote the guidelines. This exclusion was entirely proper; those physicians who benefit directly from the services rendered should be consulted for their input, but they should not be included in the specification of the best-practice guidelines. For many best practice guidelines, it is not so (Brawley and Goldberg, 2011).

Previous to 2009, the USPSTF strongly recommended that all women over the age of 50 have regular mammograms. In 2009, the USPTF changed their opinion, based on more recent studies. But, both of the bodies of studies ignored the 40% of women in their 50s and 60s who don't get mammograms, which includes about 5000 lives lost to breast cancer (Brawley and Goldberg, 2011). They recommend that some of the effort spent in developing guidelines should be redirected to research to find better tests for breast cancer. The same recommendation might apply to all medical specialties with guidelines.

Best-practice guidelines appear to be an attempt to collapse the sometimes widely varying opinions of medical practice available in the literature to form rather simplistic recommendations, and to replace intuition and common sense with evidence based on science (Brawley and Goldberg, 2011). This perception does not mean that we should not *include* evidence-based conclusions to guide best practices. Rather, we must find ways to integrate evidence-based medical knowledge with intuition and experience. And we must find better tests and treatments, rather than spend a lot of time on collapsing a lot of medical research (which may vary widely) into simplistic guidelines.

Guidelines of the American Academy of Neurology

One of the best sets of the guidelines was developed by the American Academy of Neurology (AAN), which provide at least suggestions for integrating evidence-based data and subjective judgments. Fig. 2.11 shows a set to guidelines for the treatment by anticonvulsants for painful diabetic neuropathy (PDN).

Notice the verbs "offered" and "considered." The introductory statement ("If clinically appropriate") applies to guidelines supported by all levels of evidence. These qualifiers suggest that the physician might consider these

Level of Evidence	Treatment Guidelines
Strong evidence	If clinically appropriate, (Drug-A) should be offered
Moderate evidence	Drug-B and Drug-C should be considered
Moderate evidence	Drug-D, Drug-E, and Drug-F should not be considered
Insufficient evidence	There is insufficient evidence to support use of Drug-G
Clinical context	Although Drug-H may be effective, it is potentially terotogenic, and should be avoided in diabetic women of child-bearing age.

FIGURE 2.11 AAN guidelines for the treatment of painful diabetic neuropathy. Specifications of the actual drug names are omitted. *American Academic of Neurology, http://www.aan.com.*

guidelines as inputs for the formation of treatment decisions, along with clinical observations and intuitive judgment provided by training and past experiences. This is the right way to use guidelines documents in medical treatment.

Medical records move into the digital world

Medical practitioners (in clinics, hospitals, and private practice) were mandated by the 2009 ACA to convert their records to a digital format. Initially, few data storage and management systems were structured sufficiently to serve medical data storage and retrieval. Eventually, many systems were developed to serve the need as clinics, hospitals, and private practitioners moved their records into the digital age.

Healthcare data systems

A recent report by IBM claims that 95% of healthcare IT leaders in large medical organizations state that they intend to implement cloud data storage and analysis options in the near future. Pressures on IT infrastructure brought on by the COVID-19 pandemic have greatly increased the need for cloud-based infrastructure, and for development of professional skills and security upgrades (source: https://newsroom.ibm.com/2021-01-04-IBM-Study-Majority-of-Surveyed-Companies-are-Not-Prepared-for-IT-Needs-of-the-Future-Say-U-S-and-U-K-Tech-Leaders).

There are over 70 healthcare EHR data systems available as of 2022. The pros and cons associated with 10 common healthcare data systems are presented below (abstracted from: https://www.praxisemr.com/top-ehr-vendors.html).

1. Epic
   a. Longstanding reputation, trustworthy
   b. Interoperability
   c. Third party apps to fit specific ailments
   d. Cloud host system
   e. Efficient user interface packed with shortcuts, templates, and smart phrases
   f. Patient portal
2. Praxix
   a. AI system learns from the user and gets smarter and faster with greater use
   b. Template-free model allows for utmost freedom and customization
   c. Improves medical quality
   d. Increases revenues of 30%–40%
   e. Provides legal protection
   f. Reduces charting time by 2 hours on average
   g. Lowers physician burnout and protects from stress
   h. Live technical support and comprehensive training services
   i. Patient portal
   j. DataMiner research tool allows for instant query
3. Cerner
   a. Easy navigation and organization
   b. Protects against HIPAA errors
   c. Automatic transfer of charting data using a variety of input methods

    d. Integration with third party apps and systems

    e. Patient portal

    f. Comprehensive and well-regarded support services

4. GE Healthcare

    a. Improves scalability via integration with third-party medical interfaces

    b. Increased efficiency in data entry

    c. Analyzes quality of care provided via the Quality Care dashboard

    d. Collaboration tools

    e. Customizable templates

    f. Patient portal

5. Meditech

    a. Interactive web feel

    b. Easy customization that meets specific needs to improve efficiency

    c. Well-organized interface allows for easy identification of alerts and tools

    d. Online customer support services

    e. Interoperability via Continuity of Care Documents (CCDs)

    f. Patient portal

6. eClinicalWorks

    a. Virtual visits via telemedicine application, Healow

    b. Free conversion packages

    c. Training included for small practices

    d. Patient portal

    e. Improved data streamlining via Grid cloud system

    f. Health and Wellness Tracking Data

7. Nextgen

    a. Generates patient treatment and solution plans

    b. Designed to improve patient engagement and reduce administrative burden

    c. Well-regarded training and support system

    d. Patient portal

    e. Population health management and risk stratification services

    f. Strong data analytics offerings

8. Allscripts

    a. Well-known to ease the prescribing process

    b. Longstanding reputation with much experience

    c. A plethora of EHR offerings designed to meet specific needs

    d. Streamlined administrative tasks

    e. Easy-to-use dashboard

    f. Allscripts prenatal offers tools for improving maternity care

    g. Patient engagement platform

9. Practice-fusion

    a. Well-organized system eases navigation

    b. Up-to-date advanced technology

    c. Free patient portal

    d. Library of customizable templates

    e. Continued well-regarded support services

10. GreenwayHealth

    a. Automatic health reminders for patients

    b. Comprehensive support system

    c. Minimizes administrative and clinical errors

    d. Increased productivity and efficiency

    e. Information security

    f. Offers practice analytic services

    g. Offers a patient portal

Postscript

Despite the significant problems with implementing the insurance exchanges mandated by the ACA in 2009, the Obama administration followed in the footsteps of many rulers in the past to provide medical information for the diagnosis and treatment of common people. Thus this history of written and digital medical knowledge bases has set the stage for discussions in the rest of the chapters of this book about how to analyze this impending flood of digital medical information with predictive analytics and analytical decision systems.

References

Barnett, G.O., Souder, D., Beaman, P., Hupp, J., 1981. MUMPS – an evolutionary commentary. Comput. Biomed. Res. 14, 112–118.

Bleich, H.L., 1969. Computer evaluation of acid-base disorders. J. Clin. Invest. 1969 (48), 1689–1996.

Blum, B., 1983. Mainframe, minis, and micros; past, present and future. Medcomp 1, 40–48.

Boice, N., 2020. Bills of mortality: tracking disease in early London. Lancet 395, 1186–1187. Available from: https://doi.org/10.1016/S0140-6736(20)30725-X.

Boyce, N., 2020. Bills of Mortality: tracking disease in early modern London. Lancet 395 (10231), 1186–1187. Available from: https://doi.org/10.1016/S0140-6736(20)30725-X.

Brawley, O., Goldberg, P., 2011. How We Do Harm: A Doctor Breaks Ranks about Being Sick in America. St. Martin's Press, New York, p. 303.

Breasted, J.H., 1967. Ancient Records of Egypt: Historical Documents from the Earliest Times to the Persian Conquest. U. Chicago Press.

Bryan, C.P., 1930. The Papyrus Ebers. Goeffrey Bles, 22 Suffolk St. Pall Mall, London. Available from: http://oilib.uchicago.edu/books/bryan_the_papyrus_ebers_1930.pdf.

Carlyle, T., 1836. Sartus Resartus . Available from: http://www.gutenberg.org/files/1051/1051-h/1051-h.htm.

Collen, M.F., 2012. Computer Medical Databases: The First Six Decades (1950-2010). Springer, New York, NY.

Cook, J., 2001. Dr. Simon Forman: A Most Notorious Physician. Chatto & Windus, London.

Dawson, W.R., 2010. Herodotus as a Medical Writer. Bull. Inst. Classical Stud. 33, 87–96.

Defoe, D., 1722. A Journal of the Plague Year. Louis Landa. and David Roberts. (eds), Oxford World's Classics (Oxford: Oxford University Press. 2010.), pp. xxxvii + 265. ISBN: ISBN: 978-0-19-957283-0.

Eckstein, A.M. 1987. Senate and General: Individual Decision-Making and Roman Foreign Relations, 264–194 BC University of California Press, 381 p.

Eisenstein, E.L., 1991. The Printing Press as an Agent of Change: Communications and Cultural Transformations in Early-Modern Europe, 1. Cambridge Univ. Press, p. 794.

Faulkner, R., 1969. The Ancient Egyptian Pyramid Texts, Oxford University. ISBN 0-85668-297-7, hardcover reprint ISBN 0-19-815437-2.

Gibbon, E. 1906. Decline and Fall of the Roman Empire. 12-vol. ed. John Bagnell Bury, Fred de Fau and Company, New York.

Hammond, N., 1961. Land tenure in Attica and Solon's Seisachtheia. J. Hellenistic Stud. 81, 76–98.

Johnson, E.D., 1970. History of Libraries in the Western World. The Scarecrow Press, Metuchen, NJ.

Kaplan, Robert, 2000. The Nothing That Is: A Natural History of Zero. Oxford University Press, Oxford.

Lloyd, G.E.R., 2006. Principles and Practices in Ancient Greek and Chinese Science. Variorum, Ashgate, Burlington, VT.

Miles, W.D., 1992. A History of the National Library of Medicine: The Nation's Treasury of Medical Knowledge. U. S. Government Printing Office.

Muhlberger, S. 1998. The Roman Conquest of Greece. Class syllabus for History 2055 (Ancient Civilizations), Nipissing University, Ontario, Canada. <http://www.nipissingu.ca/department/history/muhlberger/2055/l33anc.htm>.

Neugebaur, O., 1957. Exact Sciences in Antiquity. Brown University Press, Providence, RI, pp. 35–36.

Nisbet, R., Elder, J., Miner, G., 2009. Handbook of Statistical Analysis & Data Mining Applications. Academic Press, Burlington, MA.

Oppenheim, A.L., 1962. Mesopotamian medicine. Bull. Hist. Med. XXXVI, 99–108.

Polastron, Lucien X., 2007. Books On Fire: The Tumultuous Story Of The World's Great Libraries. Thames & Hudson Ltd, London, pp. 2–3.

Porter, R., 1999. The Greatest Benefit to Mankind: A Medical History of Humanity. W.W. Norton & Co., New York, p. 872.

Randell, B. (Ed.). 1982. The Origins of Digital Computers, Selected Papers, third ed. Springer-Verlag. ISBN 0-387-11319-3.

Reisner, G., 1905. The Hearst Medical Papyrus. U. California Publ. Egyptian Archeology Vol. 1.

Shanks, N.J., D. Al-Kalai. 1984. J. R. Soc. Med. 77 (1):60–65.

Sigerist, H., 1951. A History of Medicine. Oxford University Press.

Swain, S. 1996. Galen, Chapter 11. In: Hellenism and Empire: Language, Classicism, and Power in the Greek World, AD 50–250, pp. 357–379.

Unschuld, P.U., 1985. Medicine in China: A History of Ideas. Univ. Cal. Press, Berkeley.

Wellmann, M., 1901. Die Fragmente der sikelischen Arzte Akron, Philistion mtd des Diokles vonKarystos. Weidmann, BerlinKaplan, Robert. (2000). The Nothing That Is: A Natural History of Zero. Oxford University Press, Oxford, UK.

Zeller, E., 1886. Outlines of the History of Greek Philosophy. Longmans, Green, London.

Chapter 3

Bioinformatics ☆

Nephi Walton and Gary D. Miner

Chapter outline

Prelude

In Chapter 2, the history of record-keeping in Medicine was presented, along with a general introduction to the latest form of storage—the computer database. The massive amount of data available in medical databases measured in terms of volume, frequency, and resolution begs the question: "What can we learn from this ocean of data?" This chapter takes the first step towards building an understanding of the size, the nature, and some challenges we face to learn anything useful from the massive amount of medical data available today. The first task in this step, however, is to describe briefly the nature and development of the necessary processes (and some available tools) for predicting the current state (i.e., diagnosis) and future state (i.e., treatment) of human medical conditions. The general name for this technology is Predictive Analytics.

The rise of predictive analytics in healthcare

Everyone likes the concept of gazing into something like a "crystal ball" to learn what will happen in the future. Chapter 1, discusses an important element of the history of medicine and healthcare in the Middle Ages, which was

---

☆NOTE: The figures in this chapter and the chapter are Copyright by Nephi Walton, 2012, 2022.

Practical Data Analytics for Innovation in Medicine. DOI: https://doi.org/10.1016/B978-0-323-95274-3.00005-1

centered on mystical seers, who vended medical advice. In fact, the concept of the medical case book arose among those seers, seeking to document the many cases they advised. In our modern age, Science has replaced the crystal ball, but it has been based largely upon what happened in the past, generating responses that are reactive to those events, rather than proactive. Science can "look" into the future to the extent that it extends trends or events that have happened in the past. The problem is finding a way to view *new* information in a future context. But gaining new insights from old data requires the complicated analysis of many interacting factors in medicine and healthcare to generate likely scenarios that might happen in the future. This goal has eluded physicians, who are trapped in by the perceptions of their own minds. Now, we have a way to do this—with computers.

The computer has revolutionized medicine and made possible many advances that have had a tremendous impact on healthcare and the life expectancy of humans. The computer is becoming an indispensable tool in the practice of healthcare as it becomes more advanced, and as the volume of information in healthcare increases exponentially. The American Medical Informatics Association (AMIA) has formally defined Biomedical Informatics as:

> the interdisciplinary field that studies and pursues the effective uses of biomedical data, information, and knowledge for scientific inquiry, problem solving and decision making, motivated by efforts to improve human health.
>
> American Medical Informatics Association (2013).

Predictive analytics plays a key role in these efforts, and becomes even more important as we advance in this field.

In the 2002 movie Minority Report, Tom Cruise plays a cop who keeps his city crime-free by catching murderers before they have a chance to commit a crime. In principle, this is the next level in law enforcement. This notion makes a good story, because it "plucks" the heartstrings of many people, who are very interested to change what might happen in the future. The police in the Minority Report depended on reports of special "precog" people, who could "see" what the future would be, if events or actors in the present were left unchanged. How very much like the seers in the Middle Ages were the precogs of the film. The job of the police was to make the changes necessary to avoid the undesirable future (like arrest the criminal before he commits the crime). We want to take analogous actions regarding our medical healthcare. Our challenge is to find some means of precognition in the technology of the present to provide some insights about what might happen in the future. We can't change the future in such a direct manner as did the police in the Minority Report, but we can change what might happen in the future, if we can predict with reasonable accuracy what it is that might happen. This is the realm of Predictive Analytics.

Healthcare is entering an era of development that is very similar in principle to the theme in the film Minority Report. The theme developing in healthcare is focused on using predictive analytics to follow a more proactive approach to the diagnosis and treatment of disease. Homicide is also a health statistic; therefore, the theme followed in the Minority Report is of great interest to healthcare practitioners, at least in principle. In the Public Health Informatics section of this chapter, you will learn that the methodologies of predictive analytics are quite different from the way the "precog" people were used to "read" the future in the Minority Report. In contrast, healthcare is leveraging the predictive power of artificial intelligence tools to predict probabilities, rather than certainties. But, these probabilities can be high enough and accurate enough to have significant effects on the prevention of adverse consequences (e.g., sickness and death).

Moving from reactive to proactive response in healthcare

In the past, medicine has been primarily a reactive field. When we are faced with a disease, we treat it; if the pain gets worse, we alleviate it; when someone stops breathing, we resuscitate them. One of the aims in predictive analytics in healthcare is to diagnose problems at an early stage of development (or even before they occur at all), before they have had a chance to take a toll on the human body. But, the role of predictive analytics does not stop once the individual develops the disease. Another aim of predictive analytics is to guide in selecting and tailoring treatments for individuals by predicting the course of events that is likely to occur with every treatment option that is available. Of course these concepts apply not only to individuals but to populations, and by using predictive analytics we can foresee public health threats and take the necessary steps to lessen their burden or prevent them from happening at all.

Medicine and big data

Biomedical Informatics involves developing techniques to efficiently process and analyze the data producing summative results, which can then be used to improve health outcomes. One of the biggest challenges in Biomedical Informatics today is developing techniques and tools to process the immense amount of data generated in healthcare today. The amount of biomedical data available today is tremendous, and it is growing exponentially; it is becoming one of the most important sources of "Big Data" (data volumes measured in terabytes and petabytes). The critical importance of it is defined in terms of life and death of many people. Vast amounts of biomedical data are being accumulated in many forms, such as free text, radiographs, photos, gene sequences, microarrays, vital signs, and lab values. We can produce, transmit, and store more of this data than ever before, and our capacities to store it and our abilities to analyze it are increasing at staggering rates. For example, the cost of 1 MB of data storage in 1995 was over 15,000 times the cost for the same amount of storage in 2022, and the processing speed of our desktop computers for analyzing the data has increased about 30 times (from 100 MHz to over 3000 MHz) since 1995. With increasing technologies to store and process data we also increase the amount or resolution of the data we generate. The bottleneck we face now is the limitation in our ability to process and synthesize these large volumes of data.

When working with such large volumes of data, the likelihood of finding associations occurring somewhere in the data set simply by chance is quite common, and the process of finding true meaning behind data becomes extremely difficult, if not impossible (at least seemingly so). Imagine the immense number of calculations required by your brain just to perform an action as simple as throwing a wad of paper into a trash can. Light hits color sensors in your retina, which detect color, brightness, and depth. These signals are sent to the visual cortex (together with other parts of the brain), which processes every "pixel" of the digital image they form and identifies every object in your field of vision. After the trash can is identified as the target in your visual field (and the distance to it is estimated), sensory signals from your hand are transmitted to your brain, providing information about the weight, consistency, and form of the wad of paper. The brain processes millions of pattern elements in your memory formed by similar signals caused by experiences in the past to calculate the force necessary to exert on the wad of paper to make it follow the correct trajectory to the trash can. These signals and calculations are then combined to activate muscle groups, which receive visual, vestibular, proprioceptive, and tactile sensory input from thousands of neurons in the arm, shoulder, and hand to coordinate a smooth muscular action to propel the wad of paper to the trash can. Your nervous system is trained to do this through years of experience with inputs from your various sensory organs. All of these inputs must be combined together and coordinated in very complex ways to perform this apparently simple task. The time required to learn how to perform this action can be substantial, and the more experience the individual has in doing it, the smoother and more accurate is the toss. You will not get it on the first try, but each toss will get closer and closer to hitting the trash can, as your brain processes new information from each experience. This process is very similar in principle to the way predictive analytics tools learn to recognize patterns in data, example by example (i.e., row by row in the data file).

Although predictive analytics techniques are not as advanced as are analytical processes in the human brain, the capabilities of these techniques are growing continuously. They work on principles similar to those that control learning processes in the brain. Historical data are provided to the predictive analytics tool, which function as cases (or experiences) in the past, which are used to build a pattern, which is composed into an analytical model, The more historical data (or "experience") that are processed in the building (or training) of the model, the better it will perform. As time passes and new data or "experience" are available, they can be added to the training data set, and the model can be retrained. So as each year passes, your model can become increasingly accurate, as additional experience is available for the training process.

With the advent of meaningful use there were considerable financial incentives for healthcare organizations to utilize their data stores to improve patient outcomes. Predictive analytics models have played a role in meeting the goals associated with the concept of meaningful use. This chapter will give a brief overview of the biomedical informatics field and how predictive analytics can be applied to some of the key areas of informatics. The goal of this chapter is to create a stepping stone to inspire users of informatics technology to apply predictive analytics to their field in innovative and creative ways. This inspiration may lead them to create tools to improve healthcare and take on projects that will make a difference in healthcare and ultimately in people's lives.

An approach to predictive analytics projects

There are limitless possibilities when deciding on a predictive analytics project. Predictive analytics is well established in many areas of business, including: (1) customer relationship management; (2) fraud detection; (3) sales forecasting; and more recently in (4) online retail, where retailers cater ad offerings on the first web page, based on your own personal preferences for what you looked at or purchased in the past, where you live, your gender, and your age. Every person has an individual storefront showcasing products you are most likely to buy, based on predictions of what you are likely to buy. Predictive analytics are also used to manage employees and schedule their shifts automatically based on the times that buyer traffic is most likely to be the highest. For example, there appears to be a jump in traffic at a

Jamba Juice as temperatures rise (Bellcross, 2012). In the airline industry predictive analytics are used to schedule flights, and Wall Street uses these technologies extensively to manage the buying and selling of stock. It is always important to consider how and where technologies are being deployed before deploying them in our respective fields to learn from what has already been done.

In healthcare we can do similar things, including:

- Delivering treatment plans based on models of patient characteristics to optimize patient adherence and response.
- Offering additional services that a patient is likely to need and watch for additional symptoms or conditions that a patient is likely to develop.
- Scheduling of nurses, doctors, and other staff to match predicted patient volumes.
- Efficiently purchase and store medical supplies according to predicted demand.
- Prevent disease through the creation of predictive models that accurately identify individuals at highest risk and deploying preventative measures.

The predictive analytics process in healthcare

Regardless of the purpose for which predictive analytics is used, the same process of steps can be followed in the project. Fig. 3.1 shows these key steps required to tackle a predictive analytics project in medicine and healthcare. The focus of this methodology is on the directed path of operations that moves the researcher from hypothesis to solution. There are some feedback loops, which represent elements of the learning process, but the flow of operations leads to the desired end point—predictions which can be incorporated into medical decision-making. Following this methodology section are discussions of some key areas in bioinformatics where predictive analytics are being applied. There are innumerable examples in these areas, and those discussed are merely examples selected to inspire innovative thinking and encourage you to initiate projects of your own involving predictive analytics.

Process steps in Fig. 3.1

Step 1. Problem definition

Define a problem/situation for which advanced notice will change your course of action and steps can be taken to change an outcome. Initially, choose problems that will have a relatively large impact, but also for which you will have significant domain support in solving. Once you identify a problem, you may even break it up into parts and tackle a smaller piece of it before taking on the whole challenge. For example, you may be faced with the challenge to predict census at a children's hospital. There are many possible factors that can affect hospital census, therefore, one of the first tasks in solving this problem is to look at factors that have the largest impact on census. The general problem of hospital census could be broken down into various causes of admission, and it might be that the biggest driver of admissions is respiratory disease, primarily bronchiolitis. The causes of bronchitis could be further defined in terms of causal factors, the most likely of which could be RSV (RSV means Respiratory Syncytial Virus - it is a highly infectious virus that can cause respiratory illness in all age groups). Consequently, the defined problem could be to predict an RSV outbreak.

Step 2. Identify available data sources

Hospitals have a large store of data, but don't limit study to data which exist currently in the hospital. Other valuable data sources include:

- External causal data, like data on adverse weather conditions, which might drive people indoors, thereby promoting the transmission of viruses which can develop into disease outbreaks.
- Information available from local clinics and urgent care facilities such as an increase in respiratory complaints, which can be related to an increase in positive respiratory viral tests, or an increase in emergency room visits.
- Secondary data, such as a spike in medication purchases at retail stores, which can be related to the defined problem. Other secondary data might include changes in television viewing patterns when children stay home from school, or an increase in web searches for respiratory symptoms.
- Exercise your creativity, and do some research in the medical literature to identify other secondary sources of available data.

After the data sources have been identified, consider the likelihood of access to data from each one. For example, even though data for retail medication purchases are available, gaining access to that data can be quite difficult and

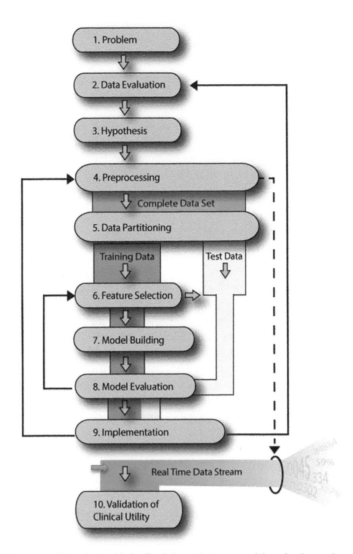

FIGURE 3.1   A predictive analytics process flow chart with feedback loops between model evaluation and preprocessing-feature selection, and an overall iteration loop between implementation, and data evaluation.

expensive. You may decide to confine yourself initially to only those data sources that are readily available, and can be accessed within your budget. Be careful not to limit yourself to just those available data sources related to known associations, because a major part of the predictive analytics process is the discovery of complex relationships that were previously unknown. Preliminary analysis might identify other useful data, which were previously unrecognized.

Step 3. Formulate a hypothesis

After you have identified the group of available data sources to use, you must formulate an hypothesis. Following the previous example, your hypothesis might be that you can predict RSV outbreaks with meteorological variables and positive viral test counts as inputs to a neural network modeling algorithm. In this case, the specific hypothesis is defined in terms of the methods proposed to test it. But, that need not be the case—the appropriate methodology can be selected later.

Step 4. Data preprocessing

Cleaning and preparing a data for predictive analytical modeling can take 70%–90% of the total project time to complete; make your plans accordingly. Data preprocessing includes:

- integration of data sets from multiple sources;
- filling of missing data elements with imputed values;

- deleting records and variables that are unusable for modeling;
- derivation of new variables to use as predictors; and
- modifications in the data structure of the input data set (e.g., balancing data sets with rare targets).

Data formats and consistencies must be checked, and corrected where necessary. For example, testing of the same lab value, such as TSH, may have different values and ranges of normal depending on the lab where it was processed. You might have to recode some variables, and derive other variables that you suspect might be predictive of the target outcome. Predictive analysis requires values in *every* row of *every* variable, or the modeling algorithm may ignore the entire row! Finally, you may need to normalize and or discretize the data to achieve better performance with your model. Discretizing means to create a separate variable for each unique value in a categorical variable (e.g., A, B, C). Remember the principle "garbage in=garbage out" and make sure your data is clean and consistent.

Step 5. Data set design

After the data set has been prepared at the data element level, you must perform several operations on the data set as a whole. These operations include:

- *Data set partitioning*. It is important to hold out a portion of your data for evaluating the accuracy of your model after it is built. Many predictive analytics algorithms will divide the input data set into two subsets used for training the testing model over many iterations through the data set. The *training set* is input to the algorithms, which evaluates the relative predictive weights associated with each variable (for neural nets), or the selected cut points in the construction of a decision tree. These parameters are used to compose the predictive model, which is used to predict the target outcome for data in the testing set after the first iteration. The predicted values are compared with the actual values for each record in the testing data set, and an overall error is calculated. One of the training parameters of the algorithm is modified slightly, based on the overall error, and the training data are input to the algorithm again. This process may go through hundreds of iterations until a specified threshold is reached (measured in terms of number of iterations performed, or a selected minimum error is reached). Thus both the training and testing sets of data are used in the training operation. Evaluation of the accuracy should be performed on a data set not used in the training operation in any way. That means that you should create a *third* partition (the validation data set) in the data partitioning process, for use in calculating prediction accuracy. Don't base prediction accuracy on the training set, or even on the testing set, because your model may be overtrained for the specific training set, and it might fail significantly on any new data set. Some algorithms create all three data sets for you, most other algorithms create only training and testing data sets, and some algorithms don't partition input data sets at all—they depend on you to do the partitioning explicitly. Therefore, know your modeling algorithm!
- *Balancing of data sets with rare targets*. If you want to model the decision to leave a company (called *churn*, or attrition), it is likely that no more than 5% of the data records will have Target value=1 (for churn), and 95% will have a Target value=0. It is very easy for a modeling algorithm to build a model that is 95% accurate, just be predicting all rows as Target=0. But that model doesn't help you predict the customers that will churn! In order to do that, you must force the algorithms to focus more on churn records, than nonchurn records. There are three ways to do that:
  o deleting enough records with Target=0 to equal the number of records with Target=1;
  o duplicating enough records with Target=1 to equal the number of records with Target=0; and
  o calculating the ones-complement of the proportion of Target=1 records (1—proportion of Target=1 records), and use that number as a weight to submit to the modeling algorithm, and treat Target=0 records analogously.

Step 6. Feature selection

In this step, we apply the principle of Occam's razor, which is essentially that if you have two competing theories that make exactly the same predictions, the simpler one is better. This is particularly important in predictive analytics with machine learning, because having too many features can lead to overfitting. A feature is the name given to a transformed variable. When a model is overfitted, it is conformed very closely to the training data set including the noise in it (meaningless or inaccurate data that has no correlation to the outcome). As a result, the model may show very poor performance on any new data it encounters. There are a number of feature selection algorithms specific to the particular methods employed in machine learning. Each of these techniques can help you select only the features with the highest correlation to your outcome, and can improve your model's performance. Aside from the problem of overfitting it makes no sense to use more data points if you can get equal or better results with a simpler model.

There is a caveat, however, that must be considered in the choice to use feature selection. Performing feature selection on your entire data set may bias your results, so make sure that you partition before performing feature selection.

If you are going to validate your model on a separate independent data set, then partitioning a third data set is not necessary. When using an independent data set, make sure it comes from the same population as the training and testing data partitions. Patient populations can differ markedly at different locations (i.e., Salt Lake City, UT vs Detroit, MI), which may introduce a significant bias to your results.

Step 7. Model building

This is where things get exciting and sometimes frustrating. There are many different predictive analytics algorithms that can be used to build your model. There are many predictive analytics software packages available currently, which contain a broad choice of modeling algorithms (https://www.ibm.com/account/reg/us-en/signup?formid=urx-19947). Popular among the choices of these packages are:

Statistica	https://www.tibco.com/products/data-science; https://www.tibco.com/products/data-science/downloads
IBM modeler	https://www.ibm.com/analytics/spss-statistics-software;
Weka	http://www.cs.waikato.ac.nz/mL/weka/;
SAS-EM	https://www.sas.com/en_us/home.html; https://welcome.oda.sas.com/login
Orange	http://ax5.com/antonio/orangesnns/;
R with RATTLE	http://www.r-project.org/;
RapidMiner	https://rapidminer.com/products/studio/
KNIME	https://www.knime.com/
Angoss	http://www.angoss.com/

Some of the modeling algorithms include: logistic regression models, time series models, decision trees, artificial neural networks, support vector machines (SVM), naïve Bayes, and k-nearest neighbors. If time and budget allow it may be useful to try several different methods and compare their results.

Ensembles of different modeling algorithms may produce more accurate models than possible with any of the constituent algorithms. Most predictive analytics packages have modeling options that permit the design of ensemble models. It is recommended that several different ensembles be tried, before selecting the single algorithm (or group of them) that works best on your data set.

Step 8. Model evaluation

After a predictive analytic model is created, that is not the end of the story. The model needs to be "evaluated" for reliability, sensitivity, and specificity. This model may have been created from small-sized datasets. Any model needs to be evaluated, but especially when the patient numbers that produced the model are small. This can be done using several methods, such as:

- Use of both a training and testing set of the data; if both subsets of data provide the about the same accuracy, then the model may be a good one—but still needs further evaluation.
- Hold out sample: part of the dataset is "held out" in a random manner, with the rest being used as the train and test sets. Then after the model is created, the "hold out" sample is run again the model to see if the same "accuracy scores" are obtained and the individual scores seem reasonable.
- Use of v-fold cross-validation; this is a process where the dataset is subsampled numerous times (10 times is commonly used in real practice); if the accuracy scores of the v-fold cross-validation are about the same as for the train, test, and hold out samples, then the model is probably quite robust.

The above list is not exclusive, as there are other additional measures that can be taken to evaluate the model.

Step 9. Model implementation

After the model is built and evaluated favorably, it can be deployed and tested in the operational systems where it will be used. This step in predictive analytics can be extremely difficult, because it may require interfacing with other systems, and collecting and analyzing current data on a daily basis, rather than working on an isolated dump of historical data. Don't put too much effort into integrating your model into the clinical workflow permanently, until you have completed step 10 below, The deployed cannot be used until it is proved to provide some clinical utility.

Step 10. Validation of clinical utility

You may spend a significant amount of time working on a prediction algorithm to predict admissions for the ER, only to find that there are no interventions that the hospital is willing or able to take to improve the outcome. You could build a very powerful predictive modeling solution, with no problem to solve. This is like designing a product that nobody wants to buy. The ability to deploy the model should be evaluated up front before you start the project. Make sure that you look at the outcome of your prediction operations in relation to actual interventions that can take place once the prediction is made. But to prove your model makes a difference with the intervention, you must have some way to compare it to existing methods, and demonstrate that your predictions actually improve healthcare.

The next operation could be labeled as Step 11: Reevaluate, add more data, and rebuild the model. As emphasized earlier, a given model is not the end of analytical modeling; it is just one step along the way. Models "age" as new data becomes available. You might be able to improve model performance significantly by adding more data reflecting local demographic or societal changes. For example, you might be able to significantly improve the accuracy of a hospital census model by adding new inputs from seven different viral outbreak models. You can also build similar models using different variables and combine their results to come up with a better estimate. Make sure and remember that every year that goes by, you have another year's worth of data for training your model, you could retrain monthly or even weekly, if you like.

Translational bioinformatics

There are a number of papers available in the literature on the use of predictive analytics in medical research, but its value can only be realized when predictions are used expressed in a form that can used successfully to impact patient care. This process is defined by AMIA as "Translational Bioinformatics" to include:

>…the development of storage, analytic, and interpretive methods to optimize the transformation of increasingly voluminous biomedical data, and genomic data, into proactive, predictive, preventive, and participatory health. Translational bioinformatics includes research on the development of novel techniques for the integration of biological and clinical data and the evolution of clinical informatics methodology to encompass biological observations. The end product of translational bioinformatics is newly found knowledge from these integrative efforts that can be disseminated to a variety of stakeholders, including biomedical scientists, clinicians, and patients.
>
> AMIA American Medical Informatics Association (2022); https://amia.org/about-amia/why-informatics/informatics-research-and-practice.

The "Tricorder" medical device used in the Star Trek TV and film productions is not a far-fetched idea in Bioinformatics; similar tools may be in use in the not too distant future. This device is used in these dramas to scan the human body, provide information about the health of the body, and make a quick diagnosis and prognosis of treatment. The human body is a complex system controlled by a complex group of interacting biological signals, of which we are becoming increasingly aware. The actions and reactions of our body are the best indicators of what is happening in the body. But, this voluminous cascade of signals can be difficult to interpret. The decision of when and where to collect these signals is problematic in itself. When your body confronts an invader organism, signals are propagated to the white blood cells that indicate the nature of the invader, where it is, and triggers the appropriate response to defend the body against the threat. If we can sense and record these signals early, we might be able to prevent a healthcare disaster. In addition to the uses of these signals as data inputs, relationships between disease and other factors (e.g., genes, proteins, and adverse healthcare events) can be combined to build powerful predictive models useful in treating the disease. The process of building such predictive models and expressing the outcomes in terms useful for diagnosis and treatment of disease is the central goal of translational bioinformatics.

Clinical decision support systems

Clinical decision support systems (CDSS) are integrated analysis and deployment systems designed to facilitate decision-making in healthcare. They combine information about the current patient with information about past diagnoses and treatments stored in a database to provide feedback or recommendations that will aid healthcare providers in the decision-making process at the point of care. The Healthcare Information and Management Systems Society (HIMSS) expands this definition to include patients as recipients of information, to permit patients to be active participants in their care. The definition of Clinical Decision Support according to HIMMS is:

>…a process for enhancing health-related decisions and actions with pertinent, organized clinical knowledge and patient information to improve health and healthcare delivery. Information recipients can include patients, clinicians and others involved in patient care delivery; information delivered can include general clinical knowledge and guidance, intelligently processed

patient data, or a mixture of both; and information delivery formats can be drawn from a rich palette of options that includes data and order entry facilitators, filtered data displays, reference information, alerts, and others.

(Osheroff et al., 2012).

CDSS provide a delivery mechanism with which we can take the results of scientific research to the bedside, to directly impact patient care.

CDSS are separated by Plato's problem, the gap between knowledge and experience.

Coauthor and Editor of this book, Robert A. Nisbet, notes: This sentence begs the question: "Why did Plato say that?." The answer is related to Plato's underlying position on the nature of being, in which he believed that the "whole" of being is greater than the sum of its parts, and that the most important parts were beyond our human senses (e.g., love and hate).

On the other hand, Aristotle, Plato's student for 20 years, believed that you could define the nature of being only in terms or human senses (e.g., what the eye could see and the hand could touch). Therefore for Aristotle, the whole of being could be defined in terms of the sum of its sensible parts. Predictive analytics combines both approaches to model realities in the world. We can reference an extended discussion of this topic in our first book by Nisbet et al. (2009).

Clinical knowledge is a *cognitive* understanding of a set of known clinical rules and principles based on medical literature, which guide our decision-making processes. Experience is an *acquired* understanding of medical outcomes gained through years of practice applying various outcomes related to various particular conditions, the majority of which cannot be learned sufficiently through reading and acquiring cognitive knowledge. This important distinction arises because there is an immense number of medical subjects in the literature that could be researched and taught. In addition, there are so many variables in the medical decision-making process that outcomes based on knowledge versus those based on experience are often discordant. It is practically impossible to teach physicians all of knowledge acquired by experience, because the environmental variables are constantly changing, so the body and nature of our experiences evolve through time, reflecting particular outcomes under specific conditions. Cognitive knowledge, however, is always associated with a limited scope of outcomes that are out of date, by necessity—there is a time-lag between subject outcomes and the reporting of them. For example, this distinction could come sharply into focus, when choosing a physician to remove your kidney in surgery; would you choose one who has mastered a surgical textbook, or one who has the experience of 300 successful operations?

CDSS's can be classified into two types of systems:

- knowledge-based support systems that are defined by a well-established set of rules that guide decisions, based on the interpretation of the medical conditions judged in the medical literature to be the best practice; and
- nonknowledge-based systems that do not use a set of defined a priori rules, but instead, uses artificial intelligence algorithms to induce the rules through machine learning methods, allowing the system to learn from hundreds or even thousands of encounters, rebuilding the "model" set of rules as environmental variables change. These systems can be based on neural networks, genetic algorithms, SVMs, decision trees, or any other machine learning technology that "learns" to recognize patterns in data sets case by case.

Hybrid clinical decision support systems

Hybrid CDSS have been developed to allow the end user to synthesize the results from both knowledge and clinical experience, and make a clinical decision based on the results of both (examples in the literature include Santelices et al., 2010).

In such a hybrid system, multiple predicted outcomes are posed for the physician, based on data input from knowledge and experience bases, and furnished with associated probabilities to permit the selection of the appropriate decision. As we continue to learn more about cognitive science, and distil this knowledge into principles, we can apply them to improve these "intelligent" systems to help us make the best clinical decisions possible at the time. This practice of continuous incorporation of patient data, cognitive knowledge, and clinical experience is referred to often as "rapid learning." Rapid learning approaches that continuously update the CDSS as new data become available provide an ability to create decision models that adapt to the availability of new treatments, interventions, and metrics (variables) that can be input to the modeling process. This paradigm is shown in Fig. 3.2.

Many CDSS provide information on drug interactions and can generate allergy alerts. Many of these alerts, however, are very basic, and don't include any information on many other factors, such as dose, time of administration, and the context in which the medications are given. Consequently, many physicians discount these warnings. On a given work day, it is very common for a physician to dismiss dozens of these warnings, as he prescribes medications in the hospital.

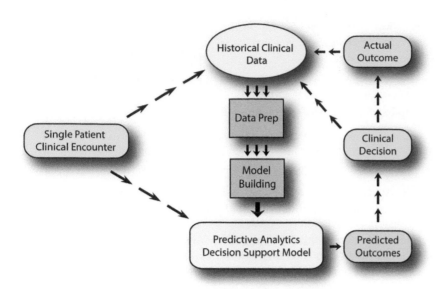

FIGURE 3.2 Illustration of the pathways in a hybrid CDSS (clinical decision support system); these have been developed to allow the end user to synthesize the results from both knowledge and clinical experience, and make a clinical decision based on the results of both.

In some instances, physicians become so used to ignoring these warnings, that they may accidently disregard an important one. It is time to make these alerts more "intelligent," by using predictive analytics to predict levels at which problems occur, and to set thresholds to control when alerts will be generated. In addition, these alerts should provide information about the effectiveness of the drug for the given clinical scenario, and suggest more effective options, if a suboptimal treatment is selected. Such a system could include analysis of a patient's antibiotic prescription history, before presenting a list of drugs for choice, or check its database for any information about the susceptibility of the patient to a bacterial invasion, if the chosen antibiotic does not provide broad enough coverage.

Consumer health informatics

The AMIA Definition of Consumer Health Informatics is:

> *the field devoted to informatics from multiple consumer or patient views. These include patient-focused informatics, health literacy and consumer education. The focus is on information structures and processes that empower consumers to manage their own health–for example health information literacy, consumer-friendly language, personal health records, and Internet-based strategies and resources. The shift in this view of informatics analyzes consumers' needs for information; studies and implements methods for making information accessible to consumers; and models and integrates consumers' preferences into health information systems.*

AMIA American Medical Informatics Association (2022).

The focus is shifted from the problem resolution to understanding by the consumers of the nature of the problems, and the availability of various solutions to fix them. This new focus centers on information structures and programs that empower consumers to manage their own health. These programs can be classified into three groups:

Patient-focused informatics

- Predicting various elective procedures associated with a given treatment.
- Predicting the level and type of information required to support a given treatment.
- Predicting treatment schedules, based on patient symptoms and basic measurements.
- Recommending various interventions to be made or precautions to be delivered to prevent future ailments.

Health literacy

- Presenting selected Internet resources to increase patient understanding of the nature of health problems, and their recommended treatments. Patients will access the Internet anyway, therefore, it is important to direct them to responsible websites, and screen out those that are inappropriate, or present questionable information.

Consumer education

This shift in the focus of medical informatics addresses the need for healthcare information perceived by consumers by providing a means for them to acquire it responsibly. Most importantly, this new focus in informatics integrates consumers' preferences into health information systems. Consumer informatics stands at the crossroads of other informatics disciplines, such as nursing informatics, public health, health promotion, health education, library science, and communication science. From this central position, consumer informatics can function as a clearing house and a provider to valuable consumer-related information to other informatics disciplines

Direct-to-consumer genetic testing

Related to consumer informatics, direct-to-consumer (DTC) genetic testing is a rapidly growing new industry, which accepts DNA samples directly from consumers. Samples are analyzed, and information regarding genetic markers, which may predispose to certain diseases, is provided. There has been considerable debate about having such options available to consumers directly, because their lack of domain knowledge and inability to understand the medical context may cause unnecessary worry and stress in individuals whose genetic markers indicate an increased risk of certain diseases. This stress can cause some people to fall prey to (or even seek) relief from opportunistic marketing schemes that offer unproven treatments, cures, or preventative options for a disease. Some of the companies that offer these genetic testing services have recognized this problem, and offer genetic counseling to anyone who uses their service. Nonetheless, there are ethical concerns about this practice by responsible physicians who question whether these tests have any positive effect on the health of the population, and that they might have an overall negative psychological effect. There is certainly a beneficial place in medicine for genetic tests, when they are properly administered. For example, testing for mutations in the BRCA ½ gene can be done by responsible healthcare professionals. People with mutations like this have a very high risk of disease, and can take preventative measures (such as prophylactic mastectomy) that have been proven to decrease mortality. Most of the thousands of other markers that are tested in DTC genetic testing are genetic variants that have minimal impact on the disease. In order to assign any practical meaning to these markers, they must be analyzed in context with the other information such as family history, other genetic markers, and patient characteristics. There is a considerable need for research on the effects of DTC genetic testing and what benefits can arise from it. Predictive analytics can play a role here also, by including many variables in the analysis, and providing probabilities of risk, which patients can manage to some degree with diet and exercise to decrease the level of risk of disease.

Use of predictive analytics to avoid an undesirable future

Tools designed to predict future health must be used in the proper context. It is known commonly that health improvements are associated with changes in diet, demographics variables, exercise, and even education level. The opportunity to use a computer to explore various options and outcomes related to modifications of certain factors in their lives could potentially lead to people make appropriate changes in their lives. Charles Dickens presented a set of similar situations to Ebenezer Scrooge in "A Christmas Carol." Instead of using a computer armed with predictive analytics programs (which, of course, didn't exist then), he faced the old miser with the Ghosts of Christmas Past, Present, and Future. He used those literary vehicles to show Scrooge how decisions and events in the past led to his present circumstances, and how, if left unchecked, they would lead inevitably to a undesirable future. Scrooge was shocked! That response led him change his present actions, in the hope of avoiding what he was convinced would happen otherwise in the future. We can bundle those "ghosts" into a predictive analytical system bring to reality the Victorian dream (and that, indeed, of everyone) to change some scary things that otherwise might happen in their future.

Consumer health kiosks

How would you like to walk up to a kiosk at a mall or drug store, and be able to get a prescription, like you can get money at an ATM? That day is not far away. In addition to prescribing medications, these kiosks could offer on the spot lab testing and vitals measurement, or reassure you that your symptoms are not life threatening. Predictive analytics will be used in these systems to predict the outcome or severity of a problem, based on the available information compared against past visits. The kiosk may even be able to tell you to see health provider, recommend that you rush to the emergency room, or do nothing, and just wait for time and nature to fix the problem.

Who uses the Internet? Nearly everybody

When people want to learn about a medical condition or a treatment, they rely on large search engines, such as Google, to find answers to their questions. Even though Google has improved over time, there still may be a significant amount of misinformation presented on the website, even on the first web page. Researching a topic takes time and effort, even with comprehensive websites like PubMed. This is particularly true for the average consumer who does not necessarily understand the terminology or context in the articles they find.

Information can be dangerous. It can lead people to spend excessive amounts of money on unproven treatments, and to neglect getting appropriate medical care, which sometimes leads to death. We could use predictive analytics to analyze internet search phrases in context of demographics, location, and search history to provide more "intelligent" search results, and report the level of understanding of the result topic by the medical community. Google is already doing some of this, but there are still significant dangers in following a Google search of information related to disease treatment. For example, criminals can take advantage of people with diseases that cannot be cured, by offering the only "cures" available.

Patient monitoring systems

A patient may be hospitalized when there is a significant risk that the patient may take a turn for the worse. Therefore, it is important to keep the patient in a controlled and monitored environment, where plenty of healthcare professionals are available should the need arise. These patients are monitored according to the level of risk assigned to the patient. For example, various levels of risk might direct that patient vitals be monitored every hour, or every 4 hours, every 8 hours, or even be continuously monitored. These measurements are necessary in order to assess the status of patient stability, and assure that there have been no changes related to the monitored values. Most hospitals have devised scoring systems to assess the patient's status and their needs to ensure receipt of the proper level of care. For example, patient status may indicate whether a patient can be watched on the general floor, or must be transferred to the intensive care unit (ICU). Despite the use of these monitoring systems patients may become unstable, and no help is available to provide the appropriate level of care for their condition. Patients may become unstable very quickly without much warning. An intern might be presenting a patient to the attending physician outside the room, and a nurse might call out a code for an emergency response, because the patient's blood pressure dropped precipitously in a period of seconds, and the patient stopped breathing. Instances like these beg the questions:

- Was there something that could have been done sooner? Probably, there was, but staff limitations precluded it.
- Could we have picked up a signal of the impending crisis quicker and taken action to prevent a near collapse and long stay in the ICU? Yes, If the nurse could be dedicated full-time to the room.
- What went wrong in this instance, was the patient not properly assessed? The assessment may have been correct at the time it was made, but patient status can change *very quickly*.
- Why not measure vitals more often on every patient?

The answer to all of these question are related to limitations in cost and resources. The more monitoring a patient receives, the more time and resources must be dedicated to that patient. This is why a stay on the general hospital ward is much less expensive than a stay in the ICU. Another common question is:

We have continuous monitoring systems available, why not put those on every patient?

This is a bad idea for many reasons, one of which is that the more equipment that is hooked up to a patient, the more you restrict their actions. This restriction may require them to stay in bed, and this is often counterproductive during rehabilitation of a patient. Another reason is excessive cost and the time you will spend contending with incidental findings and errors on monitoring devices. Because not all signals from monitoring devices are intelligently processed, they often produce erroneous measurements. It is very common in pediatrics to be called to a room to assess a patient with a low oxygen saturation, only to find the sensor dangling from the hand, or that the patient was moving so much that the machine was not picking up a good signal. Even with good signals, there is a broad range of normal responses, and you are liable to encounter values that appear to be outliers, if you monitor constantly. One particularly annoying event that physicians are commonly called to attend is the incidence of bradycardia during sleep. This condition can be completely normal, and it must be assessed in the context of the given problem and the medications the patient is taking. It is standard practice in medicine that unnecessary lab work is to be avoided, not just because of the added cost, but also because there is a 5% that you may spend hundreds or thousands of dollars chasing an "abnormal" lab value (based on the mean and standard deviation of the entire population), which is completely normal for this person.

Applications for predictive analytics in intensive care unit patient monitoring systems

In pediatrics, the commonly-used Patient Early Warning Score (PEWS) is a perfect example of an opportunity for predictive analytics to generate significant improvements in patient monitoring operations. When a pediatric patient receives a very low PEWS score (or two consecutive moderately-low scores), the patient is assessed for transfer to the ICU. When low scores occur, the transfer order is written, the ICU staff are called, and they do the transfer assessment. Based on the judgment of the ICU staff, the transfer happens, or the physicians are reassured that transfer is unnecessary. While this system works reasonably well in most cases, physicians may find that they must write PEWS exception orders, occasionally, to keep the patients on the floor despite these low scores. Why does this happen? Many other variables are not considered in the system; assessment is based on population-based metrics, and not on all characteristics of the individual case. Physicians must compensate for this weakness in the PEWS system, by overriding it. Predictive analytics can used to create models that can learn from experience and apply all the appropriate patient characteristics to make a more accurate assessment.

ICUs have electronic systems to collect high frequency measurements and closely monitor critically ill patients. In most instances, these measurements are stored in large databases, combined with information provided by other EHR systems. This situation is a prime opportunity for the design of predictive analytics projects, which compare the results from such systems to existing patient scoring systems and prognostic models. Early work on artificial intelligence in the ICU had focused on knowledge-driven techniques as described in CDSS above. More recently there has been research on data-driven methods, or experience-based artificial intelligence models. There is ample opportunity right now to implement predictive analytics systems in the ICU. It is the prime time to take advantage of the vast data stores that are available in today's ICUs, and start accounting for other characteristics of the patient that are not assessed in standard scoring methods.

Current predictions are derived from the analysis of raw signals from various monitors in the ICU. One of the key problems in these systems is the significant amount of noise in the monitors. The signals from these monitors can be affected by patient transport, patient movement, bad wiring, poor connections, and device failures, to name a few. Noise must be filtered out of signals before accurate predictions can be made. Many devices have signals that can be used in correlation with measurements to determine if the measurements are correct, such as the waveform on a pulse oximetry monitor. Some algorithms can filter out much of the noise, and thus clarify any valid signal (or the lack of it). Several of the challenges in making predictions in the ICU are to determine the points in time at which to make the measurements, and selection of the time-grain resolution on which to make predictions, considering the fact that different monitors capture information at different time resolutions. Physicians must also take into account interventions (e.g., medications given), and the status of the patient, such as bradycardia during sleep, and rise in blood pressure when a child is screaming.

Challenges of medical devices in the intensive care unit

The major challenges of current medical devices in the ICU is their relative lack of accuracy and portability. Medical devices are evolving along with informatics technology, and more accurate devices will be developed. These devices will become smaller and less sensitive to movement and less prone to errors, allowing the patient to have more mobility. This mobility will permit more frequent measurements, even permitting some patients to move away from the ICU for a period of time. As these devices evolve, predictive analytics can be progressively incorporated into them, allowing us to sense signals that may alert us to an impending crisis requiring intervention before it is too late. The majority of predictive modeling technology in use currently in the ICU is based on analytical techniques developed for classification and regression (numerical estimation), which consider measurements of variables in a time sequence as independent variables. Information can be extracted from these variables to permit modeling of a target outcome, based on changes in these variables over time. Classical time-series analysis considers only the signal present in the outcome, not the predictive signals present in the time sequence variables. See Nisbet et al. (2009) for a discussion of this subject. These time sequence analyses have been very successful in medical informatics (e.g., for the prediction of the next diabetic episode).

Apart from accuracy and portability, another of the challenges of implementing systems in the ICU is that you are dealing literally with life and death situations, in which the stakes are high and there is little room for error. Models must have a high level of accuracy and have high discriminative performance to be acceptable for use in this setting. Such models should be used not as a "crutch" to replace judgment, but as an additional aid to support it. When these models are validated against actual improvements in patient care and reduction in mortality, they may replace older methods, but not until then. Because of the challenges in this new field of technology in the ICU, there are few such systems that have been deployed in the ICU. Although many proof-of-concept studies have been done, few systems have been validated. This is an area that is ripe for the use of predictive analytics to affect patient care directly and significantly.

Public health informatics

Public Health Informatics is defined as:

> *the application of informatics in areas of public health, including surveillance, prevention, preparedness, and health promotion. Public health informatics and the related population informatics, work on information and technology issues from the perspective of groups of individuals. Public health is extremely broad and can even touch on the environment, work and living places and more.*

AMIA American Medical Informatics Association (2022).

In this discipline, the focus is shifted from individuals to groups of people, and may include many other fields that can have an effect on the health of a population. Some examples of this broad list of fields include:

- **Weather**—cold and rainy weather may cause many people to stay in enclosed spaces during periods of cold or rain. This behavior may contribute to viral outbreaks in winter months.
- **Safety features**—the design of cars, buildings, and toys may affect the general public health of the nation.
- **Architecture and layout of streets**—the way in which people are forced to move in towns and cities may contribute to accidents and congestion, which affects depression and the general state of human wellness in the vicinity.
- **Food cost and growing methods**—in large respect, we are what we eat, and type and amounts of foods that people eat can affect public health in large geographical areas. The price of food can drive people to eat food that is not nutritious and may contribute to obesity.
- **Food and sanitation standards**—local outbreaks of disease reduce the general state of public health.
- **Social programs**—great public interest has been generated about the health effects of obesity, diabetes, and sexually transmitted diseases.

All of these areas of public health provide rich sources of data for use in predictive analytics, which can provide valuable insights to programs aimed at increasing the general state of wellness in our society. Even the mining of social networks can provide data for the use of predictive analytics to compare public health problems and the status of different geographical areas.

The major problem: lack of resources

Public health departments tend to be strapped for cash always, so it is important to predict areas with the biggest problems, and prioritize the allocation of resources to those areas with the largest potential impacts. Insurance companies have begun to share their predictive analytics with healthcare providers so they can apply appropriate interventions to cut their costs. It may be possible to build similar relationships between insurance companies and departments of public health in areas where a large number of insured individuals are concentrated.

Social networks and the "Pulse" of public health

For better or worse, social media is now ubiquitous in our society. Data mined from some social media sites provide a vast resource of information about subjects that people love to share with others, even health-related issues. People are becoming increasingly connected through social media; it is hard to find someone who doesn't have a mobile device that can upload their photos, thoughts, and whereabouts into the social data cloud. This huge source of data could be tapped with predictive analytics to take the "pulse" of the general status of public health in society and suggest where it is headed. Patterns of medically related complaints can be mined at various times to provide insights about changes in patterns of disease outbreak, obesity, mental health problems, and educational needs. Social media may even be able to provide information about whether health-related interventions are working, or indicators that show an increase in wellness. This pulse of society can be related to geographical area by using GPS coordinates, and can be applied for prioritizing areas of high violence. Some police departments (e.g., in Memphis and Chicago) do this now to optimize the allocation of squad car and surveillance resources. These measures can be extended with predictive analytics to expose factors that encourage a high incidence of crime in an area. Analyses like these can be orchestrated to "diagnose" a development of conditions that might promote social unrest, depression, or other psychological conditions, and can help to design intervention measures to promote public health.

Predictive analytics and prevention and disease and injury

In recent years, an emphasis on disease prevention has arisen to compliment a prior focus on diagnosis and treatment. It is difficult (if not impossible) for a physician to discuss every preventative strategy for any disease that may befall patients during the short time allotted to a typical office visit. Through the use of predictive analytics, we can analyze large populations of people to quantify risks related to public health, and help physicians to develop intervention programs for those patients at highest risk of some ailment or medical condition.

Large companies have very significant financial incentives to prevent injuries. Companies may monitor the incidence of workplace injuries, and collect other data related to safety providing reports and real-time alerts to permit timely intervention to prevent injury. Some large companies have used predictive analytics to significantly reduce injury incidence rates by more than 60%, which in turn led to increased productivity and a decreased workers compensation fees (Schultz, 2012a,b). A research group from Carnegie Mellon University was able to build models that can predict the number of injuries on work sites with 80%–97% accuracy rates (Schultz, 2020).

Biosurveillance

Biosurveillance is a huge area in public health, primarily stemming from the national security interest and the threat of biological weapons and viral outbreaks. The importance of this field has been underscored by the recent COVID-19 pandemic. Although several syndromic surveillance systems have been installed to detect outbreaks at local and national levels, our experience with COVID has demonstrated that there is a lot more work to do (Maxmen, 2021). Many signals have been used to try and detect outbreaks, including the monitoring of healthcare encounter-related information (Lombardo et al., 2003), monitoring of purchases at large retailers to detect the purchase of medications that might signal a spreading disease (Goldenberg et al., 2002), monitoring Google searches related to symptoms, and monitoring health-related social media postings (O'Shea, 2017). These same techniques can be used to aid in hospital management and implementing public health measures when a disease outbreak is predicted, which increases public awareness, and hinders the spread of disease.

Food-borne illness

Food-borne illness outbreaks are relatively frequent and spread quickly. Predictive analytics can play an important role in predicting where the outbreak is likely to spread, and how it can be contained.

These are just a few of the applications of predictive analytics in public health. Public health problems generally involve large populations with lots of data and are ideally suited to the application of predictive analytics. This is an exciting area that presents tremendous opportunity to create predictive analytics tools that can have a large impact.

Medical imaging

Image mining is an area that is rich in opportunity for predictive analytics models. Images can be 2D, 3D, static, or moving (4D). Tools using these technologies are available for:

- Screening people for retinal macular degeneration.
- Predicting cardiovascular events by using ultrasound flow imaging to measure pressures, velocities and turbulence of flow related to the likelihood of future events.
- Finding various problems on images consisting of billions of pixels, representing enormous amounts of special data.
- CT lymph node analysis to support staging for cancer screening.

Soon, we may be able to use complex image analysis predictive models to replace more invasive means, such as biopsies, or removal of lymph nodes for tissues for staging cancer studies. We will be able to design surgeries and cater treatments with fewer invasive procedures.

Face recognition and other biometrics (e.g., eye scanning) is well-established as a component in security systems. One of the most mature technologies using image analysis is recognition of specific information in images. This technology has even been used to diagnose genetic syndromes (Srisraluang and Rojnueangnit, 2021). Similar tools are being used for cell recognition, identifying nuclei and other organelles and their features to classify tissue samples. These morphological features are often indicative of what is happening to the organism as a whole. Information from image analyses like these can be combined with physiological information to provide rich new variable combinations to use in building predictive models.

Clinical research informatics

According to the AIMA:

> *Clinical Research Informatics involves the use of informatics in the discovery and management of new knowledge relating to health and disease. It includes management of information related to clinical trials and also involves informatics related to secondary research use of clinical data. Clinical research informatics and translational bioinformatics are the primary domains related to informatics activities to support translational research.*
>
> AMIA American Medical Informatics Association (2022) (AMIA, Accessed February 9, 2022).

Clinical trials for new medications are expensive, often running into hundreds of millions of dollars. At the beginning of a trial, you should know how many patients you will be able to get, and how many are likely to drop out. Insufficient recruits can delay the trial, causing severe delays and increasing costs, and drop-outs can affect the reliability of results. Yet thousands of trials have been done, and there is an immense amount of information about these trials that is available for use. We can mine these data to increase the likelihood of better outcomes before we start the trials. Knowing what happened in the past in similar trials, we can optimize the study design before any money is invested or cancel the trial if the preliminary results appear similar to those in the past.

Intelligent search engines

Researching a topic in PubMed (or any other online medical literature source) can be quite time-consuming and difficult. Intelligent searching tools can use predictive analytics to present query results of keyword searches based on the context of the search string. One approach to do this, semantic mapping, can be incorporated into search engines to present various strands of meaning for keywords, and permit the researcher to search for what he meant, rather than just what he entered literally into the search string. These alternate search paths can give researchers more pertinent and even rather obscure results that are important, but would have been missed otherwise.

Personalized medicine

Personalized medicine is a field that has huge potential for the use of prediction and association analyses. Specific treatments can be catered, based on past experience with other patients. Using exome or full genome analysis, it will be possible soon to predict how patients will respond to various drugs and therapies. Personalized medicine systems can include information from image analysis, lab data, demographics, history of adherence to treatment, financial status, physiological signals, and other data sources to cater the best treatment to the patient based on predicted probabilities. This topic will be further discussed in the Personalized Medicine chapter of this book.

Hospital optimization

Hospital staffing, particularly nurse staffing, is a major issue in many hospitals today. A shortage of nurses can have very detrimental effects on patient outcomes, while having too many nurses on shift adds unnecessary healthcare expense, which translates directly to higher patient costs. By predicting hospital census, the scheduling of nurses with predictive analytics technology can function to increase scheduling efficiency during times of high need, while eliminating unnecessary shifts. Intelligent scheduling can be applied also to optimize availability, utilization, and storage of resources. Certain supplies or medications related to outbreaks must be available when an outbreak hits, but many of these supplies have relatively short shelf lives, which can be managed by just-in-time replenishment systems. On the other hand, having too much of these supplies with short shelf lives can increase unnecessarily the operating expenses necessary to maintain them. Many staffing tools are available that use predictive analytics in the general business world, however in medicine the stakes are higher, and the processes and relationships to staffing are very complex. This is an area that is ripe for analytics, and there are many technologies common in the business world that can be applied to the world of healthcare.

As described in the public health section above, many businesses are looking at safety measures that can be recorded in order to predict accidents, and businesses have been successful in using these measures to predict, and thereby prevent accidents from happening in the workplace. In the hospital accidents, mistakes, or changes in processes can have an even more dramatic impact, particularly in such high-stakes areas as ICUs. Incidence of morbidity and mortality can be greatly reduced by modeling outcomes, modeling various hospital measures, and comparing them with actual outcomes, while increasing patient satisfaction at the same time. This is a very broad and important area where predictive analytics can be applied to significantly improve the quality and success of healthcare.

Challenges
Data storage volumes

The size of the space and the cost necessary to store biomedical data have been significant issues in the past, but these issues are becoming increasingly pressing and important now, as very large amounts of data are generated by the existing medical systems. The prospect of storing all of the information in the entire genome of an individual is daunting enough (3 gigabytes in the Human Genome Project), but when related epigenetic effects (changes in gene expression without changes in the DNA nucleotide sequence), and temporal effects related to each gene are considered, the storage volume required for each person becomes truly gigantic (possibly, several terabytes). When you consider the data storage requirements for all the people in a hospital census, which turns over many times during a given year, the storage volume may increase into the petabytes. And that is just for one hospital for one year! We are on the brink of a monumental explosion in data volume in medicine and healthcare. The creation of advanced compression methods and algorithms to efficiently store and retrieve such information becomes paramount.

Data privacy and security

Privacy and security are also major concerns in the storage and use of any data about individuals. With the passage of the American Health Insurance Portability and Accountability Act (HIPAA) in 2003, health providers must make sure that all medical records and related information (e.g., billing records) conform to a set of standards of documentation, handling, and privacy. But, these standards are rather broad, and each state can choose the way that information is protected and made available to individuals. The problem is that there is a wide latitude among the states in regulation of patient health information. For example, Meingast et al. (2006) reported that Alabama had no general statute restricting the disclosure of patient information, while California has extensive regulations of such disclosures. This wide variability in disclosure regulations among the states, provides a high probability of leakage and misuse of patient healthcare information during transmission across stateliness. This problem is becoming exponentially worse as patient information becomes available in electronic format. It appears that we are still in the "Wild West" of information regulation in medicine and healthcare.

Meingast et al. (2006) posed some questions that remain today, even in the wake of the Patient Protections and Affordable Care Act of 2010:

- Who owns the data?
- How much data should be stored?
- Where should data be stored?
- To whom should this data be disclosed?
- Unanswered questions abound regarding where the data should be stored, who owns the data, to whom should this data be disclosed without the patient's consent?
- How should the data be secured?
    The authors suggested some solutions to these problems, which are still relevant today:
- Define clear specifications for role-based access to healthcare data, in which different rules apply to people in different usage roles.
- Define new HIPAA regulations to standardize how healthcare data can be used and transmitted between states.
- Rules to govern patient privacy in home monitoring programs.
- Policies and rules defining how data can be acquired for predictive analytical purposes, and who can have this access.

Other sources of health-related data can be scoured from the internet, which raises the question of whether or not it is ethical to use this information without an individual's permission. In one case, a medical student happened to look at the Facebook page of a certain patient, and determined that the patient might be in a high-risk situation at home. This information led to an intervention, which is likely to have saved the patient's life. Regardless of the happy ending, this example left lingering concerns about the propriety of such actions even when performed with good intention.

Standards and consistency of data

There is a considerable lack of consistency in terminology and measurements between medical practices, labs, and even within the same hospital. It is very hard to analyze data compiled from these sources and we are still quite far from a universal standard in terminology and measurement. To make things more difficult, many of the measures in medicine are extremely subjective; you could easily get four different answers from four different physicians if you asked them to characterize a murmur, for example. You also have to take into consideration the temporal aspect of a measurement and context in which it happened, which are not always recorded. We must anticipate significant challenges in these areas at the beginning of any predictive analytical project.

Interpretability of models

Physicians are primarily driven by evidence-based medicine and rely heavily on understanding the underlying mechanism of disease to deploy the correct interventions. PA models that do not enable physicians to have some understanding of the variables and methodology that contribute to the prediction or decision outcome of a model are unlikely to be widely adopted. This may evolve over time as PA and AI gain more widespread acceptance and physicians become more accustomed to this technology. This evolution, however, does not diminish the importance of the physician understanding what drives a model. Most of these systems should be designed to augment a physician's capabilities rather than replace them and in order to use the presented information in the context of a physician's clinical judgement, the physician must be able to factor in the variables driving the model.

Evidence-based guidelines and adoption of PA models

In order for PA models to be widely adopted in clinical care they will need to undergo clinical trials with published studies that demonstrate a significant impact on patient outcomes. Many PA models are developed and used at individual institutions but are never deployed outside the hospital where they were created. Publication of model results that include validation at external sites is necessary to facilitate more widespread adoption.

Portability of PA models

Perhaps one of the most challenging aspects of building PA models is developing them in a way that they can easily be deployed at other institutions. This is a huge challenge, since different hospitals may have very different demographics/characteristics than those that the original models were trained on, and the data architecture could vary to the degree that it is very difficult if not impossible to provide the same data inputs into the model. For this reason it is important to use widely adopted standards when available for your data inputs. It is possible to train your model on another institution's data if you have access to all the same standard data elements that were provided in your original model. Retraining models to adapt to differences amongst healthcare organizations is a common practice and is often required to obtain optimal performance.

Regulation of PA models

There is some suggestion that predictive analytics tools used in patient care specifically CDSSs should be regulated similar to medical devices, requiring stringent acceptance, commissioning, and quality assurance. The CDSS would have to be validated on local datasets before approval. This can be problematic for rapid learning CDSS, because they change constantly as information is gathered from patients. Methods to regulate and thereby ensure patient safety without losing the advantage of rapid learning will need to be addressed. Interestingly, coming out of the ACA Act is PCOR (Patient Centered Outcomes Research), a non-profit "contract research" agency that is committed to developing transparent networks of medical data among healthcare organizations including hospitals, clinics, and individual doctors. The goal of this, among others, is to produce CER (Comparative Effectiveness Research) to "really" determine which treatments and drugs and medical devices are working for both "groups of patients" (grouped by age, race, sex,

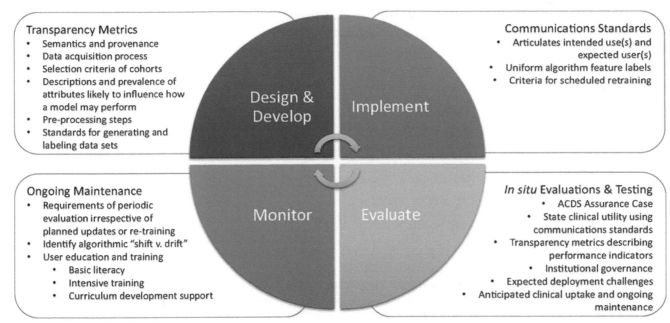

FIGURE 3.3 Policy recommendations for all stages of ACDS (*ACDS*, adaptive clinical decision support)—design and development, implementation, evaluation, and ongoing monitoring—require further development to ensure safe and effective ACDS. A concerted multistakeholder effort to identify key transparency metrics for training datasets and communications standards for AI-driven applications in healthcare is needed to understand how bias can corrupt AI-driven decision support and identify ways to mitigate such bias. Additionally, policies that standardize in situ testing and evaluation, as well as ongoing maintenance of ACDS should be established. *Fig. 3.1 from Petersen, C., Smith, J., Freimuth, R.R., Goodman, K.W., Jackson, G.P., Kannry, J., et al. (2021). Recommendations for the safe, effective use of adaptive CDS in the US Healthcare System: An AMIA position paper. J. Am. Med. Inform. Assoc. JAMIA. https://doi.org/10.1093/jamia/ocaa319.*

and other grouping factors, including genetic predisposition) and "individual patients" (which are primarily determined by DNA profiles plus other attributes). To do this will require that HIPPA laws and other regulatory processes, whether FDA or elsewhere, are worked through so that they will not be inhibitory to the development of accurate diagnostic and treatment methods (PCORI, 2022).

> **Definition: Clinical decision support systems.**
>
> CDSS are computer-based programs that analyze data within EHRs to provide prompts and reminders to assist healthcare providers in implementing evidence-based clinical guidelines at the point of care.

In January of 2021, the AMIA published a position paper on recommendations for the safe and effective use of CDS ("Clinical Decision Support") in the US healthcare system (Petersen et al., 2021). A lot of this is summarized in Fig. 3.3 (which was the one figure in their 2021 position paper).

Basically, this AMIA group suggests the need for the development of a CDSS that "trains itself" and thus "adapts its algorithms" in real time, based on new data. They refer to this as a ACDS (adaptive clinical decision system). But such a system will require new initiatives and oversight coordinated among many medical institutions. The AMIA group provided the figure (see Fig. 3.3) that they think is needed to provide a safe ACDS system, with all of the needed governance and associated standards.

For those readers interested in further details on this ACDS system, please see the reference (Petersen et al., 2021).

Summary

This is a very exciting time for predictive analytics in biomedical informatics. It is at the forefront of medical research, as we transition from reacting to disease to proactively preventing it. Predictive analytics is in its infancy in this field, and though many studies show predictions using small single-source data sets, there are few that are based on large amounts of clinical data available from many sources, such as images, lab values, physiological signals, genetics, and other patient demographics and characteristics. There are even fewer predictive analytics projects that have been incorporated into clinical practice. This situation provides abundant opportunities in almost every area of informatics for predictive analytics, and there is ample opportunity to use these tools to make a lasting difference in healthcare.

Postscript

This chapter has provided a high-level overview of the challenges of using predictive analytics to answer questions in medicine. But before we describe some aspects of personalized medicine (Chapters 6 and 7), we must consider some issues in the nature of the data storage and analysis architectures (Chapter 4) and kinds of medical data sources (Chapter 5).

References

American Medical Informatics Association. <http://www.amia.org> (accessed 10.11.13.).

AMIA (American Medical Informatics Association), <https://amia.org/about-amia/why-informatics/informatics-research-and-practice> (accessed 9.2.22–11.2.22).

Bellcross, C.A. (2012). A part-time life, as hours shrink and shift. *The New York Times*.

Goldenberg, A., Shmueli, G., Caruana, R.A., & Fienberg, S.E. (2002). Early statistical detection of anthrax outbreaks by tracking over-the-counter medication sales. Proc. Natl. Acad. Sci. USA 99 (8), 5237–5240. Available from: https://doi.org/10.1073/pnas.042117499. PMID: 11959973; PMCID: PMC122753.

Lombardo, J., Burkom, H., Elbert, E., Magruder, S., Lewis, S.H., Loschen, W., et al., 2003. A systems overview of the Electronic Surveillance System for the Early Notification of Community-Based Epidemics (ESSENCE II). Journal of Urban Health 80 (2 Suppl. 1), i32–42.

Maxmen, A., 2021. Has COVID taught us anything about pandemic preparedness? Nature 596 (7872), 332–335. Available from: https://doi.org/10.1038/d41586-021-02217-y. PMID: 34389832.

Meingast, M., Roosta, T., & Sastr, S. (2006). Security and privacy issues with health care information technology. In: Proceedings of the 28th IEEE EMBS Annual International Conference New York City, USA, August 30–September 3, 2006.

Nisbet, R., Elder, J., & Miner, G. (2009). Handbook of Statistical Analysis and Data Mining Applications 1st Edition. Elsevier-Academic Press.

O'Shea, J., 2017. Digital disease detection: a systematic review of event-based internet biosurveillance systems. International Journal of Medical Informatics 101, 15–22. Available from: https://doi.org/10.1016/j.ijmedinf.2017.01.019.

Osheroff, J. A., Levick, D. L., Saldana, L., Velasco, F. T., Sittig, D. F., Rogers, K. M., et al. (2012). Improving Outcomes with Clinical Decision Support: An Implementer's Guide, second ed. Healthcare Information and Management Systems Society, Chicago, IL.

PCORI. (February 11, 2022). Patient Centered Outcomes Research Institute. <https://www.pcori.org/> (accessed February 2022).

Petersen, Carolyn, Smith, Jeffery, Freimuth, Robert R., Goodman, Kenneth W., Purcell Jackson, Gretchen, Kannry, Joseph, et al., 2021. Recommendations for the safe, effective use of adaptive CDS in the US Healthcare System: an AMIA position paper. Journal of the American Medical Informatics Association. Available from: https://doi.org/10.1093/jamia/ocaa319.

Santelices, L.C., Wang, Y., Severyn, D., Druzdzel, M.J., Kormos, R.L., Antaki, J.F., 2010. Development of a hybrid decision support model for optimal ventricular assist device weaning. Annals of Thoracic Surgery 90 (3), 713–720. Available from: https://doi.org/10.1016/j.athoracsur.2010.03.073.

Schultz, G., 2012a. Using advanced analytics to predict and prevent workplace injuries. Occupational Health and Safety 81 (7), 88, 90-91.

Schultz, G. (July 1, 2012b). Using advanced analytics to preict and prevent workplace injuries; in occupational health and safety. <https://ohsonline.com/Articles/2012/07/01/Using-Advanced-Analytics-to-Predict-and-Prevent-Workplace-Injuries.aspx>.

Schultz, G. (June, 2020). From EHS-Today (Environmental Health & Safety). <http://www.ehstoday.com;specifically; https://info.predictivesolutions.com/hubfs/Jan%202018%20Web%20Migration/PDFs/ps-dontPDF.pdf>.

Srisraluang, W., Rojnueangnit, K., 2021. Facial recognition accuracy in photographs of Thai neonates with Down syndrome among physicians and the Face2Gene application. American Journal of Medical Genetics Part A 185 (12), 3701–3705. Available from: https://doi.org/10.1002/ajmg.a.62432. Epub 2021 Jul 21. PMID: 34288412.

Further reading

Bellcross, C.A., Page, P.Z., Meaney-Delman, D., 2012. Direct-to-consumer personal genome testing and cancer risk prediction. Journal of Cancer 18 (4), 293–302.

Cai, H., Cui, C., Tian, H., Zhang, M., Li, L., 2012. A novel approach to segment and classify regional lymph nodes on computed tomography images. Computational and Mathematical Methods in Medicine 2012, 145926. Epub 2012 Oct 31.

Cheng, S.K., Dietrich, M.S., Dilts, D.M., 2011. Predicting accrual achievement: monitoring accrual milestones of NCI-CTEP-sponsored clinical trials. Clinical Cancer Research 17 (7), 1947–1955.

Güiza, F., Van Eyck, J., Meyfroidt, G., 2012. Predictive data mining on monitoring data from the intensive care unit. Journal of Clinical Monitoring and Computing .

Improving Outcomes with Clinical Decision Support: An Implementer's Guide, second ed. HIMSS. 2011.

Isariyawongse, B.K., Kattan, M.W., 2012. Prediction tools in surgical oncology. Surgical Oncology Clinics of North America 21 (3), 439–447. viii–ix. Epub 2012 Apr 17.

Kamel Boulos, M.N., Sanfilippo, A.P., Corley, C.D., Wheeler, S., 2010. Social Web mining and exploitation for serious applications: technosocial predictive analytics and related technologies for public health, environmental and national security surveillance. Computer Methods and Programs in Biomedicine 100 (1), 16–23.

Lambin, P., van Stiphout, R.G., Starmans, M.H., Rios-Velazquez, E., Nalbantov, G., Aerts, H.J., et al., 2012. Predicting outcomes in radiation oncology-multifactorial decision support systems. Nature Reviews Clinical Oncology.

Phan, J.H., Quo, C.F., Cheng, C., Wang, M.D., 2012. Multiscale integration of -omic, imaging, and clinical data in biomedical informatics. IEEE Reviews in Biomedical Engineering 5, 74–87.

Zheng, Y., Hijazi, M.H., Coenen, F., 2012. Automated "Disease/No Disease" grading of age-related macular degeneration by an image mining approach. Investigative Ophthalmology & Visual Science pii: iovs.12–9576v1.

Chapter 4

Data and process models in medical informatics

Robert (Bob) Nisbet

Chapter outline

Prelude

Chapters 1 and 2 provided an overview of the current issues related to medical practice based on comprehensive analysis of huge quantities of available data. Chapter 3 introduced the reader to the broad landscape of medical bioinformatics practice for extracting actionable information from this huge body of data. This chapter takes the next step to building use analytical models in medicine—understanding the data elements upon which these models are built.

Chapter purpose

This chapter lays the foundation for building classification systems for diseases and maladies in medicine. It shows how present classification systems developed in history and practice to form the ICD system used now for insurance and billing purposes. Finally, it describes the OMOP-DM data model used commonly to store clinical data, from which analytical models can be built.

Introduction

It is very tempting for conscientious medical practitioners to go immediately to the available data sets, and try to make meaningful conclusions from the analysis of them. Initially, this may appear to be a reasonable approach, but it almost always turns out badly for the analyst. This common failure may prompt the conclusion that the data sets are useless to provide information for building models of acceptable accuracy for a given target variable (the entity being modeled). This conclusion may be right in some circumstances, but more often, the conclusion is wrong. The "signal" of the target variable may actually be there in the data set, but the analyst failed to find it. This is a *procedural* problem not an *informational* problem. The first step in designing a proper procedure for analyzing data is to develop an understanding of the nature of the data elements themselves. A proper understanding of data elements in a data set (e.g., clinical drug prescriptions) requires the knowledge of each data element in relation to all other data elements of its kind—in a classification system. The second step in the analytical procedure is to use the proper data processing procedures in the right

Practical Data Analytics for Innovation in Medicine. DOI: https://doi.org/10.1016/B978-0-323-95274-3.00004-X

order to build a construct for analyzing a data set (an analytical process model). This chapter will describe a common classification system and a common analytical process model used to build predictive models of medical outcomes.

Systems for classification of diseases and mortality

Individual medical maladies were described briefly in the Ebers Paprys (c. 1550 BCE), the earliest manuscript available that describes medical conditions (see Chapter 2). Several of the conditions mentioned included depression and dementia. The Hippocratic Corpus of documents (c. 430 BCE) became the standard of treatment for over 2000 years, until the rise of modern medicine in 1900 (see Chapter 1). For example, the corpus included a document named "On Injuries of the Head," which describes three modes of fractures of the head bone as contusions, breaks, and fractures. This approach worked well, as long as the same corpus was used by all medical practitioners, but it was not very comprehensive.

Bills of mortality

The first detailed records of medical maladies were the Bills of Mortality, which were weekly reports of statistics in London to monitor causes of death beginning in 1592 to 1595, and then continuously from 1603 until 1758 (Boyce, 2020). Fig. 4.1 shows an example of a Bill of Mortality for the week of February 21−28, 1664.

Notice in Fig. 4.1 the total of 393 deaths from Plague during February 21−28, 1664. At that rate, we might expect there to be about 20,000 deaths from the Plague during 1664. This was, however, only a prelude of what was to come in 1665, during which totals of Bills of Mortality amounted to slightly under 70,000. Due to travel of infected persons out of London (and dying elsewhere), the total mortality was estimated to be as high as 100,000 during what was considered to be the Great Plague year (Bell, 2001). Bills of Mortality became the first major attempt to store a historical record of causes of death.

The ICD system

The concept of the Bills of Mortality evolved into the International List of the Causes of Death. The conference on the First Revision of the International List of Causes of Death convened in Paris on August 21, 1900, adopting a detailed classification of causes of death consisting of 179 groups of causes. Various updates were made until 1940, when the Surgeon-General of the US published a list of codes for classifying diseases and injuries, in addition to causes of death. This classification effort was expanded in 1945 to the Statistical Classification of Diseases, Injuries and Causes of Death. This classification system was updated several times, until it was renamed the International Classification of Diseases ICD in 1948. The ICD was revised many times to form the ICD-11 codes in 2019, which went into effect in January, 2022 (WHO, 2018). Despite the development of extensive ICD classification system, two problems remain:

1. Different sources of medical information in different countries used different terminology to refer to the same medical condition. For example, the disease of "gout" was sometimes called "crystalline arthritis"; and
2. The system was developed primarily for the purpose of serving insurance billing needs, rather than analytical needs.

Something else was needed to provide data in the suitable format for data analytics in medicine. This is a good example of the power of data analytics in changing the very way that the medical business was conducted. A similar change happened in the late 1990s, when the concept of the Exploration Data Warehouse was proposed to serve the data needs of data mining in business (Inman et al., 1998).

The OMOP common data model

The Observational Medical Outcomes Partnership (OMOP) organization was formed in 2012 (Overhage et al., 2012). This aim of this organization was to provide a common representation of medical information to serve three primary purposes:

1. medical claims and billing systems;
2. electronic health records (EHRs); and
3. statistical analysis and analytical modeling to support data-driven decision-making.

The responsibility for the development and application of the OMOP Common Data Model (OMOP-CDM) passed to the Observational Health Data Sciences and Informatics (OHDSI) organization in 2014. Fig. 4.2 shows a broad

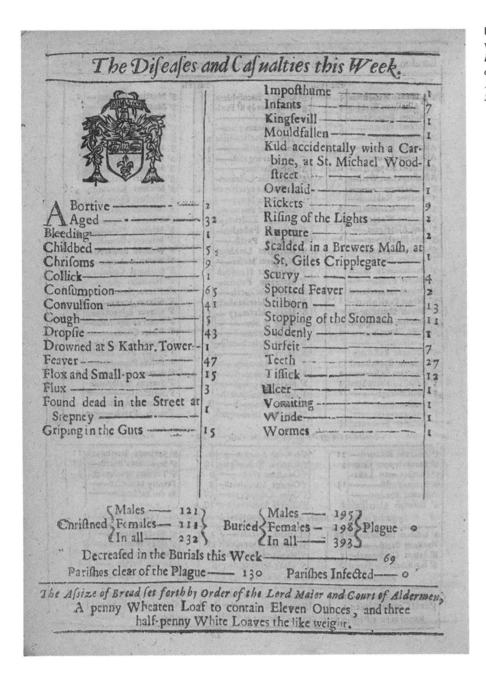

FIGURE 4.1 Bill of Mortality for the week of February 21–28, 1664 in London. *Boyce, N. 2020. Bills of mortality: tracking disease in early London. Lancet 395, 1186–1187.   https://doi.org/10.1016/ S0140-6736(20)30725-X.*

overview of how the OMOP-CDM relates to disparate observational data sources and subsequent analytical systems. The OMOP-CDM has been adopted by over 150 medical centers worldwide, including the University of California campus medical centers at San Diego, Irvine, Los Angeles, and San Francisco. At UC-Irvine, standardization in the medical outcome vocabulary permitted the development of a common Quality Measurement Engine (QME) for standardized reporting purposes. The OMOP-CDM data platform provides a consistent source of data for SQL scripts and for analytical modeling. The responsibility for the maintenance and update of the OMOP-CDM passed to the OHDSI (Observational Health Data Sciences and Informatics) organization in 2014.

Fig. 4.2 shows a high-level expression of the OMOP-CDM.

Reasons for OMOP

1. Need to understand each data element of each data source: a very difficult task.
2. Need for a standardized vocabulary across multiple data sources.

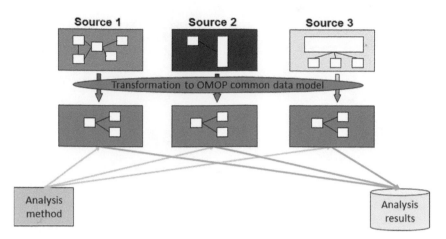

FIGURE 4.2   The relationship between disparate medical data sources and analytic data structures via initial transformations to a common representational format—the OMOP common data model. *From: https://www.ohdsi.org/data-standardization/the-common-data-model/*

3. Since US claims databases are derived off two standard forms, health Insurance Claim Form-1500 and Universal Billing form 92, one might assume the two databases would have similar content and structure; however, this is not the case. Need to provide for storage and analysis for reasons other than for billing purposes.

The Observational Medical Outcomes Partnership developed the OMOP Common Data Model (CDM) to address the standardization issue. The motivation was to enable the consistent transformation of diverse observational databases into a common vocabulary, and then use it to perform systematic analysis, including predictive analytics.

Voss et al. (2015) provides part of the value proposition for the OMOP CDM. Erica Voss elaborated on this value proposition in 2018 by listing the major elements of it (Voss, 2018).

According to Voss (2018), the CDM provides capabilities that include:

1. Cross-institutional collaborations (everyone has the same format, one code set could run for everyone).
2. One analysis program can run on multiple databases.
3. Research can be accelerated due to easy access to de-identified data that is mapped onto standard vocabularies and increased capability to do analysis due to access to free tools (like KNIME).
4. Standard OMOP Vocabularies facilitates finding of relevant codes to serve research needs.
5. Transformation of various data sets into OMOP CDM format greatly facilitates the development of an intimate knowledge of the source data elements.
6. This intimate knowledge of the destination data set facilitates quality assessment and the remediation operations during the ETL process.

The OMOP CDM provides a common data format

Medical information is stored in databases in multiple formats. In its most common format, the electronic health record (EHR) was designed to serve billing and regulatory purposes. Data in many EHR systems (e.g., Epic) are stored in data structures that are not very compatible for data analysis and decision-making. Most EHR systems store data at the granular level of the office visit by a patient to serve billing and reporting systems related to a single medical service instance. The OMOP-CDM permits the characterization (by SQL scripts) of a medical event or service provision in terms of a common medical outcome (e.g., represented by the OMOP condition code = 34.1, to indicate Scarlet fever). This common characterization of multiple outcomes among multiple EHRs for a given patient greatly facilitates further analysis of causes and treatments of Scarlet fever. Details on the OMOP CDM are available at https://ohdsi.github.io/TheBookOfOhdsi/CommonDataModel.html#fn20.

OMOP CDM architecture is patient-centric

The OMOP CDM is a dimensional model, not a relational model. Data representing various dimensions of medical issues are stored in separate tables, all linked to the Persons table. The dimensional tables store medical events for:

1. Conditions
2. Procedures

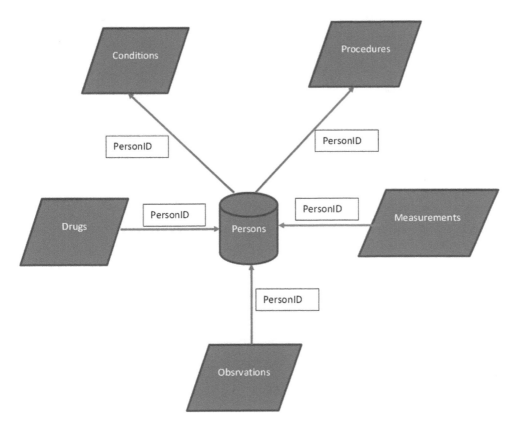

FIGURE 4.3 The dimensional OMOP model showing the patient-centric organization of information.

3. Drugs
4. Measurements
5. Observations

Fig. 4.3 shows the conceptual relationship between the five dimensions of data in the UCI colon polyp model and the central Persons table. This design is person-centric, rather than account-centric, which makes it easier to write a SQL script to gather all information of a given person ID. This architecture serves adequately to support the data access needs for reports and descriptive analytics (charts and graphs), but it does not serve all of the need of analytical modeling.

Medical data stored in the dimensional tables may exist at different levels of aggregation. For example, general demographic information in the Persons table is usually *static* (e.g., Gender and Birthdate), that is it does not change over time. Data in the five dimensional tables, however, can be very dynamic, that is many records may exist for conditions (for example) recorded in many visits to the medical center on different dates. For this reason, the records in the Persons table are organized one-to-one (one record to one patient), but the records in the dimensional tables are organized one-to-many (many records for a given patient). The numbers of OMOP codes associated with the five dimensional tables are listed in Table 4.1.

Additional data processing operations necessary to serve the analysis of OMOP data

Focus on the data representation problems in the OMOP-CDM does not, however, illustrate the huge amount of data preparation processing represented by the connecting lines between the OMOP data structure in Fig. 4.3 and analytical operations shown in Fig. 4.2. An important part of data preparation processing is to integrate all input variables for a given person to the same line (row or record) for analysis by predictive analytics algorithms.

This collapse of person data to a single line can consume a large amount of data processing. For example, Table 4.1 shows that there are 43,552 codes among the five dimensional tables that could be related to a given patient. Potentially, individual binary predictor variables (called variously "dummy" or "one-hot" variables) could be derived from each of the 43,552 codes and submitted to the modeling algorithm. A one-hot variable for a given code will contain a 1 if the code is found in any of the person records in the database, or a 0 if it is not found. While this operation can be automated in the KNIME modeling platform, many of the codes will not be relevant to colon cancer, and they will slow down

TABLE 4.1 Numbers of OMOP codes in the five dimensional tables.

Conditions	23,411
Procedures	12,197
Drugs	7863
Lab measurements	3123
Observations	81
Total	43,552

processing significantly, and just get in the way of selecting the important codes as predictors. Selecting those codes to use in deriving predictor variables (feature selection) is part of the data preparation process discussed below.

Most books on analytical data modeling proclaim that 60%–90% of project time is spent on data preparation (Nisbet et al., 2017). Sometimes, the data preparation activities are called "data wrangling." The initial task of this wrangling is to collect all data elements for a given patient on a single line (a "record") for analytical modeling (like wrangling steers for a cattle drive). For building management reports, this operation is performed often by a series of GroupBy statements in SQL.

For reporting or analysis of data at the patient level requires that many records for a given patient be grouped to the patient level. But each patient visit may be related to a different medical issue. How can all medical issues be related to a single patient in one row of data (needed for analysis at the patient level)? This problem is further complicated by the many different data representations (e.g., numbers, codes, text) for multiple data elements in each of the five dimensions of information shown in Fig. 4.3.

The Analysis Method box in Fig. 4.2 represents a large number of processes and methods necessary to support the building of an analytical model. To understand what must happen at the detailed level, we must drill down further. Before descending to the individual data preparation process level, however, the high-level process model followed in this project should be understood. Fig. 4.4 shows the most common phases of the process model followed in this project to build an analytical model, the Cross-Industry Standard Process for Data Mining (CRISP-DM).

The CRISP-DM processing model

The CRISP-DM process model for data analytics displays the order of six phases of data processing for building analytical models, and it shows also the flows and backflows of data between these phases. Analytical modeling is not a straightforward process. It involves feedbacks (backflows of data) to compose an iterative system.

CRISP-DM phases

1. Business Understanding. This process model was built initially for working with business data, but it can be applied to any operations domain, including the medical operations.
2. Data Understanding. For medical data, it is necessary initially to understand what the many OMOP codes mean, and how they relate to the medical issue to be modeled. For building the precancerous polyp model, it was necessary to review the medical literature to understand how precancerous polyps are usually diagnosed (via a colonoscopy), and how each OMOP condition and procedure code related to the colonoscopy procedure. One of the reasons for the literature review was to eliminate from the model those OMOP codes that were related to the colonoscopy procedure itself or the logical outcome of it. Many codes were eliminated in this phase of the modeling effort. For example, OMOP procedure code 45.23 (colonoscopy) was eliminated from the modeling data set.
3. Data Preparation. Most of the project time was spent performing tasks of data preparation. Major activities of data preparation include:
   a. De-identification—necessary to remove links to specific patients before the data set is moved outside the UCI Health Services IT infrastructure for analytical modeling.
   b. Data Integration—necessary to combine data elements from the central Persons table with information from the five dimensions of medical information.

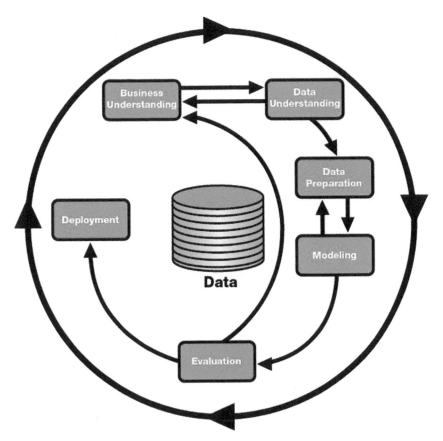

FIGURE 4.4 The cross-industry standard process for data mining (CRISP-DM) (Nisbet et al., 2018). *Nisbet, R., Miner, G., Yale, K. 2018. Handbook of Statistical Analysis and Data Mining Applications. Academic Press, Cambridge, MA, pp. 792.*

c. Data reformatting and quality analysis—necessary to convert values to a common data type.

d. Target definition—necessary to specify which categories of medical pathology should be predicted as positive (Target = 1), and those to be predicted as negative (Target = 0)

e. Data exclusions—necessary to identify and remove from the analysis those data elements that are not appropriate for modeling.

f. New variable derivation—necessary to combine existing data elements to form new variables that might be related to the Target variable.

g. Aggregation of all records to the patient level—necessary to group all data elements related to a given patient into one analysis row (record).

h. Feature selection—necessary to reduce the number of features for modeling to those features judged by some analysis method to have a high likelihood of being related to the Target variable.

i. Preliminary Modeling—necessary to decide which data preparation methods to use (e.g., which balancing method or which missing value method is best for the prepared data set). These tasks will be described in detail in the Data Preparation section as examples of those operations that must be accomplished for building models with OMOP data.

4. Modeling. Much of the rest of the project time was spent in building models with various algorithms and various data sets to identify the optimum algorithm and data set to use for the building final model. The selected modeling algorithm must be properly parameterized to build the model with the optimum accuracy (knows also as "hyperparameterization").

5. Model Evaluation. Some metrics of model performance were evaluated to select the most appropriate model for the project. The most predictive model according to one evaluation metric may not be the best model to use.

How this chapter facilitates patient-centric healthcare

This chapter provides an overview of an appropriate medical data classification and an analytical process model to help medical practitioners input patient-centric data and analyze the results in terms of the appropriate context necessary to understand results. This top-down perspective of understanding will permit the proper diagnosis and treatment of medical conditions specific to a patient expressed in terms of low-level patient data (a bottom-up perspective). Most complex systems are like this. See Chapter 14 for more information on how the merger of top-down perspectives and bottom-up perspective on natural phenomena can help the analyst to recognize and understand emergent properties in medical informatics studies.

Postscript

Chapters 1–4 presented top-down perspectives of patient-centric healthcare. The stage is now set to add to our understanding of medical informatics some bottom-up elements and issues of patient-centric data sources (Chapter 5), personalized medicine (Chapter 6), and patient-directed healthcare (Chapter 7).

References

Bell, W.G., 2001. The Great Plague in London in 1665. The Folio Society London, 256 p. (originally published in 1924).

Boyce, N., 2020. Bills of mortality: tracking disease in early London. Lancet 395, 1186–1187. Available from: https://doi.org/10.1016/S0140-6736(20)30725-X.

Inman, W., Imhoff, C., Sousa, R., 1998. The Corporate Information Factory. Wiley Computer Publishing, New York, NY, p. 274.

Nisbet, R., Miner, G., Yale, K., 2018. Handbook of Statistical Analysis and Data Mining Applications. Academic Press, Cambridge MA, p. 792.

Overhage, J., Ryan, P.B., Reich, C.G., Hartzema, A.G., Stang, P.E., 2012. Validation of a common data model for active safety surveillance research. J. Am. Med. Inform. Assoc. 19, 54–60. Available from: https://doi.org/10.1136/amiajnl-2011-000376 [PMC free article] [PubMed] [Google Scholar].

Voss, E.A., Makadia, R., Matcho, A., Ma, Q., Knoll, C., Schuemie, M., et al., 2015. Feasibility and utility of applications of the common data model to multiple, disparate observational health databases. J. Am. Med. Inform. Assoc. 22 (3), 553–564. Available from: https://doi.org/10.1093/jamia/ocu023. Epub 2015 Feb 10. PubMed PMID: 25670757; PubMed Central PMCID: PMC4457111.

Voss, E., 2018. Forum post. <https://forums.ohdsi.org/t/reasons-to-ohdsis-omop-cdm/5273>.

WHO, 2018. WHO releases not international classification of diseases (ICD 11). Press release. <https://www.who.int/news/item/18-06-2018-who-releases-new-international-classification-of-diseases-(icd-11)> (accessed October, 2021).

Further reading

OMOP CDM, Reich, C., Ryan, P.B., Belenkaya, R., Natarajan, K., Blacketer, C., 2019. OHDSI common data model v6.0 specifications <https://github.com/OHDSI/CommonDataModel/wiki>.

Chapter 5

Access to data for analytics—the "Biggest Issue" in medical and healthcare predictive analytics

Gary D. Miner

Chapter outline

Prelude

Chapter 4 described the proper data structures and processing model that are required to analyze any patient healthcare information properly. In this chapter, we will consider the vast amount of healthcare data that is now available, and how that volume is expanding exponentially. Existing problems associated with storage and formats of this huge volume of data are discussed in terms of possible solutions to permit it to serve patient-directed healthcare operations.

Size of data in our world: estimated digital universe now and in the future

In 2005 the International Data Corporation estimated the size of the "digital universe" to be about 130 exabytes (EB). And by 2017 this had expanded to 16,000 E (or 16 zettabytes=ZB). This was expected to be about 40,000 EB by 2020, or about 5200 gigabytes (GB) of data for *each* individual (Dash et al., 2019).

But just how much data does healthcare really have? Today, every second, a tremendous amount of healthcare data is being obtained by EHR (Electronic Health Records) and other places. This data is being mined, some of it in "real time," to gain information and insights. It is estimated that today 30% of the world's data volume is coming from the healthcare industry, and that by 2025 the annual growth rate of data acquisition for healthcare will reach 36% (see Fig. 5.1) (RBC-Capital Markets, 2022).

Some of the key points in all of this are that:

- healthcare is generating the world' largest volume of data;
- healthcare is pivoting (maybe, at least trying) from treatment to prevention; and
- healthcare and technology are converging to innovate wearables that measure instantly and sometimes continuously health metrics.

Practical Data Analytics for Innovation in Medicine. DOI: https://doi.org/10.1016/B978-0-323-95274-3.00008-7

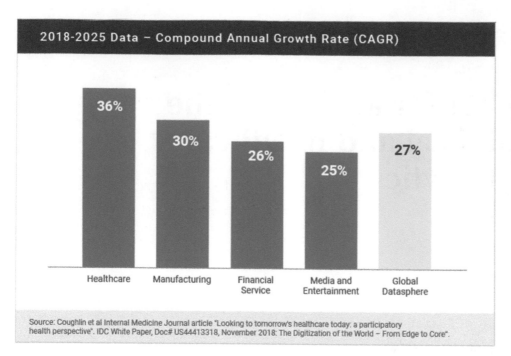

FIGURE 5.1 The growth rate in digital data acquisition, from 2018 and estimated to 2025. Healthcare clearly has the largest proportion, at 36%. *From Wiederrecht, G. Healthcare technology will, one day very soon, literally touch us all on a daily basis. <https://www.rbccm.com/en/gib/healthcare/episode/the_healthcare_data_explosion>* (accessed 05.01.22).

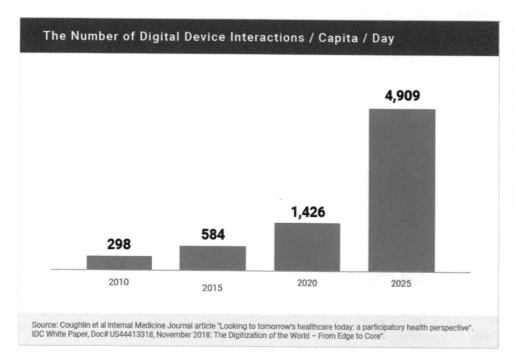

FIGURE 5.2 The number of digital interactions *each* person has per day. Data from 2010 and 2015 with estimates for 2020 and 2025. Clearly the prediction is for an exponential increase during the 3rd decade of the 21st century. *Wiederrecht, G. Healthcare technology will, one day very soon, literally touch us all on a daily basis <https://www.rbccm.com/en/gib/healthcare/episode/the_healthcare_data_explosion>* (accessed 05.01.22).

Convergence of healthcare and modern technologies

The digital revolution in healthcare is happening right now, and all around the world. We have seen this clearly during 2019 and 2020 in the COVID pandemic, where daily we get reports of the incidence and prevalence of this illness, not just from countries like the United Kingdom, Australia, and United States, but from all over the world (see Chapter 16 about visualizations of COVID-19 data for countries of the world). Instant information available to anyone almost anywhere. (See Fig. 5.2 for growth of number of human/digital interactions per day from 2005 and estimated to the future.)

Collections of biotic communities (or major life zones) classified according to dimensions of temperature and precipitation were called "biomes" by Robert Whitaker, following similar use of the term by previous ecologists (Whittaker, 1962). These lifezones could be described by the dimensions of temperature and precipitation. Likewise, the recent availability of many kinds of biological data for an individual (e.g., genetic, metabolic, and metabolomics data) has been termed the "data-ome." Wiederrecht (2022) describes the development of the data-ome as follows.

Humans are being digitized through new devices, apps and monitoring technologies, which are tracking, analyzing and storing this massive amount of data. The smart phone in your pocket is a prime example along with cloud computing, AI and the wearable on your wrist. Together, the data generated from these technologies is creating what's been dubbed as an individual's **'data-ome.'** *All of which is setting the stage for a battle to access and utilize the digital representation of the world's health and wellness....*

Source: RBC-Capital Markets (2022).

With this growth in digital interactions, the global healthcare budget will be an estimated and astounding $15 trillion by 2030. Digital health brings together a large number of industries that include genomics, ingestibles and implantables, wearables, sensors, retailers, social media, AI, analytics, clinical data, and EHR (Gopal et al., 2019; Solbah & Grünewald, 2019).

Many think that by 2030 healthcare will be centered on patients (e.g., "patient-centered" and "patient-directed") that have become enlightened enough about the healthcare industry that they are now empowered to prevent diseases instead of waiting until they must seek treatment for a disease. For this to happen, there will have to be an enablement of both data and algorithms in the healthcare system; this means that the healthcare system will have to be organized in an entirely new way. Physicians and caregivers will have to redefine their roles, and payers will have to make ways for paying for new types of healthcare modalities.

Alongside this overhaul of the healthcare system, tectonic shifts in global public- and private-sector budgets are expected: trillions of health dollars will be spent differently than they are spent today. There will be less emphasis on treatment and care, and more on prevention, diagnostics, and digital solutions, such as mobile apps, smart monitoring devices and artificial intelligence (AI)–enabled analytics tools.

From Solbah and Grünewald (2019).

Healthcare budgets will increase by 2030 but spending per patient is expected to decrease (as much as 30% maybe) because the number of people having access to good healthcare can be expected to increase but not at the rate of healthcare budgets. Most likely the big tech giants will be leaders in these changes in healthcare, instead of the big pharma, regulators, and the providers. Big tech has already entered the prevention and diagnostics markets and is also moving into biopharma's treatment markets. This will most likely impose a big challenge for biopharma companies, but at the same time can bring great opportunities for biopharma if they prepare themselves to move into diagnostics, prevention, and digital health solutions. I think this means that biopharma must reimagine themselves if they want to maintain relevance to the healthcare of the future.

Most biopharma companies seem to know this change is coming fast, since in a survey 96% agreed that the future of healthcare will be with people individually managing their own health themselves, with this involving personalized and digital methods as part of their daily lives; additionally, this survey showed:

- 68% expected this scenario to be the norm in major healthcare markets by 2030.
- 75% perceived the future of healthcare as an opportunity for biopharma if the sector is willing to disrupt itself.
- 85% said they have some or all the key elements of the future of healthcare on their corporate agenda.
- But only 25% are taking a holistic approach to addressing the challenge.

"The future of healthcare is coming, and it's not a matter of whether, but when, the sector will be disrupted" (Solbah & Grünewald, 2019).

Reasons why healthcare data is difficult to get and difficult to measure

- Much of the data is in multiple places.
- The data is structured and unstructured.
- It has inconsistent and variable definitions; evidence-based practice and new research is coming out every day.
- The data is complex.
- Changing regulatory requirements. (LeSueur, 2018).

Multiple places where medical data are found

Medical data is found in many places. Among these sources of medical data are the following:

- EMR (Electronic Medical Records) (also called EHR)
- Radiology
- And other Medical Departments in clinics and hospitals
- Pharmacy
- Hospitals
- Clinics
- Private Doctors
- Payors (the Insurance Companies, Medicare Medicaid, etc.)
- Personal Health Records
- Electronic Prescription Services (E-prescribing)
- Patient Portals
- Master Patient Indexes
- Health-Related Smart Phone Aps

If all these sources of medical data "talked to each other," e.g., were linked up so that there was interoperability among these sources of data, the use of this data might not be such a big problem. But indeed, it is just the opposite, making it a big, big issue! Different hospitals, even different departments (such as radiology, internal medicine, etc.) within a hospital may use different "Medical Record Systems"; invariable these different systems do not talk to each other.

So, the fact there are so many different places where critical medical data is stored, coupled with in most cases the different data sources cannot digitally interact with each other, has created a great difficulty for getting good data for (1) doctors easily using the data to make good diagnostic and treatment decisions, let along (2) having good data for statistical and predictive analytic modeling.

Many different formats of medical data: structured and unstructured

Data useful for data analytics and making predictions and prescriptions for doctors to use in better diagnosis and treatment decisions take on the following formats:

- Text
- Numeric
- Written on paper
- Stored digitally (electronic data)
- Pictures
- Videos
- Multimedia
- Images (radiology, etc.)
- Old handwritten medical records

And the above can be found in different formats, such as:

- Clinical data formats
- Claims data formats

An example of this is a patient with a broken arm becomes an "Image" in the clinical record but is an ICD-9 (or ICD-10) code in the claims data files.

So, I think it is easy to see that it can become a "nightmare" to gather and coordinate the data from all these sources into a single format that becomes absolutely necessary before any good data analytics can be performed and interpreted, leading to the knowledge that the medical system needs to make good, real-time diagnoses and treatment decisions.

Another problem is inconsistent definitions

- One clinic may define COPD differently than another clinic.
- Three different doctors may give three different "criteria" needed to diagnose a specific disease, like diabetes.

- And these same three doctors may give you five different ways to treat a specific disease!
- And with new research/new information continually coming out, it can change how the "doctor that reads this research" decides to treat their patients; thus, the data that is recorded, keeps changing.

This lack of standardization in defining diseases, different doctors preferred ways of treating a disease, and an "information library" that is in a continual state of flux exaggerate the difficulty in obtaining good medical data on which to perform good data analytics leading to good decisions.

Changing government regulatory requirements keep changing what data is taken and kept

Unfortunately, regulations set forth by the Federal Government and other agencies also make for additional difficulties in obtaining good data, especially good and equivalent data over time. Some of these regulatory agencies and issues are:

- Regulatory requirements from places like CMS (Center for Medical Services of the Federal Government) change and almost invariable increase.
- Reporting requirement thus are continually changing. Those from the CMS are often around things like:
  - Transparency in Pricing
  - Quality enhancements
  - Value-based medical care.

Unlike many businesses and industries where definitions and "business rules" are fixed for long periods of time even decades, in medicine and healthcare things are volatile—"a rule" (criteria of DX or treatment, for example) of today may *not* be "the best practice" tomorrow; these changes make for a great difficulty in making sense of original sources of medical data.

What are some of the benefits of using good data analytics in medical research and healthcare delivery?

Modern Data Analytics are being used successfully in most aspects of society and have been for the past 20 years. Medicine has been slow to accept. But during the remainder of this decade healthcare is predicted to rapidly implement effective data analytics and decision systems, which should lead to:

- Improved performance for operations
- Advanced and more accurate treatments for patients
- Discovering "the right treatments" for diseases
- Personalized–"person-centered" communication
- Much easier and faster (if not instant) access to key information

By the year 2030 we should see that much of this has been implemented, and "personalized medicine" has come of age.

Conclusion of Chapter 5: the importance of health care data analytics

One of the best descriptions we've seen that sums up thoroughly and concisely the value of data analytics for medicine and healthcare was stated in an online blog:

> ....We can collect all the data we want, but it doesn't do any good if we don't know what to do with that information.
>
> We need a centralized, systematic way of collecting, storing and analyzing data so we can use it to our advantage.
>
> The collection of data in health care settings has become more streamlined in recent years. Not only does the data help improve day-to-day operations and better patient care, but it can also now be better used in predictive modeling. Instead of just looking at historical information or current information, we can use both datasets to track trends and make predictions. We are now able to take preventive measures and track the outcomes.
>
> The fee-for-service style of health care is becoming a thing of the past.

(Continued)

(Continued)

There is a growing demand for patient-centric, or value-based, medical care which has led to a considerable shift towards predictive and preventive measures regarding public health in recent years. Data makes this possible.

Instead of simply treating the symptoms as they present, practitioners can identify patients at high risk of developing chronic illnesses and help to treat an issue before it surfaces. This helps to lower costs for the practitioner, insurance company and patient as the preventive treatment may help to stave off long-term issues and expensive hospitalizations.

If hospitalization is necessary, data analytics can help practitioners predict risks of infection, deterioration and readmission. This too can help lower costs and improve patient care outcomes.

Consider the impact this has had on the COVID-19 pandemic. The data being collected is analyzed in real time to understand the effects of the virus better and predict future trends so we may slow the spread and prevent future outbreaks.

Source: University of Pittsburgh: Online School of Health and Rehabilitation Services (2021)

Additionally, we can think of a couple of other conclusions from information presented in this chapter:

- All medical data formats should be standardized to permit proper aggregation and analysis of healthcare data among and between healthcare and research institutions.
- Healthcare data storage structures (databases) should be patient-centric, not condition-centric.

Some of the authors of this 2nd edition of this book have been involved in data analytics and medical research for upwards of 40 years, living through the development of modern Predictive Analytics and AI that were being researched by a few during the 1980s and 1990s, as computing power rose up to the level needed for the humongous computations required by some of the AI and ML methods, but then began to become "mainstream" in 1999 and rapidly entered the business fields in the first decade of the 21st century. Now is the time for medicine and healthcare delivery to "fully embrace" the power available in these methods and push to bring about the needed revolution in the delivery of healthcare. It has already started: we predict it will proceed rapidly during the next 5–10 years!

Postscript

The initial problem with healthcare data is very evident now—there is so much of it! And the problem is getting worse *exponentially*. A second problem is that much of the healthcare data is stored in many different formats. Integrated healthcare management in the United States will require very soon that there be some kind of central repository of healthcare information at the regional or national level. First steps have been made in the European Union for the development of such a national healthcare data repository for hospital records (EHR4CR).

Chapters 6, 7, and 8 will explore the nature and constraints of personalized healthcare. But before personalized medicine can be practiced consistently, the myriad of data sources must be standardized and committed to a central repository. A centralized system like the EHR4Dr project in Europe should be developed in the United States.

References

Dash, S., Shakyawar, S.K., Sharma, M., et al., 2019. Big data in healthcare: management, analysis and future prospects. J. Big Data 6 (54), 2019. Available from: https://doi.org/10.1186/s40537-019-0217-0. Available from: https://journalofbigdata.springeropen.com/articles/10.1186/s40537-019-0217-0#Ack1.

Gopal, G., Suter-Crazzolara, C., Toldo, L., Eberhardt, W., 2019. Digital transformation in healthcare - architectures of present and future information technologies. Clin. Chem. Lab. Med. 57 (3), 328–335. Available from: https://doi.org/10.1515/cclm-2018-0658. PMID: 30530878.

LeSueur, D., 2018. Five reasons healthcare data is difficult to measure. Hralthcatalyst. Available from: <https://www.healthcatalyst.com/insights/5-reasons-healthcare-data-is-difficult-to-measure/> (accessed 07.03.2018).

RBC-Capital Markets, 2022. The healthcare data explosion. <https://www.rbccm.com/en/gib/healthcare/episode/the_healthcare_data_explosion> (accessed 05.01.22).

Solbah, T., Grünewald, P., 2019. Driving the future of health: How biapharma can defend and grow its business in an era of digitally enabled healthcare. strategy&: Part of the PwC network. Available from: <https://www.strategyand.pwc.com/de/en/industries/health/future-of-health.html> (accessed 08.01.2019).

University of Pittsburgh, 2021. Online School of Health and Rehabilitation Services. The role of data analytics in healthcare. University of Pittsburgh Blog. <https://online.shrs.pitt.edu/blog/data-analytics-in-health-care/> (accessed 01.13.21).

Wiederrecht, G., 2022. The Convergence of Healthcare and Technology. rbc capital markets. <https://www.rbccm.com/en/gib/healthcare/episode/the_healthcare_data_explosion> (accessed 05.01.22).

Further reading

Chakraborty, M., 2021. Big data healthcare latest news: big data analytics in healthcare: possibilities and challenges. Big Data Healthcare Latest News <https://www.analyticsinsight.net/big-data-analytics-in-healthcare-possibilities-and-challenges/>.

Fagella, D., 2018. Where healthcare's big data actually comes from. <https://emerj.com/ai-sector-overviews/where-healthcares-big-data-actually-comes-from/> (accessed 05.01.22).

https://www.healthcatalyst.com/wp-content/uploads/2021/05/infographic-why-healthcare-data-is-difficult-768x602.png.

https://www.slideshare.net/healthcatalyst1/5-reasons-why-healthcare-data-is-difficult-to-measure?from_action=save.

Chapter 6

Precision (personalized) medicine

Nephi Walton

Chapter outline

Preamble

Chapter 5 described at a high level the nature of the "data-ome" of an individual patient. This chapter unpacks that concept and describes in greater detail the nature, the relevancy and the challenges associated with the use of each data type to the practice of personalized medicine. The reader might question the necessity of digging so deeply into the nature of the data that we analyze. The answer to this question is given in principle that underlies the inscription on the entrance to the Oracle at Delphi, "Know Thyself". The principle is that when you seek knowledge of anything, know yourself first.

P4 (Personalized; preventative; Predictive; and Participatory) medicine will mandate that every sector of the public and private healthcare systems rewrite their business plans over the next 10 years

Practical Data Analytics for Innovation in Medicine. DOI: https://doi.org/10.1016/B978-0-323-95274-3.00002-6

In 10- or 15-years, with the fast decline in the cost of next-generation(genetic) sequencing, we will potentially have access to the complete(human) genome and (associated) medical, molecular, cellular, and environmental data for a growing fraction of the human population in both developed and developing countries. This will afford us an unparalleled opportunity to (use predictive analytics on) these data for (developing) the predictive medicine of the future

Paraphrased from: Lee Hood, Revolutionizing medicine in the 21st Century Through Systems Approaches (Hood and Tian, 2012).

What is personalized/precision medicine?

The cartoon shown in Fig. 6.1 is from a 1971 article on personalized medicine (Gibson, 1971). The author had used the term to describe having a personalized relationship with your physician rather than seeing a battery of specialists for each condition (Gibson, 1971). This is the first article to appear in PubMed using the search term "personalized medicine." This article was written to express the fear of the medical community's departure from personalized medicine into an era of subspecialists. It is ironic that the cartoon meant to depict the medical communities moving away from a personalized format of medicine, and yet is very close to depicting personalized medicine as it is viewed today.

The first use of the term "personalized medicine" as it is used today was on April 16, 1999. This use of the term appeared in the title of an article published in the Wall Street Journal by Robert Langreth and Michael Waldholz entitled "New Era of Personalized Medicine—Targeting Drugs for Each Unique Genetic Profile" (Langreth and Waldholz, 1999a,b). The article described the formation of the single nucleotide polymorphism (SNP) consortium. Its purpose was to create a comprehensive SNP map of the human genome, which would provide the subsequent possibility of being able to develop drugs specific to a person's genetic makeup. This article was later published in the scientific literature in the journal "The Oncologist" (Langreth and Waldholz, 1999a,b; Jørgensen, 2009).

Today personalized medicine has a very broad definition that spans a number of medical fields, however, the succinct definition could be stated as "the tailoring of a treatment to an individual based on their unique characteristics." The treatment may be a medication, an exercise, a diet, a surgical procedure, or any other therapeutic measure used within the field of medicine. The treatments may be tailored to the individual by using genetic markers or other biological markers, environmental information, behavioral information, demographic information, or any other information that might change the way the patient responds to the treatment. Using these factors, physicians can determine a type of treatment, duration of treatment, what dose to give, what kind of follow-up is needed, or who is most likely to have certain side effects. With this information in hand, the appropriate monitoring system can be assigned to each patient.

At this time, however, it is not economical or realistic to think that research can be done on each individual to discover and produce medications that are appropriate for use only with that person. The President's Council of Advisors on Science and Technology accepted the succinct definition of personalized medicine presented above. The Council went on to explain further: "personalized medicine does not literally mean the creation of drugs or medical devices that

FIGURE 6.1 Cartoon on "Personalized Medicine." *Used by permisison from Gibson, W. M., 1971. Can personalized medicine survive? Can. Fam. Phys. 17 (8) 29–88.*

are unique to a patient, but rather the ability to classify individuals into subpopulations that differ in their susceptibility to a particular disease or their response to a specific treatment. Preventative and therapeutic interventions can then be concentrated on those who will benefit, sparing expense and side effects for those who will not."

Personalized medicine versus precision medicine

Personalized medicine and precision medicine are often used interchangeably, as one could argue that they are synonymous or at the least, they are overlapping. The National Research Council has stated that "personalized medicine" is an older term with similar meaning to "precision medicine." "Precision" was considered a more appropriate term than "personalized," as the latter term "could be misinterpreted to imply that treatments and preventions are being developed uniquely for each individual"; while in precision medicine, "the focus is on identifying which approaches will be effective for which patients based on genetic, environmental, and lifestyle factors." Based on these perceived differences the council preferred "precision medicine" to "personalized medicine" (MedlinePlus, 2022). The author's perspective is that "precision medicine" is synonymous with "personalized medicine" and that over time, as our knowledge increases, we will indeed be developing treatments catered to the individual as our precision increases. These terms will be used interchangeably throughout this book. While the author's preference is to use "precision medicine," "personalized medicine" has significant historical context that has defined the field.

P4 medicine

More recently, Dr. Leroy Hood of the Institute for Systems Biology has termed personalized medicine as "P4 Medicine" (Hood and Tian, 2012). He describes the four components of this medical approach with four Ps:

- The first P is for *personalized*, e.g., a medicine or treatment that takes into account a person's genetic or protein profile.
- The second P is for *preventative*, e.g., anticipating health problems and focusing on wellness, not disease.
- The third P is for *predictive*, e.g., directing appropriate treatment, and avoiding drug side effects.
- The fourth P is for *participatory*, e.g., empowering patients to take more responsibility for their health and care.

P5 to P10 medicine

Others have gone on to define further the nature of personalized medicine to include other Ps:

- **Psychocognitive** aspects referring to how individuals act to prevent, cope and react to illness; decide about different therapeutic options; interact with healthcare providers; and adhere to treatment (Gorini and Pravettoni, 2011).
- **Public**, referring to societal and population-based media formats that involve shared and open-source entities consisting of a combination of e-health, e-medicine, and telemedicine in which computers, "expert systems," and innovative online health communities, play a central role in a patient's health (Bragazzi, 2013).

While several other Ps have been suggested, such as precision, pluriexpert, and others (Pires et al., 2021; Gardes et al., 2019; Jain and 2021, 2021), these start to fall in the realm of redundancy or have poor generalization to the overall concept or personalized medicine. In the authors view, P4 medicine defines the core principles of precision/personalized medicine and the other Ps may serve as important considerations in the application of precision medicine.

Precision medicine, genomics, and pharmacogenomics

The activity of precision medicine is associated commonly with the application of pharmacological treatments to a person with a specific set of genes (a genotype). Personalized medical treatment expects that a given person will respond differently than another person, because the two people have different sets of genes, which cause different responses. The study of the specific nature of responses to drugs by people of different genetic complements is called pharmacogenomics.

Genetics and pharmacogenomics were included as components of personalized medicine when it was first described. While both of these components are crucial elements of personalized medicine, it encompasses much more, including many environmental (nongenetic) processes that occur within the human body or to the human body from external influences. The advances in genetics over the past decade have been breathtaking. Medical science has learned that the expressions of genes within each individual determine an individual's ability to fight off or contract disease and governs how they will respond to medications and treatments.

Modern gene sequencing technology can map entire genomes at rates far faster and at lower prices than previous generations of this technology. This technology has led to an onslaught of genetic data, which has raised many new possibilities, as well as substantial problems, not the least of which is the lack of a suitable infrastructure within medicine to support and process this large amount of information. This situation is changing rapidly, however, and in the near future an individual's genome sequence will be considered a vital part of everyone's medical record.

Pharmacogenomic studies have led to the development of drug therapies designed (targeted) for specific genetic complements. Many of these targeted drugs have been approved by the FDA for the treatment of disease, and many more are in the development stages. Among these targeted drugs, those designed to treat cancer, specific single gene loci genetic diseases, and infectious disease are among the most prominent treatments in use. Increasingly smaller subsets (groupings) of diseases are being identified and targeted with focused pharmacological therapies that are designed to work at the molecular level for that specific group of diseases.

Pharmacogenomics is an important application of genomic sequencing. It is well known that the same drug may have a different effect on different individuals due to both their personal genomic background and environmental living habits (Heller, 2013). People metabolize drugs at different rates based on their genetic makeup which means in those that metabolize a particular drug slower the drug stays active in the system longer, increasing the effect of the drug. In contrast, those that metabolize the drug more quickly will have less active drug circulating and therefore will have less effect. In other cases, such as with Clopidogrel, the drug may not be able to be effectively metabolized into its active form, rendering the drug ineffective (FutureMed/Singularity University, 2013; https://www.futuremedicine.com/doi/10.2217/fca-2016-0045). Because people metabolize drugs differently, drug dosages should be based on an individual's ability to metabolize the drug, similar to how we base dosages on weight now. Genetic information can be used to assign drug doses as well as reduce side effects. The FDA currently lists 72 drugs where evidence supports the use of pharmacogenetics for prescribing based on drug efficacy, side effects, or adverse outcomes (https://www.fda.gov/medical-devices/precision-medicine/table-pharmacogenetic-associations, accessed Feb 20, 2022). Additionally, based on pharmacogenomic trials, genetic tests are required by the US Food and Drug Administration (FDA) for several drugs before they can be administered to patients. Examples include the anticancer drugs cetuximab, trastuzumab, and dasatinib, and the anti-HIV drug maraviroc. The anticoagulant drug Warfarin and the anti-HIV drug Abacavir are also included in this category.

Differences among us

When considering personalized medicine, many people relate it to our genetic makeup; however, this is only a small part of the differences among us that determine how we will respond to treatments. Two people, such as identical twins, can have identical genetic makeup but if the genes are not expressed equally there will be different results. The science that describes the control of gene expression is *epigenetics*. Epigenetics is an umbrella term that covers effects of many DNA regulatory systems, including DNA methylation, histone modification, nucleosome location, and noncoding ribonucleic acid (RNA). Our genes can be considered as a set of instructions, and our epigenetic "markers" determine how those instructions are carried out. Many of these epigenetic effects related to environment influences are heritable. Therefore, what you eat, drink, and do today may affect your great grandchildren in the future (Vanhees et al., 2013). It makes perfect sense that one could alter a disease process with a genetic basis through the control of regulatory mechanisms that underlie gene expression. Pharmaceuticals used in patient care were likely having epigenetic effects before anyone truly understood the meaning of epigenetics. Today, targeted epigenetic therapies have a well-established role in hematological malignancies and the role of epigenetics in the clinical application of precision medicine will increase substantially over time, as this model of treatment moves into other areas of patient care (Ganesan et al., 2019; Brown et al., 2022).

Differences go beyond our body and into our environment

How we respond to treatment is also affected by the things we do, what we have access too, and things we touch or eat. For example, certain medications are not as effective when taken with certain foods. Some people may have limited access to certain types of food. The sociodemographic status of the individual may also affect the outcome of a treatment based on the access they have to care. Appropriate education can affect the manner and regularity with which patients will take a medication. For people who forget to take daily medications, prescription of a drug to be taken once per week might be a better choice, even though it might not be as effective as a daily dose. Certain foods we eat habitually can change our bodies, and those changes can affect our genetic expressions resulting in a different response to medications. The toxins, allergens, weather, stressors, and infections to which we are exposed in our environment can affect the state of our health, development of a disease, and response to treatments. These are just a few examples of the many variables that can affect our bodies and overall health.

Changes from birth to death

A very distinct series of genetic transcription events take place after conception, which control the development of a functional human being. These processes change over time through distinct causal pathways. The genes that are expressed in a four-celled embryo are not the same as those expressed when we are sitting in a kindergarten class learning to read. Yet all humans go through the same development phases: in utero development, infant to toddler, prepubescent to pubertal child. During each stage, important developmental milestones are achieved though distinct biological processes that determine our future functional human nature. With respect to intrinsic nature, human beings can be viewed as having a preprogrammed pathway that governs growth and development. Most humans go through these predictable milestones, unless genetic mutations or external forces (injury, illness. etc.) stop these milestones from occurring. While these milestones are somewhat predictable, the timelines or speeds that milestones are achieved and the degree to which they are achieved are heavily influenced both by our genetic makeup and our environment.

Under normal conditions, the distinct series of gene expressions control growth and maturity stages. Although tragic events or errors in gene transcription along the way may cause death before all stages are completed. It is also possible that aging and death itself is to a degree a result of preprogrammed effects determined by gene transcription. This idea has been debated for years and is often discredited, and the prevailing view remains that aging occurs as a result of the accumulation of molecular damage over time, which leads to functional and physiological decline. Single gene alterations have been shown to delay aging in model organisms, although it is thought these changes do so by improving maintenance and repair functions that slow down damage accumulation. A growing body of evidence, however, indicates that specific genetic instructions drive aging in model organisms (Jones et al., 2012; Goldsmith, 2017). Regardless of the mechanism, everyone goes through aging differently, and life span is variable, even between individuals with identical genetics. Science does know there are multiple cellular signals and genetic markers that can be used to enhance tracking of the aging process and help to predict the time of death. The ability to measure these signals and predict some of these changes may lessen the burden of aging and may prolong life.

Ancestry and disease

Historically, most disease-causing mutations and predispositions have been passed down through successive human generations. As personalized medicine becomes more prominent in our generation, the genetic and medical history in our ancestry becomes much more relevant. This information is important, because we are likely to have the same mutations and therefore the same disease type and response to treatments as those whose genes we share. These ancestral clues discovered through disease social networks will be important in the definition of disease in the future.

Gene therapies

The FDA defines gene therapy as follows "Human gene therapy seeks to modify or manipulate the expression of a gene or to alter the biological properties of living cells for therapeutic use" (Long Term Follow-Up After Administration of Human Gene Therapy Products, 2020). These therapies are delivered through plasmid DNA, viral vectors, bacterial vectors, human gene editing technology, and patient-derived cellular gene therapy products (What is Gene Therapy, FDA). At the time of writing there are 22 FDA approved gene/cellular therapies in clinical use (https://www.fda.gov/vaccines-blood-biologics/cellular-gene-therapy-products/approved-cellular-and-gene-therapy-products, accessed Feb 20 2022) with hundreds of gene therapies in clinical trials (https://patienteducation.asgct.org/gene-therapy-101/clinical-trials-process). Although extremely expensive, some costing more than a million dollars per dose, these therapies are extremely effective and in many cases curative (Gene therapies should be for all, 2021). This is ground-breaking in many ways since the majority of diseases acquired by humans that are not infectious, end up having long-term consequences for which medications are used to manage symptoms of the underlying disease without actually curing the disease itself. The first trials of gene therapies, although somewhat successful in curing the disease, brought on unintended consequences because of the lack of precision in the delivery system [PMID: 26790362]. This precision has increased markedly over time and each time a successful delivery system is developed (i.e., Luxturna for retinal disorders) the initial therapy is followed by a number of other targeted gene therapies that can take advantage of the same delivery mechanism while targeting a different gene.

It is not about just our genome

Personalized medicine expands beyond our own genome into the genome of microorganisms that inhabit our bodies. Our microbiome refers to the collective genomes of the microorganisms that reside within our bodies. These include bacteria, viruses, and microbial eukaryotes. There are 100 times more genes in our microbiome than in our own DNA. Many of these organisms are beneficial and even necessary for human health; others cause problems, and all of them affect our health. Some people have called the microbiome an "organ," because it interacts closely with the rest of the body and has considerable effects on our health. An increasing number of human diseases are being examined for correlative or causative associations with the microbiome, including, but not limited to, autoimmune disorders, cardiovascular disease, asthma, periodontal disease, obesity, and even some cancers (Weinstock, 2012; Cho and Blaser, 2012; Liu et al., 2012; Gueniche et al., 2022; Binda et al., 2022; Dietert and Dietert, 2022).

Changing the definition of diseases

In 2006 the GAIT (Glucosamine/Chondroitin Arthritis Trial) trial consisted of a randomized controlled trial of the efficacy of glucosamine and chondroitin sulfate for painful knee osteoarthritis. At the American College of Rheumatology meeting in November 2006, a large debate on this subject was conducted. Two groups of scientists argued for and against the efficacy of the drugs in treatment, using the same drug trial data from the GAIT study. Both groups considered different aspects and different subgroups of the clinical drug trial. One side was convinced of efficacy of the drugs, demonstrated in the trial data, and also by personal experience. On the surface, it might appear that both groups could not be right, yet both groups were composed of highly educated physicians using the same scientific data to debate a clinical question arrived at different conclusions. It is quite likely that both groups may have been right; they just considered different groups of people in the trial data.

Medical practice depends upon large studies of very heterogeneous groups of people to determine the most effective medications. The groups are gathered on the basis of similar symptoms, while the molecular or genetic basis of their disease may be entirely different. It is not uncommon for diseases with completely different molecular bases to be grouped together solely on the presentation of similar physical symptoms. When diseases are grouped this way, it is very difficult to find a treatment that works for everyone, because each disease is a different problem. When this approach is used, the disease is not cured, only the symptoms are alleviated. As more is learned about these diseases and how to further characterize disease processes on a molecular level, physicians and medical scientists need to rethink the definition of disease and further group diseases in order to more appropriately treat each one.

This transition in characterizing disease has become routine in the field of oncology. Molecular subtyping has now become standard practice before treatment in many different cancers (Shen, 2021; Zhao et al., 2019). It is likely that medical science will look back at the way these tumors were treated previously as rather barbaric compared to the precise therapies of the future. Physicians will no longer apply the shotgun approach to curing disease, but instead will use precision—personalized medicine to treat each disease in the most effective way.

The National Research Council (NRC) recognized this need and erected a committee to explore the feasibility and need for a new taxonomy of human disease based on molecular data. The results of the meeting of this committee were published in a report entitled "Toward Precision Medicine: Building a Knowledge Network for Biomedical Research and a New Taxonomy of Disease" (National Research Council US, 2011). This report suggests the need for an information system, which the committee termed a "Knowledge Network of disease." This network would integrate information from researchers on causes of disease with clinical information, thus allowing researchers, healthcare providers, and the public to share and update this information in real time. The system would be based on an "Information Commons," or a data repository that links layers of molecular data, medical histories, socioeconomic information, environmental information, and health outcomes to *individual patients*. The data in the Information Commons would be updated regularly by the research community and from entries in the medical records of participating patients.

Creation of such a knowledge network would have a tremendous impact on healthcare, and it would be a gold mine not only for data mining and predictive analytics for model development to predict the best way to treat an individual patient. Access to these large datasets and the latest research data can enable scientists to formulate and test hypotheses more easily and quickly to understand disease pathophysiology and assign treatments. These predictive analytic models could segregate patients further by individual characteristics (including genotypes) to create personalized treatment plans for patients. In addition, this process would define disease subtypes more accurately, which then would be included in the new taxonomies of disease. The NRC (2011) report concluded that widespread data sharing is an essential component to the creation of new disease taxonomy because of the amount of data that is necessary to redefine a disorder and create a subtype.

In order for such a knowledge network to be put in place HIPPA regulations would have to be modified to allow individual data (rather than just aggregated data) to be utilized for research and analysis. Only by using each individual's data can these reliable models for scoring new patients be developed. This can happen only through a combination of events, including individual patient's consent, de-identification of a data/case, and shared networks of data using the same format, and security controls to prevent misuse of data. Such changes in HIPPA must happen before any money is spent to develop these knowledge networks.

Systems biology

Systems biology is the study of the components, interactions, and dynamics of a biological system. In the past, most medical research was done on individual genes or proteins, without a complete understanding of the biological system of which they are a part. Advances in technology are increasing our ability to conduct research and ask questions in context of biological systems. A good analogy can be seen in the structure and function of an automobile. Individual parts (i.e., the alternator) can be studied in relationship to its ability to generate electricity. This approach is not very useful information unless we look at the alternator in the context of the whole car. A car can function well without an alternator, but it will stop functioning over time as the battery loses energy without an energy generator to replace it. A cursory analysis of the problem apart from other components of the system might lead to the conclusion that the car doesn't run, because the battery failed. The battery might be charged with an external charger, but it will fail again in the future. Our failure to make a long-term correction to the problem is based on our focus on a single element of a complex system, without an understanding of the other components that make it work properly. In contrast, an understanding of the interconnection and dynamics between the battery and the alternator may show that the cause of the problem is that the alternator that charges the battery doesn't work. This conclusion might lead to a further conclusion that the belt, which drives the alternator, may be broken. This analogy of the automobile battery can apply to many biological problems. For example, an association might be discovered between a bodily dysfunction, the lack of a protein (the "battery") related to it, and the gene (the "alternator") that generates it. Exclusive focus on the protein without consideration of its place in the biological system may obscure the true cause of the disease. Even if one has knowledge that the alternator exists. It would be difficult to figure out that the alternator caused the car to stop running because we could remove the alternator completely from the car, and yet with a fully charged battery the car would still run for a considerable amount of time. It is only over time that the effects of the nonfunctioning alternator would be apparent. We must understand the system in order to correct the problem in the best (most accurate) way. Studying the effect in the context of the biological system will permit physicians to devise long-term solutions to problems, rather than continuing to treat symptoms as they arise.

There are four key properties of systems that must be understood in order to understand how any system works (including biological systems) these are:

1. the structures or parts that compromise the system;
2. the system design;
3. the system dynamics or how the system behaves; and
4. the control mechanisms of the system.

If we can understand these key elements we can do a better job to model disease systems, and test hypotheses in intelligent and focused ways.

The study of biological systems requires teams and large collaborations of experts on each component involved in the system. Only by the formation of these collaborations and using the latest technology to layout the design, structure, and function of these networks, can the speed of scientific research be increased significantly, and focused to find solutions to problems in an efficient manner.

Predictive analytics with decisioning capabilities can and will play a crucial role in this process, creating predictions of system behavior that can be tested, validated, and placed into action by making treatment decisions for individual patients. It is crucial to create predictions of interventions needed at key positions in the system, to target accurately the pharmacological therapies appropriate for an individual. The only alternative is to follow the current medical treatment model in trying various therapies, one at a time, over a period of time, before one is found to work. In addition to predicting appropriate pharmacological therapies, this systems level understanding of the treatment can also provide clues to the understanding and prediction of side effects. The systems approach enables physicians to combine complex signals from several compartments of biological systems to create predictive models for accurate diagnosis of the disease, deploying the best treatment, and predicting accurately the best outcome.

Systems biology is still evolving and there is substantial development needed in terms of standards and technical infrastructure to accelerate this approach. The potential benefits are tremendous and there is a potential to revolutionize healthcare and our understanding and treatment of human disease (Kitano, 2002; Zupanic et al., 2020; Sordo Vieira and Laubenbacher, 2022).

Efficacy of current methods—why we need personalized medicine

Evidence-based medicine (EBM) remains the current goal for treatment in our medical system. In Chapter 7, however, evidence is presented to support the case that physicians in the United States like to think that they are using the best of EBM in their practices, but the evidence they appeal to is based on the efficacy of a therapy on a large heterogeneous group of people, which will no longer suffice as the basis for treatment standards. Physicians cannot provide treatment any longer based on the mean of the population while ignoring the outliers. Instead, they must base their treatment decisions on characteristic groups, whose individual characteristics control their response to therapy. Many of the medications used widely (i.e., antidepressants) are barely more effective than a placebo when evaluated in large studies; similar results are seen with homeopathic medications. Yet, many people have positive responses with these medications, while others do not. This diversity of responses "muddies the waters," making it difficult for the physician to prescribe the appropriate medication or treatment. If medical researchers and physicians were to break down these sample populations into subgroups, it is likely that ensuing studies would find similarities among people with similar responses, thus allowing physicians to redefine and break down the phenotype of the disease. The use of K-means and E-means clustering techniques is one way to find these subgroups of patients who response similarly.

Predictive analytics in personalized medicine

Personalized medicine presents a tremendous opportunity to develop systems for predictive analytics coupled with decisioning systems. There are so many complex relationships and variables involved in the regulation of human health that even when all the principles or relationships behind disease mechanisms are clearly understood, there will continue to be a need to refine the predictive analytics models as additional scientific knowledge is discovered, and environmental changes occur over time. To further illustrate this dynamic, refer to the example discussed in Chapter 3, in which a piece of paper is thrown into a trash can, or a basketball is thrown into a hoop. The physics that regulates the flight of the paper or the basketball is understood clearly, and the amount of force necessary to make sure the object cast falls into the goal can be judged closely. But there is large combination of environmental variables that can affect the direction of flight and the amount of force required, therefore, a static model would be impractical. Too many influences would affect the flight of the object, and it would be impossible to build a model built for every possible situation ahead of time.

The future: predictive and prescriptive medicine

Much of the medical practice in the past has been focused on reacting to problems, rather than anticipating them. Such "reactive" medicine must be replaced by *predictive and prescriptive* medicine. This is not a new concept, even Hippocrates recognized its value in ancient Greece, long before we knew anything about DNA. Hippocrates said that *"He will manage the cure best, who foresees what is to happen from the present condition of the patient"* (Adams, 1849). Yet over 2000 years later, we are only beginning to realize the full potential of that maxim, and we are just beginning to practice predictive medicine. Our conventional view of medicine focuses primarily on symptoms, which are the late manifestations of disease. Most medical practice today neglects molecular signals that may serve as harbingers hours, days, months, or even years before the onset of disease. Often, conventional medicine ignores underlying mechanisms of the symptoms also, and treats only what is physically discernable at the time the patient presents.

The future of medical practice must be focused on predicting and thereby preventing disease, rather than reacting to symptoms after the disease has already taken its toll. The human body is composed of many complex interacting systems. These systems send signals at every level in order to communicate with each other. Currently, we have the abilities to process and interpret these signals effectively; in fact, these abilities are increasing daily as medical scientists acquire new and faster technologies (i.e., "smartphone" applications). These new technologies can provide a richer landscape of data and analytical methods for building better predictive models for use in providing a near real-time scoring for individual patients for diagnosis and treatment decision-making. The human can be viewed as a complex system of "factories" communicating and shipping things around the body. Modern medicine, however, does not pay attention to this system of factories. Without this holistic treatment of such a complex system, medical treatments may cause it to become unregulated, which may overload some of the systems and cause them to fail. This rather ad hoc approach to medicine is contrasted by the highly controlled function of an actual commercial factory, where quality control workers monitor constantly all the processes that occur in the factory. If some factory process has a problem or begins operating outside of its normal range, the appropriate corrective action can be taken immediately to prevent product failures. In medicine, these failures represent misdiagnoses, improper treatments, and even death. If physicians are able and willing to harness the messages from all of these bodily systems, they will be able to correct the problems in these systems before they become overloaded, permitting the maintenance of equilibrium among bodily systems. In the future we will be able to monitor each of our organ systems through a variety of biological signatures that will tell us the current state of the organ. By analyzing the signatures of the bodily organs and systems (e.g., the circulatory system), doctors can locate the problems and begin an intervention before there are any clinical manifestations. In this way, medicine will no longer be reactive, attempting to rebuild a "factory" that has already become highly unstable and fraught with failures. Instead, medicine can become proactive and maintain systems in the human body in a healthy equilibrium. This situation has been in place in many of the commercial industries of the world for nearly 20 years. It is time that it happens in medicine, medical research, and healthcare delivery also.

Application of predictive analytics and decisioning in predictive and prescriptive medicine

There are at least four medical measures that can utilize predictive analytics to maintain our bodies in a healthy state of equilibrium:

1. Determining or predicting risk of disease and predicting best preventative measures to follow to prevent disease.
2. Monitoring health for failing organs and systems followed by applying early interventions before the body gets out of balance.
3. Early detection of disease to head it off before the patient develops clinical symptoms.
4. Predicting what our needs are, psychologically, physically, and nutritionally, and applying the appropriate therapies, nutrition, etc. to meet those needs.

When physicians and patients work together as a team to apply these measures, making interventions, as needed, patient bodily systems can be maintained in a healthy equilibrium (Fig. 6.2).

Another area for the application of predictive analytics application to healthcare delivery is to assist in the recovery of a biological system when the human body falls out of equilibrium. When this occurs either through aging, disease, or lack of sufficient nutritional, psychological, or physical support, predictive models can provide information to help move the system back into equilibrium. The nature of this equilibrium is dynamic, not static. The optimum level of any part of a complex biological system changes over time, and control systems in them must adjust to these changes. Predictive models must focus on solutions to move the functioning of the body back to the proper level for a given

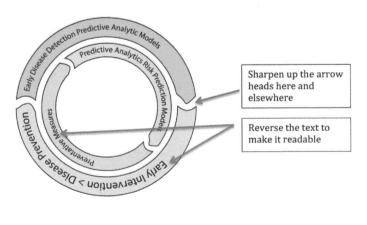

FIGURE 6.2 Four areas that can use predictive and prescriptive analytics to maintain the body in a healthy state.

time, and this level is a "moving target." Therefore, predictive modeling must be part of an adaptive process, in which the current state of the optimum solution keeps pace with the changing needs of the body.

Definitions of homeostasis will vary at different points in the human life span, and different interventions will be required at different stages of life. A good example is in our diets. Very few people can eat high-caloric junk foods like they may have done as teenagers and maintain a healthy weight. Bodily metabolic rates change with age, and opportunities for exercise may diminish. But if the caloric intake does not change, the body increases in weight. Predictive models can facilitate accurate diagnoses, provide valuable input for the prescription of appropriate therapy for the individual, monitor the response to therapy, and adjust therapy as appropriate. The opportunities include also that of patient data in intensive care units or other hospital setting, and predicting disastrous events before they occur, allowing preventative measures to be taken the events occur.

The time is right for the shift to predictive—prescriptive personalized medicine. Considerable advancements in processing of very large data volumes ("Big Data") provide us with an unprecedented ability to both understand and monitor the complexities in the function of the human body. This ability in conjunction with powerful analytical tools available today positions us at the beginning of an exciting era in which predictive analytics will play a substantial role in improving human health.

The diversity of available healthcare data

There are multiple types of data that can be used to build predictive analytics models in medicine and healthcare. It is likely that the best models will incorporate multiple data sources. Each data source provides a perspective on the pattern sought by the modeling software. Each perspective can be combined by the mathematical algorithms in a fashion analogous to the way the human brain composes the perception of depth from the separate images of the same objects viewed at slightly different angles. Mathematical algorithms can sense patterns in the combined data set that may not have been clearly defined in any single data set. Combining data from multiple sources of genetic and environment information will enable the modeling algorithms to predict the best outcome and prescribe the best intervention for a given individual.

This individual-centric approach in medical diagnosis and treatment considers factors like a person's ability to adhere to the necessary treatment when prescribing the best therapy. The use of predictive analytic numeric and text data mining capabilities combined with the abundance and diversity of data sources available provides medicine and healthcare the greatest likelihood of making a good prediction and providing an effective treatment. The best models will incorporate the best variables from among the vast array of data sources that are related to the incidence and treatment of a disease. The result is that hundreds or even thousands of variables may be analyzed using various feature selection techniques to generate the "short-list" of variables for building the model. This short list of variables is much more likely to produce a better model than will a larger list. The predictive accuracy of successive models may increase as the number of important variables decreases, as successive feature selection operations occur, at least up to a point. Beyond this "sweet spot," the predictive accuracy will decrease as the number of variables increases. Adding additional variables beyond this point functions only to add "noise" to the data set, which depreciates the accuracy of the model. This effect is likely to become increasingly important as additional open networks of healthcare data become available.

Diversity of data types available

Many types of data are available for building predictive analytical models. Data types that are critical to driving predictive analytics in precision medicine include:

1. Phenotypic data
2. Real-time physiological data
3. Imaging data
4. Genomic data
5. Transcriptomics data
6. Epigenomics data
7. Proteomics data
8. Glycomics data
9. Metabolomics data
10. Metagenomics data

11. Nutrigenomics data
12. Behavioral measures data
13. Socioeconomic status (SES) data
14. Personal activity monitoring data
15. Climatological data
16. Environmental data

Phenotypic data

This is where it all starts. We recognize traits that appear to be different from the standard population. When these traits are problematic, we call it a syndrome or disease. This is how disease has been defined for centuries. Unless there is a physical, mental, or psychiatric-psychological manifestation we don't consider a "defect" such as a genetic mutation to be a disease. Many "defects" have evolved as a means of adaptation to the environment, which offered some advantages, and therefore, they are not viewed as defects, but beneficial new characteristics. When this happens, the common perception of the norm changes. Most genetic mutations are very different from those that are imputed to cause the superpowers of characters cast in the X-Men movies. Instead, most mutations are "deleterious," in that they reduce the general state of health or vigor of the organism. A few mutations can be viewed as beneficial (or "good" for the organism). These beneficial mutations are favored in the process of natural selection, because they provide some advantage in coping with competition or changing environmental conditions.

Although there are a few scattered traits that appear to be controlled by a single gene, for most diseases, many different genes can be involved (such as eye color or hair color). For example, arthritis can be viewed as a single disease with many different manifestations. These manifestations are characterized by which joints are affected, how early the disease first appears in an individual, what medications the disease responds to, the amount of bone deformity caused, the amount of pain, the durations of pain, and related symptoms such as heart disease. The combination of these physical or physiological manifestations to a particular disease are called the "Phenotype." In the past, medicine was defined in terms of the phenotype of the disease. Medicine, however, is moving more toward defining disease by the "genotype."

Definition of the phenotype is very problematic. Defining a phenotype can be a difficult task, because there are overlapping characteristics among phenotypes in a single individual or among different individuals, who have varying levels of expression of disease. This situation is complicated further by environmental interactions and disorders caused by multiple genes. Once the phenotype is defined, however, searching for it from the clinical information based on this definition may require an extensive review of medical records, requiring the employment of text mining methods to increase the speed of retrieval. The phenotype is often included in clinical information, but, unfortunately, is often hidden in current healthcare documentation practices in large bodies of nonstandard free-form text, from which it must be derived. While medical record systems and medical coding systems have evolved to become more granular in terms of defining disease they still fall short in producing easily accessible phenotypes for research and predictive analytics models. Even where more granular ICD10 codes, for example, have been developed, the challenge of having physicians enter all the ICD10 codes that are applicable to patients remains. Medical record systems are very centered on billing and much of the documentation generated by physicians is produced to facilitate billing. For example, a patient's list of problems or phenotypic features may be defined by 30 different ICD10 codes, but a physician only needs to enter a few to generate the maximum amount of reimbursement for the visit. As most physicians are very busy, they are unlikely to do the extra work to facilitate secondary uses of the data. Physicians usually do a better job of describing these phenotypes in the unstructured data, so development of automated systems to derive structured information from clinical notes is critical to moving the field forward. As these processes evolve, the effort to define and derive the phenotype from medical records will become easier and more standardized. Unfortunately, current medical science uses a diversity of terms and expressions often to describe the same things, the choice of which is based on the experience, training, and location of service of an individual doctor or nurse.

Clinical information

Clinical information includes the total documentation for the patient encounter in the hospital or clinic. It is from this vast array of information that the phenotype is derived. It includes the physicians notes and patients' history, it also includes routine measurements such as height, weight, blood pressure, temperature, heart rate, respiratory rate, and other miscellaneous lab metrics. These metrics include all of the observations by the physician, nurse, and other healthcare staff each time a patient is seen, and any measurements taken during that patient visit. Additionally, it includes much of the information discussed below. Electronic medical record (EMR) vendors have recently begun to incorporate genetic information as structured data (Walton et al., 2020). It is likely during the next 10 years that the clinical medical record will incorporate even more information, including epigenetic and proteomic data, as the structure of EMRs evolves.

Real-time physiological data

In a hospital, patients are subjected to a barrage of measurements, particularly in intensive care units, where they are connected to continuous monitoring devices. These devices enable nurses or physicians to make an immediate intervention if a crisis arises, such as shock, heart failure, or respiratory failure. Monitoring systems like this help to save lives in hospitals and are a necessary part of any hospital system. Many times, however, it may not happen at the most opportune time using the best available resources. By the time that a nurse notices a sudden drop in blood pressure or heart rate, the damage may have occurred already. Many of these events may be predicted ahead of time, in real-time, by models which evaluate subtle changes in these physiological measurements compared against an extensive historical record to make predictions sufficiently ahead of time to prepare an appropriate intervention before significant damage occurs. Unfortunately, the current situation is that much, or all of this data gathered in real-time monitoring is discarded. If it were retained in the medical record, it could serve as a rich source for building predictive analytics models to predict such things as:

1. predicting a patient's response to treatment;
2. predicting deterioration or adverse events;
3. predicting transfer to the PICU (Patient Intensive Care Unit);
4. predicting a patients return to a particular ICU after being transferred to a regular ward bed; or even
5. predicting return to the hospital after being discharged.

Imaging data

Various forms of imaging are used to make diagnoses and to monitor patients' disease and recovery. Imaging capabilities of medical instruments are improving rapidly, allowing physicians to use less invasive high-resolution pictures to monitor what is happening in the body, often in real time. These imaging types include:

- CT scans
- X-rays
- Ultrasound
- Magnetic resonance imaging
- Functional magnetic resonance imaging
- Nuclear medicine
- Positron emission tomography
- Photoacoustic imaging
- Breast thermography
- Echocardiography
- Microscopy

It is reasonable that these images contain signals that might indicate where the disease is headed, and that they can be used to diagnose and track progression of disease to allow physicians and nurses to react sooner and to a better effect before a problem progresses. Currently, the interpretation of these images requires a trained human eye. The eye of a trained radiologist is a complex image analysis organ that can summarize quickly the image in its entirety. The brain of the radiologist contains a vast clinical knowledge base that is associated closed to what he sees in the images, as a result of many years of training while staring at these images in the context of human disease.

Despite this close linkage between the image perception and the magnificent and powerful computer of the human brain behind the eyes, it is not possible to analyze the image pixel by pixel to sense very subtle changes. This limitation becomes ever more apparent as the spatial resolution of these images increases with advances in technology. New techniques are being developed to analyze electronically and quantify pathology on these images (Wang et al., 2022; Habib et al., 2022; Liu et al., 2022; Nam et al., 2022). As these techniques mature, there will be ample opportunity to mine this vast amount of imaging data and perform analyses that are not possible using only human judgment. Data derived from common image characteristics (modalities) are currently being used for predicting prognosis in cancer and cardiovascular disease (Qi et al., 2022; Jathanna et al., 2021). Additionally, even obesity is being diagnosed more accurately and managed more effectively using these methods (Gallagher et al., 2011). While all of these imaging modalities show great potential for future predictive and prescriptive personalized medicine (Brizel, 2011; Bucciarelli-Ducci et al., 2022), there are still challenges remaining regarding clinical application, particularly around standards and integration into clinical workflows (Jathanna et al., 2021; Habib et al., 2022).

Genomic data

The cost of genome sequencing has decreased significantly over the past few years. In 2003, the cost of sequencing a human genome was $2.7 billion; this price has dropped to less than $500 dollars in 2022. While exome sequencing (ES) and panels have been the mainstay of clinical diagnosis, whole genome sequencing (WGS) is gaining momentum as a primary clinical test in this arena. What has not gained traction in the clinical space is the secondary use of all the sequencing data that is not included in the clinical report. The hope is that the decrease in cost and time required to sequence a genome will promote its use in healthcare and open up more use cases for not just diagnosis but other secondary uses of the data such as pharmacogenomics and polygenic risk scores. More universal acceptance of WGS could provide a huge amount of data for analytical modeling. Genetic data is the backbone of personalized medicine and pharmacogenomics.

DNA—the center piece of heredity and bodily differences

Genetic information is encoded in the DNA molecule (deoxyribonucleic acid), described by Watson and Crick in 1952. DNA is a double-stranded molecule, with the two strands coiled in the form of a helix (Fig. 6.3). The molecule is organized in two long chains of nucleotides, composed of pairs of deoxyribose sugar and high-energy phosphate groups, connected by pairs of nucleobases, guanine (G), adenine (A), thymine (T), and cytosine (C). Adenine always pairs with thymine, and guanine always pairs with cytosine. A specific sequence of paired nucleobases of a given length of nucleotides is called a gene. The DNA molecule is tightly coiled into a long structure called a *chromosome*. Humans have 46 chromosomes, consisting of 22 sets of homologous chromosomes and two nonhomologous sex chromosomes. The DNA content of which is inherited in approximately equal proportions from both parents.

Genes function to encode specific proteins, some of these protein function as enzymes, which act as catalysts in various chemical reactions in the body, and other of these proteins function as structural "building blocks" or function in other ways. If the specific sequence of nucleobases is disrupted by some chemical or physical disturbance, the gene will fail to encode the protein properly, and the function of that gene will be lost (or at least partially lost). This process is called a mutation. For example, dietary proteins (proteins that are eaten as food) are broken down into smaller molecules like phenylalanine, which in turn should be broken down into tyrosine, a reaction catalyzed by the enzyme

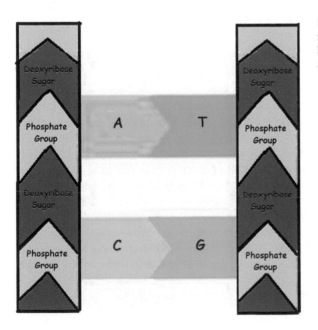

FIGURE 6.3 Structure of the DNA molecule; this shows the connection of multiple nucleotides, each one of which is composed of a deoxyribose sugar, a high-energy phosphate group, and a nucleobase. *Used by permission from DNAspecialtitle.blogspot.com.*

phenylalanine hydroxylase. One gene controls the production of phenylalanine hydroxylase, and if that gene is damaged or missing in the chromosome, the enzyme cannot be produced. This disease is called Phenylketonuria (PKU). Symptoms of PKU are severe mental retardation and spastic movements.

Many bodily processes are affected by multiple genes, therefore, a disruption in one of the genes may lead only to a decrease in the protein products, rather than a complete lack of the protein.

The combined sequence of these nucleotides among all of the 46 chromosomes composes the genotype of an individual. The genotype in combination with the unique environment of an individual, is what determines our phenotype. The expression of the phenotype comprises the physical, physiological, psychological, psychiatric, and spiritual characteristics of an individual. There are about 3 billion bases in human DNA, and more than 99% of them are the same in all people. This information is the combination of genetic sequences passed on from both parents, with the origin of each piece of the total sequence alternating from one parent or the other in a nearly random fashion, leading to the unique characterization of each individual. This unique characterization is caused by just the 1% difference in the genetic sequence. Sometimes, however, this very small genetic difference causes very large differences among individuals; small changes in DNA sequence can have large effects. Larger genetic differences will have even larger effects. For example, mice and men differ only in about 15% of their DNA; this means that men and mice are 85% identical (National Human Genome Research Institute, 2010; http://www.genome.gov/10001345).

DNA replication and mutation

The genetic system has the ability to make copies of the DNA molecules. This process is called replication. Many molecule types are necessary for this process, including several types of RNA, and the enzyme DNA polymerase. Each strand of DNA in a double helix can serve as a pattern, or template, for duplicating the sequence of nucleotides. This critical process takes place when cells divide, ensuring that the resulting cells have the same DNA as the undivided cell. A gene mutation is defined as "a permanent change in the DNA sequence that makes up a gene." The large majority of the DNA in cells of all organs and tissues in a body is the same. Some differences do arise, though, through processes of mutation.

Somatic mutations

Somatic mutations occur in body cells (not germinal cells, which control heredity) when errors occur in the DNA replication machinery. External agents (chemicals and toxic exposures) can affect this machinery and cause defects. When these defects occur the cell that contains them has a different sequence than the other cells in the body that still contain the germline sequence. These cells can then replicate themselves, perpetuating the genetic defect and sometimes even increasing the number of defects in subsequent cells if the original defect affects the cell replication and DNA replication and repair machinery. The accumulation of these defective cells can manifest in many different ways, such as tumors in oncological conditions or patches of skin discoloration in others. Sometimes, these mutations are benign, but they can be the source of cancer, and they may be implicated in other diseases as well (Poduri et al., 2013; Costantino et al., 2021). Somatic mutations are not passed on to future generations.

Germline mutations

Genetic mutations in the sexual cells of organisms (seeds in plants; gonads in animals) are inheritable. The mutations that are inherited are called "germline" mutations, and these are what confer the inheritance of human disease through subsequent generations (Genetics home Reference—NIH, 2013; http://ghr.nlm.nih.gov/). These germline mutations can happen during the recombination process in cell division in the sexual cells. These mutations are passed on via the egg or the sperm cell, which are combined to form a new DNA of the child. Once these mutations are established in the sexual cells, they can then be passed on to future generations. Some mutations will produce harmful or fatal results while others will produce no effect, an effect that is benign, or in rare cases an effect that is beneficial.

There are several types of genetic mutations that can occur from the rearrangement, deletion, or repetition of the nucleobases A, G, C, and T that form the DNA strands. When these changes happen, the resulting proteins that are produced by the gene sequence may be defective, or may have a new or altered function, or not made at all. These mutations can have a wide range of effects from making no difference at all to having minimal effect while others can be fatal. There are thousands of mutations that have been linked to human disease, physical characteristics, drug response, and other physiological properties.

Genetic testing for disease has become increasingly sophisticated and also more easily available. This testing was originally very laborious and expensive; therefore, only small sequences of DNA were examined. Today, the entire genome can be sequenced at varying resolutions. The most common genetic studies in the past few years were based on SNPs or Single-Nucleotide Polymorphisms (pronounced "snip" or "snips" for plural). SNPs are DNA sequence variations in a single nucleotide in the genome that differ between different people and cover a small span of the DNA strand. Using these relatively common variations, it is possible to detect medium to large genomic lesions. Genome-wide association studies (GWAS) involve the genotyping of hundreds of thousands or millions of SNPs, which are then analyzed in relation to a particular disease or phenotype using hypothesis-free agnostic approaches. GWAS have shown considerable success in mapping susceptibility loci for common diseases. The associations that have been identified by GWAS have identified relationships with many genes that previously would not have been thought to be good candidates as causes for particular diseases or phenotypes. Thus GWAS have added greatly to our knowledge about genes and processes involved in human disease. While the variants discovered have contributed significantly to knowledge of the disease processes, they have made only small contributions to medicine's ability to make predictions of an individual's risk for developing a particular disease; this is because the effect size of most variants is very small, and most diseases are extremely complex involving multiple genes and other complex factors. The variants identified by GWAS explain only a small proportion of the total genetic variance involved in most disorders.

Meta-analysis of GWAS involves combining the results of several GWAS studies to increase the power of the analysis with the goal of finding more statistically significant associations and novel genetic associations. This approach has become a very popular for the discovery of new genetic loci for common diseases. A large number of the genetic risk variants discovered recently have come from large-scale meta-analyses of GWAS studies (Wagner, 2013; Evangelou and Ioannidis, 2013).

Although GWAS studies have the advantage of lower cost and speed of analysis, one of their weaknesses is that they are designed only to look at common variants based on the SNPs that are analyzed, and thus skip large portions of the genome. Whole exome sequencing (WES) and WGS are rapidly replacing arrays in the research space as the price becomes more competitive. The exome is the expressed genome, which means that part of the genome that is transcribed and then translated into proteins. These newer sequencing approaches allow for a more granular understanding of the genome and allow for better targeting of the array of rare variation that underlies common disease. WGS provides sequence data on introns, or the areas of the genome that are not expressed. At one time these regions were thought of as "junk" DNA but we now know that these regions play a critical role in how, when, and where DNA is translated into proteins (Hesselberth, 2013; Shaul, 2017; Parenteau et al., 2019). While it is cheaper to sequence only the expressed portion of the genome, WGS is gaining ground. This shift is due to the decreasing cost of WGS and the increasing knowledge regarding the functional consequences of intronic DNA.

While there are many inherited Mendelian disorders that can be attributed to a variant in a single gene, finding *single* genes that define or determine the origin of complex disease is usually not possible, even with whole genome and ES. Most diseases involve complex interactions between multiple genes, other molecular factors, and the environment. This explains how disease can be passed down through families and skip generations. Because of the complexity of most diseases, robust disease prediction models must be based on multiple genes and include a multitude of other molecular and environmental variables, as outlined in this chapter. The focus of most genomic medicine has been on disease; with increased availability it will become further integrated into primary medical care playing a critical role in the assessment and promotion of wellness and health. Combining a genetic analysis with basic clinical indicators such as blood pressure, body mass index, glucose, and cholesterol measures would give the primary care physician a detailed personalized plan to prevent disease in patients at risk.

The ability to combine genetic information with clinical data is crucial for healthcare and predictive analytics. Clearly, full integration of this ability into clinical practice is needed in the future. But before a complete clinical integration can happen, datasets are vitally necessary for building predictive models that can be used for diagnosis and treatment in clinical medicine. This need has been recognized and several initiatives are in place now for gathering this data for analytical modeling and research. Below are some examples of the many notable efforts:

1. The All of US Research Program, 2015 (https://allofus.nih.gov/).
2. the Personal Genome Project (PGP) (Personal Genome Project, 2005, 2013; http://www.personalgenomes.org/).
3. the Electronic Medical Records and Genomics Network (eMERGE, 2013; http://www.genome.gov/27540473).
4. the Patient Centered Outcomes Research Institute (PCORI, 2013, 2022).
5. DIGITizE: Displaying and Integrating Genetic Information Through the EHR Action Collaborative, 2014 (https://www.nationalacademies.org/our-work/digitize-displaying-and-integrating-genetic-information-through-the-ehr-action-collaborative).
6. UK Biobank (https://www.ukbiobank.ac.uk/).

The personal genome project

The PGP is a nonprofit public repository of integrated genomic, environmental, and health datasets, whose goal is to make a wide spectrum of data about humans accessible to researchers and clinical practitioners to increase our knowledge of the human body and improve human health. This project functions to sequence all 3 billion base pairs of the DNA of individuals participating in the project, and then combines this information with their medical records and environmental exposures, and finally compares it with their phenotypic traits. This project uses open-source, open-access, and open-consent frameworks. This means that there are no constraints placed on the usage of the data. Each contributor of DNA has their own interface, which enables them to continue to contribute additional information over their lifetime from various sources including:

- Historical (longitudinal) records of status
- Medical and social history
- Environmental exposures
- Nutrition
- Lifestyle
- Physical measurements
- Blood chemistry
- Presence or absence of microbes and viruses
- Other pertinent data

This data is freely available to researchers and is a valuable resource for data mining and predictive analytics (http://www.personalgenomes.org/).

The Electronic Medical Records and Genomics network

The eMERGE Network is a national consortium organized by National Human Genome Research Institute to combine DNA biorepositories information with EMR systems for large-scale, high-throughput genetic research. There are several academic medical centers across the United States that participate in this network. The eMERGE model does not require active recruitment for the study or gathering of samples because it uses cases and controls from various EMRs, for which genetic samples have been already collected as specimens, saving time and money. This network uses this data not only for genetic research, but also to help develop some other frameworks and guidelines related to genomic medicine in the areas of ethics, legal issues, privacy, and community engagement (http://emerge.mc.vanderbilt.edu/; Katsanis and Katsanis, 2013).

The Patient-Centered Outcomes Research Institute

The Patient Centered Outcomes Research Institute (PCORI) is a new nonprofit research institute mandated by the ACA (Accountable Care Act Health Reform Laws), funded by seed money from the National Government, and to be sustained by $2 per person per year signed up for national healthcare coverage. One of the primary goals of PCORI is to develop regional and national data base networks of patient data with the goal of developing accurate predictive and prescriptive models for more accurate diagnosis and treatment of diseases on an individual basis (PCORI, 2020, 2022; http://www.pcori.org/).

Large biorepositories of genomic data with linked health-related information are critical to healthcare and research. These are especially important for rare diseases, for which pooled resources and aggregation of data from affected individuals in the country and around the world are required to generate enough cases to study the disease effectively. We are embarking on an exciting era in genomics that presents significant computational, scientific, legal, and ethical challenges. This large-scale integration and use of genomic data in healthcare systems has only just begun.

Transcriptomics data

Transcriptomics data are generated by the measurement of gene expression profiles with microarrays. The genes encoded in each cell affect the phenotype only if they are expressed. The first step in gene expression is transcription of DNA into RNA. Often, cellular mechanisms can suppress the expression of genes, and this suppression might occur only under certain conditions. Gene expression profiles provide a measure of the activity of the genes in our body. This process is very complex, as millions of genes are transcribed in different locations in our bodies at the same time.

Even though the DNA in each location of the body is the same, expression of the genes along the DNA backbone may differ significantly in different locations, because these locations (i.e., the brain and the heart) exert different influences on gene expression. These differences in gene expression provide for the different functions required in these locations. If that landscape of gene expression were not complicated enough, each cell in our body may be transcribing differently. For example, while an individual's brain and heart have the same genetic backbone, the genes that are actively being transcribed at each site are very different because these organs obviously have very different functions and don't necessarily use the same parts of the genome to execute their functions. To further complicate things, expression of genes is affected by physiological and environmental conditions, which can cause a wide variation in expression, depending on when the sample is taken. In summary, gene expression profiles vary over time, across different physiological and other environmental conditions, and also based on the location within the body. Therefore, the total bodily landscape of transcriptions (the *transcriptome*) consists of all DNA transcription molecules (i.e., messenger RNA—mRNA; transfer RNA—tRNA; small nuclear RNA—snRNA; ribosomal RNA—rRNA; and other noncoding RNA) produced in the transcription process at a specific developmental stage or physiological condition. These measurements are useful for studying diseases with focal pathological findings where expression can be measured in areas of change such as with cancer or multiple sclerosis (Sánchez-Pla et al., 2012).

Epigenomics data

The genetic information encoded in DNA is not the only thing inheritable in the human body. Through epigenetics other factors can create heritable changes in our molecular makeup; some of these include development conditions (i.e., in utero), environmental chemicals, drugs, aging, and diet. These factors can cause changes in the chromatin material (histones) that are wrapped around the double helix DNA chain to form the chromosome. These changes in chromatin are caused primarily by the addition of a methyl group to the histone that is provided by one of the modifying factors listed above. DNA methylation can affect the expression of the DNA genes, and these changes are heritable, even though the DNA base sequence is not changed. Epigenetic mechanisms play important roles in gene expression, DNA repair, and recombination. Epigenetic features are tied to specific locations on our genome and can be inherited, however they can also be modified, or erased in response to molecular signaling during development, or by environmental effects. Similar to changes in the base sequence of DNA, these epigenetic mechanisms can cause significant problems. Some defects in epigenetic regulation have been linked to human disease, including developmental defects, metabolic disorders, and cancer. Additionally epigenetic mechanisms can be associated with other diseases including but not limited to psychiatric disorders, diabetes, and asthma.

Epigenetic control of gene expression is centered primarily around the association of DNA with chromatin. This organization begins with 147 base pairs of DNA wrapped around eight different histone proteins. This assembly of DNA and histones is referred to as a nucleosome. The resultant nucleosomes are packaged tightly together into compact fibers known as chromatin. It is through this complex structure that epigenetic regulation occurs, where areas with less chromatin and hence more DNA exposed for translation are expressed, and also areas more tightly bound are silenced or not expressed. Epigenetic regulation occurs primarily through four mechanisms:

1. **The first mechanism** is the posttranslational modification of histone proteins; modifying these proteins, which form the core of the nucleosome, alters the local chromatin conformation. Changes in chromatin compaction affect the accessibility of genes for transcription. Genes in loosely packed regions, referred to as euchromatin, are more actively expressed and those genes in more tightly packed regions, called heterochromatin, are not expressed.
2. **The second mechanism** is through direct DNA modification. DNA modification, specifically methylation of cytosine, is crucial for many genetic processes including DNA imprinting, X-chromosome inactivation, and long-range silencing of genomic regions. DNA methylation is a general marker of gene silencing, and it prevents transcription by interfering with transcription factor binding. Other modified forms of cytosine, outside of methylation, have more recently been discovered. DNA methylation is typically associated with tightly condensed chromatin.
3. **The third mechanism** involves noncoding RNA molecules interacting with specific target mRNAs and triggering a cascade of events resulting in specific mRNA degradation and thus preventing transcription. Acting in concert with the previously described aspects of the epigenetic machinery, noncoding RNAs contribute to the regulation of these processes.
4. **The fourth mechanism** involves packaging the chromatin into a higher-order structure within the cell nucleus. Chromatin accessibility and nucleosome positioning within this higher order package is also an important epigenetic mechanism. It is more common to find elements that regulate gene expression in regions of the genome that are more accessible in the higher order packaging.

These epigenetic mechanisms often work together to regulate DNA expression. For example, small noncoding RNA molecules can participate in directing DNA methylation. Additionally, enhancer elements in accessible regions can even encode for small noncoding RNA molecules themselves. In summary epigenetic features primarily control DNA expression through allowing or disallowing transcriptional machinery access to DNA.

This sophisticated regulation of DNA transcription plays a critical role in many cellular processes. It is crucial in determining the type of cell (heart cell, brain cell, blood cell, skin cell, etc.) a cell will become in early development as we progress from a single fertilized cell into a complex organism. It is also an important mechanism in turning on or off genes that promote or inhibit cancer.

The *epigenome* includes the entirety of these epigenetic features throughout the genome. Compared to the more static genome, the epigenome is very dynamic and will differ depending on the tissue type, the person's age, what types of exposures the person has had, and many other factors, some yet to be discovered. This diversity of effects renders the study of the epigenome in relation to human disease much more complicated than studying the genome. The genome has a common set of DNA molecules that is fairly constant in all the cells in the body. In contrast, the epigenome must be studied in the right tissue at the right time and under the right conditions, because it can easily change based on these factors. Despite its complexity, knowledge of the epigenome is crucial to understanding human disease, therefore, significant efforts are underway to further study and better define the human epigenome (Capell and Berger, 2013; Fingerman et al., 2013; Wang and Chang, 2018).

To address the complexity of the human epigenome, the National Institute of Health (NIH) launched the Roadmap Epigenomics Project from 2007 to 2017. This project created a series of epigenome reference maps that were made available to the public. These maps include a wide array of cell lines, cell and tissue types from individuals at various developmental stages, and with various diseases and health states. These data consist of DNA methylation patterns, histone modifications, chromatin accessibility, and small RNA transcripts. The website for this project is at http://www.roadmapepigenomics.org. Links to more information, data, and tools for epigenomics can be found on the NIH common fund website (Epigenomics, 2013; https://commonfund.nih.gov/epigenomics).

Proteomic data

Following the genomic and transcriptomic processes in genetic studies, the next step is the formation of proteins, called *proteomics*. There are many molecular mechanisms that can modify the shape and function of these proteins. The science of proteomics consists of quantifying these proteins, in an effort to understand the physiological functions of these proteins in the body. The composition of proteomic processes (the proteome) varies based on the type of cell, its location, and its environment, and it can change over time. Different genes are expressed in different types of cells, so the proteins expressed in different cells can vary widely depending on the body organ, tissue, or location.

Cells are extremely complex and contain a large number of molecules that interact both within a cell and outside the cell, which enable cells to function, communicate, and perform the necessary functions to sustain life. One of the primary goals of science is to understand how these cells function and communicate. The central dogma of molecular biology describes the flow of molecular information from DNA as it is transcribed into RNA, and then translated to form proteins. Although this three-step sequence is largely true, the complete process is far more complex. There are many mechanisms by which proteins are created or modified that are not directed by information in the DNA. Some examples of these mechanisms are epigenetic marks, alternative splicing, noncoding RNAs, protein—protein interactions, and posttranslational modifications. Given the independence and many possible modifications that are possible outside the central dogma it could be argued that analysis of proteins, as the key functional entities of the cell, forms the principal level of information required to understand how cells function.

The very characteristics that give proteomics an advantage over genomics and transcriptomics in understanding cell function also complicate the study of proteins. Global protein analysis is a significant analytical challenge, considering the diversity of proteins throughout the body, the complex number of modifications, and variety of ways that they can be modified. Proteins are primarily measured with mass spectrometry, although other methods are also used. These technologies have advanced significantly during recent years. A decade ago, the sequencing and identification of an individual protein was a significant challenge. Today, the identification and quantification of nearly all proteins is now achievable in a single experiment. This huge amount of data that can now be generated has created a desperate need for computational tools to organize and mine this data (Aslam et al., 2017).

Glycomic data

Glycomics is the comprehensive assessment of glycans or sugars. Glycans are important building blocks of the four major biomolecules of life—carbohydrates, nucleic acids, lipids, and proteins. They are involved in virtually every pathophysiological condition in medicine. The analysis of these biomolecules gives medical science a better understanding of human physiology (Wells and Hart, 2013). Glycosylation is the covalent attachment of glycans to proteins or lipids. Glycosylation is the most abundant posttranslational modification of proteins. It is also the modification that provides the most structural diversity. Greater than 50% of the entire human proteome is modified with glycans. Glycosylation plays a critical role in many biological processes that affect human health including cell recognition and cell-to-cell communication (Li et al., 2013). Both cell recognition and cell-to-cell communication are critical in embryo development, immunity, autoimmune diseases, cancer, and other human diseases. Glycosylation is determined both by genetics and environmental factors, so some elements of disorders related to glycosylation are inherited but environmental effects also play a role both directly and through medication of epigenetic mechanisms. Glycans are analyzed with liquid chromatography, mass spectrometry, and capillary electrophoresis. Both quantitative and qualitative changes in glycan structures have been found in many complex diseases including cancer (Zoldoš et al., 2013).

Metabolomic data

Metabolism is the set of processes inside the cell that produce energy and cellular building blocks, such as amino acids, nucleotides, and lipids. These building blocks and the biochemical intermediate products generated during their production and utilization are referred to as metabolites. Examples of metabolites are amino acids, organic acids, sugars, fatty acids, lipids, steroids, small peptides, and vitamins. Metabolomics is the study of the complete set of metabolites of low or intermediate molecular weight reflecting physiology, developmental, or pathologic state of cell, tissue, organ, or organism. Although the way scientists conceive concepts in this field now is relatively new, even the ancient Greeks believed the concept that changes in tissue and biologic fluids could be early signs of pathology and may indicate disease. In 1506 Ullrich Pinder described the possible medical value of the smell, color, and taste of urine; all of these characteristics of urine are the result of the metabolites contained therein (Syggelou et al., 2012). Metabolite levels are a product of the sum effect of genetics, transcription, posttranscriptional regulation, and environmental processes. The measurement of cell status, health, and activity defines the physiological phenotype of the cell and its health at any given moment in time. Metabolomics is focused on comprehensive profiling of metabolites from cells and biofluids. While the genotype is relatively static, metabolomics reflects a dynamic nature of biologic systems in response to the environment and their interaction with other systems. Thus it is possible to use these measures to determine the effects on the body of hormones, drugs, food, chemicals, and any other environmental exposures.

Metabolomic studies are performed on biological fluids such as urine, plasma, cerebrospinal fluid, maternal milk, or saliva or on tissue samples from various organs in the body. For example, urine can be obtained very easily, and it contains a significant amount of information on the overall metabolic state of an individual. The unique biochemical composition of urine is affected by a complex combination of the effects of genotype, physiologic conditions, disease state, environment, nutrition, and drug or toxic elements ingested. Analysis of urine could allow for prediction of disease progression, early detection of disease, and to monitor a patient's disease progression, treatment response, or recovery.

Soon, it may be possible to develop personalized treatment regimens based on the current metabolic status of the patient, based on their static genotype, and computer readouts of the current state of their body at that moment. This facility expands the boundaries of personalized medicine to encompass any changes in these variables up to the second the patient presents. A primary goal of medicine is the creation of individualized treatment regimens that are optimized for a patient's metabolic status.

Metabolomics has an advantage over some of the other "omics" in that there is a smaller set of variables to analyze. Compared to more than 20,000 genes in genomics (disregarding the millions of possible mutations in those genes), more than 80,000 transcripts measured in transcriptomics, and the greater than 10,000,000 proteins measured in proteomics, metabolomics is concerned with fewer than 3000 metabolites. This relatively small number of variables gives us a much more manageable set of data and the measurements are a result of the activity of the other "omics" that contain a greater number of variables. These metabolites are involved in the major biochemical pathways in the human body, including glycolysis, Krebs cycle, and lipid metabolism. They are involved in signal pathways as transmitters and hormones, and they can measure specific pathological biochemical processes, such as oxidative stress. These measures provide a real-time measurement of the status of these pathways and systems (Syggelou et al., 2012; Gonzalez-Covarrubias et al., 2022).

Metagenomic data

We have a world within ourselves in the form of our microbiome. The human body is inhabited by millions of organisms living in communities working together with our own cellular environment, sometimes to our benefit, and other times to our detriment. Each of these microorganisms has their own genetic makeup, which can have a substantial effect on how we function and how our body reacts to a disease or even develops disease. The human microbiome is a complex assemblage of the microbes inhabiting many sites in the human body. The components of this microbiome include bacteria, viruses, and fungi. Today's technology allows for in-depth sequencing and analysis of these communities and their members. In the foreseeable future, measuring microbiota composition could become standard clinical practice as it may become diagnostic for some diseases or indicate increased susceptibility to others. The microbiota of a number of disease states are being examined and it is possible that we can improve or maybe even cure some disease states just by modifying the microbiome (Eloe-Fadrosh and Rasko, 2013). The microbiome has been implicated as having an effect in many diseases, including rheumatoid arthritis, autism, asthma, and cancer (Barko et al., 2018).

The NIH has recognized the importance of the microbiome in human health and disease by founding the Human Microbiome Project. Information and data are available at http://commonfund.nih.gov/hmp/ (Human Microbiome Project, 2013).

Nutrigenomics data

We are what we eat! This statement is literally true, for we get the building blocks for our body's development from the foods we ingest. Each of these foods has a unique genetic composition, and each interact differently with our body. Nutrigenomics is "the science of the effect of genetic variation on dietary response and the role of nutrients and bioactive food compounds in gene expression" (Fenech et al., 2011). What one eats can directly affect gene transcription and, therefore, it can affect overall human health and phenotype. The response to nutrients is based on the body's genetic makeup, and how the body metabolizes and interacts with these nutrients. Because the health effects of nutrition are based on the body's genetic composition, different people with different ethnicities require different types of nutrition. It is possible, therefore, that better health outcomes can be achieved through customizing nutritional intake to be consistent with the genetic composition of the body. Because dietary factors are important controllers of human health, it is not difficult to expect certain effects of McDonalds versus the Mediterranean diet, but predictions are much more complex and can be hard to make (Fenech et al., 2011; Mir et al., 2022; Felisbino et al., 2021; Marcum, 2020).

Behavioral measures data

One of the most important aspects of success of treatment and preventative measure is the patient's behavior in regard to the intervention. Will they take the medication as prescribed? Will they avoid the things that prevent recurrence? Will they make changes in their exercise or diet that prevent disease? Will they avoid risks? The answers to these and many other questions can be answered often by analyzing aspects of a patient behavior. Patient behavior is often predicted using the Theory of Planned Behavior (TPB) model (Ajzen, 1991), which attempts to take into account the many variables that can affect a patient's behavior and determine whether the patient will comply to an intervention. This test is included often in clinical data. But on a larger scale for research purposes, it can be extrapolated, to consider what and where we eat, how much we exercise, what type of job we have, how much stress we are under, how many friends we have, and how we interact with other human beings. All of these factors have effects on our physiology and overall physical and mental health. One of the assumptions of the TPB model is that a given behavior is a function of the attitude, intention, and beliefs. Such measures have been used in many health studies to predict behavior-related health outcomes, such as adherence to medication in HIV patients (Jones et al., 2012). Some of the important variables in TPB models include descriptive norms, subjective norms, self-identity, self-efficacy, locus of control, behavioral control, anticipated regret, desires and emotions, moral norms and anticipated affect, social cognition properties, prototypes and willingness, conscientiousness, and goals and their properties. The variation of the deployed TPB models acknowledges the complexity of factors underlying the ultimate behavioral choice, but nevertheless, they assume that the behavior is predicted by some combination of social cognitive factors about the behavior. Each variable that is added to the model may increase incrementally the predictive power of the TPB model but may also introduce some unexplained variance with each predictor. Meta-analyses have shown that beliefs and attitudes in the TPB model predict 39% of the behavioral intention and 27% of the actual behavior, with notably stronger prediction when behavior was based on self-reports. In order to close the gap between behavioral intention and actual behavior, Gollwitzer and Sheeran (2006) proposed the addition of implementation intention.

Socioeconomic status data

SES is a composite measure of an individual's economic and sociological standing. It is a complex assessment measured in a variety of ways that account for a person's work experience, economic and social position in relation to others, based on income, education, and occupation. SES has been a powerful determinant of health; as a general rule, wealthy people tend to be in better health than people of poorer status (Erreygers, 2013). There appears to be a significant impact of SES on a multitude of diseases, including:

- cardiovascular disease (Gershon et al., 2012);
- respiratory disease (Bashinskaya et al., 2012);
- mental health-related disorders (Businelle et al., 2013); and
- hepatocellular carcinoma (Lu et al., 2022).
  Some of the metrics of SES include:
- highest level of education attainted;
- education of parents;
- current occupation;
- net income;
- household income;
- wealth (assets, capitol); and
- other related variables.

People are separated usually into groups based on these metrics from least advantaged to most advantaged, or low, medium, or high SES (Galobardes et al., 2006). There are many complex factors in the relationship between SES and health. People with relatively few resources may not have very good access to care services, or even transportation to get healthcare. They may not have the time to focus on their health or sufficient education to realize the impact that certain elements have on their health. Priorities can vary also; one person might be trying to maintain good health, while another person is a single mother trying to maintain a family with a minimum wage job. The stress related to one's SES alone may impact their health (Businelle et al., 2013). Regardless of the mechanism, there is a strong association between SES and health. Many studies with many different diseases have found profound implications of SES on disease (Gershon et al., 2012; Bashinskaya et al., 2012).

Personal activity monitoring data

Devices that measure human movements have existed for centuries. Leonardo da Vinci was perhaps the first to conceptualize the pedometer; his design was a device worn at the waist with a long lever affixed to the thigh with a ratchet-and-gear mechanism that recorded the number of steps taken during walking. Pedometers are now integrated in to phones and watches that we use every day. Needless to say, the design of these tracking devices has improved over time, and now these devices record far more than steps, including pulse rate, respiratory rate, sleep, and can even record EKGs. Perhaps the more striking feature of today's devices (and the most important for the purposes of this book) is their ability to track and store this information over time. These devices have been used extensively for physical activity research, allowing for objective measurement of physical activity beyond what has been learned through questionnaires. They are used now for comparing a population's overall physical activity, then targeting interventions at those individuals with decreased physical activity who have a predicted higher risk of adverse outcomes associated with decreased physical activity and obesity.

Some of this data is in the public domain and can be used to produce predictive analytic models. The US government began using an accelerometer in the National Health and Nutrition Examination Survey (CDC, 2013—http://www.cdc.gov/nchs/nhanes.htm) in 2003–04. Interestingly, use of this technology found that less than 5% of US adults met the national physical activity by pedometer in contrast to 45%–51% meeting the requirement by survey. There may be some inaccuracies in the measurement of this activity but much of the discrepancy is probably also related to an individual's overestimation of their actual physical activity, thus the true value probably lies between 5% and 51%. As our technology has improved, these measures have become more accurate. Research has shown that individuals who show less activity on these devices are at increased risk of developing obesity, hypertension, dyslipidemia, and insulin resistance. There is strong evidence a wearable monitor can provide a more valid assessment of physical activity than a physical activity questionnaire.

Some devices can provide monitoring of sedentary behaviors and sleep. A sedentary activity as defined by the Sedentary Behavior Research Network as "any waking behavior characterized by an energy expenditure <1.5 metabolic equivalents while in a sitting or reclining posture." Sleep is defined and tracked differently using other devices. Some adults who engage in excessive sedentary behaviors have an increase in cardiometabolic risk factors (Bassett, 2012). In fact, one study using questionnaire data showed that time spent in two sedentary behaviors, car driving and TV watching, are highly correlated with mortality from a variety of direct causes. Significant associations have been observed between sedentary time and waist circumference, triglycerides, HDL, resting blood pressure, plasma glucose, metabolic risk, and death (Bassett, 2012). Lack of proper sleep has been associated with diabetes, cardiovascular disease, poor immunity, mood disorders, obesity, and hypertension. By tracking this information, it may be possible to predict future disease.

Personal activity monitors are built into many of the devices we use every day. These devices produce a considerable amount of data on human activity that has been missed by older methods; thus now science has the ability to mine this data to build predictive models and improve human health. Aside from personal tracking devices, there are a number of apps for mobile devices that allow tracking of activity and also calories expended. The data generated from these apps and devices are a gold mine of information for data mining and predictive analytics (Hicks et al., 2019).

Climatological data

Weather patterns have effects on health in a number of ways. Certain illnesses are more prevalent during certain weather conditions. Additionally, weather affects not only the time spent outdoors, but also the amount of time spent indoors, with increased exposure to other people. Weather conditions affect an individual's amount of physical activity also (at least for most people). Climate change is thought to have had an effect on both the incidence and global patterns of human disease. Climate-related environmental variables have been associated with cardiovascular disease, respiratory diseases, mental illness, and infections (see Fig. 6.4). The abundance and distribution in populations of

FIGURE 6.4 Climatic conditions causing noncommunicable illnesses and associated pharmaceutical treatments (please see explanations of this figure in the text).

infectious organisms in our ecosystems is dependent on climate. Infections related to these organisms account for approximately 60% of human pathogenic illnesses (Bengis et al., 2004). Climate has a considerable effect on respiratory diseases, which may exhibit exacerbated symptoms based on environmental conditions; for example, asthma: the pollen concentration in the air is dependent on weather, weather changes, pollutants trapped by inversions, and other factors (Redshaw et al., 2013). Climate can play an important role in predicting disease, especially communicable disease where weather predictions may be combined with an individual's genetic and "omic" profile to predict an oncoming disease or infection. The availability of accurate predictive and prescriptive models will allow physicians (and even individuals who monitor themselves) to manage infections effectively when these weather conditions are present.

Environmental data

Environmental data encompasses all that the body comes in contact with both externally and internally, including chemicals, food, weather, force, toxins, sensations, smells, sounds, and visualizations. Anything outside our physical body that can be introduced to the inside or affect the inside of our bodies can be an environmental stimulus. These environmental stimuli may not all be physical things but may also be situational such as in the case of stress or stressful situations. The environment regulates the amount of oxygen delivered to our lungs based on elevation, the particles that are in the air, environmental interactions, the disease exposed to, the amount of sunlight, and even things like the gravitational force on our bodies. Besides these geographical environmental variables, there are many variables that impact us daily affecting our health including chemicals, diet, lifestyle, physical and psychological stress, and infectious agents. All of these exposures can have an effect on our health and disease processes. New technologies and methodologies for assessing human exposures are being developed, which will provide opportunities to expand our knowledge base and provide useful data for building predictive models for human disease (Weis et al., 2005).

All the other OMICs

In this chapter we have covered many of the core "omics" in use today that will play important roles in predictive analytics in healthcare, but we certainly have not addressed all of them. There are a number of subomics, such as the methylome as a subset of the epigenome or the spliceome as part of the transcriptome. Other "omics" are emerging and several others exist already or are being defined at the time of this publication. Other "omics" that are emerging will also prove to play important roles in predictive analytic models, including toxicoproteomics and autoantibodyomics. Toxicoproteomics is the study of proteins expressed in response to chemical and environmental exposures, which is a part of the larger field of toxicogenomics (Wetmore and Merrick, 2004). The autoantibodyome is the compete profile of autoantibodies developed by an individual, which will be critical in the predictive management of autoimmune disease (Chen et al., 2012). All of these and other forthcoming "omics" will play important roles, and they will be important in the future for building predictive models for disease prevention and health.

The future

The future of healthcare can be related to the development of missiles in warfare. Early missiles had a predetermined target to which they were aimed. But any changes in conditions around them could set them off course, causing them to miss their target. Modern intelligent missiles still have a planned trajectory, and a target to which they are aimed, but they can adjust the trajectory and the target, following signals from sensors of their current conditions or problems that arise in flight. These adjustments have to be based on the predicted flight path, because the speed of the missile puts it in a position far from the location where the data was collected. So, all interventions must be made on a predicted trajectory. *This is very similar to the path personalized medicine will take in the future.* This basis for determining the predicted "trajectory" of healthcare for an individual will begin with genetic profiling done at birth, taking into consideration the parents' disease states and "omic" profiles. This predicted healthcare trajectory will produce a risk score to guide the nature and timing of interventions to achieve the best possible outcome. Thus our "omic" profile will then be monitored at set points in time to help guide the choice of preventative measures, and if needed make interventions to fend off or prevent disease. By maintaining the trajectory, we can minimize the risk of disease and maintain our bodies in a healthy equilibrium.

Fig. 6.5 is a conceptual interface of a possible risk scoring system for disease. The vertical dashed bars represent population-based risk matched for age and gender. The gray areas represent individual calculated risk of an individual, based on genetic makeup and background information, and the gray bar's length indicates the confidence interval.

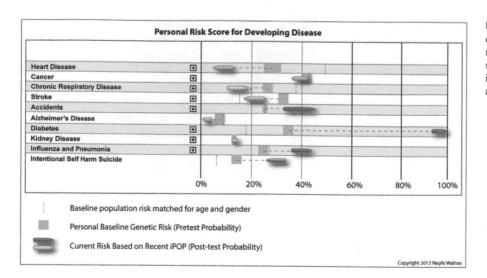

FIGURE 6.5 Personal risk score for developing disease. The direction of the marker indicates the direction your health score is trending. The color indicates green if you are below baseline risk and red is above baseline risk.

The red and green bars represent the risk, based on the individual's omics data at the time of testing. The bars are red if the risk has been elevated above their baseline, and green if it has decreased. The direction of the arrow shows the direction they are headed based on their last assessment. The length of the arrow indicates the confidence interval. The diseases shown in Fig. 6.5 are the top 10 causes of mortality in the United States. Many of these diseases could theoretically be able to be subdivided into specific types of each disease. For example, clicking on the plus sign to the right of cancer would display a list of variety of cancers with similar scores. Some of the items on the list would rely on environmental and behavioral measures, rather than upon "omic" data. For example, accidents are more likely to happen to people who have risky behaviors, although there are some genetic associations linked to risk taking that increase the likelihood of accident. An increased risk of suicide could be related to life stressors and current depressed mood, but it may be related substantially also to the presence or absence of the genes involved in major depressive disorder. The increased risk of influenza could be secondary in its effect to not having received a flu shot before the beginning of the flu season. Many measures will be required to create effective predictive models that can help maintain a healthy equilibrium.

Challenges

We face many challenges, as we integrate predictive modeling into personalized healthcare. Some of these challenges raise questions, for which we don't have answers at present.

Challenge #1

A new infrastructure must be created for healthcare. As personalized medicine based on individual characteristics becomes the norm, healthcare providers and pharmacists must be educated about these measures and how to prescribe and manage patients in a less standard manner. This transition will require also additional education in the use of complicated tools to help manage this information. Most EMR systems in their present configuration cannot even incorporate or manage structured genetic information, let alone integrate all the other variables such as metabolomics, epigenetics, and transcriptomics.

Challenge #2

Who will pay for gene sequencing of every individual for healthcare purposes? The cost of genome sequencing is declining rapidly, with a genome sequence without interpretation costing less than $500. Despite this steep decrease in price, patients with significant genetic features and disabilities continue to struggle to get payers to reimburse for genetic testing. Considering the difficulties in gaining approval for the testing of sick people, it is hard to imagine convincing insurance companies to pay for this test on an otherwise healthy individual. One argument for these tests might be that the genetic information could save money through preventative care and thus avoidance of costs associated with preventable cancer or heart disease. This argument is beginning to gain traction as the price of sequencing comes down and large sequencing initiatives start to demonstrate the financial impact of genetic screening (Rosso et al., 2017; Ramdzan et al., 2021; Verbelen et al., 2017; Catchpool et al., 2019; Koldehoff et al., 2021).

Challenge #3

Who will regulate predictive models that that directly affect patient care? How will these models be tested for clinical use? The possibility of FDA regulation for these models is a consideration, but it raises some serious concerns. If the FDA becomes involved with predictive modeling, will the costs of testing and regulation drive up the prices? Probably, it will. In addition, can we expect the FDA to be able to interpret and integrate predictive modeling properly in their deliberations? Probably, we cannot, at least in the near future. Serious questions are raised already about the regulation of genetic testing, in the wake of new offerings by many companies to provide genetic testing services directly to consumers. These and many other questions regarding genetic testing remain open, as we enter this era of predictive medicine. Here is one action plan the FDA has expressed for AI and ML software medical devices (https://www.fda.gov/medical-devices/software-medical-device-samd/artificial-intelligence-and-machine-learning-software-medical-device) for those readers that would like to explore this further.

Challenge #4

What effects will predictions of health outcome have on an individual's mental health and general overall daily anxiety levels? Knowing that there is a certain probability of developing a debilitating disease could have a significant impact on an individual's mental state. Patient response to genetic risk assessments is highly variable, depending on disease and a number of other issues. Strategies have been developed to present this information effectively, but more research is needed in this area as it develops (Lautenbach et al., 2013; Jacobs et al., 2019).

Challenge #5

Does the ability to predict someone's health outcome change his or her behavior? For years, it has been known that the effects of smoking cause cancer and cardiovascular disease, and it contributes to a myriad of other diseases. Yet people continue to smoke. Diet and exercise can also reduce your health burden significantly on a number of different fronts, yet the rate of obesity continues to rise in America. Millions of people smoke cigarettes, millions are overweight, and millions of people have high cholesterol. Regardless of the many warnings in schools and the media that these practices are associated with high risks of serious diseases, many people continue the practices. We are bombarded constantly about the benefits of healthy diet and exercise, yet we continue to get fatter, and maintain unhealthy diets. Most studies in this area suggest that genetic risk has a minimal effect on behavior and has questionable effectiveness in motivating changes in behavior (Henrikson et al., 2009; Hollands et al., 2016). On the positive side there are select studies that have shown significant behavioral changes (Carpenter et al., 2007; McCarty et al., 2018) and perhaps analyzing the differences between the disparate studies and outcomes can help us better understand how to use this information to change human behavior. A significant effort must be made to create plans for effective interventions, in order for personalized medicine to be effective in its truest form. Despite the possibility that people may not change their behavior, predicting risk does give the physician an advantage for prescribing the proper medications for people on the edge of disorders, such as diabetes or hypertension. Strong predictions by these models might induce the physician to prescribe drugs earlier than otherwise. Currently, we do not have sufficient information to determine when or how genetic risk information might motivate healthy behavior. Identifying the settings in which genetic risk can motivate healthy behavior, and identifying which people are likely to respond to this information are important goals for predictive analytics.

Challenge #6

With the possibility of building clinical risk prediction models also comes the possibility of legal liability. There has been little if any litigation that addresses the use of risk prediction models. However, as they become more widely used, the prospect of a legal liability and lawsuits will only increase. There is, however, an existing body of litigation on family history and genetics that would have some relevance to the use of risk prediction models. The possibility of a "bad prediction" is a legal concept that presents some difficulty, and it is very difficult to judge its significance at this early stage of the implementation of predictive analytic risk models. Outside of blatant misuse or absolute failure to use a required risk prediction model, there is very little legal guidance to avoid the risk for medical liability (Black et al., 2012).

Challenge #7

Many of the biological (body) specimens that are required for full "omics" analysis are not acquired easily. More effective and less invasive methods of gathering this information will be required to get a good assessment of the "omics" of the body. For example, requiring a brain biopsy and undergoing general anesthesia at every checkup is not a likely possibility considering the risk, time, and cost involved, yet some markers of neurological disease using today's technology would require this information.

Challenge #8

The technology required for personalized medicine is certainly not perfect, and many refinements and further developments are needed. For example, despite the hype about the value of WGS, with recent advances in long read sequencing we know that our traditional WGS platforms are missing a significant amount of the genome. Long-read sequencing is now commercially available but at the time of writing is too slow and too expensive for widespread clinical adoption (Murdock et al., 2022; Grosz et al., 2022; Long et al., 2022), e.g., data on just 200 patients can consume petabytes (a million gigabytes, or 1000 terabytes) of data storage.

Challenge #9

It is likely that the implementation of precision medicine into daily practice will modify the role and clinical workflow of the physician and may at times require the added services of genetic counselors to assist the patients in making their medical choices. Consideration must be given for creating a model for this type of interaction, and for determining how this process will fit into our healthcare delivery and reimbursement systems.

Challenge #10

The scientific disciplines covered by personalized predictive medicine are numerous and the tremendous amount of data and computing power required to employ them is unprecedented. New devices and techniques must be developed to collect and analyze this data. The complete implementation of personalized predictive medicine will require the concerted work of teams of scientists and engineers from many fields, along with guidance from ethical and legal professionals to develop guidelines for implementing this technology. The task is daunting, but also very exciting.

Challenge #11

Racial and ethnic bias are serious concerns for predictive analytics models. This is particularly true for precision medicine where the genetic variation that underlies our racial differences can become inadvertently associated to outcomes without a scientific basis and may result in disparate treatments based on race. Historically there have been differences that have been useful to know about as they do impact differences in risk and care (Stepler et al., 2022; Bienzle et al., 2022); however; care must be taken to ensure that these differences have a scientific basis and don't exacerbate existing disparities. One of the challenges related to this in precision medicine is that our genetic reference database is based largely on people of European ancestry (accessed Feb 27, 2022: https://www.pbs.org/newshour/science/genetic-research-has-a-white-bias-and-it-may-be-hurting-everyones-health). This means that most predictive models that are based on genetic information will inherently be catered to this population as variants that cause disease in this population may not be seen in other populations and disease-causing variants in populations with poor representation may not even be documented. There are many efforts underway to improve the diversity of our reference genome that will help us alleviate these problems, but there is still considerable work to be done (https://www.genome.gov/about-genomics/policy-issues/Genomics-Health-Disparities).

Challenge #12

The development of standards around data and implementation and evaluation of predictive analytics models is critical to precision medicine. To enable large-scale deployment of precision medicine predictive analytics models requires standards-based architectures for the data that go into the model as well as the results that come out of the model. There must also be standards by which these models are evaluated and standards to which they must be developed in order to be considered for deployment. These areas are efforts in progress (Amarasingham et al., 2016), but these must become more mature for wide-scale deployment and use of predictive analytics models in precision medicine.

Challenge #13

Predictive Analytics (PA) models that use large amounts of biological data are inherently difficult to interpret. It is important that models maintain some level of transparency so that physicians have an idea of what drives them in order to appropriately weigh the model's recommendations in the context of the patient's presentation and clinical care. Until we achieve extremely high levels of performance in these models it is critical that they are built to augment a physician's judgement and not replace it. Adoption of PA models is likely to be significantly hampered by lack of transparency as physicians are unlikely to adopt technologies that they cannot understand or trust.

Postscript

In this chapter, we have taken a deep dive into the ocean of many types of patient data, which can be used to characterize the diagnosis and treatment of individual patients. The primary challenges before us to make patient-directed healthcare happen involve the development of entirely new service and IT structures to support it. This development must proceed within a structure of guidelines for the implementation of predictive models that affect patient healthcare. The legal landscape must be rebuilt to provide protections like the General Data Protection Regulation (GDPR) in the EU. State-specific implementation of GDPR-like regulations have been passed in some states (e.g., the Consumer Data Protection Act by Virginia in 2021). The next step in the implementation of personalized medicine (described in this chapter) is to integrate predictive analytics into patient-directed decision-making, which is described in Chapter 7.

References

Adams, F., 1849. The genuine works of Hippocrates. W. Wood and Company, New York.

Ajzen, I., 1991. The Theory of Planned behaviour. Org. Behav. Hum. Decis. Process. 50, 179–211.

Amarasingham, R., Audet, A.M., Bates, D.W., Glenn Cohen, I., Entwistle, M., Escobar, G.J., et al., 2016. Consensus statement on electronic health **predictive analytics**: a guiding framework to address challenges. EGEMS (Wash. DC) 4 (1), 1163. Available from: https://doi.org/10.13063/2327-9214.1163. eCollection 2016. PMID: 27141516.

Aslam, B., Basit, M., Nisar, M.A., Khurshid, M., Rasool, M.H., 2017. Proteomics: technologies and their applications. J. Chromatogr. Sci. 55 (2), 182–196. Available from: https://doi.org/10.1093/chromsci/bmw167. Epub 2016 Oct 18. PMID: 28087761.

Barko, P.C., McMichael, M.A., Swanson, K.S., Williams, D.A., 2018. The gastrointestinal microbiome: a review. J. Vet. Intern. Med. 32 (1), 9–25. Available from: https://doi.org/10.1111/jvim.14875. Epub 2017 Nov 24. PMID: 29171095; PMCID: PMC5787212.

Bashinskaya, B., Nahed, B.V., Walcott, B.P., Coumans, J.V., Onuma, O.K., 2012. Socioeconomic status correlates with the prevalence of advanced coronary artery disease in the United States. PLoS One 7 (9), e46314.

Bassett, D.R., 2012. Device-based monitoring in physical activity and public health research. Physiol. Meas. 33 (11), 1769–1783.

Bengis, R.G., Leighton, F.A., Fischer, J.R., Artois, M., Morner, T., Tate, C.M., 2004. The role of wildlife in emerging and re-emerging zoonoses. Rev. Sci. Technol. Oie. 23, 497–511.

Bienzle, U., Sodeinde, O., Effiong, C.E., Luzzatto, L., 2022. Glucose 6-phosphate dehydrogenase deficiency and sickle cell anemia: frequency and features of the association in an African community. PMID: 1174693.

Binda, C., Gibiino, G., Coluccio, C., Sbrancia, M., Dajti, E., Sinagra, E., et al., 2022. Biliary diseases from the microbiome perspective: how microorganisms could change the approach to benign and malignant diseases. Microorganisms 10 (2), 312. Available from: https://doi.org/10.3390/microorganisms10020312. PMID: 35208765.

Black, L., Knoppers, B.M., Avard, D., Simard, J., 2012. Legal liability and the uncertain nature of risk prediction: the case of breast cancer risk prediction models. Public Health Genomics 15 (6), 335–340. Available from: https://doi.org/10.1159/000342138. Epub 2012 Sep 12.

Bragazzi, N.L., 2013. From P0 to P6 medicine, a model of highly participatory, narrative, interactive, and "augmented" medicine: some considerations on Salvatore Iaconesi's clinical story. Patient Prefer. Adherence 7, 353–359.

Brizel, D.M., 2011. Head and neck cancer as a model for advances in imaging prognosis, early assessment, and posttherapy evaluation. Cancer J. 17 (3), 159–165.

Brown, L.J., Achinger-Kawecka, J., Portman, N., Clark, S., Stirzaker, C., Lim, E., 2022. Epigenetic therapies and biomarkers in breast cancer. Cancers (Basel) 14 (3), 474. Available from: https://doi.org/10.3390/cancers14030474. PMID: 35158742; PMCID: PMC8833457.

Bucciarelli-Ducci, C., Ajmone-Marsan, N., Di Carli, M., Nicol, E., 2022. The year in cardiovascular medicine 2021: imaging. Eur. Heart J. ehac033. Available from: https://doi.org/10.1093/eurheartj/ehac033. Epub ahead of print. PMID: 35259251.

Businelle, M.S., Mills, B.A., Chartier, K.G., Kendzor, D.E., Reingle, J.M., Shuval, K., 2013. Do stressful events account for the link between socio-economic status and mental health? J. Public Health (Oxf.) .

Capell, B.C., Berger, S.L., 2013. Genome-wide epigenetics. J. Invest. Dermatol. 133 (6), e9. Available from: https://doi.org/10.1038/jid.2013.173.

Carpenter, M.J., Strange, C., Jones, Y., Dickson, M.R., Carter, C., Moseley, M.A., et al., 2007. Ann. Behav. Med. 33 (1), 22–28. Available from: https://doi.org/10.1207/s15324796abm3301_3. PMID: 17291167.

Catchpool, M., Ramchand, J., Martyn, M., Hare, D.L., James, P.A., Trainer, A.H., et al., 2019. Goranitis I. Genet. Med. 21 (12), 2815–2822. Available from: https://doi.org/10.1038/s41436-019-0582-2. Epub 2019 Jun 20. PMID: 31222143.

CDC, 2013. <http://www.cdc.gov/nchs/nhanes.htm>.

Chen, R., Mias, G.I., Li-Pook-Than, J., Jiang, L., Lam, H.Y., Chen, R., et al., 2012. Personal omics profiling reveals dynamic molecular and medical phenotypes. Cell 148 (6), 1293–1307.

Cho, I., Blaser, M.J., 2012. The human microbiome: at the interface of health and disease. Nat Rev Genet. 13 (4), 260–270. Available from: https://doi.org/10.1038/nrg3182. PMID: 22411464; PMCID: PMC3418802.

Costantino, I., Nicodemus, J., Chun, J., 2021. Genomic mosaicism formed by somatic variation in the aging and diseased brain. Genes (Basel) 12 (7), 1071. Available from: https://doi.org/10.3390/genes12071071. PMID: 34356087; PMCID: PMC8305509.

Dietert, R.R., Dietert, J.M., 2022. Using microbiome-based approaches to deprogram chronic disorders and extend the healthspan following adverse childhood experiences. Microorganisms 10 (2), 229. Available from: https://doi.org/10.3390/microorganisms10020229. PMID: 35208684.

Eloe-Fadrosh, E.A., Rasko, D.A., 2013. The human microbiome: from symbiosis to pathogenesis. Annu. Rev. Med. 64, 145–163.

Epigenomics, 2013. <http://www.ncbi.nlm.nih.gov/epigenomics/>.

Erreygers, G., 2013. A dual atkinson measure of socioeconomic inequality of health. Health Econ. 22 (4), 466–479.

Evangelou, E., Ioannidis, J.P., 2013. Meta-analysis methods for genome-wide association studies and beyond. Nat. Rev. Genet. 14 (6), 379–389.

FDA, What is gene therapy. <https://www.fda.gov/vaccines-blood-biologics/cellular-gene-therapy-products/what-gene-therapy>.

Felisbino, K., Granzotti, J.G., Bello-Santos, L., Guiloski, I.C., 2021. Nutrigenomics in regulating the expression of genes related to type 2 diabetes mellitus. Front. Physiol. 12, 699220. Available from: https://doi.org/10.3389/fphys.2021.699220. PMID: 34366888; PMCID: PMC8334860.

Fenech, M., El-Sohemy, A., Cahill, L., Ferguson, L.R., French, T.A.C., Shyong Tai, E., et al., 2011. Nutrigenetics and nutrigenomics: viewpoints on the current status and applications in nutrition research and practice. J. Nutrigenet Nutrigenomics 4 (2), 69–89. Published online 2011 May 28.

Fingerman, I.M., Zhang, X., Ratzat, W., Husain, N., Cohen, R.F., Schuler, G.D., 2013. NCBI Epigenomics: What's new for 2013. Nucleic Acids Res. 41 (D1), D221–D225.

FutureMed/Singularity University, 2013. <http://singularityu.org/2012/11/01/futuremed-scheduled-for-february-4-9-2013-at-singularity-university/; http://singularityu.org/tag/futuremed/>.

Gallagher, D., Shaheen, I., Zafar, K., 2011. State-of-the-art measurements in human body composition: a moving frontier of clinical importance. Int. J. Body Compos. Res. Author manuscript; available in PMC.

Galobardes, B., Shaw, M., Lawlor, D.A., Lynch, J.W., Davey Smith, G., 2006. Indicators of socioeconomic position (part 2). J. Epidemiol. Community Health 60, 95–101.

Ganesan, A., Arimondo, P.B., Rots, M.G., et al., 2019. The timeline of epigenetic drug discovery: from reality to dreams. Clin. Epigenet 11, 174. Available from: https://doi.org/10.1186/s13148-019-0776-0.

Gardes, J., Maldivi, C., Boisset, D., Aubourg, T., Vuillerme, N., Demongeot, J., 2019. Maxwell®: an unsupervised learning approach for 5P medicine. Stud. Health Technol. Inf. 264, 1464–1465. Available from: https://doi.org/10.3233/SHTI190486. PMID: 31438183.

Gene therapies should be for all, 2021. Nat. Med. 27, 1311. <https://doi.org/10.1038/s41591-021-01481-9>.

Genetics home Reference—NIH, 2013. <http://ghr.nlm.nih.gov/>.

Gershon, A.S., Dolmage, T.E., Stephenson, A., Jackson, B., 2012. Chronic obstructive pulmonary disease and socioeconomic status: a systematic review. COPD 9 (3), 216–226.

Gibson, W.M., 1971. Can personalized medicine survive? Can. Fam. Physician 17 (8), 29–88.

Goldsmith, T.C., 2017. Externally regulated programmed aging and effects of population stress on mammal lifespan. Biochem. Mosc. 82, 1430–1434. Available from: https://doi.org/10.1134/S0006297917120033.

Gollwitzer, P.M., Sheeran, P., 2006. Implementation intentions and goal achievement: a meta-analysis of effects and processes. Adv. Exp. Soc. Psychol. 38, 69–119.

Gonzalez-Covarrubias, V., Martínez-Martínez, E., Del Bosque-Plata, L., 2022. The potential of metabolomics in biomedical applications. Metabolites 12 (2), 194. Available from: https://doi.org/10.3390/metabo12020194. PMID: 35208267; PMCID: PMC8880031.

Gorini, A., Pravettoni, G., 2011. P5 medicine: a plus for a personalized approach to oncology. Nat. Rev. Clin. Oncol. 8 (7), 444.

Grosz, B.R., Stevanovski, I., Negri, S., Ellis, M., Barnes, S., Reddel, S., et al., 2022. J. Peripher. Nerv. Syst. Available from: https://doi.org/10.1111/jns.12485Online ahead of print. PMID: 35224818.

Gueniche, A., Perin, O., Bouslimani, A., Landemaine, L., Misra, N., Cupferman, S., et al., 2022. Advances in microbiome-derived solutions and methodologies are founding a new era in skin health and care. Pathogens. 11 (2), 121. Available from: https://doi.org/10.3390/pathogens11020121. PMID: 35215065.

Habib, A.R., Kajbafzadeh, M., Hasan, Z., Wong, E., Gunasekera, H., Perry, C., et al., 2022. Artificial intelligence to classify ear disease from otoscopy: a systematic review and meta-analysis. Clin. Otolaryngol. Available from: https://doi.org/10.1111/coa.13925. Epub ahead of print. PMID: 35253378.

Heller, F., 2013. Genetics/genomics and drug effects. Acta Clin. Belg. 68 (2), 77–80.

Henrikson, N.B., Bowen, D., Burke, W., 2009. Does genomic risk information motivate people to change their behavior? Genome Med. 1 (4), 37.

Hesselberth, J.R., 2013. Lives that introns lead after splicing. Wiley Interdiscip. Rev. RNA.

Hicks, J.L., Althoff, T., Sosic, R., Kuhar, P., Bostjancic, B., King, A.C., et al., 2019. Best practices for analyzing large-scale health data from wearables and smartphone apps. NPJ Digit. Med. 2, 45. Available from: https://doi.org/10.1038/s41746-019-0121-1. PMID: 31304391; PMCID: PMC6550237.

Hollands, G.J., French, D.P., Griffin, S.J., Prevost, A.T., Sutton, S., King, S., et al., 2016. BMJ 352, i1102. Available from: https://doi.org/10.1136/bmj.i1102. PMID: 26979548.

Hood, L., Tian, Q., 2012. Systems approaches to biology and disease enable translational systems medicine. Genomics Proteom. Bioinforma. 10 (4), 181−185.

Human Microbiome Project, 2013. <http://commonfund.nih.gov/hmp/>.

Jacobs, C., Patch, C., Michie, S., 2019. Eur. J. Hum. Genet. 27 (4), 511−524. Available from: https://doi.org/10.1038/s41431-018-0310-4. Epub 2018 Dec 20. PMID: 30573802.

Jain, K.K., 2021. Textbook of Personalized Medicine, thir ed., p. 2.

Jathanna, N., Podlasek, A., Sokol, A., Auer, D., Chen, X., Jamil-Copley, S., 2021. Diagnostic utility of artificial intelligence for left ventricular scar identification using cardiac magnetic resonance imaging-a systematic review. Cardiovasc. Digit. Health J. 2 (6 Suppl.), S21−S29.

Jones, G., Hawkins, K., Mullin, R., Nepusz, T., Naughton, D.P., Sheeran, P., et al., 2012. Understanding how adherence goals promote adherence behaviours: a repeated measure observational study with HIV seropositive patients. BMC Public Health 12, 587.

Jørgensen, J.T., 2009. New era of personalized medicine: a 10-year anniversary. Oncologist 14 (5), 557−558.

Katsanis, S.H., Katsanis, N., 2013. Molecular genetic testing and the future of clinical genomics. Nat. Rev. Genet. 14 (6), 415−426.

Kitano, H., 2002. Systems biology: a brief overview. Science 295 (5560), 1662−1664.

Koldehoff, A., Danner, M., Civello, D., Rhiem, K., Stock, S., Müller, D., 2021. Value Health 24 (2), 303−312. Available from: https://doi.org/10.1016/j.jval.2020.09.016. Epub 2021 Jan 5. PMID: 33518037.

Langreth, R., Waldholz, M., 1999a. New era of personalized medicine: targeting drugs for each unique genetic profile. Oncologist 4 (5), 426−427.

Langreth, R., Waldholz, M., 1999b. Genetic mapping ushers in new era of profitable personal medicines. The Wall Street Journal <https://www.wsj.com/articles/SB924225073307249185>.

Lautenbach, D.M., Christensen, K.D., Sparks, J.A., Green, R.C., 2013. Communicating genetic risk information for common disorders in the era of genomic medicine. Annu. Rev. Genomics Hum. Genet. 14, 491−513.

Li, F., Glinskii, O.V., Glinsky, V.V., 2013. Glycobioinformatics: current strategies and tools for data mining in MS-based glycoproteomics. Proteomics 13 (2), 341−354.

Liu, B., Faller, L.L., Klitgord, N., Mazumdar, V., Ghodsi, M., Sommer, D.D., et al., 2012. Deep sequencing of the oral microbiome reveals signatures of periodontal disease. PLoS One 7 (6), e37919.

Liu, M.H., Zhao, C., Wang, S., Jia, H., Yu, B., 2022. Artificial intelligence-a good assistant to multi-modality imaging in managing acute coronary syndrome. Front. Cardiovasc. Med. 8, 782971. Available from: https://doi.org/10.3389/fcvm.2021.782971. PMID: 35252367; PMCID: PMC8888682.

Long, J., Sun, L., Gong, F., Zhang, C., Mao, A., Lu, Y., et al., 2022. Gene 822, 146332. Available from: https://doi.org/10.1016/j.gene.2022.146332. Online ahead of print. PMID: 35181504.

Long Term Follow-Up After Administration of Human Gene Therapy Products; Guidance for Industry, January 2020. <https://www.fda.gov/media/113768/download>.

Lu, W., Zheng, F., Li, Z., Zhou, R., Deng, L., Xiao, W., et al., 2022. Front. Public Health 10, 741490. Available from: https://doi.org/10.3389/fpubh.2022.741490. eCollection 2022. PMID: 35252078.

Marcum, J.A., 2020. Nutrigenetics/nutrigenomics, personalized nutrition, and precision healthcare. Curr. Nutr. Rep. 9 (4), 338−345. Available from: https://doi.org/10.1007/s13668-020-00327-z. PMID: 32578026.

McCarty, C.A., Fuchs, M.J., Lamb, A., Conway, P., 2018. Optom. Vis. Sci. 95 (3), 166−170. Available from: https://doi.org/10.1097/OPX.0000000000001188. PMID: 29424826.

MedlinePlus, 2022. National Library of Medicine (US) <https://medlineplus.gov/genetics/understanding/precisionmedicine/precisionvspersonalized/>.

Mir, R.A., Nazir, M., Sabreena, Naik, S., Mukhtar, S., Ganai, B.A., et al., 2022. Utilizing the underutilized plant resources for development of life style foods: putting nutrigenomics to use. Plant. Physiol. Biochem. 171, 128−138. Available from: https://doi.org/10.1016/j.plaphy.2021.12.038. Epub 2022 Jan 3. PMID: 34998100.

Murdock, D.R., Rosenfeld, J.A., Lee, B., 2022. Annu. Rev. Med. 73, 575−585. Available from: https://doi.org/10.1146/annurev-med-042120-014904. PMID: 35084988.

Nam, D., Chapiro, J., Paradis, V., Seraphin, T.P., Kather, J.N., 2022. Artificial intelligence in liver diseases: improving diagnostics, prognostics and response prediction. JHEP Rep. 4 (4), 100443. Available from: https://doi.org/10.1016/j.jhep.2022.100443. PMID: 35243281; PMCID: PMC8867112.

National Human Genome Research Institute, 2010. <http://www.genome.gov/10001345>.

National Research Council (US) Committee on A Framework for Developing a New Taxonomy of Disease, 2011. Toward Precision Medicine: Building a Knowledge Network for Biomedical Research and a New Taxonomy of Disease. National Academies Press, US, Washington, DC.

Parenteau, J., Abou Elela, S., Introns, 2019. Good day junk is bad day treasure. Trends Genet. 35 (12), 923−934. Available from: https://doi.org/10.1016/j.tig.2019.09.010. Epub 2019 Oct 25. PMID: 31668856.

PCORI, 2013. Comprehensive Inventory of Research Networks. Clinical Data Research Networks, Patient-Powered Research Networks, and Patient Registries. Patient Centered Outcomes Research Institute. PDF. Accessed: <https://www.pcori.org/assets/2013/06/PCORI-Comprehensive-Inventory-Research-Networks-061213.pdf>.

PCORI, 2020. Patient Centered Outcome Research Institute: Project Summary. PCORnet: Clinical Research Network (CRN). Accessed: <https://www.pcori.org/research-results/2020/insight-clinical-research-network>.

PCORI, 2022. Patient Centered Outcome Research Institute: Improving methods for conducting patient-centered outcomes research. Accessed: <https://www.pcori.org/funding-opportunities/announcement/improving-methods-conducting-patient-centered-outcomes-research-2022-standing-pfa>.

Personal Genome Project, 2005, 2013. <http://www.personalgenomes.org/>.

Pires, I.M., Denysyuk, H.V., Villasana, M.V., Sá, J., Lameski, P., Chorbev, I., et al., 2021. Mobile 5P-medicine approach for cardiovascular patients. Sensors (Basel) 21 (21), 6986. Available from: https://doi.org/10.3390/s21216986. PMID: 34770292; PMCID: PMC8587644.

Poduri, A., Evrony, G.D., Cai, X., Walsh, C.A., 2013. Somatic mutation, genomic variation, and neurological disease. Science 341 (6141), 1237758.

Qi, T.H., Hian, O.H., Kumaran, A.M., Tan, T.J., Cong, T.R.Y., Su-Xin, G.L., et al., 2022. JBCR; Ai3. Multi-center evaluation of artificial intelligent imaging and clinical models for predicting neoadjuvant chemotherapy response in breast cancer. Breast Cancer Res. Treat. Available from: https://doi.org/10.1007/s10549-022-06521-7. Epub ahead of print. PMID: 35262831.

Ramdzan, A.R., Manaf, M.R.A., Aizuddin, A.N., Latiff, Z.A., Teik, K.W., Ch'ng, G.S., et al., 2021. Int. J. Env. Res. Public. Health 18 (16), 8330. Available from: https://doi.org/10.3390/ijerph18168330. PMID: 34444091.

Redshaw, C.H., Stahl-Timmins, W.M., Fleming, L.E., Davidson, I., Depledge, M.H., 2013. Potential changes in disease patterns and pharmaceutical use in response to climate change. J. Toxicol. Env. Health B Crit. Rev. 16 (5), 285−320.

Rosso, A., Pitini, E., D'Andrea, E., Massimi, A., De Vito, C., Marzuillo, C., et al., 2017. Ann. Ig. 29 (5), 464−480. Available from: https://doi.org/10.7416/ai.2017.2178. PMID: 28715059.

Sánchez-Pla, A., Reverter, F., Ruíz de Villa, M.C., Comabella, M., 2012. Transcriptomics: mRNA and alternative splicing. J. Neuroimmunol. 248 (1-2), 23−31.

Shaul, O., 2017. How introns enhance gene expression. Int. J. Biochem. Cell Biol. 91 (Pt B), 145−155. Available from: https://doi.org/10.1016/j.biocel.2017.06.016. Epub 2017 Jul 1. PMID: 28673892.

Shen, J.P., 2021. Artificial intelligence, molecular subtyping, biomarkers, and precision oncology. Emerg. Top Life Sci. 5 (6), 747−756. Available from: https://doi.org/10.1042/ETLS20210212. PMID: 34881776; PMCID: PMC8786277.

Sordo Vieira, L., Laubenbacher, R.C., 2022. Curr. Opin. Biotechnol. 75, 102702. Available from: https://doi.org/10.1016/j.copbio.2022.102702. Online ahead of print. PMID: 35217296 Review.

Stepler, K.E., Gillyard, T.R., Reed, C.B., Avery, T.M., Davis, J.S., Robinson, R.A.S., 2022. ABCA7, a genetic risk factor associated with Alzheimer's disease risk in African Americans. J. Alzheimers Dis . Available from: https://doi.org/10.3233/JAD-215306. Online ahead of print. PMID: 35034901.

Syggelou, A., Iacovidou, N., Atzori, L., Xanthos, T., Fanos, V., 2012. Metabolomics in the developing human being. Pediatr. Clin. North. Am. 59 (5), 1039−1058.

Vanhees, K., Vonhögen, I.G., van Schooten, F.J., Godschalk, R.W., 2013. You are what you eat, and so are your children: the impact of micronutrients on the epigenetic programming of offspring. Cell Mol. Life Sci.

Verbelen, M., Weale, M.E., Lewis, C.M., 2017. Pharmacogenomics J. 17 (5), 395−402. Available from: https://doi.org/10.1038/tpj.2017.21. Epub 2017 Jun 13. PMID: 28607506.

Wagner, M.J., 2013. Rare-variant genome-wide association studies: a new frontier in genetic analysis of complex traits. Pharmacogenomics 14 (4), 413−424.

Walton, N.A., et al., 2020. Pilot implementation of clinical genomic data into the native electronic health record: challenges of scalability. ACI Open. 4 (02), e162−e166.

Wang, K.C., Chang, H.Y., 2018. Epigenomics: technologies and applications. Circ. Res. 122 (9), 1191−1199. Available from: https://doi.org/10.1161/CIRCRESAHA.118.310998. PMID: 29700067; PMCID: PMC5929475.

Wang, L., Chang, L., Luo, R., Cui, X., Liu, H., Wu, H., et al., 2022. An artificial intelligence system using maximum intensity projection MR images facilitates classification of non-mass enhancement breast lesions. Eur. Radiol.

Weinstock, G.M., 2012. Genomic approaches to studying the human microbiota. Nature. 489 (7415), 250−256.

Weis, B.K., Balshaw, D., Barr, J.R., Brown, D., Ellisman, M., Lioy, P., et al., 2005. Personalized exposure assessment: promising approaches for human environmental health research. Environ. Health Perspect. 113 (7), 840−848.

Wells, L., Hart, G.W., 2013. Glycomics: building upon proteomics to advance glycosciences. Mol. Cell Proteom. 12 (4), 833−835.

Wetmore, B.A., Merrick, B.A., 2004. Toxicoproteomics: proteomics applied to toxicology and pathology. Toxicol. Pathol. 32 (6), 619−642.

Zhao, L., Lee, V.H.F., Ng, M.K., Yan, H., Bijlsma, M.F., 2019. Brief. Bioinforma. 20 (2), 572−584. Available from: https://doi.org/10.1093/bib/bby026.

Zoldoš, V., Horvat, T., Lauc, G., 2013. Glycomics meets genomics, epigenomics and other high throughput omics for system biology studies. Curr. Opin. Chem. Biol. 17 (1), 34−40.

Zupanic, A., Bernstein, H.C., Heiland, I., 2020. Systems biology: current status and challenges. Cell Mol. Life Sci. 77 (3), 379−380. Available from: https://doi.org/10.1007/s00018-019-03410-z. Epub 2020 Jan 13. PMID: 31932855.

Further reading

Altelaar, A.F., Munoz, J., Heck, A.J., 2013. Next-generation proteomics: towards an integrative view of proteome dynamics. Nat. Rev. Genet. 14 (1), 35−48.

Angell, H.K., Gray, N., Womack, C., Pritchard, D.I., Wilkinson, R.W., Cumberbatch, M., 2013. Digital pattern recognition-based image analysis quantifies immune infiltrates in distinct tissue regions of colorectal cancer and identifies a metastatic phenotype.mBr. J. Cancer.

Arnold, R.M., Forrow, L., 1990. Rewarding medicine: good doctors and good behavior. Ann. Intern. Med. 113 (10), 794−798.

Bastien, R.R.L., Rodríguez-Lescure, Á., Ebbert, M.T.W., Prat, A., Munárriz, B., Rowe, L., et al., 2012. PAM50 breast cancer subtyping by RT-qPCR and concordance with standard clinical molecular markers. BMC Med. Genomics 5, 44. Published online 2012 October 4.

Budin, F., Hoogstoel, M., Reynolds, P., Grauer, M., O'Leary-Moore, S.K., Oguz, I., 2013. Fully automated rodent brain MR image processing pipeline on a Midas server: from acquired images to region-based statistics. Front. Neuroinform 7, 15.

Charlab, R., Zhang, L., 2013. Pharmacogenomics: historical perspective and current status. Methods Mol. Biol. 1015, 3−22.

Chen, R., Snyder, M., 2013. Promise of personalized omics to precision medicine. Wiley Interdiscip. Rev. Syst. Biol. Med. 5 (1), 73−82.

Cho, I., Blaser, M.J., 2012. The human microbiome: at the interface of health and disease. Nat. Rev. Genet. 13 (4), 260−270.

Clegg, D.O., Reda, D.J., Harris, C.L., Klein, M.A., O'Dell, J.R., Hooper, M.M., et al., 2006. Glucosamine, chondroitin sulfate, and the two in combination for painful knee osteoarthritis. N. Engl. J. Med. 354 (8), 795−808.

de Magalhães, J.P., 2012. Programmatic features of aging originating in development: aging mechanisms beyond molecular damage? FASEB J. 26 (12), 4821−4826.

Emmett, M.R., Kroes, R.A., Moskal, J.R., Conrad, C.A., Priebe, W., Laezza, F., et al., 2013. Integrative biological analysis for neuropsychopharmacology. Neuropsychopharmacology.

Fanos, V., Antonucci, R., Atzori, L., 2013. Metabolomics in the developing infant. Curr. Opin. Pediatr.

Glubb, D.M., Paugh, S.W., van Schaik, R.H., Innocenti, F., 2013. A guide to the current web-based resources in pharmacogenomics. Methods Mol. Biol. 1015, 293−310.

Gomez-Casati, D.F., Zanor, M.I., Busi, M.V., 2013. Metabolomics in plants and humans: applications in the prevention and diagnosis of diseases. Biomed. Res. Int. 2013, 792527.

Gregson, R.A., 2013. Decision making with complex nonlinear systems: inference and identification in the context of DS22q11.2. Nonlinear Dyn. Psychol. Life Sci. 17 (2), 173−181.

Healthy People 2020, 2013. <http://www.healthypeople.gov/2020/>.

Healthy People, 2020. Framework: The Vision, Mission, and Goals of Health People 2020 <http://www.healthypeople.gov> (accessed 19.09.13).

Heidrich, A., Schmidt, J., Zimmermann, J., Saluz, H.P., 2013. Automated segmentation and object classification of ct images: application to in vivo molecular imaging of avian embryos. Int. J. Biomed. Imaging 2013, 508474.

Hindorff, L.A., MacArthur, J., Morales, J., et al., 2013. National Human Genome Research Institute. A catalog of published genome-wide association studies. <http://www.genome.gov/gwastudies> (accessed 14.09.13).

Hochberg, M.C., The Great Debate: Perspectives on Glucosamine and Chondroitin Sulphate Moderator. University of Maryland, Baltimore, MD.

Hood, L., Balling, R., Auffray, C., 2012. Revolutionizing medicine in the 21st century through systems approaches. Biotechnol. J. 7 (8), 992−1001.

Kuhl, P.K., Coffey-Corina, S., Padden, D., Munson, J., Estes, A., Dawson, G., 2013. Brain responses to words in 2-year-olds with autism predict developmental outcomes at age 6. PLoS One 8 (5), e64967.

Ng, S.C., Bernstein, C.N., Vatn, M.H., Lakatos, P.L., Loftus Jr, E.V., Tysk, C., et al.,Epidemiology and Natural History Task Force of the International Organization of Inflammatory Bowel Disease (IOIBD) 2013. Geographical variability and environmental risk factors in inflammatory bowel disease. Gut 62 (4), 630−649.

Patel, C.J., Sivadas, A., Tabassum, R., Preeprem, T., Zhao, J., Arafat, D., et al., 2013. Whole genome sequencing in support of wellness and health maintenance. Genome Med. 5 (6), 58.

Pavlopoulos, G.A., Oulas, A., Iacucci, E., Sifrim, A., Moreau, Y., Schneider, R., et al., 2013. Unraveling genomic variation from next generation sequencing data. BioData Min. 6 (1), 13.

Priorities for Personalized Medicine, 2008. Report of the President's Council of Advisors on Science and Technology. OCLC Digital Archive <http://www.ostp.gov/galleries/PCAST/pcast_report_v2.pdf>.

Ramos, E., Doumatey, A., Elkahloun, A.G., Shriner, D., Huang, H., Chen, G., et al., 2013. Pharmacogenomics, ancestry and clinical decision making for global populations. Pharmacogenomics J.

Richardson, S.R., O'Malley, G.F., 2022. Glucose 6 Phosphate dehydrogenase deficiency. In: StatPearls. Treasure Island, FL: StatPearls Publishing. <https://www.ncbi.nlm.nih.gov/books/NBK470315/>.

Ridaura, V.K., Faith, J.J., Rey, F.E., Cheng, J., Duncan, A.E., Kau, A.L., et al., 2013. Gut microbiota from twins discordant for obesity modulate metabolism in mice. Science 341 (6150), 1241214.

Song, J.W., Lee, J.H., 2013. New morphological features for grading pancreatic ductal adenocarcinomas. Biomed. Res. Int. 2013, 175271.

StatSoft Decisioning System. 2012. 2013 <http://www.statsoft.com/products/statistica/decisioning-platform/>.

Uddin, L.Q., Supekar, K., Lynch, C.J., Khouzam, A., Phillips, J., Feinstein, C., et al., 2013. Salience network-based classification and prediction of symptom severity in children with autism. AMA Psych. 1−11.

Vaidyanathan, G., 2012. Redefining clinical trials: the age of personalized medicine. Cell 148 (6), 1079−1080.

Vermeersch, K.A., Styczynski, M.P., 2013. Applications of metabolomics in cancer research. J. Carcinog. 12, 9.

Wade, C.H., Tarini, B.A., Wilfond, B.S., 2013. Growing up in the genomic era: implications of whole-genome sequencing for children, families, and pediatric practice. Annu. Rev. Genomics Hum. Genet. 14, 535−555.

Wang, Z., Liu, X., Yang, B.Z., Gelernter, J., 2013. The role and challenges of exome sequencing in studies of human diseases. Front. Genet. 4, 160.

Chapter 7

Patient-directed healthcare

Linda A. Miner

Chapter outline

Practical Data Analytics for Innovation in Medicine. DOI: https://doi.org/10.1016/B978-0-323-95274-3.00019-1

Prelude

The need for patient-centric data structures (discussed in Chapters 4 and 5) and the great promise of using many new data sources for personalizing medical diagnosis and treatment beg the question of how can we make this happen? A follow-on question is how can we do this without compromising the quality of healthcare? This chapter dives into the many issues related to incorporating the patient directly into their diagnosis and treatment as the primary enabling factor in personalizing medicine.

Empowerment in patient-directed medicine

Something exciting is happening much more now than in the year of the first edition. We've been going through a quiet revolution in medicine—from a time in which patients went to their doctors seeking all the answers to now when they take much information with them to the doctor. One can Google just about anything on one's cellphone and research studies are just a click away, even if one does not have access to the old PubMed to access online journals. As of the middle of May, 2020, PubMed became PubMed.gov.

I was able to enter into the search box, "Ivermectin as treatment for covid" and found this URL: https://pubmed.ncbi.nlm.nih.gov/?term=ivermectin+as+treatment+for+covid, with 10 articles listed on the first page. Considering that ivermectin has been a controversial topic, it was surprising to see so many. Even more surprising, when looking at various articles, there were often potential flaws or lack of randomization, which rendered conclusions moot. One example of such was a JAMA article by López-Medina and López (2021), found not on PubMed but by searching further to find results of clinical trials mentioned (this one claiming to be randomized), appeared to be hopelessly flawed, with both groups, placebo and treatment, inadvertently receiving ivermectin and with no checking of blood levels to see if the groups differed; it was little wonder that the conclusion found no results. At any rate, anyone can now access the PubMed articles. But as always, reader beware.

Chapter 6 discusses personalized medicine that is patient-centered, while this chapter discusses *patient-directed* medicine. There is certainly overlap between these two chapters, but there are distinctions. Patient-centered (personalized) medicine views the patient from the point of view of the provider. Providers attempt to increase patient satisfaction and good medical outcomes. Patient-directed healthcare is from the patient's point of view, wherein the patients assume more responsibility for their own good health outcomes. The two emphases should complement one another. This chapter asks how patients can take more responsibility for their own healthcare, and how might predictive analytics (PAs) be useful to patients as they take more responsibility.

Enter the word "patient" into a Google search string, and these definitions emerge:

- Adjective: able to wait without becoming annoyed or anxious
- Noun: a person receiving or registered to receive medical treatment
- Synonyms: adjective. Uncomplaining-long suffering-enduring-tolerant; noun.

Self-monitoring, N of 1 study

Contrast the above with how Gary (Dr. Gary Miner, organizer of this book) took charge of his own eye care and conducted an N of 1 study (discussed later in the chapter and in Chapter 11). He was attempting to follow his doctors' recommendations but when the course of disease looked like blindness in his left eye, he took charge. He arranged to purchase from one of his wonderful doctors, Dr. Ellen, a self-monitoring device for taking his own eye pressure, to help him manage his recovery from his latest operation. Please see Gary's eye chapter, Chapter 11. Using his home device, which was rare at the time, he was able to develop line graphs of his pressures under various conditions. Using these, he was able to work *with* his doctors as the doctors worked with one another for a good result (see the coauthors of his chapter). Gary along with his doctors saved his remaining eyesight. If they could have started earlier, he would have saved even more of his eyesight.

Gary first kept records of his intraocular pressure (IOP) under various conditions. In this way, he was able to analyze what seemed to work or not work when following the orders of his doctors. Fig. 7.1 shows one of his information collections:

The graph in Fig. 7.2 demonstrates an example of a resultant line figure of the outcome (predicted) variable, eye pressure:

Possible research: Patients who self-monitor could be contrasted to patients who do not self-monitor to predict course of disease. Additional prediction (independent) variables besides self-monitoring variable, could be interest in self-monitoring (if they are consistent with the task, regularity of adding values to their portal), and various self-knowledge values, which could indicate independence and motivation.

Measurement time ▾	Device type	Device name	IOP (OD)	Quality OD	IOP (OS)	Quality OS	Controls
30/07/2021 ⏰ 09:16	Icare HOME	OMEG iCare 01	-		11	Excellent	☐ Exclude
1-2 MINUTES after L-EYE MASSAGE							
30/07/2021 ⏰ 09:14	Icare HOME	OMEG iCare 01	14	Excellent	-		☐ Exclude
30/07/2021 ⏰ 09:13	Icare HOME	OMEG iCare 01	-		20	Excellent	☐ Exclude
~80 minutes AFTER l-EYE MASSAGE							
30/07/2021 ⏰ 09:13	Icare HOME	OMEG iCare 01	15	Excellent	-		☐ Exclude
30/07/2021 ⏰ 08:25	Icare HOME	OMEG iCare 01	-		18	Excellent	☐ Exclude
35 minutes AFTER L-eye Massage							
30/07/2021 ⏰ 08:25	Icare HOME	OMEG iCare 01	14	Excellent	-		☐ Exclude
30/07/2021 ⏰ 07:48	Icare HOME	OMEG iCare 01	-		16	Excellent	☐ Exclude
1 minute AFTER l-EYE MASSAGE							
30/07/2021 ⏰ 07:48	Icare HOME	OMEG iCare 01	18	Excellent	-		☐ Exclude
30/07/2021 ⏰ 07:47	Icare HOME	OMEG iCare 01	-		21	Excellent	☐ Exclude
30 minutes AFTER AM Strong Cup of Coffee							
30/07/2021 ⏰ 07:47	Icare HOME	OMEG iCare 01	18	Excellent	-		☐ Exclude

FIGURE 7.1 Partial data collection form for IOP in July, 2021.

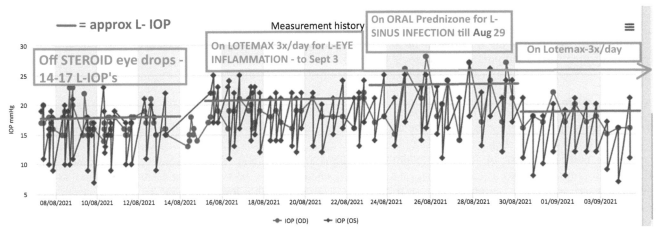

FIGURE 7.2 One of Gary's line graphs of his eye pressure—the data in this graph occurred in August and beginning of September 2021.

Traditionally, patients waited; they sat in waiting rooms, waited for the medical people to dispense medicine to them. Patients were long-suffering, enduring pain, waiting for someone else to serve them, to save them. Patients assumed that the doctors knew best. Most patients do what doctors tell them to do, otherwise they might fear that they could get worse or even die. One thing that patients often did not know was that their doctors didn't always know. This was so evident during the 2020 pandemic in which, at first, patients were quickly put onto ventilators, which, as it turned out, given oxidative stress, may have contributed to the inability to get air into the alveoli, and many died as a result. When patients are suddenly hurt, acutely ill, or are unconscious and in the hospital, of course they may be totally dependent on medical personnel. However, if they are able and perhaps before things get bad, patients can keep their own data, detect patterns so that they and their doctors are able to observe more, and to learn more about their conditions. There is a resulting self-reliance that helps them find their own best paths.

Patients are becoming empowered, and the trend is increasing. However, patient empowerment *does not* mean that patients have no need for doctors. Doctors have studied medicine for many years before they see patients. Then, in their clinics, they see many patients and learn what to expect and not to expect. This real-life experience is invaluable. There is no way that patients should ignore the established medical community. However, there are so many ways in which patients can and should take more responsibility for themselves—there are many more online resources at one's fingertips and medical grade at-home medical devices, such as EKG machines to monitor fibrillation or Gary's ICare machine, allowing glaucoma patients to monitor their own eye pressures [again, see Dr. Gary Miner's Chapter 11 on glaucoma and the self-development of a predictive algorithm]. Also see the example of an individual's self-development of a tailored weight management algorithm, found in the tutorials (F1 on page 388) of the first edition of this book (found on this book's companion webpage).

Research questions

This chapter discusses the following questions and more, all helping to define the limits and promises of patient-directed medicine. All of which could be good topics for research, hence the red outlined box.

- What are some of the productive areas for patient-directed medicine?
- What do patients look like when they are sharing in the responsibilities and when they are directing their own care?
- How are patients changing the ways in which medicine is practiced?
- What are some methods patients could use as they better direct their own healthcare?
- What are the responsibilities of patients for maximizing their own health? What are the constraints and limits of patient empowerment?
- How has the concept of "patient data" changed for researchers?
- And something that has changed everything: In what ways has the COVID-19 pandemic changed the role of patients—who does and who should make decisions, particularly when the answers are unknown? (A sign in one of Gary's doctors' entries stated something like this: "We don't care if the CDC says it's okay now. If you come in, you wear a mask!" No choice there. Sometimes decisions are made for the patient, of course.)

Rather than simply looking at what medical professionals can do for patients, this chapter looks at what patients might do for themselves and possible areas for PAs.

The responsible patient

Formerly passive patients are sensing that the best healthcare for them is to step out of that waiting role. Many patients are taking more control of their medical destinies. Today, many patients seek information before they go to the doctor. They watch YouTube videos of doctors explaining symptoms, possible diagnoses, and treatments. They watch as other patients with similar symptoms tell their stories, what happened to them, best treatments, drug trials, and so on. They look up research articles as per above and go to their medical clinic with a lot of information. One might think that at least the very old sit by passively, but they have more time to investigate if they are retired. The roles of doctor (and all medical people) and "patients" have changed. Again, one thing that patients may be discovering is how little their doctors know, and they wonder what research the doctors are using for their decisions. Some physicians welcome a knowledgeable patient, others don't.

First, what do patients look like when they are sharing the responsibility for their care and when they are taking responsibility for themselves? It might be useful to develop a patient responsibility index paper and pencil instrument that could be used in future prediction studies. Second, possible PA research concerning the responsible patient: after defining and identifying responsible patients (such as the developed index), one could look at basic health indices such as BMI, diabetes, arthritis, asthma, and so on to see if responsibility could be predicted from those health issues. Conversely, health outcomes could be predicted by factors of responsibility (found in the responsibility index from above).

Patients changing how medicine is practiced

Sometimes patients think they know more than they do know and can make themselves nearly impossible to work with. They can also be totally wrong. Patients will need to use common sense people skills if they are to assume more responsibility for themselves, without alienating the medical community. They need to learn to work with their doctors and to sense when to say something and when not—in other words, how to work in a team. By the same token, the medical community would be wise to realize that patients are changing and accept patients as partners in medical care.

Some do, but some do not. If a physician assumes the old position of a god, he or she may not have patients in the future, as patients increasingly are seeking doctors who collaborate with them.

Politics now seem to play a role in medicine. We must go back to honest scientific investigation. Patients as well as physicians really do need to understand research—what makes it good and what makes it bad—so that political persuasions have nothing to do with working to determine truth. Potential profits also should have nothing to do with understanding the science, whether that understanding, and those answers bring in a lot of money or only a little. Preference should not be given to the most expensive medicines and treatments if older, less expensive ones will do just as well, given research, always research. Patients would do well to try to determine if proponents of certain procedures or medicines have anything to gain from their use. It is important to search for peer reviewed, excellent research and not simply going to someone's opinion on a blog. If going to a blog or video one must make sure the blogger's comments are based upon sound research.

Training courses could be placed on portals for patients to access—the courses could be on communication skills with physicians, and/or basic scientific principles and how to know when a research article is "good." Then studies could be conducted on health outcomes based on patients' knowledge, and communication skills. Other research could be conducted predicting medical community opinions of patients, with patient communication and research skills as independent variables.

Patient empowerment versus compliance

Yes, patients are changing the way medicine is practiced but is the way, right? How much should they be involved and what decisions should be their own? Abject compliance or total control of themselves—how much of each? The following is a good study to consider when making those decisions.

A study in China came up when searching for status of patient empowerment (Lu and Zhang, 2019; https://www.ncbi.nlm.nih.gov/pmc/articles/PMC6535977/). The study was found because the authors mentioned decision-making preferences and self-determination theory. Their article, however, was on patient compliance rather than empowerment, in other words, totally trusting everything said or done to the patients, and assuming the doctors knew everything. Just to understand cultural differences, it was noted that physician *expertise could not be controlled* but that the outcomes for patients depended on patient compliance to what the doctor said. Under the section on "Patient Compliance," they said, "Given that we cannot control the degree of physicians' professionalism, we pay increasing attention to patient compliance to improve the curative effects."—to which I said, Whaaat? In other words, the doctor could be incompetent but doing as the doctor says, even if wrong, would lead to better outcomes for the patient.

The need for empowerment likely depends on the patient's cultural norms—at times presenting limitations, regardless of those norms. However, certainly, for the past 10 years or so, the patient empowerment idea has grown in western countries. Patients feeling the need to depend more on self puts blind compliance in question. If a doctor is threatened by a patient's knowledge of the latest research, it might be time to find a doctor who appreciates an informed patient so they can have rational and informed discussion.

Collaboration between patients and the medical community

Patient involvement

Involvement and empowerment are closely related. If a patient feels no personal power, that person is less likely to become involved in decision-making, regardless of the attempts to the contrary on the part of the medical community. With positive personal power, the patient will be more apt to participate, even if the system does not encourage participation. The best situation is developed when both patient and medical community believe that patients should actively participate in their own healthcare as much as possible, including participation in positive lifestyle habits to increase health-taking responsibility for self. In other words, there should be a collaboration between patients and the medical community. Some aspects of this topic are listed below.

- *Patient involvement.* Cvengros et al. (2009) noted that there were inconsistencies in various studies concerning the importance of patient involvement in achieving good outcomes. In some studies patient involvement seemed to add to good outcomes, and in others it was not so. Cvengros et al. hypothesized that congruence between physicians and patients was most important for increasing good outcomes. The authors thought that the mismatches were what was driving the inconsistencies in the literature. They called congruence, "symmetry."

Admittedly, Cvengros et al. had many limitations that affected the generalizability of their study, but it certainly provides some variables to test in a PA study and can speak to the ways in which medicine may be changing because of patient empowerment. In such studies, matching characteristics between physician and patient might be examined concerning:
1. desire for information sharing;
2. decision-making;
3. behavioral involvement;
4. emotional support needed and given; and
5. patient satisfaction measures needed and provided.

Patient involvement in medical education

One research group in Canada (Moreau et al., 2019) wondered if patients could be beneficially involved in the assessment of medical residents. They used two phases—phase one was surveying residency program directors and phase two was contacting the respondents and interviewing them. The purpose was to find out how many programs used patients (or intended to use them) to help assess the skills of residents. Sixty-three out of the 462 surveys sent to program directors responded (less than 14% response rate). Most of those (64% of the returnees) did not know if patients would be involved, 20.8% intended to involve patients, and 16.8% did not involve patients nor were they thinking of it. So then, when they went out to interview, they found that of the program directors in phase one, 72.7% did not use patients to help assess residents, 15.2% (five) did use patients to help assess residents, and 12.1% did not know, even at that point, if they involved patients. These were program directors who likely had a large part in helping to design the residency programs, so it would be reasonable to think they might know if they involved patients in assessment. The kind of programs Moreau et al. were mainly concerned with were "Competence By Design" (CBD) programs and the program directors would be in charge of them. CBD programs required multiple kinds of assessments of residents, and patients were one aspect. Obviously, from the lack of patient involvement, patients were deemed not important. Patients could have helped assess communication skills, perceptions of skills and abilities of the residents, and some intangibles that most doctors, it seems, should want such as positive relations with patients. This study is listed in the references below, but the reader might like going directly to it now, on the Internet (https://www.ncbi.nlm.nih.gov/pmc/articles/PMC6445318/).

Not being involved in such a study likely would not impact those patients who didn't know they should have been asked; however, the long-term effects could mean that future physicians could be oblivious to how they came across to patients. Failing to involve patients represented missed opportunities for increasing effectiveness.

Patients could be helpful in developing and maintaining resident training. It should be done in such a way as to be nonthreatening to overworked residents. However, knowing what patients were gaining from visits could be quite enlightening. Perhaps attendings should volunteer first. It could be quite useful to know if patients were able to say what was bothering them without being interrupted in the first 10 seconds, for example.

Limitations of patient involvement

- *Hindrances to patient involvement.* Another literature review stated that empowerment was identified as a key factor in improving healthcare (Davis et al., 2007). In the article, the authors stated that the United Kingdom National Patient Safety Agency identified that international indicators of patient involvement were key factors in improving health services and particularly improving safety. Safety is an important goal, as stated in Chapter 1, p. 13, concerning goals of The Joint Commission for reducing harm. Davis et al. (2007) identified factors that contributed to higher rates of involvement in patient healthcare (factors predicting high involvement):
1. Demographics: patients being younger, female, and highly educated.
2. Emotions and coping: having active coping skills (high resilience).

3. Severity of illness: varied by individual patient (and a variable that might be more conducive to personalized PA, rather than population analysis)—some people wanted to get more involved the sicker they got while others wanted someone else to take over.

4. Illness symptoms: again, another area of individualized response on variables such as individual response to symptoms, type of treatment, and individually perceived impact of participation on outcomes.

5. Prior experience and beliefs about those incidents: how the individual viewed previous experiences with the medical community would impact the individual's willingness or demands to participate,

6. Knowledge and beliefs of the attending physician: how the physician believed patient involvement would help or hinder. For example, in the Davis et al. (2007) study, 58% of the physicians thought that patients were somewhat often or very often partially responsible for medical errors that were committed. Generally, physicians viewed patient involvement positively. However, if matching patients to physicians concerning involvement is important then PA studies might examine congruence and not simply physician preference.

7. Physician—patient interactions: when interactions were positive, patient involvement was higher. When patient/physician interactions were negative, patient involvement was not encouraged and was lower.

Possible PA Research: Intuitively, it seems that positive interactions, positive self-esteem, knowledge, and patient involvement in decision-making (as much as is feasible) would all score positively for assuring good medical outcomes. However, more research should be done in this area to understand the impact on medical outcomes of patient involvement, even in the long haul. Medical outcomes could include mental as well as physical well-being.

Evidence supporting patient involvement

Some evidence that supports the hypothesis of better outcomes with more patient involvement includes:

- *Patient involvement in therapy.* Arnetz et al. (2004) noted that in physical therapy, better outcomes accompanied goal setting that involved both therapist and patient.
- *Patient involvement in medication compliance.* Later, Arnetz et al. (2010) looked at cardiac patients from 11 Swedish hospitals. The patients were given questionnaires on their cardiovascular symptoms, their medication compliance, their participation in cardiac rehabilitation, and their achievement of secondary preventive goals (p. 298). The authors conducted follow-ups on the patients 6–10 weeks after their hospitalizations for myocardial infarctions. There was not a consistency between involvement and all final outcomes. Some associations were found between more involvement and fewer cardiovascular symptoms upon follow-up. Other outcomes were nonsignificant. (Of course, use of multiple significance tests should be regarded as suspect, or possibilities, but not out and out trusted.)

From the first edition of the text, there has been continued support for involvement of patients in their care. Involvement includes decision-making in medical care (Susanti et al., 2020; Hyde et al., 2017; Zizzo et al., 2017; Eliacin et al., 2015; Manias et al., 2019; Probst et al., 2018; Dehghan et al., 2018; Ibrahim et al., 2019).

One interesting article concerned public and patient involvement in Indonesia for mental healthcare. According to Susanti et al. (2020), basic human rights were traditionally violated for the mentally ill who were hospitalized, "including arbitrary and prolonged hospital detention, involuntary treatment and tens of thousands of people being illegally chained up ("pasung") in unsanitary conditions, both in the community and in hospital settings" (p. 378). Some of these violations continued to present even after an initiative in 1993 to improve mental healthcare. In this study, the authors used focus groups to begin getting more involvement from both patients and the public, while caring for people with psychoses. The challenges and barriers were discussed, as was the collaboration with the UK in this important work. If patients and the public are not involved, things often continue as they always have been, without anyone even thinking about how things might become better—except for those who must suffer under awful conditions, that is.

Zizzo et al. (2017) conducted a study in Montreal with Parkinson patients who were in the early stages. As part of their data collection, they used the Autonomy Preference Index (Morandi et al., 2017) to determine desire for participation and desire for communication. They said that patient involvement was a complex situation, while supporting shared decision-making but felt that especially communicating and staying in communication were very important. The complexity of the disease meant that the situation was different for each person, making communication especially crucial. Not surprisingly, patient characteristics did vary in the study, but in examining the data most people wanted to make their own major medical decisions, tended not to go along with their doctor's advice if they disagreed, thought they should be making their own decisions regarding hospitalization care, and should make everyday medical decisions. On the other hand, as illness progressed, patients thought the doctors should make more of the decisions and that doctors should determine how many and when checkups should occur. So, the takeaway in the Zizzo et al. study is that patients, even with such a debilitating disease, very likely wanted to be empowered and involved until the disease progressed greatly. Again, communication was very important.

The Eliacin et al. (2015) VA study involved 54 interviews with veterans who were mental health outpatients. The qualitative study used thematic analysis (the eyeball technique rather than a more quantitative analysis such as text mining) on the interviews. They reported in their abstract that the desire to be involved depended on perceptions of the patient–physician relationship, including themes such as "fear of being judged, perceived inadequacy, and a history of substance abuse" (par. 4). One imagines that a trusting relationship between doctor and patient might vary both with patient experiences (PTSD, types of injuries, circumstances, need for disability, and so on) and the military physician personality, rank, and tenure. Trust was a theme visited many times in the articles mentioned below.

DeRosa et al. (2019) brought an important component into the empowered, involved patient—that of the role of librarians and other information professionals. Patients do not have to go only to the Internet for information. They may find that fully functioning librarian professionals are just as close as their local library, including state and sometimes private college/university libraries close to them. The DeRosa study focused on "patient- and family-centered care" (PFCC) programs and initiatives provided by the Institute for Patient- and Family-Centered Care (2020, https://www.ipfcc.org/). However, most librarians would be willing to help someone searching for information on diseases, clinical trials, and other resources. In the discussion, DeRosa et al. stated:

> As health care decision-making requires informed patients, it is not surprising that the included PFCC cases utilized interventions to empower their patients. Librarians were noted to contribute to shared decision-making through consulting directly with patients, improving the health literacy of patients by teaching them how to evaluate health information, and delivering information to both providers and patients with the goal of fostering collaborative communication (p. 319).

Manias et al. (2019) conducted another qualitative study (thematic analysis) reviewing the literature examining how families help older family members deal with medications across transitions of care (such as staff shift changes in hospitals, nursing facilities, and even home health workers, or leaving the hospital for home, etc.). In examining the 23 articles that met the criteria for inclusion, the authors mentioned difficulties of families in being heard, but rather, family members could be instrumental in helping with the proper administration of medications, with watching out for adverse events, beneficial outcomes, and in helping to answer questions of the elderly patient. Transition of care related to hospitalizations, moving to care facilities, going home, changes in medication, and other such transitions. Family members often found that the medical staff was simply too busy to explain what they needed to know—such as what adverse effects they should watch out for. One main theme, involving 14 of the articles was "Participating in Decision-Making" and subthemes were, "medication decisions occurring on admission to hospital, those occurring during transfers and at discharge, those that happened after discharge home and to aged care facilities, and characteristics of health professionals and family members in fostering participation in decision making" (par 1 in the section). For the very important role that families of the elderly ill were playing, they seemed to be met with quite a lot of frustration. This frustration was due to the medical community being swamped, due to delays in transfers, not being listened to, and due to not getting the information they truly needed for their vital task. A researcher of PAs might find a gold mine of important studies contained in the ideas of this qualitative study. This study represents, in my opinion, the best of what a qualitative study is good at—providing areas that need quantitative research that can result in meaningful changes in healthcare.

A study by Probst et al. (2018) was secondary to an earlier multicenter study (Anderson et al., 2014) which randomly assigned patients to either a study program on decision-making or to standard care. In the Probst et al. study, video and audio recordings were analyzed using the OPTION-12 scale which measured the degree to which clinicians engaged patients in decision-making (Elwyn et al., 2005), and the score of each from the first study's Control Preferences Scale (CPS). The CPS was a Likert-scaled questionnaire eliciting strongly agree to strongly disagree answers to questions about how much autonomy and decision-sharing the patient desires. The lower the score, the more autonomous was the patient; and the higher the score, the more the clinician controlled. They also measured health literacy using two instruments.

They had information on each patient such as gender, age. The authors worked hard at controlling, explaining exclusion criteria, reliability, and validity. The study could be repeated with different analyses, in fact, a predictive study. They could have conducted a PA study that analyzed individuals and would come up with a predictive algorithm that would be able to predict individual autonomy and desire for shared decision-making.

With all these data on each patient (perfect for data mining), they analyzed the data using multiple regression and p values between various groups and displayed five big tables of differences and p values.

Family-wise statistical errors

Looking at the Probst et al. (2018) study as an example, it was very traditional, but also quite inconclusive, given the increased possibility of type-one errors or family-wise errors.[1]

Here is a brief explanation:

Basically, a type one error is the probability of rejecting a null hypothesis when we should not have. If we are using a 0.05 level of significance and a probability is less than 0.05, we might think the difference between two groups, for example, is significant. However, if we are conducting multiple tests on the same data, we can run into one or more type one errors. If we have a huge table of comparisons and just pick out the ones that are significant at less than a probability of 0.05, we likely are making that type of mistake. The probability equation for a type 1 error is the following: where α_{FWE} is the probability of rejecting a null hypothesis when we should not, α_{EC} is the probability level we think we are using to make the decision, and the exponent c is the number of comparisons we are making. $\alpha_{FWE} \leq 1 - (1 - \alpha_{EC})^C$. Using this test, if we are using the cutoff of 0.05 and run, say four comparisons, then the result would be, $1 - (1 - 0.05)^4 = 1 - (.95)^4 = 1 - 0.8145 = 0.1855$.

So instead of thinking that we are 95% certain the differences are real, we are only 81% certain. If one has an entire table of comparisons, you can see that one cannot be at all certain that any one difference is significant, much less pick out those that are less than 0.05 and call them significant.

Communication and trust

Trust between patient and physician is crucial for patients with life-threatening diseases. And trust is increased in a climate of both two-way communication and when the healthcare providers provide information and time to discuss (Dehghan et al., 2018; Ibrahim et al., 2019).

Trust, information, and spending time led to a *treatment partnership*, increasing life satisfaction. As doctors actively engaged patients in conversations while providing information customized for the patient, more patient involvement was elicited (Ibrahim et al., 2019).

Vaughan and Tinker (2009) also stressed the importance of trust and communication by health professionals to vulnerable populations in a pandemic.

Communication and trust during the pandemic

The people of the United States certainly experienced many opportunities for communication during the COVID-19 pandemic concerning the elderly with comorbid conditions and the importance of staying sheltered. Oddly, by sheltering, it seemed that when people did emerge before vaccinations were developed, they were quite at risk and were among the first to become victims of opening the country. Vaughan and Tinker (2009) noted that if information was inconsistent, it was more difficult to maintain trust, especially among the vulnerable and at-risk populations. Certainly, healthcare organizations such as the WHO and the CDC should have consistent communications—did masks really protect? Did it matter what kind of masks? Many people asked why all people were quarantined instead of just the most vulnerable or just the sick. Instead, ill patients were sent from hospitals and into nursing homes, where many sickened and died. Children who were deemed not at risk were kept out of school, causing parents who could not work from home to lose their jobs. It was during this time of confusion that the predictive algorithms were not very useful as they overpredicted deaths, frightening the population further. CDC officials said one thing one week and then the opposite the next week. Physicians were not listened to when they were using drugs such as Plaquenil which appeared to be saving people, used at first symptoms. Studies were contradictory—at times, data was faked. Some thought "Big Pharma" seemed to have a vaccination agenda; others called those views "conspiracy theory." The world was in a muddled state of mutual distrust. At this writing, we continued in our confusion. Needing data and answers, we often found confusion and conflicting research studies. One such example of muddlement, when asked in a FOIA for documents related to its

---

1. Remember the problem with multiple p value tests from Chapter 2, first edition, p. 25.

vaccine development, Pfizer asked to have 55 years to fully share those documents. Many of those needing answers would be dead by then.

Concepts of communication and risk are important during pandemics and might lead one to some excellent topics for PA projects. Pandemic topics may show themselves in the future, in *predicting which individuals are most vulnerable* to disease, and for determining the *best treatment options for individuals*.

A study of the conversations between physicians and patients in the exam room highlighted the involvement or lack thereof that tends to occur. Patients expect their doctors to diagnose, but do not want them to make the wrong diagnosis (Gill et al., 2010). Gill et al. spoke of preemptive resistance in which the patient had already decided on possible diagnoses, reporting symptoms that agreed with those diagnoses, and then providing counterexamples when the physicians suggested something else. What could go wrong here? Meanwhile, the physicians were trying to collect data, make hypotheses, and come up with good diagnoses. The study showed the dynamic interactions that occur in a situation like this. The authors wondered if the patients might exert undue influence on the physicians, causing imprecision in diagnosis. On the other hand, they maintained that such interactions may broaden the possibilities that the doctor might consider. The role of patient seemed to be evolving.

Possible PA Research: Videotaped patient–doctor interviews could be the basis of an interesting predictive study of these interactions. The use of text mining using learning techniques and any pertinent hard data would enhance the analysis. The correct diagnosis could be the dependent variable, degree of participation could serve as an independent variable along with the text analysis.

Collaboration and limitations

Older articles related to the collaborative nature of the patient/physician relationship ranged from wondering if too much patient input could derail doctors (Gill et al., 2010) to thinking that doctors should listen more to patients (Cocksedge and May, 2005). Others spoke of legal aspects of patient communication and patient involvement such as for parents of children or for end-of-life situations (Dworkin, 2003).

- *Importance of patient autonomy* Who should be the one to decide when the patient is incompetent or underage? Dworkin (2003) provided much legal expertise on patient autonomy, explaining the rights of parents to make decisions for their children. Also included in this study was much information on patient autonomy in general. Dworkin told of an old, evidently well-known lawsuit against a surgeon who did not awaken the patient from anesthesia before completing the same operation on her right ear as had been done on the left ear. The patient sued.
- *End-of-life situations* Likewise, end-of-life choices have special legal considerations (Dworkin, 2003).

How patient-directed medicine works using predictive analytics

Privacy concerns can hinder research

Informed consent and Health Insurance Portability and Accountability Act (HIPAA) laws increase patient autonomy, control over self-information, and participation. However, those very privacy provisions can frustrate well-intentioned predictive research.

In 2009, the Institute of Medicine studied the effects of HIPAA on research and stated that HIPAA regulations made it more difficult to recruit patients in studies which then likely caused selection biases, required more paperwork for IRBs, cost more as a result, and caused more errors because data had to be totally deidentified. The increase in rules and regulations including differences in IRBs made it much more discouraging to attempt studies and multicenter studies. Selection biases can reduce generalizability of studies, especially non-PA studies. Because predictive studies use learning algorithms that do not depend on normal curves, some of these problems are not as important in PA studies, which do not necessarily depend on representative samples.

PA studies do, however, greatly depend on the accuracy of the variables, such as correct diagnoses, and accuracy of any independent variables, say blood tests, lifestyle, or perhaps genomic markers. The outcome variable also needs to be accurate—if death versus survival, one must know that outcome.

Predictive analytics for patient-directed research

The researcher is predicting to an individual not to a group, but because of the Health Insurance Portability and Accountability Act (HIPAA), the researcher often must depend upon volunteers, which could not be assumed to be a random sample of some disease, say a type of cancer. To develop a predictive model, as described in Chapter 2, within the data of that group of volunteers, however, random subgroups are formed—a training group, a testing group, and a holdout group for validating the prediction model. Those groups must be randomly selected. So, we are supposing that the only data the researcher has for a certain cancer are those of volunteers. It would be better if the researcher had data, or a large random sample, from *all* victims of that cancer, but the reality is that one is fortunate to have data from volunteers, or from those of a certain hospital(s) provided through an IRB process—still not a random group from all such patients, but useful, nonetheless. First, the prediction model is developed using the training data. The model is then tested on the testing group, and finally validated on a third group, the holdout data, all from those original volunteers.

The model would likely be published or patented and sold. Next, the model would be used for people at large who had the disease, perhaps in hospitals to predict death or survival, depending on that person's measured variables used by the model. It is possible that the model would not be as accurate for that person as it was for the original volunteers. There could be some intervening variable that would change the outcome to invalidate the prediction (say, an internal fear of losing control) which might be an internal characteristic that could have some systemic effect influencing the trajectory of the disease. However, for the model to be used by a "nonvolunteer" whatever variables predicted the outcome, the PA model would have to be measured on the nonvolunteer, potentially providing more data to the researchers (if the IRB would allow it) and an opportunity to fine-tune the model to predict to a greater number of individuals. **Prediction accuracy (percent of correct predictions)** and not traditional between groups, p values, is the desired outcome of the processes of PAs.

Patents are mentioned because there is often a desire for finding algorithms that can be sold. It is likely that such a motivation can be a deterrent to patient health—it almost seems that people patent every little thing in the hopes that some patent will result in great wealth. Researchers may also have hidden ties to manufacturers, which would likely invalidate objectivity in research. The desire to "cash in" can cause older medications, which might be effective for new diseases, to be overlooked or discouraged, because they are inexpensive, unless they can be reformulated to have a new patent for added expense. Even worse, is the thought that researchers could be motivated to create new infective (pathogenic) organisms, patent those organisms, and then create antidotes, say, in the form of vaccines, to sell. What might people be willing to do for billions of dollars? For science to work, these attitudes and practices are detrimental.

Regardless of patents and possible greed, which can hinder good medical outcomes and research, if a good prediction model is created and published as an inexpensive app, the use of such gives patients more precise information concerning the outcome of their malady—diagnosis, treatment, or other outcome prediction. Such information gives the patient more power to make good decisions, hence, more autonomy.

Moreover, informed patients can do the research themselves if they have a reliable measurement tool to track their progress. When devices for self-monitoring are developed, such as Gary's ICare, eye pressure machine, glucose monitors, or a home EKG machine, patients are better able to stay on top of chronic illnesses. Patients could write up their own case study, or even include other patients from their support groups—not research likely to occur, but certainly not out of the realm of possibility.

Patient involvement is generally crucial during the end-of-life. Patients and the families of patients generally want to be informed and involved, at least in the United States. Other countries may differ. It is a good plan for patients to have directives in case of incapacitation. A good site for learning more about advance planning is the one by authors Sedini et al. (2021). This website reviews what is known and as in the title, attempts to be up to date. Explanations are given for various types of advance care along with websites for finding documents such as living wills and durable power of attorney. Such resources can help to empower patients and family members for end-of-life decisions.

Parents must be included in the decisions concerning their children, in terms of withdrawing or withholding treatment, or in terms of whether the child will receive treatment or vaccines. There must be clear and timely communication and exchanges of information (Giannini et al., 2008). Schools can dictate to parents, for example, but parents can withdraw their children from schools. Honest exchanges of information and concerns must be a part of the decisions.

Cultures and decisions

Mo et al. (2012) found no consistent pattern for support among patients in Korea and recommended that information and decision-making should be individualized according to the needs, preferences, and culture of the patient. PA would be good for such studies. Some cultures don't tell a loved one if the person has a terminal disease—the doctor tells the family. In such a culture, patients may not want to be making decisions. See below the Singapore study (Chan and Goh, 2000).

Legal euthanasia occurs in some other countries, such as in Belgium (Van Wesemael et al., 2009). Decisions in those countries need specialized physicians who are called in to consult. Certainly, culture and religious beliefs would influence such decisions.

In the United States it is frequently assumed that patients need to know their diagnoses to take more control of their care. This view is not necessarily held worldwide. In some countries, doctors believe that patients should not be involved in decisions. In Singapore, many doctors believed that their patients were not capable of rational thought, because they were ill, and should not always be included in the diagnosis (Chan and Goh, 2000). Instead, family members were included, especially if the patient did not want to, or was too incapacitated to take the medical advice. A similar finding concerning the sharing of prognosis was uncovered in Kazakhstan by Shinkarenko et al. (2007). Physicians and patients preferred that the patient not be informed of a terminal illness diagnosis. Rather, the families of the patient were to be told the diagnosis. Not telling continued as of 2018, with doctors telling the family 98% of the time before telling the patient, and of those, 82% of the families deciding not to tell the patient (Wang et al., 2018).

Possible PA Research: Physician/caregiving variables (in cultures in which patients often make decisions), might predict patient desire for involvement in their medical care, particularly those variables that match patient characteristics with physicians ("symmetry hypothesis"). Need for symmetry might also require more communication skills of the physician. Researchers could monitor physician listening behaviors for use in predictive studies.

Patient outcomes might be predicted from patient characteristics such as resiliency locus of control, and personal efficacy. Another good study might examine this matching process in end-of-life decisions. Personalized medicine using predictive analysis would be especially useful in end-of-life decisions.

Coordination of care and communication for patient-directed healthcare

Patient care depends upon good communication and coordination. Increased use of specialists has also increased the opportunities for miscommunications or lack of communication among the specialists, the primary care physicians (PCP), and patients. Goldhill (2013) illustrated this situation very well with the example of his friend, "Bill," a wealthy cofounder of an investment firm, who began having breathing problems during a ski trip. (Likely today, Bill would be tested for COVID.) This event prompted a journey to specialists who disagreed with one another and did not coordinate with each other. He saw four pulmonologists, a hematologist, and a cardiologist. Diagnoses from this suite of specialists included lymphoma, of which he was told that he would die very soon, a congenital heart defect, and lung diseases of various kinds. Bill was shocked at the lack of coordination between doctors about their diagnoses, and how they communicated with him. The specialists seemed unable to think beyond their specialties, and they could not generate alternate diagnoses. Bill's wealth could not generate "more coordinated service." "It was shocking how much work he had to do himself, from hounding doctors for appointments, making sure their test results were communicated to other specialists, to wading through mountains of online data to determine the best treatment options" (Goldhill, 2013).

Goldhill's passage brought back memories of helping to coordinate one of the first Alzheimer's drug trials in the late 1980s. Patient records were replete with instances of the older people taking medications that interacted negatively with one another. (This was before pharmacies or hospitals kept track of patient drug interactions.) Physicians who called the Alzheimer clinic often said, "I never knew so and so was taking that!" Each patient had at least three or four physicians and often each one prescribed a different medication. Often a medication from one doctor caused symptoms that another doctor then unknowingly treated with a different, confounding medication. The need for coordination has a long history. The patients in the study were anything but self-directed. They and their caregivers were positively swamped with trying to cope with a horrible disease. Those patients needed a medical advocate. How is it with patients and their families today? Do they continue to simply go from doctor to doctor, not questioning the medications or reasons for those medications? Or do they question, read, and decide?

The situation likely has not changed over the years since. Patients surely would feel more confidence in a coordinated system between PCPs and specialists. As practices become busier (increased panel size), coordination becomes ever more difficult, which was born out in the research study by Mohr et al. (2019). People have not become less busy and so coordination among one's doctors is still a concern for patients. Patients need to be proactive in ensuring coordination. Patient advocates could still be one answer. Regardless, patients demanding coordination among medical personnel will cause shifts in how medicine is practiced. Again, please refer to Chapter 11 on glaucoma.

Possible PA Research: Certainly, disease type, severity of disease, and incapacitation level must be considered in any predictive model involving patient-directed medical care. The point at which a patient needs an advocate would be a good subject for research. Using positive outcome as the predicted/dependent variable, predictors/independent variables could be the amount of patient involvement and the kinds of patient involvement, plus the attitudes of the doctors involved, concerning patient involvement.

Communication skills in the medical setting

Communication studies

Patients go to their doctors for medical advice. In the delivery of this advice, patients must listen to their doctors, and the doctors must listen to their patients. Effective communication is a two-way street. Studies point to various topics of importance in doctor—patient communication.

- Physician Tune-out. Cocksedge and May (2005) analyzed doctor—patient consultations by examining nonverbal cues and behaviors, along with the verbal exchanges. The authors found that some physicians might simply tune out their patients' ideas by using the strategies listed in Table 7.1.

In an Advisory Board daily briefing (2018), it was said that most doctors think they are good listeners. However, according to the source, in fact, they are not. They interrupted and allowed only 11 seconds for the patient to speak before doing so.

- *What patients want.* Two authors investigated what patients want and who they consider to be a good doctor; in a qualitative study patients evaluated a physician as "good" if he or she listened to them (Kliems and Witt, 2011). In addition, patients can tell when their doctors assume what the problems are, and jump to conclusions about a diagnosis without considering patient concerns (Kliems and Witt, 2011; Cocksedge and May, 2005).

TABLE 7.1 Methods of tuning out the patient (Cocksedge and May, 2005, p. 1003).

Tuning out strategies
Deferring listening to another time
Reassuring
Changing the subject
Interrupting
Nodding appropriately but allowing the mind to wander
Using body language (such as standing up, a closed posture)
Reducing sympathy
Being directive
Making a plan

- *Listen to what patients say.* Again, much like the Advisory Board briefing above, according to Kliems and Witt (2011) German doctors spent 11 − 24 seconds in the introductions and 7.6 minutes interviewing the patient. Some HMOs in the United States required the physician to spend an average of 7−8 minutes per patient. According to National Public Radio, the average patient was allowed to speak only 12−15 seconds before the physician interrupted (Varney, 2012).
- Evidently things had gotten a bit worse since the above studies were published. Another source found that the patients were able to speak for only 11 seconds—with specialists worse than primary doctors (Singh Ospina et al., 2019).

Many patients do not consider the problem of misdiagnoses when they visit their physicians. Fig. 7.3 lists the steps patients can take to help prevent diagnostic errors. Patients should take the initiative to minimize diagnostic errors if they value their own health. Patients must remember that physicians are human; they have limitations. They get tired, they must interact with many people, and just like the rest of the world, they would like to get away and go home at the end of the day. At the end of a day, physicians can feel absolutely exhausted by providing care for all the people in their practice that day. Seldom do they have the luxury of spending hours (even days) with a team of junior colleagues debating various diagnoses (as is often portrayed on TV shows such as "House").

In research articles the key ingredient concerning good communication seemed to be the PCP in that the PCP is the hub of communication (Pham et al., 2009; Zaggocare, 2018). In the context of medical homes (explained below), family doctors were to coordinate a team of specialists and medical practitioners to provide the required care for their patients. Playing a more central role in medical practice would place even greater burdens on the schedules of PCPs. Generally, PCPs have a high patient load and work long hours for much less pay than their specialist counterparts. They have the greatest responsibility for good communication but have the least amount of time to do so. The direction of a patient's own care can become more difficult and more imperative by the characteristics of their specific healthcare system. It is likely that patients will need to take more responsibility in facilitating communication and overcoming barriers in the future. Patients must ask questions, follow up interactions with physicians, and conduct research for themselves as suggested in Fig. 7.4. They will have to do so with great tact and skill, to avoid being considered a "difficult" patient (van Nostrum and Elaine, 2008). In the video, the character, Elaine, reads in her own chart that at one time a nurse commented that she was "difficult." Elaine tried to contest the label only to be judged by the physician as more

What Patients Can Do | Steps you can take to prevent or detect diagnostic errors

TELL YOUR STORY WELL Communicate symptoms and timing carefully.	**DON'T ASSUME** no news is good news: Follow up if you don't hear back after a test or appointment.	**ASK YOUR DOCTOR** these questions:	• Will the tests you are proposing change the treatment plan?	• When should I except to see my test results?
KEEP ACCURATE RECORDS of symptoms and when they started.	**ENCOURAGE** your doctors to think broadly.	• Can you review my primary concerns and symptoms? • How confident are you of the diagnosis?	• Are there findings or symptoms that don't fit your diagnosis? • What else could it be?	• What resources can you recommend for me to learn more about the diagnosis?
MAKE SURE YOU KNOW your test results.	**KNOW** that there may be uncertainty and that the initial diagnosis is only a working diagnosis.	• What further tests might be helpful to make you more confident?	• Can you facilitate a second opinion by providing me with my medical records?	Source: BMJ Quality & safety The Wall Street Journal

FIGURE 7.3   Preventing Diagnostic Errors (The Wall Street Journal, 2013, p. R2).

FIGURE 7.4   The difficult patient (van Nostrum and Elaine, 2008). Available to view at https://www.youtube.com/watch?v=pyossoHFDJg.

difficult. To escape the label she engaged other physicians, only to find the pronouncement following her from practice to practice. Her rash remained untreated.

(The reader should spend a few minutes reviewing this important video. Very humorous, yet not so far-fetched that one cannot identify. The reader should also remember that the disorder most often misdiagnosed is infection, and infections can lead to death.)

Newman-Toker et al. (2019) estimated that between 40,000 and 80,000 misdiagnoses lead to death each year (p. 228). Citing a 2003 study, it was estimated that as many as 12 million diagnostic errors occur each year, with a third of these lead to "serious permanent damage" or "immediate or inevitable death" (p. 228).

The Newman-Toker et al. (2019) study examined "well characterized" malpractice claims for a period of 10 years. Of the 55,377 malpractice suits, 11,529 (or 21%) were claims due to diagnostic errors. Just among the diagnostic errors, 7379 of the 11,529 (or 34%) were "high severity" cases (p. 230). Vascular events, infections, and cancers comprised over 74% of the high severity misdiagnosis cases. In terms of the misdiagnoses, about half originated from primary doctors and half among specialties (specialists such as oncologists, cardiologists, infection specialists, as well as surgical services, diagnostic and other specialty service providers, plus a few unspecified).

One wonders if specialists might be called in sooner to help combat the half of diagnostic errors made by general doctors. How soon are specialists called in when those three diagnostic areas are suspected? If a PCP or hospitalist encounters someone with a laceration from a rafting accident, for example, with an infection that doesn't seem to respond to infused antibiotics, how soon is a specialist called in? How often does sepsis occur in such cases before a specialist is called, and how many deaths occur as a result? A good PA study could hold sepsis spread, or death versus survival, as the dependent variable and time until specialist was called as one of the independent variables.

Barriers to productive communication

There are many barriers hindering PCPs from becoming more involved with patient healthcare in communicating with diverse specialists. The following list describes some of these barriers.

- *Physician workload.* The first barrier is the sheer number of other physicians with which the PCP must coordinate. The authors observed that a PCP might deal with hundreds of other physicians (usually specialists) in the treatment of his or her patients. The number of communications and the time to perform them will continue to be a barrier to effective coordination among specialists treating a patient. Patients can help by making sure messages are relayed properly between doctors. This might mean patients should transport reports personally to facilitate proper communication. Heavy workloads can lead to physician burnout, which certainly can hinder communication, particularly if physicians decide to leave the profession. Berg (2019) quoted, "'The tide has not yet turned on the physician burnout crisis,' said AMA President Barbara L. McAneny, MD. 'Despite improvements in the last three years, burnout levels remain much higher among physicians than other US workers, a gap inflamed as the bureaucracy of modern medicine interferes with patient care and inflicts a toll on the well-being of physicians.'" (Berg, 2019, par. 4).
- *Personal characteristics as a barrier.* Personal characteristics of patients can become either barriers to or facilitators of good communication and coordination between medical providers and patients living with cancer (Kahana et al., 2009). This study showed that negative self-evaluation and various disabilities were among those demographics and personal characteristics of patients that significantly inhibited communication. Widowed or divorced women were less likely to get help. If patients had relatively low income or chronic illness, they were even less likely to get coordinated medical help. On the other hand, the study showed that patients had better informal support systems if the family of origin had strong norms of family obligation (p. 179). Such support aided communication and coordination.

Another study agreed that personal characteristics of patients can either hinder or help communication (Kahana et al., 2009). Proper communication was facilitated also when patients had access to various social resources to organize support. Optimism, resourcefulness, faith, prayer, and spirituality likewise facilitated communication. Also, Kahana et al. (2009) gave some advice for treatment of patients in wheelchairs, which insisted that doctors speak to them, as well as to the person pushing the wheelchair. The patient should be able to talk with the doctor directly. The authors mentioned the value of marshaling resources on the Internet and social networks, such as Facebook and on smartphones. Table 7.2 (from Kahana et al., 2009) shows developmental, age-related issues related to care-giving, most of which represent problems in communication and coordination.

TABLE 7.2 Life-course specific care-getting issues.

Challenges to care-getting	Childhood	Adolescence	Young adulthood	Midlife	Old age
Personal barriers	Trust limited to parents as protectors	Embarrassed to disclose symptoms	Fearful of appearing weak and needy	Conflict between generativity needs and care getting	Threatened by dependency when asking for help
	Unskilled in articulating needs	Negotiation skills not fully developed	Help seeking threatens independence	Illness-related loss of self-esteem limits help seeking	Unassertive in communicating needs
Social barriers	Wants to limit parental distress	Fears stigma from peers	Job/family responsibilities inhibit care-getting	Job/family responsibilities inhibit care-getting	Patient may also be a caregiver
Barriers to getting medical care	Separation from parents during treatment	Only assent required for treatment	Lack of support from partner in dealing with healthcare system	Lack of support from family in dealing with healthcare system	Comorbidities interfere with getting responsive cancer care
Facilitators of care getting	Skillful parental advocacy and support	Strong peer support	Emotional support from partner	Emotional support from family	Advocacy by family
	Supportive school environment	e-literacy; self-advocacy	e-literacy; self-advocacy	e-literacy; self-advocacy	Patient–physician–family partnership
Issues of access/availability to formal and informal care-getting	Lack of direct communication with healthcare providers	Conflict with parents may inhibit medical care-getting	Availability of partner or close friends	Availability of family members friends/neighbors	Availability of spouse/adult child friends/neighbors
	Older sibling availability	Peer "shield" availability	Parent/work role strain inhibits access	Parent/work role strain inhibits access	Access to physician

Source: From Kahana, E., Kahana, B., Wykle, M., Kulle, D., 2009. Marshalling social support: a care-getting model for persons living with cancer. J. Family Soc. Work. 12 (2), 168–193. https://doi.org/10.1080/10522150902874834.

Tavakoly Sany et al. (2020) trained physicians on communication skills and found that good patient outcomes increased because of the training. Their controlled study of hypertensive patients was conducted in Iran.

- *Poor or no insurance leads to less care.* One study found that the children receiving the highest amount of good coordination and communication were those with insurance, and the well insured children were less ill than others in the study (Tippy et al., 2005). There were no significant differences related to any other factor. This outcome was a strong indictment for the 2005 medical community and suggested that money may have been an important factor in receiving adequate coordinated care and excellent communication patterns. Under various capitation plans, physicians were not rewarded for taking the more difficult cases, because difficult diagnoses and treatment took more time. Comorbidities (health problems related to other health problems) required more time also, causing greater expense for their treatment. The type of insurance and payment types may have impacted communication by subtly influencing the physician, as mentioned above. For example, the physician may have had time limits per patient interview, and possibly unreimbursed comorbidities might not have been discussed as a result.

As of 2019 (Collins et al., 2019), more people had healthcare insurance than they did before the Accountable Care Act, but more of them were underinsured. People experienced high out-of-pocket expenses and high deductibles. Insurance companies pulled out of the market, making for less competitive pricing.

Donald Trump campaigned on the promise of repealing and replacing Obamacare. The repeal failed but there were some changes that eventually emerged through executive orders. Some of the provisions were allowing the purchase of insurance across state lines, allowing more companies to form their own Association Health Plans, allowing more short-term plans, and expanding employers' abilities to offer Health Reimbursement Arrangements (Wikipedia, last (ed.) June 11, 2020). The individual mandate penalty (but not the mandate) was eliminated in 2017 when the Tax Cuts and Jobs Act was signed into law (Healthsherpa, 2018). President Biden, also through executive action, attempted to restore the Obama plan and reverse Trump whenever possible.

Possible PA Research: All the barriers listed above are good candidates for future PAs research in patient-directed medicine. Predictive analytical models could be built to determine if good medical outcomes could be predicted by the degree of match between doctor and patient in various trains, including communication expectations. Education models for both patients and physicians might result from such research. Other studies might examine communication styles of physicians/support staff and interruptions as predictors of good outcomes. PA studies could look at predicting medical outcomes based on type of insurance and/or decisions to have insurance.

- *Lack of patient-centered collaborative programs.* Medical homes plan for integrated care is conceptually a good way to increase communication among all entities. Medical homes centered on patients, have comprehensive service programs, and use teams of medical providers. According to the Patient-Centered Primary Care Collaborative (PCPCC, 2013), there are five elements of a medical home:
1. Collaborative (patients and care providers form a team)
2. Comprehensive (prevention to treatment of chronic illnesses; all-inclusive physically and mentally)
3. Coordinated (organized and active communication among entities)
4. Accessible (better hours, shorter waiting times)
5. Quality and safety (commitment through use of IT, informed decision making)

Patients selecting their best models of care

Medical homes

Medical homes expect that physicians will lead a team, in which decisions result from a group effort guided by government mandated standards (Massina, 2013). The current role of the physician in medical homes must change from being totally responsible for diagnoses and treatments to one of partnering with patients and the rest of the medical team. The goal will be for more communication and coordination of care. Patients will be grouped into "homes" to provide consistent care over time. Physicians should experience some relief in burden but ineffective EMRs could reduce this relief, and they may experience more electronic "paperwork" to document all the required information.

van den Berk-Clark et al. (2018) conducted a review of the literature and metaanalysis concerning patient-centered medical homes, particularly for low-income patients. They found moderate improvements in healthcare outcomes, particularly for addictions and for diabetes. This author could not find in the literature if the numbers of health homes were expanding or retracting. Most of the research seemed to have ended about 2015, possibly coinciding with government grant funding drying up by around 2014.

However, one organization, the Primary Care Development Corporation (PCDC, 2020) is actively assisting medical homes (800 as of May 8, 2020). PCDC helps in many ways, teaching, one-on-one coaching, helping to organize resources and time, sometimes funding, and especially encouraging patients to take active roles in their own health. It was quite encouraging to read the article on their achievements.

The integrated healthcare delivery system model

Another model for increasing communication between the disparate parts of healthcare is the integrated healthcare delivery systems (IHCDS) or HMO. Kaiser Permanente uses a form of this system (Bevan and Janus, 2011; Kaiser Permanente, 2013, 2020). Kaiser Permanente integrates health plans, physicians, and medical teams to achieve positive outcomes for the patients. Kaiser Permanente incorporates sophisticated information technology to achieve smooth transitions between clinics, primary care, specialists, and hospitals, all with the patient at the center (Kaiser Permanente, 2013).

Bevan and Janus (2011) wondered why the IHCDS service model had not been used in England, and why it was not in more widespread use in the United States, considering that communication and coordination is reportedly increased by its use. In England, patients go to a general practitioner first, who then sends them to specialists as needed. Bevan

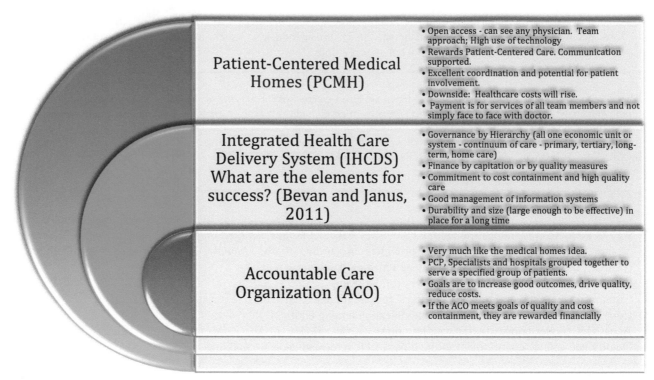

FIGURE 7.5 PCMH, IHCDS, and ACO comparisons.

and Janus thought that overall care would be better for patients in an IHCDS model. The IHCDS model is contrasted with medical homes and accountable care organizations in Fig. 7.5. Note that communication and coordination are prime objectives.

Comparison with accountable care organization

Refer Fig. 7.5.

Direct pay/direct care model

One wonders in looking at the types of traditionally leaning care delivery above, where would the people fit who do not choose to have an insurance plan? A fourth category could be added—that of the Direct Pay/Direct Care Model (Miner et al., 2019, pp. 242–245), in which patients "subscribe" to a medical clinic and all basic medical needs are addressed without outside insurance. The patient would also need catastrophic medical insurance. Increasing numbers of physicians are switching to the Direct Care Model. Communication in such a model would greatly depend on the PCP and the medical team. The PCP would control the number of patients in the practice and knowledge of the patient would follow—the fewer the patients the more communication would be possible. If there were choices concerning numbers of patients, those patients might even choose their doctor depending on how many patients there were in the practice. If one could not speak during a visit, that patient might want to go to a doctor that had the time to listen more.

In other countries, the more common use of EMR systems and treatment teams increases the efficiency and coordination of care (Schoen et al., 2006), therefore, the trend in the United States towards referral to medical homes may be a very good move.

Examining the elements of the care giving model, such as medical homes, ACO, IHCDS, or Direct Pay would be good sources of variables for predicting communication effectiveness, patient satisfaction, as well as medical outcomes.

Consumerism and advertising in patient-directed healthcare

Advertising to patients

We seem to be in a consumer-driven age. Advertising by doctors and hospitals has become a norm. Sometimes, patients are unduly influenced because of all the medical advertising they see. Direct to Consumer Advertising (DTCA) in medicine has greatly proliferated in the past 15 years, and its impact was thought at one time to be negative for patients (Lexchin, 1999; Mackert et al., 2010). Media influences public perceptions, and when the media aligns its products with medicine, scientific research, clinical tests (or anything that even sounds scientific), it appears that the influence increases. A spokesperson in a white coat or even an actual physician in a commercial may command attention. "Diseases" that at one time were unheard of, such as dry eyes, erectile dysfunction, OAB (over active bladder), or LOW T, are referred to commonly in the media, and the curatives are as familiar as gum or brands of orange juice. Billboards advertise operations as though they were cars, as in Fig. 7.6 (adapted from a picture of an actual billboard—note one of the authors on the billboard).

Billboards today advertise all kinds of "Health Improvement Procedures"; doctors and specialist hospital groups compete with one another to get patients. (Billboard by LM, and she's rather proud of it.) Gibson and Singh (2010) relayed one heartbreaking example of a young boy with pectus excuvatum (funnel chest). His condition was relatively mild. The condition inhibited his day-to-day activities only slightly. The boy was actually very healthy but may have experienced a somewhat better quality of life if the problem were fixed. His parents saw advertised on a billboard a new "less invasive" procedure to fix the problem. The ad seemed so enticing, indicating that the operation would fix the problem simply, and it would be easy on the patient. At the initial visit, the surgeon gave the parents a single study showing impressive statistics; he gave them also other educational material on the condition. The parents were intelligent and well-educated; they had the ability to make good judgments. They elected to have their son's chest "fixed." They trusted what the doctor said, and they wanted to believe that the procedure would be easy on their son. The result, though, was catastrophic. Cascading problems and the unavailability of a senior surgeon on the weekend when the surgery was performed led to complications from a perforated duodenal ulcer that were fatal (Gibson and Singh, 2010).

On the other side of targeted advertising, physicians are often pressured by patients to prescribe the newest medicines advertised in the media. Pirisi (1999) found that some physicians feel pressured into prescribing the medications that prompt patients to visit them. Family physicians were more likely than internists to succumb to the wishes of the patients. The problem with this response pattern is that physicians know that patients value the prescription of the medicine, and so they treat the desire, not the medical need. A prescription in hand seems to validate the office visit, for some reason.

Television is a commonly used avenue for the marketing medical devices and medication. Dramatic effects of hushed tones are employed often at the end of a glorious advertisement, along with music to generate a feeling of relaxed interest in the viewer. Images of nature and restful bathtubs are used to couch possible side effects soothingly expressed in the background. Disclaimers like, "...may cause shortness of breath, liver damage, increased chances of heart attack, stroke, and even death. If you experience any of these please contact your doctor immediately." Some disclaimers state that, "If you are pregnant or likely to become pregnant, you should not take ...." That disclaimer would apply to most women between the ages of 15 and 50, but they want very much to insulate the viewer from drawing that conclusion.

FIGURE 7.6 Billboard advertisement.

It seems that psoriasis is one of the most prevalent diseases in the United States if one is simply to watch television ads (print media as well). While the disease affects only about 3% of the population (Rachakonda et al., 2014), commercials abound. However, if one considers that the treatments cost $40,000 to $50,000 per year and they do not *cure* the disease, only treat the symptoms, then the 8–9 million potential customers seem totally worth all the advertisements.

Research studies related to advertising and consumerism

- *Advertising does drive medical consumerism.* Datti and Carter (2006) found that advertising seemed clearly to influence patients to visit their physicians. Datti and Carter hypothesized that older people would be more susceptible than younger people to the "direct to consumer advertising" (DTCA). Surprisingly they did not find an effect of age in their findings. They did find, however, that about a third of all the people exposed to DTCA visited their doctors to ask for prescriptions of advertised medications. And about 69% of those received it. There were 2601 people in the study. This means that about one fourth of the people so exposed to the advertising, received the advertised medicines—a boom for pharmaceutical companies.
- *People are consumer oriented.* Salgo (2006) stressed that we tend to be consumer-oriented; we pay for goods and services, and we expect good results. Doctors are also beginning to view patients as consumers or paying customers. There is a dark side to seeing patients as consumers, though. As physicians move more people through the healthcare system, they make more money (Salgo, 2006). It is not simply Medicare that requires physicians to see more patients in less time. The combined effect of this consumerism in healthcare is to move physicians toward the goal of making money, rather than caring for people.
- Prospective patients want to go to hospitals that they have seen advertised on television (Tongil and Diwas, 2020), if they had insurance to select the hospital. People with lesser forms of insurance were not as persuaded by television advertising. For people with little or no insurance the choices were limited significantly.

Privacy of prescription data. Is it private?

Gilbert (2011) cited a Supreme Court case (Sorrell v. IMS Health) on the question of whether a law could be enacted to prohibit prescription histories being sold to drug companies.

Regardless of how we feel about it, PAs have been and are being used to target advertising. For example, in the above court case (Gilbert, 2011), PAs could be used to predict which elements should be included in strong sales pitches to doctors, using prescription histories, for prescribing new medications.

In the end, the law enacted to restrict the sales was found by the Supreme Court to violate the First Amendment of free speech. This sort of data fed, targeted marketing is used also by internet social media and search engines to push medications. As early as 2020, many people felt that our smartphones, televisions, and computers were listening in and algorithms picked out interests, expressed needs, and so on. Advertising was soon to follow. Facebook and Google were not the only data miners among media, and it is not just the patients who are targeted but physicians as well. In their own defense, drug companies claim that they do not want to waste the doctors' time with talking about medicines of no interest to them. Data mining will be used by an increasing number of elements of our society in the future. By 2021, algorithms were easily determining what information was "disinformation" and those sites were pulled down. Many questions remained about who determined the algorithms and whether such was limiting free speech. Privacy of information, it seemed, was a thing of the past. Large amounts of data may be had by huge companies, owning monopolistic platforms, and evidently those entities were wealthy enough to be listening and watching constantly. Data miners might need to find employment within the huge companies or hospitals to work on projects.

Fortenberry et al. (2010) found that billboards were effective in encouraging patients to frequent specific medical practices. They suggested billboard rotation frequency, revealed that the use of corner tabs was unproductive, and suggested ways that digital billboards could be used more effectively.

PAs tools can be used to save lives, save time, and save money. PA can also be used to serve pure profit motives, with the focus on the paying consumer, or to serve political ends. Power and wealth are huge drivers. True scientists, it is hoped, will instead be motivated by the search for truth for the betterment of humanity.

Patients diagnosing themselves amid targeted advertising

In the United States (and in some other countries), people may go to their doctors with lists of orders for advertised medicines and Google searches of possible diagnoses. We are becoming consumer-oriented in our healthcare thinking, and we expect to be served; television, Facebook, and Twitter tell us what the "truth" is, and we want it! If you want to know what five foods should be eliminated to decrease belly fat, simply look at the right-hand side of your Facebook page, as in Fig. 7.7 (assuming, of course, one is overweight and has been searching for such topics, setting off the algorithm).

Many people use Google to search for their symptoms, sometimes in response to an advertisement they have seen, and are then targeted by the Internet data miners. Patients are becoming increasingly more self-directed and empowered, while at the same time, easier prey for advertisers than before data were collected from social media.

> Possible PA Research: A good PA study could be to gather information on what diagnoses patients have looked up before they visit their doctors, and then see if the final diagnosis could be predicted from them. Other studies could look at how the targeted marketing affects the patient/physician interactions and final diagnoses. Which patients would benefit most with the best outcomes with what kind of care—self-empowered, taking charge versus forms of more physician control of their medical care?

Some forms of advertising are rather indirect, and they aren't obvious to the patient. Years ago, physicians' office staff welcomed the drug reps who came in loaded with "goodies" of pens, note pads, handy items that would be useful in an office, with the company logo written on them. This practice is no longer allowed; however, the representatives still try to influence in any way possible. The pharmaceutical reps still make appointments with physicians to tell them about the latest advancement in medicines. Some medical practices will allow the sales reps to bring lunches for the entire staff. When introducing new appliances for operations, the reps will even go into the operating rooms to help the physician use the new device properly.

The Federal "Sunshine Act" (enacted February 1, 2013) mandated that all physicians must start recording all payments and gifts received from medical device companies and pharmacies, beginning in August 2013 (Fisher, 2017). In the first edition, this author thought it would be interesting to determine the effects of the Sunshine Act on the purchases of devices and drugs. Though not a PA study, Guo et al. (2017) attempted to assess changes in prescribing behaviors of physicians because of states' level of transparency. Before 2013, it was up to states as to how much would be reported and to whom. The authors compared the prescribing behaviors in Massachusetts, which was a "transparent" state due to Open Payment Law in 2009, to surrounding states that did not have such laws. They examined differences in prescribing behaviors before and after the law, and Massachusetts to the other surrounding states (Connecticut, New Hampshire, Rhode Island, and New York). They tried to match cities that were close to the borders in each case to reduce the variability between the locations of the physicians. What they found was that there was a reduction of prescriptions (of the three types they looked at—statins, antidepressants, and antipsychotics) overall, and a reduction of branded prescriptions. Evidently, exposure had an effect. This study made this author think of a study by one of her undergraduates in which physicians were characteristically late to surgeries. The student was a surgical nurse. Her

Beyond Diet

Learn about the 5 Foods You Should NEVER
Eat for Weight Loss

5 Foods You Must Not Eat
http://www.beyonddiet.com/

Click to Watch Our Video
Presentation, and Start Losing Today!

WEIGHT LOSS

Like · Comment · Share · 👍 6,126 💬 941 📝 3,590 · ☀ · Sponsored

👍 Like Page   FIGURE 7.7 Ad from Facebook.

before and after study involved a simple posting of the arrival times of the physicians—resulting in a marked reduction of tardiness. Operations started on time thereafter. To see some up to date reporting requirements of the Sunshine Act, please see the reference, Appzen (2020), and to see dates when various reports are due, see Porzio Aggregate SpendID (2021). One must sign into the Porzio site for additional information.

> Possible PA Research: A study could be done predicting device purchases from reps who either do or do not go in the operation room. Such a study was done by another undergraduate of Southern Nazarene University-Tulsa. The results might be quite surprising. See Tutorial Z in the first edition of this book, in which the presence of device representative did seem to influence sales. Additional studies could use sales as the outcome variable predicted from various drug rep attempts at advertising and helping physicians.

Patients making use of technology and advertising for good or for bad

Patients are also targeting doctors and medications using data. Lagasse (2019) spoke of patients becoming savvy with respect to technology, targeting doctors whom the patients think will be the best providers, who will use their insurance, who perhaps will provide off-hours virtual exams, and allow online appointments. Patients find evaluations by other patients and basically shop out doctors.

Perhaps advertising increases awareness of possible medications. But how much does the public know about the clinical trials conducted to validate their use? Do people in general understand even the basics of the concept of a double-blind random design? Are those designs being used by drug companies? The answer is that most patients do not know how drug trials are conducted. Most people assume all medicines approved for use by the FDA have been rigorously researched and tested sufficiently to assure their safety and effectiveness. If patients knew that a new medicine had been tested for only 6 weeks (or even 6 months, or a year), would they be willing to take it? What happened to the rollout of COVID vaccinations that were developed so quickly that the only way they could be used was by using the "emergency" label in which drug companies were not liable for any damage. Were there not skeptics among the population? People in general want to believe that their medications and devices are safe to use. Do people conduct as much research prior to a knee replacement, as is done commonly prior to a car replacement? Do people visit one car lot and automatically believe everything the salesperson says? All these questions beg the same answer: No! Patients should be equally concerned about finding the right knee replacement procedure (and the surgeon to cut them open), as they are with finding the right car to buy. Patients need to find out everything they can concerning their own care—patient-directed healthcare! Certainly, patients need not to believe that vaccinations are the "mark of the beast," but they do need to make their own decisions and then suffer their own consequences resulting from their decisions. Directing one's own healthcare implies the responsibility of finding all relevant information and doing the work that accompanies self-direction.

Patients must assume increasing responsibilities to prevent illness, to monitor costs, and to be accountable for their own health. For nonemergency care, the patient who practices self-directed healthcare will be concerned about costs of tests, how many tests are run, and will learn the latest research on the risks and benefits of various treatments. In other words, patients must become well-informed consumers of healthcare services.

Comparison shopping is another stratagem patients can practice to increase the quality of their healthcare and lower its cost. One area of comparison shopping is for imaging services. Many doctor clinics have imaging capacities, where this imaging could be done. Currently, government law allows hospitals to charge more for diagnostic tests than it allows doctors to charge. Imaging at hospitals may be encouraged by hospitals because it is much more lucrative than it is for a physician in his own office. For example, the cost for a shoulder MRI in Tulsa in 2013 ranged from a low of $600 to a high of $4000 (New Choice Health, 2013). Patients should learn about the price differentials before they choose a relatively high-priced procedure. New Choice Health offered transparent cost comparisons in featured cities. Resources such as this one could assist patients seeking information on costs. Armed with this information, a patient could inform the PCP about it, who might be able to arrange for a test at the least expensive facility and receive the same information. In the shoulder imaging example, the full cost of the cheapest alternative was less than the copay of some of the others.

Possible PA Research: Areas of consumerism could certainly be ripe for predictive studies:

1. Attitudes of patients and physicians toward new medicines, and vaccinations
2. Public understanding of basic research design.

3. How influenced are physicians by patient pressure?
4. Hospitals researching the influence on purchases of the presence of the company representatives during operations
5. Medical outcomes due to "direct to customer" advertising
6. Influence of customer price shopping on those services shopped for
7. A study of drug trials—lengths of time studies run, numbers of adverse events
8. One could even mine the texts of the long pharmaceutical inserts in medications.

Patient payment models and effects on self-directed healthcare

Proper healthcare in the United States (and in some other countries) means having insurance so that we can visit a doctor for little or no out-of-pocket expense. Those with insurance are not charged as much for a given healthcare service as those without insurance. As Goldhill (2013) pointed out, we don't count the costs we pay in insurance premiums and monies we pay the government. We also do not count the costs we are deferring to the future in terms of national debt.

When asked, "Do you have healthcare?" We think insurance. We are frightened if we don't have "healthcare" and feel vulnerable, helpless, and exposed. For those who had no health insurance, the Affordable Care Act (ACA) came not a moment too soon. The huge downside of ACA was that costs could increase for many with the addition of previously uninsured people and those who thought they were healthy and didn't need expensive coverage simply accepted the tax penalty and didn't sign up. Companies fled the markets of some states leaving only one or two "choices" for those who wanted to sign up.

Later, with the election of Donald Trump, healthcare became ever more political and, as of 2021, the left and the right fought for their interpretations of healthcare, while with huge quantitative easing, the national debt rose astronomically. At the time of this writing, it was difficult to determine what an empowered patient could do in terms of using insurance companies. However, below is one suggestion that does not depend on what the government does or does not do.

The Direct Care/Direct Pay method (also called Subscription Medicine) of basic healthcare is one solution, if only one could find a practice that used the subscription method (Subscription Medicine, 2015). Increasing numbers of physicians are moving to such a plan for their clinics. A family of four might pay $80 per month for each adult and $50 a month for each child (fees depend on what the doctor charges). No insurance is filed for routine care. Visits, emails, online visits, generic medicines, and basic tests are all supplied with the subscription. The patients would need catastrophic insurance for exigencies such as surgeries, chemo, or hospitalizations. Many physicians are attracted to the subscription model because not only is there more control over one's own health by patients, but physicians gain control as well. Physicians can decide on how many patients they want in their practices and do not have to file with insurance companies. If they should accept too many patients, then patients naturally receive less care and might want to find a different practice with fewer patients. The model can save thousands of dollars per year per family. The clinic saves money because it no longer must deal with many different insurance companies, and it would have a steady income.

Research topics on the importance of healthcare insurance, whether that be a subscription to a clinic or traditional types of insurance, include:

- *Uninsured people receive less treatment.* Doyle, Jr. (2005) researching outcomes from automobile accidents found that uninsured patients received 20% less treatment than those insured patients.
- *Reimbursement amount versus Number of patients, particularly considering COVID hospitalizations.* In 2014, it was expected that people who were within a specified level of poverty were to be eligible for Medicaid, under the provisions of the ACA (APHA, 2013). It was assumed that all those people being added would lead to many more services provided. However, White (2012) found that the number of services was more closely tied to the amount of insurance reimbursed per patient than the number of patients. According to White, "either the use of managed care tools or the relatively low reimbursement rates, or both—may have limited the utilization effect of the coverage expansion" (p. 978). However, according to Daly (2020) at the height of the pandemic, hotspot hospitals were to get $50,000 per COVID-19 patient admission. That was to be a reduction, from the earlier, $77,000 per patient. Lest we think that is a lot to be receiving, each COVID patient required more specialized care that cost more. It would be interesting to study the actual effects of a pandemic on hospital financials and staffing.
- *Length of enrollment can be important.* On the other hand, Howell and Trenholm (2006) compared outcomes for undocumented children enrolled in an insurance program for a year with those who just entered the program, finding that those who were enrolled longer were healthier. One wonders, in 2022, how our country might deal with undocumented children with the huge influx of such, but it is hoped and expected that compassion will rule.

- *Public versus private insurance differences.* Einarsdóttir et al. (2013) studied babies from over 1000 Australian women in public and private insurance groups and found that babies in the public insurance group were more likely to have a lower APGAR score and were given less specialized care than were babies of mothers with private insurance. Thankfully, there was no difference in resuscitation rates. The authors were troubled by the increased risk of respiratory morbidity, with infants of publicly insured women having a lower likelihood of admission to specialized care programs.

Contrast the above study with the more recent findings that some uses of ventilators were associated with increased morbidity for COVID patients at the beginning of the pandemic, regardless of the type of insurance. The difference reflects a difference in understanding of a new disease, and likely due to the nature and progression of oxidative stress, endothelial cell dysfunction and thrombosis in the lungs, particularly in obese patients. Over the course of the pandemic, physicians learned that the disease had phases and each phase needed different treatment. Additionally, pneumonias may have been of two types in COVID-19. The first type (L) involved a lack of profusion into the blood (low ventilation to perfusion ratio). In this type L, the lungs are not full of fluids and move freely, but the oxygen is not getting to the blood. This type L likely was not expected and maybe not looked for at first, when little was understood concerning this novel disease. The second, more what hospitals were used to, is the acute respiratory distress syndrome, type H. It is possible that patients first presented to the hospitals with the L type and then as the disease progressed, fluids could accumulate making the patients look like the more expected type H in terms of lung function. Treatment, particularly ventilation, should be different for these two types as explained by Dr. Roger Seheult (2020b, #53). The point is that treatments can vary by insurance type, but also by what is understood about novel diseases.

Burden of healthcare—predicting the future

Predicting life and death

When people get sick, most don't think about the enormous burden they may be placing on future generations, caused by requested medical scans and/or procedures—unless they must pay for the procedures by themselves. Increases in technology and in medical diagnostic activity lead to increases in the use of them. Physicians may use new technology as a protection against litigation. Lawsuits increase costs, technology increases costs, and covering more people increases costs. The public learns to expect that the level of medical treatment will increase with severity of disease, which leads to increasing costs, at rates far exceeding inflation, and even when the rest of the economy suffers recession. The result is that medical costs continue to rise precipitously, apparently without question—if insurance pays for it. Often when a family member is dying, the family wants to go the whole distance, resuscitating the loved one with any state-of-the-art equipment available. End-of-life care, in fact, costs upward of 25% of all Medicare costs (Ashish, 2018).

The case study in Chapter 17 shows the steps taken with the Loma Linda Hospitals data (Chandnani et al., 2018) to predict life or death in patients with disseminated intravascular coagulopathy based on pediatric ICU admissions. Three factors were identified as important predictors. Such studies can help make decisions about what kinds of treatments are used and potentially could lower costs. Patient-directed medicine should include any relevant studies concerning outcomes so that patients can make informed decisions.

> Ashish (2018) wrote that it would be good if an accurate prediction model could be devised as to expected death. Having extra time with one's family is precious, and it would be quite helpful to know when interventions will be fruitless or fruitful. Lord (2020) gives good advice in her blog on helping patients make good end-of-life decisions and provides questions that might be used in helping the patient (and/or family) make those decisions. PA research could be vitally helpful for doctors, patients, and their families.

Goldhill (2013) suggested that we can "follow the money," when discussing the impacts of healthcare costs. In US dollars (adjusted for inflation) the United States Hospital Insurance budget increased from $893 million in 1966 to $208,419 million (over 208 billion) in 2012 (Office of Management and Budget, 2013, Table 4.1). Thus a 233-fold increase occurred over the past 46 years. This rate of increase was not anticipated by those who made projections at the inception of the budget (Sanders, 1965, 1967; Myers, 1967, 1994).

According to the US Census (US Pop 1900−2022; NPG, 2013), the US population increased from an estimate of 198,712,056 in 1967 to 313,914,040 in 2012 and would climb even further to over 315,000,000 in 2013. The population ratio between 2013 and 1967 is about 1.58 (not even twice as large). These figures show that about $4.49 per person was budgeted for healthcare in 1967, which rose to about $661.65 per person in 2013. According to the Bureau of Labor Statistics (2013), $1.00 in 1966 would be worth $7.18 in 2013. Multiplying the per person 1967 budget $4.49 times 7.18 yields $32.23, which is what would be budgeted in 2013 to match the cost of 1967. Though, what happened was that healthcare costs increased 20.53 times (661.65/32.23) in 46 years (corrected for inflation). The situation was even worse in 2021 and one could hardly look at the spending bills being proposed in Washington D.C.

A greater than 20-fold increase seems a bit exorbitant over a 46-year period. If the current trend continues, we might have thought, who would be able to afford health insurance? Not very far into the future, it could be likely that the only people able to afford healthcare in the United States will be the super wealthy. Who knows what will be "reimagined" for healthcare? This prospect behooves us to become more involved in costs. We should determine which costs we should bear, and which we should not. Should we expect insurance to cover the cost of band aids, or for only catastrophic events?

The Congressional Budget Office (2020) reported April 15, 2020, that the nation expected to receive $3.5 trillion in revenues and expected the outlay to be $4.4 trillion. This, of course, was when the pandemic had just started. Of that, likely underestimated, budget, Medicare would spend at least $644 billion, while Medicaid $409 billion. Again, we should not worry about the National Debt, because the Federal Bank is able to print unlimited amounts of money (quantitative easing) each night to satisfy any financial needs. (That statement was facetious, regardless of how true that fact is as of this writing in 2022.)

Misapplication of treatment increases costs

Another factor of cost increase is the application of various treatments to the wrong people. Often, it is difficult for clinicians to know for certain if the intended treatment is appropriate; they certainly don't want to err on the side of sending sick people away, or worse, of sending sick, highly contagious people to live with highly vulnerable, heretofore unexposed people.

In the first edition of this book, in Case Study I, (pp. 544−557) we described a model that predicted the need for hospitalization for emergency patients complaining of chest pain. One model was 100% sensitive (correctly identifying patients that did have a subsequent heart attack) but was less specific (correctly identifying those who could safely go home). A different model was 100% specific and less sensitive. By combining both models they were able to correctly identify those who needed to be sent home from those who needed to be kept, resulting in hospital savings (Alekseiv and Harris, 2013).

Possible PA Research: Again, related to predicting when someone will die, PAs may help by predicting more accurately when treatments will do no further good, and when to use a high-priced technology, the patient is beyond a certain cutoff. Thus PA can help to keep costs down by correctly targeting appropriate applications. Such models might help future research on how to help the individual. Sometimes the predictors that "work" in the predictive algorithm, might have a causative effect on outcome and can be the basis for controlled research that ends up causative, thus saving more lives.

Another possible research topic: How do the traditional insurance approaches compare (outcome-wise) to nontraditional approaches such as subscription medicine, also called Direct Pay, described earlier?

Models of insurance—predicting the best for individuals

Insurance plans are many and diverse. Various coverage models that at one time were quite different in terms of structure and function now seem to be blending. All plans are created for cost containment and increases in quality, and their creators do not want to lose money.

The most basic categories of insurance are the Indemnity Model (traditionally Blue Cross/Blue Shield), and Managed Care (HMOs such as Kaiser and Humana). Point of Service is a kind of hybrid between HMO and Indemnity, and Direct Care is a new model, gaining in popularity. Fig. 7.8 compares some of the choices that patients might have within each basic model.

Even the basic breakdown in Fig. 7.8 does not distinguish the types completely. Within each type are subtypes, which tend to overlap with other models. Commonalities are more prevalent than differences among many insurance companies. Payment authorization often varies between companies. Sometimes payment authorization can differ even

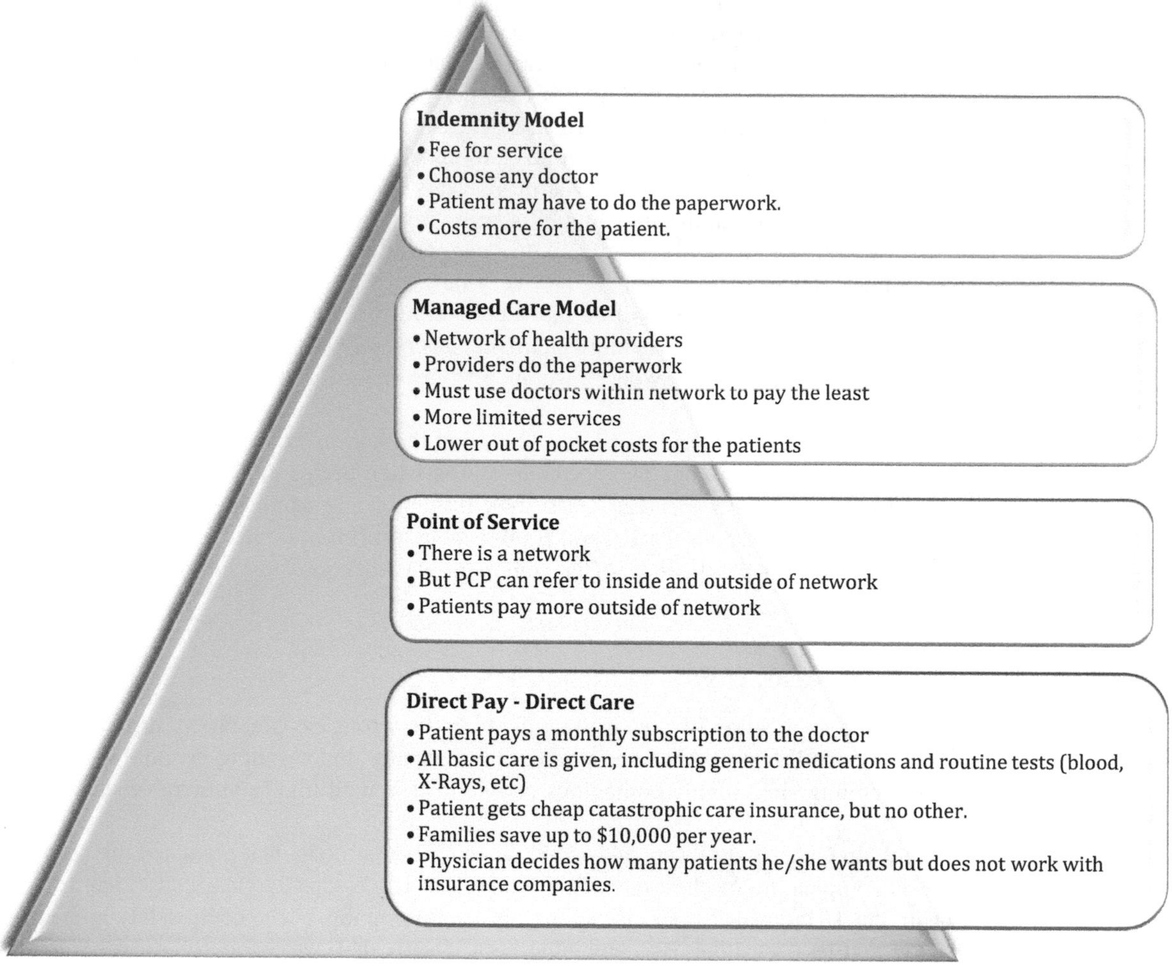

Indemnity Model
- Fee for service
- Choose any doctor
- Patient may have to do the paperwork.
- Costs more for the patient.

Managed Care Model
- Network of health providers
- Providers do the paperwork
- Must use doctors within network to pay the least
- More limited services
- Lower out of pocket costs for the patients

Point of Service
- There is a network
- But PCP can refer to inside and outside of network
- Patients pay more outside of network

Direct Pay - Direct Care
- Patient pays a monthly subscription to the doctor
- All basic care is given, including generic medications and routine tests (blood, X-Rays, etc)
- Patient gets cheap catastrophic care insurance, but no other.
- Families save up to $10,000 per year.
- Physician decides how many patients he/she wants but does not work with insurance companies.

FIGURE 7.8   Models of insurance (and no insurance) types.

within the same company. The plan most in the control of both the physician and the patient is the Direct Care/Direct Pay Model as described above.

One example of this distinction in the life of Kevin Dwyer was portrayed on the Today Show on July 16, 2013. Kevin was in the latter stages of cystic fibrosis. At 40 years of age, Kevin was older than the average age of survival of the disease, but he continued to plan for his life, including his recent engagement. He and his sister, Martha, had researched all medications, drug trials, looked at interactions, mode of action, and so on. They were extremely well-informed and self-directed. They found a new drug, Kayldeco, that they thought could work. Kevin and his doctor wrote to Kevin's insurance company. The drug was expensive ($25,000) per month. The drug company reviewer turned down Kevin's repeated requests (three appeals) stating that the drug was not approved by the FDA and was not deemed "medically necessary." What was so extraordinary about the story was that Kevin's sister also had the disease, at about the same stage and had been approved for the drug. Kevin by this time was close to death. Martha was seemingly quite well as she had been on the drug for some time. Martha's company was the same as Kevin's. The two experienced the "luck of the draw" when their requests went to different reviewers. The Today Show was able to tell Kevin that he would be covered. The insurance company relented after at least two sources had reported on the discrepancy.

Patients opting for the Direct Pay Model surely would have to be capable of making their own decisions. For example, in selecting their catastrophic insurance, they would want to know their family histories. PAs could be quite helpful to patients in this model as they determined what kind of catastrophic insurance to purchase by helping them predict what their own health likely would be, in the future.

Surely, these patients would be most keenly aware of what made for a healthy lifestyle. They would have to be diligent in directing their own care, negotiating catastrophic insurance for themselves. This option would truly be self-directed healthcare, while partnering with their PCP.

Workers in companies that supply employee insurance may not be given any choice of insurance plans unless they seek their own insurance outside their companies. Employers may let employees choose between a PPO and HMO. Retirees who were used to having their places of work make insurance decisions, could become shocked at the varieties of choices open for them (beyond the basic provisions of Medicare, Part B and Part D). Even Medicare can seem scary, with so many regulations, forms, and decisions to make. Patients must become informed to make good choices for themselves. They might be able to convince their employers to pay for their Direct Care/Direct Pay Model and the catastrophic insurance, as it would certainly be a less expensive option for the business.

> Possible PA Research: PA studies could predict medical outcomes from variables of reimbursement decisions, types of insurance, healthcare costs. Patient-driven healthcare is not simply a nicety but could save one's life. As stated above, PA could be quite helpful to the individual trying to predict future healthcare needs for themselves and their families. Here is where patient portals could be used if there were predictive models that the patient could review and try while subscribing.

Seven years ago, this author wrote a blog, Seven Ways PAs Can Improve Healthcare (Miner, 2014). In revising this chapter, I ran into the blog again. The reasons continue to have validity today. If patients have good predictive studies to examine, it certainly could make their decisions easier. Good predictive studies could be invaluable to all healthcare professionals who work with patients as they struggle over difficult diagnoses and treatments plans.

Research assisting patients in self-education and decisions

What sources should patients turn to when directing their healthcare? Researching medical conditions can be quite enlightening and interesting (and sometimes frightening). But one thing is certain, a patient cannot make informed healthcare decisions without knowledge about its elements. Genetic information will become increasingly important in self-directed healthcare. What is done with that information will be important also, and it will be very interesting. Patients can also keep data related to various circumstances on their condition to help guide them (as can be seen in Dr. Gary Miner's tutorial on glaucoma, Chapter 11).

As stated above, knowledge of the results of PA studies may help patients make better informed decisions for themselves. Sometimes they make their own decisions that differ from their physicians' preferences for them. Gill et al. (2010) conducted a study that is a good example of how patients really can direct their own care when given information. The Gill et al. study consisted of 89 breast cancer patients, some of which were diagnosed with the Oncotype DX genetic test. Results of the test combined with a PAs model (for those who were not tested) were used to recommend chemotherapy and hormonal treatments to 23 patients. In this article on personalized medicine, the authors talked about a breast cancer genetic test called Oncotype DX, which when used with PA, can produce a prediction score for recurrence which can be helpful in deciding whether to have chemotherapy. If a low score, the decision was easy and there should be no chemo. A high score meant chemotherapy. In the middle, though, one needed to use clinical judgment and hunches. (It would have been better if an accurate individual prediction model could have been developed.) The oncologists had recommended chemotherapy (and hormones) for 42 of the 89 patients when those oncologists did not know the results of the Oncotype DX test. After the results they narrowed to just 23 patients that they thought needed chemo. Most of the patients followed their doctor's advice. However, seven of the women in the study did not follow their doctor's advice and opted for other treatments or observation.

The Gill et al. (2010) study did not address the reasons for the noncompliance. However, the new test did seem to reduce anxiety over choices for both the doctors and the patients and the knowledge from the study did allow the patients to make their own decisions, regardless of what their doctor recommended. PAs need not be 100% accurate to provide benefit in decision-making; however, the higher the percentage of "hits," or correct predictions, the better. The Gill et al. (2010) study is a good example of how patients really may need to direct their own care when given information because the information is not specifically for them as individuals.

Quite a few genes for breast cancer have since been identified and Raman et al. (2013) provided a long list of them. In addition, other cancers and even some forms of Alzheimer's have been studied and their various genes greatly increase predictive accuracy for treatments.

Raman, Avandano and Chen (2013) provided one-page summaries of the genetic tests for cancers. Patients can be extremely well-versed on the research in a disease or disorder. As previously stated, patients can even bring information to a physician in a collaborative partnership which informs the physician. In contrast physicians can also educate their patients in the office or even in seminars which they open to the public. Such educational seminars can at the same time advertise the physician to the public (Van Doren and Blank, 1992).

Patient self-responsibility: highlight on obesity

Ultimately, patients are responsible for much of their own health, and they should educate themselves. Some conditions appear very unexpectedly, but others are suspected in advance. Similarly, patients might know that a parent died early of heart disease, or that other members of the family developed Alzheimer's disease. Patients cannot remain ignorant of vital information relating to their health, but should meet the challenges that face them, by whatever is possible to ameliorate the debilitating effects our genes have on our health and well-being.

Our current epidemic of obesity highlights the need for patient-directed healthcare. Information is needed both by patients and their care providers. Such information might be part of what physicians teach their patients, perhaps by using patient portals (discussed below). On the other hand, it may be necessary for patients to bring up the topic and becoming more involved themselves.

> Possible PA Research: Often doctors are reluctant to talk about weight with their patients. The reasons are unclear but determining the reasons for reluctance to discuss one of the major health problems in the United States (as well as other countries) would be a great topic for a PA study in personalized medicine. The problem of obesity is complicated, but we need to know how much of the problem is related to lack of knowledge about food as the patient chooses what and how much to eat, how much is related to the financial ability of patients to purchase high-grade foods, how much is related to psychology, or genetics, or other factors.

Percent of obesity

Estimates at first writing were that up to 60% of the American public was either overweight or obese. Obesity in children was endemic and getting worse. If patients are to direct their own medical care, they really need to start with doing something about their choices leading to obesity. Obesity is quickly becoming our number one health issue in the United States and in other countries as well. For example, obesity rates have grown in the United States from 12% in 1990, and 23% in 2005, to over 30% in 2013 (Menifield et al., 2008) (see Fig. 7.9). Oxidative stress related to obesity can adversely affect the endothelial cells in blood vessels, increasing risk of strokes and heart attacks. Obesity affects how insulin is used in the body and increases the risk of type 2 diabetes as well as other pathologies.

By examining Fig. 7.9, one sees the problem was not simply the United States, although the US led the world in obesity. Mexico was said to have outstripped the United States (though people should probably keep their clothes on in

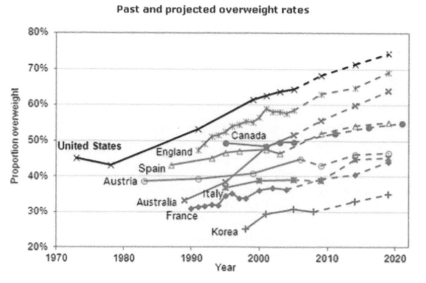

FIGURE 7.9 Projected rates of overweight over time (CBS, 2013).

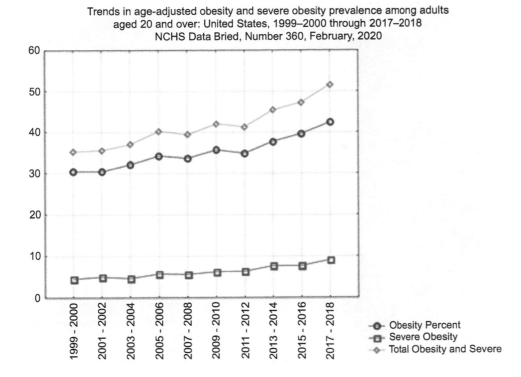

Trends in age-adjusted obesity and severe obesity prevalence among adults aged 20 and over: United States, 1999–2000 through 2017–2018 NCHS Data Bried, Number 360, February, 2020

- Obesity Percent
- Severe Obesity
- Total Obesity and Severe

FIGURE 7.10 Age-adjusted (controlling for differences in population age distributions) obesity rates in the United States, 1999 to 2018 (NCHS, 2020). *Graph by Linda Miner, Data from NCHS Data Brief Number 360, 2020. Data brief 360. Prevalence of obesity and severe obesity among adults: United States, 2017–2018. <https://www.cdc.gov/nchs/data/databriefs/db360_tables-508.pdf#page=4>.*

both countries!) (CBS, 2013). We are not far from the prediction, but the graph in Fig. 7.10 is frightening, as we are becoming both more obese and more greatly obese.

More recently, the situation did not seem to be improving. Fig. 7.9 predicted about 75% overweight in the United States by 2020. In 2021, we seemed to be at least 70% obese and overweight according to Gulbin (2021), so we continue to have a problem. We are not far from the prediction, and the following graph (Fig. 7.10) is frightening, as we are becoming increasingly not only obese but greatly obese. The COVID pandemic of 2020 did not help, with many people eating more while being bored and possibly anxious at home. They also may not have been getting enough exercise at least in 2020. By 2021 people were beginning to move around more and some were horrified at their sizes and vowed to do something about it.

Possible PA Research: Education concerning nutrition and exercise may be needed and, likely, the earlier the better. Variables for a study of obesity could be "knowledge of good nutrition, socioeconomic status, dietary consumption, age, the availability of fitness facilities, employee sponsored fitness and training facilities, insurance company sponsored fitness facilities, incentives to join health and fitness programs, wellness programs, and lifestyle choices, i.e., time spent watching television, engaging in low-level exercise, etc." (Menifield et al., 2008, p. 87.) Researchers could also focus on programs that advertise such as NOOM, Nutrisystem, Weight Watchers, GOLO, and others. Is it helpful to have systems send the food to eat? Does weight rebound upon completion? Or they could try the relatively new intermittent fasting (De Cabo and Mattson, 2019; Fung, 2016) reportedly having many positive health effects besides weight loss.

Distribution of obesity in the United States—costs and related diseases

Geographic location of obesity might be predicted from factors such as how much nutrition education is emphasized in various parts of the country. Fig. 7.11 shows the geographical distribution of obesity (Bendix and Lutton, 2020).

If one compares the map of Fig. 7.11 with that of the first edition (Fig. 14.8 in the first edition) it is evident that the Mississippi River Valley is not doing well at all, with all those states gaining (literally). These are self-reported obesity findings from telephone interviews and of adults. Because only up to 35% of adults called admitted to obesity, one also wonders if perhaps the Mississippi River Valley has more truth tellers than other parts of the country. Or perhaps other estimates of obesity are incorrect. My bet is that people did not accurately report over the phone. Regardless, obesity seems to be expanding along with our waistbands. It is an epidemic all its own!

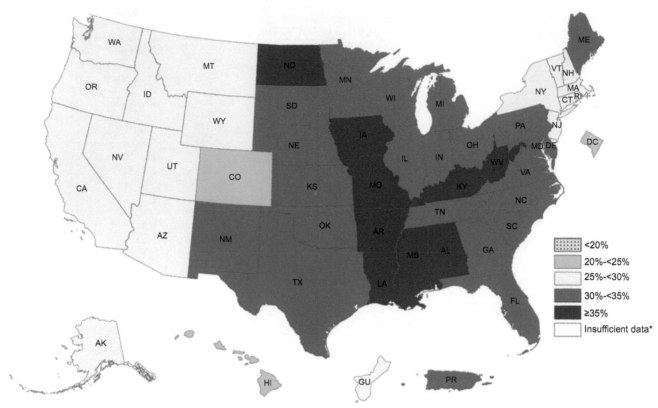

FIGURE 7.11   Overall obesity (CDC, 2018).

Costs of Obesity. Obesity increases costs in healthcare and productivity. In 2011 costs attributable to obesity were about $73.1 billion for the United States (Yang and Nichols, 2011). Those costs included medical costs, loss of productivity while at work, and absenteeism. The Affordable Care Act (PPACA) allows employers to reduce the health care premiums of the nonobese by as much as 30%.

Obesity Contributes to the Development of Diseases. Obesity is linked to many diseases and disorders such as coronary heart disease, hypertension, type 2 diabetes, osteoarthritis and damage of load-bearing joints, gallstones, stroke, asthma, and high cholesterol (Bidgood and Buckroyd, 2005; Yang and Nichols, 2011). Are all people aware of these facts? Nutrition programs can also include the consequences of weight gain, in the same way that smoking education has informed children of the dangers of smoking.

Bendix and Lutton (2020) stated that the healthcare costs related to obesity accounted for more than $1.5 trillion in 2018. Let's see what the actual increase comes to when one takes inflation into account.

Using the US Bureau of Labor Statistics calculator (https://www.bls.gov/data/inflation_calculator.htm), $73.1 billion dollars in 2011 would have inflated to $83.6 billion dollars in 2018. So, dividing $1.5 trillion by $83.6 billion would account for 17.9 times increase in actual costs. And the costs have only increased since 2018. Bendix and Lutton (2020) provided the top 10 diseases, numbers of people affected, and the 2016 costs for each, in their article. The top 10 diseases were dyslipidemia (too much lipid in the blood), hypertension, osteoarthritis, chronic back pain, type 2 diabetes, asthma and COPD, gall bladder disease, coronary heart disease, stroke, and congestive heart failure.

Cascading effects on sleep of obesity

As people enlarge, the incidence of sleep apnea increases but sleep apnea does not necessarily cause obesity. Rather there could exist subtypes of obesity, with excessive daytime sleepiness perhaps related to metabolic disturbances and stress (Vgontzas et al., 2008). Short hours of sleep were associated with significant weight gain in a prospective weight study of 7027 Finlandians (Lyytikäinen et al., 2011). Further, sleep deprivation of only one night increased ghrelin (feed me hormone) levels (Schmid et al., 2008). Schmid et al. cited other studies in which leptin (don't feed me) levels decreased with longer periods of sleep deprivation.

An excellent resource is the Harvard T.H. Chan School of Public Health (2020), which is likely kept up to date as current research comes out. In the section prevention of obesity sleep was discussed. The section included the information as above and the importance of getting enough sleep to keep appetite at bay. People who sleep only five hours per night were much more likely to eat—they had more time to eat for one thing and they were more likely to eat out, have irregular eating patterns, and more snacking.

> Perhaps they were confusing sleep deprivation for hunger. (Might be a good topic for research.) New studies need to be done to understand the mechanisms of hunger and insomnia, especially PA studies for individuals.

People should be self-interested enough to take charge of their weight and live a lifestyle that supports good health. However, achieving weight loss and avoiding weight gain is more complicated than patients deciding to take charge. It is likely that most large persons think about weight every day and want to be thinner. Patients and physicians need to stay abreast of the newest research in obesity and researchers truly need to find answers. Education about the root causes of obesity could be quite helpful in avoiding it for those who are not yet obese, such as children. On the other hand, parents worrying constantly about their child's weight can cause the child to become obsessed with food and eating and may contribute to bulimia and anorexia. It is a hard issue and a very individual issue—what works for one person might not for another. For this reason, PA studies would be helpful.

If we could conquer obesity in our country, our medical costs would no doubt decrease. Researchers tend to view obesity as any other disease from population statistics. Doctors traditionally look at patients' symptoms and the collection of those symptoms gives rise to diagnoses. By considering population statistics, physicians tend toward eliminating or reducing the symptoms without regard for the individual characteristics of the patient (Haiech and Kilhoffer, 2012). By looking at groups, the individual patient does not occupy the central position—hence, again, there is need for PAs that predict to the individual. Patients taking charge of their own care can help physicians focus on the individual but PA algorithms would be useful in determining what would work best for individuals.

Obesity, cholesterol, statins, and patient-directed healthcare

As stated above, obesity is related to quite a few diseases, including the tendency to have high cholesterol (Bidgood and Buckroyd, 2005; Yang and Nichols, 2011; NIH, 2021). We have heard for a few years now that cholesterol was not the cause of heart attacks and that statins might not be good for people. Statins are among the most prescribed medications there are, and tens of millions of people take them in the United States, often obese people. Statins have been the "gold standard" for decreasing cholesterol. There is a lot of money going to the drug companies for that medication alone and at least a trillion dollars in sales since the inception of statins (Kendrick, 2020). In 2004, the US National Cholesterol Education Program lowered the threshold as to what was considered "high" cholesterol. At that time, according to Demasi (2018), eight of the nine committee members making the recommendation had direct ties to pharmaceutical companies selling the statins. Of course, with the new threshold, more statins were prescribed.

Kendrick (2020) discussed the research on statins and the actual mechanisms that reduce cardiovascular disease, quoting a paper by DuBroff et al. (2020), "Considering that dozens of [randomized controlled trials] of LDL-cholesterol reduction have failed to demonstrate a consistent benefit, we should question the validity of this theory" (par. 19). Bowden and Sinatra (2012) and Gundry (2021) also questioned the link between cholesterol, as traditionally measured, and cardiovascular risk. As traditionally measured, cholesterol is generally reported as overall measure, high density lipoprotein, and low density lipoprotein. They believed that all the particles should be measured and reported, and they believed it is only the small, dense particles that contribute to heart disease, along with inflammatory markers.

> An excellent PA study would be to measure all 13 or 14 particles and then develop a predictive algorithm for individuals and risk for heart disease.

Until such predictive studies are done and considering that statins can have negative effects on muscles, energy levels, kidneys, and so on, sometimes leading to death, perhaps the empowered patient should do a good search on the research and may decide for themselves whether to take the drugs any longer but use something like aspirin to thin the blood (as thinning the blood may be the major reason that statins have any effect at all).

The need for N of 1 studies

Sometimes, patients can better direct their own healthcare and patients need to wake up to the fact that they really need to take more responsibility for ensuring their own health. There are times when people absolutely need to rely on the medical community—times that they will need emergency care, operations, and treatments such as chemotherapy and life-saving medications. It's not that doctors and other medical personnel are not needed, they always will be. But patients absolutely need to have a role in their own health. They need to eat healthy foods, exercise, read up on all medications given to them by doctors, search out treatments when they have diseases and so on. Sitting back and having doctors "fix them" will not be as fruitful as participating and working at it in addition to the input of their doctors. Patients could participate in PAs by independently keeping track of their own data, or if there were predictive algorithms on portals, patients could input their data and generate their own prediction (called N of 1 studies).

Haiech and Kilhoffer (2012) offered a new paradigm for education and personalized medicine: In this new paradigm the patient, rather than the symptoms, becomes the center of the physician's thoughts. Patient centeredness can result in a complex integration of elements from the ecosystem, such as viral attack, xenobiotics, and individual genetic characteristics. (See Chapter 6.) PA studies allow finely tuned, homogeneous stratification of the patient populations in which the elements aid in the customization of treatment (and in some instances, the patients could input data on portals to generate those customized treatments). The process truly *personalizes medicine* (Haiech and Kilhoffer, 2012, p. 299). Louca (2012) agreed that the potential of personalized medicine requires researchers to find new methods of data analysis given the huge data streams of genomic data and the need for translating the findings into medical practice (p. 211). Truly personalized medicine creates more knowledge for and about the individual patient, but one should not forget about patient-directed medicine, in that he or she can make quite a contribution with N of 1 self-study (see below) as opposed to the medical community conducting N of 1 studies. Both have advantages. Truly, with self-knowledge, and gathering data on self, as in N of 1 studies, the patient becomes more involved in care.

Cerrato and Halamka's (2020, pp 134−135) HIMSS book agreed that N of 1 studies were in the future of personalized care of patients. Extending to patient-directed N of 1 makes a lot of sense.

> A good idea! The authors suggested N of 1 studies could be good vital in clinical decisioning, and could be aggregated for conducting randomized trials. Kravitz et al. (2008) outlined how such "N of 1 aggregated" trials could be conducted using a random, double-blind design.

Neal and Kerckhoffs (2010) presented workflow for Patient Specific Modeling used in tailoring treatments and optimizing for individual patients. Their blood flow models were 3D and patient specific. Imaging from magnetic resonance imaging, computed tomography, ultrasound, and image processing were used to generate models of computational fluid dynamics. According to Kravitz et al. (2008), by using their modeling methods and advanced PAs other researchers have been able to predict aneurism ruptures (pp. 112−113) for individuals, such as abdominal aortic aneurysms and cerebral aneurysms.

There are fundamentally two parts of predictive medicine comprising the part that *predicts what subgroup* the patient will fall into, and the *individualized characteristics* of the patient that help to fine-tune treatment modalities. Involving the patient in collecting and analyzing his or her own data can accomplish the fine-tuning and becomes, as Haiech and Kilhoffer said, "truly personalized." Both the patient and the physician learn the patient.

N of 1 study examples

Weight Loss Examples. In the first edition of this book, tutorials E and F centered on weight management demonstrating the use of PA methods are now found on the Companion Web Page. Tutorial E examines data from many patients, while tutorial F1 and 2 comprise N of 1 examples, which allow an individual to self-educate in a way that was quite new to the field. This kind of study is what was meant concerning patients using a portal themselves to generate predictions—given the predictive algorithm on the portal.

- E. Obesity—Group: predicting medicine and conditions that achieved the greatest weight loss in a group of obese/morbidly obese patients.
- F1. Obesity—Individual: predicting best treatment for an individual from portal data at a clinic.
- F2. Obesity—Individual: automatic binning of continuous variables and weight of evidence to produce a better model than the "hand binned" stepwise regression model of tutorial F1

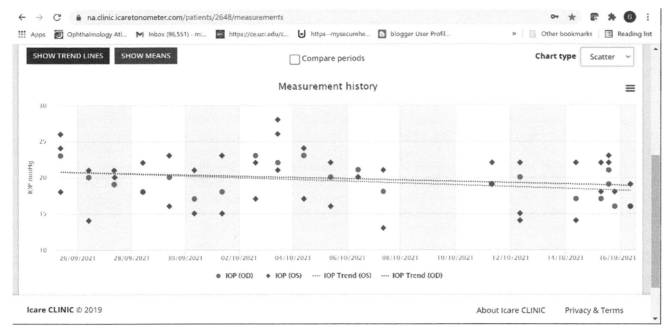

FIGURE 7.12 Gary Miner—iCare scatter plot with trend lines—example of N of 1 research.

Glaucoma Example (Chapter 11): As described above, author Dr. Gary Miner, was able to track his IOP at home for his glaucoma and by producing graphs, was able to work with his doctors to enhance the operation, called the Ahmed procedure, that he had on his left eye. Fig. 7.12 shows another such graph. A video has been uploaded onto YouTube of Gary learning to use the machine, "Gary Trains on the iCare Machine" (iCare, 2022). The URL is https://youtu.be/6nakad5Y3HY (Right-Brain, Ink, 2022).

Chapter 11 explains all that happened and emphasizes that he was able to collect the data because he had an in-home instrument. This instrument allowed him to obtain his IOP several times a day and send high readings (over 20) to his doctors.

With such data, Gary analyzed the conditions which produced the lowest IOP (holding IOP as the dependent variable and the readings from different conditions as his independent, predictor variables).

Pravettoni and Gorini (2011) warned that in our efforts at individualizing predictions in genomic research, we should be careful not to leave out the psychocognitive aspects of the patients. Patients need to be empowered in the process, the knowledge from which comes the power to transform their lives. Again, one of the differences between personalized medicine and patient-directed medicine is the point of view. In the former, the direction is from the doctor to the patient and the latter is directed by the patient. Both are important.

Data scientists could make a fortune—development of apps and artificial intelligence for phones and PC application

Possible PA Research: Some smart data scientists who also know IT, or who may partner with IT persons, are going to develop N of 1 type apps for smartphones and make a lot of money helping people develop their own prediction model for best weight loss or for other self-monitoring activities.

Besides data scientists developing apps for various ailments, patients can educate themselves concerning triggers and behaviors and PAs can help the process in an N of 1 type personal study without the use of formal PA. People find it difficult to lose weight and maintain weight loss. Likely there are methods for weight loss that work best in populations. Individuals working with their physicians might benefit from PA by gathering repeated data on their own individual patterns because often what works for a group simply will not work for an individual struggling on one's own.

Physicians and patients may collaborate through patient portals. The portals would collect the data available to both. Data from many individuals could be combined through those same portals, moving the emphases from individual N of

1 studies to a wider PA study. Again, data scientists could make a fortune developing apps for portals. Or smart patients could develop their own and even sell their algorithms over YouTube. They could test various methods and keep their own data, testing advertised weight loss programs, or self-programs such as intermittent fasting (Fung, 2020—this is only one of many YouTube videos on intermittent fasting emphasizing the many health benefits of fasting. These benefits include reducing blood pressure, correcting insulin resistance and even Type II diabetes, and reducing cardiovascular disease.) Berry (2021) is another excellent resource for learning about intermittent fasting, which is becoming quite popular and is free.

Again, PAs could be taught to individual patients through portals as N of 1 studies. The process of keeping track of one's data (blood pressure, eye pressure, as in Dr. Gary Miner's case, weight, EKG readings, depression level, or any other measurable outcome) might be too much for some. Other patients would really enjoy and learn from the experience. Patient portals could also be equipped with short courses that patients could take to help them understand studies that they read and to understand the information the physician supplied. Ideally, patients would learn some basic research methods and physicians would communicate transparently the numbers involved. Researchers could then assess the effectiveness of such efforts.

Patient portals

The use of patient portals is a good way of communicating records and messages from doctors, labs, and so on to the patient and to help demonstrate compliance to standards. Portals should also be considered as a means of helping patients to self-educate, as mentioned above. Much more now than at first writing, clinics and hospitals are using patient portals. The use and quality of portals varies widely. It is common to find portals that are used for little more than setting appointments. Others allow for limited dialog between patients and their doctors. Ideally, portals would allow patients to add their own data and information that were used both by patients and clinicians. Patient portals can document medical outcomes by treatment and would make for excellent predictive research. There is much hope that portals will also empower patients to be more informed concerning their own individual care (Ammenwerth et al., 2012). Individual care could store data for patients as they conduct their own N of 1 studies.

Lofty ideals of increased communication and using patient portals for patient education are often met with the reality of time constraints and lack of personnel and energy. At the very least, patients should be able to see their information online. They can better direct their own healthcare with more information. They could access test results, doctors' notes, and education concerning any condition. There might also be links to specific diseases and online research articles. However, there are practices in which the medical community barely has the time to fill the patient portals with anything but the next appointment and perhaps the lab results of blood tests. The dream meets reality, but improvements are being made. Ultimately, it is possible that the time spent setting up and maintaining the portals could save time in future interactions, particularly if there were a way for patients to add their own data, or to collect data from patients' at-home devices. With Gary's ICare machine, when he entered data at home, values if over a predetermined level were automatically emailed to the doctor. In that way, the physician was aware of increased pressure, along with the patient, but was not bothered by too much data coming to him.

One very interesting study used PAs to find resources for burn patients in the Netherlands. Doupi and van der Lei (2002) explored importing information from patient profiles via EMR systems and then linking them to relevant HTML documents. This study was written over 19 years ago. By now, healthcare IT specialists should be much more able to achieve such results.

This author personally found in one of her portals and in the doctor's notes, information on her operation using canned paragraphs that could be copied and pasted into the portal, which was great, but also were not checked to see if they matched what happened. For example, the copied narration said there were staples used to close the wound but glue was used, which then made her wonder if there were other things that did not match. Nonetheless, the operation was a success. And information was better than no information.

No doubt there is a variety in how portals are utilized by both patients and medical personnel. Beth Israel, in setting up its patient portals, invited the doctors (PCPs) to participate (Walker et al., 2011). Those PCPs who participated were able to withhold a note if they felt a note would harm a patient. But for the most part, the notes were shared by participating PCPs. Patients were enthusiastic about seeing their doctor's notes. Some physicians were enthusiastic, and some were not. Most (61%−81%) of the participating physicians thought the open notes were a good idea but only 16%−33% of nonparticipating physicians from the three sites thought patient notes were a good idea.

Dendere et al. (2019) reviewed the literature concerning portals. They noted the literature they examined attested to the worth of portals in a variety of ways. The authors listed benefits, problems, and patient preferences concerning patient portals. A variety of functions were used by portals with patients often wanting to be able to communicate with healthcare professionals. Although the authors stated that portals could increase adherence to medication regimens, they found some studies reported increases in good outcomes and other studies did not. It is likely that the variety of uses for portals and the differences in what was put into the portals in terms of time and effort accounted for those differences in outcomes. This would be an area worthy of investigation.

Patient education is one area in which portals can be used, as well as communication between healthcare providers and patients. Portals represent a huge potential for communication, for patient education, and for patient empowerment. If physicians thought that patients would not understand the patients' own medical information, there could also be links to statistical information and patients could educate themselves as to terminology and research methods. One study recommended that statistics be taught to all children so that when children became adults, all patients could understand concepts such as absolute risks, mortality rates, and natural frequencies (Gigerenzer et al., 2007). Mini courses comprising concepts such as those could be incorporated within patient portals so that patients could educate themselves. One area of information could be definitions, advantages, and disadvantages of various kinds of research so patients would understand more of what they read in the literature and might be better able to evaluate the worth of articles. They might also be better able to understand clinical trials if that information were provided.

> Possible PA Research: PA might predict patient outcomes based on physician participation in sharing notes within patient portals, or in setting up mini courses for patients. Effects of patient education on outcomes could also be researched, given the kinds of portals patients were encountering. In addition, patients could conduct their own N of 1 studies and researchers could set up standards for those patients so that the studies could be combined into larger studies. This could be done easier using specialists' practices in which the range of maladies is reduced.

Alternatives and new models

Medical tourism

At first writing, medical tourism was on the rise and insurance companies were willing to pay for operations and treatments for patients who were willing to travel to other countries where costs were not as great (Gibson and Singh, 2010). The pandemic made a dent in medical tourism but did not wipe it out.

Companion Global Healthcare is a company connected at arm's length to Blue Cross and Blue Shield. Blue Cross of South Carolina (2013) invited its members to consider Companion Global Healthcare.

From the 2013 Global Healthcare website, enticing its members to consider international medicine:

"Members traveling abroad will receive:

- Surgical services
- Travel arrangements, including flights and hotels
- VIP transfer services from airports to hospitals
- Passports and visas (if necessary)
- Coordination of medical services
- While abroad, members will also receive: Hospitality services unavailable in many US hotels

Choice of different room types—from private to deluxe accommodations International staff with on-site interpreter services for multiple languages Significant cost savings" (Blue Cross of South Carolina, 2013).

At Companionglobalhealthcare.com, one found the header as shown in Fig. 7.13.

Note the "hip replacement" tab. Or how about this—if people could not afford to have their teeth whitened in the United States, they could fly to India to have it done while they got their root canals and crowns as well. Prices generally included the airfare and the accommodations. The only thing patients may have missed were family and friends by their side.

Companies, not simply insurance companies were sending employees abroad for operations—to the tune of about one and a half million Americans a year (Evans, 2020). The reason was the companies were saving tons of money by doing so. One story recounts a knee operation in Costa Rica in which the patient not only did not pay a penny, but instead, got a bonus for agreeing to the operation out of the country. The company saved about half the cost or nearly $20,000 on the deal.

The 2020 pandemic certainly changed medical tourism. Many people elected not to have nonessential surgeries and lockdowns changed just about everything. Incredibly, the medical tourism market in 2020 exceeded $10 billion.

FIGURE 7.13 Come travel with us! (Blue Cross of South Carolina, 2013).

The Medical Tourism Index (2020) provides information on medical tourism for those interested. The five top destinations from the index in 2020 to 2021 were Canada, Singapore, Japan, Spain, and the United Kingdom. And according to Global Market Insights (2021), as the pandemic winds down, the market is expected to rise to over $37 billion.

Where could it go wrong?

Forced human organ harvesting was reported to have been big business in China as they had set a goal of being the largest transplant country by 2020. As the world became aware of the horrid practice, calls for reform were made. However, by 2019, the practice was reported to have continued regardless of the statements of reform from China (PR Newswire, 2017). At the time of this writing, though, as China denied harvesting, there were continued allegations of enslavement and mistreatment of over a million Uyghurs. Forced organ harvesting was not out of the question (BBC News, 2021).

Certainly, as recently as 2019, it was purported that there was continuing forced organ harvesting in China, particularly among the Falun Gong believers, and Uyghur Muslims, at an international genocide conference, which included eyewitnesses (PR Newswire, 2019; Health Europa Quarterly (HEQ), 2020).

According to Wendy Rogers, Professor in Clinical Ethics, forced organ harvesting was still going on in 2020, and in fact had accelerated. Fong Gong and Uyghur practitioners were a major focus, though she also heard of Christians and other groups, "considered enemies of the state," (HEQ, 2020, par. 19), being captured for later harvesting. Given such a rationalization for murder, it's also big business and a major source of income. In addition, academic papers were being published which said the organs were donated by "volunteers" but those were reportedly lies (HEQ, 2020, par. 24).

It is hoped that the reports are not true. However, the empowered patient cannot be naïve, in our world today.

Alternative screenings

One company touts noninvasive diagnostic screening tests (Life Line Screening, 2013). Set up in churches generally, asymptomatic "patients" are queued one at a time to receive the four or five tests that are given. They are later alerted to any abnormalities; information they may take to their doctors.

One family doctor in Washington D.C. (Common Sense, 2011) stated that the Life Line Screening tests are unproven. It is conceivable that this kind of searching out problems could cause more harm than good in that given false positives, people could end up with unnecessary treatment.

On the other hand, Rasmussen et al. (2007) found that in primary care appropriate screening was useful in prolonging life without any increase in total costs. Their randomized study followed patients for 6 years. It might be good for patients to suggest to their doctors that they be screened for cardiovascular disease. One of the screens in the Rasmussen study was BMI and the physicians scheduled regular follow-up appointments helping patients to improve their cardiovascular risk odds. Follow-ups with patients might be a good idea for PCPs to do in general, given that obesity is so common. Often, after telling a patient that he or she is obese, there is no follow-up. Providing information on methods that help, such as intermittent fasting, or conducting weekly weigh-ins might be helpful. It was thought at first writing that the situation might change, because obesity was being listed as a disease with the possibility of more attention from the medical community. Evidently, that thought was not to be, with obesity ever increasing. There was no research yet to back this thought up, but it seemed to me in talking with people and observations on social media, that the first year of the pandemic served to fatten people but then the second year became a time for some to take charge and slim down. One can hope.

Self-diagnostic kits

In the future, private companies might spring up to provide low-cost diagnostic alternatives to patients at the patients' requests. Patients who are concerned or curious could find out for themselves. People already are using drugstore screenings for pregnancy, blood sugar levels, HIV, urinary tract infections, high blood pressure, so why not for strep, ultrasounds (well, that might be a bit difficult!), COVID, which in 2021, finally had inexpensive at-home test kits that were sensitive enough to tell if someone was positive and contagious. With self-diagnostic kits, patients could take care of routine problems and leave the more difficult diseases to the physicians. However, the quality of kits needs to be scientifically evaluated. Patients need to research, to carefully evaluate the quality of home tests, and if there is any question, visit their physician for checking the results.

Patient-directed does not mean the patient is the doctor. The adage about having a fool for a doctor comes to mind. Patient-directed means patients enter a partnership with the medical community, always trying to improve their health by researching the best ways to take care of themselves, adopting a healthy lifestyle, by monitoring themselves, and by consulting with the medical community.

Here is an example from a patient who had degenerative disk disease (DDD), and so said the patient's orthopedic surgeon. (Note, patient consulted with a doctor, before launching out on patient's own.) The doctor had nothing to offer, however. So, the patient figured it was up to self to figure out if something could be done. A common condition of getting old, at times the patient would "throw their back out" and be rather incapacitated. When bad, the patient wondered if there should be an operation. Instead, the patient began researching, looking up research, and looking at YouTube. There, the patient found Bob and Brad, "the world's most famous physical therapists on the internet (in their own opinion, of course)." Naturally, there are many fine physical therapists on the internet with remedies and moves to alleviate pain. But they happened onto Bob and Brad. They had many videos on DDD, and since using their techniques for sitting, standing, and walking, the patient's back has not "gone out." Bob and Brad have over 3000 videos, with over 390 million views and 4.28 million subscribers. Here is a link to their website: https://www.youtube.com/c/BobBrad. This example is not telling you what to do—you should always check with your doctor before launching on an unknown. The point is, we can also be working on ourselves. We don't always have to run to a doctor. If we are to take care of ourselves, it is incumbent on us to make sure we stay up with recommended shots, screenings, and visits to our PCPs. We just don't need to run to the doctor for everything. We can lose weight on our own, we can exercise on our own if our PCP says it is important. We can eat what we know is healthy or look it up if we don't have a clue. We can stop behaviors that we know are unhealthy, such as smoking or drinking. We can direct, for a large part, our own healthcare.

Possible research—These self-tests, again, could be quite useful in N of 1 studies. Researchers could aggregate the like studies into larger predictive studies with outcome as the dependent variable. One other area of study could be the reliability and accuracy of the various self-tests. Inaccurate results could easily lead a patient astray. Building a list of accurate self-tests would be an invaluable service to the public at large. Someone could write a book concerning interpretation of various tests or sell to practices educational materials for patient portals.

An alternative to traditional insurance

If patients are to take more initiative in self-care, they will have to start thinking more like consumers and health insurance may need to change accordingly. Often people with insurance perceive that if someone else is paying the bills, they really don't care how much procedures cost. If they are paying themselves, then they have a different attitude. Consumer-driven health plans (CDHP) may be an answer (Goldhill, 2013). These plans would cover catastrophic events, would have high deductibles for day-to-day medical care, but also would have lower premiums. Those low premiums could be coupled with depositing the difference into their health savings accounts (HSA), perhaps saving up to $8000 per year in insurance costs. A good site for an HSA reference is Publication 969 from the Internal Revenue Service (IRS, 2020), which details the regulations and tax-exempt opportunities. Goldhill also pointed out that low premiums from CDHP could allow more people to afford insurance and they would pay for themselves rather than increasing the burden on government funds. Generally, the lower income people tend to be younger and just starting out. They also tend to be healthier and would not need to pay deductibles since they would not be accessing service as much. Goldhill did not talk about children, however. Health expenses for children are high, and perhaps plans should consider covering children regardless. Subscription prices tend to be low for a family with children—anywhere from $100 to $200 per month for a family of four, contrasted with thousands per month for regular insurance. Recent increases in child care tax credits for families surely will help. Families could get up to $3600 per year (IRS, 2021). The healthcare consumer would no doubt be interested in shopping around to find less expensive procedures and care if contributing to an HSA. That cost comparison shopping should eventually reduce all costs.

Doctors striking out on their own

If more physicians opt for the no insurance subscription, Direct Pay/Direct Care model, also called direct primary care (DPC) for practicing medicine, then market supply and demand could keep prices down. Doctors could decide how many patients they wanted to take in, giving them more control over their stress levels, plus they would not have to deal with multiple insurance companies.

As previously stated, the Direct Pay method allows doctors to stop working with insurance companies. In 2012, 9539 physicians had opted out of Medicare (Beck, 2013). Subscription-based PCP doctors are increasing in numbers and there is even help to patients seeking such a physician/practice on the Internet (AAPS, 2020). In 2021, The Direct Primary Care Coalition (TDPCC, 2021) reported "Today, about 1,200 DPC practices in 48 states provide peerless access to great primary care to over 300,000 American patients" (par. 1).

According to Beck (2013), doctors who no longer take insurance will rather depend on subscriptions to their practices, perhaps charging what Medicare would pay them. Some have begun making house calls now that they have extra time from filling out all the required paperwork. Patients who opt for a CDHP plan as above would surely appreciate a concierge-type personal doctor. They would not have to rely on insurance either, except for catastrophic coverage. ACA did demand everyone have insurance but quite a few young people found it less expensive to take a penalty rather than to sign up (Weaver and Radnofsky, 2013). But then, with the Trump administration, the individual mandate was eliminated, making more room for a Direct Pay model of healthcare payment. As of this writing, the Biden administration had not achieved any significant changes or improvements. However, there is still time.

Interesting research might concern the well-being of physicians and other medical personnel between the traditional insurance system or the subscription system. Doctors could decide how many patients they would like to take on. If they wanted smaller workloads, they could decide on that. They could also decide how long they would like to talk with individuals without having a patient quota to maintain. Perhaps they would feel more in control and less stressed. Their staff would not have to deal with a multitude of insurance companies, each with its own rules. Life might be better for all. It would be interesting to find out if this were all true. Or would doctors feel more pressure? Perhaps a researcher could predict to the individual physician.

Other possible PA Research: Using good/bad outcomes as the dependent variable, predictor variables could be amount of collaboration between patients and medical community, whether traditional insurance was used or using the Direct Pay method, amount of research patients engage in, use of private screening "roundups," successful attempts at weight loss, and variables such as dietary choices and sleep patterns.

Alternative ways of knowing about ourselves—genomic predictions

As Chapter 10 tells us, gene research has and will become increasingly important for predictions in all diseases.

The future is now, given PA techniques, as researchers are cataloging genes, gathering warehouses of data, and making just such predictions as were imagined many years ago. (Refer to Chapter 10 for more information on genetic predictions.) Aspinall (2009, p. 527) provided perspective:

One hundred years ago, if a patient came in with lethargy, bruising, and night sweats, the best that the physician could say is you have a disease of the blood. Twenty years later we understood leukemia and lymphoma were different. Twenty years after that, we knew chronic, acute, indolent, and aggressive. What happened next created the initial work on personalized medicine—we can now quantify 90 or perhaps as many as 150 different genotypes for leukemia and lymphoma. What are the other areas that we might fruitfully participate?

Angelina Jolie made her decision concerning a mastectomy, because of her genetic propensity. Of course, we all have individual genetic endowments and some of those genes supply us with likelihoods of diseases. Even given those endowments, to a large extent, patients control their outcomes. They may know that they have a family history of heart disease. If so, they could control their diets and exercise to ameliorate the genetic potential. Equally negative decisions also serve to influence their fates. Patients could exert more precise control over their destinies by making proper modifications if their genes were mapped and if they knew what probabilities accompanied their futures. They would need support of data and information.

It would be wonderful if researchers could build a cooperative national database. Efforts at building a cooperative database and cooperative research efforts have already begun (Kawamoto et al., 2009; Lerena et al., 2007). The Clinical Genomic Database (CGD, 2021) is one such cooperative database. Predictive research could follow such databases.

Some concerns

Complete gene mapping is expensive, but if there is a specific test ordered by a physician, generally insurance will pay for it. Personalized medicine can potentially save money by reducing the number of drugs that a patient needs (Aspinall, 2009). In addition, costs for gene testing have declined over time, as we read in Chapter 9, and likely will become much more affordable in the future. As we learned in Chapter 6, genomic research holds great promise for determining possible outcomes. Saturday, August 24, 2013, Gary and I were privileged to help host the family reunion following the Tulsa Alzheimer's walk. This family was one of the families we tested 30-some years prior (Miner et al., 1989) and children of some of the members apparently had about a 95% chance of inheriting their family's Alzheimer gene. Some of the discussion revolved around getting life and long-term care insurance before taking the gene test. After the gene test, they knew that they would not be able to purchase insurance. Many issues are involved in gene testing, and one is the issue of who would be privy to the information.

Barring a genetic map, people can still examine family trees and note what diseases everyone had. By collecting their own data from their families, they can draw tentative conclusions about what fates await them—unless they consider lifestyle changes for some diseases. Information on genetics could be another item of education posted in patient portals. People are already swabbing their cheeks and sending their DNA off to be analyzed for paternity, to determine if they are "part Native American," or where their ancestors came from, why not to get their partial or full gene maps? In addition to having information in portals, patients could get a gene map app for their phones to analyze their genome and predict future diseases in the privacy of their own homes or at a restaurant with friends. As mentioned in Chapter 1, there are some potential problems with knowing one's genome. The Wall Street Journal called the problem, "patients living in limbo" (Marcus, 2013). If one learns he or she has a chance of acquiring a disorder or disease, what kind of chance is it? Are there percentages that are known? Is the test able to give a time frame? Are there environmental activities that can prevent the expression of the gene? PAs has the potential of moving people out of "Limbo Land."

We know, given the genetic data on individuals, that PA research can go forward. The question is, do people want to have tests done by sharing their genetic information? Lord knows, all our other information is out there on various public sites. Do we also want to include our genetic data?

Here are some types of genetic research that can be done and some benefits: Personalized medicine using genomic information can streamline and personalize treatments, potentially saving money. The studies below emphasize personalization.

Diagnosing tumor types for individuals: *Baek et al. (2008) used microarray technology to analyze best practices for individuals, to diagnose accurately, and specifically for tumor types. They used predictive analytic methods such as boosted trees, feature selection, including cross-validation methods. Although it seemed they were not satisfied with the accuracy of the algorithms developed, they did feel hopeful that these methods could be trained to increase accuracy and have "good potential for selecting gene sets" that could potentially benefit in prescribing specific and targeted treatments for individuals.*

Using organizational charting for single nucleotide polymorphism (SNPs) data: Holger et al. (2008) suggested methods for organizing huge amounts of genetic data used in genetic personalized medicine. They used diabetes as a model and presented organizational charts. They also mentioned a very interesting website for formulating and organizing data, Galaxy (2013). An excellent short basic "101" tutorial among other tutorials was given at the Galaxy site (Galaxy, 2013). The site is free, and one can set up an account there to develop data sets. The tutorial demonstrates determining within one of the free data sets the highest number of SNPs on a particular nucleotide sequence (or exon). This was done by first generating a SNPs database, then an exons database, merging the two, and then using an algorithmic "tool" to count the number of SNPs on the exon.

On the other hand, there are some downsides of genetic testing. One is the possibility of creating expensive orphan drugs that would respond to the individual needs. Friedman (2012) raised the thought that personalized medicine might increase the prices of medications by essentially creating those "orphan" drugs, which are drugs needed by only a very small number of individuals. Before thinking that pharmaceutical companies are simply greedy, think that it is incredibly expensive to develop drugs. Orphan drugs are created by some companies for rare diseases. PAs could determine subgroups that respond to the orphan niches thus increasing the numbers that could use the orphans. Friedman feared either "mini monopolies" of pharmaceutical companies would have the ability to command a high price, or on the other side, and equally feared, governments would regulate prices. Increasing the subgroups using PAs could alleviate this problem as well.

Psychiatry and genetic testing are also being combined to construct predictive modeling (Mitchell et al., 2010). Part of the discussion of the authors was the possibility that genetic modeling concerning psychiatric diseases might increase the stigma of mental disorders.

Many issues of privacy and dealing with insurance companies certainly would follow predictive research related to genetics (Hoffman, 2019). Hoffman discussed the legal and ethical considerations of predicting health using artificial intelligence, genetics, and other predictive methods in her North Carolina Law Review article.

A good source of information concerning genomic research is the NIH National Human Genome Research Institute (NHGRI, 2020), mentioned above for the genomics database. There one can many other resources related to genetic research to see if any of the databases would be good to use for PA.

Possible PA Research: Patients and their doctors could access sites such as NHGRI, and the Galaxy site in developing their own personalized studies (https://usegalaxy.org/).

Predictive analytics for patient decision-making

In 2011 then IHI second-year resident, Dr. Diane Watson, researched ankle injuries and using PAs found that two questions asked instead of six of the guidelines would provide more predictability of whether someone needed an X-ray for an ankle injury (Watson et al., 2011) than would the answers to all six. Two questions were faster than six and the interpretation was smoother with a better chance of being accurate. As mentioned above, Alekseiv and Harris (2013), also with IHI, provided their tutorial on a prediction study of who needed to be hospitalized after entering the ER with chest pain. With 100% sensitivity and 100% specificity by combining two predictive models, their study had the potential of saving up to 75% of the hospital costs (see Tutorial I, first edition of book). Studies such as these, help doctors make decisions for patients to personalize medicine. Patients also need to make decisions for themselves.

However, predicting individual outcomes also supports patient decisions. As expected, researchers are finding ways to predict to individuals in genetics (Baek et al., 2008; Blokzijl et al., 2010; Colombet et al., 2004; Galas and Hood, 2009). Patients need to be included in the process and help to make decisions for themselves.

Evers et al. (2012) emphasized that interdisciplinary and "translational" views of medicine are what is needed. They purported the importance of participatory healthcare in which centers involved patients. Grzymski et al. (2020) used population screening in the Healthy Nevada Project and found people who were carriers of three genetic diseases, hereditary breast and ovarian cancer, Lynch syndrome, and familial hypercholesterolemia, who otherwise may not have known it if they hadn't gone for screening. Self-monitoring and going for exams after such information would be a responsibility of patients.

Concerning patient involvement in PAs, as explained earlier, fine-tuning responses is possible when an individual tracks his or her own data over time. Self-tracking is what Dr. Daniel Kraft (Gagliani, 2023) called the "quantified self," or N of 1 studies, as defined above. This might have been the first time the term "N of 1 studies" was coined. Kraft said that for patients to be a part of collecting their own data using apps of various kinds, physicians need to be incentivized to using such data. At first writing, most physicians were not compensated for answering emails or conducting Skype sessions. However, after the pandemic struck, many physicians offered online sessions instead of having patients go into the clinics, and payments were determined. Perhaps physicians could be incentivized with some form of monetary compensation.

Physician incentives for getting patients to add data to their portals would ensure that patients' data would be used thus, providing for more patient involvement. Communication between patients and physicians exists when patients are included in such tasks and with communication, increased chances for knowledge. Increased communication and involvement may also increase the effectiveness of treatments. Getting our smart medical devices, such as smart watches, patient EKG machines, and so on, to connect to our portals could ensure automatic data entry.

Connectivity

Social media has a potential for increasing patient control. Data miners can use Twitter, Facebook, and Google searches to predict outbreaks of illnesses such as flu and colds. Cooperation among the social giants allow targeted contact monitoring. However, patients can also connect to others through social media, finding those who have their symptoms to see what they have found out and how they have been treated. Support groups form spontaneously on the Internet for every disorder imaginable. Although physicians may dread our lists, patients can also fairly well rule out certain disorders even before seeking medical help. There are other sources for connecting patients more directly with their care and giving them more opportunities to have a say in their care. More and more hospitals and clinics are introducing computerized portals for patients to view and make additions.

Controlling some diseases by searching research on one's own

When patients further research what they hear and read on the Internet, they may find ways of combatting diseases without even needing doctors. And again, this is *not* to say that people should skip checking with their doctors, but only that they might find new methods of controlling symptoms such as eliminating statins and gaining energy. Some physicians, like Dr. Roger Seheult (Seheult, 2020a, # 69) on the YouTube MedCam offer free courses for anyone who wants to learn. It was there that it can be learned, for example, that N-Acetylcysteine (NAC) and Quercetin can reduce inflammation, particularly in the epithelial cells in the lungs and could be good medications to take if one contracted COVID.

The supplements, taken prophylactically reduced one patient's asthma symptoms by reducing the eosinophil count, virtually eliminating constant nasal gland infections. The patient felt better and virtually eliminated the need for taking antibiotics, which was always worrisome as the more antibiotics are used, the less effective they may become. The patient did not need to take an expensive medication to reduce eosinophils, as had been recommended by one allergist. In addition, the patient obtained relief from the patient's hand-bruising that was happening so often if the patient even gently hit a hand on something. Perhaps the cell walls of the blood vessels were strengthened. The cell walls could have been compromised should the patient had acquired COVID with its cytosine storms. For an interesting exposition of leukocyte endothelial cell adhesion, please look at Granger and Senchenkova (2010) in NIH books.

In addition, this patient had experienced high eosinophils in the blood, which were thought to contribute to allergy-induced asthma, and which were reduced to normal after being on the two supplements. At least in mouse models, NAC also reduced the neutrophil and eosinophil counts in the bronchoalveolar lavage fluid (Eftekhari et al., 2013). Oxidative stress is a problem both in COVID and in asthma. NAC, and Quercetin as well, may be helpful for both diseases. Randomized controlled studies should be conducted.

As indicated above, part of the process of inflammation involves leukocytes adhering to endothelium cells in blood vessels (Granger and Senchenkova, 2010). Most people can fight off the COVID virus, but in some people there occurs a hyperinflammatory cascade in which the lungs are injured, and fluids leak out eventually causing death. This inflammatory type of increase is also present with people who have asthma (Rodriguez and Veciana, 2020).

Angiotensin II is involved in the inflammation process as it allows for increases oxidative stress and inflammation. Please see MedCram number 61 on YouTube that explains the cascade role of Ang II (Seheult, 2020c, #61). And then for how NAC can be helpful, please watch MedCram number 69 (Seheult, 2020a, #69).

It is interesting that NAC has been sold as a dietary supplement for decades but as of the summer of 2021, the FDA has asked that it no longer be advertised as a supplement, and Amazon was no longer selling it. One wonders if plans are being made by pharmaceutical companies to produce a medication much like NAC to be sold for higher prices.

NAC could be quite effective in treating many diseases that involve mitochondrial dysfunction, such as fatigue, and chronic diseases such as asthma, COPD, and pulmonary fibrosis (Nicolson, 2014). RCT studies need to be done. Studies that could be done include diseases such as Huntington's disease, 3-NP-induced mitochondrial dysfunctions (Sandhir et al., 2012), neurodegeneration in X-linked adrenoleukodystrophy, and perhaps the prevention of the mitochondria-specific effects of very long chain fatty acids (Zhou et al., 2021). Two interesting podcasts were created by Berk (2016, 2017) who cited Brown et al. (2013) in the 2016 podcast. In the podcast, Professor Berk proposed that NAC could be beneficial pharmacotherapy for psychiatric illnesses such as depression, schizophrenia, and drug addiction (also see Trivedi and Deth, 2015, particularly concerning epigenetic status and gene priming in drug addiction), and OCD. NAC crosses the blood−brain barrier and increases glutathione, plus it repairs mitochondria. This seems like a wonder drug, and we would need to see all the RCT studies on all the claims. However, when Dr. Seheult (2020a, #69) explained the role of NAC in reducing oxidative stress, it certainly seemed reasonable.

Quercetin is another over-the-counter supplement that has been found in mouse models to reduce the cellular action of eosinophils, the action found in allergic rhinitis and asthma (Sakai-Kashiwabara and Asano, 2013). The authors concluded that quercetin might be useful in combatting the effects of eosinophils in allergic asthma. (The same fate as NAC may also befall Quercetin, however.) The above patient took Quercetin and NAC during the pandemic, and in general, to enhance the patient's immune system and reduce possible oxidative stress. It was surprising to find that the measured eosinophil level had reduced to a normal level. The medication that the patient was to take to do the same was around $40,000 per year and the patient's copay would have been about $4000 per year. Because NAC and Quercetin were potentially good candidates for treating several diseases, it seems they are now garnering attention from the pharmaceutical industry and the FDA is deciding that perhaps they should not be sold as supplements (Alliance for Natural Health, 2021). As of this writing, both were available from health food sites. However, NAC was no longer available on Amazon.

Portals, evidence medicine, and gold standards in predictive analytics

At first writing Beth Israel Deaconess Medical Center (BIDMC, 2013) was home to over 300 clinical trials. BIDM is committed to research and in-patient information/education. BIDMC boasts using all the "gold standards" but also boasts of conducting much research. Today, BIDMC has over 2500 active research studies covering 52 areas of research! It actively matches patients with researchers, as a member of the national ResearchMatch group, and BIDMC also conducts clinical trials so one need only go online to try to find one that matches (Beth Israel Deaconess Medical Center, 2020). Many computerized patient portals now provide documents and reports from hospitals and clinicians. The BIDMC portals also provides reminders for patients and clinicians. These reports reminded the physicians of gold standards from evidence-based medicine, which can not only affect mortality but also financials.

Perhaps PA capabilities could be woven into portals for doctors today who understand machine learning. Then instead of the gold standard for populations, they could find best practices for individual patients. Reminders can help the patients take better care of themselves, but particularly if patients had the responsibility for their own personal data. [Please be reminded of the two individual patient obesity tutorials in the first edition—F1 and F2 (pp. 388−445 and 446−461)]. Portals could be a means of incentivizing physicians to interact with patients and to tailor treatments to individual patients given their data. Such a concept could be used by residents fulfilling their research requirements. Portals could be a means for collecting data for analysis—both for forecasting and for PAs.

Data can be gathered and analyzed and sent to governmental agencies that require transparency. BIDMC (2013) not only sent data out but brought them back again, plus additional data, using a structure, process, and outcome framework. The data gathered from HIMSS, the Hospital Consumer Assessment of Healthcare Providers and the Systems survey, were matched to their data by using CMS hospital identifiers (Angst et al., 2012) and additional research was possible. Not only did the hospital use data for research but they included the patients in a very personal sense using its

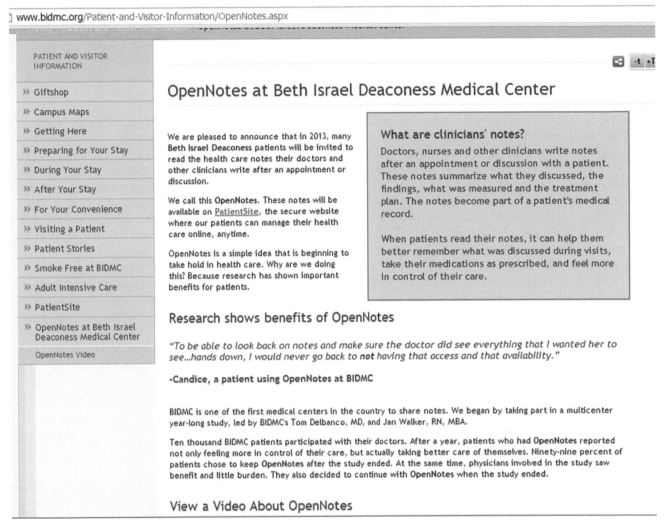

www.bidmc.org/Patient-and-Visitor-Information/OpenNotes.aspx

PATIENT AND VISITOR
INFORMATION

» Giftshop

» Campus Maps

» Getting Here

» Preparing for Your Stay

» During Your Stay

» After Your Stay

» For Your Convenience

» Visiting a Patient

» Patient Stories

» Smoke Free at BIDMC

» Adult Intensive Care

» PatientSite

» OpenNotes at Beth Israel
   Deaconess Medical Center

OpenNotes Video

OpenNotes at Beth Israel Deaconess Medical Center

We are pleased to announce that in 2013, many Beth Israel Deaconess patients will be invited to read the health care notes their doctors and other clinicians write after an appointment or discussion.

We call this OpenNotes. These notes will be available on PatientSite, the secure website where our patients can manage their health care online, anytime.

OpenNotes is a simple idea that is beginning to take hold in health care. Why are we doing this? Because research has shown important benefits for patients.

What are clinicians' notes?

Doctors, nurses and other clinicians write notes after an appointment or discussion with a patient. These notes summarize what they discussed, the findings, what was measured and the treatment plan. The notes become part of a patient's medical record.

When patients read their notes, it can help them better remember what was discussed during visits, take their medications as prescribed, and feel more in control of their care.

Research shows benefits of OpenNotes

*"To be able to look back on notes and make sure the doctor did see everything that I wanted her to see...hands down, I would never go back to **not** having that access and that availability."*

-Candice, a patient using OpenNotes at BIDMC

BIDMC is one of the first medical centers in the country to share notes. We began by taking part in a multicenter year-long study, led by BIDMC's Tom Delbanco, MD, and Jan Walker, RN, MBA.

Ten thousand BIDMC patients participated with their doctors. After a year, patients who had OpenNotes reported not only feeling more in control of their care, but actually taking better care of themselves. Ninety-nine percent of patients chose to keep OpenNotes after the study ended. At the same time, physicians involved in the study saw benefit and little burden. They also decided to continue with OpenNotes when the study ended.

View a Video About OpenNotes

FIGURE 7.14   Beth Israel's OpenNotes on Patientsite (BDIMCb, 2013).

patient portal called "Patientsite." See Fig. 7.14. Surely, we are more sophisticated in portals now in 2020 than in 2012, and so we are only limited by our imagination as to what the medium can provide.

Patientsite at Beth Israel

Patientsite of the Beth Israel Deaconess Medical Center, is an online portal for patients to manage their healthcare. Patients can view their records securely 24 hours a day. They may make amendments to those records and communicate with their doctors and other healthcare personnel. The site was available for PCs, tablets, and smartphones,

The site was to add "open notes" (BDIMCb, 2013) allowing patients to read and comment on whatever notes are shared among medical personnel. As of 2021, Open Notes is an ongoing enterprise. One can find information about Open Notes at the Beth Israel site (BIDMC, 2021; Open Notes, 2021).

Cleveland clinic

Cleveland Clinic is known for innovation (producing the first face transplant surgery, for example). Cleveland Clinic, as well as Beth Israel, went online with its patients (Monegain, 2013). Cleveland started its patient portal, MyChart, in June of 2013 and planned to bring everything in patients' records securely to them by the end of 2013. Patients have access to lab reports, medication lists, summaries of visits, preventive care advice, X-ray reports, physician notes, and so on. One thing provided was a delay in results so that information would first be reported by the physician. In this way, patients would not have to find out they have a serious diagnosis via the Internet.

By 2020, most hospitals and clinics used patient portals, though some are better than others. Portals have the potential of taking some burdens off physicians and staff, though likely not initially.

Body computing

Dr. Leslie Saxon, USC cardiologist, presented a TED talk describing her Body Computing Conferences of 2008 and 2009 (Saxon, 2010). She defined the traditional model of patient/physician interaction or the "nurturing physician" image with these words, "I'm in a room with a patient that's anxious and that's unclothed, and many don't come with their mothers. And if with their mothers their mothers are more nervous than they are. And [I'm] trying to impart some information to them in that very tense environment and then locking that information up within me, which then is very difficult to access. Never letting it [that information] go anywhere or vetting it or letting it get expressed or get further iterated is a big mistake" (transcribed from Ted talk). The 2021 Body Computing Conference stream may be found here: 2021 Virtual Body Computing Conference—USC Center for Body Computing (uscbodycomputing.org).

Saxon mentioned implantable defibrillators, fully networked, which increase sudden heart attack survival rates from 2% to 99%. Saxon has devised several electronic devices that allow a patient to broadcast information. For example, there are body tattoos that broadcast heart rates wirelessly into smartphone apps. Apple has produced a phone/watch that tracks blood pressure, heartbeat, EKGs, number of steps taken per day, and monitors sleep and so on. Saxon clarified that technology that is seemingly cold and impersonal can allow for more compassionate care and better outcomes.

> Possible PA Research: Let patients own their data: (Cohen, 2021). So why not give them the data? If they have the data, why not make individual predictions from those data?

Diagnostic apps

The Wall Street Journal predicted the increased use of online symptom checkers and said that it was good for patients to search out possible diagnoses (Landro, 2013). That was a good prediction as increasing numbers of patients are using various body tools and apps, plus online resources to find possible diagnoses. They can then bring to their physicians their thoughts about diagnoses, and their reasons for ruling out some of them. Physicians, particularly residents, may also rely on various online helps when deciding on diagnostic tests. The lists that the patients bring to their physicians could be helpful in improving diagnoses.

A good article to help with the process of integrating apps into the clinic is by Gordon et al. (2020). In the article they suggest a framework for prescribing apps in clinical care. They looked at regulations, validation, payment, workflow, and education. They suggested "digital formularies" that would work much like traditional formularies.

Patrick Soon-Shiong (in Fineman, 2013) described his "medical information highway" that allows a patient's doctor to submit a patient's tissue to a site that quickly determines how the patient's mutated DNA is speaking to the cancer cells. This information is submitted to a database in the cloud. The doctor can then determine the specific nanotherapy for that individual (Fineman, 2013). Astonishing as this revelation may be, more astounding is the idea of patients sending their own samples and receiving individualized treatment plans from the cloud.

> Possible PA Research: Researchers could use diagnostic apps, body computing, and use of such in portals as predictor variables. If patients are involved in adding their own data to patient portals, such as bp, weight, steps per day, and so on, would such efforts increase compliance, increase good outcomes, etc.? A great topic to research, in general, whether PA or not.
>
> Examining the connections between asthma and COVID as it concerns oxidative stress, epithelial cell degradation and possible treatments for both using agents, such as NAC and others, might provide answers for individuals—what "cocktail" works for one patient may not for another. Comorbidities, stage of disease, perhaps genomic indicators, and so on could be predictors along with the agents, and the individual outcome could be the dependent variable.

All the ideas about technology and self-diagnosis are wonderful if one has access to and knowledge of online resources and can purchase the various self-diagnosis gadgets. Without help, all of that can leave out people with limited resources of time and money. Ideas and funds to help vulnerable populations are needed and that need will continue as technology increases. Research could be done to see how to use medical clinics to improve access for patients in need. Perhaps a loan library of devices could be set up so that patients could self-monitor. Or better yet, medical insurance companies could provide digital medical technology to patients. Early diagnosis and the self-monitoring of chronic conditions could reduce medical costs and cover the costs of such.

Chapter conclusion

PAs reaches all people, even without their knowledge. People encounter that long arm in every click as they surf the Internet or speak to someone on their smartphones. It is not unusual to have a conversation with one's spouse in front of one's smartphone and find the topic of that conversation in one's Facebook page as an advertisement, for example. We are constantly being monitored and data are being sent to a cloud from many of our smart devices. The appetite for data among data miners knows no boundary and will only increase in the future. It is not surprising to realize that companies monitor employees via electronic devices measuring heretofore private domains, such as heart rates and measures of stress, as they say what to whom at meetings (The Wall Street Journal, 2013). Privacy issues are already a concern as we become ever more connected to monitoring devices that connect to our bodies—perhaps even nanodevices that will reside inside our bodies. What social pressures will exist in the future to "voluntarily" submit to employer's requests for data? Often no one asks us if we want our data collected. How about medical data? Who will decide? How about vaccinations, when drug companies have no legal responsibility for harm that they may cause? How will patients participate in the future? What will be the rules for participation? Will HIPPA disappear in the demand for data "for their own good"? How invasive will our government become? These are questions that were facing many people as of this writing. Patients were confronting ethical dilemmas concerning privacy versus a kind of utilitarian health or individual privacy versus public health. At times, they had to decide between their jobs and their reasons for not wanting to vaccinate. Deeply divided, others could not understand why anyone would reject something that would keep them from disease. Would it be constitutional for the government to demand that everyone take a vaccine to protect the public regardless of the reason the person had to not vaccinate, such as already having innate immunity after contracting COVID? Will heretofore free people be told they must shelter and stay home? Or to wear masks when the masks provided no protection? These were times in which there was a definite struggle for some, for compliance with something they did not want. Empowered patients may have limits, particularly in times of pandemics. At times, individuals may have to give up some personal liberty for the good of the whole. Who will determine when?

Patients of the future may have a greater responsibility for directing their own healthcare and in some areas, and they may be told what they will do in other areas. What will be the role of the government or the pharmaceutical industry? All the areas from above will impact patient-directed care, including patient empowerment and involvement, communication and coordination, education and self-education, consumerism, payment options, and finally all the new advances from genetics and technology. Certainly, PAs will have an important place and researchers today remain on the cusp of exciting applications. PA research in patient-directed medicine holds great promise for increasing good outcomes while reducing costs. PA also leaves us with ethical issues that need to be addressed.

At least at present, however, patients must depend upon themselves to be smart about their own healthcare, and to make decisions about what they will and will not do. They need to take responsibility for their own health by creating a healthy lifestyle for themselves. They cannot simply trust that their government or even that their doctors will always know what to do and will communicate with one another. They may have to make decisions about what they will and will not do.

Patients must take responsibility for their *involvement*, must be sure that *communication* is going both directions, need to become better *consumers*, especially where *costs and payments* are concerned. They need to *educate* themselves to maintain health as much as possible and need to actively gain awareness of *changes in technology now and in the future*. (See Fig. 7.15.) They need to *direct* their own healthcare by *taking responsibility* for their lifestyle. They know what to do to make for a healthier future, for the most, but will they do it? Patient-directed healthcare is a life and death matter, not to be taken lightly.

Ultimately, the terms empowered-patient and patient-directed healthcare mean being responsible for oneself, collaborating with professionals, and making one's ultimate decisions about one's healthcare. For noncatastrophic ailments,

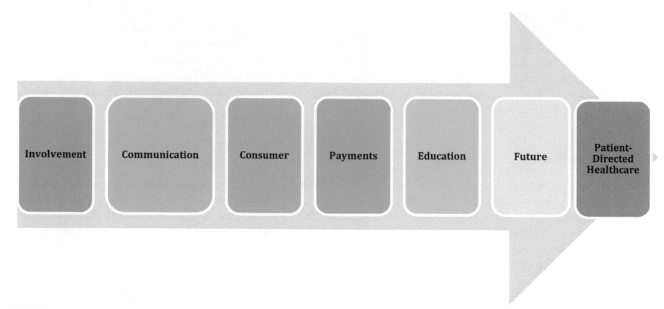

FIGURE 7.15   Areas of patient-directed healthcare.

patient-directed means honestly assessing oneself and stepping up to do whatever one can to achieve better health outcomes, whether eating healthy foods, exercising, stopping bad behaviors, *whatever we know we need to do*, so that we can stop running to doctors to fix things that we have broken.

If we all would start doing this, our general health would likely improve, and healthcare costs could decrease.

Postscript

Some of the benefits that can accrue from patient-directed healthcare include:

1. It may provide better control of patient data privacy.
2. It can lead to reduced costs by limiting optional procedures.
3. It can facilitate greater patient response to medical treatment programs.
4. It promotes a sense of patient empowerment, which can generate greater patient satisfaction about their medical treatment.
5. It can discover patterns and treatments missed by clinicians.
6. It can be wrong if patients do not collaborate with their physicians.

If patients participate more directly in decision-making for their healthcare, how does that involvement interact with regulatory measures? We will explore that subject in the next chapter.

References

Advisory Board, 2018. 87% of doctors say they're great at listening. (But the evidence does not support that). <https://www.advisory.com/daily-briefing/2018/12/14/listening>.

Alekseiv, S., Harris, A., 2013. Detection of stress induced ischemia in patients with chest pain after "rule-out ACS" protocol. In: Paper Presented to In His Image Scientific Assemblage and Retreat, 6/27/13. Case Study I in Miner, et al., 2015, pp. 544–557.

Alliance for Natural Health, 2021. Why the FDA is attacking NAC supplements. <anh-usa.org>.

Ammenwerth, E., Schnell-Inderst, P., Hoerbst, A., 2012. The impact of electronic patient portals on patient care: a systematic review of controlled trials. J. Med. Internet Res. 14 (6), 37. Available from: https://doi.org/10.2196/jmir.2238.

Anderson, R.T., Montori, V.M., Shah, N.D., et al., 2014. Effectiveness of the chest pain choice decision aid in emergency department patients with low-risk chest pain: study protocol for a multicenter randomized trial. Trials 15, 166.

Angst, C.M., Devaraj, S., D'Arcy, J., 2012. Dual role of IT-assisted communication in patient care: a validated structure-process-outcome framework. J. Manag. Inf. Syst. 29 (2), 257–292.

APHA, 2013. American Public Health Association: Medicaid expansion. <http://www.apha.org/advocacy/Health+Reform/ACAbasics/medicaid.htm>.

Appzen, 2020. Sunshine Act Compliance: All You Need to Know. AppZen.

Arnetz, J.E., Almin, I.I., Bergström, K.K., Franzén, Y.Y., Nilsson, H.H., 2004. Active patient involvement in the establishment of physical therapy goals: effects on treatment outcome and quality of care. Adv. Physiother. 6 (2), 50−69.

Arnetz, J., Winblad, U., Höglund, A., Lindahl, B., Spångberg, K., Wallentin, L., et al., 2010. Is patient involvement during hospitalization for acute myocardial infarction associated with post-discharge treatment outcome? An exploratory study. Health Expect. 13 (3), 298−311. Academic Search Premier, EBSCOhost, viewed 11 July 2013.

Ashish, K.J., 2018. End of Life Care Not, End of Life Spending. JAMA Health Forum. <https://jamanetwork.com/channels/health-forum/fullarticle/2760146>.

Aspinall, M., 2009. Personalized medicine and pathology. Arch. Pathol. Lab. Med. 133 (4), 527−531.

Association of American Physicians and Surgeons (AAPS), 2020. Find a direct payment/cash friendly practice (or list your practice). <https://aapsonline.org/direct-payment-cash-friendly-practices/>.

Baek, S., Moon, H., Ahn, H., Kodell, R.L., Lin, C., Chen, J.J., 2008. Identifying high-dimensional biomarkers for personalized medicine via variable importance ranking. J. Biopharm. Stat. 18 (5), 853−868. Available from: https://doi.org/10.1080/10543400802278023.

BBC News. 2021. Who are the Uyghurs and why is China being accused of genocide?, June 21. Available at: https://www.bbc.com/news/world-asia-china-22278037.

Beck, M., 2013. More doctors steer clear of medicare. Wall Str. J. 247 (24), A1.

Bendix, J., Lutton, L., 2020. The true cost of obesity, *Medical Economics*, February 24. Available at: https://www.medicaleconomics.com/view/true-costs-obesity?page=4.

Berg, S., 2019. New Survey Shows Decline in Physician Burnout. AMA Physician Health. <https://www.ama-assn.org/practice-management/physician-health/new-survey-shows-decline-physician-burnout>.

Berk, M., 2016. Glutathione, oxidative stress and N-acetylcysteine (NAC) in psychiatric disorders—Prof Berk, October 30. Available at: https://www.youtube.com/watch?v=tvel1kOuUO4&t=6s.

Berk, M., 2017. Applications of N-acetylcysteine (NAC)—from addiction to autism by Prof Berk, March 2. Available at: https://www.youtube.com/watch?v=s945zr6P_rs.

Berry, K.D., 2021. Fasting doesn't help weight loss? (Dr. Jason Fung Wrong?): (91) Fasting doesn't help weight loss? (Dr Jason Fung Wrong?) Intermittent fasting—2021, June 20. Available at: https://www.youtube.com/watch?v=qyYHRce0vL4&t=2s.

Beth Israel Deaconess Medical Center BIDMC, 2020. Clinical trials and research studies. <https://www.bidmc.org/research/clinicaltrials#:~:text=With%20approximately%202%2C500%20active%20research,human%20body%20and%20mind%20work> (accessed 13.08.20.).

Bevan, G., Janus, K., 2011. Why hasn't integrated health care developed widely in the United States and not at all in England? J. Health Polit. Policy Law 36 (1), 141−164. Available from: https://doi.org/10.1215/03616878-1191135.

BDIMCb, 2013. Open notes. <http://www.bidmc.org/Patient-and-Visitor-Information/OpenNotes.aspx>.

Bidgood, J., Buckroyd, J., 2005. An exploration of obese adults' experience of attempting to lose weight and to maintain a reduced weight. Counsel. Psychother. Res. 5 (3), 221−229. Available from: https://doi.org/10.1080/17441690500310395.

BIDMC, 2013. Beth Israel Deaconess Medical Center: we never forget that before you are a patient you are a person. <http://www.bidmc.org/About-BIDMC.aspx>.

BIDMC, 2021. OpenNotes and PatientSite. BIDMC of Boston.

Blokzijl, A.A., Friedman, M.M., Pontén, F.F., Landegren, U.U., 2010. Profiling protein expression and interactions: proximity ligation as a tool for personalized medicine. J. Intern. Med. 268 (3), 232−245. Available from: https://doi.org/10.1111/j.1365-2796.2010.02256.x.

Blue Cross of South Carolina, 2013. Companion global healthcare: because it matters how you're treated. <http://www.southcarolinablues.com/members/discountsaddedvalues/companionglobalhealthcare.aspx>.

Bob and Brad Channel: <https://www.youtube.com/c/BobBrad>.

Bowden, J., Sinatra, S., 2012. The Great Cholesterol Myth: Why Lowering your Cholesterol Won't Prevent Heart Disease. Fair Winds Press, ISBN 9781610586344.

Brown, R.M., Kupchik, Y.M., Kalivas, P.W., 2013. The story of glutamate in drug addiction and of N-acetylcysteine as a potential pharmacotherapy. JAMA Psychiatry 70 (9), 895−897. Available from: https://doi.org/10.1001/jamapsychiatry.2013.2207. PMID: 23903770.

Bureau of Labor Statistics, 2013. CPI inflation calculator. <http://www.bls.gov/data/inflation_calculator.htm>.

CBS, 2013. Mexico takes title of "Most Obese" from America. <http://www.cbsnews.com/8301-202_162-57592714/mexico-takes-title-of-most-obese-from-america/>.

Cerrato, P., Halamka, J., 2020. Prinventing Clinical Decision Support: Data Analytics, Artificial Intelligence and Diagnostin Reasoning, HIMSS. Routledge, Taylor & Francis Group, Boca Raton, FL.

Chan, D., Goh, L., 2000. The doctor-patient relationship: a survey of attitudes and practices of doctors in Singapore. Bioethics 14 (1), 58.

Chandnani, H., Goldstein M., Khichi, M., Miner L., Tinsley C., 2018. Prediction model for mortality in patients with disseminated intravascular coagulopathy based on pediatric ICU admission: three factors identified. In: 9th World Congress on Pediatric Intensive and Critical Care, 2018, Singapore. PICC8-O690.

Clinical Genomic Database, 2021. Advancing Human Health Through Genomics Research: National Human Genome Research Institute. Clinical Genomic Database. <nih.gov>.

Cocksedge, S., May, C., 2005. The listening loop: a model of choice about cues within primary care consultations. Med. Educ. 39, 999−1005. Available from: https://doi.org/10.1111/j.1365-2929.2005.02264.x. https://www.advisory.com/daily-briefing/2018/12/14/listening.

Cohen, J., 2021. The path to better health: Give patients their data. Issues in Science and Technology. Patients Should Own Their Health Data. <https://issues.org/>.

Collins, S.R., Bhupal, H.K., Doty, M.M., 2019. Health insurance coverage eight years after the ACA: Fewer Uninsured Americans and Shorter Coverage Gaps, But More Underinsured. The Commonwealth Fund: Health Insurance Coverage Eight Years After the ACA: Fewer Uninsured Americans and Shorter Coverage Gaps, But More Underinsured. <https://www.commonwealthfund.org/>.

Colombet, I., Bura-Rivière, A., Chatila, R., Chatellier, G., Durieux, P., 2004. Personalized vs. non-personalized computerized decision support system to increase therapeutic quality control of oral anticoagulant therapy: an alternating time series analysis. BMC Health Serv. Res. 427–428. Available from: https://doi.org/10.1186/1472-6963-4-27.

Common Sense Family Doctor, 2011. Common sense thoughts on health and conservative medicine from a family doctor in Washington, DC: preventive health screenings that are hardly a life line <http://commonsensemd.blogspot.com/2011/02/preventive-health-screenings-that-are.html>.

Cvengros, J.A., Christensen, A.J., Cunningham, C., Hillis, S.L., Kaboli, P.J., 2009. Patient preference for and reports of provider behavior: impact of symmetry on patient outcomes. Health Psychol. 28 (6), 660–667. Available from: https://doi.org/10.1037/a0016087.

Daly, R., 2020. Hospitals in COVID-19 Hotspots to Receive $10 Billion More in Federal Aid. Healthcare Financial Management Association. <hfma.org>.

Datti, B., Carter, M.W., 2006. The effect of direct-to-consumer advertising on prescription drug use by older adults. Drugs Aging 23 (1), 71–81.

Davis, R.E., Jacklin, R., Sevdalis, N., Vincent, C.A., 2007. Patient involvement in patient safety: what factors influence patient participation and engagement? Health Expect. 10 (3), 259–267. Available from: https://doi.org/10.1111/j.1369-7625.2007.00450.x.

De Cabo, R., Mattson, M.P., 2019. Effects of intermittent fasting on health, aging, and disease. N. Engl. J. Med. 381 (26), 2541–2551. Available from: https://doi.org/10.1056/NEJMra1905136.

Dehghan, H., Keshtkaran, A., Ahmadloo, N., Bagheri, Z., Hatam, N., 2018. Patient involvement in care and breast cancer patients' quality of life—a structural equation modeling (SEM) approach. Asian Pac. J. Cancer Prev.: APJCP 19 (9), 2511–2517. Available from: https://www.ncbi.nlm.nih.gov/pmc/articles/PMC6249441/.

Demasi, M., 2018. Statin Wars: how Big Pharma infiltrates governments and the medical profession, *Michael West Media Independent Journalists*, January 23. Available at: https://www.michaelwest.com.au/statin-wars-how-big-pharma-infiltrates-governments-and-the-medical-profession/.

Dendere, R., Slade, C., Burton-Jones, A., Sullivan, C., Staib, A., Janda, M., 2019. Patient portals facilitating engagement with inpatient electronic medical records: a systematic review. J. Med. Internet Res. 21 (4), e12779. Available from: https://doi.org/10.2196/12779Full article can be found at this URL. Available from: https://www.ncbi.nlm.nih.gov/pmc/articles/PMC6482406/.

DeRosa, A.P., Baltich Nelson, B., Delgado, D., Mages, K.C., 2019. Involvement of information professionals in patient-and family-centered care initiatives: a scoping review. J. Med. Library Assoc. 107 (3), 314–322. Available from: https://ezproxy.snu.edu:2292/10.5195/jmla.2019.652. https://www.ncbi.nlm.nih.gov/pmc/articles/PMC6579588/.

Doupi, P.P., van der Lei, J.J., 2002. Towards personalized Internet health information: the STEPPS architecture. Med. Inform. Internet Med. 27 (3), 139–151. Available from: https://doi.org/10.1080/1463923021000014149.

Doyle Jr.J.J., 2005. Health insurance, treatment and outcomes: using auto accidents as health shocks. Rev. Econ. Stat. 87 (2), 256–270. Available from: https://doi.org/10.1162/0034653053970348.

DuBroff, R., Malhotra, A., de Lorgeril, M., 2020. Hit or miss: the new cholesterol targets. BMJ Evid.-Based Med. Available from: https://doi.org/10.1136/bmjebm-2020-111413Published Online First. Available from: https://ebm.bmj.com/content/early/2020/07/20/bmjebm-2020-111413.

Dworkin, R.B., 2003. Getting what we should from doctors: rethinking patient autonomy and the doctor-patient relationship. Health Matrix: J. Law-Med. 13 (2), 235.

Eftekhari, P., Hajizadeh, S., Raoufy, M.R., Masjedi, M.R., Yang, M., Hansbro, N., et al., 2013. Preventive effect of N-acetylcysteine in a mouse model of steroid resistant acute exacerbation of asthma. EXCLI J. 12, 184–192.

Einarsdóttir, K., Haggar, F.A., Langridge, A.T., Gunnell, A.S., Leonard, H., Stanley, F.J., 2013. Neonatal outcomes after preterm birth by mothers' health insurance status at birth: a retrospective cohort study. BMC Health Serv. Res. 13 (1), 1–7. Available from: https://doi.org/10.1186/1472-6963-13-40.

Eliacin, J., Salyers, M.P., Kukla, M., Matthias, M.S., 2015. Factors influencing patients' preferences and perceived involvement in shared decision-making in mental health care. J. Ment. Health 24 (1), 24–28. Available from: https://pubmed.ncbi.nlm.nih.gov/25279691/.

Elwyn, G., Hutchings, H., Edwards, A., et al., 2005. The OPTION scale: measuring the extent that clinicians involve patients in decision-making tasks. Health Expect. 8, 34–42.

Evans, B., 2020. US companies saving money by sending employees overseas for healthcare, *The Denver 7 ABC*, February 14. Available at: https://www.thedenverchannel.com/news/national/us-companies-saving-money-by-sending-employees-overseas-for-healthcare.

Evers, A.M., Rovers, M.M., Kremer, J.M., Veltman, J.A., Schalken, J.A., Bloem, B.R., et al., 2012. An integrated framework of personalized medicine: from individual genomes to participatory health care. Croatian Med. J. 53 (4), 301–303. Available from: https://doi.org/10.3325/cmj.2012.53.301.

Fineman, H., 2013. Meet Patrick Soon-Shiong, the LA billionaire reinventing your health care. Available at: <http://www.huffingtonpost.com/2013/12/01/patrick-soon-shiong_n_4351344.html>.

Fisher, J.H., 2017. The Sunshine Act Laws Blog. <https://protectingpatientrights.com/blog/the-sunshine-act/>.

Fortenberry Jr.J.L., Elrod, J.K., McGoldrick, P.J., 2010. Is billboard advertising beneficial for healthcare organizations? An investigation of efficacy and acceptability to patients. J. Healthc. Manag. 55 (2), 81–96.

Friedman, Y., 2012. Will personalized medicine be a driver for widespread price controls? J. Commer Biotechnol. 3–4. Available from: https://doi.org/10.5912/jcb.559.

Fung, J., 2016. The Obesity Code: Unlocking the Secrets of Weight Loss. Greystone Books, Vancouver/Berkeley, ISBN: 1771641258.

Fung, J., 2020. Intermittent fasting—Life changing. Available at: YouTube.

Gaglani, S., 2023. Interview with FutureMed Executive Director, Dr. Daniel Kraft. MedGadget. <https://www.medgadget.com/2013/10/interview-with-futuremed-executive-director-dr-daniel-kraft.html>.

Galas, D.J., Hood, L., 2009. Systems biology and emerging technologies will catalyze the transition from reactive medicine to predictive, personalized, preventive and participatory (P4) medicine. Interdisc. Bio Cent. (1), 1−4. Available from: https://doi.org/10.4051/ibc.2009.2.0006.

Galaxy, 2013. Galaxy 101: the very first tutorial you need. <https://main.g2.bx.psu.edu/>.

Giannini, A., Messeri, A., Aprile, A., Casalone, C., Jankovic, M., Scarani, R., et al., 2008. End-of-life decisions in pediatric intensive care. Recommendations of the Italian Society of Neonatal and Pediatric Anesthesia and Intensive Care (SARNePI). Pediatr. Anesth. 18 (11), 1089−1095. Available from: https://doi.org/10.1111/j.1460-9592.2008.02777.x.

Gibson, R., Singh, J.P., 2010. The Treatment Trap. Ivan R. Dee, Chicago.

Gigerenzer, G., Gaissmaier, W., Kurz-Milcke, E., Schwartz, L.M., Woloshin, S., 2007. Helping doctors and patients make sense of health statistics. Psychol. Sci. Public. Interest. 8 (2), 53−96. Available from: https://doi.org/10.1111/j.1539-6053.2008.00033.x.

Gilbert, S., 2011. Medicine that's a little too personalized. Hastings Cent. Rep. 41 (4), 49.

Gill, V., Pomerantz, A., Denvir, P., 2010. Pre-emptive resistance: patients' participation in diagnostic sense-making activities. Sociol. Health Illn. 32 (1), 1−20. Available from: https://doi.org/10.1111/j.1467-9566.2009.01208.x. http://wiley.com.

Global Market Insights, 2021. Industry trends. Medical tourism market trends l Growth Outlook 2021 − 2027. <gminsights.com>.

Goldhill, D., 2013. Catastrophic Care: How American Health Care Killed My Father—And How We Can Fix It. Alfred A. Knoff, New York.

Gordon, W.J., Landman, A., Zhang, H., et al., 2020. Beyond validation: getting health apps into clinical practice. NPJ Digit. Med. 3, 14. Available from: https://doi.org/10.1038/s41746-019-0212-zhttps://www.nature.com/articles/s41746-019-0212-z#citeas.

Granger, D.N., Senchenkova, E., 2010. Inflammation and the Microcirculation. Morgan & Claypool Life Sciences, San Rafael, CA, Chapter 7, Leukocyte−endothelial cell adhesion. https://www.ncbi.nlm.nih.gov/books/NBK53380/.

Grzymski, J.J., Elhanan, G., Morales Rosado, J.A., et al., 2020. Population genetic screening efficiently identifies carriers of autosomal dominant diseases. Nat. Med. Available from: https://doi.org/10.1038/s41591-020-0982-5.

Gulbin, S., 2021. How fat is America? An overview of obesity statistics. In: Diet & Weight Loss. Livin3.

Gundry, S., 2021m The #1 killer—and they're treating it ALL wrong. Episode 173, *Dr. Gundry Podcast*, Sept. 14. Available at: https://youtu.be/sxH07tV2B6c.

Guo, T., Sriram, S., Manchanda, P., 2017. 'Let the Sun Shine in': The Impact of Industry Payment Disclosure on Physician Prescription Behavior. April 17. <https://ssrn.com/abstract=2953399>.

Haiech, J., Kilhoffer, M.-C., 2012. Personalized medicine and education: the challenge. Croatian Med. J. 53 (4), 298−300. Academic search premier, EBSCOhost (accessed 16.04.13.).

Harvard T.H. Chan School of Public Health, 2020. Obesity prevention source. <https://www.bls.gov/data/inflation_calculator.htm>.

Health Europa Quarterly (2020) Forced organ harvesting: "I'm going to China, they're shooting my donor", January 30. Available at: https://www.healtheuropa.eu/im-going-to-china-theyre-shooting-my-donor/97063/.

Healthsherpa, 2018. Is health insurance mandatory in the United States?. <https://www.healthsherpa.com/blog/health-insurance-mandatory-united-states/>.

Hoffman, S., 2019. What genetic testing teaches about predictive health analytics regulation. In: North Carolina Law Rev. 98, p. 123. <https://scholarship.law.unc.edu/nclr/vol98/iss1/5>.

Holger, M., Hogan, J., Kel, A., Kel-Margoulis, O., Schacherer, F., Voss, N., et al., 2008. Building a knowledge base for systems pathology. Brief. Bioinforma. 9 (6), 518−531. Available from: https://doi.org/10.1093/bib/bbn038.

Howell, E.M., Trenholm, C., 2006. The effect of new insurance coverage on the health status of low-income children in Santa Clara County. Health Serv. Res. 42 (2), 867−889. Academic Search Premier, EBSCOhost (accessed July 15.07.13.).

Hyde, C., Dunn, K.M., Higginbottom, A., Chew, G.C.A., 2017. Process and impact of patient involvement in a systematic review of shared decision making in primary care consultations. Health Expect. 20 (2), 298−308. Available from: https://ezproxy.snu.edu:2292/10.1111/hex.12458.

Ibrahim, F., Sandström, P., Björnsson, B., Larsson, A.L., Drott, J., 2019. "I want to know why and need to be involved in my own care...": a qualitative interview study with liver, bile duct or pancreatic cancer patients about their experiences with involvement in care. Support. Care Cancer 27 (7), 2561−2567. Available from: https://pubmed.ncbi.nlm.nih.gov/30430301/.

iCare Home, March, 2022. Welcome to the new iCare. iCare products and services for modern and reliable ocular diagnostics. <https://www.icare-world.com/us/product/icare-home/>.

Institute for Patient- and Family-Centered Care, 2020. Transforming Health Through Partnerships. <https://www.ipfcc.org/> (accessed July 28, 2022).

Institute of Medicine (US) Committee on Health Research and the Privacy of Health Information: The HIPAA Privacy Rule, 2009. In: Nass, S.J., Levit, L.A., Gostin, L.O. (Eds.), Beyond the HIPAA Privacy Rule: Enhancing Privacy, Improving Health Through Research. National Academies Press (US), Washington, DC5, Effect of the HIPAA Privacy Rule on Health Research. Available from: Publication 969 (2020), Health Savings Accounts and Other Tax-Favored Health Plans l Internal Revenue Service. Available from: irs.gov.

IRS, 2020. Publication 969 (2020), Health savings accounts and other tax-favored health plans. <https://www.irs.gov/publications/p969#en_US_2020>.

IRS, 2021. Advance Child Tax Credit Payments in 2021. Internal Revenue Service. <irs.gov>.

Kahana, E., Kahana, B., Wykle, M., Kulle, D., 2009. Marshalling social support: a care-getting model for persons living with cancer. J. Family Soc. Work. 12 (2), 168−193. Available from: https://doi.org/10.1080/10522150902874834.

Kaiser Permanente, 2013. Kaiser Permanente's Integrated Health Care Model. <http://mydoctor.kaiserpermanente.org/ncal/facilities/region/santarosa/area_master/about_us/health_care_model/>.

Kaiser Permanente, 2020. Our Model. Kaiser Permanente.

Kawamoto, K., Lobach, D.F., Willard, H.F., Ginsburg, G.S., 2009. A national clinical decision support infrastructure to enable the widespread and consistent practice of genomic and personalized medicine. BMC Med. Inform. Decis. Mak. 9 (1), 1−14. Available from: https://doi.org/10.1186/1472-6947-9-17.

Kendrick, M., 2020. Home Page: Cholesterol lowering has no impact. <https://drmalcolmkendrick.org/2020/08/05/cholesterol-lowering-has-no-impact/> (August 5, 2020).

Kliems, H., Witt, C.M., 2011. The good doctor: a qualitative study of German Homeopathic Physicians. J. Alt. Complement. Med. 17 (3), 265−270. Available from: https://doi.org/10.1089/acm.2010.0158.

Kravitz, R.L., Duan, N., Niedzinski, E.J., Hay, M., Subramanian, S.K., Weisner, T.S., 2008. What ever happened to n-of-1 trials? Insiders' perspectives and a look to the future. Milbank Q. 86 (4), 533−555. Available from: https://doi.org/10.1111/j.1468-0009.2008.00533.x.

Lagasse, J., 2019. Consumers Are Increasingly Self-Empowered in Their Care Decisions—And Placing a Growing Premium on Convenience. Healthcare Finance. <https://www.healthcarefinancenews.com/news/consumers-are-increasingly-self-empowered-their-care-decisions-and-place-growing-premium> (Nov. 5, 2019).

Landro, L., 2013. A better diagnosis. Wall Str. J. 247 (19).

Lexchin, J.J., 1999. Direct-to-consumer advertising: impact on patient expectations regarding disease management. Dis. Manag. Health Outcomes 5 (5), 273−283.

Life Line Screening, 2013. Life line screening: the power of prevention. <http://www.lifelinescreening.com/>.

Lerena, A., Michel, G., Jeannesson, E., Wong, S., Manolopoulos, V.G., Hockett, R., et al., 2007. Third Santorini conference pharmacogenomics workshop report: "Pharmacogenomics at the crossroads: what else than good science will be needed for the field to become part of personalized medicine?". Clin. Chem. Lab. Med. 45 (7), 843−850. Available from: https://doi.org/10.1515/CCLM.2007.182.

Lóópez-Medina, E., López, P., 2021. Effect of ivermectin on time to resolution of symptoms among adults with mild COVID-19: a randomized clinical trial. JAMA .

Lord, L., 2020. The Value of Understanding Patient Preferences in Treatment Plans and Protocols. Cipher Health <https://cipherhealth.com/the-value-of-understanding-patient-preferences-in-treatment-plans-and-protocols/>.

Louca, S., 2012. Personalized medicine - a tailored health care system: challenges and opportunities. Croatian Med. J. 53 (3), 211−213. Available from: https://doi.org/10.3325/cmj.2012.53.211.

Lu, X., Zhang, R., 2019. Impact of physician-patient communication in online health communities on patient compliance: cross-sectional questionnaire study. J. Med. Internet Res. 21 (5), N.PAG. Available from: https://ezproxy.snu.edu:2292/10.2196/12891.

Lyytikäinen, P., Rahkonen, O., Lahelma, E., Lallukka, T., 2011. Association of sleep duration with weight and weight gain: a prospective follow-up study. J. Sleep. Res. 20 (2), 298−302. *PsycINFO*, EBSCOhost (accessed 17.07.13.).

Mackert, M., Eastin, M.S., Ball, J.G., 2010. Perceptions of direct-to-consumer prescription drug advertising among advanced practice nurses. J. Med. Mark. 10 (4), 352−365. Available from: https://doi.org/10.1057/jmm.2010.26.

Manias, E., Bucknall, T., Hughes, C., Jorm, C., Woodward-Kron, R., 2019. Family involvement in managing medications of older patients across transitions of care: a systematic review. BMC Geriatrics 19 (1), 95. Available from: https://ezproxy.snu.edu:2292/10.1186/s12877-019-1102-6. https://bmcgeriatr.biomedcentral.com/articles/10.1186/s12877-019-1102-6.

Marcus, A.D., 2013. Genetic testing leaves more patients living in limbo. *The Wall Street Journal*, D1, November 19.

Massina, J., 2013. Obamacare transforms med school. Modern Healthcare. <http://www.modernhealthcare.com/article/20130204/INFO/302049986>.

Medical Tourism, 2020. The Medical Tourism Index 2020−21.

Menifield, C.E., Doty, N., Fletcher, A., 2008. Obesity in America. ABNF J. 19 (3), 83−88.

Miner, L.A., 2014. Seven Ways Predictive Analytics Can Improve Healthcare. Elsevier Connect: <https://www.elsevier.com/connect/seven-ways-predictive-analyticscan-improve-healthcare>.

Miner, G.D., Richter, R.W., Valentine, J.L., Miner, L.A. (Eds.), 1989. Alzheimer's Disease: Molecular Genetics, Clinical Perspectives and Promising New Research. Marcel Dekker, Inc, New York.

Miner, G.D., Miner, L.A., Dean, D., 2019. Healthcare's Out Sick—Predicting a Cure—Solutions that Work. Routledge Taylor & Francis Group, New York.

Mitchell, P.B., Meiser, B., Wilde, A., Fullerton, J., Donald, J., Wilhelm, K., et al., 2010. Predictive and diagnostic genetic testing in psychiatry. Psychiatr. Clin. North. Am. (1), 225−243. Available from: https://doi.org/10.1016/j.psc.2009.10.001. 20159347.

Mo, H., Shin, D., Woo, J., Choi, J., Kang, J., Baik, Y., et al., 2012. Is patient autonomy a critical determinant of quality of life in Korea? End-of-life decision making from the perspective of the patient. Palliat. Med. 26 (3), 222−231.

Mohr, D.C., Benzer, J.K., Vimalananda, V.G., Singer, S.J., Meterko, M., McIntosh, N., et al., 2019. Organizational coordination and patient experiences of specialty care integration. J. Gen. Intern. Med. 34 (Suppl. 1), 30−36.

Monegain, B., 2013, July. Cleveland Clinic Opens EMR to Patients. Healthcare IT News 10 (7), HIMSS, 20. Available from: https://www.healthcareitnews.com/news/cleveland-clinic-opens-emr-patients?topic=08,18.

Morandi, S., Golay, P., Vazquez-Montes, M., Rugkasa, J., Molodynski, 2017. Factorial structure and long-term stability of the Autonomy Preference Index. Psychological Assessment. PMID: 27124100. Available from: https://doi.org/10.1037/pas0000327.

Moreau, K., Eady, K., Jabbour, M., 2019. Patient involvement in resident assessment within the competence by design context: a mixed-methods study. Can. Med. Educ. J. 10 (1), e84−e102.

Myers, R.J., 1967. What would "Medicare" cost?: comment. Am. Risk Insurance Assoc. 34 (1), 141−147.

Myers, R.J., 1994. How Bad were the original actuarial estimates for Medicare"s hospital insurance program? Actuary 6−7.

NCHS Data Brief Number 360, 2020. Data brief 360. Prevalence of obesity and severe obesity among adults: United States, 2017−2018. <https://www.cdc.gov/nchs/data/databriefs/db360_tables-508.pdf#page=4>.

Neal, M., Kerckhoffs, R., 2010. Current progress in patient-specific modeling. Brief. Bioinforma. 11 (1), 111−126. Available from: https://doi.org/10.1093/bib/bbp049.

New Choice Health, 2013. Your healthcare marketplace. <http://www.newchoicehealth.com/Home>.

Newman-Toker, D.E., Schaffer, A.C., Yu-Moe, C., Nassery, N., Saber Tehrani, A.S., Clemens, G.D., et al., 2019. Serious misdiagnosis-related harms in malpractice claims: The "Big Three"—vascular events, infections, and cancers. Diagnosis 6 (3), 227−240. Available from: https://doi.org/10.1515/dx-2019-0019.

NHGRI, 2020. Online research resources. <https://www.genome.gov/10000375/online-research-resources>.

Nicolson, G.L., 2014. Mitochondrial dysfunction and chronic disease: treatment with natural supplements. Integr. Med. (Encinitas, Calif.) 13 (4), 35−43.

NIH, 2021. Overweight and Obesity. NHLBI, NIH.

NPG, 2013. Negative population growth: facts and figures. <http://www.npg.org/facts/us_historical_pops.htm>.

Office of Management and Budget, 2013. Historical Tables. Table 2.4 composition of social insurance and retirement receipts and of excise taxes: 1940−2018. <http://www.whitehouse.gov/omb/budget/HistoricalsPCDC> (accessed 05.05.13).

Open Notes, 2021. OpenNotes—Patients and clinicians on the same page.

PCPCC, 2013. About us: the leading national coalition dedicated to advancing the patient-centered medical home <http://www.pcpcc.org/about>.

Pham, H.H., O'Malley, A.S., Bach, P.B., Saiontz-Martinez, C., Schrag, D., 2009. Primary care physicians' links to other physicians through medicare patients: the scope of care coordination. Ann. Intern. Med. 150 (4), 236-W:40.

Pirisi, A., 1999. Patient-directed drug advertising puts pressure on US doctors. Lancet 354 (9193), 1887.

Porzio Aggregate SpendID, 2021. US State and Local Aggregate Spend Reporting Requirements. HCP & HCO Aggregate Spend Reporting Requirements | State & Local.

PR Newswire US, 2017. The biggest cover-up in China: forced live organ harvesting-DAFOH says prominent healthcare organizations fall for fake news of pledged transplant reforms, August 10.

PR Newswire US, 2019. DAFOH discusses role of forced organ harvesting as a tool of religious persecution and genocide in China, September 23.

Pravettoni, G., Gorini, A., 2011. A P5 cancer medicine approach: why personalized medicine cannot ignore psychology. J. Eval. Clin. Pract. 17 (4), 594−596. Available from: https://doi.org/10.1111/j.1365-2753.2011.01709.x.

Primary Care Development Corporation, 2020. PCDC marks 800th patient-centered medical home <https://www.pcdc.org/patient-centered-medical-home/?creative=377231586957&keyword=pcmh%20model&matchtype=b&network=g&device=c&gclid=EAIaIQobChMI1p3Y-JyH6wIVla_ICh1ZtwxHEAAYASAAEgKCAvD_BwE>.

Probst, M.A., Tschatscher, C.F., Lohse, C.M., Fernanda Bellolio, M., Hess, E.P., 2018. Factors associated with patient involvement in emergency care decisions: a secondary analysis of the chest pain choice multicenter randomized trial. Acad. Emerg. Med. 25 (10), 1107−1117. Available from: https://ezproxy.snu.edu:2292/10.1111/acem.13503. https://pubmed.ncbi.nlm.nih.gov/29904986/.

Rachakonda, T.D., Schupp, C.W., Armstrong, A.W., 2014. Psoriasis prevalence among adults in the United States. J. Am. Acad. Dermatol. 70 (3), 512−516. Available from: https://ezproxy.snu.edu:2292/10.1016/j.jaad.2013.11.013.

Raman, G., Avendano, E.E., Chen, M., 2013. Update on Emerging Genetic Tests Currently Available for Clinical Use in Common Cancers [Internet]. Agency for Healthcare Research and Quality (US), Rockville, MDAppendix A, One-page summaries of the genetic tests for cancers. Available from: https://www.ncbi.nlm.nih.gov/books/NBK285334/.

Rasmussen, S.R., Thomsen, J.L., Kilsmark, J., Hvenegaard, A., Engberg, M., Lauritzen, T., et al., 2007. Preventive health screenings and health consultations in primary care increase life expectancy without increasing costs. Scand. J. Public. Health 35 (4), 365−372. Available from: https://doi.org/10.1080/14034940701219642.

Right-Brain, Ink, 2022. Gary Trains on the ICare Machine. Author, Tulsa, OK. Available from: https://youtu.be/6nakad5Y3HY (Private—only those with the URL can view.).

Rodriguez, C., Veciana, C., 2020. Asthma and COCID-19: The Eosinophilic Link. QEIOS <https://www.qeios.com/read/5IY4IF>.

Sakai-Kashiwabara, M., Asano, K., 2013. Inhibitory action of quercetin on eosinophil activation in vitro. Evid.-Based Compl. Altern. Med. 2013, . Available from: https://doi.org/10.1155/2013/127105Article ID 127105, 7 pages.

Salgo, P., 2006. The Doctor Will See You for Exactly Seven Minutes. The New York Times. <http://www.nytimes.com/2006/03/22/opinion/22salgo.html?_r=2&>.

Sanders, B.S., 1965. What would Medicare cost? J. Risk Insurance 32 (4), 579−594.

Sanders, B.S., 1967. What would Medicare cost?: Author's reply. J. Risk Insurance 34 (1), 148−166.

Sandhir, R., Sood, A., Mehrotra, A., Kamboj, S.S., 2012. N-Acetylcysteine reverses mitochondrial dysfunctions and behavioral abnormalities in 3-nitropropionic acid-induced Huntington's disease. Neurodegener. Dis. 9 (3), 145−157. Available from: https://doi.org/10.1159/000334273. 22327485. nih.gov.

Saxon, L., 2010. Body Computing and Network Communications. TEDx Talks. <http://www.youtube.com/watch?v=bl9do6mGq1M>.

Schmid, S.M., Hallschmid, M., Jauch-Chara, K., Born, J., Schultes, B., 2008. A single night of sleep deprivation increases ghrelin levels and feelings of hunger in normal-weight healthy men. J. Sleep. Res. 17 (3), 331−334. Available from: https://doi.org/10.1111/j.1365-2869.2008.00662.x.

Schoen, C., Osborn, R., Phuong Trang, H., Doty, M., Peugh, J., Zapert, K., 2006. On the front lines of care: primary care doctors' office systems, experiences, and views in seven countries. Health Aff . Available from: https://doi.org/10.1377/hlthaff.25.w55525w555-w571.

Sedini, C., Biotto, M., Crespi Bel'skij, L.M., Moroni Grandini, R.E., Cesari, M., 2022. Advance care planning and advance directives: an overview of the main critical issues. Aging Clin Exp Res. 34 (2), 325–330. Available from: https://doi.org/10.1007/s40520-021-02001-y. Epub 2021 Oct 15. PMID: 34655048; PMCID: PMC8847241.

Seheult, R., 2020a. MedCram COVID 19 Update #69. NAC supplementation and COVID 19, #69. <https://www.youtube.com/watch?v=Dr_6w-WPr0w>. November 5.

Seheult, R., 2020b. MedCram COVID 19 Update #53. Coronavirus pandemic update 53: anticoagulation; can mechanical ventilation make COVID 19 worse?. <https://www.youtube.com/watch?v=o8aG63yigjA>, April 10.

Seheult, R., 2020c. MedCram COVID 19 Update #61. Blood clots and strokes in COVID 19, # 61 <https://www.youtube.com/watch?v=22Bn8jsGI54>, April 27.

Shinkarenko, A., Shinkarenko, I. Miner, L.A., Miner, G.D., 2007. Truth telling about terminal diagnoses in Kazakhstan. In: Paper Presented at the 25th Annual Scientific Assembly and Retreat, IHI Family Medical Residency Program, Western Hills, June.

Singh Ospina, N., Phillips, K.A., Rodriguez-Gutierrez, R., et al., 2019. Eliciting the patient's agenda-secondary analysis of recorded clinical encounters. J. Gen. Intern. Med. 34, 36–40. Available from: https://doi.org/10.1007/s11606-018-4540-5.

Subscription Medicine, 2015. On demand healthcare for everyone. <https://www.ariasystems.com/blog/subscription-medicine-on-demand-healthcare-for-everyone/>.

Susanti, H., James, K., Utomo, B., Keliat, B., Lovell, K., Irmansyah, I., et al., 2020. Exploring the potential use of patient and public involvement to strengthen Indonesian mental health care for people with psychosis: a qualitative exploration of the views of service users and carers. Health Expect. 23 (2), 377–387. Available from: https://ezproxy.snu.edu:2292/10.1111/hex.13007. https://pubmed.ncbi.nlm.nih.gov/31782266/. full text. https://onlinelibrary.wiley.com/doi/epdf/10.1111/hex.13007.

Tavakoly Sany, S., Behzhad, F., Ferns, G., et al., 2020. Communication skills training for physicians improves health literacy and medical outcomes among patients with hypertension: a randomized controlled trial. BMC Health Serv. Res. 20, 60. Available from: https://doi.org/10.1186/s12913-020-4901-8.

TDPCC, 2021. About Us | direct-primary-care. <dpcare.org> (accessed 08.08.21).

The Congressional Budget Office, 2020. The Federal Budget in 2019: an infographic <https://www.cbo.gov/publication/56324>.

The Wall Street Journal, 2013. Journal report leadership: information technology, Dow Jones & Company, October 21, R1–R6.

Tippy, K., Meyer, K., Aronson, R., Wall, T., 2005. Characteristics of coordinated ongoing comprehensive care within a medical home in Maine. Matern. Child. Health J. 913–921. Available from: https://doi.org/10.1007/s10995-005-4747-z.

Tongil, K., Diwas, K.C., 2020. The impact of hospital advertising on patient demand and health outcomes. Mark. Sci. 39 (3), 612–635. Available from: https://ezproxy.snu.edu:2292/10.1287/mksc.2019.1153. https://pubsonline.informs.org/doi/pdf/10.1287/mksc.2019.1153.

Trivedi, M.S., Deth, R., 2015. Redox-based epigenetic status in drug addiction: a potential contributor to gene priming and a mechanistic rationale for metabolic intervention. Front. Neurosci. 8, 444. Available from: https://doi.org/10.3389/fnins.2014.00444. 25657617. PMCID: PMC4302946.

US Population by Year 1900–2022. US Resident PopulationUS Resident Population <https://www.multpl.com/united-states-population/table/by-year>. (Accessed July, 2022).

van den Berk-Clark, C., Doucette, E., Rottnek, F., Manard, W., Prada, M.A., Hughes, R., et al., 2018. Do patient-centered medical homes improve health behaviors, outcomes, and experiences of low-income patients? A systematic review and *meta*-analysis. Health Serv. Res. 53 (3), 1777–1798. Available from: https://doi.org/10.1111/1475-6773.12737.

Van Doren, D.C., Blank, K.M., 1992. Patient education: a potential marketing tool for the private physician. J. Health Care Mark. 12 (1), 71–77.

van Nostrum, Elaine, May 21, 2008. A Difficult Patient – Seinfeld Television Series (May 21, 2008). (66) A Difficult Patient – YouTube. <http://www.youtube.com/watch?v=pyossoHFDJg>, May 21.

Van Wesemael, Y., Cohen, J., Onwuteaka-Philipsen, B.D., Bilsen, J., Distelmans, W., Deliens, L., 2009. Role and involvement of life end information forum physicians in euthanasia and other end-of-life care decisions in Flanders, Belgium. Health Serv. Res. 44 (6), 2180–2192. Available from: https://doi.org/10.1111/j.1475-6773.2009.01042.x.

Varney, S. 2012. What's Up, Doc? When Your Doctor Rushes Like the Road Runner. Shots, Health News from NPR. <http://www.npr.org/blogs/health/2012/05/24/153583423/whats-up-doc-when-your-doctor-rushes-like-the-road-runner>.

Vaughan, E., Tinker, T., 2009. Effective health risk communication about pandemic influenza for vulnerable populations. Am. J. Public. Health 99 (Suppl. 2), S324–S332. Available from: https://doi.org/10.2105/AJPH.2009.162537. https://www.ncbi.nlm.nih.gov/pmc/articles/PMC4504362/pdf/AJPH.2009.162537.pdf.

Vgontzas, A.N., Bixler, E.O., Chrousos, G.P., Pejovic, S., 2008. Obesity and sleep disturbances: meaningful sub-typing of obesity. Arch. Physiol. Biochem. 114 (4), 224–236. Available from: https://doi.org/10.1080/13813450802521507.

Walker, J., Leveille, S.G., Long, N., Vodicka, E., Darer, J.D., Dhanireddy, S., et al., 2011. Inviting patients to read their doctors' notes: patients and doctors look ahead. Ann. Intern. Med. 155 (12), 811–819.

Wang, H., Zhao, F., Wang, X., Chen, X., 2018. To tell or not: the Chinese doctors' dilemma on disclosure of a cancer diagnosis to the patient. Iran. J. Public. Health 47 (11), 1773–1774.

Watson, D., Rylander, E., Miner, L.A., Miner, G.D., 2011. Ottawa guidelines for ankle X-rays; an incidence Study at Family Medical Care. In: Paper Presented at the 29th. Annual Scientific Assembly and Retreat, IHI Family Medical Residency Program, Western Hills, OK, June.

Weaver, C., Radnofsky, L., 2013. New Health-Care Law's Success Rests on the Young: Will young and healthy give up disposable income to pay for insurance? The Wall Street Journal. Available from: https://www.wsj.com/articles/SB10001424127887324263404578613700273320428.

White, C., 2012. A comparison of two approaches to increasing access to care: expanding coverage vs. increasing physician fees. Health Serv. Res. 47 (3pt1), 963–983. Available from: https://doi.org/10.1111/j.1475-6773.2011.01378.x.

Wikipedia, 2020. Executive order 13813. <https://en.wikipedia.org/wiki/Executive_Order_13813#:~:text=The%20Executive%20Order%20Promoting %20Healthcare,of%20the%20Obama%20Administration%20is>.

Yang, Y., Nichols, L.M., 2011. Obesity and health system reform: private vs. public responsibility. J. Law, Med. Ethics 39 (3), 380–386. Available from: https://doi.org/10.1111/j.1748-720X.2011.00607.x.

Zaggocare, 2018. What are doctors doing to reduce diagnostic errors?. <https://zaggocare.org/doctors-reduce-diagnostic-errors/>.

Zhou, J., Terluk, M.R., Orchard, P.J., Cloyd, J.C., Kartha, R.V., 2021. *N*-acetylcysteine reverses the mitochondrial dysfunction induced by very long-chain fatty acids in murine oligodendrocyte model of adrenoleukodystrophy. Biomedicines 9 (12), 1826. Available from: https://doi.org/10.3390/ biomedicines9121826. http://nih.gov. 34944641. PMCID: PMC8698433.

Zizzo, N., Bell, E., Lafontaine, A.-L., Racine, E., 2017. Examining chronic care patient preferences for involvement in health-care decision making: the case of Parkinson's disease patients in a patient-centred clinic. Health Expect. 20 (4), 655–664. Available from: https://ezproxy.snu.edu:2292/ 10.1111/hex.12497. https://pubmed.ncbi.nlm.nih.gov/27624704/.

Chapter 8

Regulatory measures—agencies, and data issues in medicine and healthcare

Gary D. Miner

Chapter outline

NOTE: *This Chapter 8 in this 2nd Edition is a very short abstraction of 1st Edition Chapters 4—12; NOTE: the original 1st Edition Chapters 4—12 are available for readers on the Elsevier companion web page for this 2nd Edition.*

Prelude

Capture and storage of health information is provided by a number of organizations described in previous chapters. The effective coordination and communication of this information, however, is necessary to use this information to guide proper healthcare. Several organizations exist to provide some information consistency and regulatory structure within which this information can be used effectively. This chapter will describe the regulator landscape for healthcare as of 2022.

Introduction

In the first edition of this book we provided extensive chapters on several regulatory agencies that exist to help maintain a consistency in health information. In addition we examined "quality control" statistical measures (root-cause analysis, Six-Sigma), evidence-based medicine (EBM), and EHR (electronic health records). In order to devote space to other important concepts in AI and Machine Learning in this 2nd Edition we omitted each of these chapters, but provide two things in their place: (1) a summary of these here; and (2) the original 1st Edition Chapters are provided for the interested reader on this book's Elsevier companion web page.

HIMSS and Organizations that Develop Health Information Technologies (HIT) Standards (summary of Chapter 4 in 1st Edition; for the complete 1st Edition Chapter 4, please see the Elsevier companion web page for this book's 2nd Edition.).

It seems obvious that healthcare could be improved if all the healthcare entities coordinated and communicated better; communication would improve if HIT were standardized around best practices. There are organizations dedicated to improved communication and improved healthcare technology, including the Healthcare Information and Management Systems Society, the American National Standards Institute (ANSI), and the National Coordinator for Health Information (ONC). Whereas ANSI and ONC are standards organizations, the HIMSS not-for-profit organization is broader and

Practical Data Analytics for Innovation in Medicine. DOI: https://doi.org/10.1016/B978-0-323-95274-3.00021-X

concentrates on "optimal use" of IT. HIMSS has over 50,000 international members, who for the most part work in health-care fields. HIMSS enjoys the membership of not-for-profit organizations and corporate members, totaling over 770 organizations (HIMSS, 2013).

Electronic Medical Records: Analytics Best Hope (summary of Chapter 5 in 1st Edition: for the complete 1st Edition Chapter 5, please see the Elsevier companion web page for this book's 2nd Edition.)

NOTE: EMR (Electronic Medical Records) and EHR terminology is used interchangeably in some of the literature, yet others define EMR and EHR precisely, with different, but overlapping, functions; you will find this difference discussed in Chapter 26 (1st Edition: The Future of Global Healthcare, by Gary Miner).

As early as 1965, it was recognized that, "The need for better means of recording and retrieving data for medical records has become increasingly apparent" (Baird and Garfunkel, 1965). For the most part, that statement is as relevant today as it was then. The article goes on to describe a project at St. Christopher's Hospital for Children in Philadelphia, in which certain elements of the medical charts were "recorded via electronic data processing equipment", and then transcribed for the patients' charts. The data were entered via "mark−sense" cards which were manually coded, and then fed into a machine to place a punch in the appropriate column. The card could then be read by a computer, the data tabulated and reports produced. Many advantages were appreciated by users, including legible notes, chronological summaries, statistical reports, and control of utilization of drugs and lab tests. Ultimately, the project was ended due primarily to cost factors. The authors concluded that the ability to provide information in "real time" via a computer "will be associated with considerable expense, with administrative difficulties, and with a need for intensive educational effort." After reading this article almost 50 years later, a present-day physician or administrator who has had any significant experience with electronic medical systems will feel a sense of "déjà vu" (or "paramnesia," using the medical term).

What is an electronic medical records?

In simplest terms, an EMR is a digital reproduction of the information commonly found in a paper medical chart. It is usually locally housed (or remotely hosted but controlled) by the provider or institution, and it may have a degree of interoperability with other electronic systems (e.g., lab, billing) within the institution. This describes the vast majority of systems in use today.

Similar to but broader in scope than the EMR, the EHR, as defined by HIMSS, is a historical electronic record (historical instances of patient health treatment information) generated by one or more encounters in any care delivery setting. Included in this information are patient demographics, progress notes, problems, medications, vital signs, past medical history, immunizations, laboratory data, and radiology reports. The EHR automates and streamlines the clinician's workflow. The EHR has the ability to generate a complete record of a clinical patient encounter. Included in the EHR are other supporting care-related activities gathered either directly or indirectly via some interface, including systems for evidence-based decision support, quality management, and outcomes reporting.

The potential benefits of EHR systems include:

- real-time on-site and remote access to patient information;
- electronic prescription ordering with automatic checking for drug interactions;
- electronic ordering and posting of lab results with some simple analysis, including trending of values;
- enhance charge capture through communication with the billing system;
- clinical decision support, which could prevent or minimize errors in treatment, and provide reminders for needed lab immunications or other procedures;
- clear space used previously for paper medical records;
- decrease or eliminate the need for transcription of information;
- provide reports to facilitate quality improvement;
- provide legible notes; and
- provide valuable data for research on population health and other studies.

Today there are literally hundreds (over 1700 vendors, to be exact, have surfaced since the Rand Report of 2005; not all of these are in business today) of different EMR products with widely differing interfaces and functionality. Many of these systems are "niche" products, dedicated to specific functions (i.e., Pharmacy, Radiology, ER, PT/Rehab, ICU). It is difficult to compare one system to another directly, even though interoperability appears to be the intended design of Meaningful Use mandates. Many of the more established companies do not even provide online demonstrations without the presence of a salesperson. There is no direct interoperability between systems, and most have their own proprietary data structures. This can pose a problem or even be a deterrent for changing products due to the very real potential for losing all or significant amounts of patient data accumulated in the current system because it may not be portable to the new system.

Interoperability of EMR systems is still, as of the time we write this in 2022, one of the biggest problems in easy access of patient data among a patient's various specialist doctors. It is also the biggest issue preventing the availability of good data for making effective predictive models and thus providing the basis for the best medical decisions by a patient and their doctors.

If you are interested in the historical understanding of the development of EMR over the years, we suggest you go to d the 1st Edition Chapter 5 on EMR, which is available to you, the reader, on the Elsevier companion web page for this 2nd Edition.

Open Source EMR and Decision Management Systems: (summary of Chapter 6 in 1st Edition: for the complete 1st Edition Chapter 6, please see the Elsevier companion web page for this book's 2nd Edition.)

Over the past 40 years, open source EMR systems have been among the leading systems in the development and evolution of medical record systems. The focus in this chapter is on free and open-source medical software systems and their potential to develop global EMR systems that are integrated with predictive analytic models and advanced decision management systems. The term "free" software means there is no charge for its use. The term "open-source" means that you have access to the source code: most open source software is also free software. Some free software products do not provide access to the source code.

It can be argued that only commercial software vendors can provide the level of testing and validation necessary to support efficient use of software applications for business and industrial use. The validity of that argument is drawn into question by the success of the Linux and MySQL applications common in many businesses and industries today. Even in the EMR field, comprehensive open source hospital systems like World Vista (see below) have been developed with the close cooperation of the Veterans Administration. The success of VistA (and several other open source EMR applications) has demonstrated that reliable open source software can be developed and supported at a tiny fraction of the cost of commercial solutions. Some reasons for selecting open source EMR solutions include:

1. High cost of commercial EMR systems
2. Maintain control, rather than cede healthcare operations to a software vendor
3. Complexity of operations, tight schedules, and the need for specific customizations in healthcare organizations require considerable software configuration
4. Companies like Medsphere provide implementation services for open source EMR software (http://www.medshpere.com)
5. Open Source software is used commonly in other commercial applications
   a. Linux in a large number of industries
   b. MySQL in many businesses

Five of the best open source electronic medical records systems for medical practices

Several open source EMR systems are targeted for medical practices and clinics. Highlighted below are a few of the leading open source EMR systems that can be used globally and have potential to support analytic models.

1. **The OSCAR EMR System**—OSCAR is the only widely deployed open source EMR system used in Canada. The name "OSCAR" is an acronym for "Open Source Clinical Application Resource."
2. **OpenEMR**—OpenEMR is an open source EMR and practice management application suite that includes patient billing and a patient portal. The website (http://www.open-emr.org/) describes the OpenEMR community and provides full documentation, demos, and software downloads.
3. **Peace Corps (PC)-EMR**—The PC back at the time we wrote the 1st Edition of this book (2012—13) was planning on developing their own open EMR system. However, in 2013—14 they adopted the OpenEMR SYSTEM maintained by EnSoftek (https://www.ensoftek.com/wp2015/?case-studies = peace-corps-ehr-pcmedics) to implement and maintain it (Maduro and Herr, 2013). This was probably because by that time OpenEMR was one of the most widely implemented EMR systems in the world with more than 15,000 installs around the globe and translations into 19 languages. In 2020, however, the Peace Corp temporarily discontinued services because of COVID, and as we write this chapter we are not sure of the status of the PC today, although they have made several announcements that they plan to return to service (https://www.peacecorps.gov/news/library/peace-corps-takes-steps-return-service-overseas/).

4. **ZH Healthcare: a Medical Records System used in India: A cloud-based implementation of the OpenEMR—** We understand that this started out as a partnership between HP and ZH Healthcare in India (https://www.prweb.com/releases/2012/7/prweb9699462.htm). It is done as a Health IT as a Serevice, committed to making healthcare records affordable to everyone (https://zhhealthcare.com/). ZH Healthcare is a global health IT innovator and developer of blueEHR, one of the most powerful EHR, EMR, and Revenue Cycle Management solutions used by patients, physicians and healthcare providers, health systems, and health IT companies in over 100 countries. Today it is available in 35 languages (https://blueehr.com/secure-cloud-based-architecture-disaster-recovery-business-continuity/). Their address is: 33/2165, Zaiham Tower, Vennala, P.O, Kochi, Kerala 682028, India (https://zhhealthcare.com/); however, their headquarters today appears to be in Virginia (https://www.cbinsights.com/company/zh-health-care). Their LinkeIn page (https://www.linkedin.com/company/zhhealthcareinc/) proclaims that they make health IT (EMR, etc) affordable and accessible to everyone—see https://blueehr.com/about-us/ which also lists a Bethesda, MD address. So today they seem to be located in several offices around the world and easily accessible.

5. **OpenMRS by Partners in Health**—In 2001, Partners In Health designed and deployed its first web-based EMR system. PIH's initial EMR system was implemented to manage the treatment of patients with multidrug-resistant tuberculosis at PIH's sister organization in Peru, Socios En Salud. Later, the Medical Informatics team adapted this EMR to support PIH's HIV treatment programs in rural Haiti. In both countries, the original PIH EMR has proved to be a valuable clinical and program management tool.

(If you, the reader, are interested in more details of some of the above Open Source EMR, please see the 1st Edition Chapter 6, located on the Elsevier companion web page for this book's 2nd Edition.)

EBM: (summary of Chapter 7 in 1st Edition; for the complete 1st Edition Chapter 7, please see the Elsevier companion web page for this book's 2nd Edition.)

EBM refers to the practice of medicine from the primary basis of empirical evidence gathered from cases in the past. It is distinguished from the historical practice of medicine primarily in its emphasis on objective elements of decision-making (data, analyses, and conclusions), rather than subjective elements (intuition, judgment, and "informed" guesses). EBM appears to be the current "rage" in professional and regulatory circles to such an extent that historical medical practice based on "best" practices appears be in danger of being shoved aside. The subject role of the physician should not be discarded, though. Elements of objective data and subjective judgment should be combined into a holistic view of the medical reality the physician is attempting to diagnose and treat. EBM is a necessary (and indeed the primary) basis for medical decision-making. But, these objective elements must be combined with subjective judgment for the physician to treat the person, not just the disease or disability.

But what about Predictive Analytics and EBM? EBM is yet to be proven in itself as a better way to practice medicine. So far, there are no controlled trials proving the superiority of the EBM approach. It is plausible, however, to theorize that using the tools of analytics, it may be possible to gather and analyze data in ways that allow more rapid and complete determination of the best evidence. Even so, there will still be the challenge of translating that knowledge to action at the bedside—the human understanding of the analytics and other factors will remain part of the decision process.

(For those readers wishing more information on EBM, please see the complete 1st Edition Chapter 7, on the Elsevier companion web page for this book's 2nd Edition.)

ICD 10 (International Classification of Disease Version 10): (summary of Chapter 8 in 1st Edition: for the complete 1st Edition Chapter 8, please see the Elsevier companion web page for this book's 2nd Edition.)

Efforts to classify the causes of morbidity and mortality are not new. The first international classification, the International List of Causes of Death, was introduced by the International Statistical Institute (ISI) in 1893 at the ISI World Congress in Chicago. The ICD has been revised a number of times and published in a series of editions to reflect advances in health and medical science over time (Weigel and Lewis, 1991; Bebbington, 1992; Innes et al., 1997; Topaz et al., 2013; Organization, 2013a).

Rise of the international classification of disease

In 1948, the then new World Health Organization (WHO) took over scrutiny of the classification and published the 6th version. Coincident with this transition, the International Classification of Disease (ICD) assumed its present name. ICD-6 was the first edition that incorporated morbidity (illness) as well as mortality. The WHO Nomenclature Regulations were adopted in 1967, and mandated that organizational member countries use the most current ICD revision to compile their mortality and morbidity statistics (Organization, 2013a; Mulvihill, 2011; Maguire, 2012).

The ICD is now in its tenth revision. ICD-10 was approved in 1990 by the Forty-third World Health Assembly (Organization, 2013a,b). There are well over 20,000 scientific articles that have cited ICD-10 metrics. ICD has been translated into 43 languages (Organization, 2013b). Approximately 120 countries worldwide use the classification to report their morbidity and mortality statistics. The ICD has become the global health information standard for mortality and morbidity statistics (Brouch, 2000; Innes et al., 2000; ICD-10, 2005; Dimick, 2008; Organization, 2013c). ICD is increasingly relevant not only in clinical care but in research settings, and it has been used to define diseases processes and patterns. New applications have taken ICD-10 into the realm of healthcare management as well as outcomes research. Importantly, in the new paradigm of accountable care, it has been used to effectively allocate resources (Dimick, 2012a; Hartman et al., 2012; Kostick, 2012). Mortality data are reported by more than 100 countries. As a major indicator of health outcomes, ICD-10 helps to monitor morbidity and mortality rates. These metrics can be used to gauge success in reaching Millennium Development Goals. $3.5 billion or roughly 70% of the world's health expenditures are allocated using ICD reimbursement tools (Organization, 2013c; Implementation of the International Statistical Classification of Diseases and Related Health Problems, 1997; Bowman and Scichilone, 2013). ICD-10 includes better healthcare metrics, which facilitate data tracking and permit a more comprehensive view of history and treatments. This more complete and flexible system of data tracking can promote better decision-making in both clinical and financial areas, leading to more equitable reimbursements (Baldwin, 2010; Danzig, 2010).

(For those readers wishing more information on ICD, please see the complete 1st Edition Chapter 8, on the Elsevier companion web page for this book's 2nd Edition.)

Meaningful Use—The new (in 2013–14) buzz word in medicine—Today (in 2022) this term is rarely used, as its "principles" have been, for the most part, adopted in healthcare services over the past 7 years: (summary of Chapter 9 in 1st edition: for the complete 1st Edition Chapter 9, please see the Elsevier companion web page for this book's 2nd Edition.)

As it pertains to healthcare, meaningful use is the new nirvana. This term, "meaningful use," has been bantered around enough that we "think" we know what it means when we see it. Nevertheless, there is a dichotomy of thought as to what the best path to achieve a "meaningful product" that embodies the use of the term. That said, most medical professionals are clear in their need to adapt to this new pathway but are less clear in terms of what this means for them for the near and distant future.

For example:

- Is it sufficient to accept the hospital's recommendation for a specific MIS (medical information system), HIS (health information system) or EMR product that will comply with meaningful use despite the fact that the software is not designed for the particulars of the office or specialty practice?
- What if the software vendor no longer meets certain objectives?
- Does the physician practice lose the opportunity to participate in certain incentive-based programs?
- Further down the line, will the physician be punished if the office uses software that no longer carries the "meaningful use" imprimatur?

There needs to be a defined path to achieve certain concrete objectives designed to meet the new mandates. Moreover, there needs to be a certain guarantee that compliance to this process will be rewarded in a manner that is neither arbitrary, nor capricious.

Today, in the early part of 2022, most of the goals of meaningful use (Phase I, Phase II, and Phase III) have been more-or-less implemented. However, one in particular, the *patient portal*, is now a part of most healthcare services, so that the clinicians involved can document that they have implemented this aspect of meaningful use, but the author writing this finds that few of the patient portals really work well: some health services have excellently designed patient portals but do not know how to use them, or just do not use/respond to the patient on the portal. Other health services make it a goal to respond to a patient portal patient inqjuiry within 24 or 48 hours, and do so. But this author has yet to see a patient portal where one can attach a pdf or word-doc or other document, and instead has to get the doctor's email address and email documents. The doctors have the same issues, in that on most patient portals they cannot attach pdf or other attachments to portal messages. However, in most cases, most portals allow the patient to see their lab test results in one form or another, and in many cases download a pdf of these.

(For those readers that wish to learn more about the Meaningful Use topic, please see the complete 1st Edition Chapter 9 on the Elsevier companion web page for this book's 2nd Edition.)

The Joint Commission (TJC) (Formerly called **JACHO = Joint Commission on Accreditation of Healthcare Organizations):** (summary of Chapter 10 in 1st edition: for the complete 1st Edition Chapter 10, please see the Elsevier companion web page for this book's 2nd Edition.)

Ernest Codman was a Boston surgeon who practiced in the early part of the 20th century. He was an acknowledged advocate of hospital reform, a pioneer in the field of outcomes management in patient care, and the first American physician to track patient outcomes past their initial hospitalizations in a systematic manner. Using "End Result Cards," Dr. Codman kept basic demographic data on every patient. Diagnosis, treatment, and outcome were codified. Each patient was tracked on for at least 12 months to determine long-term outcomes. The "end results system" created an opportunity to identify clinical misadventures. These clinical misadventures would be used as the basis for improving the care of future patients by focusing attention on process improvement. Dr. Codman felt that both clinical information and process improvement should be made public so that patients could make an informed choice of physicians and hospitals. These efforts contributed to the founding of the American College of Surgeons Hospital Standardization Program (Roberts et al., 1987; Pearre, 1955; McCleary, 1977). The Joint Commission on Accreditation of Hospitals (JCAH) was created by merging the Hospital Standardization Program with like quality measures contributed by the American College of Physicians, the American Hospital Association, and the American Medical Association (Roberts et al., 1987; Pearre, 1955).

From its founding in 1951, the JCAHs developed processes that were largely targeted towards developing the metrics for the certification process and analysis of what constituted grounds for corrective action (JOINT accreditation commission is proposed, 1951; BYLAWS of the Joint Commission on Accreditation of Hospitals, 1952). In 1965 the federal government selected the organization's accreditation process as the one that would be used to qualify a hospital for the Medicare Conditions of Participation (i.e., accreditation at the federal level) (JOINT accreditation commission is proposed, 1951; BYLAWS of the Joint Commission on Accreditation of Hospitals, 1952; JOINT Commission accredits 199 hospitals, 1953; JOINT Commission on Accreditation of Hospitals, 1955; ACCREDITATION of hospitals; some policies and actions of the joint commission in regard to who may do surgery, consultations, chiropodists, and many other matters, 1955; Carroll, 1971). Although this process had been in place for 45 years, as of July 15, 2010, Section 125 of the Medicare Improvements for Patients and Providers Act of 2008 changed the certifying agency to the Centers for Medicaid and Medicare Services. During the transition period, a methodology involving corporation between the two agencies was worked out to avoid a delay or mismanagement of the accreditation process (Card and Lehmann, 1987; Medicare and Medicaid programs; recognition of the Joint Commission on Accreditation of Healthcare Organizations standards for home care organizations—HCFA, 1993; Jost, 1994).

At the time of its first inception, TJC was known as the JCAH. Subsequently, TJC was known as the Joint Commission on Accreditation of Healthcare Organizations. It was only more recently that the organization transitioned to its current identity as "The Joint Commission" (Roberts et al., 1987; Pearre, 1955; McCleary, 1977; Tousignaut, 1977; Joint Commission, 1988; Franko, 2002). The organization is currently based in Oakbrook Terrace, Illinois but maintains a virtual presence across the United States through numerous satellite networks of reviewers. The organization mission statement defines TJC's function "to continuously improve healthcare for the public, in collaboration with other stakeholders, by evaluating healthcare organizations and inspiring them to excel in providing safe and effective care of the highest quality and value" (Rich, 1996; Flanagan, 1997).

(For those readers that wish to learn more about the Joint Commision topic, please see the complete 1st Edition Chapter 10 on the Elsevier companion web page for this book's 2nd Edition.)

Root-Cause Analysis, Six-Sigma and overall Quality Control and Lean Concepts: (summary of Chapter 11 in 1st edition: for the complete 1st Edition Chapter 11, please see the Elsevier companion web page for this book's 2nd Edition.)

Six Sigma

Six Sigma aims to achieve quality by reducing variation in processes. Six Sigma aims to reduce error rates such that six standard deviations around the mean of a process fit between the upper and lower tolerances. For example a manufacturer producing lock barrels, had to make sure the diameters of the barrels were right. If they were too large they would not fit the assembly of the lock and if they were too small, they also would not fit. The upper and lower control limits ideally include six standard deviations on each side of the mean to result in no more than 3.4 bad parts in 1,000,000 opportunities for error (99.9997% accuracy) (Harry and Crawford, 2004). Of course, that very stringent standard is often modified, depending upon the analysis of the process for a company. There is a curve of diminishing returns and if the return on investment (ROI) is less with more precision, then less precision is tolerated until the ROI diminishes. (Finding that point of diminishing returns would be an opportunity for predictive analytics.) Harry and Crawford (2004) used an airline example. Baggage arrivals are on about a four-sigma process, whereas passenger arrival (fatality rate) is on a six sigma process. "To put that in perspective, a person is about 1800 times more likely to get to his destination safely than his luggage is" (p. 8).

Early in the 20th century in the United States, more error was tolerated in production. If a company owned a majority of the market share then if a production method produced 92% good parts and 8% bad parts, there was little concern. It was easier to do what had always been done and not modify systems. The bad parts were simply discarded or reworked. If a company has the market, then it does not have to worry too much about sending out bad parts, as long as it is making a comfortable or increasing profit. If energy is abundant, the companies did not care if energy was wasted. If air was plentiful, then a little pollution wouldn't hurt—it would simply blow away. If, on the other hand, other companies start competing with better products, as did Japan after WWII, with the help of D. Edward Deming, then suddenly American companies started losing the market share. Japan was using Six Sigma methodology (interestingly a US invention) and the United States was not (Carson and Carson, 1993). Scarcity of natural resources and governmental regulations further propelled companies to reduce waste and to streamline processes. One question as we apply quality concepts to healthcare: what model has been used in the past and what will constitute the future? And what happens if a government takes over healthcare?

Deming's famous 14 points (Anderson et al., 1994) have emphasized continuous improvements in quality through excellence in leadership, training, and trust and elimination of the fear of making mistakes, of quotas, and of barriers to quality. He ended by saying that transformation is everyone's job from the top to the bottom of an organization.

Quality control

Total Quality Management (TQM) is sometimes used synonymously with Six Sigma because quality management generally combines the standardization of Six Sigma with the processes of DMAIC (defining, measuring, analyzing, improving, and controlling) (Chang, 2009). However, TQM and DMAIC are but two of the methodologies of Six Sigma. Levels of Six Sigma in terms of amount of training and in levels of savings to the company are generally green belt and black belt. Information on certification, levels, and amounts of money saved by an organization may be found on the Six Sigma website (Six Sigma, 2013). The various methodologies of Six Sigma may be found on the Six Sigma website as well.

Quality attainment in medical settings, including hospitals, aims to provide the faster service while increasing quality care. If patients are seen in a timely fashion, with accurate diagnoses and treatments, then the healing can begin sooner and costs can be saved. Faster and more accurate processes lead to serving more patients within a time interval and increasing income, while avoiding money suckers such as hiding and fixing errors and paying out lawsuits. Faster and more accurate processes also mean increasing patient satisfaction and even more important, helping people to live longer by increasing health.

Lean concepts for healthcare: the lean hospital as a methodology of Six Sigma

Lean hospitals practice DMAIC. Lean clinics practice DMAIC. As processes are examined, inefficient and noncontributing processes are lopped off (surgically removed), resulting in a cleaner and leaner organization. Mazur, McCreery, and Rothenberg (2012) discussed the implementation of lean processes in hospitals during the first year of implementation. The authors identified two major processes, single loop and double loop learning models in "the detection and the elimination of error" (p. 11). First loop issues are generally dealt with first. The first loop finds errors and attempts to correct them. The deep values or root issues are not dealt with at first. After the first and immediate issues (symptoms) have been solved and put into the "control" mode, the deep roots and values issues are dealt with. In the second loop the hospital reflects about what its values are, directions it wants to move, and what those directions mean. The authors stated that it is important to help employees get to the second level of reflection because it is at that level at which a deeper understanding and commitment to the lean process occurs.

Many lessons can be learned from an institution that involved the entire institution in quality improvement, such as Mount Sinai in 2000 (Chassin, 2008). Such a large undertaking had to have top-down support, and it did, but the author warned that hospital settings are not exactly like manufacturing plants and that top-down mandates do not always match the subsystem culture but that teamwork was needed. Deming no doubt would agree even for manufacturing facilities. Much time, effort, training, and money no doubt were given to such an undertaking, but in the end, the efforts paid off. Many projects were undertaken in the data-driven environment. One thing they learned was to depend on data rather than on people's perceptions of how they thought things were going. They carefully measured baselines and then found the metrics that would ensure reliability of measure. Use of control charts aided in their collection and analysis. Some of their improvements were a "28% reduction in excess dosing in patients with kidney dysfunction, 52 minute reduction in turnaround time, 700% improvement in tracking of patient controlled analgesics, 90 minute reduction in bed turnover time, 91% defect reduction in cardiac stents, AICD, pacemakers, 85% defect reduction in chemotherapy revenue capture, and 90% defect reduction in OB/GYN revenue enhancement" (pp. 51—52). The last three provided savings and revenue of over $7 million. Definitely, Six Sigma paid off for them in terms of money but no doubt it paid off in terms of increasing good patient outcomes. The 1st Edition Chapter 12 of this book (please see Elsevier companion web page for that chapter) discusses the many aspects of the lean hospital movement and provides excellent examples.

Root cause analysis

Root cause analysis (RCA) is one of the tools used within Six Sigma (Six Sigma, 2013). The "five whys" is a method of discovering root causes. Continuously asking why can help drive the discovery of reasons for errors (Six Sigma, 2013; Connelly, 2012). The most fundamental reason for the failure or the error is the root cause (Dattilo and Constantino, 2006).

"The five whys" is a technique used many times when human interactions are involved in the processes from which the problem or error came. For example, when the patient in Chapter 1, Why this book?—What we want to accomplish with this second edition of our first "big green book" died because the computer froze, it was important to ask why it happened.

- Why did the patient die? Because the operation could not continue.
- Why could the operation not continue? Because the computer screen froze.
- Why did the computer screen freeze? Because the computer had not been powered down for many days of use.
- Why was the machine not powered down? Because those running the surgeries did not know the computer needed to be shut down and it was faster simply to hibernate the machine.

After determining the root cause of the problem, the problem could be solved and someone was designated to power down the computer after operations. Problem discovery for healthcare systems often is more complex than the previous example in which many causes are found to interact with an event. Connelly (2012) suggests involving all stakeholders in an interdisciplinary team is important for drilling down to causes. The team needs to answer what happened, how did it happen, why did it happen and what should be done to prevent its reoccurrence (p. 316).

(For those readers that wish to learn more about Root-Cause Analysis, Six Sigma and overall quality control and lean concepts topics, please see the complete 1st Edition Chapter 11 on the Elsevier companion web page for this book's 2nd Edition.)

Lean Hospital Examples: (summary of Chapter 12 in 1st edition: for the complete 1st Edition Chapter 12, please see the Elsevier companion web page for this book's 2nd Edition.)

Increasingly hospitals embrace lean concepts. They do so for many reasons: to ensure best practices, to avoid litigation, for accreditations standards and better patient outcomes, because they see the principles working so well in industry for the bottom line, and because the current perception is that the best hospitals practice lean principles. These are but some of the reasons. Lean practices that incorporate predictive analytics are becoming the future of medicine.

When lean concepts first started hitting hospitals at the turn of the millennium, Alan Mulally, at that time the CEO of Ford Motor Company, who had previously been in top leadership positions at Boeing where he had used lean principles, started applying this to the Henry Ford Hospitals. Mullally's previous position with Boeing was as head of commercial planes. He had worked at Boeing for 38 years before moving to Ford where he also used lean principles, such as "working together" (Hoffman, 2012). His leadership at Boeing achieved lean production. Mulally was also a major reason Boeing pulled up its nose after an economic dive. He joined Ford in 2006 as CEO and Ford's stock steadily rose (Hoffman, 2012), in part by using lean concepts (although it should be noted that Henry Ford was a very early creator of lean concepts when he opened his first plant using assembly lines and standard practices in 1913; Strouse, 2008). Fig. 8.1 shows Mulally at a news conference in New York in December 2012.

Henry Ford Hospitals and Virginia Mason Hospital

What is interesting about Mulally's career trajectory is that the Henry Ford Health System adopted a lean philosophy, taking its training from Henry Ford Production Systems (Zarbo, 2013). Zarbo announced the 2014 training sessions on the hospital website for "physicians, nurses, technologists, pathologists, residents, directors, managers, administrators, quality and medical officers" (p. 1). In his announcement, Zarbo stated, "This approach to LEAN is based upon Deming's management principles of leading and practicing in a culture that uses manufacturing-based work rules and process improvement tools derived from the Toyota Production System" (p. 1). The sessions were also to focus on the

FIGURE 8.1 Alan Mulally at a meeting in New York in December, 2012 (Norihiko and Ingrassia, 2013). *From Norihiko, S, Ingrassia, P., Oct. 18, 2013. Ford's Mulally won't dismiss Boeing, Microsoft speculation <http://www.reuters.com/article/2013/10/18/us-ford-mulally-idUSBRE99H15P20131018>, p. 1.*

problem-solving method, PDCA, which stands for plan, do, check, and act. So, using an outside entity, healthcare improved for this hospital.

When Virginia Mason Hospital (in the Seattle, Wasington area) decided it wanted to overhaul itself in 2001, it noticed Boeing's strides. Boeing had been using Toyota Motor Company as its model. Toyota was transformed by Taiichi Ohno who developed many of the kaizen methods (Kaplan, 2012). Interestingly Ohno developed his ideas by observing supermarkets in the United States and seeing how the customers could select anything they wanted without delay.

And so, two major hospitals were influenced by industry, and in particular, by two motor companies

(For those readers that wish to learn more about the Lean Hospital Movement, including many more examples, please see the complete 1st Edition Chapter 12 on the Elsevier companion web page for this book's 2nd Edition.)

Postscript

Within the boundaries of healthcare regulation, much work can be done to develop the use of digital information to guide personalized healthcare. But, before some examples of the use of predictive analytics in healthcare are presented in Chapters 11–17, some distinctions should be made among the types of digital data available to healthcare analysts. Chapter 9, Predictive analytics with multiomics data will discuss the topic of metabolomics, in which various metabolites present in bodily effluents (e.g., blood, saliva, and breath) can be analyzed and related to many dysfunctions and diseases of the body.

References

Accreditation of hospitals; some policies and actions of the joint commission in regard to who may do surgery, consultations, chiropodists, and many other matters. Ohio Med 51, 168–170, 1955.

Anderson, J.C., Rungtusanatham, M., Schroeder, R.G., 1994. A theory of quality management underlying the deming management method. Acad. Manag. Rev. 19 (3), 472–509. Available from: https://doi.org/10.5465/AMR.1994.9412271808.

Baird, H.W., Garfunkel, J.M., 1965. Electronic data processing of medical records. N. Engl. J. Med. 272 (23), 1211–1215.

Baldwin, G., 2010. Is ICD-10 the industry's wake up calls? Health Data Manag. 18, 46–48. 50, 52.

Bebbington, P., 1992. Welcome to ICD-10. Soc. Psychiatry Psychiatr. Epidemiol. 27, 255–257.

Bowman, S., Scichilone, R., 2013. ICD-10-CM/PCS part of a global network of information standards. J. AHIMA/Am. Health Inf. Manag. Assoc. 84, 52–53.

Brouch, K., 2000. Where in the world is ICD-10? J. AHIMA/Am. Health Inf. Manag. Assoc. 71, 52–57.

BYLAWS of the Joint Commission on Accreditation of Hospitals, 1952. Bull. Am. Coll. Surg. 37, 238–244.

Card, W.F., Lehmann, R., 1987. An overview of the methodology used by the Joint Commission to evaluate Medicare-certified HMOs. QRB. Qual. Rev. Bull. 13, 415–417.

Carroll, W.W., 1971. Joint Commission myth—(and the reality.). AORN J. 14, 37—41.

Carson, P., Carson, K.D., 1993. Deming vs traditional management theorists on goal setting: Can both be right? Bus. Horiz. 36 (5), 79.

Chang, J.L., 2009. Six Sigma and TQM in Taiwan: an empirical study of discriminate analysis. Total. Qual. Manag. Bus. Excell. 20 (3), 311—326. Available from: https://doi.org/10.1080/14783360902719675.

Chassin, R., 2008. The Six Sigma Initiative at Mount Sinai Medical Center. Mt. Sinai J. Med. 75 (1), 45—52. Available from: https://doi.org/10.1002/msj.20011.

Connelly, L.M., 2012. Root cause analysis. MEDSURG Nurs. 21 (5), 316-313.

Danzig, C., 2010. Billing & coding. ICD-10: experts fear executive attention lags. Hosp. Health Netw./AHA 84, 19.

Dattilo, E., Constantino, R.E., 2006. Root cause analysis and nursing management responsibilities in wrong- site surgery. Dimens. Crit. Care Nurs. 25 (5), 221—225.

Dimick, C., 2008. ICD-10 postcards. Canadians, Australians share experiences with ICD-10 implementation. J. AHIMA/Am. Health Inf. Manag. Assoc. 79, 33—35.

Dimick, C., 2012a. Welcome to ICD-10 university: lessons learned from the ICD-10 summit. J. AHIMA/Am. Health Inf. Manag. Assoc. 83, 38—41.

Flanagan, A., 1997. Ensuring health care quality: JCAHO's perspective. Joint Commission on Accreditation of Healthcare Organizations. Clin. Therap. 19, 1540—1544.

Franko, F.P., 2002. The important role of the Joint Commission. AORN J. 75, 1179—1182.

Harry, M.J., Crawford, J., 2004. Six Sigma for the little guy. Mech. Eng. 126 (11), 8—10.

Hartman, K., Phillips, S.C., Sornberger, L., 2012. Computer-assisted coding at the Cleveland Clinic: a strategic solution. Addressing clinical documentation improvement, ICD-10-CM/PCS implementation, and more. J. AHIMA/Am. Health Inf. Manag. Assoc. 83, 24—28.

HIMSS Analytics, 2013. History <http://www.himssanalytics.org/about/history.aspx> (accessed 28.08.13).

Hoffman, B.G., 2012. American Icon: Alan Mulally and the Flight to Save Ford Motor Company. Random House, Inc, New York.

ICD-10: the Canadian experience, 2005. J. AHIMA/Am. Health Inf. Manag. Assoc. 76, 62—63.

Implementation of the International Statistical Classification of Diseases and Related Health Problems, 1997. Tenth Revision (ICD-10). Epidemiol. Bull. 18, 1—4.

Innes, K., Hooper, J., Bramley, M., DahDah, P., 1997. Creation of a clinical classification. International statistical classification of diseases and related health problems—10th revision, Australian modification (ICD-10-AM). Health Inf. Manag.: J. Health Inf. Manag. Assoc. Aust. 27, 31—38.

Innes, K., Peasley, K., Roberts, R., 2000. Ten down under: implementing ICD-10 in Australia. J. AHIMA/Am. Health Inf. Manag. Assoc. 71, 52—56.

Six S, 2013. Determine the root cause: The 5 whys. <http://www.isixsigma.com/tools-templates/cause-effect/determine-root-cause-5-whys/> (accessed 11.4.13.).

JOINT accreditation commission is proposed, 1951. Mod. Hosp. 76, 70.

JOINT Commission accredits 199 hospitals, 1953. Bull. Am. Coll. Surg. 38, 159. passim.

JOINT Commission on Accreditation of Hospitals, 1955. J. Am. Med. Assoc. 157, 1614—1615.

Joint Commission: who, when, why?, 1988. Disch. Plan. Update 8, 9—10.

Jost, T.S., 1994. Medicare and the Joint Commission on Accreditation of Healthcare Organizations: a healthy relationship? Law Contemp. Probl. 57, 15—45.

Kaplan, G.S., 2012. Waste not: the management imperative for healthcare. J. Healthc. Manag. 57 (3), 160—166.

Kostick, K.M., 2012. Coding diabetes mellitus in ICD-10-CM: improved coding for diabetes mellitus complements present medical science. J. AHIMA/Am. Health Inf. Manag. Assoc. 83, 56—58. quiz 59.

Maduro, R.A., Herr, C., November, 2013. U.S. Peace Corps Adopts OpenEMR for use in 77 Countries World-wide <https://www.openhealthnews.com/hotnews/us-peace-corps-adopts-openemr-use-77-countries-world-wide>.

Maguire, N., 2012. ICD-10-CM diagnosis coding (October 1, 2014). J. Med. Pract. Manag.: MPM 27, 393—394.

Mazur, L., McCreery, J., Rothenberg, L., 2012. Facilitating lean learning and behaviors in hospitals during the early stages of lean implementation. Eng. Manag. J. 24 (1), 11—22.

McCleary, D., 1977. Joint Commission on Accreditation of Hospitals—twenty-five years of promoting improved health care services. Am. J. Hosp. Pharm. 34, 951—954.

Medicare and Medicaid programs; recognition of the Joint Commission on Accreditation of Healthcare Organizations standards for home care organizations--HCFA, 1993. Final notice. Fed. Register 58, 35007—35017.

Mulvihill, L., 2011. The national transition from ICD-9 to ICD-10. J. Registry Manag. 38, 100—101. quiz 108—109.

Norihiko, S., Ingrassia, P., Oct. 18, 2013. Ford's Mulally won't dismiss Boeing, Microsoft speculation. <http://www.reuters.com/article/2013/10/18/us-ford-mulally-idUSBRE99H15P20131018>.

Organization, W. H., 2013a. History of the development of the ICD. <http://www.who.int/classifications/icd/en/HistoryOfICD.pdf>.

Organization, W. H, 2013b. Home Page of the World Health Organization. <http://www.who.int/classifications/icd/en/>.

Organization, W. H., 2013c. International Statistical Classification of Diseases and Related Health Problems 10th Revision. <http://apps.who.int/classifications/icd10/browse/2010/en>.

Pearre, A.A., 1955. The history and organization of the Joint Commission on Accreditation of Hospitals. Md. State Med. J. 4, 698—700.

Rich, D.S., 1996. Meeting Joint Commission requirements for competence assessment. Am. J. Health-System Pharm 53 (726), 729.

Roberts, J.S., Coale, J.G., Redman, R.R., 1987. A history of the Joint Commission on Accreditation of Hospitals. JAMA: J. Am. Med. Assoc. 258, 936—940.

Strouse, R., 2008. Adopting a lean approach. EE: Eval. Eng. 47 (4), 56–60.

Topaz, M., Shafran-Topaz, L., Bowles, K.H., 2013. ICD-9 to ICD-10: evolution, revolution, and current debates in the United States. Perspect. Health Inf. Manag./AHIMA, Am. Health Inf. Manag. Assoc. 10, 1d.

Tousignaut, D.R., 1977. Joint Commission on Accreditation of Hospitals' 1977 standards for pharmaceutical services. Am. J. Hospital Pharm. 34, 943–950.

Weigel, K.M., Lewis, C.A., 1991. Forum: in sickness and in health—the role of the ICD in the United States health care data and ICD-10. Top. Health Rec. Manag. 12, 70–82.

Zarbo, R. Henry Ford Health System Pathology and Laboratory Medicine: On site lean training. Henry Ford System Training Sessions. <http://www.henryford.com/body.cfm?id = 50135> (accessed 11.7.13).

Further reading

Ahima, 2005. U.S. must adopt ICD-10-CM and ICD-10-PCS. J. AHIMA/Am. Health Inf. Manag. Assoc. 76 (26), 28.

Bowman, S., 2008. Why ICD-10 is worth the trouble. J. AHIMA/Am. Health Inf. Manag. Assoc. 79, 24–29. quiz 41-22.

Conn, J., 2006. House passes IT bill deadline set for ICD-10 implementation. Mod. Healthc. 36, 12.

Dimick, C., 2012a. 'Don't slow down': an ICD-10 summit wrap-up. J. AHIMA/Am. Health Inf. Manag. Assoc. 83 (52–56), 58.

Dimick, C., 2012b. ICD-10 delay impacts all sectors of healthcare: industry attempts to answer the question 'what now'? J. AHIMA/Am. Health Inf. Manag. Assoc. 83, 32–37.

Kloss, L., 2005. The promise of ICD-10-CM. Health Manag. Technol. 26 (48), 47.

Schwend, G., 2007. Expanding the code. The methodical switch from ICD-9-CM to ICD-10-CM will bring both challenges and rewards to healthcare. Health Manag. Technol. 28 (12), 14.

Chapter 9

Predictive analytics with multiomics data

Robert A. Nisbet

Chapter outline

Prelude

Medical science and practice is becoming increasingly data-driven. These data are being derived from studies of an increasing number of biological processes generating many different kinds of data. The challenge for today is developing effective ways to combine those data into compatible data sets that are properly prepared for analysis. This chapter will focus largely on the proper integration and preparation of diverse data sets necessary to support diagnostic analysis and the development of effective treatment protocols.

Introduction to multiomics

Multiomics is a group of information entities that includes all of the individual functions of each biological process participating in the overall function of the cell. The primary categories of multiomics were introduced as "Omics," and described in Chapter 13, 1st Edition, to include genomics, epigenomics, transcriptomics, proteomics, glycomics, and metabolomics. The set of characteristic components of each of these Omics categories compose the "Ome," such that the metabolome consists of all of the metabolites present in across all instances of it. Similarly, the genome is composed of genetic elements (e.g., genes), the proteome is composed of proteins, and the transcriptome is composed of genetic transcription elements like transfer-RNA. Each of the Omics relates to multiple tissues, to each other (according to the arrow connections), and to multitissue networks, as shown in Fig. 9.1. Areas of secondary interest described in Chapter 13, 1st Edition, include metagenomics, glycomics, and nutrigenomics. Microbiomics data (composing the microbiome) have become very important as a basis of many studies during the last 5 years, particularly regarding bacteria (e.g., fecal transplants).

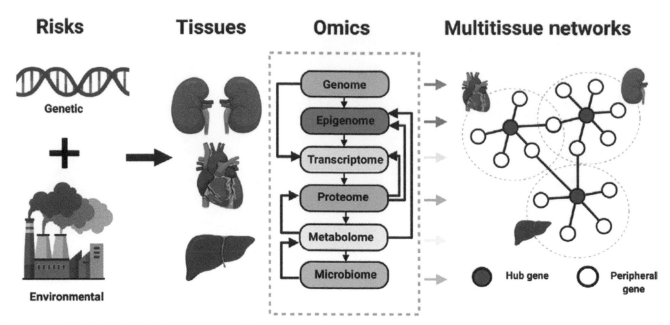

FIGURE 9.1 Relationship of Omics groups to multiple tissues and tissue networks. *Source: From Yang, X., 2020. Multitissue multiomics systems biology to dissect complex diseases. Trends Mol. Med. 26 (8), 718–728.*

Genomics

The human genome contains 30,000—35,000 genes and encodes for nearly 100 trillion cells in the human body (Guttmacher and Collins, 2002). The first four Omics shown in Fig. 9.1 (genomics, epigenomics, transcriptomics, and proteomics) are sometimes referred to as genomics layers. Metabolomics and microbiomics are focused on molecules and organisms not directly related to genetic processes. The biological and chemical processes related to each of these genomics layers has been studied in hundreds of books and papers over the last 15 years. Out of these studies has emerged recognition of the need for integration of data for all of these Omics layers and for those of many others.

Multiomics

The term "multiomics" has been used by many investigators in the last 5 years to refer to that combined body of information. Yang (2020) described multiomics not only in terms of the included cellular functional entities, but also in terms of the interaction between them (see Fig. 9.1). Because of the strength of some of these interactions, Yang (2020) views them within the context of systems biology, which includes all aspects of structure and function of the biological cell. This view is crucial to permit the study of all aspects of diagnosis and treatment of diseases. For example, the strong interaction between cardiovascular disease (CVD) and its associated risk factor, type-2 diabetes, requires the integration of information from both data domains to guide development of effective treatment strategies for both conditions.

Fig. 9.2 shows a graphical representation of the interactions between genome-wide association studies (GWAS), CVD, and various biological functions.

Despite many decades of extensive studies, many complex diseases remain without an effective treatment. Yang (2020) studied CVD and its associated risk factor, type-2 diabetes. He explains that one of the reasons for the lack of effective treatment is the many interacting genetic risk factors identified by GWAS, shown in Fig. 9.2. Another reason is the diversity of environmental factors (e.g., diet, exercise, pollutants, and pathogens) which have a significant effect on the development of CVD. These interacting factors compose a network of relationships that includes molecular entities also (e.g., proteins). One of these proteins, Caveolin 1, is encoded by the CAV1 gene, and it is a part of a large interacting network involving over 30 connections to genes and molecules (see Fig. 9.2). This body of interactions was named by Sanchez et al. (1999) as the *interactome*.

Most previous studies of cause and effect for various diseases like CVD have followed the *reductionist approach*, which have been effective only at showing important aspects of individual genes, proteins, or other molecules (Fang and Casadevall, 2011; Mazzocchi, 2012). These studies have been largely successful in exposing the functions of

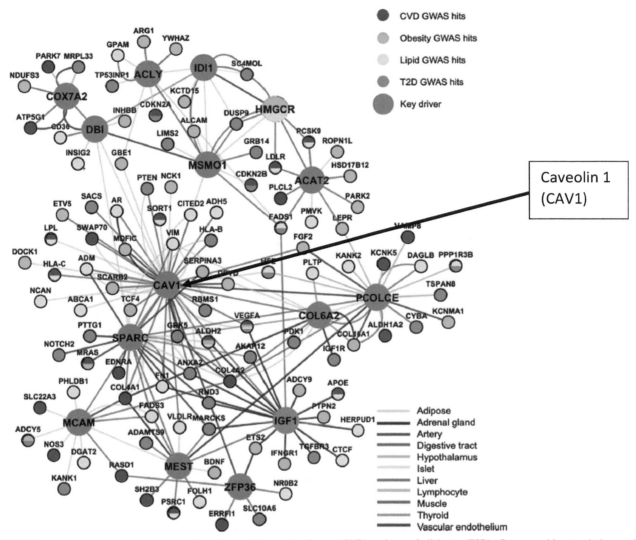

FIGURE 9.2 A cross-tissue gene networks shared by cardiovascular disease (CVD) and type 2 diabetes (T2D). Genome-wide association studies (GWAS) of candidate genes of various cardiometabolic diseases are network nodes indicated with different levels of gray. Key driver or hub genes are large nodes. An example of this connectedness is provided by the interrelationships between the protein Caveolin (CAV1) and over 30 genes or molecules (Shu et al., 2017). *Source: From Yang, X., 2020. Multitissue multiomics systems biology to dissect complex diseases. Trends Mol. Med. 26 (8), 718–728 (with permission from Shu, L., K. Chan, G. Zhang, T. Huan, Z. Kurt, Y. Zhao, et al. (2017) Shared genetic regulatory networks for cardiovascular disease and type 2 diabetes in multiple populations of diverse ethnicities in the United States. PLoS Genet. 13, e1007040).*

individual genes or biological molecules. Single-cell or single-tissue analysis has many advantages in comparison with multitissue approaches, including sampling accuracy and library construction. It is difficult, however, to study the highly interactive effects of diseases on multiple body tissues. Therefore, Yang (2020) described a more comprehensive approach, referred to as *multitissue multiomics systems biology*.

Multiomics systems biology

Hundreds of papers and books have accumulated during the last 5 years, which describe the use of relationships between various diseases and each (or combinations) of these Omics. This chapter will not try to review the work done in each of these study areas. Rather, this chapter will focus primarily on the integration, data preparation, and analytics operations using data from some examples among the Omics. This new collective discipline has been accorded the name, *multiomics*. We will consider the subject of multiomics within the context of systems biology.

The many interacting components of cause and effect in multiple tissues constitute a multidimensional complexity that should be treated systemically, not just individually. Single-cell approaches are valuable in elucidating the function and expressions of single genes, but they are not efficient to describe polygenic disease effects involving multiple genes. Disease effects are defined not only by multiple gene expressions affecting many tissues, but also by the micro- and macroenvironment in which they are embedded. This analysis framework can be applied to many different medical issues and diseases.

Basic analytics operations in multiomics

This chapter will focus on the basic operations for analyzing multiomics data, which include:

1. Data integration
2. Data preparation
3. Analysis methods
4. Knowledge discovery from analytics results

Some results of various studies will be presented in passing, but this chapter will focus on the data processing operations required for multiomics analysis.

Multiomics data integration

Integrating data from multiple sources apart from a common molecule or function is challenging (Zhang et al., 2021). Different multiomics sources may provide data on different scales or time intervals. Even when these features are consistent, the simple concatenation of these measurements onto the data record as separate columns (variables) increases data dimensions and contributes additional noise. Independent feature selection from each experiment may miss the interactions among variables of different multiomics components. These effects can significantly reduce the accuracy of models based on the integrated data set. Multiomics data should be integrated in ways that preserve any functional relationships between multiomics components.

Kedaigle and Fraenkel (2018) describe a number of ways to integrate multiomics data from a functional perspective:

1. Correlation analysis between different molecules was used by Noushmehr et al. (2010) to identify a mutation related to a specific epigenetic state.
2. Mapping multiomics information to known functions (Fabregat et al., 2018; Mi et al., 2016) can provide a basis for different preparation protocols among the omics data sets. These methods, however, are limited by the incomplete knowledge of molecular pathways.
3. Inference of networks and functions among the multiomics data. Habermann et al. (2015) used different types of networks as their underlying model. The CVD network shown in Fig. 9.2 must be coordinated with other disease networks to define the appropriate transformations for each. Only in this way can interference effects be assessed and built into treatment protocols.
4. The omics integrator tool (Tuncbag et al., 2016). The underlying networks and input data types in this tool are flexible to match different experimental situations. Methods for measuring accuracy (e.g., Specificity) and robustness of models generated from the integrated data are included. This tool provides an intuitive interface, so that neither computational nor modeling expertise is needed in order to use the tools effectively.

Kedaigle and Fraenkel (2018) suggest that this integration framework can be expanded in the future to encompass temporal and spatial information. Temporal data, for example, can show trends of change in processes over time that must be built into process models. Temporal data for a given analysis variable (longitudinal data) can be analyzed for lag effects on the relationships between multiomics data and medical outcomes.

Multiomics data preparation

Medical data in general can require significant data preparation operations. These general data preparation operations include: (1) feature selection and dimensionality reduction; (2) feature engineering; and (3) bias removal. Detailed discussions of feature selection and engineering operations are summarized in (Nisbet et al., 2017). A more detailed discussion of bias removal is appropriate for this chapter, because so many analysis problems and failures are caused by bias. For example, Murray et al. (2008) found that over half of the studies indexed in PubMed and MedLine between 2002 and 2006 suffered so severely from methodological bias errors that the probability of their results being correct was no more than that of flipping a coin.

Methodological bias

The two most common sources of bias in medical analyses are *algorithm bias* and *data bias*.

Algorithm bias is caused by characteristics of the modeling algorithm that are not appropriate for a given application of it. For example, many parametric statistical algorithms (e.g., linear regression) assume that data distributions are normal (bell shaped), effects of each predictor variable are completely independent of those of any other variable; and the combined effects of the predictor variables is linear (they plot as a straight line). Among machine learning (ML) algorithms, artificial neural nets suffer from the assumption that a local minimum error is a surrogate for the global minimum error location on the multivariate decision surface. Ignorance of these assumptions can lead to the use of algorithms that are not suited properly for a given analysis.

Data bias is also a major source of analysis error. Analysts know the old maxim "Garbage in ... garbage out," but mistakes are made often in attempts to remove the "garbage." Sometimes, the most efficient thing to do is to delete the "offending" rows or variables (e.g., to remove possible effects of missing values). Often, data bias problems can be removed by operations other than removing data. Data bias problems include: (1) sampling bias (e.g., nonrepresentative sampling of smokers vs. nonsmokers); (2) set bias (e.g., a 2015 Google facial recognition algorithm classified African-Americans as gorillas, because too few African-Americans were in the training set); and (3) data segmentation errors (e.g., unconcern for smoking habits can cause significant prediction errors of some diseases and cancers).

Data bias error can also be caused by improper balances in the data set caused by: (1) unrepresentative negative examples; and (2) imbalance of data sets with rare target values.

Unrepresentative negatives

Before building a model of a medical condition, the presence or absence of this condition must be properly characterized. Representative instances of positive medical conditions are often easier to characterize than representative instances without the medical condition (the negative condition). It may be necessary to eliminate some unrepresentative examples of the negative condition before modeling through the process of *positive unlabeling* (Fusilier et al., 2015). This process begins by treating all the negatives as truly negative and training the classifier. Next, using the classifier, score the unknown class and eliminate those for whom the classification is least reliable. Retrain with the positives against that truncated negative set, and repeat as necessary until some stopping condition is met (Fusilier et al., 2015; Liu et al., 2003). Another approach involves bagging the unknown into distinct subsets and evaluating each subset for training (Liu et al., 2003).

Imbalance of data sets with rare target variables

ML classification algorithms learn case-by-case, the way our human brains do it. The number of cases in each class presented to the algorithm must be nearly equal to avoid causing a learning bias. There are three common approaches for preventing this learning bias.

1. Undersampling. This approach reduces the number of rows of the common class to equal the number of rows of the rare class. The approach does create the balance of positive and negative examples, but it discards some of the richness in the data to characterize the common class appropriately.
2. Oversampling. This approach increases the number of rows of the rare class to equal the number of rows of the common class. This process is accomplished by oversampling the existing common class rows to a sufficient extent to equal the number of common class rows.
3. Synthetic row generation. The Simple Minority Oversampling Technique (SMOTE) algorithm creates synthetic rows by performing a nearest-neighbor analysis on a set of rare category rows to select variable values from other rows most similar to the set of rare category rows used. The SMOTE technique is the most widely used method of reducing bias due to imbalanced data sets in classification problems.

Data preparation issues specific to particular omics data sets

In addition to these generic data preparation operations, each multiomics data type may require several specific data preparation and preprocessing operations prior to integration with data from other types of multiomics data before they can be related to medical outcomes. You must understand fully what form of data is output from the sensors associated with the physical analysis of each data type. For example, very different data preprocessing steps are required to use data obtained from microarray analysis, gene sequencing analysis, and mass spectrometer (Ms) analysis. Each of these physical analyses may require very specific data preprocessing steps before modeling can begin.

Microarray data

Data from microarray analysis may require one or all of the following data preprocessing steps.

1. Noise reduction contributed by various factors (fluorescence of glass slides, dyed molecules, etc.).
2. Normalization is necessary to account for different environmental conditions in which the multiomics data were gathered, and different calibrations of analysis tools used in each of them.
3. Probe set summarization must be consistent across probe sets used in various studies. Different probe sets used in different microarray analyses at different times must be summarized in a consistent manner to permit mathematical analysis.

Gene sequencing data

Some common data preparation operations for genomic data include:

1. Removing unknown sources of technical variation. Nonbiological variation can be minimized by careful control of technical aspects of processing, such as maintaining uniform distribution of replicates across batches analyzed at different times. This control, however, may not be sufficient to prevent nonbiological variation in cases involving different amounts of dissociation or sorting of samples for single-cell RNA sequencing. In these cases, covariate analysis can be used to remove batch terms.
2. Starting quantities of single-cell RNA may vary between cells, and contributes to the total variation. Normalization of RNA length of data across batches can remove some of this variation.
3. Other cell-to-cell variation can be removed by scaling each cell with a size (length) factors, instead of one used in batch processing, or employing spike-in controls to account for spectra peaks introduced from nonbiological sources.

Mass spectrometer data

Liquid chromatography or gas chromatography operations require ionization of input substances to produce adsorption spectra for further analysis with a Ms to determine the specific molecular weights of various ions produced. Yu et al. (2019) reported a series of such data preparation operations of metabolomics gas chromatograph-mass spectrometer (GCMS) data, including EIC extraction, peak detection, peak marking and annotation, peak alignment and registration, peak filling, and statistical analysis of the Ms data.

Peak extraction from the chromatogram

Peaks associated with instrumental noise are removed if the ratio between the peak and the lowest value is less than 10, or the number of successive scans is less than 10. Specific processing steps include:

Peak detection

Many peak detection algorithms are available based on calculation of the 2nd derivative value of the signal (analogous to the slope of the line). These methods calculate the 2nd derivative as the representative of the slope of the curve at each point along the curve, and find the greatest slope value. Other heuristic relationships can be used to identify the highest peaks. The simplest heuristic method is to difference the data (calculate the difference between each data point and its previous value in the time-series as a surrogate to the slope), and find the greatest positive difference.

1. Derivative methods

    These methods calculate the 2nd derivative as the representative of the slope of the curve at each point along the curve, and find the greatest slope value. Other heuristic relationships can be used to identify the highest peaks. See Sboner et al. (2007) for a review of simple peak-finding methods.
2. Wavelet transform methods

    Wee et al. (2008) use a wavelet proportionate to the 1st derivative of a Gaussian function to find peaks in electrophoresis signal data. This approach pairs sequences of maxima (peaks) and minima (troughs) to define peaks in the signal data set.

3. Momentum-based methods

   This approach uses the momentum parameter of a neural network algorithm to find peaks in a signal data set (Harmer et al., 2008). The routine that is employed in neural network algorithm processing to find the global minimum error is used to find a succession of troughs in conjunction with nearby peaks in the signal data.

4. Differencing methods

   One of the data preprocessing operations used in time-series analysis (the "I" in ARIMA algorithms) calculates the difference between each data point and its previous value in the time-series. These differences integrate information on the signal at any time with the previous value of the signal in the time-series. This preprocessing step can also be used to find peaks (maximum differences) in relation to nearby troughs (Palshikar, 2009).

Peak merging and annotation

Peaks with a small side (relatively abrupt in change) can be merged with the neighboring peak. Candidate peaks. Clustering of isotropic ions will identify the MH^+ ions (element + hydrogen atoms) suitable for further analysis.

Peak alignment

Each peak is marked by four elements, namely, elution time, high-resolution m/z value, chromatogram number, and sample number. All detected peaks are sorted in ascending order according to high-resolution m/z values.

Peak registration

Peaks are grouped iteratively based on high-resolution m/z values and retention times until each group is characterized by only one peak.

Peak filling

Sometimes, a peak for a given registered group is missing in a sample. The missing peak is calculated by summing the ion intensities and m/z tolerance of the registered component. (See Yu et al. (2019) for details.)

Analysis methods

Two groups of analysis methods have been used extensively for modeling of multiomics data: (1) parametric statistical analysis; and (2) ML analysis.

Statistical analysis methods

Statistical analysis methods require that the analyst be aware of the assumptions of the standard parametric approach (called the "Parametric model") and their possible effects on bias in the results: (1) data sets must be normally (bell-shaped) distributed (some specialized algorithms can accommodate other data distribution formats); (2) the relationship between the dependent variable and predictor variables must be linear (following a straight-line); and (3) predictor variables must all be independent in their effects on the dependent variable. The problem with many real-world data sets is that all of these assumptions are often violated significantly. The amount of bias in the analytical results is proportional to the combined effect of violation of these three assumptions. Analysts who use parametric statistical methods must be prepared to defend any level of bias generated in their results. Murray et al. (2008) found that over half of the cancer trials papers indexed in the MedLine and PubMed between 2002 and 2006 were so flawed in either experimental design or choice of the analysis algorithms that results were little better than random choice.

Commonly used statistical methods include Analysis of Variance (ANOVA), Linear Regression, Piece-wise Linear Regression, and Principal Components Analysis. Yu et al. (2019) use parametric statistical modeling methods (ANOVA and Piece-wise Linear Regression) to discover relationships with various predictor ions. Several of these steps, (e.g., iterative clustering to ID a single peak) is performed to conform the data set input to the statistical modeling algorithms as much as possible to linear relationships.

Machine learning methods

Most machine learning methods do not suffer from the three assumptions of the Parametric model. The prepared data set can be submitted to a modeling algorithm to identify relationships between the registered peak values and any

TABLE 9.1 Data preparation steps prior to statistical analysis.

Data Analysis Steps in AntDAS2	
EIC extraction	1. Dividing mass spectral space into a number of sub-spaces with space width equal to 0.1* *m/z* tolerance. 2. Countering number of ions in each sub-space and clustering ions. 3. Constructing pseudo EICs based on ions in each cluster. 4. Removing bad EICs.
Peak Detection	1. Baseline correction for each EIC by a local-minimal value-based strategy. 2. Pseudo peak detection by using the developed peak window extending algorithm. 3. Instrumental noise estimation for each EIC by using a moving window smoothing strategy. 4. Filtering out peaks with low signal-to-noise ratios and performing peak merging. 5. High-resolution *m/z* marking for each peak.
Peak annotation	Clustering isotopic ions, adduct ions, and isotopic adduct ions, like $[M+H]^+$, $[M+1+H]^+$, $[M+2+H]^+$, $[M+3+H]^+$, $[M+H-H_2O]^+$, $[M+H-2H_2O]^+$, $[M+K]^+$, $[M+Na]^+$, $[M+NH_4]^+$, etc.
Peak Alignment	Using peak retention time and high-resolution *m/z* information in combination with dynamic strategy for peak alignment
Peak Registration	Using the developed adaptive network searching algorithm to group peaks corresponding to the same compound
Peak Filling	Integrating signals for samples which cannot detect peaks in the corresponding elution range
Statistical Analysis	Unsupervised methods: HCA, PCA Supervised methods: ANOVA, PLS

Source: After Yu Y.-J., Zheng, Q.-X., Zhang, Y.-M., Zhang, Q., Zhang, Y.-Y., Liu, P.-P., et al., 2019. Automatic data analysis workflow for ultra-high performance liquid chromatography-high resolution mass spectrometry-based metabolomics. J. Chromatogr. A 1585:172–181.

medical outcome. When ML algorithms are used, these data preparations steps might be eliminated, because ML algorithms [e.g., Random Forests (RF) or Deep Learning Neural Nets] are much better for finding strongly nonlinear patterns in data. ML algorithms do, however, have several assumptions of their own: (1) classification algorithms require the same number of both classes of the dependent variable. If the dependent variable has two classes (0 and 1), both classes must be "balanced," that is both classes must have the same number of examples, or bias will result in the ML processing. Decision trees for regression-type problems (regression trees) assume that the values are normally distributed in their ranges, which they may not be.

Table 9.1 shows a diagrammatic representations of the steps in the data preparation of genomic data provided by Yu et al. (2019) for parametric statistical analysis. These steps were automated to form the automatic data analysis workflow (AntDAS2) methodology.

Data conditioning

Before analysis with statistical or ML algorithms, data sets must be conditioned to conform as far as possible to the assumptions of the algorithms. The mathematics operations of regression algorithms assumes that all predictor variables have the same range, but this is often not so. Therefore, a common data conditioning operation performed is *normalization*. For ML classification models, the data set must be *balanced* for submission to the classification algorithm.

Normalization

Even for ML algorithms, data normalization is often necessary, to account for the different calibrations and machine settings between batches, particularly when multiple omics data types are used. Two forms of normalization are used commonly: (1) min/max; and (2) z-score. The min/max normalizes data in each variable range separately, assigning a number between 0 and 1.0. the z-score method normalizes data in each variable to range between $-\infty$ and $+\infty$, using the z-value transform, but where 99.5% of the values range between -3 and $+3$ units. This range is related to the range of standard deviation units away from the mean value for 99.5% of the values in a normal statistical distribution. The z-score normalization transformation works best for parametric statistical methods and for ML methods that use linear regression in processing operations (e.g., regression tree algorithms like RFs for Regression).

Data set balancing

There are two general approaches to balancing data sets for classification modeling in which one (or more) of the classes are relatively rare: (1) by undersampling the common class(es) to use the same number of cases for each class; and (2) oversampling the rare class to equal the number of the common class(es).

The undersampling approach discards data from the common class which might be necessary to define properly the common class, therefore, many analysts prefer to oversample the rare class(es). One popular oversampling method is the SMOTE, in which synthetic cases are constructed from existing cases by a nearest-neighbor analysis for values for each variable.

Data preprocessing tools in multiomics

Table 9.2 shows examples of various data preprocessing applications available for several multiomics types.

Multiomics analytical methods

A large number of analytical methods have been applied to multiomics data during the last 20 years.

Open source tools for multiomics analytics

Bioinformatics includes all mathematical and statistical analyses of biological data. Most of these biological applications include studies of genomics and pharmaceutical data. Many of these tools are applicable to other multiomics data types. Gururaj and Pavithra (2020) discuss the two commonly used coding packages for Perl (Bioperl) and Python (Biopython). Biopython is one of the most common bioinformatics packages built on the Python language, available for

TABLE 9.2 Statistical preprocessing tool applications available for multiomics (Vasileiou et al., 2019; Yi et al., 2016).

Genomic	Transcriptomic	Epigenomics	Proteomic	Metabolomic
• GMAP	• GMAP	• MACS	• OpenMS	• OpenMS
• BWA	• BWA	• SISSRs	• Mzmine 2	• Mzmine 2
• STAR	• STAR			
• GATK	• HTSeq			
• SAM Tools	• BED Tools			
	• RSEM			
	• Cufflinks			
	• Defuse			
	• TopHat-Fusion			
	• Trans-ABySS			
	• Trinity			
	• Scripture			

Source: From Vasileiou, P., Magiorkinis, G., Lagiou, P., Gorgoulis, V. 2019. Analytical methods for systems medicine. Ch. 7 In: Translational Systems Medicine and Oral Disease. Sonis, S., A. Villa. 458 p. See also Yi, L., Dong, N., Yun, Y., Deng, B., Ren, D., Liu, S., et al., 2016. Chemometric methods in data processing of mass spectrometry-based metabolomics: a review. Anal. Chim. Acta 914:17–34.

Linux, Windows, and Mac. Gururaj and Pavithra (2020) provide detailed instruction for installation of the package. Some of the capabilities of Biopython include:

- Parsing bioinformatics files using Python data structures
- Support for numerous file formats
- Includes processing of online data
- Provides interfaces for common bioinformatics data processing programs
- Provides tools for performing common operations
- Connects with BioSQL.

Machine learning tools in multiomics analytics

Shastry and Sanjay (2020) discuss some common ML algorithms used in Bioinformatics analysis in three categories: (1) filter methods; (2) wrapper methods; and (3) embedded methods.

Filter methods

Filter methods include various heuristic logic used to describe some aspect of a data set. These methods include Chi-squared, F-score, and Spearman's Correlation Coefficient. These methods are used primarily in the process of data sleuthing and data profiling.

Wrapper methods

Wrapper methods focus primarily on feature selection. Algorithms employed include various classification methods used by Classification and Regression Trees (CART), ID3, and C4.5. Problems involving classification operations have used RF, radial basis function methods, and kernel methods, like those provided in some Support Vector Machine (SVM) algorithms. Complex classification methods many require deep learning neural networks with convolutions (CNN).

Embedded methods

Embedded methods predict numeric outcomes. Classical statistical methods used since the 1920s to predict numerical outcomes include linear and nonlinear regression algorithms and several multivariate nonparametric methods like the Kruskal—Wallis test. Examples of ML algorithms used to predict numeric outcomes include neural networks, regression trees, and SVM.

Deep learning for genome and epigenome analysis

Kelley et al. (2016) trained a convolutional neural network on one-hot-encoded DNA-sequencing data to predict chromatin accessibility signal. One-hot coding (or dummy variable coding) assigns a 1 or a 0 to a new variable for each class in the categorical variable, depending on whether or not that class is present in the variable for a given data row. This technique converts textual information to numeric information for analysis with many algorithms (e.g., logistic regression). Zhang et al. (2021) explained that chromatin accessibility cannot be explained comprehensively by DNA sequences, but the potential effect of noncoding mutations on chromatin accessibility can be determined. Kelley et al. (2016) described Basset, an open source package to apply CNNs to learn the functional activities of DNA sequences. Basset was used to predict that an increase in the binding of the critical genome architecture protein CTCF presupposes the development of the vitiligo disease polymorphism rs4409785.

Other deep learning models specifically aim to identify the causal mutations in disease.

DeepSEA, for example, trains a convolutional network on 919 epigenomic features encompassing chromatin accessibility, histone modification, and transcription factor binding (Zhou and Troyanskaya, 2015). One DeepSEA model identified the role of previously unknown driver mutations in autism spectrum disorders (Zhou et al., 2019).

Focus on metabolomics

During the previous decade, many studies of multiomics data have been performed for purposes of cancer and disease diagnosis and management. Whole books could be composed easily from the recent research studies of each of type of multiomics data. This section will drill down a little deeper into one of those types, metabolomics, and discuss how ML methods have been applied to predict the presence of various types of cancer or disease from the analysis of metabolomics data. In addition, an example will be described in general for cancer diagnosis. A detailed tutorial of this example is available on this book's web page.

Prediction of pancreatic and lung cancer from metabolomics data

Metabolites have been defined as molecules less than 1500 Daltons molecular weight that are produced endogenously by cellular metabolism (Fiehn, 2002). Cells altered by cancer or disease produce additional metabolites or significantly higher levels of existing metabolites. The quantitative analysis of these metabolites can be related to the presence/absence of the cancer or disease in a patient.

After cancer and disease metabolites are emitted, they dissolve into organic fluids in the cell and bloodstream. Many of these volatile organic compounds (VOCs) are subject to being ionized prior to being chromatographed in a GCMS. The number of ions in these VOCs can be counted in representative absorption peaks and related to normal and abnormal (diseased or cancerous) conditions. Specific compounds from which the ions are derived can be identified from standard tables. The presence of significantly higher than normal ion counts for a compound from a diseased or cancerous subject can tag them as markers for the presence of cancer. Various of these marker compounds have been used as predictors of cancer presence for lung cancer, breast cancer, liver cancer, pancreatic cancer, and colon cancer.

VOCs have been used to predict cancer presence with high accuracy in hundreds of studies during the last 10 years. For example, Nisbet (2020) analyzed metabolites in blood plasma from a group of lung cancer, pancreatic cancer, and normal patients. A GCMS system was used to count the number of ions of various molecular weights flowing through the system. Output data were preprocessed with a fast Fourier-transform to convert data from the time domain to the frequency domain. A Python program was used to further preprocess the GCMS output data to generate the data sets for building the analytical models in the KNIME analytical platform. All data rows were randomized and data values were normalized with a z-score function. Missing values were imputed appropriately, and the data set input to the Random Forest modeling algorithm was balanced with the SMOTE technique. Finally, important features were selected iteratively with a backward feature elimination technique using a Random Forest modeling algorithm. Fig. 9.3 shows the KNIME workflow used for developing the multicategorical model.

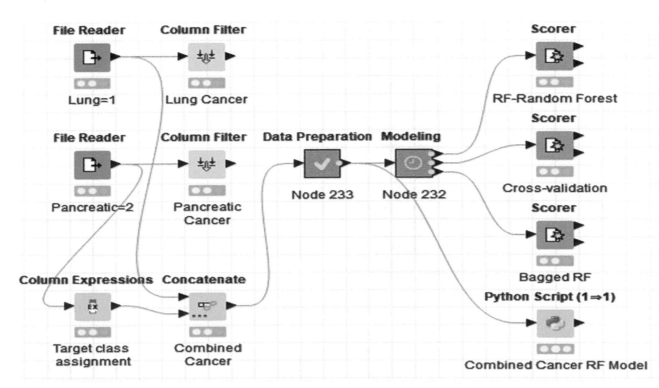

FIGURE 9.3   KNIME workflow for developing random forest models with both KNIME and Python.

Results showed a perfect model to predict the correct cancer state of each row. How could this be? Past experience has shown that perfect models are usually the result of either mistakes in algorithm specification or variable definition. For example, the most common cause of a perfect model is using a target variable that is in some way related to one of the predictor variables *by definition*! This error generates a "leak from the future" (Nisbet, 2009).[1] These perfect models were not the results of leaks from the future, but rather because the ion counts of specific metabolites for both groups of cancer patients, range completely above those for normal patients. The modeling algorithm had no trouble producing a perfect model. Scouring of the literature uncovered four other cases where modeling of VOCs for cancer prediction produced perfect models (see Gordon et al., 1985; Westhoff et al., 2009; Turki and Wei, 2018; Qian et al., 2018). This example is presented as a full tutorial on this book's web page.

Postscript

The 1960s brought an increasing awareness of the connectedness among many organisms on Earth. The previous 50 + years in the sciences of botany and zoology were dedicated to understanding how individual species or closely related groups of species were generated, how they grew, and what killed them. Scientists began to realize that they could not understand very deeply how and why plants and animals live and die as they do without understanding how they compose a system of ecological relationship—the *ecosystem*. Extending from 1964–74, the International Biological Program (IBP) sought to understand the systems-level nature of biological dynamics on the Earth. Much research funded through the National Science Foundation focused on systems-level processes in groups of related ecosystems called *biomes*. The author of this chapter was one of the IBP funded researchers in the Desert Biome. The International Geosphere-Biosphere Program (1987–2015) extended research to integrate the biophysical and geophysical aspects of Earth's biomes, focusing on global changes, particularly climate change. It is time for multiomics studies to follow a similar path of systems-level development to understand the relationships and contributions among the "Omes" for the purpose of improving medical diagnosis and treatment regimes. Should we call for an International Multiomics Program?

References

Fabregat, A., Jupe, S., Matthews, L., Sidiropoulos, K., Gillespie, M., Garapati, P., et al., 2018. The reactome pathway knowledgebase. Nucl. Acids Res. 46, D649–D655.

Fang, F.C., Casadevall, A., 2011. Reductionistic and holistic science. Infect. Immun. 79, 1401–1404.

Fiehn, O., 2002. Metabolomics—the link between genotypes and phenotypes. Plant. Mol. Biol. 48, 155–171.

Fusilier, D.H., Montes-y-Gómez, M., Rosso, P., Guzmán Cabrera, R., 2015. Detecting positive and negative deceptive opinions using PU-learning. Inf. Process. Manag. 51, 433–443. Available from: https://doi.org/10.1016/j.ipm.2014.11.001.

Gordon, S.M., Szldon, J.P., Krotoszynski, B.K., Gibbons, R.D., O'Neill, H.J., 1985. Volatile organic compounds in exhaled air from patients with lung cancer. Clin. Chem. 31 (8), 1278–1282.

Gururaj, T., Pavithra, A.P., 2020. Open-source software tools for bioinformatics. In: Srinivasa, K., Siddesh, G.M., Manisekhar, S.R. (Eds.), Statistical Modelling and Machine Learning Principles for Bioinfomatics Tools, and Applications. Springer Nature Singapore Pte Ltd, pp. 75–94.

Guttmacher, A.E., Collins, F.S., 2002. Genomic medicine-A primer. N. Engl. J. Med. 347, 1512–1520.

Habermann, B., Villaveces, J., Koti, P., 2015. Tools for visualization and analysis of molecular networks, pathways, and -omics data. Adv. Appl. Bioinforma. Chem. 8, 11.

Harmer, K., Howells, G., Sheng, W., Fairhurst, M., Deravi, F., 2008. A peak-trough detection algorithm based on momentum. In: Proceedings—1st International Congress on Image and Signal Processing, CISP 2008 4, pp. 454–458. Available from: https://doi.org/10.1109/CISP.2008.704.

Kedaigle, A., Fraenkel, E., 2018. Turning omics in therapeutic insights. Curr. Opin. Pharm. 42, 95–101. Available from: https://doi.org/10.1016/j.coph.2018.08.006.

Kelley, D.R., Snoek, J., Rinn, J.L., 2016. Basset: learning the regulatory code of the accessible genome with deep convolutional neural networks. Genome Res. 26, 990–999.

Liu, B., Dai, Y., Li, X.L., Lee, W.S., Philip, Y., 2003. Building text classifiers using positive and unlabeled examples. In: ICDM 2003, Third IEEE International conference on data mining, November 2003.

Mazzocchi, F., 2012. Complexity and the reductionism-holism debate in systems biology. Wiley Interdiscip. Rev. Syst. Biol. Med. 4, 413–427.

Mi, H., Poudel, S., Muruganujan, A., Casagrande, J.T., Thomas, P.D., 2016. PANTHER version 10: expanded protein families and functions, and analysis tools. Nucl. Acids Res. 44, D336–D342.

Murray, D., Pals, S., Biltstein, J., Alfano, C., Lehman, J., 2008. Design and analysis of group-randomized trials in cancer: a review of current practices. J. Natl Cancer Inst. 100, 483–491.

---

1. See Chapter 20 by John Elder.

Nisbet, R. 2020. Modeling cancer detection with volatile organic compounds. Presented at Predictive Analytics World Conference, Las Vegas, June, 2020.

Nisbet, R., Elder, J., Miner, G., 2009. Handbook of statistical analysis and data mining applilcations. Academic Press, San Diego, CA, p. 822.

Nisbet, R., Miner, G., Yale, K., 2017. Handbook of statistical analysis and data mining applications. Academic Press, San Diego, CA, p. 792.

Noushmehr, H., Weisenberger, D.J., Diefes, K., Phillips, H.S., Pujara, K., Berman, B.P., et al., 2010. Identification of a CpG island methylator phenotype that defines a distinct subgroup of glioma. Cancer Cell 17, 510–522.

Palshikar, G., 2009. Simple algorithms for peak detection. <https://www.researchgate.net/publication/228853276_Simple_Algorithms_for_Peak_Detection_in_Time-Series>.

Qian, K., Wang, Y., Hua, L., Chen, A., Zhang, Y., 2018. New method of lung cancer detection by saliva test using surface-enhanced Raman spectroscopy. Thorac. Cancer 9, 1556–1561.

Sanchez, C., Lachaize, C., Janody, F., Bellon, B., Röder, L., Euzenat, J., et al., 1999. Grasping at molecular interactions and genetic networks in Drosophila melanogaster using FlyNets, an Internet database. Nucl. Acids Res. 27 (1), 89–94. Available from: https://doi.org/10.1093/nar/27.1.89. PMID: 9847149. PMCID: PMC148104.

Sboner, A., et al., 2007. Simple methods for peak and valley detection in time series microarray data. In: McConnell, P., Lin, S.M., Hurban, P. (Eds.), Methods of Microarray Data Analysis V. Springer, Boston, MA. Available from: https://doi.org/10.1007/978-0-387-34569-7_3.

Shastry, K., Sanjay, H.A., 2020. Machine learning for bioinformatics. In: Srinivasa, K., Siddesh, G.M., Manisekhar, S.R. (Eds.), Statistical Modelling and Machine Learning Principles for Bioinfomatics Tools, and Applications. Springer Nature Singapore Pte Ltd, pp. 25–40.

Shu, L., Chan, K., Zhang, G., Huan, T., Kurt, Z., Zhao, Y., et al., 2017. Shared genetic regulatory networks for cardiovascular disease and type 2 diabetes in multiple populations of diverse ethnicities in the United States. PLoS Genet. 13, e1007040.

Tuncbag, N., Gosline, S.J.C., Kedaigle, A., Soltis, A.R., Gitter, A., Fraenkel, E., 2016. Network-based interpretation of diverse high- throughput datasets through the omics integrator software package. PLoS Comput. Biol. 12.

Turki, T., Wei, Z., 2018. Boosting support vector machines for cancer discrimination tasks. Comp. Biol. Med. 101 (2018), 236–249.

Vasileiou, P., Magiorkinis, G., Lagiou, P., Gorgoulis, V., 2019. Analytical methods for systems medicine. Ch. 7 In: Translational Systems Medicine and Oral Disease. Sonis, S., A. Villa. 2019, 458 p.

Wee, A., Grayden, D., Zhu, Y., Petkovic-Duran, K., Smith, D., 2008. A continuous wavelet transform algorithm for peak detection. Electrophoresis 20, 4215–4225. Available from: https://doi.org/10.1002/elps.200800096, Wiley Online Library.

Westhoff, M., Litterst, P., Freitag, L., Urfer, W., Bader, S., Baumbach, J., 2009. Ion mobility spectrometry for the detection of volatile organic compounds in exhaled breath of patients with lung cancer: results of a pilot study. Thorax 64, 744–748.

Yang, X., 2020. Multitissue multiomics systems biology to dissect complex diseases. Trends Mol. Med. 26 (8), 718–728.

Yi, L., Dong, N., Yun, Y., Deng, B., Ren, D., Liu, S., et al., 2016. Chemometric methods in data processing of mass spectrometry-based metabolomics: a review. Anal. Chim. Acta 914, 17–34.

Yu, Y.-J., Zheng, Q.-X., Zhang, Y.-M., Zhang, Q., Zhang, Y.-Y., Liu, P.-P., et al., 2019. Automatic data analysis workflow for ultra-high performance liquid chromatography-high resolution mass spectrometry-based metabolomics. J. Chromatogr. A 1585, 172–181.

Zhang, L., Karimzadeh, M., Welch, M., McIntosh, C., Wang, B., 2021. Analytics methods and tools for integration of biomedical data in medicine. Artificial Intelligence in Medicine. Academic Press, pp. 113–129, Ch. 7. Available from: https://onlinelibrary.wiley.com/doi/full/10.1002/elps.200800096.

Zhou, J., Troyanskaya, O.G., 2015. Predicting effects of noncoding variants with deep learning-based sequence model. Nature Methods 12, 931–934. Available from: https://doi.org/10.1038/nmeth.3547.

Zhou, J., Park, C.Y., Theesfeld, C.L., et al., 2019. Whole-genome deep-learning analysis identifies contribution of noncoding mutations to autism risk. Nat. Genet. 51, 973–980.

Further reading

Koul, N., Manvi, S., 2020. Machine-learning algorithms for feature selection from gene expression data. In: Srinivasa, K., Siddesh, G.M., Manisekhar, S.R. (Eds.), Statistical Modelling and Machine Learning Principles for Bioinfomatics Tools, and Applications. Springer Nature Singapore Pte Ltd, pp. 151–164.

Koul, N., Manvi, S.S., Mi, H., Poudel, S., Muruganujan, A., Casagrande, J.T., et al., 2016. PANTHER version 10: expanded protein families and functions, and analysis tools. Nucl. Acids Res. 44, D336–D342.

Chapter 10

Artificial intelligence and genomics

Nephi Walton and Gary D. Miner

Chapter outline

Prelude

Artificial Intelligence (AI) refers broadly to the theory and development of computer systems to function in ways that normally require human intelligence. In medical informatics, the target behavior usually involves decision-making to support medical diagnosis or treatment. Genomics is a type of data composed of genetic information (e.g., genes) that can be related to conditions requiring diagnosis and treatment. This chapter discussed tools and methodologies for analysis of genomic data with AI tools and methodologies.

How do we enable the clinical application of artificial intelligence in genomics?

AI is used in many of the tools that analyze genetic data prior to its direct clinical application. However, there is little direct clinical application of AI with genomics at the time of writing (January/February of 2022). Genomics data presents a great opportunity for application of AI while also presenting significant challenges.

The challenges: The large dimensionality of genomic data makes it good target for AI applications. However, the relatively small amount of genomic data in relation to the dimensionality creates a problem for training.

The other major challenge is the lack of knowledge about the function of all the genes, let alone the millions of variants of those genes. We still do not understand the clinical impact of many of the variants in even our most studied genes.

Training of AI on models on genomic data is also limited by the quality of phenotype data available in the EMR.

Genomics fast moving field—and now ready for artificial intelligence to have an impact

Genomics is a fast-moving field and quickly developing into a field where AI can have significant impact and application. The decrease in the price of sequencing the genome from $300 million to less than $500 has resulted in a rapid increase in sequencing as well as increasing adoption for clinical application. The more WGS (Whole Genome

Practical Data Analytics for Innovation in Medicine. DOI: https://doi.org/10.1016/B978-0-323-95274-3.00014-2

Sequencing) testing we do, the more useful its clinical application becomes. As an example we now know there are more than 1500 genes that contribute to epilepsy. This increase in knowledge of the genetic basis of epilepsy has been largely due to the increasing use of exome sequencing. Despite the fact that we now know these genes cause epilepsy, we know very little about the prognosis, treatment, or penetrance of these gene specific disorders. This is because we have not done enough testing to build large cohorts of each cause and study its epidemiology, pathogenesis, and management. As we increase testing we find more patients with each gene and begin to better understand how to manage them. As we learn how to manage epilepsy more specifically based on gene, genetic testing becomes even more valuable and more frequently ordered and knowledge continues to increase cyclically as we do more sequencing.

Need to open existing large datasets to more researchers

Opening up existing large datasets to more researchers is another way to rapidly increase our knowledge of genetics. While many large genomic datasets exist, the access to such datasets is usually restricted to a few at specific organizations. These organizations do not have the manpower to fully utilize these datasets or ask all the relevant questions that can lead to further scientific discovery. Large national efforts such as the England Genomics 100,000 genomes project and the All of US Research Program seek to open this data up to more researchers and will ultimately result in large increases in knowledge around genomics.

Successful artificial intelligence models will be ones that use smaller and manageable portions of the human genome

A better understanding of biological pathways and genetic interactions will allow for the creation of AI models that utilize a smaller and perhaps more manageable portion of the genome. Human genetics is complex, especially when you consider that every cell in your body regardless of structure and function contains the same genetic code. This is possible because these genes are transcribed or expressed differently in different cell types and different locations in the body. Therefore, understanding the effects of a specific genetic variant on one system does not necessarily lead us to an understanding of the function in a different organ system or type of cell. This variability in expression is well exhibited in many complex diseases that have many different manifestations in different organ systems resulting from the change in a single variant. In addition to the complexity of when where and how genes are expressed there are extensive genetic interactions in the body particularly when you look at genes that are all part of complex pathways. Often a single mutation is not what drives disease but rather a series of mutations in multiple genes that through an additive or interactive effect can cause disease. AI models can take advantage of our knowledge of biological pathways to create models that rely on a smaller subset of data directly related to the pathway of interest.

Polygenic risk scores

Polygenic risk scores are a prime area for application of genomics and AI. AI models can combine genetic data from polygenic risk scores with phenotypic information about the patient to create more accurate risk models allowing for earlier intervention for those at high risk of disease. Better standardization, well-defined regulatory guidelines, and more implementation science research is needed to help drive clinical adoption of such models. Efforts are underway to standardize the application and reporting of polygenic risk scores (Wand et al., 2021). This is important work, and more work should be done to integrate these models into the EHR and clinical workflow. There are hundreds if not thousands of polygenic risk scores that have been created for many different diseases however the clinical application and utility of such scores is severely lacking. AI has the chance to improve clinical adoption through its potential to improve on simple polygenic risk score models but could also hinder adoption through its lack of transparency. Physicians are less apt to adopt the decisions of clinical decision support models if they do not understand the underlying principles that drive the results of the model.

Artificial intelligence models cannot replace but must augment physicians diagnosis and treatment decisions

AI models that augment a physician's abilities are more likely to succeed than those that attempt to replace them. Clinical judgement is very important in medicine, and it is unlikely that we would be able to develop AI models that capture all of the complex data that surrounds a patient interaction. Physicians commonly encounter uncommon scenarios that rely not on experience with a given situation but rather a full understanding of all of the physiological/biological processes that are taking place at a given moment. It would be difficult if not impossible to train a machine learning algorithm on all the potential scenarios that a clinician might encounter. Too much reliance on a AI can actually be dangerous as physicians may began to utilize less clinical judgement and could in fact lose some of this clinical judgement in the process. Similar to how many of us use GPS so commonly that we would be lost without it or that none of us remember phone numbers anymore. It will be in the very distant future that AI can replace the clinical judgement that physician learns through years of training. Even some of most established and perhaps most advanced AI technologies have significant error, such as voice to text; a misspelling, however, does not present as near the problem as when error is introduced in a life-or-death situation with a patient. The importance of the human augmentation of AI is accentuated by the examples of Tesla accidents, this also illustrates the potential catastrophic effects of relying too much on AI and deferring or ignoring one's own human judgement.

Since genetic differences underlie many of our ethnic and racial differences. Models that utilize genomics and AI models are very susceptible to racial and ethnic bias and could easily lead to disparities in care. This is particularly true where our knowledge of the genomics of many racial and ethnic subgroups is extremely limited.

The magnitude of the data required to train genomics and AI models is substantial and the concern with such large repositories of data is the lack of privacy. Detailed medical histories and phenotypic features make genomics and AI models more useful. However, gathering such large data sets leads to significant privacy concerns.

Governance—balance between rapid approval of models and ensuring no human harm

Medicine is very driven by clinical guidelines and clinical trials. Though regulation of AI models in medicine is still evolving it is likely that widescale adoption of such models would require clinical trials that prove the effect of such models in clinical care. While regulation can be cumbersome it is also at time necessary to ensure that we do not adopt technologies that are harmful to patients. With rapid increases in technology, we need to develop regulatory models for the implementation of new technologies that do unnecessarily delay their implementation when there could be significant clinical benefit. The balance between rapid approval and ensuring no harm is a delicate one.

EHR and integration of artificial intelligence into clinical workflows

Integration of AI models into the EHR and clinical workflow is essential to generalized adoption. Physicians work under very tight time constraints and are not apt or likely to switch between applications or add additional things to their workflows that consume additional time that will take time away from the patient. The most successful AI models will be well integrated and actually decrease the time required for the physician to successfully execute and encounter. The integration into the EHR can be particularly difficult since it often requires the cooperation of EHR vendors who are likely to have other priorities. The implementation of appropriate clinical decision support around these models is needed to ensure adoption and use.

Physicians are largely uneducated about AI, its limitations, benefits, applications, and underlying principles. While it is not expected that a physician would be able to build an AI model, a basic understanding of how such models work would be useful as part of their medical curriculum as these technologies start to become integral to patient care. Physicians are already required to have training in understanding research studies, clinical trials, and the underpinnings of evidence-based clinical guidelines. Adding some basic understanding of the priniciples underlying AI may not be too big of a lift and may generate more acceptance of such models when they are introduced clinically.

What would an artificial intelligence and genomics integration look like?

An illustration of some of the components involved with using AI (Artificial Intelligence) with genomic data has been showed in Fig. 10.1.

Real-world examples of artificial intelligence and genomics modeling systems emerging in 2022

Although significant clinical application examples of AI with Genomics have not developed over the past 5 years, regardless of all the hype that such would happen, I think you can see from what we presented above in this chapter, that the application of AI with Genomics was not ready to happen because of the former lack of sufficiently fast real-time computing power to

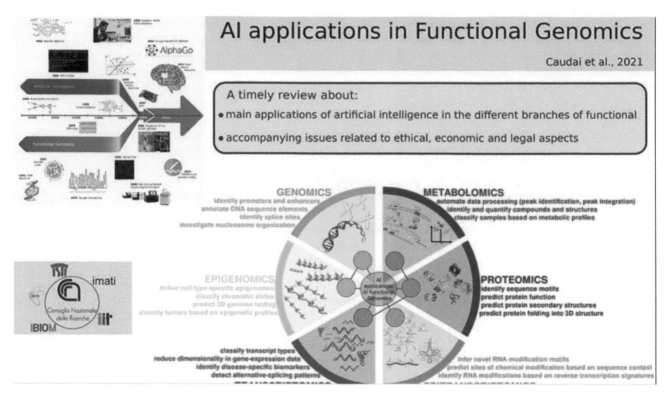

FIGURE 10.1 An illustration of some of the components involved with using artificial intelligence with genomic data. *From https://www.sciencedirect. com/science/article/pii/S2001037021004311.*

make it happen. But today this is changing. There are minimally two different research centers where this will begin to change in 2022:

1. The SINGAPORE "Digital Platform" project. The NUS (National University of Singapore) and NUHS (National University Health System) in 2018 announced their efforts in the "A Glimpse Into Hospital of the Future" press release (https://www.nuhs.edu.sg/sites/nuhs/NUHS%20Assets/News%20Documents/NUHS%20Corp/Media%20Releases/ 2018/060718%20Media%20release_Glimpse%20into%20hospitals%20of%20the%20future.pdf). This announcement was about their "DISCOVERY AI" platform that they had built and were continuing to tweak and enhance.

   "DISCOVERY AI platform: The platform is a production level, secure sandbox for building AI tools from huge multidomain medical databases. It is a platform that is able to anonymously link multiinstitution datasets yet share data securely and equitably between clinicians, researchers and data scientists. The platform is fortified by a proprietary blockchain technology coupled with enterprise-grade security. Access to this platform is unified through a cluster-wide governance policy. Multiple AI tools have been built on this platform and are integrated with the existing EMR system to alert clinicians directly. This platform and associated AI technologies are developed in-house with proprietary technology from the NUS School of Computing and partners in other institutes of higher learning. The DISCOVERY AI platform has been test-bedded at the National University Hospital (NUH) and will be officially launched on 6 July 2018. Health" (from: https://www.nuhs.edu.sg/sites/nuhs/NUHS%20Assets/News%20Documents/NUHS%20Corp/Media%20Releases/ 2018/060718%20Media%20release_Glimpse%20into%20hospitals%20of%20the%20future.pdf).

   By the end of 2021 it was not clear if the NUHS was finally ready for real-time integration of GENOMICS and AI (Thomas, 2021 and BioSpectrum-Asia Edition, 2021); however, it was clear that they had added a new "hardware/software" tool called NVIDIA DGX A100 system, which was given the name ENDEAVOUR PLATFORM, to work with the DISCOVERY AI PLATFORM at incredibly increased computing speeds, such that real time data-streaming allowed physicians instant access to important information needed to making real-time patient diagnoses and treatment decisions. Even though the new NVIDIA DGX A100 has just gone live, NUHS is already looking forward to the next generation of the system to tackle the expected growth in datasets and speed needed to process those data. The Singapore group has invested in programs that look at genomics using not gigabytes, but terabytes of genomics data per day. More computing power is needed to handle this, which may require the addition of more NVIDIA DGX A100 units to get the GPUs needed.

But the way to do this may have already been mapped out by the Oxford Nanopore Technologies genome sequencing technologies which have deploying the NVIDIA technology in a variety of genomic sequencing platforms to develop AI tools that improve speed and accuracy in genomic analysis. Using the Cambridge-1 supercomputer should improve genomic accuracy and allow faster turnaround time for genomic analysis. (Ciborowski, 2021). NVIDIA already in early 2022, working with Meta Works, is building "massive supercomputing power" by putting 760 of the NVIDIA DGX A100 systems into a "Super-Super-Computer" as its compute nodes. This packs a total of 6080 NVIDIA A100 GPUs linked on a NVIDIA Quantum 200 Gb/s InfiniBan network to deliver 1895 petaflops of TF32 performance; and in a second phase later in 2022 they plan to expand the GPUs from 6080 to 16,000, which they believe will deliver 5 exaflops of AI performance (Boyle, 2022), Such supercomputing power should go a long way toward making genomics data available in real time for clinical use.

2. The Mayo Clinic "Digital Platform" project. On April 14, 2021 the Mayo Clinic (Rochester, Minnesota; also Scottsdale, Arizona, and Jacksonville, Florida) announced their launch of a digital technology platform that would revolutionize healthcare delivery. At that time they called it "The Remote Diagnostics and Management Platform" (RDMP) (Anastasijevic, 2021). The purpose of this new platform is to provide the next generation of clinical decision tools to help clinicians make faster and more accurate diagnoses and provide truly continuous care to patients. This RDMP provides what we can call "event-driven medicine," for example, insights in the right context, at the right time.

The Mayo Clinic launched two new companies, with partners, to support this newly created digital platform (RDMP). One of these new companies is called Anumana, Inc.; its purpose is to develop and commercialize AI-enabled algorithms. Anumana will initially focus on developing "state-of-the-art" Neural Network analytics applied to the Mayo Clinic's billions of pieces of heart health data. There are many heart conditions, including silent arrhythmias, where evidence-based treatments already exist that can prevent heart failure, stroke, or death, but the key is to detect these before symptoms develop. The new RDMP digital analytics should be able to provide analytic modeling that can make these predictions prior to symptoms.

The second of these new companies associated with Mayo is Lucem Health, Inc. Lucem Health's purpose is to collect, orchestrate, and curate data from any device...yes any device! This is to provide a platform for connecting remote patient telemetry devices with the AI-enabled algorithms in the Mayo digital platform, for generating needed information to make real-time diagnoses and treatment decisions (Anastasijevic, 2021).

In a January 2022) announcement, Mayo also continued to reveal more details of their "Digital Platform," with a blog post discussing genomics, proteomics, and biomarkers in general (Halamka and Cerrato, 2022). The overriding purpose of the Mayo Digital Platform is to develop a "New Standard of Care" that integrates data from multiple sources, including phenotype, genotype, imaging, digital pathology, and signals from patient telemetry; doing so will contribute greatly, for the first time, in providing real personalized diagnosis and personalized disease treatment. Today data is available from a long list of "omics," like genomics, proteomics, metabolomics, and microbiomics (see Chapter 9 of this book for further details of the "omics"). These omics can be coupled with medical imaging to develop deeper, more person-centered views of what lies within the person, and thus a much better understanding of how to handle a disease situation for the individual person.

Aspects of this Mayo Digital Platform are also discussed or referenced in Chapter 11 (Glaucoma Practical Application), and Chapter 22 (The latest in Predictive and Prescriptive Analytics), and the final Chapter 26 (21st Century Healthcare: Getting the System that Meets Global Needs).

Conclusions

Genomics and AI have seen a lot of hype over the past 5–7 years with regard to what it can do for better healthcare delivery. However, the computing power needed to handle the tremendous amount of genomic data in real time has not been available until recently. Thus we should see the beginning of some real advances in 2022 and the next few years.

Postscript

The challenges and considerations discussed in this chapter may move some researchers to avoid genomic analysis, because data is not available even now, or the analysis of it is so complicated and computationally intensive. These problems may also characterize the incorporation of any of the "omics" data into medical research. This situation may change, however, in the immediate future. The exciting promise that "omics" data provide will drive an increasing number of medical organizations to enable this data to be used to personalize medical diagnosis and treatment during the rest of the 21st century.

References

Anastasijevic, D., April 14, 2021. Mayo Clinic launches new technology platform ventures to revolutionize diagnostic medicine. <https://newsnetwork.mayoclinic.org/discussion/mayo-clinic-launches-new-technology-platform-ventures-to-revolutionize-diagnostic-medicine/>.

BioSpectrum-Asia Edition Dec 1, 2021. NUHS builds AI production platform using NVIDIA DGX A100 for better healthcare predictions <https://www.biospectrumasia.com/news/45/19420/nuhs-builds-ai-production-platform-using-nvidia-dgx-a100-for-better-healthcare-predictions-.html>.

Boyle, C. January 24, 2022. Meta works with NVIDIA to build massive ai research supercomputer <https://blogs.nvidia.com/blog/2022/01/24/meta-ai-supercomputer-dgx/>.

Ciborowski, J. July 6, 2021. NVIDI launches UK's most powerful supercompter, for research; in AI and healthcare <https://nvidianews.nvidia.com/news/nvidia-launches-uks-most-powerful-supercomputer-for-research-in-ai-and-healthcare>.

Halamka, J.,Cerrato, J. January 17, 2022. Precision medicine, omics, and the black box <https://www.mayoclinicplatform.org/2022/01/17/precision-medicine-omics-and-the-black-box/?linkId=149537359&s=03>.

Thomas, W., Nov 24, 2021. NUSH builds new architecture for AI-based patient care. CDO-Trends <https://www.cdotrends.com/story/16041/nuhs-builds-new-data-architecture-ai-based-patient-care?refresh=auto>.

Wand, H., Lambert, S.A., Tamburro, C., Iacocca, M.A., O'Sullivan, J.W., Sillari, C., et al., 2021. Improving reporting standards for polygenic scores in risk prediction studies. Nature. 591 (7849), 211−219. Available from: https://doi.org/10.1038/s41586-021-03243-6. Available from: 33692554. Available from: https://pubmed.ncbi.nlm.nih.gov/33692554/.

Further reading

Ahmed, A.A.A., Donepudi, P.K., Asadullah, A.B.M., 2020. Artificial intelligence in clinical genomics and healthcare. Eur. J. Mol. Clin. Med. 7 (11). Available from: https://ejmcm.com/article_5590_a17522791fa369a82505aff76274826d.pdf.

Prologue to Part II

Part II of this 2nd edition is found entirely on this book's COMPANION WEB PAGE (URL: https://www.elsevier.com/books-and-journals/book-companion/9780323952743).

In Part II we get into the core of effective learning hands-on examples in the form of step-by-step tutorials and case studies:

1. **Case studies**. Through these examples, you, the reader, can get a good idea of the parameters that need to be considered in various domains, and if you wish to create a simple data set yourself based on variables explained in the case study, you can also work through these types of examples. The reasons that data sets are not supplied for these case studies primarily reside in current HIPPA regulations.
2. **Tutorials**. In effect, these are case studies with data. Using the data sets supplied in the accompanying Elsevier companion web page for this book, you can follow step-by-step and work through these examples so that you can gain the skills in using effective analytic and decisioning methods effectively. These tutorials are done in various predictive analytic software.

(For the commercial software you, the reader, will need to work through the links provided on the companion web page for the various vendors to see what is currently available as either a free trial version or a low-price academic version. For the free software, like R and KNIME, links are also provided to help you, the reader, download this software.)

Part II

Practical step-by-step tutorials and case studies

The following tutorials are available on the companion web page (https://www.elsevier.com/books-and-journals/book-companion/9780323952743):

1st Edition_TUTORIAL A Case Study: Imputing Medical Specialty Using Data Mining Models

1st Edition_TUTORIAL B Case Study: Using Association Rules to Investigate Characteristics of Hospital Readmissions

1st Edition_TUTORIAL C Constructing Decision Trees for Medicare Claims Using R and Rattle

1st Edition_TUTORIAL D Predictive and Prescriptive Analytics for Optimal Decisioning: Hospital Readmission Risk Mitigation

1st Edition_TUTORIAL E Obesity Group: Predicting Medicine and Conditions That Achieved the Greatest Weight Loss in a Group of Obese/Morbidly Obese Patients

1st Edition_TUTORIAL F1 Obesity Individual: Predicting Best Treatment or an Individual From Portal Data at a Clinic

1st Edition_TUTORIAL F2 Obesity Individual: Automatic Binning of Continuous Variables and WoE to Produce a Better Model Than the "Hand Binned" Stepwise Regression Model

1st Edition_TUTORIAL G Resiliency Study for First and Second Year Medical Residents

1st Edition_TUTORIAL H Medicare Enrollment Analysis Using Visual Data Mining

1st Edition_TUTORIAL I Case Study: Detection of Stress-Induced Ischemia n Patients with Chest Pain "Rule-Out ACS" Protocol

1st Edition_TUTORIAL J1 Predicting Survival or Mortality for Patients with Disseminated Intravascular Coagulation and/or Critical illnesses

1st Edition_TUTORIAL J2 Decisioning for DIC

1st Edition_TUTORIAL K Predicting Allergy Symptoms

1st Edition_TUTORIAL L Exploring Discrete Database Networks of TriCare Health Data Using R and Shiny

1st Edition_TUTORIAL M Schistosomiasis Data from WHO

1st Edition_TUTORIAL N The Poland Medical Bundle

1st Edition_TUTORIAL O Medical Advice Acceptance Prediction

1st Edition_TUTORIAL P Using Neural Network Analysis to Assist in Classifying Neuropsychological Data

1st Edition_TUTORIAL Q Developing Interactive Decision Trees using Inpatient Claims (With SAS Enterprise Miner)

1st Edition_TUTORIAL R Divining Healthcare Charges for Optimal Health Benefits Under the Affordable Care Act

1st Edition_TUTORIAL S Availability of Hospital Beds for Newly Admitted Patients: The Impact of Environmental Services on Hospital Throughput

Prologue to Part III

Part III is a complete new part added to this 2nd edition. Part III contains seven chapters that are practical application examples of using data analytics, including predictive analytics, to find new information in data from which healthcare providers can make more accurate medical diagnoses and treatments decisions.

These seven examples involve the following medical situations:

1. Making the best decisions on glaucoma eye disease diagnosis and treatment via data taken by the patient at home
2. Making the best decisions on diuretics drugs to use in kidney disease by the patient taking data at home
3. Creating a predictive model to make better treatment decisions for preterm infants at risk of BPD (bronchopulmonary dysplasia)
4. Making earlier and better decisions on colon cancer using predictive analytic modeling
5. Making earlier and better diagnosis of pancreatic and lung cancer from metabolic data obtained from breath analysis
6. Following COVID-19 incidence and prevalence data around the world to make better decisions about treating an epidemic
7. Making better diagnosis and treatment decisions about DIC for pediatric ICU patient admissions.

Part III

Practical application examples

Chapter 11

Glaucoma (eye disease): a real case study; with suggested predictive analytic modeling for identifying an individual patient's best diagnosis and best treatment

Gary D. Miner, Linda A. Miner and Billie Corkerin[1]
[1]C.P.O.T.; I-CARE HOME Trainer; OMEG, Jenks, OK, United States

Chapter outline

Prelude

Other case studies in this book present several examples of the use of predictive analytics in the practice of personalized medicine. This chapter relates one man's struggle with a serious eye disease (glaucoma), and his efforts in finding the appropriate treatment regime for it. The alternative action is for him to wait for blindness to descend upon him. As you read this account, consider that a successful outcome of any predictive model of glaucome represents much more to the patient than just the provision of an example of personalized healthcare, but the preservation of his vision of the world around him. Data to build this model are not currently available; therefore, the author presents the design for a hypothetical model, which readers can use as a template for use in the future when data to serve are available.

Why this chapter in this book?

This book's primary message is that healthcare and medicine need to fully engage in the "delivery revolution" that has already started. Several terms are being used for this revolution: "Personalized Medicine," "Person-Centered Medicine," and even "Patient-Directed Medicine" and healthcare. In this chapter, we are presenting Gary's 12-year-plus encounter with an eye disease, glaucoma. Gary Miner, project director for this book, has had to endure operations, difficult treatment and numerous physicians in our efforts at finding ways to save his eyes The pathway of diagnosis and treatment over the past 12 years that he has experienced clearly speaks to the "lack of person-centered" care, and directs the ways needed in the future. It is much easier for readers to *grab into the essence* and *importance* of a concept when they can hear about or read about a personal experience. Thus, our presenting Gary's story should help all readers to clearly understand the importance, and the urgency of doing everything possible to get medicine and healthcare delivery to this next *Gold Standard of Person-Centered Medicine*. This glaucoma story should also help the reader understand the importance of patients being involvement in their own care - and at times, directing their own healthcare.

This chapter presents how still, in 2022, medicine is *not* being practiced as a patient-centered endeavor, and thus shows clearly *why* medicine *has to change* in 2022 and the years ahead and is probably why so many of the Internet Blog writers on healthcare topics (Taranto, 2021; Gupta, 2021) chose to make predictions in December 2021 that 2022 would be the *Year of Personalized Medicine Revolution*—the real beginning of what has to happen.

How serious is glaucoma? Why do we need to watch for it?

How long does it take to go blind from glaucoma?

> On an average, untreated glaucoma takes around 10−15 years to advance from early damage to total blindness. With an IOP (Intraocular Pressure) of 21−25 mmHg it takes 15 years to progress, an IOP of 25−30 mmHg around 7 years, and pressure more than 30 mmHg takes 3 years. *(Neoretina Eye Care Institute, 2018, par. 1).*

What is a normal eye pressure?

Eye pressure is measured in millimeters of mercury (mm Hg). Normal eye pressure ranges from *10−21 mm Hg.* Ocular hypertension is an eye pressure of greater than 21 mm Hg (Seltman, 2020).

Characteristics of glaucoma disease

According to Alicea (2021), most blindness is caused by glaucoma. Patients' primary care physicians and optometrists are the medical providers most able to catch the disease, as glaucoma is not felt or noticed by the patient in the first stages. These medical providers should know the patient's medications and conditions and risks that most lead to glaucoma. One of the risks, the patient should know, and that is of a family history of glaucoma, particularly in one or more blood parent.

There are two general types of glaucoma—open angle and closed angle (see Fig. 11.2). The disease, with its increased pressure inside the eye, affects the optic nerve and retinal nerve resulting in damage that, if untreated, will cause permanent loss of sight in the affected eye(s). The only thing that will stop the damage is lowering the pressure. This eye pressure is called "intraocular pressure" (IOP). IOP comes from the amount of fluid (aqueous humor) in the eye versus the amount that can escape the eye though the natural tubing of the eye (which is meant to take away excess fluid).

Patients recognized as at risk for glaucoma should be referred to an ophthalmologist who will conduct a thorough exam and will monitor the success of the treatments or, if necessary, recommend treatments (Alicea, 2021). For example, patients often start out with various drops designed to reduce IOP, that are prescribed by their general OD or general ophthalmologist. However, if various drops do not help, then other strategies are recommended, and a glaucoma specialist ophthalmologist is needed for making these diagnostic and treatment decisions.

Risk factors and treatment

People most likely to get glaucoma are older, nonwhite, and have a family history of glaucoma (Stein et al., 2021). First degree relatives of someone with glaucoma have a lifetime risk of 22%, according to the authors. Average risk for the disease is 2.3%. So, the likelihood of acquiring glaucoma for someone with a parent with glaucoma is ten times the normal population. Stein et al. estimated heritability to be around 70%. They stated that developing a predictive algorithm would be quite useful. This would help to predict those people whose glaucoma would progress, help to determine expected visual loss, and even to find methods of stabilizing the IOP of the individual.

Methods of stabilization included glaucoma drops such as prostaglandin analogs, alpha agonists, and topical beta-blockers (for open angle glaucoma). For closed angle glaucoma, they mentioned topical alpha-agonists, carbonic anhydrase inhibitors (either topical or oral), and beta blockers (Stein et al., 2021).

When drops and medications no longer work then either noninvasive, minimally invasive, or surgically invasive operations are tried. These include laser trabeculoplasty, surgical trabeculoplasty—trabeculotomy, XEN-GEL Stint, and Ahmed valve shunt for open angle glaucoma. For closed angle glaucoma various other laser and surgical options are available.

The bottom line is that the longer the pressure is above normal in one's eye, the more likely is vision loss.

Further in this chapter the reader will find a a description of Gary's patient journey with glaucoma, detailing some of the treatments and expectations of various attempts at controlling his IOP in both eyes, but primarily the left eye. His case was intractable, but not hopeless.

Note to readers: The following few pages will go into topics like the anatomy of the eye, and some specific tests used in diagnosing glaucoma. This is presented here for the general reader who does not have knowledge of these things. This information is not easily obtainable from any one source and it has taken considerable time to summarize it into a short background summary of eye anatomy, physiology, and related medical terminology related to glaucoma. If you, the reader, are familiar with these things, feel free to skip down to the section of this chapter called: "Case Study: Gary's Glaucoma Progression (from about 2010 to 2022)." From this point on in the chapter, Gary will be speaking in first person whenever he is describing his own case. Of particular note, is his ever increasing participation in his own care. Linda credits his increased participation and his final and continuing N of 1 study as being extremely important in the saving of his eye. She believes this cannot be over stated. Without his involvement and efforts, she believes his eye would have lost sight by now.

Basic anatomy of the eye and relation of physical structure to glaucoma disease

Fig. 11.1 shows the parts of the eye that are important to aqueous fluid movement; this figure shows a normally functioning eye. Of particular interest here is the normal fluid flow pathway.

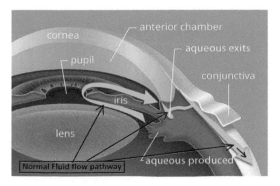

FIGURE 11.1 Parts of the eye of particular interest in aqueous fluid outflow from the eye where the IOP can become elevated above 21 mmHg if the flow is hindered in any way. *From Spraybary, A., Kelley, S. March, 2021, updated Jan, 2022. Aqueous humor. All about vision. <https://www.allaboutvision.com/eye-care/eye-anatomy/aqueous-humor/>, Modified by author.*

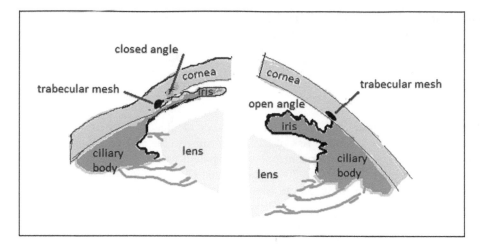

FIGURE 11.2 Illustration of open and closed angles with respect to glaucoma. Note the difference in the angle and the amount of space between the iris and the cornea. Drawing by co-author Linda Miner. *Adapted from Santra Dhali, R., Munshi, S. (2017). Emerging glaucoma therapeutics. Int. J. Basic. Clin. Pharmacol. [S.l.], 4, 4, 606–612, 2017. ISSN 2279-0780. <https://www.ijbcp.com/index.php/ijbcp/article/view/808>. http://doi.org/10.18203/2319-2003.ijbcp20150360.*

What is glaucoma?

In general, glaucoma symptoms are usually high pressure inside the eye that damages the optic nerve which can result in death of optic neurons (neurons that "see") and thus can result in permanent vision loss, either partially or total vision loss (Weinreb et al., 2014; Tsai, 2021). But more specifically, glaucoma is a group of diseases characterized by increased intraocular pressure resulting in damage to the optic nerve (inner retina and axons), retinal ganglion cells, and nerve fibers. It can be a combination of ocular conditions that result in damage to the optic nerve head, eventually causing a loss of the visual field. Most forms of glaucoma follow the classic triad of (1) increased intraocular pressure, (2) optic nerve damage, and (3) a loss of side (peripheral) vision. (Or it can result in loss of "central vision" with side vision (peripheral vision) fairly well maintained, as is the case of the person of this case study). In summary, glaucoma is a symptomatic condition of the eye in which the I.O.P. exceeds the "acceptable tolerances" needed to maintain a healthy eye, thus the affected eye eventually becomes visually dysfunctional. There are two main types of glaucoma: open angle glaucoma, and closed angle glaucoma.

Please see Fig. 11.2 for an illustration of a side-view cross section of the eye and location of many of the parts of the eye being discussed in this chapter.

A good layman's definition of open angle glaucoma is the following:

Chronic open angle glaucoma is a painless condition which causes damage to the optic nerve at the back of your eye and can affect your vision. The person affected is not aware they have it: it is detected by an optician or eye doctor. It is usually caused by an increase in pressure within your eye (Starr, 2018).

For the patient, the symptoms are usually "silent," developing without the person aware, until the person becomes aware of a significant loss of vision, maybe peripheral vision, and particularly noticed in night driving. But if it progresses, other symptoms can be experienced, as shown in the next Fig. 11.3.

What is the normal pressure (IOP) in the eye?

In general, a "normal" pressure range is between 10–21 mm Hg. Some people can have a higher IOP which is "normal for them" and thus live a long life without ever encountering visual field loss. But this is unusual, so eye doctors always want to see a range of 10–21, and when the IOP gets to 22 they want to start monitoring it closely (which today, at this stage when first encountered means checking the IOP every 3–6 months, maybe even once a year).

What causes a rise in intraocular pressure above the norm of 10–21?

Simply stated, a rise in IOP above the norm can be caused by (1) an *increase* in formation of the aqueous fluid within the eye, and/or (2) a *decrease* in the drainage of the aqueous fluid within the eye to the outside of the eye ball. Decreased drainage of fluid from the eye can be caused by one or several things:

Symptoms of Open-Angle Glaucoma

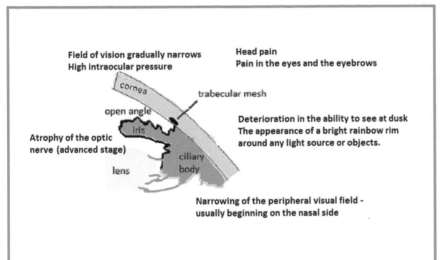

Field of vision gradually narrows
High intraocular pressure

Head pain
Pain in the eyes and the eyebrows

cornea

trabecular mesh

open angle

iris

Deterioration in the ability to see at dusk
The appearance of a bright rainbow rim
around any light source or objects.

Atrophy of the optic
nerve (advanced stage)

ciliary
body

lens

Narrowing of the peripheral visual field -
usually beginning on the nasal side

FIGURE 11.3 Various symptoms that a person may experience as glaucoma (if untreated) can present . Drawing by co-author Linda Miner. *Adapted from: Oday Alsheikh, Braverman-Terry-OEI. (2018). Glaucoma: the silent thief of vision. <https://drstuartterry. com/2019/08/30/glaucoma-the-silent-thief-of-vision/>.*

1. Improper development of the angle of AC

> The anterior chamber (AC) is the aqueous humor-filled space inside the eye between the iris and the cornea's innermost surface, the endothelium. Hyphema, anterior uveitis, and glaucoma are three main pathologies in this area.

From: Cassin, B.; Solomon, S. (1990). Dictionary of eye terminology. Gainesville, Fla: Triad Pub. Co. ISBN 978−0−-937404−33-1)
Narrow configuration of the angle of A.C Cassin and Solomon, 1990)

2. Obstruction of the trabecular mesh work (particulate material)
3. Peripheral anterior synechiae (PAS)

> Peripheral anterior synechiae (PAS) refers to a condition in which the iris adheres to the angle. PAS can develop in various ocular conditions, including ocular inflammation, a posttraumatic condition, after cataract surgery, or with an iris bombe in a pupillary block glaucoma.

From Lee et al., (2006).

4. Plasmoid aqueous condition

> Plasmoid aqueous indicates **a high degree of inflammation and flare**. If left untreated, it can lead to uveitic glaucoma and vision loss. Investigation of underlying systemic disease or infection in anterior uveitis may be pursued based on clinical findings and patient history.) *(Yaseen, 2020).*

> We should understand here some things about the aqueous humor (AC) of the eye, since it is critical to IOP: the AC functions to maintain IOP, but it also functions in providing nutrients to the cornea and lens (as these two parts are avascular, e.g., do not have a blood supply), and additionally functions by removing waste from the cornea and lens.

The illustration in Fig. 11.4—a complete anatomical cross-section of the eye—is included here for the reader's convenience in examining and understanding all of the parts of the eye.

5. Pupil block (lens/seclusio pupillae)

> What causes pupillary block glaucoma? Pupillary block is the most common mechanism leading to acute angle-closure glaucoma, and it occurs when the flow of aqueous humor from the posterior chamber to the anterior chamber is obstructed by a functional block between the pupillary portion of the iris and the lens.*(Flowers et al., 1996).*

> Seclusio pupillae occurs when the synechiae extend 360 degrees around pupillary border. Peripheral anterior synechiae (PAS) may lead to secondary angle-closure glaucoma if they fuse circumferentially and are typically found inferiorly. PAS are more often found superiorly when associated with primary angle closure *(Pittner et al., 2021).*

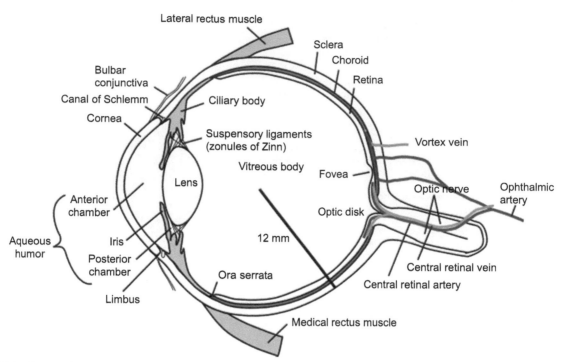

FIGURE 11.4 Cross section of the eye. The aqueous humor is illustrated at the lower left part of this diagram; please note the position of the iris and the lens with regard to the anterior chamber and thus how the aqueous humor fluid bathes both the iris and lens, providing them with nutrients since the lens and iris do not have a blood supply (Feher, 2012). *From Feher, J., (2012). Quantatum human physiology. Elsevier and accessed in Science Direct under: Aqueous Humor <https://www.sciencedirect.com/topics/engineering/aqueous-humor)> <https://doi.org/10.1016/C2009-0-64018-6>.*

FIGURE 11.5 Increase outflow resistant in trabecular meshwork or in Schlemm canal: aqueous humor (*green*) drains through the trabecular meshwork into Schlemm's canal. *(Drawing by co-author Linda Miner; Adapted from: Dudek, K. September 12, 2014. Glaucoma slide presentation. <https://www.slideserve.com/krysta/glaucoma >.)*

In summary for this section of Chapter 11: There are three main types of glaucoma: (1) primary congenital glaucoma; (2) primary open angle glaucoma (PAOG), usually caused by high pressure (IOP) but it can also be associated with either "low pressure" or "normal pressure," and usually simply termed *open angle glaucoma*; and 3) primary angle closure glaucoma (PACG), usually simply termed *closed angle glaucoma*. See Figs. 11.2 and 11.3 above for diagrammatic illustrations of open angle and closed angles.

Pathophysiology of glaucoma

Glaucoma's high IOP pressure can be caused by several things but usually it is an interference in the outflow of fluid from the front part of the eye. It can also be an "increase" in fluid to the eye, which happens at the back of the eye; interestingly some new "Minimally invasive" laser surgery methods are beginning to be used in an attempt to lower this inward flow, but this is done in the surgical center as it requires getting the laser to point at the very back of the eye.

The interference with flow can be due to degenerative changes in (1) the trabecular network inside the eye, or (2) the Schlemm's canal inside the eye, and/or (3) exit channels for fluids to leave the eye. Fig. 11.5 below illustrates an up close diagram of these parts of the eye.

Diagnosis of glaucoma

Visual field testing and other exams: VF (visual field) testing is essential to following a glaucoma patient; both the "central field" and the "peripheral field" must be followed, but additional tests are also needed to fully diagnose glaucoma. The following must be determined: (1) early detection with funduscopy; (2) observation and follow-up over time (a few months or more); (3) measuring IOP frequently not only with the Goldman Gold Standard, but by using different methods for verification; (4) a CDR evaluation; and (5) photography of the disc of the eye.

What is a "funduscopic" exam? A funduscopy is an exam that uses a magnifying lens and a light to check the fundus of the eye (back of the inside of the eye, including the retina and optic nerve). With it, the doctor can see the retina (which senses light and images), the optic disk (where the optic nerve takes the information to the brain), and blood vessels.

What is a CDR evaluation? The *cup-to-disc ratio* (often notated CDR) is a measurement used in ophthalmology and optometry to assess the progression of glaucoma. The optic disc is the anatomical location of the eye's "blind spot," the area where the optic nerve and blood vessels enter the retina (Fernandez-Granero et al., 2017).

What is involved with "photography of the disc of the eye"? Also called "fundus photography," it is able to provide a *picture of the retina*, the retinal vasculature (blood vessels), and the optic nerve head, where retinal blood vessels enter the eye. It can also show drusen, abnormal bleeding, scar tissue, and areas of atrophy. (Herndon, 2019; Spaeth et al., 2016; Dudek, 2014).

Illustrations/photo of eye

Several illustrations of a cross-section of the eye are presented in Figs. 11.6 and 11.7. These show both normal fluid flow within and out of the eye, and abnormal flow that is usually present in glaucoma. Fig. 11.6 shows normal flow; Fig. 11.7 shows abnormal flow.

Management of open angle glaucoma: The first thing is to attempt to reduce the IOP levels of the eye (or both eyes if both are affected). This is usually done by liquid eye drops; there are several kinds of eye drops, each having different modes of operation, but all with the goal of reducing the IOP pressure. However, if the defects (loss of vision) in the "visual field tests" are progressing, or if the eye drops stop working in lowering the IOP to an acceptable level (usually acceptable = 11−17 or so), then various surgical treatments need to be taken. These can be either minimally invasive or invasive surgical methods. Laser treatments (which do not involve a "cutting" of the eye) that can make changes inside the eye via the laser beam are the usual first choice; sometimes they work, sometimes they do not. If they do not work, then invasive surgical methods need to be seriously considered.

"Minimally invasive" surgeries can be invasive

Laser trabeculoplasty (either argon laser or diode laser trabeculoplasty) uses *a very focused beam of light to treat the drainage angle of the eye* (Healthwise Staff, 2020). If this surgery is successful, it makes it easier for fluid to flow out of the front part of the eye, decreasing pressure in the eye. There are two types of laser trabeculoplasty:

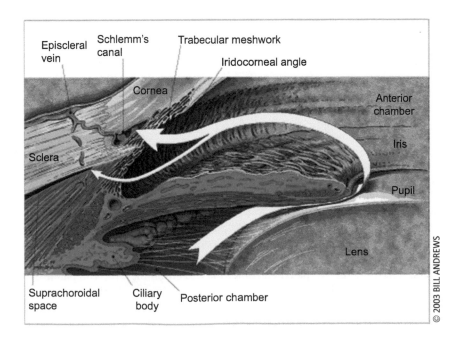

Episcleral vein
Schlemm's canal
Trabecular meshwork
Iridocorneal angle
Cornea
Anterior chamber
Iris
Sclera
Pupil
Lens
Suprachoroidal space
Ciliary body
Posterior chamber
© 2003 BILL ANDREWS

FIGURE 11.6 Normal aqueous outflow from the eye. *(free slideshare; adapted from: <https://www.slideserve.com/krysta/glaucoma>.)*

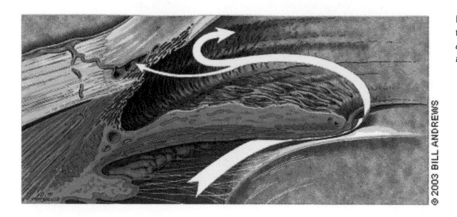

FIGURE 11.7 Abnormal aqueous outflow from the eye (present in glaucoma). *(Free slideshare; adapted from: <https://www.slideserve.com/krysta/glaucoma>.)*

1. Argon laser trabeculoplasty (ALT).
2. Selective laser trabeculoplasty (SLT).

SLT uses a lower-power laser than ALT does (Healthwise Staff, 2020).

If laser trabeculoplasty does not work, then a newer minimally invasive cutting surgery can be done, such as the *ABiC procedure that "reams out" the Schlemm's canal* allowing fluid to again flow through this canal (Khaimi, 2015). (Wife, Linda, called this the roto rooter procedure.)

I, Gary, had this done by the inventor (Khaimi, 2015) of one of the newer types of this process, at Dean McGee Eye Institute in Oklahoma City; in my case it kept the IOP down in the 16−18 range for 4 years, but then rather rapidly during the year 2020, COVID's first year, stopped working, as the IOP bounced up and down between 16 and into the mid-20s.

Invasive surgical treatments

Trabeculectomy

Trabeculectomy is *a type of glaucoma surgery performed on the eye that creates a new pathway for fluid inside the eye to be drained.* This is an outpatient procedure performed in the operating room. When performing trabeculectomy, an eye surgeon creates a flap in the sclera—the white part of the eye—underneath the upper eyelid. Underneath this flap, a pathway is created to allow fluid to drain, which lowers eye pressure. The flap is then placed loosely back down against the sclera to protect the drainage pathway, and the entire surgery site is covered up using the outermost covering of the eye (the conjunctiva). This process results in a small blister of fluid, called a bleb, on top of the eye's surface but underneath the eyelid. The fluid inside the blister gets absorbed into blood vessels and does not cause excess tearing. The bleb is hidden by the upper eyelid and generally not noticeable by other people. Trabeculectomy does not restore vision that has already been lost. The purpose of the procedure is to prevent further loss of vision due to glaucoma (Johnson, 2022).

In the long term, trabeculectomy has been proven to have a high success rate. It is estimated that *90% are successful,* with two-thirds of individuals no longer needing eye drop medication to control the condition afterward. Approximately 10%−12% of people who receive a trabeculectomy will require a repeat procedure (Gotter, 2018).

Trabeculotomy

A surgical procedure aimed at lowering intraocular pressure by unblocking the entry of Schlemm's canal to ease the outflow of aqueous humur (https://medical-dictionary.thefreedictionary.com/trabeculotomy). Trabeculotomy is a procedure more often done on children who are suffering from glaucoma. In the trabeculotomy, the surgeon will create a flap in the conjunctiva, the clear covering over the white of the eye and lining the eyelids, and the sclera, the white of the eye. The surgeon will then identify the drainage system canal and insert an instrument to open the canal wall to help the fluid inside the eye drain better. The scleral and conjunctival flaps then will be closed with stitches. These stitches will dissolve on their own and do not need to be removed. An eye patch and shield is placed on the eye. The patch will be removed by the surgeon in the office the following day (University of Pittsburgh Medical Center, 2021).

And the more modern newer surgeries put in a "shunt" to release the aqueous fluid to outside of the eye, instead of "making a new drainage tube" out of the trabecular network" (as in the trabeculectomy):

XEN-GEN SHUNT—The XEN gel stent is a *surgical implant designed to lower high eye pressure* in open angle glaucoma patients where previous surgical treatment has failed and/or medications alone were insufficient (also known as refractory glaucoma) (from AqueSys, Inc., 2019; Fea et al., 2020).

Both XEN gel stent implantation and trabeculectomy show similar proportions of surgical success and of complications and are therefore both recommendable for clinical routine. However, trabeculectomy seems to be more effective in lowering intraocular pressure than the XEN implantation (from Wagner et al., 2020).

What does the XEN-gel stint look like? What is its size?

The appearance of XEN-gel stint is the "black line" has been described in Figs. 11.8–11.10.

Another way to look at the XEN gel shunt, so that it is clear, is illustrated in Fig. 11.11, where we see where the "bleb" is created by the surgeon on the outside of the eye (underneath the eyelid) which includes an outlet port that releases excess fluid outside of the eye, less lowering IOP pressure.

Ahmed valve shunt. What does the Ahmed valve shunt look like?

The appearance of Ahmed valve shunt before being implanted in the eye has been described in Fig. 11.12.

The Ahmed valve shunt is placed in the eye, with the larger/wider fluid collecting part placed back underneath the surface of the eye, and the drainage tube leading toward the iris of the eye (Figs. 11.13 and 11.14):

Long-term results of using Ahmed valve shunts for glaucoma

How long does the Ahmed glaucoma valve last?

Research studies are ongoing, but success rates for aqueous shunt procedures are within the 60%–80% range *after 5 years*. However, every patient's eye anatomy and stage of glaucoma is different (from: Eye physicians of Long Beach, 2022).

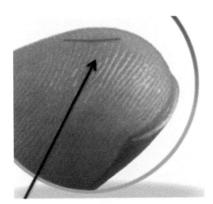

FIGURE 11.8 The XEN-gel stint is the "black line" seen in the illustration above, shown resting on a person's finger. It is about the size of an eyelash (6 mm long with internal diameter of 45 μm (Costello Eye Physicians & Surgeons, 2019). *From Dudek, K. September 12, 2014. Glaucoma slide presentation. <https://www.slideserve.com/krysta/glaucoma>; Costello Eye Physicians & Surgeons. (2019). Glaucoma treatment and surgery: XEN 45 gel implant. <https://costelloeye.com/minimally-invasive-glaucoma-surgery/xen-45-gel-implant/>.*

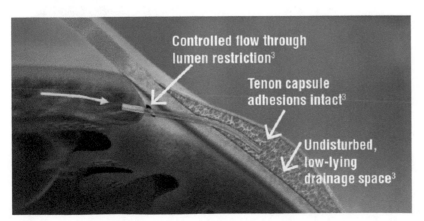

FIGURE 11.9 The XEN-gel stint tube can be seen going from the open angle at the front of the eye to a drainage space out to the side of the eye. This illustration is highly magnified as this "tube" is only the size of an eyelash. *(Free glaucoma slides, adapted from: <https://www.slideserve.com/krysta/glaucoma>.)*

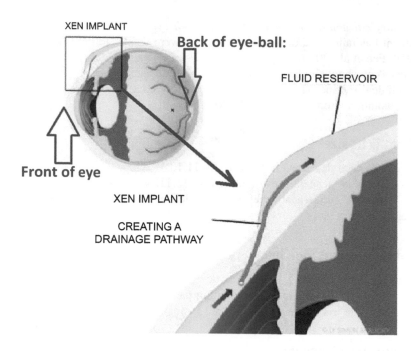

XEN IMPLANT

Back of eye-ball:

FLUID RESERVOIR

Front of eye

XEN IMPLANT

CREATING A
DRAINAGE PATHWAY

FIGURE 11.10 Exact placement of the XEN gen stint––shunt. It is placed between the "front chamber of the eye" and a "flud bleb" created on the surface of the eye out to the side underneath the eyelids (Skalicky, 2022). *From Eye Surgery Associates; Skalicky, S., (2022). Glaucoma implant surgery. <https://www.drsimonskalicky.com.au/glaucoma-specialist/xen-implant/>.*

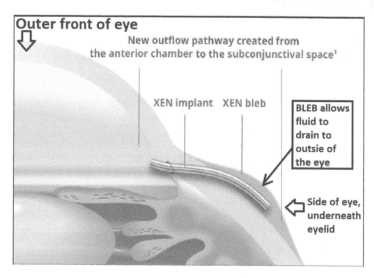

Outer front of eye

New outflow pathway created from
the anterior chamber to the subconjunctival space[1]

XEN implant   XEN bleb

BLEB allows
fluid to
drain to
outsie of
the eye

Side of eye,
underneath
eyelid

FIGURE 11.11 The XEN implant and the XEN bleb, illustrating how this shunt can allow excel fluid to drain out of the eyeball to the exterior of the eye (underneath the eyelid) through a port ("hole") created by the surgeon and known as a "bleb" (Allergan, 2019). *From Allergan. August, (2019). What is XEN? <https://www.xenglaucomaimplant.com/en/hcp/whatisxen, and modified by author Gary Miner in 2022>.*

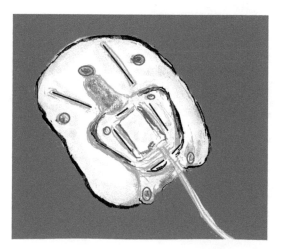

FIGURE 11.12 What the Ahmed valve shunt looks like before being implanted in the eye. *(Drawing by co-author Linda Miner; adapted from: Labtician Ophthalmics, Inc. 2019. The Ahmed glacoma valve. <https://www.labtician.com/product/ahmed-glaucoma-valve/>.)*

Ahmed Placement in Eye

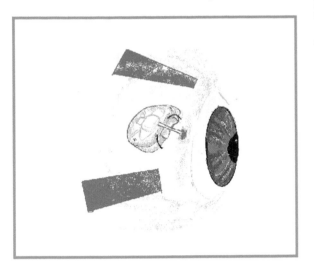

FIGURE 11.13 Diagrammatic representation of how the Ahmed valve is paced in the eye. *(Drawing by co-author Linda Miner; adapted from: Glaucoma Associates of Texas, 2022. What are glaucoma drainage implants? <https://www.glaucomaassociates.com/incisional-glaucoma-surgery/glaucoma-drainage-implant-surgery/>.)*

Ahmen Valve Bleb

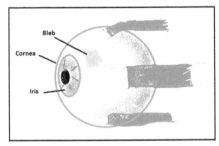

FIGURE 11.14 Positioning of the Ahmed valve shunt. Usually from the outer-upper part of the eye with the tube leading toward the iris. But can be placed from the outer-lower part of the eye with the tube also leading toward the iris. This surgery is one that can be redone, if it does not work the first time (but it usually does, up to 80%−90% effective); or a second Ahmed valve can be put in a second quadrant of the eye if the first one does not reduce the pressure sufficiently, which can happen in about 5%−10% of cases. *(Drawing by co-author Linda Miner; adapted from: Vision and eye Health, 2022. Glaucoma tube shunt <https://www.vision-and-eye-health.com/glaucoma-tube-shunt.html>.)*

A group of patients with refractory glaucoma (e.g., glaucoma that was difficult to control), and after various other treatments, including other surgeries, were finally treated with the Ahmed Valve Shunt. It was found: "Complete and qualified success rates were *31.5% and 46.0%,* respectively, at the end of follow-up. The most common complications were encapsulated cyst formation in 51 eyes (41.1%), complicated cataract in nine eyes (7.25%), recessed tube in eight eyes (6.45%), tube exposure in six eyes (4.8%), and corneal touch in six eyes (4.8%)" (from Elhefney et al., 2018).

Fluid flow in the two main types of glaucoma

Open angle

The open angle glaucoma has been described in Fig. 11.15.

Closed angle

The Closed angle glaucoma has been described in Fig. 11.16.

Photography of eye—looking at fundus in the diagnosis of glaucoma

Fig. 11.17 is further clarified and made understandable by Fig. 11.18:

Case study: my (Gary's) glaucoma progression (from about 2010 to 2022)

First medical treatment—Eye Drops only—to control IOP treatments, ∼ *2009−13* (but eye drop therapy continued on throughout up to present day). When it was discovered, during a routine eye glass exam with my optometrist in the late

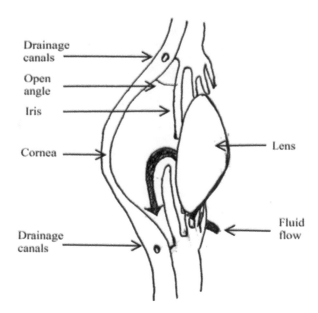

Drainage canals

Open angle

Iris

Cornea

Drainage canals

Lens

Fluid flow

FIGURE 11.15 Open angle glaucoma. The flow of fluid is into the area in front of the lens, but it apparently does not get out through the "drainage canals" (Schlemm canal) easily. This is the most seen type of glaucoma. It is also referred to as wide-angle glaucoma. It arises due to partial blockage in the drainage canal by which pressure increases slowly as fluid is not able to drain properly as seen in the figure (**Gupta et al.**). *PHOTOS and specifically from: Gupta, A., Soman, V., Bhardwaj, S. Performance improvement in detection of glaucoma in human eyes. Copyright 20xx IEEE...so a IEEE J.*

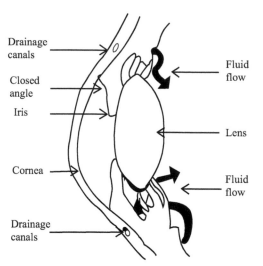

Drainage canals

Closed angle

Iris

Cornea

Drainage canals

Fluid flow

Lens

Fluid flow

FIGURE 11.16 Closed angle glaucoma (also called angle closure glaucoma). It is also referred to as acute angle glaucoma and is caused by sudden and complete blockage of aqueous humor drainage. The pressure rises suddenly leading to a speedy loss of vision. It is caused because of the narrow drainage angle as well as a thin and hanging down iris. *PHOTOS and specifically from: Gupta, A., Soman, V., Bhardwaj, S. Performance improvement in detection of glaucoma in human eyes.*

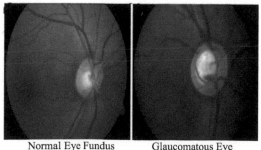

Normal Eye Fundus Image

Glaucomatous Eye Fundus Image

FIGURE 11.17 Using this fundus image of the eye, elaborate data processing is done on this image, looking at (1) the optical cup and disc areas (CDR = cup disc ratio) and (2) blood vessel segmentation ("structure−formation"); from this a SVM ("support vector machine" algorithm; a predictive analytic method) classifier is used which gives a binary result, either "positive glaucoma" or "negative glaucoma." *PHOTOS and specifically from: Gupta, A., Soman, V., Bhardwaj, S. Performance improvement in detection of glaucoma in human eyes.*

2000s that my IOP (intra optic pressure), especially left eye, was above 21, and in the 26−28 range, ophthalmologists were consulted. At this point various IOP lowering eye drops were tried, with varying results. Most lowered the pressure, but many had allergic side effects, causing inflammation of the eyes. Some IOP eye drops caused more

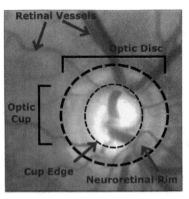 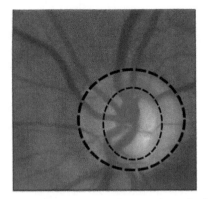 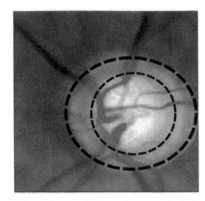

FIGURE 11.18 This illustrates the optic nerve head (ONH) in the fundus images. The optic disc (OD), optic cup (OC), and neuroretinal rim are labeled. The images show a healthy eye on the left side, and an initial state of glaucoma in the middle, and an advanced stage of glaucoma on the right. The relative expansion of the OC and the thinning of the neuroretinal rim are noticeable as this also has an effect on the retinal blood vessels. *PHOTOS and specifically from: Abdullah, F., Imtiaz, R., Madni, H. A., Khan, H. A., Khan, T. M., Khan, M. A. U. et al. 2021. A review on glaucoma disease detection using computerized techniques. Digital Object. Identifier https://doi.org/10.1109/ACCESS.2021.3061451. IEEE ACCESS: 9, 2021 Received February 2, 2021, accepted February 15, 2021, date of publication February 23, 2021, date of current version March 10, 2021.*

inflammation (reddening, plus swelling—inflammation) and others less. At one point it was thought that the preservatives in the eye drops may be the cause (as this had been determined in the scientific literature to often be the case), so at one point sterile drops without preservative were tried. These sterile drops came in individual ampules, each for one day's use and then tossed; these were more expensive (not always covered by insurance), and did not seem to have that much effect, so eventually we settled on the "usual drops," like "Generic-COSOPT" and some of the other's in their generic forms whenever possible. During this period of time we started spending part of our year living on the East Coast, so this meant having a set of opthamologists (eye medical doctors) in two places, the Midwest and the East Coast.

During this time period (2009—13) the glaucoma appeared to be an open angle glaucoma—but the story gets more complicated (see the next section).

My Atlanta, GA IRIDOTOMY—IRIDOPLASTY, 2014. In the 2013—14 time frame, the IOP started to "bounce around a bit," so my East Coast ophthalmologist sent me to a glaucoma specialist, that soon discovered that the "angle was closing" and the glaucoma disease was becoming a closed angle glaucoma case.

So my glaucoma was not simply either an open angle glaucoma or a closed angle glaucoma, but a more complicated one, having characteristics of both types.

Closed angle glaucoma can become very serious. At some point one can rapidly experience excruciating headaches. At this point emergency treatment is needed. Hopefully this treatment can be provided in time to prevent complete loss of vision in an eye.

So the Atlanta glaucoma specialist immediately scheduled what appeared to be a combination of laser peripheral iridotomy plus laser peripheral iridoplasty. This is an in-office procedure where a laser is used to "cut a v-shaped" hole out of the iris, thus preventing the complete closure of the eye's angle, and additionally hopefully giving an outlet for fluid drainage. Another goal of this laser procedure is to hopefully reduce the need for eye drops (or as many eye drops) and additionally prolong the time before invasive glaucoma surgery is needed (Chen et al., 2017).

However, this glaucoma specialist doctor was very disappointed when I returned for a follow-up visit a few weeks later to find that the IOP had not been reduced—she acted "dumb-founded" that the procedure had not worked—so the next thing was to add more eye drops, to see if this would help. But additional eye drops did not provide significant changes in the IOP.

My SLT (Selective Laser Trebeculoplasty), 2014: Selective laser trabeculoplasty, or SLT, is a form of laser surgery that is used to *lower intraocular pressure in glaucoma*. It is used when eye drop medications are not lowering the eye pressure enough or are causing significant side effects. It can also be used as initial treatment in glaucoma (Francis, 2020).

This laser treatment is "selective" in that not all of the trabecular network is a target, but instead only parts of this fluid draining network. The purpose is to "open up" some of the trabecular network in an attempt to increase fluid outflow.

Results may not be seen for up to 1–3 months after the SLT procedure is done. If successful, the effect will generally last between 1 and 5 years, and in some cases, longer than that. If it does not last at least 6–12 months, it is usually not considered successful.

In my case, there did not seem to be any significant effect. Thus it was repeated a second time, but again no significant effect was evident. (This SLT is a "cold laser" causing little side scarring of tissue, and thus can be used repeatedly; another type of laser, an older method, is the argon laser which can be used only a couple or three times.)

My Minimally Invasive surgery, ~2015/2016, ABiC (Schlemm Canal "clean out/opening"): Since the SLT did not work, and the IOP was bouncing around to a higher value on some of the doctor visits, the next thing to consider was a "minimally invasive eye surgery" (Khaimi, 2015; Klink et al., 2011).

(To watch Dr. Khaimi perform ABiC, visit https://eyetube.net/videos/phaco-cp-mak-mov; https://eyetube.net/videos/abic-procedure-reduces-iop; https://eyetube.net/spotlight/quantel-medical/interventional-glaucoma-roundtable)

This ABiC eye surgery process is illustrated in Fig. 11.19.

> The key difference between ABiC and traditional canaloplasty is that, in the former, no tensioning suture is required to maintain the IOP reduction. Of 106 patients treated in a recent case series, there was a total average decrease in IOP of 35% and in glaucoma medications of 100% 6 months postoperatively compared to baseline (M.A.K., unpublished data, 2014–15) *(Khaimi, 2015).*

At the time I had this procedure done (year 2016) there was only about 3–4 years of follow-up data after surgery available to indicate how effective and for how long this microinvasive ABiC surgery would be, although a large percentage of patients had retained low IOPO (in the mid-teens level, that is, 13–17 mmHg range). This ABiC process did keep the IOP levels down in the mid-teens for about 3 years.

In me, however, the IOP began to start increasing and "bouncing around" after 3 years, and during the 4th year got to the point where something additional had to be done. This was the year 2020 (the COVID year) isolating us on the East Coast for months, where my eye center refused to measure his IOP on a monthly (or even weekly) basis, but instead insisted on the usual 3–4 month intervals. Even when I requested that he come in weekly and have the technicians measure the IOP (something that only takes about 90 seconds) he was refused, even though CMS has a medical code for letting technicians do this on a "walk-in basis" (however, it does not reimburse the eye center as much as if the ophthalmologist took the readings); in fact at this point, this East Coast eye center had even stopped having the technicians take the IOP prior to having the ophthalmologist take it (not sure if this was the big medical center's decision from headquarters, or a local doctor's decision). At all other eye centers on both the East Coast and the Midwest it was my experience that the IP was taken twice on a doctor's visit, initially by the nurse-technician, and again by the doctor, so this seems to be the "standard of care" in this field.

BUT:

This "standard of care" is not frequent enough, as in my case in which I ended up losing about another 1/3 of my L-eye's vision during this 2020 year, all since IPO exams/readings were not routinely available. If IOP had been taken weekly, this new sudden loss of vision likely could have been prevented.

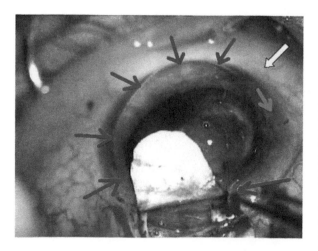

FIGURE 11.19 Canaloplasty of the Schlamm canal of the eye. This photograph shows the microcatheter being advanced through the Schlemm canal. The yellow arrow is pointing at the beacon tip of the microcannula visible through the sclera. The red arrow is pointing to the entry point of the microcatheter. The gold arrow is pointing to the direction the microcannula is advancing (and will advance until it has traversed the entire 360 degrees of this Schlemm canal). And the blue arrows show where this microcannula has already traversed This process is done to "ream out" this canal so that fluid can again flow through it; a "gel-like substance" is pushed into this canal as the microplastic hollow tube is withdrawn; this "gel like substance" holds open the canal and is gradually absorbed by the body's systems, with the idea that the canal will stay open (at least for a period of time—1 year? 2 years? 3 years? Or more?) (Lewis et al., 2009). *Diagram reframed with arrows and labels added from: Lewis, R.A., von Wolff, K., Tetz, M., Koerber, N., Kearney, J.R., Shingleton, B.J. et al. (2009). Canaloplasty: circumferential viscodilation and tensioning of schlemm canal using a flexible microcatheter for the treatment of open angle glaucoma in adults. J. Cataract Refract. Surg. 35:814–824, Published by ELSEVIER; https://new-glaucoma-treatments.com/wp-content/pdf/Canaloplasty%20Two-year%20Study%20Results.pdf.*

Self-monitoring intraocular pressure by the patient for more accurate DX and treatment decisions

I had been concerned with the need for patients taking their own IOP measurements, and starting in October of 2019 had been working with my Midwest doctor to get a patient "self IOP measuring" device. One that was like a contact lens, but with a measuring device and transmitter built into it, seemed a good option; this device sent a continuous signal to a receiving computer so that the IOP values for all times of the day could be evaluated. This device was developed in Germany and is called the "Triggerfish contact lens." This "Triggerfish contact lens sensor" is designed to provide continuous 24-hour recording of ocular dimensional changes, which are related to IOP. However, the signal sent by this Triggerfish lens is not the IOP values; thus the signals have to be translated into IOP values at the computing center receiving the signal, and then relayed to the patient. However, it is not quite that simple; the patient also has to wear a recorder around their neck to capture the signals from the device. Also a self-adhesive small antenna has to be pasted to the skin close to the eye to transmit the Triggerfish's lens signal to the recorder. But that is not the end; then the patient has to schedule a visit with their doctor where the clinic staff downloads the information stored on the recorder (around the patient's neck), before the patient can be given any of this information. So this process is not in real time. Additionally, people must not bath, shower, or swim while wearing this device, as there is danger of shock! In addition to curtailing cell phone usage, as it might interfere with the Triggerfish's signal. (NICE (National Institute for Health and Care Excellence), 2014; (Figs. 11.20 and 11.21)

But after several months of attempting to get this device, we finally discovered that it had been taken off the market.

i-CARE home device for patient home monitoring of intraocular pressure values

We will not go into the different types of "tonometry" (ways of measuring IOP values) in this chapter, but for those readers with interest, the following reference provides a good overview (Piltz-Seymour and Tai, 2021).

Instead, we will only look at patient HOME IOP measurement devices in this chapter.

The Icare HOME tonometer device has been available to European glaucoma patients since 2014, and is now available to patients in the United States. It uses a disposable probe to measure eye pressure and can be used up to six times a day. The home tonometer takes six rapid IOP measurements, then calculates eye pressure, and stores it in the device's memory (Glaucoma Research Foundation, 2020).

Here is a video of how the i-CARE HOME intraocular pressure measuring device is used: https://youtu.be/bXgEoQV0orM

Several other "Tonometers" for home use are now available as of 2021, but all have a fairly high price tag (and at this time, it is unlikely that any insurance plan will pay for a patient's use of these devices):

1. The "intelligent" tonometer: Icare ic100 ($1950.00)
2. i-CARE HOME device: ($1000.00 in this 2020 reference, but I discovered that it was available for a minimum of $2000 cost to the patient if ordered through a doctor's clinic, or $2600 if ordered directly from the company by the patient but only with a doctor's prescription) (Figs. 11.22 and 11.23).

FIGURE 11.20 The SENSIMED Triggerfish contact lens, for continuous measurement of ocular dimensional changes which correlate to IOP eye pressure. *Adapted from Valtronic, 2021. Smart contact lens—a novel solution for the management of glaucoma <https://valtronic.com/portfolio-items/smart-contact-lens/>.*

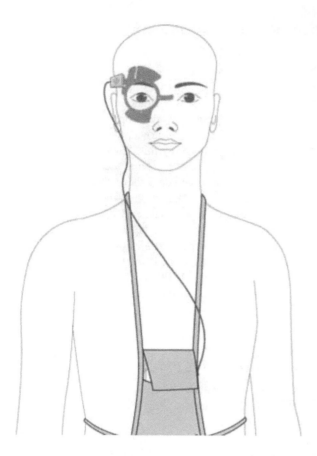

FIGURE 11.21   The Trigerfish contact lens IOP measuring device with all of its accoutrements: antenna taped to the skin close to the eye (or a wire running down with signal option); recorder of data held in a pouch worn around the neck (and do not take a shower or go swimming, as you may get an electrical shock). *Adapted from Valtronic, 2021. Smart contact lens—a novel solution for the management of glaucoma <https://valtronic.com/portfolio-items/smart-contact-lens/>.*

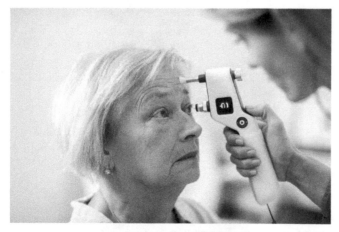

FIGURE 11.22   The "intelligent" Tonometer- Icare ic100. *From: https://www.ophthalmetryoptical.com/blog/tonometer-for-home-use/.*

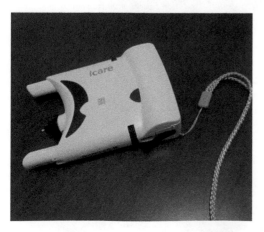

FIGURE 11.23   i-CARE HOME device. *(Photo taken by authors of this book, Gary and Linda Miner.)*

For those readers interested, here is a YouTube video of Gary being trained by Billie Corkerin (Guest Author). in how to use the i-CARE HOME device: https://youtu.be/6nakad5Y3HY—Title: *Gary Trains on the iCare Machine (filmed and edited by coauthor Linda Miner).*

1. A versatile and accurate tonometer: Reichert tono-pen avia ($1,590.00) (refurbished, on Internet, January 18, 2022: $3195)
2. A credible and reliable tonometer: Reichert tono-pen XL ($1,290.00) ($2705 refurbished on Internet January 18−2022)
3. Noncontact tonometer- Keeler PSL One ($2,990.00) ($5101 on internet Jan 18−2022) (Figs. 11.24, 11.25 and 11.26).
4. Balanced and comfortable tonometer: Reichert PT-100 (New cost on Ebay January 18, 2022 $4,995; preowned $2995) (Fig. 11.27).
5. New Israeli startup device called IOPerfect: it is an AI-based contactless IOP device from an Israeli startup called Ophthalmic Sciences (Fig. 11.28).

A new device in the treatment of glaucoma has been introduced by Opthalmic Sciences. IOPerfect works by using artificial intelligence (AI) in a virtual reality-type headset which allows telediagnosis and remote monitoring.

It can be self-administered, and patients can wear the headset to monitor their IOP (fluid pressure inside the eyes) at home. It is noninvasive and is without the need for eye drops or constant calibration.

The test is performed in less than 3 minutes, and is unaffected by corneal thickness. Its algorithm provides reliable AI-based image processing analysis of vascular pressure response.

The IOPerfect device applies mild controlled air pressure within the headset while microvideo cameras capture the difference in response to pressure of internal versus external eye blood vessels.

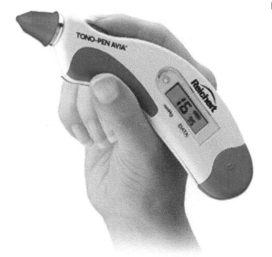

FIGURE 11.24 Reichert tono-pen avia. *(From free stock photos).*

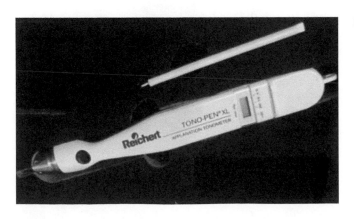

FIGURE 11.25 Reichert tono-pen XL. *(From free stock photos).*

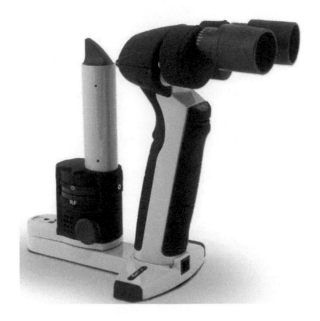

FIGURE 11.26 Keeler PSL One tonometer. *(From free stock photos).*

FIGURE 11.27 Reichert PT-10. *(From free stock photos).*

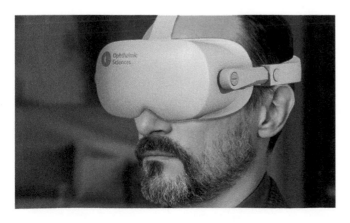

FIGURE 11.28 IOPerfect, new AI-based contactless IOP device from an Israeli startup called "Ophthalmic Sciences." *(Figure Credits to Ophthalmic Sciences; From Keating, F. Dec 8, 2021. Israeli startup unveils AI device for measuring eye fluid pressure in glaucoma treatment. Healthcare iT News. <https://www.mobihealthnews.com/news/emea/israeli-startup-unveils-ai-device-measuring-eye-fluid-pressure-glaucoma-treatment>.)*

The data is uploaded to a secure cloud where analysis takes place. The results can be accessed by medical staff for interpretation (Keating, 2021).

As others are stating

We are now moving away from the days when only professionals could perform tonometry on glaucoma patients. Patients can now have a tonometer for home use to monitor intraocular pressure without the doctor's presence. It would be best if you (the patient) kept on testing your eye pressure because it can fluctuate rapidly before your next visit to the doctor. *(Smith, 2021)*.

Being able to measure eye pressure at home can be beneficial for patients for whom going to the doctor is a burden—for example, those who can't drive themselves, or who have physical limitations. And of course, during the coronavirus pandemic, there are a lot of people who don't want to come to the clinic or doctor's office to get their pressure checked. But there's another good reason for being able to check one's pressure at home.

If someone has glaucoma, or is suspected of developing glaucoma, and their condition is stable—meaning they've been seen multiple times in clinic and their eye pressure has been in a safe range and their visual field test and optic nerve scan findings are stable—then they'll typically be seen once every six to twelve months to have their pressure measured.

Whenever you're measuring pressure, then, you're only getting a snapshot of what is a very dynamic process. For that reason, measuring eye pressure more frequently—including at different times during the day—could provide a much better picture of the overall intraocular pressure." *(From Johns Hopkins Medicine, 2020)*.

Further treatment - my late March/early April 2021—Argon LASER "Trebeculoplasty" to again, as a "last ditch effort," see if this would reduce IOP. It did not, thus I went to the another "more modern" eye care group for shunt surgery (as the this first glaucoma specialty group only did trebeculoplasTY-trebulotomy surgically)

In late 2020 it became obvious that something "more serious" was going on in my left eye, as in the fall of 2020 the IOP as measured by the very few "infrequent" times that my local East coast eye clinic was agreeable to measuring my IOP, was finally showing some values in the higher 20s and one about 31. So, after his local eye doctor prescribed one more eye drop for the month of December, 2020 to January 2021 to see if this would work, and it did not, additional consults were needed with a glaucoma specialist. I was assigned to a glaucoma specialist during February and March 2021; this specialist tried two more eye drops as one more last effort to reduce IOP, but instead of helping the IOP to go down, these just inflamed the eye further (as I was allergic to many/most of the usual IOP-reducing eye drops). Then as a last ditch effort we tried argo laser 'trabeculoplasty" once again (remember SLT trebeculoplasty had been done about 6 years earlier, but did not work) to see if it would open up the fluid pathways, thus lowering the IOP; again this failed. This medical eye group only did the most invasive type of glaucoma surgery, the true "surgical trabeculoplasty/trabeculotomy." This type of glaucoma surgery has been used for years, but it does have major risks, one being creating too low of an iOP (e.g., lower than 8 mmHg), which is also not good for the eye's health.

I really wanted the more modern "shunt/valve" surgeries (that had been around for 10−20 years, anyway, with good results), so this meant searching out a good medical eye group that did all types of glaucoma surgeries. An excellent one was found, and a superb young surgeon who "did them all" was selected. This of course delayed the needed surgery by about 4−6 weeks, but I judged this to be the "best option" *(unfortunately, the local eye doctor who had referred Gary to the first local glaucoma specialist, had not informed us that this group did not do all of the current glaucoma surgeries, and also did not inform us that they still only had Argon Laser Trabeculoplasty, as the Argon can only be used 2 or 3 times; whereas the modern SLT Laser can be used many many times)*. Again, bad medical advice (=bad diagnostic and treatment decisions) from our local eye clinic, which only further delayed the needed surgery, all the time I was losing more and more retinal cells.

My invasive surgery—2021—XEN-gel shunt and later Ahmed valve shunt

So luckily we found a superb glaucoma surgeon, one of the younger generation trained in all glaucoma surgeries and not afraid to do them, and luckily got an appointment within 2 weeks! At the appointment Gary went through the usual tests, like visual field testing, photographs of the eye, and IOP measurements. But we discussed surgery immediately, not wanting to waste time. The L-eye was very inflamed from all of the eye drops prescribed (four different ones from previous eye doctors). We discussed the XEN-gel stint, the Ahmed valve shut, and also the old-fashioned full surgical trabeculoplasty—trabeculotomy. The XEN-gel stint was suggested as the "best to try," and if that did not work the Ahmed valve shunt. A newer type of laser, but requiring the operating room, where the laser was placed at the back of the eye with the goal of lowering the fluid flow into the eye, was also discussed; but as the surgeon pointed out, this is new and the effective rate was not high, more of a "long shot" at this point, but something to think about.

The XEN-gel stint was selected. The surgery was originally scheduled 3 weeks away, at the end of May, 2021, but the surgery scheduler discovered a spot had just opened the next week, so XEN-gel stint was scheduled for May 5. The surgery went well (a 15—30 minute procedure), and the next day at follow-up with the eye clinic's follow-up opthamologist (located closer to where we lived, so more convenient) the IOP was about 12 so it appeared to be working. However, 3 days later the IOP was up in the 30s even 40s, as apparently all of the eye inflammation had surrounded this small (eyelash like in size) tube and pushed it shut. So the doctors put me on heavy oral steroids and diuretics to rapidly bring down the IOP values. However, the XEN-gel stint did not open, so emergency surgery was scheduled on May 22 to put in the Ahmed valve shunt (this valve is wider, larger, so can more easily pull out the excel eye fluids; also it has a "cap" built into it so that it will not let the pressure go below 8 IOP units, which is a nice added feature).

Ahmed valve shunt surgery: This surgery, scheduled as an "emergency" on May 22, went well. The next day at the follow-up with the follow-up glaucoma ophthalmologist, the first thing this doctor said when he looked at the eye, and said in an 'exclamatory voice tone: "*It is perfectly placed!*" The pressure was also down where we wanted it, about 11 that day. Steroid eye drops, taken every 2 hours for the first week, were prescribed, as it is important to keep the new "exit port," called the Bleb, open, thus preventing it from healing over. The routine eye drops that I'd taken for some years ("Generic-COSOPT" = Dorzolamide—Timolol) were also continued, as backup insurance. Because of the nature of this surgery, it can take some weeks or months to fully heal and for the IOP levels to stabilize. Thus I had routine eye doctor visits every 3 days at first, and then every week over the first month following this surgery, and then every 2—3 weeks for a couple of months, and then visits were spaced out between 1—2 months. It took about 6 months, with trials of various hard and soft steroids, and different IOP lowering eye drop combinations, to come up with a regimen and a stabilization of the IOP at a 15—18 level.

One of the reasons for it taking 6 months was that patient my "genetics" made me very sensitive to the steroid eye drops; these steroid drops would increase IOP more or less, depending on the steroid used (*plus the steroid eye drop also had the uncomfortable side effect of inducing frequent trips to the bathroom for urination—strong steroids every 30—60 minutes, and soft steroids every 60 minutes to 4 hours—hours being more reasonable!*). So it took many combinations of "steroid/eye drop" until we found one that did not increase the IOP drastically plus eliminated the frequent bathroom trips.

My use of i-CARE HOME device to measure my own IOP on routine basis (June 26, 2021 onward). Data collection and analysis to determine "the right treatment" that in hindsight "should have been the standard of care" from 2010 onward.

While undergoing these surgeries with the East coast eye clinics, I was still was in touch with his Midwest Eye doctor (his doctor who had overseen all of his glaucoma treatments over a 10-year period) to obtain a device that I could use at home to measure my own IOP. (In fact, I was in touch with the Midwest doctor before the surgeries, as I considered coming back to the Midwest for the surgery but decided that the surgery could be done sooner on the East Coast). The Midwest doctor had been trying to get me a home IOP device for over a year at that point, but the one we were after apparently was "taken off the market," and thus we were forced to find another option.

We came up with the i-CARE HOME IOP measuring device in mid-May. Unfortunately, at that time, a patient could not get one of these i-CARE HOME devices by themselves but had to go through a doctor. So, my Midwest eye doctor got i-CARE HOME to ship me the device, and then he sold it to me, so that I owned it and could use it for several years. (A device like this should be something easily obtained by the patient at a "reasonable cost," but of course we were still working with the "old model of healthcare delivery" as the "new model of patient-centered, patient-directed" is just beginning to invade the healthcare market!) The device cost $2000 obtained through the doctor (however, I could have gotten it for $800 additional, e.g., $2800, if he'd gotten a doctor's prescription for it and then ordered directly from the i-CARE HOME company—a Finnish company). *Reasonable, makes sense, right?*

My Midwest eye doctor assigned his research technician to "learn the device" (she had already in a previous position, taught patients how to use other similar products that this CARE HOME company produced) and then had her teach it to me at their offices in the Midwest the last week of June 2021. This "teaching session" occurred, was successful, and I have been using the i-CARE HOME device ever since. The IOP readings obtained by the device are fed, through a short cable, into either the patient's smartphone or computer, and when this is done, they go directly to the i-CARE HOME computer center web page, which both the patient and the doctor can access. In fact, my Midwest eye doctor had it set to give him an email alert every time an IOP reading of 20 or greater was obtained. I was his first patient using this; he wanted to work through its operation, as he was thinking towards the future in making it an option with any of his glaucoma patients. He is on the "right track," as having the patient take IOP readings frequently is essential to correctly diagnosing glaucoma and making good treatment decision for a patient's glaucoma BEFORE too much retinal cell death (irreversible damage) is done; the need for this will be explained, and obvious, in the last part of this chapter, where we show what the i-CARE HOME was able to do in coming up with the best treatment plan following the surgery, for me. Coming up with this effective treatment plan would have been impossible if only the IOP values from doctor visits were used in making decisions. The need for frequent measurements cannot be overstated. Obtaining only one reading per six months in the doctor's office should be considered substandard treatment. There is a great need for patients gathering their own frequent data.

My—Data presentations *(tests and data obtained by technicians at the eye clinics)*: (Figs. 11.29 and 11.30).

OD (from Latin: "oculus dextrus") = right eye; OS (from Latin "oculus sinister") = left eye (Fig. 11.31).

OD (from Latin: "oculus dextrus") = right eye; OS (from Latin "oculus sinister") = left eye (Fig. 11.32).

Both right (OD) and left (OS) eyes are illustrated in thee diagram below: (Fig. 11.33).

Both right (OD) and left (OS) eyes are illustrated in thee diagram below: (Fig. 11.34).

OD (from Latin: "oculus dextrus") = right eye; OS (from Latin *"oculus sinister"*) = left eye (Fig. 11.35).

OD (from Latin: "oculus dextrus") = right eye; OS (from Latin *"oculus sinister"*) = left eye (Fig. 11.36).

Increased night-time urination frequency was an unpleasant side-effect of my using steroid eyedrops

A consideration that was of concern for me, was that these "very low dose" steroid eye drops triggered frequent urination; this is a rare side effect, but one that patient I, sadly, experienced. The strong steroid Durozol induced urination urges sometimes even less than 1 hour apart, with an average of about 75 minutes. Since the steroid eye drop was prescribed to be taken in the evening before bedtime, this induced frequent trips to the bathroom during "sleep time," upsetting good sleep, thus both an unpleasant side effect, but also a more major side effect in that good sleep was impacted.

The softer steroid Prednisolone induced fewer bathroom trips, but still about every 90 + minutes. Unfortunately, the Fluoromethholone, the soft inexpensive steroid that we had hoped to reduce this urination frequency, worked about the same as the Prednisolone. The only thing that worked well was either "no steroid" (which was probably not a good idea after eye bleb type of surgery) or LOTEMAX SM 1x per day. The LOTEMAX SM 1 time per day in the evening (in left eye only, of course) reduced the nightime urination interval to between 3—5 hours, which was acceptable (see Fig. 11.37 for graphic details).

Is increase in "urination frequency" a common side effect of use of "steroids in eye drops"?

In general, use of steroids, especially if taken orally or via injection, can increase urinary frequency (Kapla, TB; August 6, 2021). But do the small amounts of steroid in steroid eye drops generally cause increased urination frequency? Looking at the literature found in eye journals and ophthalmological sources seemed to indicate "no," as this increased urination frequency was not listed among the numerous possible common side effects. But recently a Mayo Clinic Medical website suggested otherwise, as it said: "Talk to your doctor right away if you have more than one of these symptoms while you are using prednisone steroid eye drop, including . . . increased thirst or urination (Mayo Clinic, 2021) and another source said that ophthalmic prednisone can increase urination and thirst (Drugs.com, 2022)." (Mayo Clinic, 2021). The Mayo Clinic article was primarily addressing prednisone taken orally or by injection, and not in form of eye drops; yet it did give an indication that steroids may induce increased. The Drugs.com article spoke of ophthalmic prednisolone. The Drugs.com article (April 30, 2022) does discuss ophthalmic prednisolone.

In general, it appears that "increase in urination" is listed in the literature as "something that has not been thoroughly studied and thus the incidence is not known"; yet there are references to its possibility (Mayo Clinic, 2022c; EveryDayHealth, 2022).

However, it appears that "increased urination" may only be found when a "low dose" of steroid is given; there appears no clear-cut research on low-dose steroid eye drops and increased urine production as a side effect. But the study by Lui et al. in the Journal of Cardiovascular Pharmacology (2015), conducted with heart patients, using a randomized controlled trial, showed that low-dose predinozone increased urination frequency; however, this effect was not seen with medium or high doses of steroid.

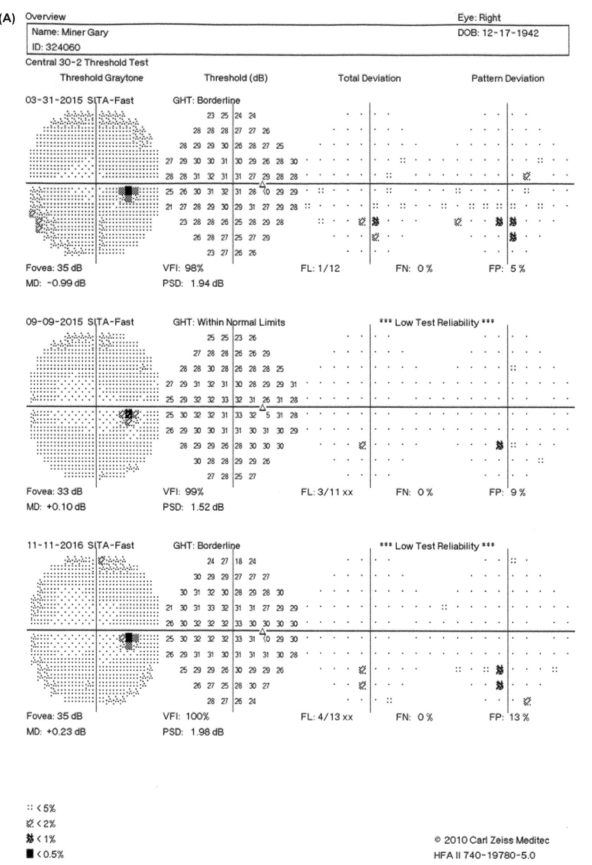

FIGURE 11.29 (A)–(C). Visual field tests, right eye, showing March 2015 through September 2021 progression. Little change is seen in the right eye. Visual field tests is one of a couple of tests that are very important in really understanding if there is progression of the Glaucoma disease; thus this becomes a very important test to use along with the actual IOP values in predicting the future progression of the disease for any one individual.

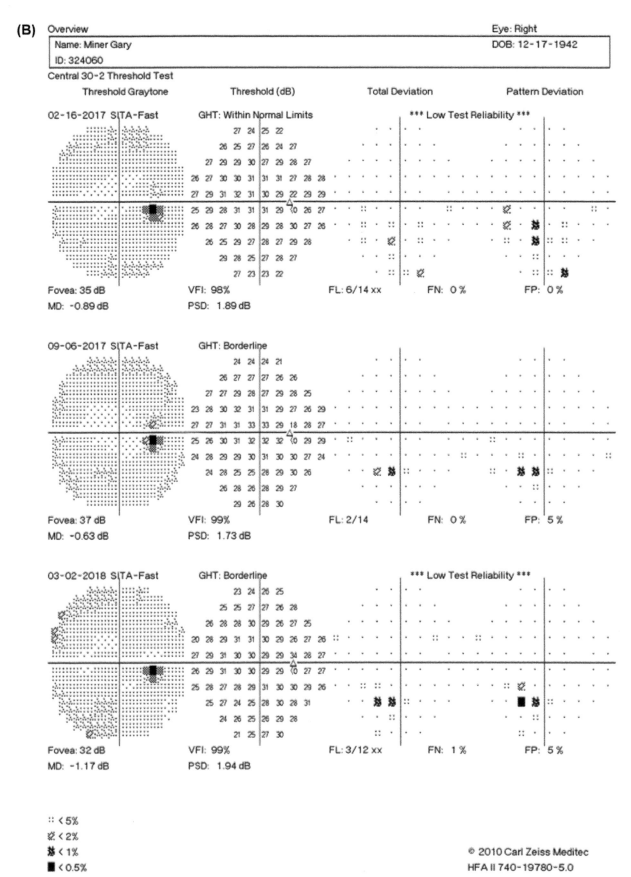

FIGURE 11.29 Continued.

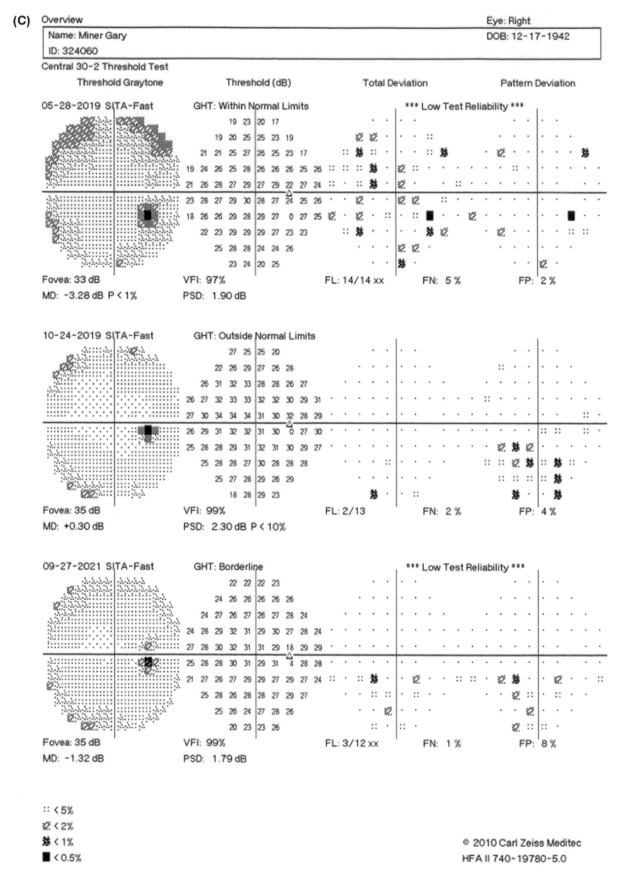

FIGURE 11.29 Continued.

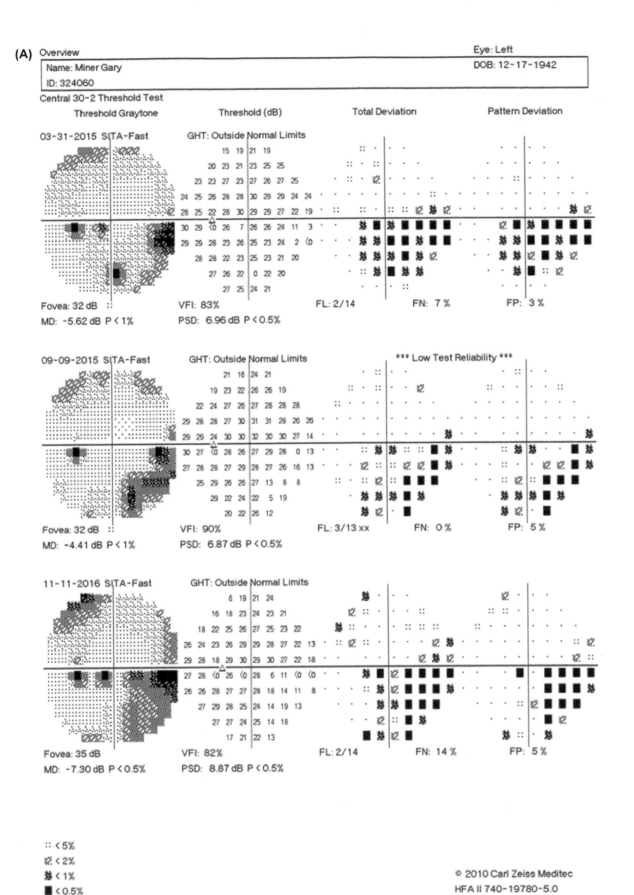

FIGURE 11.30 (A)–(C). Left eye. Visual field test, left eye, showing March 2015 through Sept, 2021 progression. The left eye is the eye affected by severe glaucoma in this case study, so this is the one to focus on in this chapter. A lot of change is seen in this left eye; this is observed by the "dark areas" in the visual field. These dark areas show the areas of the visual field where the retinal neurons have died, thus are not functioning; vision from these areas is lost forever.

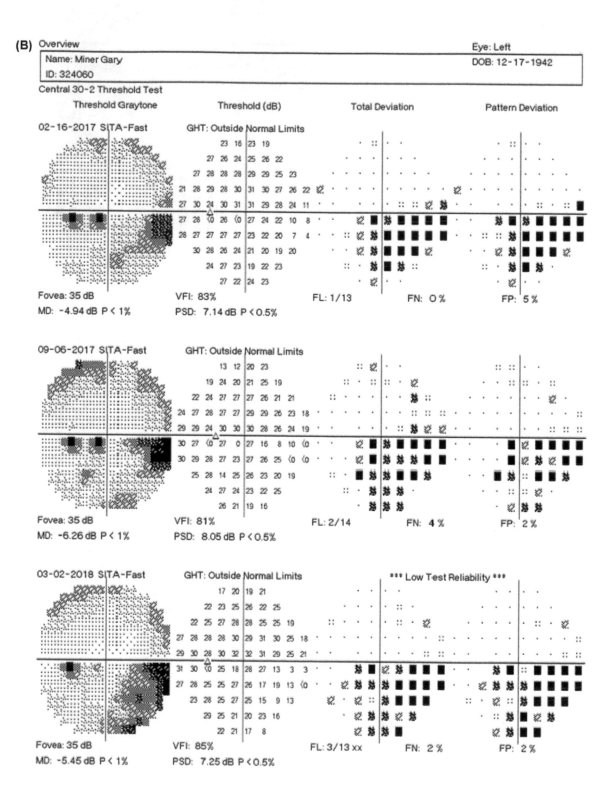

(B) Overview   Eye: Left

Name: Miner Gary   DOB: 12-17-1942
ID: 324060

Central 30-2 Threshold Test

FIGURE 11.30   Continued.

(C) Overview

Eye: Left

| Name: Miner Gary | DOB: 12-17-1942 |
| ID: 324060 | |

Central 30-2 Threshold Test

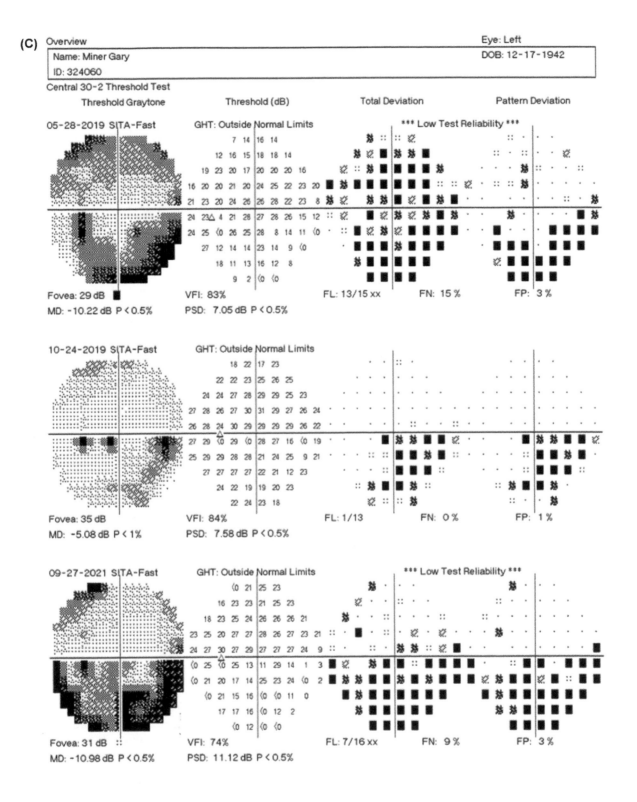

Fovea: 29 dB

MD: -10.22 dB P < 0.5%

VFI: 83%

PSD: 7.05 dB P < 0.5%

FL: 13/15 xx   FN: 15 %   FP: 3 %

Fovea: 35 dB

MD: -5.08 dB P < 1%

VFI: 84%

PSD: 7.58 dB P < 0.5%

FL: 1/13   FN: 0 %   FP: 1 %

Fovea: 31 dB

MD: -10.98 dB P < 0.5%

VFI: 74%

PSD: 11.12 dB P < 0.5%

FL: 7/16 xx   FN: 9 %   FP: 3 %

:: < 5%
⊘ < 2%
❊ < 1%
■ < 0.5%

© 2010 Carl Zeiss Meditec

HFA II 740-19780-5.0

FIGURE 11.30   Continued.

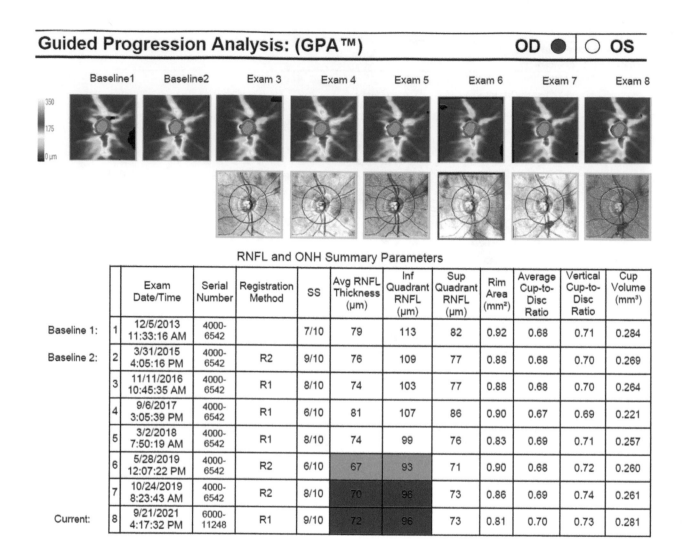

Guided Progression Analysis: (GPA™)   OD ● | ○ OS

Baseline1 Baseline2 Exam 3 Exam 4 Exam 5 Exam 6 Exam 7 Exam 8

RNFL and ONH Summary Parameters

		Exam Date/Time	Serial Number	Registration Method	SS	Avg RNFL Thickness (µm)	Inf Quadrant RNFL (µm)	Sup Quadrant RNFL (µm)	Rim Area (mm²)	Average Cup-to-Disc Ratio	Vertical Cup-to-Disc Ratio	Cup Volume (mm³)
Baseline 1:	1	12/5/2013 11:33:16 AM	4000-6542		7/10	79	113	82	0.92	0.68	0.71	0.284
Baseline 2:	2	3/31/2015 4:05:16 PM	4000-6542	R2	9/10	76	109	77	0.88	0.68	0.70	0.269
	3	11/11/2016 10:45:35 AM	4000-6542	R1	8/10	74	103	77	0.88	0.68	0.70	0.264
	4	9/6/2017 3:05:39 PM	4000-6542	R1	6/10	81	107	86	0.90	0.67	0.69	0.221
	5	3/2/2018 7:50:19 AM	4000-6542	R1	8/10	74	99	76	0.83	0.69	0.71	0.257
	6	5/28/2019 12:07:22 PM	4000-6542	R2	6/10	67	93	71	0.90	0.68	0.72	0.260
	7	10/24/2019 8:23:43 AM	4000-6542	R2	8/10	70	96	73	0.86	0.69	0.74	0.261
Current:	8	9/21/2021 4:17:32 PM	6000-11248	R1	9/10	72	96	73	0.81	0.70	0.73	0.281

Likely Decrease

Possible Decrease

FIGURE 11.31 Patient Garys right ("good") eye: Guided progression analysis from baseline (first time test done) in December, 2013, over eight total times eye scan taken over several years, through September 2021.

So a "good working hypothesis" is low-dose steroid eye drops may indeed cause increased urination frequency, in at least some patients, even if rare. A good overview of the systemic side effects of eye drops with illustrations of the routes of administration is presented in the journal Clinical Ophthalmology in 2016. One illustration from this paper is probably of value to our understanding in this chapter (Farkouh et al., 2016), where it clearly indicates that over 50% of any eye drop does NOT penetrate into the eye, but instead goes throughout the body with the potential of creating side effects on various body organs and systems (see Fig. 11.38).

Suggested absorbsion pathway of Loetmax SM; Helping to determine best treatment

Lotemax liquid was the form for this eye drop, until recently, when LOTEMAX SM took over as the "form of choice." What is the difference between them? LOTEMAX SM delivers the drug in a submicron particle size for faster drug

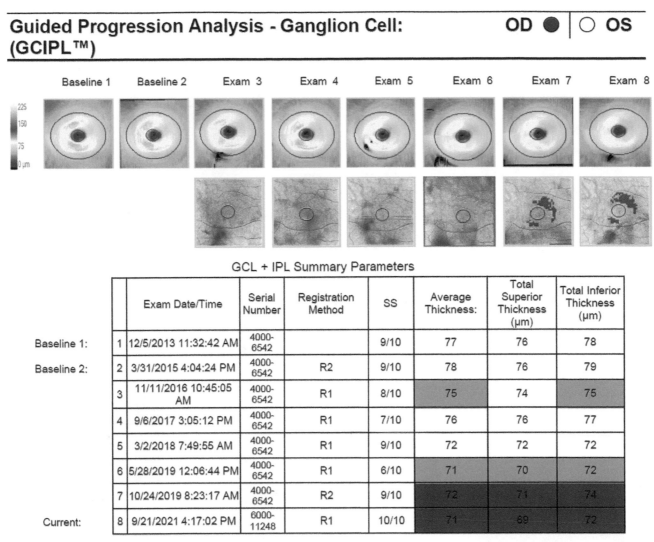

Guided Progression Analysis - Ganglion Cell: (GCIPL™)   OD ● | ○ OS

GCL + IPL Summary Parameters

		Exam Date/Time	Serial Number	Registration Method	SS	Average Thickness:	Total Superior Thickness (µm)	Total Inferior Thickness (µm)
Baseline 1:	1	12/5/2013 11:32:42 AM	4000-6542		9/10	77	76	78
Baseline 2:	2	3/31/2015 4:04:24 PM	4000-6542	R2	9/10	78	76	79
	3	11/11/2016 10:45:05 AM	4000-6542	R1	8/10	75	74	75
	4	9/6/2017 3:05:12 PM	4000-6542	R1	7/10	76	76	77
	5	3/2/2018 7:49:55 AM	4000-6542	R1	9/10	72	72	72
	6	5/28/2019 12:06:44 PM	4000-6542	R1	6/10	71	70	72
	7	10/24/2019 8:23:17 AM	4000-6542	R2	9/10	72	71	74
Current:	8	9/21/2021 4:17:02 PM	6000-11248	R1	10/10	71	69	72

FIGURE 11.32 Patient Garys right ("good") eye: Guided progression analysis, focusing on the "ganglion cell." Analysis from baseline (first time test done) in December, 2013, over eight total times eye scan taken over several years, through September 2021.

dissolution within the tears of the eye. It also is reported to provide twofold greater penetration to the aqueous humor of the eye compared to the older formulation Lotemax (Cavet et al., 2015). Comparing LOTEMAX SM to Prednisolone, Lotemax rapidly undergoes hydrolysis in the eye's anterior chamber to become an inactive derivative, thus while its strength may be a little less than Prednisolone Acetate 1%, it is less likely to increase the eye's IOP (Cakanac, 2015).

So this adhering of LOTEMAX SM drug to the surface of the eye giving better absorption, even two times greater penetration into the aqueous humor, and thus presumably less of this steroid going out systemically to other parts of the body, may be the reasons why LOTEMAX SM works the best for Gary.

Predictive analytic modeling possibilities

There are no really good "predictive analytic models" that have been discovered, yet, to *fully predict* what is the "best diagnosis" and "what is the best treatment" for an individual person (patient) that has been discovered to be suffering from glaucoma. There are some studies on selected aspects, selected eye tests, that have used SVM (Support Vector Machines, a specific "Machine Learning" method) and some other ML (Machine Learning) and "Deep Learning" methods (Thompson et al., 2020; Mursch-Edimayr et al., 2020; Burgansky-Eliash et al., 2005; Faizan Abdullah et al., 2021), but none that have produced an *overall* model that is "patient-centered."

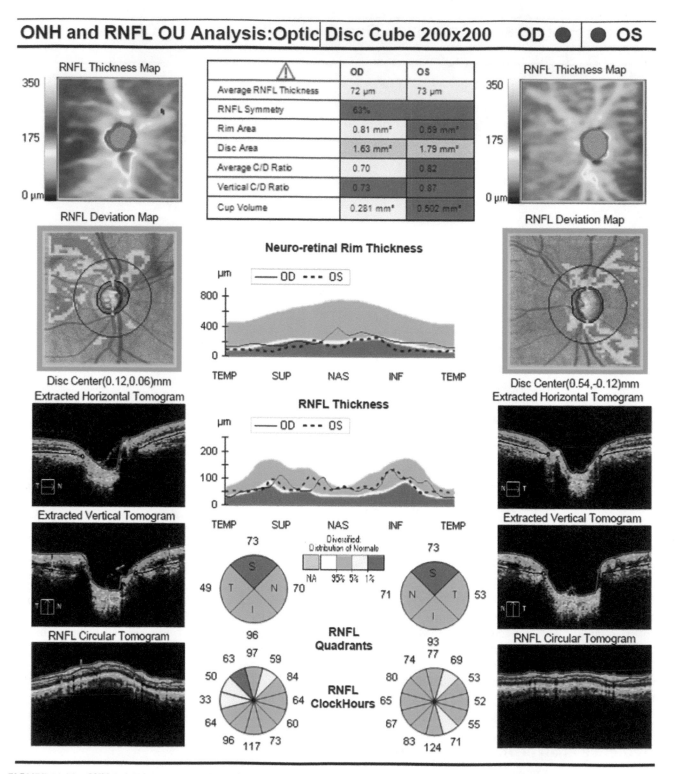

FIGURE 11.33   OHN and GNFL OU Analysis: optic disc cube. Done for both eyes, as shown here, in September 2021

It is expected that over 111.8 million people (worldwide) will have glaucoma by 2040 (Tham et al., 2014). And most people developing glaucoma do not know they have it, as the "symptoms are silent," so usually by the time it is detected there already is some retinal cell death—irreversible damage to the eyesight (Budenz et al., 2013; Hennis et al., 2007; Weinreb et al., 2014; Harwerth et al., 2004; and Harwerth et al., 1999). Thus early detection of glaucoma

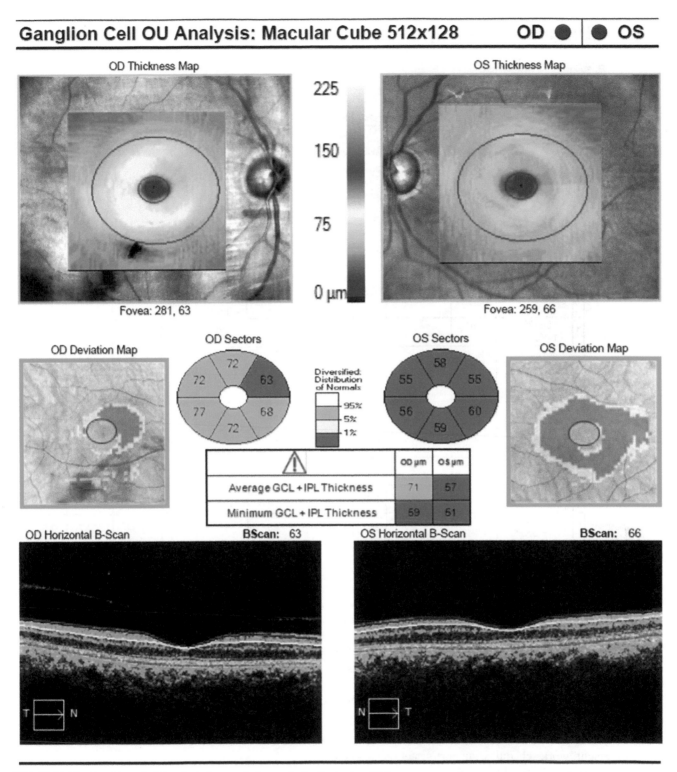

FIGURE 11.34 Ganglion cell OU analysis done on both eyes during September, 2021.

is important and may be improved by introducing novel approaches for screening, diagnosis, and detection of change over time.

If I, Gary, were to design a predictive model for glaucoma, the following predictor variables (also called "independent variables" in traditional statistics) should be included in a beginning data set used to make an overall

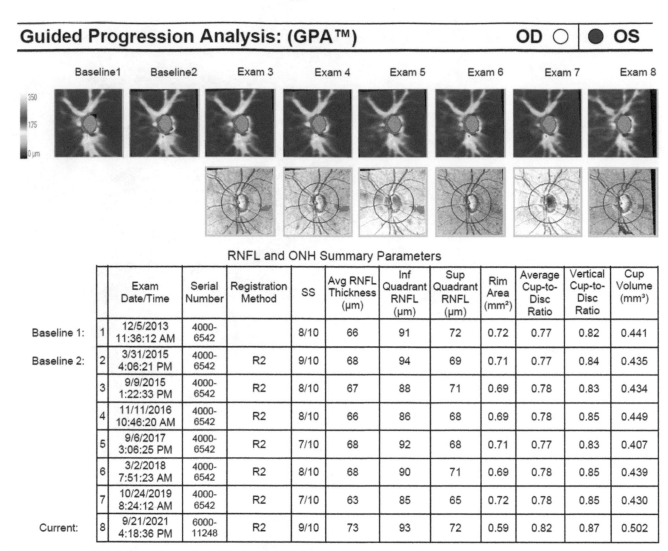

Guided Progression Analysis: (GPA™)   OD ○ | ● OS

RNFL and ONH Summary Parameters

		Exam Date/Time	Serial Number	Registration Method	SS	Avg RNFL Thickness (µm)	Inf Quadrant RNFL (µm)	Sup Quadrant RNFL (µm)	Rim Area (mm²)	Average Cup-to-Disc Ratio	Vertical Cup-to-Disc Ratio	Cup Volume (mm³)
Baseline 1:	1	12/5/2013 11:36:12 AM	4000-6542		8/10	66	91	72	0.72	0.77	0.82	0.441
Baseline 2:	2	3/31/2015 4:06:21 PM	4000-6542	R2	9/10	68	94	69	0.71	0.77	0.84	0.435
	3	9/9/2015 1:22:33 PM	4000-6542	R2	8/10	67	88	71	0.69	0.78	0.83	0.434
	4	11/11/2016 10:46:20 AM	4000-6542	R2	8/10	66	86	68	0.69	0.78	0.85	0.449
	5	9/6/2017 3:06:25 PM	4000-6542	R2	7/10	68	92	68	0.71	0.77	0.83	0.407
	6	3/2/2018 7:51:23 AM	4000-6542	R2	8/10	68	90	71	0.69	0.78	0.85	0.439
	7	10/24/2019 8:24:12 AM	4000-6542	R2	7/10	63	85	65	0.72	0.78	0.85	0.430
Current:	8	9/21/2021 4:18:36 PM	6000-11248	R2	9/10	73	93	72	0.59	0.82	0.87	0.502

FIGURE 11.35   Guided progression analysis of the left eye (OS).

patient-centered glaucoma best treatment/best diagnosis model, as all of these could be related to developing and to the ultimate severity of glaucoma. NOTE: that some of the variables listed below may seem "completely out-of-the-box," but at this stage of development of a good model, we just do not know very well what might be important, thus in predictive analytic modeling we throw into the mix as many variables as possible, even though they seem, to our "common senses," to be completely off base. When the AI and ML modeling is done, the learning process figures out what is important and what is not, and will end up with a small list of most important variables, probably between three and 10 in number.

Predictor variable possibilities for developing a good ML-AI model to predict aspects of glaucoma:

1. Age of patient
2. Life-style of patient (make a scale of "lifestyle"), which might include such things as indoor/outdoor, community involvement, work schedules, support system, and so on.
3. Sex of patient
4. Genetic race of patient (essential as different races have different genetic alleles at gene sites that may be very important, since glaucoma appears to be inherited in about 70% of cases)
5. Line of latitude where patient is living (and has lived most of life)
6. Line of longitude where patient is living (and has lived most of life)
7. Occupation
8. Activity level (on a scale)

Guided Progression Analysis - Ganglion Cell: (GCIPL™)

OD ○ | ● OS

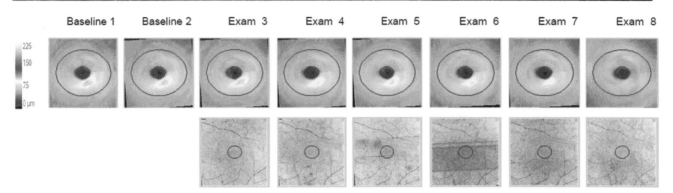

| | Baseline 1 | Baseline 2 | Exam 3 | Exam 4 | Exam 5 | Exam 6 | Exam 7 | Exam 8 |

GCL + IPL Summary Parameters

		Exam Date/Time	Serial Number	Registration Method	SS	Average Thickness:	Total Superior Thickness (µm)	Total Inferior Thickness (µm)
Baseline 1:	1	12/5/2013 11:35:29 AM	4000-6542		9/10	63	57	70
Baseline 2:	2	3/31/2015 4:06:48 PM	4000-6542	R2	10/10	63	57	69
	3	9/9/2015 1:22:55 PM	4000-6542	R2	9/10	63	57	69
	4	11/11/2016 10:47:06 AM	4000-6542	R2	10/10	62	56	67
	5	9/6/2017 3:07:04 PM	4000-6542	R2	9/10	64	59	69
	6	3/2/2018 7:50:58 AM	4000-6542	R2	9/10	62	58	67
	7	10/24/2019 8:24:46 AM	4000-6542	R2	9/10	62	58	66
Current:	8	9/21/2021 4:18:18 PM	6000-11248	R2	10/10	57	56	58

FIGURE 11.36 Guided progression analysis—ganglion cell. Left (OS) eye, taken September, 2021.

9. Where most of waking time is spent (outdoor light or indoor artificial light)
10. If waking time mostly spent outdoors, are sunglasses worn? Or eyes fully exposed to sunlight
11. Right-eye IOP (mmHg), median (with range—highest and lowest value)
12. Right-eye IOP (mmHg) as average (+/− 95% confidence limit)
13. Left-eye IOP (mmHg), median (with range—highest and lowest value)
14. Left-eye IOP (mmHg) as average (+/− 95% confidence limit)
15. Medications (# of different types of IOP lowering eye drops)
16. VA = visual acuity score
17. SE (D = diopters) = spherical equivalent
18. SAP MD (dB = decibels) = standard automated perimetry with mean deviation
19. SAP PSD (dB = decibels) = standard automated perimetry with pattern standard deviation
20. FDT MD (dB = decibels) = frequency doubling technology with mean deviation
21. FDT PSD (dB = decibels) = frequency doubling technology with pattern standard deviation
22. SAPrgc (× 1000 cells) = SAP-derived estimate of total number of retinal ganglion cells
23. OCTrgc (× 1000 cells) = OCT-derived estimate of total number of retinal ganglion cells
24. WRGC (× 1000 cells) = weighted number of retinal ganglion cells based on Optical coherence tomography (OCT) and Standard automated perimetry (SAP) measurements
25. CSFI (%) = combined structure-function index

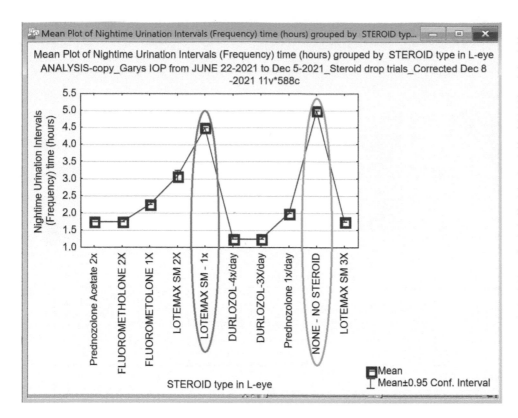

FIGURE 11.37 Steroids in the form of eye drops induced frequent urination, a uncommon side effect, apparently. Thus we had to balance this side effect with the desired goal of L-eye IOP of 15. Fortunately, the soft steroid LOTEMAX SM 1x/day gave the best L-eye IOP, which also gave a nighttime (sleeping) urination interval of about 4.5 h, almost as good as when "no steroids" were tested giving a nighttime urination interval of 5 h. The strongest eye steroid, DURLOZOL at either $3 \times$/ day or $4 \times$/day (the only times tested in this study) gave the highest urination frequency of 1.25 h (=75 min); and DURLOZOL induced highest L-eye IOP elevation to an "unacceptable level" over a long time period, but needed for the first 3—4 weeks following Ahmed valve shunt surgery for healing of the surgical incisions.

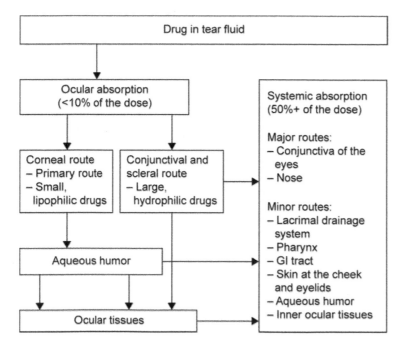

FIGURE 11.38 Possible absorption pathways of a drug administered to the eye: overview of the ocular absorption pathways of ocular drugs. Note that over 50% of the eye drops are not absorbed by the eye, but instead go throughout the body systematically, thus these eye drops can have side effects on other organs and systems of the human body.

26. Glaucoma "specialist" rating of patient's possible glaucoma Dx (on scale of 1 (lowest—no glaucoma)—10 (highest probability of glaucoma)
27. Glaucoma "ophthalmologists" rating of patient's possible glaucoma Dx (on scale of 1 (lowest—no glaucoma) to 10 (highest probability of glaucoma)
28. T2 hyperintense signal in optic nerve
29. abnormality of cerebrospinal fluid (CSF) flow dynamics

30. Chronically enlarged lateral and 3rd ventricles of the brain *(note: No one has every created a Gaussian distribution of the range of expected human 3rd ventricle size. Mayo Clinic Platform, leveraging it's Discover product, will do that in 2022* (Halamka and Cererato, 2022)
31. Possible hyperdynamic CSF flow within the lateral and 3rd ventricles of the brain
32. proximal aqueductal narrowing in the brain
33. MRE brain stiffness pattern
34. asymmetric stiffness right versus left (of brain)
35. OMICS variables:
    a. liver enzyme variant called CYP2C9*2
    b. Glial fibrillary acidic protein (GFAP)
    c. retinol-binding protein 4 (RBP4).
        (references for T2 optic nerve signal, CSF, brain, and omics variables above: Halamka and Cererato, 2022)
    d. And a multitude of other possible omics markers of aging spanning genomics, transcriptomics, proteomics, metabolomics, integromics, microbiomics, in the tables published in Wu et al. (2021) and Montaner et al. (2020).

The predictor variables in the above list having to do with omics may turn out to be more important than one might imagine initially. This is because some of those listed have to do with aging; and aging is associated with glaucoma, thus there may be a connection.

Variables being Predicted (also called "dependent variables" in "traditional statistics") that need to be considered in the interests of the patient:

1. How long (time: weeks, months, etc?) will it be before significant optic nerve cells die and vision begins to be lost?
2. What is the best IOP reducing eye drop for specific person "X"?
3. Would ABiC minimally-invasive eye surgery be helpful to patient "X"? For how long?
4. Would XEN-gel Stint or Ahmed valve shunt (invasive eye surgery) be best for patient "X"?

Some published research suggests that predictive analytics are useful in the glaucoma field. Specific areas where AI, ML, and deep learning have been used in glaucoma are:

1. screening with:
   a. *Fundus Photography* (FP),

and:

1. diagnosis and detection of glaucoma progression with
   a. OCT; and
   b. SAP (Thompson et al., 2020).

Some studies used SVMs to enhance the detection of glaucoma damage from data obtained from imaging (Burgansky-Eliash, et al., 2005; Bowd et al., 2008; Shigueoka et al., 2018). ML methods other than SVM have also been explored over the past 20 years (Bowd et al., 2008; Chan et al., 2002; Kim et al., 2017; Sample et al., 2005; Goldbaum et al., 2012; Yousefi et al., 2016; Yousefi et al., 2018; Goldbaum, 2005). *However, these methods have not been widely incorporated into clinical practice.* Yet these "traditional ML methods" may provide the best solutions to the problems of accurate diagnosis and also provide the best treatment plans for an individual patient (Thompson et al., 2020).

"Deep learning" belongs to a class of ML algorithms that use "representation learning" (illustrated in Fig. 11.39 above). These algorithms can learn features (or "representations") from data automatically. A benefit of deep learning is that the user does not need to be fully knowledgeable of the subject matter; however, these "learned features" may be difficult to understand or explain, thus just like regular neural networks models, they are like a "black box." Nevertheless, one type of deep learning network called a *convolutional neural network* has been used in computer vision applications, including face recognition and also self-driving cars. This type of learning requires training a model, then testing a model, and then validating a model; for this three separate data sets area used, one for each phase of developing an accurate model. For glaucoma, one should be able to train a model to identify patterns of the visual field such that a glaucoma diseased eye can be distinguished from a normal healthy eye. This basic approach has been used to find change in the visual fields over time (Yousefi et al., 2016; Wang et al., 2020; Sample et al., 2004; Goldbaum, 2005; Goldbaum et al., 2005; Christopher et al., 2018).

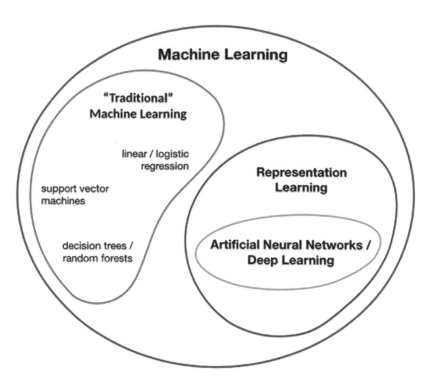

FIGURE 11.39 Fig. 11.1 from Thompson, et al. (July 2020), showing the relationship between "Traditional Machine Learning" algorithms and the newer so-called "Representational Learning" algorithm. Representational Learning is the stuff of "Artificial Neural Networks" and the newer "Deep Learning" methods; the Deep Learning are basically Neural Networks put together in newer, more complicated patterns or architectures.

Even visual field tests can now be automated with artificial intelligence—machine learning methods

In a 2018 study using ML analytic methods detected eyes in which the visual field defects were progressing eyes earlier than other methods consistently; and in fact the ML detected more slowly progressing eyes sooner and better than other methods:

> The time to detect progression in 25% of the eyes in the longitudinal data set using global mean deviation (MD) was 5.2 (95% confidence interval, 4.1−6.5) years; 4.5 (4.0−5.5) years using region-wise, 3.9 (3.5−4.6) years using point-wise, and 3.5 (3.1−4.0) years using machine learning analysis. The time until 25% of eyes showed subsequently confirmed progression after 2 additional visits were included were 6.6 (5.6−7.4) years, 5.7 (4.8−6.7) years, 5.6 (4.7−6.5) years, and 5.1 (4.5−6.0) years for global, region-wise, point-wise, and machine learning analyses, respectively. *(Yousefi, 2018).*

Archetype patterns analysis: Wang et al. (2019) used reliable visual fields (VFs) to develop 16 archetype patterns and then used those archetypes in their artificial intelligence method for their study of 12, 217 eyes of 7360 patients. Wang et al. (2019) agreed that the Archetype method can be a useful way to evaluate and understand the progression of patterns in visual field studies...

For those readers of this book that are interested in the visual patterns obtained with these archetype methods, please consult this reference for excellent diagrams of the method (Wang et al, 2019).

In a 2022 study (Montesano et al., 2022) the "Humphrey field analyzer" (HFA) data set and its value to glaucoma diseases understanding were described; this is an "Open Source" data set provided by the Department of Ophthalmology at the University of Washington, Seattle. This open access VF data set serves as a source of raw data for investigations such as VF behavior, clinical comparisons to trials, and development of new ML algorithms. This University of Washington glaucoma data set is available for use as an "open source of raw data" at the following site: https://github.com/uw-biomedical-mL/uwhvf.

> Again, we do not have the space to describe in this chapter all of the terminology and meanings of the University of Washington Humphrey visual field analyzer, but for those readers interested in understanding this more thoroughly, references are supplied above and in the References section at the end of this chapter.

My IOP measurements following Ahmed valve surgery and how taking these measurements every day by patient provided critical information needed to understand how to continue to treat the glaucoma eye condition in the months following this eye surgery, and also in the years ahead:

The next graphs are from the i-CARE HOME IOP output. Rather than discussing all of these graphs in the text, please read the captions of the graphs, as these point out items of interest to focus on; you the reader will be able to see the graph and caption in the "same view" and thus should be able to rapidly comprehend the meaning. (As mentioned earlier in the chapter, this, this YouTube URL presents a very short video showing me being trained in the use of the i-CARE HOME device: https://youtu.be/6nakad5Y3HY—Title: *Gary Trains on the iCare Machine, video filmed and edited by coauthor Linda Miner.*). Billie Corkerin is the trainer in the film.

> SEPTEMBER 3, 2021: As of September 3, 2021, my L-eye IOP appeared to be running in the "new normal range" a little more than 3 months following Ahmed valve shunt surgery. This "new normal" was in the 14–17 range when no steroid eye drop was used, but slightly higher bumping into the low 20s when steroid was used. But on September 3 using i-CARE HOME device, I got R-eye 15 and L-eye 17 in the AM prior to the eye doctor appointment (East coast) where the Dr got 16 for both eyes using the Goldman IOP, the so-called "Gold Standard" method. So it seemed like we were getting to the place we wanted to be.

But more "steroid/IOP lowering eye drops" trials were in store for the next few months.

However, this East coast doctor wanted the following for the next 6 weeks, adding stronger steroid, because of the past week's "eye inflammation," using an IOP lowering eye drop that was not compatible with my "phenotype–genotype," and it would be 6 weeks until the next doctor visit, thus this stronger steroid added "insurance":

Thus the PRESCRIPTION for next 6 weeks:

1. *Prednizolone acetate E1%, 2x day l eye*
2. *Generic COSOP, 2x/day in both eyes*

Again, the doctor primarily wanted this stronger steroid to make sure the L-eye did not get inflamed and also that the bleb did not heal over (Fig. 11.42).

At a September 22, 2021 eye doctor appointment at the Midwest doctor the Goldman IOP readings were R-20 and L-20; I had gotten, using the i-CARE HOME device, R-20 and L-21 earlier in the day, so the i-CARE HOME and Goldman IOP measuring devices appeared to be "in synch." The high IOP readings seemed to indicate that the stronger

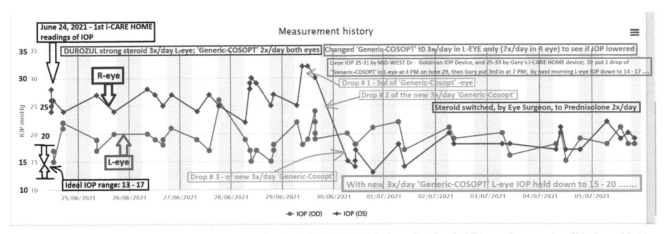

FIGURE 11.40 This graph shows the first 2 weeks of using the i-CARE HOME device, after Gary's Midwest doctor and staff had provided it to him on June 22, and taught him how to use it. On June 11 (3 weeks after Ahmed valve shunt surgery) the very strong steroid, Durozul eye drop, was dropped from 4 × /day to 3 × /day (it had been at every 2 h waking hours the first week after surgery); But then the East coast eye surgeon (via email communication) changed this to the less strong steroid Prednisolone 2 × /day on July 2 while patient was in the Midwest, after being alerted to the high IOP values obtained the last week of June. Conclusion here: Taking "Generic-COSOP" 3 × /day (instead of the usual drug manufacturer's recommendation of 2 × /day) seemed to bring down the L-eye IOP from the high 20s/low 30s to the teens and low 20s. It also appears that switching from the very strong steroid eye drop Durozul to the lower steroid strength Prednisolone brought the R-eye IOP levels down a bit (throughout the use of steroid in the L-eye, that always appeared to be a small cross-over effect to the R-eye). Not all people respond with higher IOPs to eye drop steroids, but some patients do get increased IOP levels with steroid use.

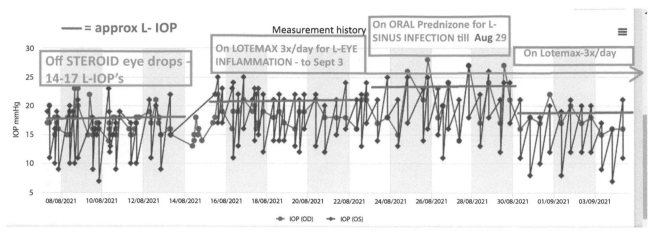

FIGURE 11.41 Approximate IOP levels from August 8, 2021 to September 3, 2021, under four different medication regimens. The period of August 8–14, when Gary was off any type of eye drop steroid gave the lowest consistent L-eye IOPs since the May 20 surgery; the approximately 14–17 mmHg IOP of L-eye was where we wanted it. (NOTE: in the above chart, all of the low L-eye IOP values, those 13 and lower, were all caused by a "Shunt value working check" where the L-eye was massaged, to push out fluid from the shunt through the bleb—when the IOP value measured immediately after this massage goes down several points, we know that the bleb/shunt port is open and working; so these value have to be disregarded in looking at the real IOP values.).

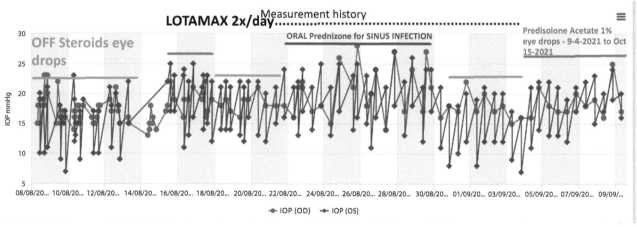

FIGURE 11.42 This figure demonstrates that when the Prednizolone was added 2x/day in early September, it again raised the L-eye IOP values. (Prednisolone acetate 1% is being taken 2 × /day for the Sept 4, 2021 to October 15, 2021 6 week period).

Prednisolone was probably causing this, and even though it was applied only to the L-eye, it was leaking to the R-eye affecting this eye's IOP also.

Thus we all agreed there was a need to try to find a steroid that is "lower in power" and will NOT raise IOPs significantly, since Gary is a "fairly high responder to increase in IOP from steroid." Both the Midwest doctor and the East coast follow-up doctor agreed that it is a "better way to keep Bleb open/functioning (and not healing/scarring over)," in the long run, and thus continued use of a steroid, in the smallest dose possible, that works for Gary, is what we needed; However, the East coast glaucoma specialist eye surgeon came from a different school of thought that believed that eye massage was the best way to go, and that Gary could even use the 30-second massage up to four times each day. The eye massage "pushed out fluid" through the bleb, thus the IOP should be lower immediately after eye massage. We, Gary and Linda, both trained in medical epidemiology and public health, felt the use of the eye massage should be useful. So another "balancing act" was needed by Gary to accomplish this. Gary decided to use L-eye massage just once a day, as a compromise, taking an i-CARE HOME IOP reading immediately after massage to see if the IOP had gone down from a reading taken before massage; this seemed like a good way to keep tabs on whether or not the bleb was open and working, and Gary's intuition told him that the once a day eye massage would not (should not) have a bad effect on retinal neurons, since the medical opthalamologist "jury" was out on this, anyway (see below).

Some eye doctors are concerned that massage can lead to damage; the "story/conclusion" is out on this, but many think that the eye massage brings the IOP pressure up high, maybe 40, 50, 60, during the massage process, thus this may be damaging/cause retinal cell loss etc., but there are arguments on both sides of this today in the glaucoma field (Could this may mean that I should be "judicious" in the use of massage, do it gently, or something else? The main reason I have been doing the eye massage once a day is to get the IOP immediately after, which is lower, thus showing that the BLEP is working, the massage pushed fluid out of the eye. To my thinking this is a way to "keep tabs on the bleb" when one is not able to actually "see the bleb" as the Dr does when he examines the eye at the doctor's office, since Eye Doctors do not like to see patients in 3-month intervals.

Several i-CARE IOP graphs are presented below, at various time intervals over a 6-month period. We will not say too much about them in the text, but readers who are interested can study the captions under each figure to understand the value and importance of patient generated data for use by the patient and their doctors in making future treatment decisions (Figs. 11.43 and 11.44).

At an October 15, 2021 East coast eye doctor appointment, after using the Predizolone 2 × /day in L-eye for 6 weeks, the technician got IOPs of R-15 and L-16; the doctor followed up with a repeat Goldman Device IOP readings

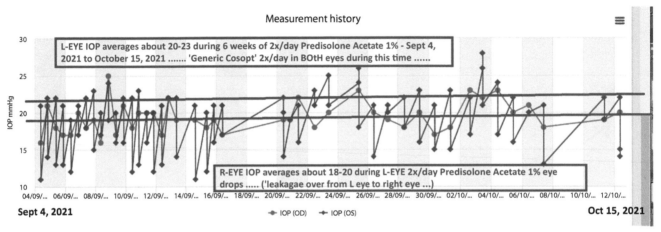

FIGURE 11.43 Gary's IOP from Sept 4, 2021 to October 15, 2021, while on a 2x/day intermediate strength eye drop steroid called Prednisolone acetate 1%; IOP lowering eye drop Generic-COSOPT was only used 2 × /day during this period (the manufacturer's recommended dosage).

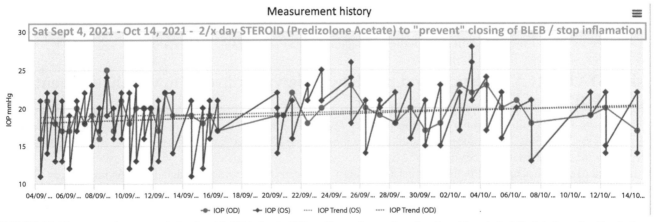

FIGURE 11.44 Sept 4, 2021 through October 14, 2021 Gary's IOP readings, when undergoing the trial period of Prednisolone 2 × /day in L-eye only, and Generic-COSOPT 2 × /day in both eyes. Trend Lines are included in above chart, but note that the L-eye trend line is lower than normal daily reality, as it includes in its calculation the very low (under 14 IOP) measurements made immediately after L-eye massage (which lowers the IOP, as this massage pushes out fluid from the Ahmed valve shunt tube, lowering the IOP for a short period of time). In reality, the L-eye IOPs were reading in the 20s all too often on this medication regimen.

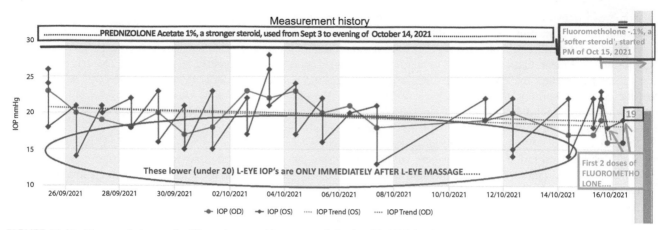

FIGURE 11.45 Fluorometholone, a "soft" eye drop steroid, was started October 15, 2021 for Gary. Immediately, after a few uses, it appeared that this soft steroid was not increasing significantly the IOP.

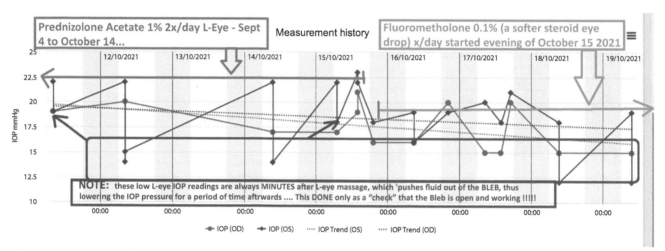

FIGURE 11.46 After a longer period of time on the soft eye steroid, Fluorometholone, it was obvious that the IOP was lower than it had been for the Prednisolone 2 × /day steroid. *So it is pretty clear that we have a balancing act to work with here, and must find the "happy mediums" between the type of steroid used and the type of IOP lowering eye drop used.*

of R-15 and L-15. There was no indication of eye inflammation so we decided to go with lower strength steroid drop trials during the next 2 months, until the middle of December, 2021 (Figs. 11.45 and 11.46).

Using STATISTICA statistical and predictive analytic software to visualize patient Gary's IOP data

The data used for the following analysis came from the few times that my eye doctor in-office visits used the Goldman IOP device for obtaining IOP values, and additionally most of the IOP data came from me taking my own IOP values daily using the i-CARE HOME device (Fig. 11.47).

In the above graph, Durlozol is the strongest steroid; it was used in the days immediately following surgery, initially every 2 hours for a few days and then 4 × per day for a couple of weeks. Prednisolone is considered a strong steroid, but lower in strength than Durlozol. LOTEMAX SM is a soft steroid; we switched to it after discovering the prednisolone was upping the IOP more than we liked. However Lotemax SM is a "messy gel" and Gary did not like the inefficiency of its application, which wasted a lot of it, and it was costing the patient considerably more (4−5 times as much), so after initial use for about 3 months following surgery, we tried to find another soft eye drop steroid that was of "water consistency" (like most IOP drops are) instead of "heavy syrup consistency" gel. Fluorometholone was the one we discovered to fit both the category of "soft steroid" and "low generic price" (other soft steroid generics were

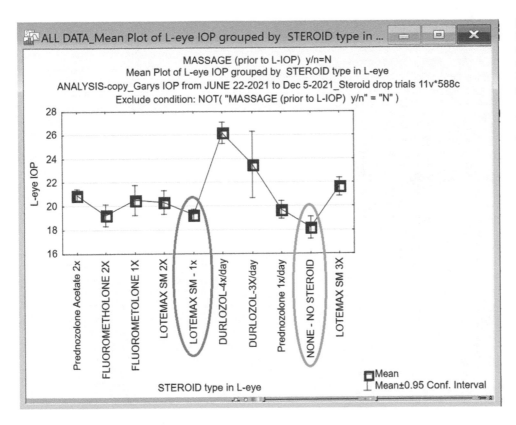

FIGURE 11.47 L-IOP under various "steroid eye drop" treatments during a 6-month period of time (June−December, 2021). Only the LOTEMAX SM 1 time/day treatment gave a L-IOP average of 18.75 that was close to our acceptable goal of 15; When no steroids were used the L-IOP was a bit closer to the goal, at about an average of 18.2, and had greater variability (as you can see with the ERROR BARS in the graph below) in our data than LOTEMAX SM 2x/day.

running in the $500−$800 per month, e.g., per 1 r-10 mL small vial; these older generics should have been at low price, as their "sister" Fluorometholone was at about $11/month, but you, the reader, "know the story" here) (Fig. 11.48).

Now that the eye steroid drop that appeared to be the "best choice" for me was selected, but the IOP still needed to be brought down a few points. I decided in mid-December to go with the three doses per day of "Generic-COSOPT," as his Midwest eye doctor had suggested 6 months previously, and with a short 1 week trial last June discovered that indeed the 3x/day of "Generic-COSOPT" did bring down the IOP a few points. Use of "Generic-COSOPT" was started on Christmas Day, Dec 25, 2021 (Fig. 11.49).

Another way of looking at this, using 3 weeks of the 3 times per day of "Generic-COSOPT" eye drops in both eyes, is to take about 4 weeks back from when the 3x/day was started, thus having 4 weeks of 2x/day data, as seen in Fig. 11.50 and Fig. 11.51).

DOSE OF "Generic-COSOPT" (=Dorzolamide−Timolol)—is three times a day OK?

COSOPT (or its generic form: Dorzolamide−Timolol) has been used for many years at a suggested dosage of two times per day (eye drops). However, recently some medical research has suggested that three times a day is doable.

Dosing

The dose of this medicine will be different for different patients. Follow your doctor's orders or the directions on the label. The following information includes only the average doses of this medicine. If your dose is different, do not change it unless your doctor tells you to do so.

The amount of medicine that you take depends on the strength of the medicine. Also, the number of doses you take each day, the time allowed between doses, and the length of time you take the medicine depend on the medical problem for which you are using the medicine. *(Mayo Clinic, 2022b)*

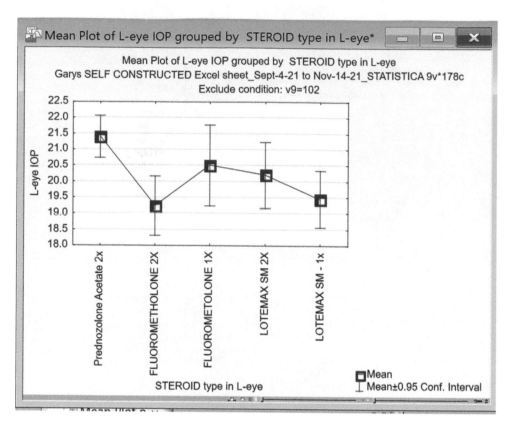

FIGURE 11.48 Effect of different eye drop steroids on IOP level of the L-eye. Although Fluorometholone at 2 doses a day gave the lowest IOP values, it induced frequent urination, thus the LOTEMAX-SM 1 dose a day (almost identical in IOP level) was selected in early December, 2021 to be the "steroid of choice" going forward. The IOP was still higher than desired, so the next goal was to adjust the IOP lowering eye drops.

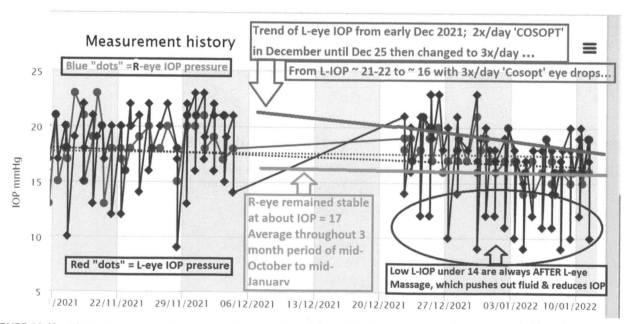

FIGURE 11.49 After 1–2 week trials of various combinations of "steroid and IOP-reducing eye drops" over a period of 5 months, it appeared that the best L-IOP was averaging in the 18–21 range (goal closer to 15; and under 20 for sure). So went back to the 3x/day "Generic-COSOPT" in both eyes in late December. The above graph shows 20 days (almost 3 weeks) of being on the 3x/day regimen; it clearly shows that finally the L-eye IOP was brought down to about 15–17 average. Additionally, the R-eye IOP stayed closer to the 15 goal, ranging from 15–17 most days.

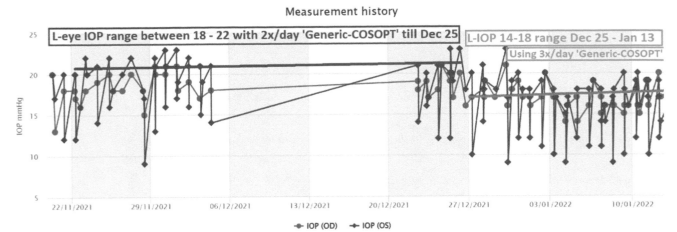

FIGURE 11.50 Almost 2 months of data on using either 2 times per day "Generic-COSOPT" or upping it to 3 times per day during the last 3 weeks of this trial. One can see that the 3x/day clearly brought the L-eye IOP down about 3–4 points at most readings (BTW; all of the low readings, below 14, above for the L-eye (red dots) are only immediately after massage was performed on the l-eye; this massage pushes out fluid through the Ahmed valve drainage tube, thus temporarily lowering the IOP; this is also a good test to take regularly to ensure that the valve drainage tube is working, thus that the bleb is open.)

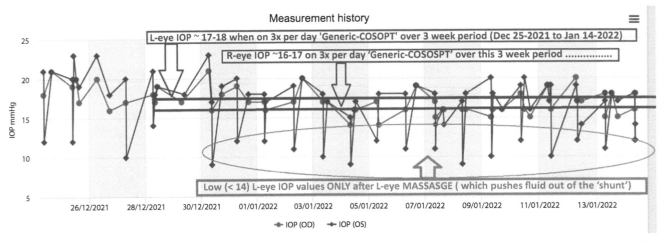

FIGURE 11.51 Three full weeks on 3x per day of "Generic-COSOPT" for both eyes, finally kept the IOPs below the 20 range (graph by Gary Miner on January 14, 2022, from i-CARE HOME IOP data readings and graphs). (BTW; all of the low readings, below 14, above for the L-eye (red dots) are only immediately after massage was performed on the L-eye; this massage pushes out fluid through the Ahmed valve drainage tube, thus temporarily lowering the IOP; this is also a good test to take regularly to insure that the valve drainage tube is working, thus that the bleb is open.)

After very long search, finally one article from Pub-Med was found from 2001 comparing Dorzolamide given $3 \times$/day and Timolol $2 \times$/day: Heijl et al. (1997) conducted a 6-month randomized control study of 184 patients with glaucoma. They used three drops a day of Dorzolamide, and two drops a day of Timolol successfully, as it lowered IOP without systemic adverse effects.

Gavidia (2021) reported on Pakravan et al. (2021) explaining that in the first month of trial using two drops of timolol-dorzolamine, there was a significant drop in IOP, but after adding another drop in the second month there was another significant drop. In the original study, the patients were monitored closely for any cardiovascular changes and none were found. The final paragraph of the Gavida article stated the possible concerns of Pakravan et al. (2021):

> Although we did not observe any change in BP by increasing the dose of timolol, aggravated nocturnal hypotension remains a concern for administering timolol-dorzolamide 3 times a day, said researchers. Moreover, we only followed our patients for 2 months, and it is possible that a longer follow-up and more exposure to timolol could result in inadvertent cardiac and respiratory adverse events.

Another source gave the following information on dosage, just for Dorzolamide (only one of the two compounds in COSOPT):

1. *For ophthalmic dosage form (eye drops):*

2. *For glaucoma or hypertension of the eye:*

3. *Adults and teenagers—Use one drop in the affected eye three times a day.*

4. *Children—Dose must be determined by your doctor. (Mayo Clinic, 2022d).*

The same source reported the following on dosage of the Timolol compound only, as an eye drop?

1. *For ophthalmic solution dosage form (eye drops):*

2. *Adults and children 2 years of age and older—Use 1 drop in the affected eye two times a day. Your doctor may need to adjust your dose as needed.*

3. *Children younger than 2 years of age—Use and dose must be determined by your doctor. (Mayo Clinic 2022)*

Another reference suggested the following dosage for using both Dorzolamide and Timolol together as an IOP reducing eye dropo:

This medication is for use in the eye(s), usually one drop in the affected eye(s) 2 times a day, or as directed by your doctor.... *(WebMD, 2022b).*

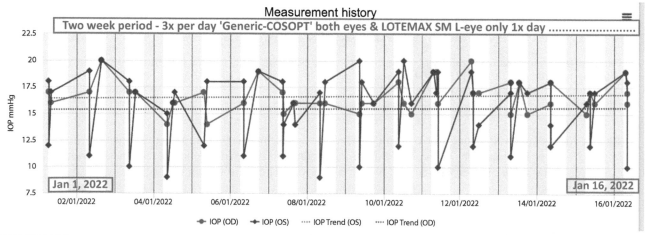

FIGURE 11.52 i-CARE HOME IOP readings for both R and L eyes taken over a 2-week period from Jan 1, 2022 through Jan 16, 2022. "Generic-COSOPT" was used 3 times per day (8 am; 2 pm, and 8 pm) during this time. Generally, over the past 10 years or so, the recommendation for dosage of this eye drop is 2 × per day; however the Dorzolamide part of this eye drop had been tested years ago (over 20 years ago) and it was shown that 3 × day dosage was OK; recently, in 2021, some research papers have been published showing that 3 × a day of the "Generic-COSOPT is OK and the Timodol does not appear to have any bad effect on heart functioning. Blood pressure and pulse rate were taken daily over this 2-week period, and Gary's values of BP and Pulse remained the same as the previous 2.5 months when "Generic-COSOPT" was only taken 2 × day (8 am and 8 pm daily). Above we see a stale trend line for both R and L eye IOPs over this 2-week period, which is what we wanted, and had been attempting to get, trying various eye drop and various hard and soft steroids over a 6-month period.

TABLE 11.1 Average of both R and L eye IOP during a 2-week period where 3× per day of "Generic-COSOPT" was used. Right eye has an average of IOP = 16.6, and Left eye an average IOP = 17.3. These IOP values are in the range we desire, thus this regimen of LOTEMAX SM 1x day in L-eye, and "Generic-COSOPT" in both eyes 3× per day seems to be the "sweet spot in eye meds" for Gary.

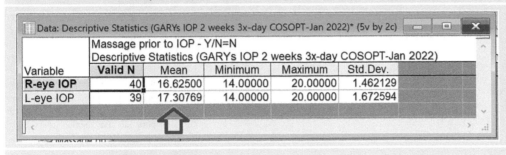

| | Massage prior to IOP - Y/N=N | | | | |
| | Descriptive Statistics (GARYs IOP 2 weeks 3x-day COSOPT-Jan 2022) | | | | |
Variable	Valid N	Mean	Minimum	Maximum	Std.Dev.
R-eye IOP	40	16.62500	14.00000	20.00000	1.462129
L-eye IOP	39	17.30769	14.00000	20.00000	1.672594

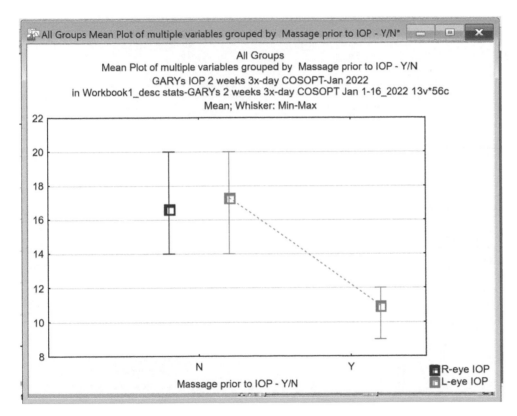

FIGURE 11.53 The range of IOP values in both R and L eyes during the 2-week 3× day "Generic-COSOPT" trial medication period. Note that the range in IOP values was between 14 and 20 for both eyes; additionally, IOP taken immediately after L-eye massage showed that that range was 9−12 (indicating that the bleb drainage port was open and working).

The above research references also reported on prescribing "Generic-COSOPT" (=Dorzolamide−Timolol) three times per day and found that it was well tolerated in glaucoma patients, thus Gary's Midwest eye doctor's advice seems reasonable, based on these studies.

The latest IOP data/graphs (and last to be included in this chapter) from Gary during JANUARY 1−16, 2022, obtained as this book's authors were finishing the writing of this book: (Fig. 11.52, Table 11.1 and Figs. 11.53, 11.54 and 11.55)

Patient measure of IOP as the next "Gold Standard of Healthcare Delivery" for Glaucoma. Essential to "CATCH" the critical times when either medical or surgical treatments are needed:

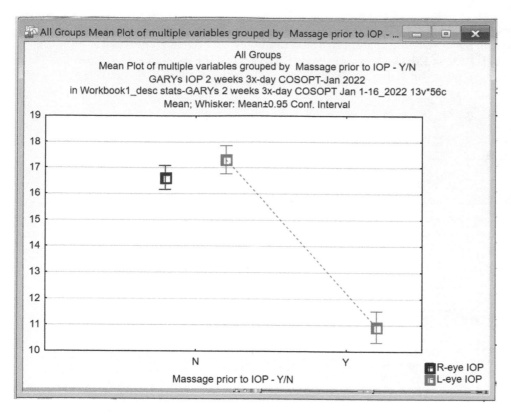

FIGURE 11.54 95% Confidence intervals for both the R and L eye IOP values showing that the "deviation" is tight with a range of about 1 IOP point around their means. Additionally, in the lower right is the 95% CI for the L-eye IOP taken immediately after L-eye massage, which causes fluid to be pushed out of the Ahmed valve shunt tube to the outside of the eye, thus lowering the IOP for a short period of time, but importantly being a test to ensure that the outlet port has not healed over but is open and working.

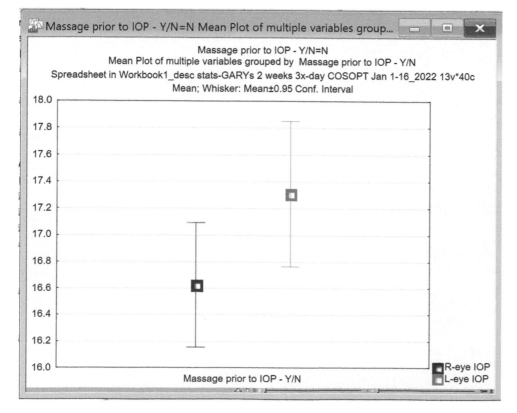

FIGURE 11.55 Close up of the means and 95% confidence intervals for both the R and L eye IOP values, during the 2-week trial period where "Generic-COSOPT" was used 3 times per day (8 am; 2 pm; and 8 pm) along with the soft steroid LOTEMAX SM 1 × day in evening (this soft steroid is used to ensure that the eye does not become inflamed, and more importantly that the bleb drainage shunt created during the Ahmed valve shunt surgery does not heal over—this drain must remain open to the outside of the eye to keep the IOP values in our "sweet spot" of 14−18 mmHg).

Future possible treatments for glaucoma

According to the Glaucoma Research Foundation (GRF) (2021), there are over 80 million people in the world with glaucoma and over 30 million glaucoma patients in the United States. Glaucoma is incredibly difficult to treat, as it tends to progress regardless of the treatment. What are new ideas and treatments that might provide some hope?

The GRF article mentioned the possibility of a twice yearly injection that Georgia Tech is working on, described further down. GRF hoped this injection, when developed, could replace eye drops while controlling IOP for patients (GRF, 2021). Another promising study reported by the GFR in the same 2021 article involved genes and the discoveries of 44 new gene loci to add to the already located 83 loci. The hope is that gene therapies might be developed after causal genes are discovered.

A possible "gene therapy" of the future is being studied at Harvard Medical School where researchers have restored vision in mice by changing retinal eye cells to recapture their youthful gene function. If this could be applied to humans, it would indeed be the first advance at reversing glaucoma. This is probably a long shot, but worth following (Glaucoma Research Foundation, 2021).

Additionally, a new "direct selective laser trabeculoplasty (DSLT)" has been developed by the Israeli start-up Belkin Laser. It has undergone a nonrandomized clinical trail to assess its safety and ability to reduce IOP. It appears that it is both safe and does reduce IOP in open angle glaucoma, thus providing an option to the current SLT, when it gets approved for clinical use (GRF, 2021).

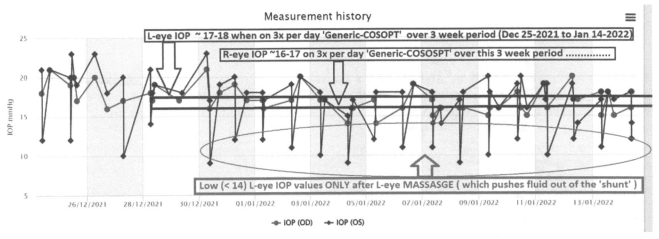

FIGURE 11.56 Raw data from the i-CARE HOME on Gary's left and right eyes over a 3-week period when only LOTEMAX SM was used 1 time per day in the L-eye, and the IOP lowering eye drop "Generic-COSOPT" was used 3 times per day in both eyes. These values were "eye-balled" as approximately L-eye average of 17−18 IOP and R-eye average of 16−17 IOP.

TABLE 11.2 Average value of both left and right eye IOP for Gary using "optimal regimen" of LOTEMAX SM soft steroid in L-eye (1 × day, evening) and "Generic-COSOPT" (Dorzolamide−Timolol) 3 × day over a 3-week trial. R-eye average is 16.9 + /−1.6, with a minimum value of 14 and a maximum value of 21; L-eye average is 17.8 + /- 1.9, with a minimum value of 14 and a maximum value of 23 observed during this period. High values of 20, 21, and 23 were only observed on the first couple of days of this 3-week trial period; as the regimen of these two eye drops "kicked in" the IOP values were always under 20. So a better representation of this is to drop off week 1 from the analysis, and just use the Jan 1−Jan 16, 2022 2-week period, as shown in Fig. 11.59.

Variable	Valid N	Mean	Minimum	Maximum	Std.Dev.		
	Descriptive Statistics (Spreadsheet in Garys 3X-DAY GENERIC-COSOPT Dec 25-2-21 to Jan 15-2022)						
R-eye IOP	49	16.91837	14.00000	21.00000	1.605158		
L-eye IOP	49	17.79592	14.00000	23.00000	1.999787		

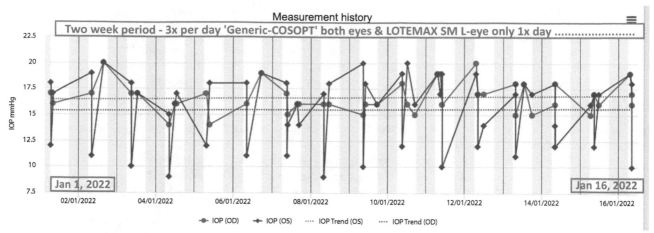

FIGURE 11.57   i-CARE HOME IOP readings for both R and L eyes taken over a 2 week period from Jan 1, 2022 through Jan 16, 2022. "Generic-COSOPT" was used 3 times per day (8 am; 2 pm, and 8 pm) during this time. Generally, over the past 10 years or so, the recommendation for dosage of this eye drop is 2× per day; however the Dorzolamide part of this eye drop had been tested years ago (over 20 years ago) and it was shown that 3× day dosage was OK; recently, in 2021, some research papers have been published showing that 3× a day of the "Generic-COSOPT" (Dorzolamide−Timolol) is OK and the Timodol does not appear to have any bad effect on heart function. Blood pressure and pulse rate were taken daily over this 2-week period, and Gary's values of BP and pulse remained the same as the previous 2.5 months when "Generic-COSOPT" was only taken 2× day (8 am and 8 pm daily). Above we see a stale trend line for both R and L eye IOPs over this 2-week period, which is what we wanted, and had been attempting to get, trying various eye drops and various hard and soft steroids over a 6-month period.

TABLE 11.3 Average of both R and L eye IOP during a 2 week period where 3× per day of "Generic-COSOPT" was used. Right eye has an average of IOP = 16.6, and left eye an average IOP = 17.3. These IOP values are in the range we desire, thus this regimen of LOTEMAX SM 1× day in L-eye, and "Generic-COSOPT" in both eyes 3× per day seems to be the "sweet spot in eye meds" for Gary.

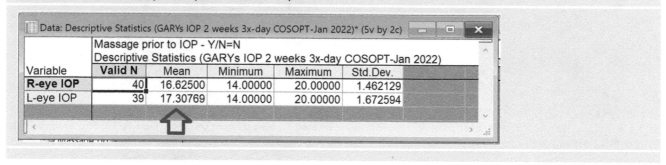

Additionally, the FDA in the Unite States has approved Durysta Biomatoprost implant, a biodegradable sustained-release implant to reduce IOP in open angle glaucoma, offering another treatment possibility for patients that have a difficult time complying with daily eye drops (GRF, 2021).

FINAL IOP levels for Gary upon finding "optimum mix of steroid and IOP eye drops"

The Raw data from the i-CARE HOME on Gary's left and right eyes over a 3-week period has been described in Fig. 11.56.

Below in Table 11.2 we see the actual descriptive statistics of this three-week regimen, giving L-eye average of 17.8 IOP, and R-eye average of 16.9 IOP, showing that this statistician's (author Gary) "eye ball" guesstimate was right in line with the actual calculated statistical averages (Fig. 11.57, Table 11.3 and Figs. 11.58−11.60).

Again as we presented in the introduction paragraph to this chapter, the personal example presented above shows how still, in 2022, medicine is not being practiced as a "patient-centered" endeavor, and thus shows clearly why this

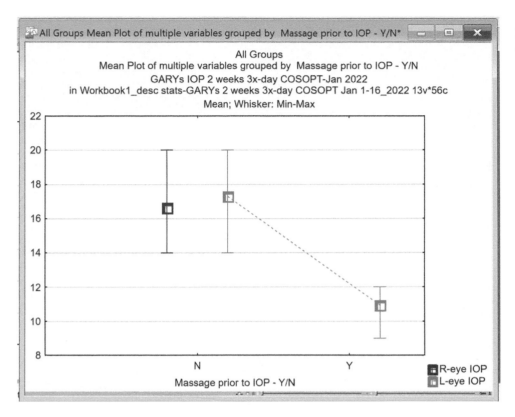

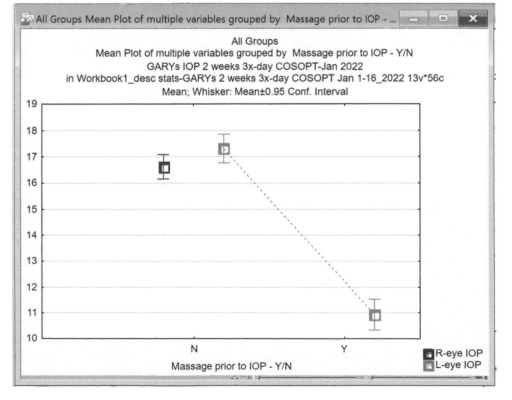

FIGURE 11.58 The RANGE of IOP values in both R and L eyes during the 2 week 3x day "Generic-COSOPT" trial medication period. Note that the range in IOP values was between 14 and 20 for both eyes; additionally IOP taken immediately after L-eye massage showed that that the range was 9–12 (indicating that the bleb drainage port was open and working).

FIGURE 11.59 95% Confidence Intervals for both the R and L eye IOP values showing that the "deviation" is tight with a range of about 1 IOP point around their means. Additionally, in the lower right is the 95% CI for the L-eye IOP taken immediately after L-eye massage, which causes fluid to be pushed out of the Ahmed valve shunt tube to the outside of the eye, thus lowering the IOP for a short period of time, but importantly being a test to ensure that the outlet port has not healed over but is open and working.

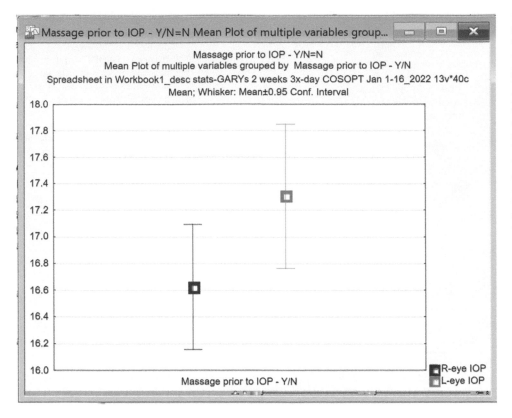

FIGURE 11.60 Close up of the means and 95% confidence intervals for both the R and L eye IOP values, during the 2-week trial period where "Generic-COSOPT" was used 3 times per day (8 am; 2 pm; and 8 pm) along with the soft steroid LOTEMAX SM 1x day in evening (this soft steroid is used to ensure that the eye does not become inflamed, and more importantly that the bleb drainage shunt created during the Ahmed valve shunt surgery does not heal over—this drain must remain open to the outside of the eye to keep the IOP values in our "sweet spot" of 14–18 mmHg).

has to change in 2022 and the years ahead. This is probably why so many of the Internet Blog writers on healthcare topics chose to make predictions in December, 2021 that 2022 would be the "Year of Personalized Medicine Revolution"—the real beginning of what has to happen.

A need is best presented by a personal story, as readers can easily "grab its meaning and value," gaining both an intellectual and emotional understanding of the need. Thus this personal story in this chapter is really the "crux of this book"—the main message that the world and the medical establishment has to embrace, understand, and deploy in the years ahead.

Ending conclusions, based on Gary's own eye care healthcare delivery experiences of just the past 2 years (fall 2019—December 2021):

Two Groupings of "bad practices" and "incorrect diagnosis & treatment decisions":

GROUP I Concerns:

We, between our two geographic locations of residence, worked with six eye doctors over this latest 2-year period, eventually leading to Ahmed valve shunt surgery on his L-eye. The following narration and graphs show the two groupings, again, from Gary's perspective.

1. Midwest eye doctor, who had overseen all of my glaucoma eye disease progression for the previous 10 years, and continued to follow me, keeping me abreast of what "were the next steps"; we had considered XEN-gel stint or Ahmed valve shunt as two of three "next possibilities" in the fall of 2019, when it appeared that the L-eye IOP might be increasing and needed frequent measuring. We even began at this point to find a "patient IOP measurement device" that was workable (unfortunately, several months later as we researched this, it was discovered that our first choice appeared to have "been taken off the market" for unknown reason; and all other options were "tremendously expensive," not covered by healthcare insurance).

2. East coast eye doctor, who had been following me since about 2013, in our small-town vacation residence; this eye clinic was regional, serving a good proportion of the state, and had ophthalmologists and optometrists of various specialties, but none specialized in glaucoma surgery. By the end of year 2019, it was obvious that I needed L-eye IOP measurements at least two times per month, if not more frequently, to fully assess if the IOP was increasing or "bouncing up and down," thus indicating invasive surgery (shunt type surgery), since additional IOP-reducing eye drops were not working; however this clinic refused to do this, saying the regular "every 3 months" IOP check was

sufficient. When I balked at this, and suggested that the clinic's technical support people take the measurements, he was refused that (even though I pointed out that there was a CMS payment code for such, although it did not pay the clinic as much for this service). This could have been handled as a "walk-in service," as it only takes a few seconds to take the Goldman IOP readings (or optionally, a simple handheld "Tonopen" could have been used, and simply, easily, and rapidly handled as a "treatment window" in the doctor's waiting room, just like routine allergy shots are handled at the Allergy & Asthma Doctor's offices). But no, complete refusal (*whose interests were being respected here? The provider? or the patient? or the payor?*).

During the year 2020 I got my IOP readings done at the clinic, but only about every 3 months, as "per their protocol." Near the end of this year, in late September, the IOP finally showed "too high" in the L-eye, so this ophthalmologist wanted to try a couple more types of eye drops, before recommending consultation with a local glaucoma surgeon. These were tried over the next 3 months, but the IOP did not improve; plus, it was obvious to me that the L-eye was not seeing as well, and then the visual field test confirmed that, so the ophthalmologist referred me to a Glaucoma specialist/surgeon about 30 miles away. (But in retrospect, and as we've indicated early in this paper, where an understanding of the VF tests is presented, visual field tests should have been conducted by this Regional Eye Clinic at least two, if not three times during the months of 2019. Note: this was the first COVID year, so what part this played in a clinic not following "best practices" is unknown.

3. Glaucoma eye doctor closer to the State's largest city:

This young Glaucoma specialist seemed extremely concerned, thorough, and appeared competent, and "patient concerned"— she appeared to be of the "younger generation of patient-centered doctors" and checking out this doctor's training and credentials we were very satisfied. This doctor wanted to prescribe two more, newer, presumably higher potency, IOP lowering eye drops to use for 2 weeks and then IOP check; at this 2-week IOP check nothing had changed, the L-eye was still in the mid-to-upper 20s; so we tried a "last ditch" effort using argon laser trabeculoplasty (they only had an argon laser, which can only be used 2−3 times; whereas an SLT laser can be used many times, but this Eye group not having an SLT made me question whether or not they were "current in the eye field"). It takes about 2−3 weeks to see if the argon laser treatment is effective; 3 weeks later the answer was" no," no lowering of the L-eye IOP. So now it was urgent that we select a surgery to obtain fluid drainage out of the eye. The only surgery this young doctor did was the "old-fashioned true surgical trabeculoplasty/trabeculotomy," which I did not want (too many risks, including ending up with too low IOP, which is also dangerous). So she had the clinic's partner and founder glaucoma surgeon, who did all of the Glaucoma types of surgeries, see me.

4. *Glaucoma eye doctor's glaucoma partner*: This doctor was the most arrogant doctor patient I had ever encountered in his entire life—and obviously "old school." He insisted that patient I had already lost 90% of his eyesight (not true!) in L-eye, gave us (Gary and Linda) a long, patronizing speech, and told us emphatically that his partner, who only did the true surgical trabeculoplasty−trabeculotomy, would do the surgery. *AND if we did not take his advice we were doing the "Wrong Thing," mind you!* At that point I knew that this was NOT the level of medical care and treatment I needed—taking time to get an evaluation at this clinic had been a "complete mistake."

So, with this unneeded delay of about 6 weeks, during which all the time more L-eye optic neurons were dying, the search for a modern competent glaucoma surgeon continued!

5. *Local friends' recommendation glaucoma specialist:* Luckily, friends who had suffered from glaucoma, and had lived in the region for many years, gave us an excellent recommendation. A glaucoma specialist in an eye group having offices in suburbs all around the state's largest city and 40 miles away. I was able to get an appointment within 10 days. As it turned out, this ophthalmologist was excellent, had 20 years of experience doing trabeculoplasty−trabeculotomy, with patients of 20 years ago still having good IOP control with their blebs open and working. But this doctor had stopped doing glaucoma surgeries, as the trabeculoplasty was the only one he did, and he felt the new generation, well-trained in all the modern glaucoma surgeries, was the way to go. So, he was able to get me in to see this eye group's most acclaimed glaucoma surgeon the very next week.

6. *This sixth doctor in my glaucoma treatment search, was a young glaucoma surgeon, who did "all of the modern glaucoma surgeries, and did them well,"* but was located a bit further away, 50 miles, close to "City center." Fifty miles may seen like a "short distance" but because of the built-up nature of the area and no room to build new roads, it took between 2−2.5 hours in heavy traffic for this 50-mile drive (each way—thus 5 hours of driving plus appointment time = full day affair), not what an older retired couple wants! But we did it. This glaucoma surgeon was great. The surgery schedule was booked up for about a month or 6 weeks, but a spot opened the next week and

the surgery scheduler got me into that spot. So only 3 weeks had elapsed from leaving the "old-fashioned outdated" eye clinic and getting surgery at this place.

However, this approximately 18-month journey, to finally get diagnosed with IOP levels that need surgery, and to get the surgery, had resulted in another 1/3 loss of my L = eye vision.

In conclusion, I think we can easily see that this search to find the "right treatment" should not be necessary where there have been good studies leading to development of the best treatment standards for eye healthcare clinics. The lack of this, as I found out, shows clearly that *this field is not "scientifically stabilized."*

A *"Gold Standard"* that is accurate and reliable and patient-centered *is yet to be achieved* in "eye care health clinics"—*we have not yet arrived!*

But there are additional concerns:
GROUP II Concerns:

1. Relying on eye doctor IOP measurements (recommended as *infrequently* as yearly, maybe bi-yearly for glaucoma patients), the IOP "got out of control" before the eye doctor's exams "discovered it," *resulting in an additional 1/3 loss of vision of the L-eye.*

2. After final Ahmed valve shunt surgery, 2 months after the surgeon would not listen to me, when she wanted to add Bimodinopine eye drops—She was told this causes intense eye inflammation because I am allergic to them (we discovered this at this time of initial visit with this eye surgeon, as I had been on Brimonidine for 4 weeks prior at that visit, with a bad L-sinus infection being treated with Prednisone and antiobiotics)—but she did not answer my portal question asking to not use Brimonidine. So not wanting to be a "difficult patient," I followed the doctor's orders, *which when used caused a second very bad L-sinus infection with its associated problems . . . resulting in 5 more months of "trial and error"* with different "steroid/IOP drops" to find the combination that worked best for me *(should not have taken that long, but after 2 months following surgery, the Eye Doctors like to go on patients being seen every 6 weeks, and then every 2 months, and then every 3 months). Listening to the patient, and following the evidence in the patient's record (where Brimonidine was listed as an allergic eye drop for me), could have cut off about 4 of these 5 months, so that we'd have had the "optimum combination" about 3 months out from surgery.*

3. The mid-December 2021 eye doctor follow-up visit, where the Goldman IOP device got a "22" reading for R-eye (not our eye of concern), and the *doctor saying that I "had to be on another eye drop, wanting Vysulta (which again in the records of this eye clinic, showed me to be "allergic" causing eye inflammation in me) or have surgery for this R-eye, was completely "poor judgment"* to say the least, but really rearing into "bad Dx and treatment Decision," as this R-eye had an average IOP value of about 16–17 (ideal—where we wanted it) over the past 6-month period, with minimum values of 13 and this mid-Dec "22" being a max (and not seen in the days previously, nor in the 3 weeks after) and truly a "fluke" reading-day.

4. *In retrospect, I should have been put on "self IOP measurement" back in year 2020,* as this would have caught the development of the "bouncing up and down" of the L-eye IOP, *thus a decision to do Shunt valve surgery would have happened about a year earlier, preventing most of the additional 1/3 loss of vision.* (Patient self-measurement devices are available, from those eye practices on the "cutting edge," as a rental device from the doctor, for example; or the patient could have come in weekly as a "walk in" for the Eye Tech to do the Goldman IOP readings, which only take a few seconds, and for which there is an CMS Dx-Payment code. . ... But all of this was refused by the local East coast regional eye clinic . . .. Why?).

Solutions needed:
1. When patients present high IOP initially (e.g., greater than 21, especially if high 20s or higher) the patient should have regular (weekly or more often) IOP checks for 3–4 weeks, to determine what is really going on, from which an *average IOP value with maximum and minimum and 05% confidence intervals* can be calculated, thus indicating if further treatment, probably surgery, is needed

2. The clinical staff and the ophthalmologists need to (1) *heed closely patients reports* of how they respond to eye drops, and (2) also *read/know what is in the patients clinical record regarding reactions* to any previously taken IOP-reducing eye drops, and also any reactions to "steroid eye drops" (if they had been used previously).

IF the doctors had taken the above two considerations into account, and acted upon them, I, likely, would not have lost an additional 1/3 of my L-eye vision. From which we can conclude:

Who/what was of most concern here? The patient? Or the provider/payor side of healthcare?

Single Statement Summary: Eye care healthcare delivery needs to be "revamped" for the 21st century with (1) digital self-monitoring of IOP, etc., and (2) data analytics that monitor treatments and tests in "real time," thus alerting the doctors and clinical staff what is of concern when a patient has a visit, whether in person or via televisit.

Bottom Line: The "Gold Standard" has changed!

How does this chapter fit into the preferred "Patient—People-Centered" "Patient-Directed" models compared to the "Provider-Centered" and "Payor-Centered" models of Healthcare Delivery?

Synopsis: The first part of this chapter covers a lot of the "current delivery methods" which are more payor-centered and provider-centered. During the treatment received by the author, the basis of the "case study" in last part of this chapter, a new patient-centered mode of eye care delivery was discussed, which will need to become the "Gold Standard of Care" in the future, if we want to have excellence and cost-effectiveness in healthcare.

External review of this Chapter 11 glaucoma story by a person with the perspective of one who has analyzed health insurance data on thousands of doctors and hundreds-of-thousands of patients over a long career—see following "text box":

Another perspective on the patient—provider and payor axes:

I read most of the background information and also your personal story. My interpretation of your personal story is:

1. glaucoma is incompletely understood, there are lots of "treatments" that are ineffective (and have serious side effects);
2. medical professionals vary widely in competence and some are just jerks;
3. the provider network is fragmented and disorganized (same as usual); and
4. you had a particularly difficult case and it is hard to know whether it should have been diagnosed and treated faster.

Unfortunately I don't think this this story is unique, as I've seen similar cases over the years.

What do I conclude from this?

For routine medical issues, the system (albeit disorganized and inefficient) works, and generates successful outcomes in most cases. Medicare Advantage programs are supposed to help you navigate this process. Usually it works. (But, the patient needs to be involved.)

For rare, nonroutine medical issues, it is a crap-shoot—find the right doc and the right treatments and you can have a positive outcome. But, I suspect most people with these unusual cases don't find the right doc and the right treatment.

I am not sure what this means for patient-centered (individualized) care. It seems to me that I got practically every possible glaucoma treatment available, but it took time to work through the possibilities. That is the way HMO models are trained to work—you start with the simpler, less expensive procedures before you jump to the complicated expensive procedures. It has been pointed out to me that this approach can actually result in higher costs, including prolonging pain and discomfort for the patient. But you have to have a very good diagnostician to start with, who will move quickly to the treatment (albeit the expensive procedure) and will successfully treat the condition.

Postscript

Some readers may have similar stories of the practice of generalized medicine based on experience and general trends and averages. The central "flaw" in the use of trends and averages as a basis for healthcare is that individual patients do not have *average* problems (it may be that no patients do). Rather, individual patients have *individual* problems and require *individual* healthcare. This distinction points to the major premise of this chapter: we are on the cusp of a paradigm shift in healthcare—the *Year of the Personalized Medicine Revolution*.

References

Abdullah, F., Imtiaz, R., Madni, H.A., Khan, H.A., Khan, T.M., Khan, M.A.U., et al., 2021. A review on glaucoma disease detection using computerized techniques. Digital Object. Identifier 9. Available from: https://doi.org/10.1109/ACCESS.2021.3061451.

Alicea, J., January 13, 2021. Glaucoma review summarizes current diagnostic and treatment practices <https://www.hcplive.com/view/glaucoma-review-summarizes-current-diagnostic-treatment-practices>.

Allergan. August, 2019. What is XEN?; <https://www.xenglaucomaimplant.com/en/hcp/whatisxen>, and modified by author Gary Miner in 2022).

AqueSys, Inc. 2019. What is the XEN GEL Stint? <https://www.xengelstent.com/XENGelStent#>

Bowd, C., Hao, J., Tavares, I.M., et al., 2008. Bayesian machine learning classifiers for combining structural and functional measurements to classify healthy and glaucomatous eyes. Invest. Ophthalmol. Vis. Sci. 49, 945—953 [CrossRef] [PubMed].

Budenz, D.L., Barton, K., Whiteside-de Vos, J., et al., 2013. Prevalence of glaucoma in an urban West African population: the Tema Eye Survey. JAMA Ophthalmol. 131, 651—658 [CrossRef] [PubMed].

Burgansky-Eliash, Z., Wollstein, G., Chu, T., Ramsey, J.D., Glymour, C., Noecker, R.J., et al., 2005. Investig. Ophthalmol. Vis. Sci. 46, 4147−4152. Available from: https://doi.org/10.1167/iovs.05-0366. Available from: https://iovs.arvojournals.org/article.aspx?articleid = 2182263.

Cakanac, C. May 15, 2015. Don't hold back: Topical steroids in clinical practice. <https://www.reviewofcontactlenses.com/article/dont-hold-back-topical-steroids-in-clinical-practice>.

Cassin, B., Solomon, S., 1990. Dictionary of Eye Terminology. Triad Pub. Co, Gainesville, FL, ISBN 978-0-937404-33-1.

Cavet, M.E., Glogowski, S., DiSalvo, C., Richardson, M.E., 2015. Ocular pharmacokinetics of submicron loteprednol etabonate ophthalmic gel 0.38% following topical administration in rabbits. Invest. Ophthalmol. Vis. Sci. 56 (7), 1524.

Chan, K., Lee, T.W., Sample, P.A., Goldbaum, M.H., Weinreb, R.N., Sejnowski, T.J., 2002. Comparison of machine learning and traditional classi-fiers in glaucoma diagnosis. IEEE Trans. Biomed. Eng. 49, 963−974 [CrossRef] [PubMed].

Chen, S., Lv, J., Fan, S., Zhang, H., Xie, L., Xu, L., et al., 2017. Laser peripheral iridotomy vs laser peripheral iridotomy plus laser peripheral irido-plasty in the treatment of multi-mechanism angle closure: study protocol for a randomized controlled trial. Trials 18 (1), 130. Available from: https://doi.org/10.1186/s13063-017-1860-4. Available from: https://www.ncbi.nlm.nih.gov/pmc/articles/PMC5356270/.

Christopher, M., Belghith, A., Bowd, C., et al., 2018. Performance of deep learning architectures and transfer learning for detecting glaucomatous optic neuropathy in fundus photographs. Sci. Rep. 8, 16685 [CrossRef] [PubMed].

Drugs.com (April 30, 2022). Prednisolone Ophthalmic Side Effects. Web site: Prednisolone ophthalmic Side Effects: Common, Severe, Long Term - Drugs.com.

Costello Eye Physicians & Surgeons. 2019. Glaucoma treatment and surgery: XEN 45 gel implant <https://costelloeye.com/minimally-invasive-glau-coma-surgery/xen-45-gel-implant/>.

Dudek, K. September 12, 2014. Glaucoma slide presentation. <https://www.slideserve.com/krysta/glaucoma>.

Elhefney, E., Mokbel, T., Samra, W.A., Kishk, H., Mohsen, T., El-Kannishy, A., 2018. Long-term results of Ahmed glaucoma valve implantation in Egyptian population. Int. J. Ophthalmol. 11 (3), 416−421. Available from: https://doi.org/10.18240/ijo.2018.03.11. Available from: https://www.ncbi.nlm.nih.gov/pmc/articles/PMC5861231/.

Elze, T., Pasquale, L.R., Shen, L.Q., Chen, T.C., Wiggs, J.L., Bex, P.J., 2015. Patterns of functional vision loss in glaucoma determined with arche-typal analysis. J. R. Soc. Interface 12, 20141118.

EveryDayHealth. 2022 <https://www.everydayhealth.com/multiple-sclerosis/treatment/side-effects-ms-steroid-treatment/> <https://www.rnib.org.uk/sites/default/files/GP%20factsheet%20-%20Steroids%20and%20the%20eye.pdf>.

Eye physicians of Long Beach. 2022. What is the Ahmed Valve <https://www.eyephysiciansoflongbeach.com/services/glaucoma-long-beach/ahmed-valve/>.

Eye Surgery Associates. Skalicky, S. 2022. Glaucoma implant surgery <https://www.drsimonskalicky.com.au/glaucoma-specialist/xen-implant/>.

Farkouh, A., Frigo, P., Czejka, M., 2016. Systemic side effects of eye drops: a pharmacokinetic perspective. Clin. Ophthalmol. (Auckland, N.Z.) 10, 2433−2441. Available from: https://doi.org/10.2147/OPTH.S118409PubMed. Available from: https://www.ncbi.nlm.nih.gov/pmc/articles/PMC5153265/.

Fea, A.M., Durr, G.M., Marolo, P., Malinverni, L., Economou, M.A., Ahmed, I., 2020. XEN® gel stent: a comprehensive review on its use as a treat-ment option for refractory glaucoma. Clin. Ophthalmol. (Auckland, N.Z.) 14, 1805−1832. Available from: https://doi.org/10.2147/OPTH.S178348. Available from: https://www.ncbi.nlm.nih.gov/pmc/articles/PMC7335291/. Available from: 32636610.

Feher, J. 2012. Quantatum human physiology. Elsevier and accessed in Science Direct under: aqueous humor <https://www.sciencedirect.com/topics/engineering/aqueous-humor>; <https://doi.org/10.1016/C2009-0-64018-6>.

Fernandez-Granero, M.A., Sarmiento, A., Sanchez-Morillo, D., Jiménez, S., Alemany, P., Fondón, I., 2017. Automatic CDR estimation for early glau-coma diagnosis. J. Healthc. Eng. 2017, 5953621. Available from: https://doi.org/10.1155/2017/5953621. Available from: https://www.ncbi.nlm.nih.gov/pmc/articles/PMC5723944/.

Flowers, C.W., Reynolds, D., Irvine, J.A., Heuer, D.K., 1996. Pupillary block, angle-closure glaucoma produced by an anterior chamber air bubble in a nanophthalmic eye. Arch. Ophthalmol. 114 (9), 1145−1146. Available from: https://doi.org/10.1001/archopht.1996.01100140347021.

Francis, B.A. October 1, 2020. Selective laser trebeculoplasty. glaucoma research founation <https://www.glaucoma.org/treatment/selective-laser-tra-beculoplasty-10-commonly-asked-questions.php>.

Gavidia, M. August 31, 2021. Study finds increased dosage of timolol-dorzolamide fixed combination safe, effective in open angle glaucoma. AJMC. <https://www.ajmc.com/view/study-finds-increased-dosage-of-timolol-dorzolamide-fixed-combination-safe-effective-in-open angle-glaucoma>.

Glaucoma Associates of Texas. 2022. What are glaucoma drainage implants? <https://www.glaucomaassociates.com/incisional-glaucoma-surgery/glaucoma-drainage-implant-surgery/>.

Glaucoma Research Foundation. 2021. https://glaucoma.org/.

Glaucoma Research Foundation. August 10, 2020. Home IOP measurements and continuous IOP monitors <https://www.glaucoma.org/treatment/home-iop-measurements-and-continuous-iop-monitors.php>.

Goldbaum, M.H., 2005. Unsupervised learning with independent component analysis can identify patterns of glaucomatous visual field defects. Trans. Am. Ophthalmol. Soc. 103, 270−280 [PubMed].

Goldbaum, M.H., Sample, P.A., Zhang, Z., et al., 2005. Using unsupervised learning with independent component analysis to identify patterns of glau-comatous visual field defects. Invest. Ophthalmol. Vis. Sci. 46, 3676−3683 [CrossRef] [PubMed].

Goldbaum, M.H., Lee, I., Jang, G., et al., 2012. Progression of patterns (POP): a machine classifier algorithm to identify glaucoma progression in visual fields. Invest. Ophthalmol. Vis. Sci. 53, 6557−6567 [CrossRef] [PubMed].

Gotter, A. 2018. Trabeculectomy: what you should know. Healthline: <https://www.healthline.com/health/trabeculectomy>.

Gupta, A. Dec 14, 2021. The future of health: three healthcare trends for 2022. Forbes Business Council. <https://www.forbes.com/sites/forbesbusi-nesscouncil/2021/12/14/the-future-of-health-three-healthcare-trends-for-2022/?sh = 4d64ff64541c>.

Gupta, A., Soman, V., Bhardwaj, S. Performance improvement in detection of glaucoma in human eyes. Accessed August 10, 2022. https://www.scribd.com/document/406322079/Glaucoma-IEEE-Paper.

Halamka, J., Cerrato, P. Jan 17, 2022. Precision medicine omics, and the black box: mayo clinic platform <https://www.mayoclinicplatform.org/2022/01/17/precision-medicine-omics-and-the-black-box/?linkId = 149537359&s = 03>.

Harwerth, R.S., Carter-Dawson, L., Shen, F., Smith 3rd, E.L., Crawford, M.L., 1999. Ganglion cell losses underlying visual field defects from experi-mental glaucoma. Invest. Ophthalmol. Vis. Sci. 40, 2242−2250 [PubMed].

Harwerth, R.S., Carter-Dawson, L., Smith Barnes III, E.L., Holt, W.F., Crawford, M.L.J., 2004. Neural losses correlated with visual losses in clinical perimetry. Invest. Ophthalmol. Vis. Sci. 45, 3152−3160 [CrossRef] [PubMed].

Healthwise Staff. August 31, 2020. Laser trabeculoplasty for glaucoma. University of Michigan Health. <https://www.uofmhealth.org/health-library/hw155165>.

Heijl, A., Strahlman, E., Sverrisson, T., Brinchman-Hansen, O., Puustjärvi, T., Tipping, R., 1997. A comparison of dorzolamide and timolol in patients with pseudoexfoliation and glaucoma or ocular hypertension. Ophthalmology. 104 (1), 137−142. Available from: https://doi.org/10.1016/s0161-6420(97)30348-0. Available from: 9022118. Available from: https://pubmed.ncbi.nlm.nih.gov/9022118/.

Hennis, A., Wu, S.Y., Nemesure, B., Honkanen, R., Leske, M.C., Barbados Eye Studies G., 2007. Awareness of incident open angle glaucoma in a population study: the Barbados Eye Studies. Ophthalmology 114, 1816−1821 [CrossRef] [PubMed].

Herndon J.R. February 2019. What is fundus photography? Health Union <https://maculardegeneration.net/fundus-photography>.

Johnson, T.V. 2022. What is a trabeculectomy?; John Hopkins Medicine: <https://www.hopkinsmedicine.org/health/wellness-and-prevention/trabeculectomy>.

Keating, F. Dec 8, 2021. Israeli startup unveils AI device for measuring eye fluid pressure in glaucoma treatment. MobiHealthNews: Healthcare iT News <https://www.mobihealthnews.com/news/emea/israeli-startup-unveils-ai-device-measuring-eye-fluid-pressure-glaucoma-treatment>.

Khaimi, M.A. 2015, November−December. Ab Interno canoloplasty: a minimally invaive and maximally effective glaucoma treatment. Glaucoma Today in Surgical Pearls. <https://glaucomatoday.com/pdfs/gt1115_surgpearls.pdf>.

Kim, S.J., Cho, K.J., Oh, S., 2017. Development of machine learning models for diagnosis of glaucoma. PLoS One 12, e0177726 [CrossRef] [PubMed].

Klink, T. MD, E. Panidou, MD, B. Kanzow-Terai, MD, J. Klink, MD, W G. Schlunck, MD, and F. J. Grehn, MD; 2011; Are there filtering blebs after canaloplasty?; J. Glaucoma; http://www.glaucomajournal.com; https://www.semanticscholar.org/paper/Are-There-Filtering-Blebs-After-Canaloplasty-Klink-Panidou/872b1df27fe528d6439dcff108008bbf1a345c37

Labtician Ophthalmics, Inc. 2019. The Ahmed glacoma valve. <https://www.labtician.com/product/ahmed-glaucoma-valve/>.

Lee, J.Y., Kim, Y.Y., Jung, H.R., 2006. Distribution and characteristics of peripheral anterior synechiae in primary angle-closure glaucoma. Korean J. Ophthalmol.: KJO 20 (2), 104−108. Available from: https://doi.org/10.3341/kjo.2006.20.2.104. Available from: https://www.ncbi.nlm.nih.gov/pmc/articles/PMC2908823/. Available from: 16892646.

Lewis, RA, MD, K. von Wolff, MD, M. Tetz, MD, N. Koerber, MD, J. R. Kearney, MD, B. J. Shingleton, MD, et al.; 2009; Canaloplasty: circumfer-ential viscodilation and tensioning of Schlemm canal using a flexible microcatheter for the treatment of open angle glaucoma in adults. J Cataract Refract Surg 2009; 35:814−824, <https://new-glaucoma-treatments.com/wp-content/pdf/Canaloplasty%20Two-year%20Study%20Results.pdf>.

Mayo Clinic (retrieved Nov. 15, 2021). Drugs and Supplements: Corticosteroids Ophthalmic Route Side Effects. Web site: Corticosteroid (Ophthalmic Route) Side Effects - Mayo Clinic.

Mayo Clinic. 2022b. Dorzolamide and Timolol (Ophthalmic Route) <https://www.mayoclinic.org/drugs-supplements/dorzolamide-and-timolol-oph-thalmic-route/proper-use/drg-20061826>.

Mayo Clinic. 2022c. <https://www.mayoclinic.org/drugs-supplements/prednisolone-oral-route/side-effects/drg-20075189?p = 1>.

Mayo Clinic. 2022d <https://www.mayoclinic.org/drugs-supplements/dorzolamide-ophthalmic-route/proper-use/drg-20063524>.

Montaner, J., Ramiro, L., Simats, A., Tiedt, S., Makris, K., Jickling, G.C., et al., 2020. Multilevel omics for the discovery of biomarkers and therapeu-tic targets for stroke. Nat. Rev Neurol. 16 (5), 247−264. Available from: https://doi.org/10.1038/s41582-020-0350-6Epub 2020 Apr 22. PMID. Available from: 32322099.

Montesano, G., Chen, A., Lu, R., Lee, C.S., Lee, A.Y., 2022. UWHVF: a real-world, open source dataset of perimetry tests from the humphrey field analyzer at the University of Washington. Transl. Vis. Sci. Technol. 11, 2. Available from: https://doi.org/10.1167/tvst.11.1.1. Available from: https://tvst.arvojournals.org/article.aspx?articleid = 2778219.

Mursch-Edlmayr, A.S., Ng, W.S., Diniz-Filho, A., Sousa, D.C., Arnould, L., Schlenker, M.B., et al., 2020. Artificial intelligence algorithms to diag-nose glaucoma and detect glaucoma progression: translation to clinical practice. Transl. Vis. Sci. Technol. 9, 55. Available from: https://doi.org/10.1167/tvst.9.2.55. Available from: https://tvst.arvojournals.org/article.aspx?articleid = 2770923.

Neoretina Eyecare Institute. October 22, 2018. How long does it take to go blind from glaucoma? <https://neoretina.com/blog/faqs/how-long-does-it-take-to-become-blind-from-glaucoma/>.

NICE (National Institute for Health and Care Excellence). Nov 19, 2014. The SENSIMED Triggerfish contact lens sensor for continuous 24-hour recording of the ocular dimensional changes in people with or at risk of developing glaucoma. <https://www.nice.org.uk/advice/mib14/chapter/technology-overview>.

Alsheikh O, Braverman-Terry-OEI. 2018. Glaucoma: the silent thief of vision <https://drstuartterry.com/2019/08/30/glaucoma-the-silent-thief-of-vision/>.

Pakravan, M., Beni, A.N., Yazdani, S., Esfandiari, H., Mirshojaee, S., 2021. Efficacy and safety of timolol-dorzolamide fixed-combination three times a day vs two times a day in newly diagnosed open angle glaucoma. J. Drug. Assess. 10 (1), 91–96. Available from: https://doi.org/10.1080/21556660.2021.1967642.

Piltz-Seymour, J., Tai, T. Y. T. December 23, 2021. IOP and tonometry. American Academy of Ophthalmology <https://eyewiki.aao.org/IOP_anInsd_Tonometry>.

Pittner, A., Sharpe, R.A.; Chelnis, A., Sharpe, R.A., Palestine, A., Gonzales, J., et al. Aug 3, 2021. Synechiae. American Academy of Opthalmology—EyeWiki; <https://eyewiki.aao.org/Synechiae#>.

Sample, P.A., Chan, K., Boden, C., et al., 2004. Using unsupervised learning with variational bayesian mixture of factor analysis to identify patterns of glaucomatous visual field defects. Invest. Ophthalmol. Vis. Sci. 45, 2596–2605 [CrossRef] [PubMed].

Sample, P.A., Boden, C., Zhang, Z., et al., 2005. Unsupervised machine learning with independent component analysis to identify areas of progression in glaucomatous visual fields. Invest. Ophthalmol. Vis. Sci. 46, 3684–3692 [CrossRef] [PubMed].

Santra Dhali, R., Munshi, S., 2017. Emerging glaucoma therapeutics. Int. J. Basic. Clin. Pharmacol. 4 (4), 606–612. Available from: https://doi.org/10.18203/2319-2003.ijbcp20150360. Available from: https://www.ijbcp.com/index.php/ijbcp/article/view/808.

Seltman, W. July 21, 2020. Ocular hypertension <https://www.webmd.com/eye-health/occular-hypertension>.

Shigueoka, L.S., Vasconcellos, J.P.C., Schimiti, R.B., et al., 2018. Automated algorithms combining structure and function outperform general ophthalmologists in diagnosing glaucoma. PLoS One 13, e0207784.

Spaeth, G.L., Rahmatnejad, K., Zeng, L. June, 2016. Is there still a role for optic isc photography? New is not necessarily better <https://glaucomatoday.com/articles/2016-may-june/is-there-still-a-role-for-optic-disc-photography>.

Starr, O. July 2018. Chronic open angle glaucoma <https://patient.info/eye-care/acute-angle-closure-glaucoma/chronic-open angle-glaucoma>.

Stein, J.D., Khawaja, A.P., Weizer, J.S., 2021. Glaucoma in adults—screening, diagnosis, and management: a review2021JAMA 325 (2), 164–174. Available from: https://doi.org/10.1001/jama.2020.21899.

Taranto, B. Dec 20, 2021. The growing power of digital health: 6 trends to watch in 2022. <https://techcrunch.com/2021/12/20/the-growing-power-of-digital-healthcare-6-trends-to-watch-in-2022/>.

Tham, Y.C., Li, X., Wong, T.Y., Quigley, H.A., Aung, T., Cheng, C.Y., 2014. Global prevalence of glaucoma and projections of glaucoma burden through 2040: a systematic review and *meta*-analysis. Ophthalmology 121, 2081–2090 [CrossRef] [PubMed].

Thompson, A.C., Jammal, A.A., Medeiros, F.A., 2020. A review of deep learning for screening, diagnosis, and detection of glaucoma progressionJuly 2020Transl. Vis. Sci. Technol. 9, 42. Available from: https://doi.org/10.1167/tvst.9.2.42. Available from: https://tvst.arvojournals.org/article.aspx?articleid = 2770356.

Tsai, J.C. 2021, August, 20. High eye pressure and glaucoma. Glaucoma Research Foundation. High Eye Pressure and Glaucoma | Glaucoma Research Foundation. <https://www.glaucoma.org/gleams/high-eye-pressure-and-glaucoma>.

University of Pittsburgh Medical Center—Children's Hospital. 2021. <https://www.chp.edu/our-services/ophthalmology/ophthalmology-patient-procedures/trabeculotomy>.

Vision and eye Health. 2022. Glaucoma tube shunt. <https://www.vision-and-eye-health.com/glaucoma-tube-shunt.html>.

Wagner, M., Schuster, A K-G., Emmerich, J., Chranopoulos, P., Hoffmann, E.M. April 20, 2020. Efficacy and safety of XEN®—implantation vs. trabeculectomy: data of a "real-world" setting <https://journals.plos.org/plosone/article?id = 10.1371/journal.pone.0231614>; <https://doi.org/10.1371/journal.pone.0231614>

Wang, M., Pasquale, L.R., Shen, L.Q., et al., 2018. Reversal of glaucoma Hemifield test results and visual field features in glaucoma. Ophthalmology 125, 352–360.

Wang, M., Tichelaar, J., Pasquale, L.R., et al., 2020. Characterization of central visual field loss in end-stage glaucoma by unsupervised artificial intelligence. JAMA Ophthalmol. 138, 190–198 [CrossRef].

Wang, M., Shen, L.Q., Pasquale, L.R., Petrakos, P., Formica, S., Boland, M.V., et al., 2019. An artificial intelligence approach to detect visual field progression in glaucoma based on spatial pattern analysis. Investig. Ophthalmol. Vis. Sci. 60, 365–375. Available from: https://doi.org/10.1167/iovs.18-25568.

WebMD. 2022b. Dorzolamide-timolol ophthalmic eye details. <https://www.webmd.com/drugs/2/drug-7707/dorzolamide-timolol-ophthalmic-eye/details>.

Weinreb, R.N., Aung, T., Medeiros, F.A., 2014. The pathophysiology and treatment of glaucoma: a review. JAMA 311 (18), 1901–1911. Available from: https://doi.org/10.1001/jama.2014.3192.

Wu, L., Xie, X., Liang, T., Ma, J., Yang, L., Yang, J., et al., 2021. Integrated multi-omics for novel aging biomarkers and antiaging targets. Biomolecules 12 (1), 39. Available from: https://doi.org/10.3390/biom12010039Published online 2021 Dec 28. Available from: https://www.ncbi.nlm.nih.gov/pmc/articles/PMC8773837/.

Yaseen, W., 2020. Plasmoid aqueous with hypopyon in Anterior Uveitis. American Academy of Optometry <https://www.aaopt.org/detail/knowledge-base-article/plasmoid-aqueous-with-hypopyon-in-anterior-uveitis>.

Yousefi, S., Balasubramanian, M., Goldbaum, M.H., et al., 2016. Unsupervised Gaussian mixture-model with expectation maximization for detecting glaucomatous progression in standard automated perimetry visual fields. Transl. Vis. Sci. Technol. 5, 2 [CrossRef] [PubMed].

Yousefi, S., Kiwaki, T., Zheng, Y., Sugiura, H., Asaoka, R., Murata, H., et al., 2018. Detection of longitudinal visual field progression in glaucoma using machine learning. Am. J. Ophthalmol. 193, 71–79. Available from: https://doi.org/10.1016/j.ajo.2018.06.007Epub 2018 Jun 18. PMID. Available from: 29920226. Available from: https://iovs.arvojournals.org/article.aspx?articleid = 2723059.

Further reading

Belghith, A., Bowd, C., Medeiros, F.A., Balasubramanian, M., Weinreb, R.N., Zangwill, L.M., 2015. Learning from healthy and stable eyes: a new approach for detection of glaucomatous progression. Artif. Intell. Med. 64, 105–115 [CrossRef] [PubMed].

Biosphere. 2019. Bausch + Lomb Announces FDA Approval Of LOTEMAX® SM (loteprednol Etabonate Ophthalmic Gel) 0.38% for the treatment of postoperative inflammation and pain following ocular surgery. <https://www.biospace.com/article/releases/bausch-lomb-announces-fda-approval-of-lotemax-sm-loteprednol-etabonate-ophthalmic-gel-0−38-percent-for-the-treatment-of-postoperative-inflammation-and-pain-following-ocular-surgery/>.

Goldbaum, M.H., Sample, P.A., Chan, K., et al., 2002. Comparing machine learning classifiers for diagnosing glaucoma from standard automated perimetry. Invest. Ophthalmol. Vis. Sci. 43, 162–169.

Grigorian, R.A., Shah, A., Guo, S., May, 2007. Comparison of Loteprednol Etabonate 0,5% (Lotemax) to Prednisonolone Acetate 1% (Falcon) for inflammation treatment following cataract surgery. <https://iovs.arvojournals.org/article.aspx?articleid = 2383347>.

Kaplan, T.B., August 6, 2021. EveryDay health. <https://www.everydayhealth.com/multiple-sclerosis/treatment/side-effects-ms-steroid-treatment/>.

Karimian, F., Faramarzi, A., Fekri, S., Mohammad-Rabie, H., Najdi, D., Doozandeh, A., et al., 2017. Comparison of loteprednol with fluorometholone after myopic photorefractive keratectomy. J. Ophthal. Vis. Res. 12 (1), 11–16. Available from: https://doi.org/10.4103/2008-322X.200161. Available from: https://www.ncbi.nlm.nih.gov/pmc/articles/PMC5340049/.

Liu, C., Zhao, Q., Zhen, Y., Zhai, J., Liu, G., Zheng, M., et al., 2015. Effect of corticosteroid on renal water and sodium excretion in symptomatic heart failure: prednisone for renal function improvement evaluation study. J. Cardiovasc. Pharmacol. 66 (3), 316–322. Available from: https://doi.org/10.1097/FJC.0000000000000282. Available from: 25992918. Available from: https://pubmed.ncbi.nlm.nih.gov/25992918/.

Mayo Clinic. 2022a. Fluorometholone (ophthalmic route: description and brand names). <https://www.mayoclinic.org/drugs-supplements/fluorometholone-ophthalmic-route/proper-use/drg-20060781?p = 1>.

Minigh, J. 2007. in xPharm: the comprehensive pharmacology reference. <https://www.sciencedirect.com/topics/pharmacology-toxicology-and-pharmaceutical-science/fluorometholone>.

Spraybary, A., Kelley, S., March, 2021, updated Jan, 2022. Aqueous humor. All about vision. <https://www.allaboutvision.com/eye-care/eye-anatomy/aqueous-humor/>.

Web MD. 2022a. Fluorometholone suspension drops (final dose forms): uses, side effects, and more. <https://www.webmd.com/drugs/2/drug-12354-185/fluorometholone-ophthalmic-eye/fluorometholone-suspension-ophlemmthalmic/details>.

Wikipedia #1. April 11, 2022. Anterior chamber of eyeball. <https://en.wikipedia.org/wiki/Anterior_chamber_of_eyeball#>.

Chapter 12

Using data science algorithms in predicting ICU patient urine output in response to diuretics to aid clinicians and healthcare workers in clinical decision-making

Anna J.C. Russell-Toner, MComm(Statistics)
The Boss Lady, The Data-Shack, York, United Kingdom

Chapter outline

Prelude

This case study was part of a larger study at Beth Israel Deaconess Medical Center in Boston. The study was designed to minimize the time to determine the proper diuretic treatment program for patients with acute kidney failure using a home-based medical device. The models presented in this case study optimize the diuretic treatment program design to maximize urine output. This type of predictive model can be effectively integrated with personalized medical programs by incorporating inputs and analyses from both clinical and home-based operations.

The first three-fourths of this paper goes through a "step-by-step" process of how the scientific researchers thought and decided upon each step of their study; this is tedious to go through, but is presented here for the reader that wants to learn "how to think about developing predictive analytic model" and understand the steps involved. For the reader that already knows how to do this, or is not interested at this time, you can skip down to near the end of this paper where two real patients are discussed. These patient examples show how the predictive analytic model coupled with treatment step can allow the individual patient to provide some of the data themselves to help make the best clinical decision on best treatment. This is "patient-centered medicine" in practice.

Practical Data Analytics for Innovation in Medicine. DOI: https://doi.org/10.1016/B978-0-323-95274-3.00027-0

Introduction

This study was originally started based on research done for a customer in the healthcare space to assess the feasibility of using an algorithm (ML or AI type analytic model) to shorten the duration of time it takes to determine if a patient's urine output goals will be met (with or without diuretics), Thus a physician or other healthcare worker could make a fast and more accurate decision, and potentially life-saving, decision, on continuing treatment strategies. This was intended to support diuresis for deresuscitation of patients in the ICU, by making it easy to initiate and monitor best practices for diuretic dosing, UO (urine output) goal setting, and UO monitoring with ICU care teams. This research was expanded on, after closure of the customer external project, and is still in progress internally to study the prevalence of acute kidney injury or mortality on these patients. This chapter focuses on the data science algorithms (Machine Learning and AI type methods) we tested for feasibility, given the data available, and the conclusions made.

The data we used to develop the models comes from the Medical Information Mart for Intensive Care (MIMIC-IV), which is a large US-based critical care database which has been integrated and deidentified. It contains a comprehensive clinical dataset containing all the patients admitted to the ICUs of Beth Israel Deaconess Medical Center in Boston, MA, from June 1, 2001, to October 31, 2012. There were 53,423 distinct hospital admissions for adult patients (aged 16 years or above) admitted to the ICUs during the study period.

A literature review was done on acute kidney injury and related topics and a summary of the results from this has also been published in this chapter. This is followed by some algorithm and model outputs and decisions, based on various options considered and given the nature of the data available.

This chapter also discusses the selection of a *Champion Predictive Model and Algorithm*, based on model performance, and finishes with some conclusions and recommendations for further work.

In summary, we found that data science algorithms can provide significant new information, and thus inform healthcare workers so they can predict with increasing levels of accuracy, what patient outcomes are likely to be, thus providing individualized patient treatment plans. We also discovered that we could make these good predictions with much less data than was previously anticipated.

Outputs and conclusion from a literature review

The papers and articles studied in the literature review (literature reviewed is available in the "References" section below), gave us an exceptionally good overview and base for choosing a useful research data set, led us to choosing appropriate variables, and confirmed our approach and choice of algorithms to be used for building the models.

The data used

Source of data

Data were extracted from the MIMIC-IV, which is a large US-based critical care database which has been integrated and deidentified. It contains a comprehensive clinical dataset containing all the patients admitted to the ICUs of Beth Israel Deaconess Medical Centre in Boston, MA, from June 1, 2001, to October 31, 2012. There were 53,423 distinct hospital admissions for adult patients (aged 16 years or above) admitted to the ICUs during the study period.

Since our study was an analysis of a third-party anonymized publicly available database, we needed to get approval for its use. Approval was obtained by the relevant research team members. This included a short exam and submission to the hospital's review board, which was then approved and access to the data provided.

Data demographics

Google Big Query was used to extract this data and load it into an internal SQL database.,
The following inclusion criteria were used for patients:

1. Over 18 y/o
2. One of the following (depending on which has the highest number of patients)
   a. 3 × consecutive hours of UO (urine output) data over a 12-hour period
   b. 6 × UO data points over a 24-hour period

Additional criteria

1. Patients with more than one diuretic administration.
2. Variables indicating an internal urinary catheter were included.
3. Patients included are either recipients of IV administration or med bolus.
4. Serum creatinine (SCr) and various other variables were included to determine their effect (if any) on urine output.

Any other exclusions, which were model/algorithm specific, are described in the section on Algorithm Outputs and Decisions (Figs. 12.1–12.6).

Further demographics are described in the Image below:

(A)

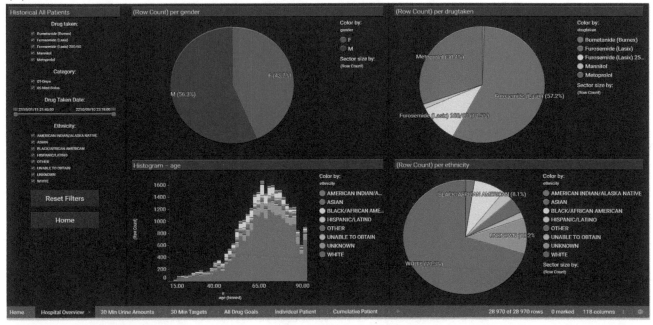

(B)

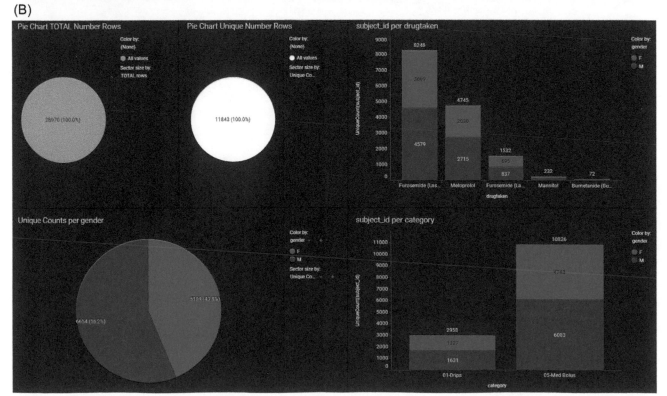

FIGURE 12.1 A and B. Data demographics for the data used in this study.

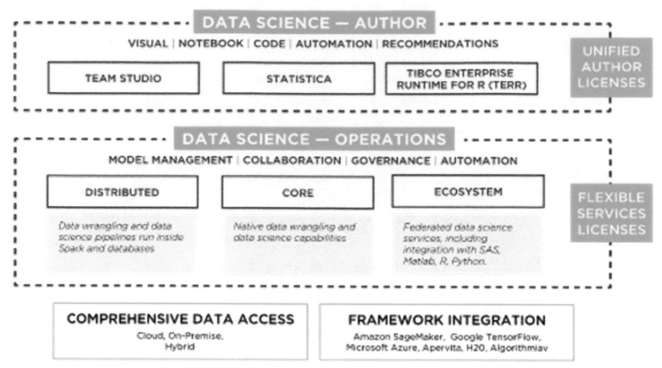

FIGURE 12.2   The "Data Science Plan of Operation" used for this study.

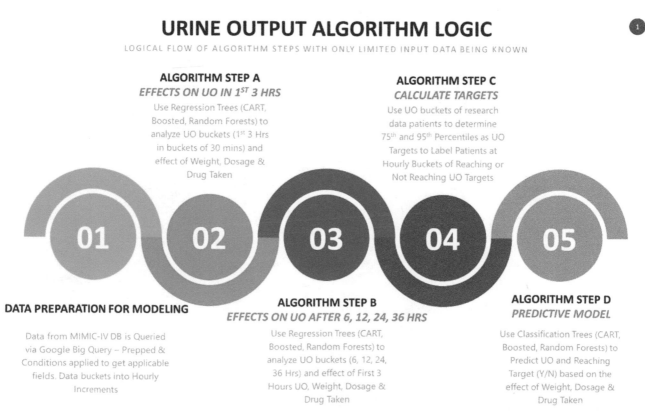

FIGURE 12.3   The logic of the algorithms used to measure urine output (UO = Urine Output).

(A)

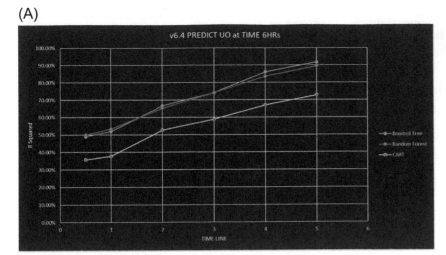

FIGURE 12.4 A and B. R-squared predicting hour 6, hourly after 1–6 hours of data collected.

(B)

<u>DEPENDENT</u> **CUM 6HRs UO Summary**

	Basic V plus Independent Variables below			v6.4
	TIME	**max time interval**	**Method**	**R^2**
Independent	0-0.5HRs	0.5	Boosted Tree	48.93%
		0.5	CART	35.69%
		0.5	Random Forest	49.90%
	until 1-1.5HRs	1	Boosted Tree	51.74%
		1	CART	37.63%
		1	Random Forest	53.29%
	until 2-2.5HRs	2	Boosted Tree	66.64%
		2	CART	52.68%
		2	Random Forest	65.27%
	until 3-3.5HRs	3	Boosted Tree	74.04%
		3	CART	58.76%
		3	Random Forest	74.14%
	until 4-4.5HRs	4	Boosted Tree	86.08%
		4	CART	66.98%
		4	Random Forest	83.52%
	until 5-5.5HRs	5	Boosted Tree	91.67%
		5	CART	72.77%
		5	Random Forest	89.54%

Technology used

For this research, the following technology has been used:
   TIBCO Data Science Workbench

Data science author

Workbench/Statistica provides a rich user interface that allows data scientists to create advanced analytic workflows using 16,000 functions. Users can also seamlessly integrate Python, R, and other nodes within the pipelines. Additionally, users can create parameterized workspaces that can be invoked via Spotfire.

Data science operations

Core services are used with Statistica. These services provide model management, platform management, scoring, and governance capabilities for Statistica https://www.tibco.com/tibco-data-science-overview.

(A)

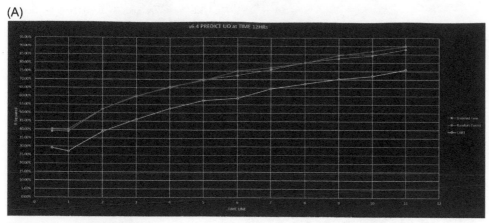

FIGURE 12.5   A and B. R-squared predicting hour 12, hourly after 1−12 hours of data collected.

(B)

DEPENDENT   **CUM 12HRs UO Summary**

Basic V plus Independent Variables below			
			v6.4
TIME	max time interval	Method	R^2
0-0.5HRs	0.5	Boosted Tree	39.05%
	0.5	CART	29.19%
	0.5	Random Forest	40.41%
until 1-1.5HRs	1	Boosted Tree	39.01%
	1	CART	27.30%
	1	Random Forest	40.34%
until 2-2.5HRs	2	Boosted Tree	52.35%
	2	CART	38.97%
	2	Random Forest	52.28%
until 3-3.5HRs	3	Boosted Tree	60.13%
	3	CART	46.20%
	3	Random Forest	59.79%
until 4-4.5HRs	4	Boosted Tree	65.15%
	4	CART	52.68%
	4	Random Forest	65.65%
until 5-5.5HRs	5	Boosted Tree	69.95%
	5	CART	57.52%
	5	Random Forest	69.37%
until 6-6.5HRs	6	Boosted Tree	72.33%
	6	CART	58.74%
	6	Random Forest	75.00%
until 7-7.5HRs	7	Boosted Tree	75.89%
	7	CART	64.51%
	7	Random Forest	77.07%
until 8-8.5HRs	8	Boosted Tree	80.21%
	8	CART	67.47%
	8	Random Forest	80.08%
until 9-9.5HRs	9	Boosted Tree	82.66%
	9	CART	70.35%
	9	Random Forest	84.51%
until 10-10.5HRs	10	Boosted Tree	84.37%
	10	CART	72.15%
	10	Random Forest	87.05%
until 11-11.5HRs	11	Boosted Tree	87.98%
	11	CART	75.86%
	11	Random Forest	89.86%

(A)

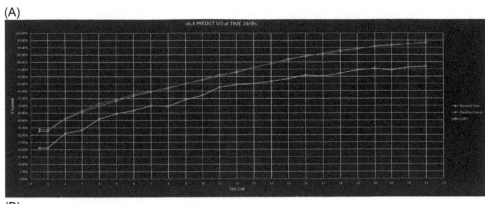

FIGURE 12.6 A and B. R-squared predicting hour 24, hourly after 1–24 hours of data collected.

(B)

<u>DEPENDENT</u> **CUM 24HRs UO Summary**

Basic V plus Independent Variables below		v6.4	
TIME	max time interval	Method	R^2
0-0.5HRs	0.5	Boosted Tree	32.64%
	0.5	CART	21.13%
	0.5	Random Forest	33.94%
until 1-1.5HRs	1	Boosted Tree	32.62%
	1	CART	21.12%
	1	Random Forest	33.69%
until 2-2.5HRs	2	Boosted Tree	40.80%
	2	CART	31.08%
	2	Random Forest	41.20%
until 3-3.5HRs	3	Boosted Tree	45.72%
	3	CART	33.26%
	3	Random Forest	46.74%
until 4-4.5HRs	4	Boosted Tree	49.92%
	4	CART	40.92%
	4	Random Forest	51.61%
until 5-5.5HRs	5	Boosted Tree	53.31%
	5	CART	44.52%
	5	Random Forest	54.39%
until 6-6.5HRs	6	Boosted Tree	56.73%
	6	CART	46.88%
	6	Random Forest	57.97%
until 7-7.5HRs	7	Boosted Tree	59.43%
	7	CART	50.13%
	7	Random Forest	60.05%
until 8-8.5HRs	8	Boosted Tree	62.41%
	8	CART	49.77%
	8	Random Forest	61.87%
until 9-9.5HRs	9	Boosted Tree	64.95%
	9	CART	54.15%
	9	Random Forest	65.26%
until 10-10.5HRs	10	Boosted Tree	68.12%
	10	CART	57.19%
	10	Random Forest	67.58%
until 11-11.5HRs	11	Boosted Tree	71.24%
	11	CART	62.76%
	11	Random Forest	70.10%
until 12-12.5HRs	12	Boosted Tree	72.78%
	12	CART	64.41%
	12	Random Forest	73.71%
until 13-13.5HRs	13	Boosted Tree	76.22%
	13	CART	65.14%
	13	Random Forest	76.34%

FIGURE 12.6 A and B. R-squared predicting hour 24, hourly after 1–24 hours of data collected.

	14	Boosted Tree	78.75%
until 14-14.5HRs	14	CART	66.75%
	14	Random Forest	78.97%
	15	Boosted Tree	82.26%
until 15-15.5HRs	15	CART	68.50%
	15	Random Forest	81.29%
	16	Boosted Tree	83.81%
until 16-16.5HRs	16	CART	70.97%
	16	Random Forest	84.53%
	17	Boosted Tree	85.93%
until 17-17.5HRs	17	CART	70.38%
	17	Random Forest	84.91%
	18	Boosted Tree	88.10%
until 18-18.5HRs	18	CART	72.25%
	18	Random Forest	87.12%
	19	Boosted Tree	89.27%
until 19-19.5HRs	19	CART	74.55%
	19	Random Forest	88.91%
	20	Boosted Tree	90.77%
until 20-20.5HRs	20	CART	75.31%
	20	Random Forest	91.01%
	21	Boosted Tree	91.32%
until 21-21.5HRs	21	CART	74.29%
	21	Random Forest	91.99%
	22	Boosted Tree	92.49%
until 22-22.5HRs	22	CART	76.42%
	22	Random Forest	92.40%
	23	Boosted Tree	93.07%
until 23-23.5HRs	23	CART	76.90%
	23	Random Forest	93.67%

FIGURE 12.6   Continued.

TIBCO Spotfire Immersive, smart, real-time insights for everyone

TIBCO Spotfire software enables everyone to explore and visualize new discoveries in data through dashboards and advanced analytics. Spotfire analytics delivers capabilities at scale, including predictive analytics, geolocation analytics, and streaming analytics. Spotfire mods allow building tailored analytic apps rapidly, repeatedly, and to scale (https://www.tibco.com/products/tibco-spotfire).

Algorithm outputs and decisions

Algorithm version 1

Introduction

The image below describes the logic of the algorithm built as "Version 1."
   This algorithm is aimed at doing the following:

1. *Step 1: DATA PREPARATION FOR MODELING*
   a. Google Big Query was used on the MIMIC-IV databases to obtain the following variables:
      i. Patient Weight
      ii. Diuretic Received

    iii. Date/Time of Administering the Diuretic
    iv. Dose of Diuretic Received
    v. Urine Output Records per patient
    vi. Date/Time of Urine Output

The Date-Time of Urine Output was then used and Pivoted to create Urine Output Values bucketed in hourly Increments up to 36 hours, for demographics and visualization. For modeling of Urine Output however, we are using cumulative values per hour for up to 24 hours.

b. For Dose of Diuretic Received, we normalize this variable in 2 ways:
i. To control for whether a med bolus or type of IV was used, we apply a formula where we normalize for the effect of either, by dividing the total dosages into optimal time brackets.
ii. To control for how many times a diuretic was administered in previous consecutive hours and the effect of this, we included the number of previous doses into the model. We found it to have a negligible effect. We have therefore pooled this into the normalization above, as a weighting, in case it may have a small effect for certain patients.

1. *Step 2: ALGORITHM STEP A—EFFECTS ON URINE OUTPUT IN FIRST 3 HOURS*
    Use Regression Trees (CART, Boosted, Random Forests, Neural Networks) to analyze Urine Output buckets (first 3 hours in hourly buckets—Cumulative UO) and the effect of weight, dosage, and drug taken. Results of this are published below.

2. *Step 3: ALGORITHM STEP B—EFFECTS ON URINE OUTPUT AFTER 6, 12, 24 HOURS*
    Use Regression Trees (CART, Boosted, Random Forests, Neural Networks) to analyze Urine Output buckets (6, 12, 24 hours—Cumulative UO) and the effect of first 3 hours Urine Output, in relation to weight, dosage and drug taken. Results of this are published below.

3. *Step 4: ALORITHM STEP C—CALCULATE URINE OUTPUT TARGETS*
    Use hourly Urine Output buckets of research data patients to determine 75th and 95th percentiles as Urine Output Targets to label patients at hourly buckets of reaching or not reaching Urine Output targets. Results of this are published below.

4. *Step 5: ALGORITHM STEP D—PREDICTIVE MODEL*
    Use Classification Trees (CART, Boosted, Random Forests, Neural Networks) to predict Urine Output and reaching target (Yes/No) based on the effect of previous hours Urine Output (Cumulative hourly), including weight, dosage and drug taken.

Algorithm selection

Algorithms are selected depending on the business/research questions one wants answered, as well as the structure of the data available to answer these questions.

Selection criteria—algorithm Steps A and B

In the case of Steps A and B of the algorithms above, we are looking to predict a continuous (Numerical) dependent/output variable (Urine Output in mL), from a mix of continuous (Numerical) and categorical independent/input variables. Our options for algorithms here, would be linear/nonlinear regression models, trees models or neural networks. As linear/nonlinear regression models rely heavily on normal distributions and can be skewed heavily in this case, and many of the variables tested for our analyses had nonnormal distributions, this was ruled out at the outset. It needs to be noted, that one could use regression techniques, but it would mean significant data transformations back and forward to normalize, which is why we opted out of using them.

For that reason, we used simple C&RT (CART—Classification and Regression Trees), more advanced and modern boosted trees and random forests, as well as regression with neural networks. With the sample size within the hourly urine buckets, they proved to be too sparse for the neural networks to be effective, so that was removed from our modeling options.

This left us with the CART, Boosted Trees, and Random Forest models/algorithms—which given the data structures and sample sizes available in buckets, performed well.

In the case of steps A and B, with a continuous output, the type of CART Tree used is a regression tree (if our output variable had been categorical or binary, we would have used a classification tree). A description of each of these will follow.

Selection criteria—algorithm Step C

Algorithms are selected depending on the business/research questions one wants answered, as well as the data available to answer these questions. In the case of step C for the algorithms above, we are looking to calculate a target (to simulate the "target" a clinician might set) for us to predict whether a particular patient is "Yes"—likely to reach this target, or "No"—not likely to reach this target at the respective Urine Output time buckets.

To calculate these "Targets," we looked at the population of patients available, based on the Selection Criteria above. Given these criteria and population, we calculated for each Urine Output time bucket, a target based on the 75th and 95th percentiles of Urine Output our population of patients were likely to reach. We thus had scientifically created simulated "Targets" to be able to illustrate how our algorithms perform in predicting a Yes or No, as per above.

As an example: To calculate the 75th Percentile of Urine Output in mL at hour 3, one would find the Urine Output Value which 75% of patients had reached in the research population at hour 3. This would then be used as the simulated "Target" (which would normally be entered by the clinician). We did the same for the 95th percentiles at each hourly time bucket. This gave us the option of using either the 75th or 95th percentiles as targets, depending on how sensitive we wanted the model/algorithm to be.

This allowed us to code our data with a Yes or No to use as an output/dependent variable to train our algorithm for step D—given the same input variables used in Step A and B.

Selection criteria—algorithm Step D

In the case of step D of the algorithms we are looking to predict a categorical dependent/output variable (Yes or No—for the Patient reaching the Simulated Target calculated above), from a mix of continuous (Numerical) and categorical independent/input variables. Our options for algorithms here, would be Logistic Regression Models, Trees Models, or Neural Networks. As basic regression models rely heavily on normal distributions and can be skewed heavily in this case, and many of the variables tested for our analyses had nonnormal distributions, this was ruled out at the outset. It needs to be noted, that one could use regression techniques, but it would mean significant data transformations back and forward to normalize, which is why we opted out of using them.

For that reason, we used simple C&RT (CART—Classification and Regression Trees), more advanced and modern boosted trees and random forests, as well as regression with neural networks. With the sample size within the hourly urine buckets, they proved to be too sparse for the Neural Networks to be effective, so that was removed from our modeling options as well.

This left us with the CART, boosted trees, and random forest models/algorithms—which, given the data structures and sample sizes available in buckets, performed well.

In the case of step D, with a categorical output, the type of CART Tree used is a classification tree (if our output variable had been continuous (Numerical), we would have used a regression tree). A description of each of these follows.

Short description of each algorithm

Regression in general and classification and regression trees

The general purpose of multiple regression (the term was first used by Pearson, 1908) is to analyze the relationship between several independent/input or predictor variables and a dependent/output or criterion variable. The computational problem that needs to be solved in multiple regression analysis is to fit a straight line (or plane in an **n**-dimensional space, where **n** is the number of independent variables) to data points.

Classification/Regression trees are used to predict membership of cases or objects in the classes of a categorical dependent variable/or values of a continuous dependent variable, from their measurements on one or more predictor variables. This includes techniques for computing binary *classification trees* based on univariate splits for categorical predictor variables, ordered predictor variables (measured on at least an ordinal scale), or a mix of both types of predictors. It also has options for computing classification trees based on linear combination splits for interval scale predictor variables.

Boosted trees

The concept of boosting applies to the area of predictive modeling, to generate multiple models or classifiers (for prediction or classification), and to derive weights to combine the predictions from those models into a single prediction or predicted classification.

A simple algorithm for boosting works like this. Start by applying some method (e.g., a tree classifier such as C&RT) to the learning data, where each observation is assigned an equal weight. Compute the predicted classifications and apply weights to the observations in the learning sample that are inversely proportional to the accuracy of the classification. In other words, assign greater weight to those observations that are difficult to classify (where the misclassification rate is high), and lower weights to those that are easy to classify (where the misclassification rate is low). In the context of C&RT, for example, different misclassification costs (for the different classes) can be applied inversely proportional to the accuracy of prediction in each class. Then apply the classifier again to the weighted data (or with different misclassification costs) and continue with the next iteration (application of the analysis method for classification to the reweighted data).

Boosting will generate a sequence of classifiers, or trees where each consecutive classifier in the sequence is an "expert" in classifying observations that are not well classified by those preceding it.

Note that boosting can also be applied to learning methods that do not explicitly support weights or misclassification costs. In that case, random subsampling can be applied to the learning data in the successive steps of the iterative boosting procedure, where the probability for selection of an observation into the subsample is inversely proportional to the accuracy of the prediction for that observation in the previous iteration (in the sequence of iterations of the boosting procedure).

For our Boosted Trees Algorithm specifically, the general idea is to compute a sequence of (very) simple trees, where each successive tree is built for the prediction residuals of the preceding tree. This method will build binary trees, such as partitioning the data into two samples at each split node. Now suppose that you were to limit the complexities of the trees to three nodes only (actually, the complexity of the trees can be selected by the user): A root node and two child nodes, for instance, a single split. Thus, at each step of the boosting (algorithm), a simple (best) partitioning of the data is determined, and the deviations of the observed values from the respective means (residuals for each partition) are computed. The next 3-node tree will then be fitted to those residuals, to find another partition that will further reduce the residual (error) variance for the data, given the preceding sequence of trees.

It can be shown that such additive weighted expansions of trees can eventually produce an excellent fit of the predicted values to the observed values, even if the specific nature of the relationships between the predictor variables and the dependent variable of interest is very complex (nonlinear in nature). Hence, the method of gradient boosting—fitting a weighted additive expansion of simple trees—represents a very general and powerful machine learning algorithm.

Detailed technical descriptions of these methods can be found in Friedman (1999a, b) as well as Hastie, Tibshirani, & Friedman (2001); see also Nisbet, Elder, & Miner (2009).

Random forests

The Random Forest Algorithm used is an implementation of the so-called Random Forest classifiers developed by Breiman. The algorithm is also applicable to regression problems. A Random Forest consists of a collection (ensemble) of simple tree predictors, each capable of producing a response when presented with a set of predictor values. For classification problems, this response takes the form of a class membership, which associates (classifies) a set of independent (predictor) values with one of the categories present in the dependent variable. Alternatively, for regression problems, the tree response is an estimate of the dependent variable given the predictors.

A Random Forest consists of an arbitrary number (ensemble) of simple trees, which are used to vote for the most popular class (classification), or their responses are combined (averaged) to obtain an estimate of the dependent variable (regression). Using tree ensembles can lead to significant improvement in prediction accuracy (i.e., better ability to predict new data cases).

The response of each tree depends on a set of predictor values chosen independently (with replacement) and with the same distribution for all trees in the forest, which is a subset of the predictor values of the original data set.

For classification problems, given a set of simple trees and a set of random predictor variables, the Random Forest method defines a margin function that measures the extent to which the average number of votes for the correct class exceeds the average vote for any other class present in the dependent variable. This measure provides us not only with a convenient way of making predictions, but also with a way of associating a confidence measure with those predictions.

A note on variable importance: Generally, in predictive algorithms, the input variables are not equivalent, that is, often few variables have greater effect on the target variable than the others. Thus, it is useful to learn the relative influence (importance) of each predictor in predicting the response.

For a given decision tree, we compute predictor importance by summing over all nodes in the tree—the drop (*delta*) in node impurity and expressing these sums relative to the largest sum found over all predictors, i.e., the most important variable.

Detailed technical descriptions of these methods can be found in Friedman (1999a, b) as well as Hastie, Tibshirani, and Friedman (2001); see also Nisbet, Elder, and Miner (2009).

Algorithm results

Summary of R-squared values projected to predict UO at hour 6, 12, and 24

The summaries below were created to evaluate the performance of the respective models to predict unknown UO from hour 1 onwards, and the value of R-squared, as it increases with more learning.

How is R-squared used and interpreted

The *R-squared* value can be interpreted as the proportion of variability around the mean for the dependent variable, that can be accounted for by the respective model. In terms of our Urine Output and the Predictions below, we are showing the accuracy of the Respective different Models, to be able to predict Urine Output at each respective hourly bucket, given the Input (Independent variables we used—in this case Patient Weight, Diuretic Received, Date/Time of Administering the Diuretic, Dose of Diuretic Received, Urine Output Records per patient). Each hourly prediction is dependent only on what we would know for each patient at that slice in time, but given that time bracket, how confident we are to give a clinician the prediction at hour 6. In layman's terms—and as an example below—if we were looking at between hour 3.5 to 4, we are approaching an R-squared of .80 or 80%. This means that at this point, we can explain 80% of the variability in UO with the input variables/data and what the Urine Output at hour 6 is likely to be. Of course, the closer we get to hour 6 and knowing the Urine Output of preceding hours, this figure starts approaching 90%.

To evaluate and investigate the example used above, for any other hours or R-squared values, full details are published below.

Detailed prediction results

Hour 1 results for predicting hour 6 The Cumulative Urine Output for each hourly bucket is used to predict the Cumulative Urine at hour 6, given that we only know the Cumulative Urine Output in the first 30 minutes—1 hour, Patient Weight, Diuretic (Drug Taken) and Dosage Amount.

Table 12.1 shows a model/algorithm comparison at this bucket in time:

TABLE 12.1 Model—algorithm comparisons at hour 1 of cumulative urine output.

Model Quality Summary												
1 Model name	2 Data na	3 Sourc	4 Data us	5 Mean error	6 Absolute mea	7 Mean squared	8 Root Mean squar	9 R squar	10 Adjusted R sc	11 Sum of square	12 Sum Squared Re	13 Valid
1 Cum6-6.5HRS - IndepUO1HR - Boosted Reg		Data		-1.6809	265.047	136589	369.58(	0.51742	0.51742	729526	151172	534
2 Cum6-6.5HRS - IndepUO1HR - Random Fore		Data		0.43777	255.647	132216	363.61	0.53287	0.53287	706168	151172	534
3 Cum6-6.5HRS - IndepUO1HR - C&RT Standa		Data		-0.000000000001	294.684	176534	420.15	0.37629	0.37629	942869	151172	534

The best performing model is the Random Forest Regression with an $R^2 = 53.28\%$ accuracy, meaning, that at hour 1, we can predict the Urine Output at hour 6 with around 53% accuracy.

Table 12.2 shows the predictor importance, which means, by impact size as described in the algorithm details above, in descending order of importance, how each predictor contributes Urine Output:

TABLE 12.2 Predictor importance on how each predictor variable contributes to urine output.

C&RT - Standard			BOOSTED TREE			RANDOM FOREST		

	Predictor importance 1 (v22 Dependent>0) Dependent variable: hour 1-1.5 urine amount Options: Continuous response, Tree number 1			Predictor importance (v22 Dependent>0) Response: hour 1-1.5 urine amount			Predictor importance (v22 Dependent>0) Response: hour 1-1.5 urine amount	
	Variable rank	Importance		Variable Rank	Importance		Variable Rank	Importance
druptaken	100	1.000000	dosage_amt_normalized	100	1.000000	dosage_amt_normalized	100	1.000000
dosage_amt_normalized	93	0.925315	drugtaken	91	0.914077	drugtaken	87	0.874709
input_weight	9	0.088887	input_weight	73	0.729526	input_weight	50	0.499284

For example, for the Random Forest, the Dosage Amount has an importance value of 100, relative to Drug taken at 87 and Input Weight at 50.

Hour 3 results for predicting hour 6 The Cumulative Urine Output for each hourly bucket is used to predict the Cumulative Urine at hour 6, given that we only know the Cumulative Urine Output in the first 30 minutes−3 hours, Patient Weight, Diuretic (Drug Taken) and Dosage Amount.

Table 12.3 shows a model/algorithm comparison at this bucket in time:

TABLE 12.3 Cumulative urine output at hour 3 predicting for hour 6.

	Model Quality Summary												
	1 Model name	2 Data na	3 Sourc	4 Data us	5 Mean error	6 Absolute mea	7 Mean squared	8 Root Mean squar	9 R squar	10 Adjusted R sc	11 Sum of square	12 Sum Squared Re	13 Valid
1	CUM6-6.5HRS - IndepUO3HRs - Boosted Reg	Data			2.4780	170.540	56919.	238.577	0.74035	0.74035	849860	327310	149
2	CUM6-6.5HRS - IndepUO3HRs - C&RT Stand	Data			0.0000000000000	197.98	90393.	300.655	0.58764	0.58764	134966	327310	149
3	CUM6-6.5HRS - IndepUO3HRs - Random For	Data			-1.8263	163.31	56690.	238.098	0.74139	0.74139	846452	327310	149

The best performing model is the Random Forest Regression with an $R^2 = 74.13\%$ accuracy, meaning, that at hour 3, we can predict the Urine Output at hour 6 with around 74% accuracy (Table 12.4).

TABLE 12.4 Predictor importance at hour 3 of urine output predicting for hour 6.

Table 12.4 shows the predictor importance, which means, in descending order of Importance, how each predictor contributes to Urine Output:

For example, for the Random Forest, the Drug Taken has an importance value of 100, relative to Dosage Amount at 65 and Input Weight at 61.

Hour 6 results or predicting hour 12 The Cumulative Urine Output for each hourly bucket is used to predict the Cumulative Urine at hour 12, given that we only know the Cumulative Urine Output from the first 30 minutes−12 hours, Patient Weight, Diuretic (Drug Taken), and Dosage Amount.

Table 12.5 below shows a model/algorithm comparison at hour 6 for predicting hour 12 of Urine Output:

TABLE 12.5 Hour 6 urine output predicting for hour 12.

	Model Quality Summary									
	1 Model name	2 Mean err	3 Absolute mean	4 Mean squared	5 Root Mean square error	6 R square	7 Adjusted R squ	8 Sum of squared	9 Sum Squared Regr	10 Valid
1	CUM12-12.5HRS - IndepUO6HRs - C&RT Standard	127.459	317.702	219028.	468.005	0.563063	0.563063	3592729	8222541	164(
2	CUM12-12.5HRS - IndepUO6HRs - Boosted Regres	-1.11250	266.971	138744.	372.48	0.723220	0.723220	2275833	8222541	164(
3	CUM12-12.5HRS - IndepUO6HRs - Random Forest	-0.472939	286.507	165457.	406.764	0.66993	0.66993	2713996	8222541	164(

The best performing model is the Boosted Regression Tree with an $R^2 = 72.32\%$ accuracy, meaning, that at hour 6, we can predict the Urine Output at hour 12 with around 72% accuracy.

Hour 8 results or predicting hour 12 The Cumulative Urine Output for each hourly bucket is used to predict the Cumulative Urine at hour 12, given that we only know the Cumulative Urine Output from the first 30 minutes−12 hours, Patient Weight, Diuretic (Drug Taken), and Dosage Amount.

Table 12.6 below shows a model/algorithm comparison at hour 6 for predicting hour 12 of Urine Output:

TABLE 12.6 Hour 8 urine output predicting for hour 12.

Model Quality Summary												
1 Model name	2 Data na	3 Sourc	4 Data us	5 Mean er	6 Absolute mea	7 Mean square	8 Root Mean squa	9 R squa	10 Adjusted R s	11 Sum of square	12 Sum Squared Re	13 Valid
1 CUM12-12.5HRS - IndepUO8HRs - C&RT Stan	Data			19.602	241.19	134082	366.17	0.72587	0.72587	241871	882353	180
2 CUM12-12.5HRS - IndepUO8HRs - Boosted Re	Data			0.45322	216.73	95342.	308.77	0.8050	0.8050	171987	882353	180
3 CUM12-12.5HRS - IndepUO8HRs - Random Fc	Data			-2.045	214.20	104692	323.56	0.7859	0.7859	188855	882353	180

The best performing model is the Boosted Regression Tree with an $R^2 = 80.50\%$ accuracy, meaning, that at hour 6, we can predict the Urine Output at hour 12 with around 81% accuracy.

Hour 12 results or predicting hour 24 The Cumulative Urine Output for each hourly bucket is used to predict the Cumulative Urine at hour 24, given that we only know the Cumulative Urine Output from the first 30 minutes—23 hours, Patient Weight, Diuretic (Drug Taken), and Dosage Amount.

Table 12.7 below shows a model/algorithm comparison at hour 12 for predicting hour 24 of Urine Output:

TABLE 12.7 Hour 12 urine output predicting for hour 24.

Model Quality Summary												
1 Model name	2 Data na	3 Sourc	4 Data us	5 Mean error	6 Absolute me	7 Mean squar	8 Root Mean squ	9 R squa	10 Adjusted R	11 Sum of squa	12 Sum Squared R	13 Valid
1 CUMM24HRS - IndepUO12HRs - C&RT	Data			-0.0000000000	628.94	80635	897.97	0.6441	0.6441	276016	775567	34
2 CUMM 24HRS - IndepUO12HRs - Booste	Data			10.594	576.67	61685	785.39	0.7277	0.7277	211148	775567	34
3 CUMM 24HRS - IndepUO12HRs - Randc	Data			-5.689	550.93	59574	771.84	0.7370	0.7370	203924	775567	34

The best performing model is the Random Forest Regression with an $R^2 = 73.70\%$ accuracy, meaning, that at hour 12, we can predict the Urine Output at hour 24 with around 74% accuracy.

Hour 18 results or predicting hour 24 The Cumulative Urine Output for each hourly Bucket is used to predict the Cumulative Urine at hour 24, given that we only know the Cumulative Urine Output from the first 30 minutes—23 hours, Patient Weight, Diuretic (Drug Taken) and Dosage Amount.

Table 12.8 below shows a Model/Algorithm comparison at hour 18 for predicting hour 24 of Urine Output:

TABLE 12.8 Hour 18 urine output predicting for hour 24.

Model Quality Summary												
1 Model name	2 Data nan	3 Source	4 Data usa	5 Mean error	6 Absolute mean	7 Mean squared	8 Root Mean square	9 R squar	10 Adjusted R sq	11 Sum of squared	12 Sum Squared Reg	13 Valid
1 CUMM24HRS - IndepUO18HRs - C&RT Standar	Data			0.0000000000004	508.492	628832	792.989	0.72246	0.72246	2152492	775567	342
2 CUMM 24HRS - IndepUO18HRs - Boosted Regre	Data			4.90174	376.997	269701	519.330	0.88096	0.88096	923188	775567	342
3 CUMM 24HRS - IndepUO18HRs - Random Fores	Data			6.49113	356.729	291926	540.302	0.87115	0.87115	999265	775567	342

The best performing model is the Random Forest Regression with an $R^2 = 88.09\%$ accuracy, meaning, that at hour 18, we can predict the Urine Output at hour 24 with around 88% accuracy.

Algorithm step C: calculation of targets As explained above for Step C for the algorithms above, we were looking to calculate a target (to simulate the "Target" a clinician might set) for us to predict whether a particular patient is "Yes"—likely to reach this target or, "No"—not likely to reach this target, at the respective Urine Output time buckets.

To calculate these "Targets," what we have done is to look at the population of patients available, based on the selection criteria above. Given these criteria and population, we had calculated for each Urine Output Time Bucket, a target based on the 75th and 95th percentiles of Urine Output that our population of patients were likely to reach. We thus had created simulated "Targets" to be able to illustrate how our algorithms may perform in predicting a Yes or No, as per above.

As an example: To calculate the 75th Percentile of Urine Output in mL at hour 3, one would find the Urine Output value which 75% of patients had reached in the research population at hour 3. This would then be used as the simulated "Target" (which would normally be entered by the clinician). We did the same for the 95th percentiles at each hourly time bucket. This gave us the option of using either the 75th or 95th percentiles as Targets, given how sensitive we wanted the model/algorithm to be.

This allowed us to code our data with a Yes or No—to use as an Output/Dependent Variable to train our algorithm for Step D—given the same input variables used in Step A and B (Table 12.9).

TABLE 12.9 Descriptive statistics for the population of patients available for this part of the study.

Variable	Descriptive statistics									
	Valid N	Mean	Median	Minimum	Maximum	Lower quartile	Upper quartile	Percentile 5%	Percentile 95%	Std. Dev.
0–1 Total Cum	28970	27.698	0.000	0.000000	2050.00	0.000	0.000	0.0000	185.000	87.399
0–2 Total Cum	28970	85.281	0.000	0.000000	2500.00	0.000	100.000	0.0000	450.000	178.456
0–3 Total Cum	28970	160.654	25.000	0.000000	3825.00	0.000	230.000	0.0000	680.000	261.258
0–4 Total Cum	28970	247.873	140.000	0.000000	4300.00	0.000	350.000	0.0000	890.000	322.166
0–5 Total Cum	28970	344.039	225.000	0.000000	5100.00	82.000	485.000	0.0000	1085.000	374.531
0–6 Total Cum	28970	446.778	320.000	2.000000	5980.00	150.000	610.000	49.0000	1270.000	416.919
0–7 Total Cum	28970	544.603	405.000	2.000000	7780.00	210.000	740.000	75.0000	1465.000	469.996
0–8 Total Cum	28970	636.260	495.000	2.000000	8030.00	270.000	855.000	105.0000	1630.000	512.738
0–9 Total Cum	28970	732.933	580.000	2.000000	8330.00	325.000	990.000	130.0000	1850.000	566.852
0–10 Total Cum	28970	832.161	675.000	2.000000	9230.00	380.000	1115.000	160.0000	2050.000	621.879
0–11 Total Cum	28970	930.954	760.000	4.000000	9680.00	435.000	1250.000	180.0000	2250.000	679.509
0–12 Total Cum	28970	1034.473	860.000	4.000000	10130.00	500.000	1385.000	210.0000	2460.000	733.582
0–13 Total Cum	28970	1143.945	965.000	4.000000	11450.00	570.000	1530.000	230.0000	2650.000	788.008
0–14 Total Cum	28970	1257.772	1082.500	4.000000	13750.00	645.000	1680.000	260.0000	2860.000	839.829
0–15 Total Cum	28970	1367.942	1190.000	4.000000	15000.00	720.000	1815.000	285.0000	3055.000	893.732
0–16 Total Cum	28970	1472.148	1290.000	4.000000	15000.00	785.000	1950.000	310.0000	3250.000	943.798
0–17 Total Cum	28970	1573.870	1390.000	4.000000	15250.00	850.000	2085.000	330.0000	3450.000	994.878
0–18 Total Cum	28970	1672.839	1480.000	4.000000	15250.00	915.000	2220.000	355.0000	3650.000	1045.873
0–19 Total Cum	28970	1767.429	1565.000	4.000000	15470.00	970.000	2345.000	375.0000	3840.000	1096.555
0–20 Total Cum	28970	1855.344	1650.000	4.000000	15520.00	1023.000	2460.000	395.0000	4025.000	1143.998
0–21 Total Cum	28970	1944.528	1727.500	4.000000	15535.00	1075.000	2580.000	415.0000	4200.000	1194.099

(Continued)

TABLE 12.9 (Continued)

Variable	Descriptive statistics									
	Valid N	Mean	Median	Minimum	Maximum	Lower quartile	Upper quartile	Percentile 5%	Percentile 95%	Std. Dev.
0–22 Total Cum	28970	2032.301	1807.000	4.000000	15575.00	1125.000	2705.000	426.0000	4385.000	1246.947
0–23 Total Cum	28970	2117.930	1882.000	4.000000	15635.00	1172.000	2830.000	445.0000	4575.000	1300.041
0–24 Total Cum	28970	2204.988	1960.000	4.000000	15635.00	1218.000	2948.000	458.0000	4765.000	1353.311

Algorithm step D: predicting if patients would reach targets (yes/no)—target at 75th percentile As described above, here we are looking to predict a Yes or No—for the patient reaching the simulated target calculated above, from a mix of continuous (Numerical) and categorical independent/Input variables.

With criteria described above we decided on the CART, Boosted Trees, and Random Forest models/algorithms—which, given the data structures and sample sizes available in buckets, were performing well.

Predicting a yes/no for hour 24 at hour 6, 12, and 24

Hour 6. Table 12.10 below shows a model/algorithm comparison at hour 6 for predicting hour 24 of Urine Output:

TABLE 12.10 Model–algorithm comparison for predicting a Yes/No for hour 24 based on urine output of hour 6.

Model Quality Summary								
	1 Model name	2 Data name	3 Source	4 Data usage	5 Misclassification error rate	6 Chi-square statistic	7 G-square statistic	8 Valid N
1	6-6.5 HRS - Boosted ClassificationTrees		Data		0.326373626	909.888186	4345.48852	5460
2	6-6.5 HRS - Random Forest Classification		Data		0.338461538	985.415695	4538.33742	5460
3	6-6.5 HRS - C&RT Standard ClassificationTrees		Data		0.385164835	6012.04921	7616.15717	5460

The best performing model is the Boosted Classification Tree with a misclassification error rate of 32.63%, which means that around 67% of the time it predicts a Yes or No accurately versus around 33% predicting it inaccurately.

Hour 12. Table 12.11 below shows a model/algorithm comparison at hour 12 for predicting hour 24 of Urine Output:

TABLE 12.11 Model–algorithm comparison for predicting a Yes/No for hour 24 based on urine output of hour 12.

Model Quality Summary								
	1 Model name	2 Data name	3 Source	4 Data usage	5 Misclassification error rate	6 Chi-square statistic	7 G-square statistic	8 Valid N
1	12-12.5 HRS - Boosted ClassificationTrees		Data		0.366680195	1082.21262	4524.71878	4928
2	12-12.5 HRS - Random Forest Classification		Data		0.355925325	1107.29602	4423.80397	4928
3	12-12.5 HRS - C&RT Standard ClassificationTrees		Data		0.458198052	42408.3847	14142.0819	4928

The best performing model is the Random Forest Classification with a misclassification error rate of 35.59% which means that around 64% of the time it predicts a Yes or No accurately versus around 36% predicting it inaccurately.

Hour 24 +. Table 12.12 below shows a model/algorithm comparison at hour 24 for predicting hour 24 of Urine Output:

TABLE 12.12 Model–algorithm comparison for predicting a Yes/No for hour 24 + based on urine output of hour 24.

Model Quality Summary								
	1 Model name	2 Data name	3 Source	4 Data usage	5 Misclassification error rate	6 Chi-square statistic	7 G-square statistic	8 Valid N
1	24+ HRS - Boosted ClassificationTrees		Data		0.390538429	3366.71197	7870.74695	7039
2	24+ HRS - Random Forest Classification		Data		0.361699105	3083.32135	7278.4868	7039
3	24+ HRS - C&RT Standard ClassificationTrees		Data		0.425770706	33691.1309	16332.4872	7039

Best performing model is the Random Forest Classification with a misclassification error rate of 36.16%, which means that around 64% of the time it predicts a Yes or No accurately versus around 36% predicting it inaccurately (Table 12.13).

The detailed misclassification matrix for each of the above is shown below:

TABLE 12.13 A, B, and C. Detailed Misclassification Matrix for each of the three time intervals for STEP D of the algorithm method used in this paper.

(A)

6 Hours

(B)

12 Hours

(C)

24 Hours

Algorithm step D: predicting if patients would reach targets (yes/no)—target at 95th percentile For Classification Type Models, to evaluate model accuracy, one looks at the misclassification error rate, as well as the confusion matrix, which shows the number and percentage of patients/cases which are correctly and incorrectly classified in each output class. In our case these are either Yes (a patient will reach their Urine Output Target), or No (a patient will not reach their Urine Output Target).

Hour 6, 12, and 24 +

Hour 6. Table 12.14 below shows a model/algorithm comparison at hour 6 for predicting hour 24 of Urine Output:

TABLE 12.14 A model/algorithm comparison at hour 6 for predicting hour 24 of Urine Output.

	Model Quality Summary							
	1 Model name	2 Data name	3 Source	4 Data usage	5 Misclassification error rate	6 Chi-square statistic	7 G-square statistic	8 Valid N
1	6-6.5 HRS - Boosted ClassificationTrees		Data		0.288803089	172.599004	897.344453	1295
2	6-6.5 HRS - Random Forest Classification		Data		0.30965251	194.511862	969.82223	1295
3	6-6.5 HRS - C&RT Standard ClassificationTrees		Data		0.388416988	403.321936	1323.82135	1295

The best performing model is the Boosted Classification Tree with a misclassification error rate of 28.88% which means that around 71% of the time it predicts a Yes or No accurately versus around 29% predicting it inaccurately.

Hour 12. Table 12.15 below shows a model/algorithm comparison at hour 12 for predicting hour 24 of Urine Output:

TABLE 12.15 A model/algorithm comparison at hour 12 for predicting hour 24 of Urine Output.

	Model Quality Summary							
	1 Model name	2 Data name	3 Source	4 Data usage	5 Misclassification error rate	6 Chi-square statistic	7 G-square statistic	8 Valid N
1	12-12.5 HRS - Boosted ClassificationTrees		Data		0.329135181	194.157853	952.09712	1191
2	12-12.5 HRS - Random Forest Classification		Data		0.340890008	249.251967	1018.56329	1191
3	12-12.5 HRS - C&RT Standard ClassificationTrees		Data		0.240134341	169.128169	712.461067	1191

The best performing model is the Standard CART Classification Tree with a misclassification error rate of 24.01% which means that around 76% of the time it predicts a Yes or No accurately versus around 24% predicting it inaccurately.

Hour 24 +. Table 12.16 below shows a model/algorithm comparison at hour 24 for predicting hour 24 of Urine Output:

TABLE 12.16 A model/algorithm comparison at hour 24 for predicting hour 24 of Urine Output.

	Model Quality Summary							
	1 Model name	2 Data name	3 Source	4 Data usage	5 Misclassification error rate	6 Chi-square statistic	7 G-square statistic	8 Valid N
1	24+ HRS - Boosted ClassificationTrees		Data		0.384615385	763.504261	1869.76714	1716
2	24+ HRS - Random Forest Classification		Data		0.367132867	428.146624	1609.34139	1716
3	24+ HRS - C&RT Standard ClassificationTrees		Data		0.264568765	309.701163	1159.03571	1716

The best performing model is the Standard CART Classification Tree with a misclassification error rate of 26.45% which means that around 74% of the time it predicts a Yes or No accurately versus around 26% predicting it inaccurately (Table 12.17).

The Detailed Misclassification Matrix for each of the above is shown below:

TABLE 12.17 Detailed Misclassification Matrix for each of the above Tables 12.14. 15. and 17. is shown below in A, B, and C part of Table 12.17.

(A)

6 Hours

Color maps of predicted category frequencies relative to the total observed class frequency for Target 95th 6-6.5 (325)
Color band of percentages = 0% >0 and <10 % 10% 20% 30% 40% 50% 60% 70% 80% 90% 100%

(Actual)	1 No (Predicted) 6-6.5 HRS - Boosted ClassificationTrees	2 Yes (Predicted) 6-6.5 HRS - Boosted ClassificationTrees	3	4 No (Predicted) 6-6.5 HRS - Random Forest Classification	5 Yes (Predicted) 6-6.5 HRS - Random Forest Classification	6	7 No (Predicted) 6-6.5 HRS - C&RT Standard ClassificationTrees	8 Yes (Predicted) 6-6.5 HRS - C&RT Standard ClassificationTrees
No	78.67%	21.33%		75.12%	24.88%		73.11%	26.89%
Yes	36.42%	63.58%		37.04%	62.96%		50.77%	49.23%
	Overall Accuracy = 71.12%			Overall Accuracy = 69.03%			Overall Accuracy = 61.16%	

(B)

12 Hours

Color maps of predicted category frequencies relative to the total observed class frequency for Target 95th 12-12.5 (380)
Color band of percentages = 0% >0 and <10 % 10% 20% 30% 40% 50% 60% 70% 80% 90% 100%

(Actual)	1 No (Predicted) 12-12.5 HRS - Boosted ClassificationTrees	2 Yes (Predicted) 12-12.5 HRS - Boosted ClassificationTrees	3	4 No (Predicted) 12-12.5 HRS - Random Forest Classification	5 Yes (Predicted) 12-12.5 HRS - Random Forest Classification	6	7 No (Predicted) 12-12.5 HRS - C&RT Standard ClassificationTrees	8 Yes (Predicted) 12-12.5 HRS - C&RT Standard ClassificationTrees
No	69.26%	30.74%		56.25%	43.75%		59.29%	40.71%
Yes	35.06%	64.94%		24.54%	75.46%		7.51%	92.49%
	Overall Accuracy = 67.09%			Overall Accuracy = 65.91%			Overall Accuracy = 75.99%	

(C)

24 Hours

Color maps of predicted category frequencies relative to the total observed class frequency for Target 95th 24+ (4870)
Color band of percentages = 0% >0 and <10 % 10% 20% 30% 40% 50% 60% 70% 80% 90% 100%

(Actual)	1 No (Predicted) 24+ HRS - Boosted ClassificationTrees	2 Yes (Predicted) 24+ HRS - Boosted ClassificationTrees	3	4 No (Predicted) 24+ HRS - Random Forest Classification	5 Yes (Predicted) 24+ HRS - Random Forest Classification	6	7 No (Predicted) 24+ HRS - C&RT Standard ClassificationTrees	8 Yes (Predicted) 24+ HRS - C&RT Standard ClassificationTrees
No	82.15%	17.85%		72.81%	27.19%		55.66%	44.34%
Yes	59.02%	40.98%		46.22%	53.78%		8.61%	91.39%
	Overall Accuracy = 61.54%			Overall Accuracy = 63.29%			Overall Accuracy = 73.54%	

Algorithm accuracy, sensitivity and specificity for the classification trees above

For regression algorithms, model accuracies are expressed as R-squared values, as described above. Of course, the higher the R-squared value, the better the input variables are at explaining the variability in the output variable, in our case, Urine Output. Like traditional regression models, the variable importance values, shows the respective importance (effect) of each input variable on the output variable.

For classification type models, to evaluate model accuracy, one looks at the misclassification rate, as well as the confusion matrix, which shows the number and percentage of patients/cases which are correctly and incorrectly classified in each output class. In our case these are either Yes (a patient will reach their Urine Output Target), or No (a patient will not reach their Urine Output Target).

A further evaluation criteria for classification type models, and specifically to evaluate sensitivity and specificity, an ROC curve can be used to evaluate the goodness of fit for a binary classifier. It is a plot of the true positive rate (rate of events that are correctly predicted as events) against the false positive rate (rate of nonevents predicted to be events) for the different possible cut-points.

An ROC curve demonstrates the following:

1. The trade-off between sensitivity and specificity (any increase in sensitivity will be accompanied by a decrease in specificity).
2. The closer the curve follows the left border and then the top border of the ROC space, the more accurate the test.
3. The closer the curve comes to the 45-degree diagonal of the ROC space, the less accurate the test.

ROC curve for algorithm Step D: predicting if patients would reach targets (yes/no)—target at 75th percentile

ROC curves are illustrated below in Fig. 12.7A−C, with the following Sensitivity/Specificity area under the curve for each at *hour 6*:

1. Boosted Classification Tree—0.726
2. Random Forest Classification—0.712
3. CART—0.641

ROC curves are illustrated below I Fig. 12.8A−C, showing the following Sensitivity/Specificity area under the curve for each, at *hour 12*:

1. Boosted Classification Tree—0.693
2. Random Forest Classification—0.699
3. CART—0.598

ROC curves are illustrated below in Fig. 12.9A−C, showing the following Sensitivity/Specificity area under the curve for each at *hour 24*:

1. Boosted Classification Tree—0.657
2. Random Forest Classification—0.690
3. CART—0.645

ROC curve for algorithm step D: predicting if patients would reach targets (yes/no)—target at 95th percentile

ROC curves illustrated below in Fig. 12.10A−C, shows the following Sensitivity/Specificity area under the curve for each of the three algorithms, respectively at *hour 6*:

1. Boosted Classification Tree—0.768
2. Random Forest Classification—0.752
3. CART—0.668

The ROC Curves in Fig. 12.11A−C as illustrated below, show the following Sensitivity/Specificity area under the curve for each algorithm, at *hour 12*:

1. Boosted Classification Tree—0.745
2. Random Forest Classification—0.721
3. CART—0.772

The ROC Curves illustrated below in Fig. 12.12A−C, show the following Sensitivity/Specificity area under the curve for each, at *hour 24*:

(A)

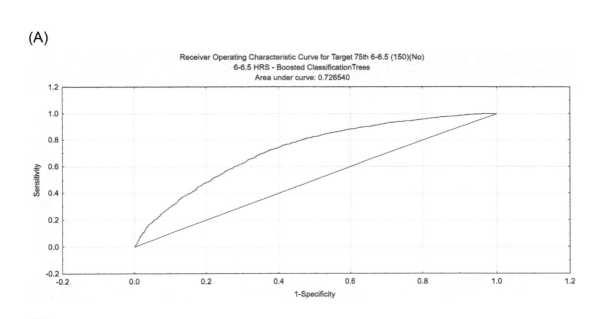

(B)

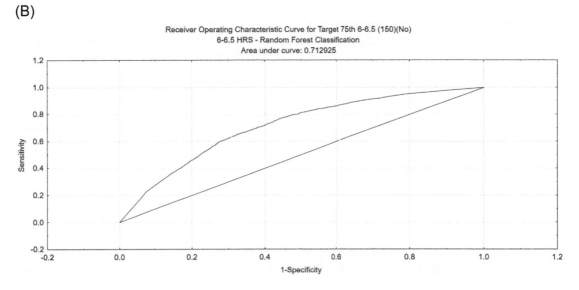

(C)

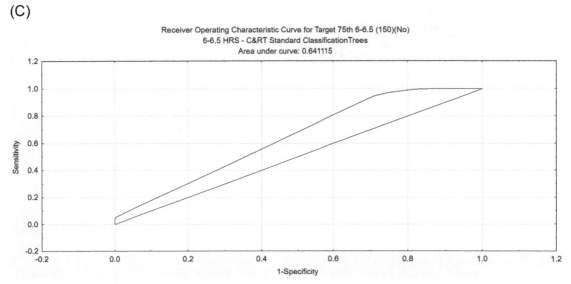

FIGURE 12.7   These figures show the ROC curves predicting if patients would reach targets (yes/no)—target at 95th percentile, showing the Sensitivity/Specificity area under the curve for each of the three algorithms, at hour 6.

(A)

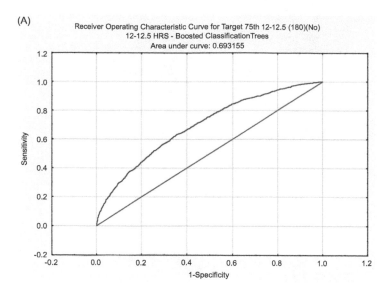

(B)

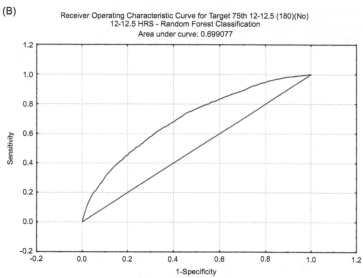

(C)

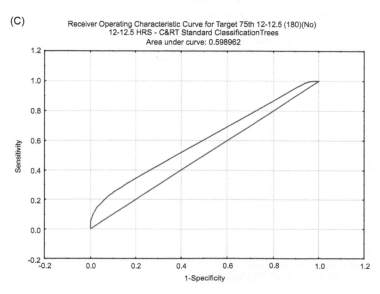

FIGURE 12.8 Sensitivity/Specificity area under the curve for each algorithm, at hour 12.

(A)

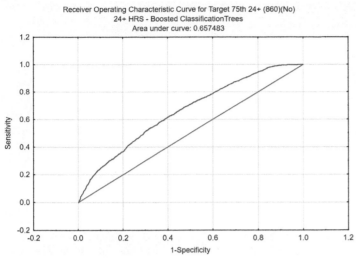

(B)

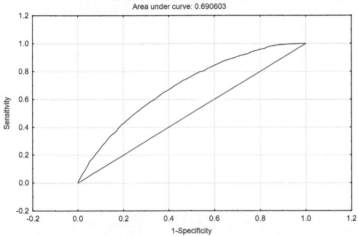

(C)

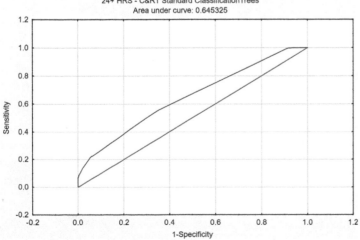

FIGURE 12.9 Sensitivity/Specificity area under the curve for each algorithm at hour 24.

(A)

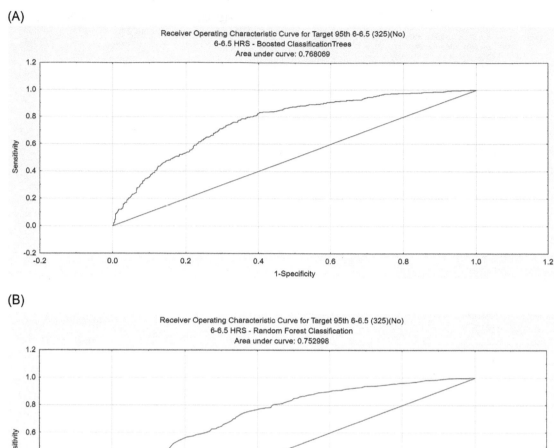

(B)

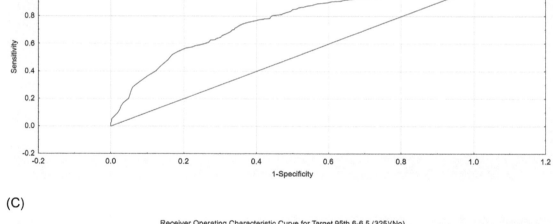

(C)

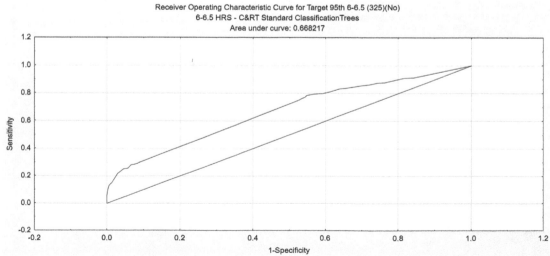

FIGURE 12.10 Predicting if patients would reach targets (yes/no)—target at 95th percentile ROC curves illustrated below in Fig. 12.10AC, shows the following Sensitivity/Specificity area under the curve for each of the three algorithms at hour 6.

(A)

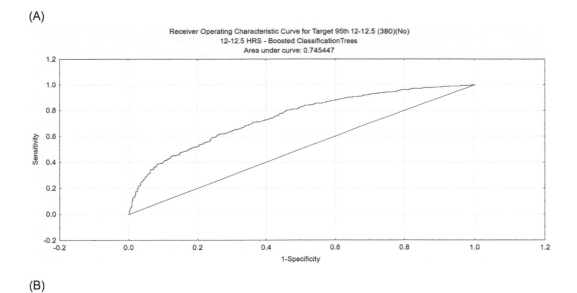

(B)

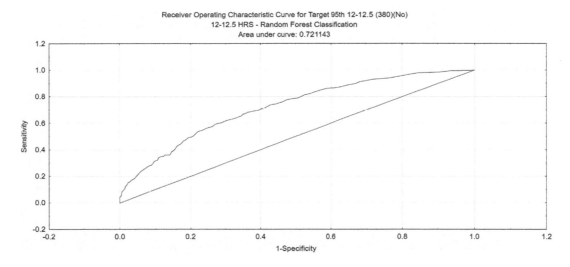

(C)

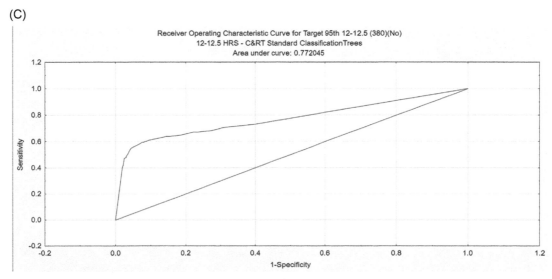

FIGURE 12.11 Sensitivity/Specificity area under the curve for each algorithm, at hour 12.

(A)

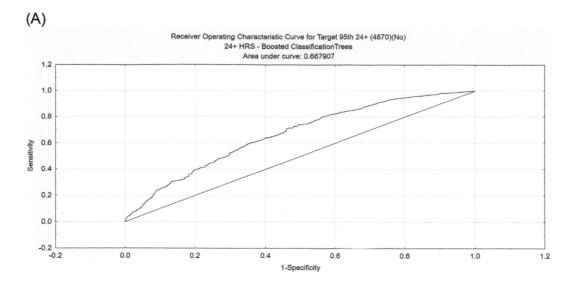

(B)

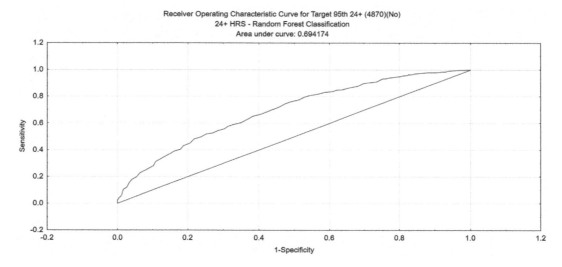

(C)

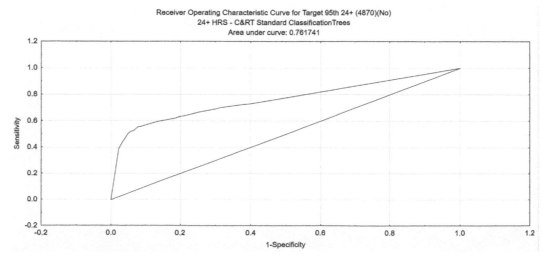

FIGURE 12.12 Sensitivity/Specificity area under the curve for each, at hour 24.

1. Boosted Classification Tree—0.667
2. Random Forest Classification—0.694
3. CART—0.761

Algorithm version 2

Introduction

The image below describes the logic of the algorithm built as Version 2. As opposed to Version 1, in Version 2 we have added all other possibly important variables which could be used in future from other data sources, to see how they could enhance the algorithm. Its logic is depicted and explained below (Fig. 12.13):

1. *Step 1: DATA PREPARATION FOR MODELING*

    Using the criteria as described in the data demographics section above, data was obtained from the MIMIC-IV database using Google Big Query; the following variable were used:

    a. Patient Weight
    b. Diuretic Received
    c. Date/Time of Administering the Diuretic
    d. Dose of Diuretic Received
    e. Urine Output Records per patient
    f. Date/Time of Urine Output

We then used the Date-Time of Urine Output and pivoted this to create Urine Output Values bucketed into hourly increments up to 36 hours, for Demographics and Visualization. For modeling of Urine Output, however, we used cumulative figures per hour for up to 24 hours.

For Dose of Diuretic received, we normalized this variable in two ways:

a. To control for whether a med bolus or IV was used, we applied a formula which normalized for the effect of either, by dividing the Total Dosages into optimal time brackets.

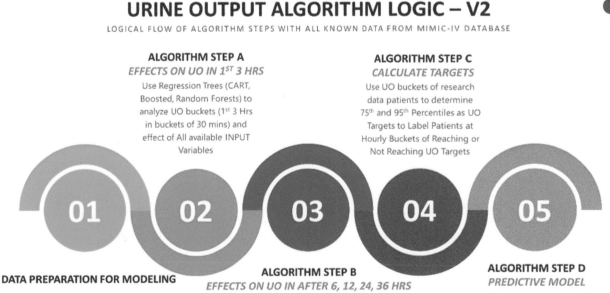

FIGURE 12.13 Version 2 algorithm logic for urine output.

b. To control for how many times a diuretic was administered in previous consecutive hours and the effect of this—we included the number of previous doses into the model. We found it had a negligible effect. We have therefore pooled this into the normalization, as a weighting, in case it might have a small effect for certain patients.

This data set is then augmented by adding the following variables, as shown in Table 12.18:

TABLE 12.18 This data set is then augmented by adding the following variables.

Variable	Descriptive Statistics (Dependent>0 v80>0 and v69>0)					
	Mean	Std.Dev	Minimum	Maximum	N	No.cases Missing
age	67.3725	14.3267	18.0000	91.000	3423	0
PH (dipstick)	6.3006	4.6372	5.0000	104.000	519	2904
Hematocrit (whole blood - calc)	61.1155	42.9807	18.0000	238.000	329	3094
Glucose finger stick (range 70-100)	649.0783	655.2216	3.0000	5316.000	2809	614
Total Bilirubin	3.4234	8.2977	0.0000	145.000	1110	2313
Sodium (whole blood)	238.9167	150.4712	125.0000	944.000	300	3123
Potassium (serum)	7.8629	4.1390	3.0000	45.000	3275	148
Creatinine (serum)	2.5280	2.6420	0.0000	35.000	3271	152

Data were run through a data reduction algorithm called "Feature Selection," which, in large datasets uses a trees regression type techniques, and by calculating an F-value (the size of contribution of each variable to the variance in the dependent variable) as well as a p-value (for what proportion of the population is the F-value significant), we used this to determine the optimum variables to include in our analysis.

The results are shown below in Table 12.19:

TABLE 12.19 "Feature Selection," which is used to to determine the optimum variables to include in predictive modeling data analysis.

	Best predictors for continuous dependent var: hour 24+ urine amount (v68 Dependent>0 - Feature)	
	F-value	p-value
age	15.36840	0.000000
Creatinine (serum)	10.38013	0.000000
drugtaken	8.87780	0.000000
dosage_amt_normalized	8.11271	0.000000
PH (dipstick)	5.49882	0.004133
Hematocrit (whole blood - calc)	3.19225	0.002305
Glucose finger stick (range 70-1	2.85248	0.008898
Total Bilirubin	2.36913	0.037086
Sodium (whole blood)	1.77625	0.182819
Potassium (serum)	1.71651	0.100128

2. *Step 2: ALGORITHM STEP A—EFFECTS ON URINE OUTPUT IN FIRST 3 HOUR*

This used regression trees (Boosted, Random Forests, and Neural Networks) to analyze UO buckets (first 3 hours in Hourly Buckets—Cumulative UO) and the effects of all available input variables. Results of this are shown below. Please note that CART has been left out of this algorithm, as it kept failing with the additional variables.

3. *Step 3: ALGORITHM STEP B—EFFECTS ON URINE OUTPUT AFTER 6, 12, 24, 36 HOURS*

This used regression trees (CART, Boosted, Random Forests) to analyze UO buckets (6, 12, 24, 36 hours in Hourly Buckets—Cumulative UO) and the effects of first 3, 6, and 12 hours cumulative UO, as well as all other available input variables. Results of this are shown below. Please note that CART has been left out of this Algorithm, as it kept failing with the additional variables.

4. *Step 4: ALORITHM STEP C—CALCULATE URINE OUTPUT TARGETS*

   This used hourly urine output buckets of data from patients to determine 75th and 95th percentiles as Urine Output targets to label patients at hourly buckets of reaching or not reaching Urine Output targets. Results of this are shown below.

5. *Step 5: ALGORITHM STEP D—PREDICTIVE MODEL*

This used classification trees (Boosted, Random Forests, and Neural Networks) to predict cumulative UO and reaching target (Y/N) at cumulative UO buckets (6, 12, 24 hours) and the effect of first 3, 6, and 12 hours Cumulative UO, as well as all other available input variables. Results of this are shown below. Please note that CART has been left out of this algorithm, as it kept failing with the additional variables.

Algorithm selection

Algorithms are selected depending on the business/research questions one wants answered, as well as the structure of the data available to answer these questions.

   Selection criteria are the same as previously.

Short description of each algorithm

Algorithm descriptions are done in detail in section "Short description of each algorithm" above.

Algorithm results

Summary of R-squared values projected to predict UO at hour 6, 12, and 24

The summaries below were created to evaluate the performance of the respective models to predict unknown UO from hour 1 onwards, and the value of R-squared, as it increases with more learning.

How is R-squared used and interpreted

The *R-squared* value usage and interpretation is described in detail in section "Algorithm results" above.

   For this model—and as an example below—if we were looking at hour 3, we are at an R-squared of .80 or 80%. This means that at this point, we can explain and predict 80% of the variability with the input data available and what the Urine Output at hour 6 is likely to be. Of course, the closer we get to hour 6 and knowing the Urine Output of preceding hours, this passes 90%.

   To evaluate and investigate the example used above, for any other hours or R-squared values, see the full details presented below in Figs. 12.14—12.16.

Detailed prediction results

Hour 1 results for predicting hour 6 The Cumulative Urine Output for each hourly bucket is used to predict the Cumulative Urine at hour 6, given that we know the Cumulative Urine Output in the first 30 minutes— 1 hour, Patient Weight, Diuretic (Drug Taken), and Dosage Amount as well as all the added variables outlined above.

   Table 12.20 shows a model/algorithm comparison at this bucket of time:

TABLE 12.20 Cumulative Urine Output for each hourly bucket is used to predict the Cumulative Urine at hour 6.

Model Quality Summary												
1	2	3	4	5	6	7	8	9	10	11	12	13
Model name	Data na	Sourc	Data us	Mean er	Absolute mea	Mean square	Root Mean squa	R squa	Adjusted R s	Sum of squar	Sum Squared R	Valid
1 Cum6-6.5HRS - IndepUO1HR - Boosted R		Data		5.2628	258.74	129806	360.28	0.64159	0.64159	541810	151172	41
2 Cum6-6.5HRS - IndepUO1HR - C&RT Sta		Data		6.996113	226.21	100511	317.03	0.99913	0.99913	130668	151172	1
3 Cum6-6.5HRS - IndepUO1HR - Random F		Data		5.972	248.27	125210	353.85	0.6542	0.6542	522629	151172	41

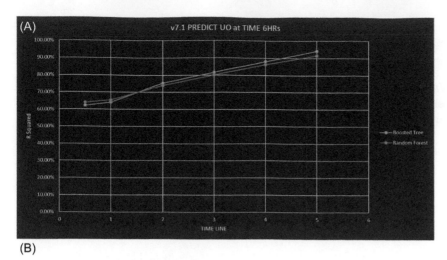

(A)

(B)

FIGURE 12.14 A and B. R-squared predicting hour 6, hourly after 1–6 hours of data collected.

DEPENDENT CUM 6HRs UO Summary

Basic V plus Independent Variables below			v7.1
TIME	max time interval	Method	R^2
0-0.5HRs	0.5	Boosted Tree	62.21%
	0.5	Random Forest	63.95%
until 1-1.5HRs	1	Boosted Tree	64.16%
	1	Random Forest	65.43%
until 2-2.5HRs	2	Boosted Tree	75.45%
	2	Random Forest	74.10%
until 3-3.5HRs	3	Boosted Tree	81.60%
	3	Random Forest	80.10%
until 4-4.5HRs	4	Boosted Tree	88.00%
	4	Random Forest	86.30%
until 5-5.5HRs	5	Boosted Tree	94.17%
	5	CART	99.97%
	5	Random Forest	91.55%

The best performing model is the Random Forest Regression with an $R^2 = 65.42\%$ accuracy, meaning, that at hour 1, we can predict the Urine Output at hour 6 with around 65% accuracy.

Table 12.21 shows the predictor importance, which means, in descending order of Importance, how each predictor contributes to Urine Output:

TABLE 12.21 The predictor importance, in descending order of Importance, showing the rank of how each predictor variable contributes to Urine Output.

BOOSTED TREE

	Predictor importance (Depe Response: 0-6 Total Cumm	
	Variable Rank	Importance
0-1 Total Cumm	100	1.000000
0-0.5 Total Cumm	69	0.691868
drugtaken	69	0.687525
dosage_amt_normalized	37	0.369611
age	36	0.362291
input_weight	30	0.298237
Creatinine (serum)	14	0.142913
Potassium (serum)	14	0.142040
Glucose finger stick (range 70-100)	9	0.087538

RANDOM FOREST

	Predictor importance (De Response: 0-6 Total Cum	
	Variable Rank	Importance
0-1 Total Cumm	100	1.000000
0-0.5 Total Cumm	52	0.523020
drugtaken	36	0.355534
dosage_amt_normalized	22	0.219823
age	14	0.135544
input_weight	10	0.098525
Potassium (serum)	6	0.064982
Creatinine (serum)	6	0.061756
Glucose finger stick (range 70-100)	4	0.039755

(A)

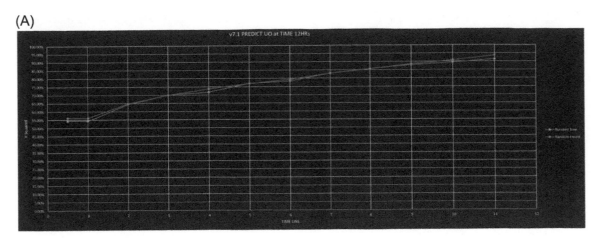

(B)

<u>DEPENDENT</u>   **CUM 12HRs UO Summary**

Basic V plus Independent Variables below			v7.1
TIME	max time interval	Method	R²
0-0.5HRs	0.5	Boosted Tree	54.28%
	0.5	Random Forest	56.12%
until 1-1.5HRs	1	Boosted Tree	54.34%
	1	Random Forest	56.10%
until 2-2.5HRs	2	Boosted Tree	64.46%
	2	Random Forest	64.84%
until 3-3.5HRs	3	Boosted Tree	69.98%
	3	Random Forest	69.94%
until 4-4.5HRs	4	Boosted Tree	71.79%
	4	Random Forest	73.69%
until 5-5.5HRs	5	Boosted Tree	77.11%
	5	Random Forest	76.94%
until 6-6.5HRs	6	Boosted Tree	78.37%
	6	Random Forest	79.35%
until 7-7.5HRs	7	Boosted Tree	83.01%
	7	Random Forest	83.11%
until 8-8.5HRs	8	Boosted Tree	85.71%
	8	Random Forest	85.61%
until 9-9.5HRs	9	Boosted Tree	87.99%
	9	Random Forest	88.51%
until 10-10.5HRs	10	Boosted Tree	89.46%
	10	Random Forest	90.77%
until 11-11.5HRs	11	Boosted Tree	91.36%
	11	Random Forest	93.69%

FIGURE 12.15   A and B. R-squared predicting hour 12, hourly after 1−12 hours of data collected.

Hour 3 results for predicting hour 6 The Cumulative Urine Output for each hourly bucket is used to predict the Cumulative Urine at hour 6, given that we only know the Cumulative Urine Output in the first 30 minutes−3 hours, Patient Weight, Diuretic (Drug Taken), and Dosage Amount, as well as all the added variables outlined above.

Table 12.22 below shows a model/algorithm comparison at this Bucket in Time:

(A)

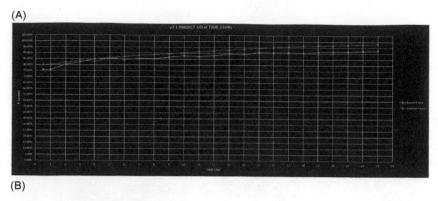

FIGURE 12.16 A and B. R-squared predicting hour 24, hourly after 1–24 hours of data collected.

(B)

DEPENDENT **CUM 24HRs UO Summary**

	Basic V plus Independent Variables below		v7.1	
Independent	TIME	max time interval	Method	R^2
	0-0.5HRs	0.5	Boosted Tree	79.86%
		0.5	Random Forest	80.18%
	until 1-1.5HRs	1	Boosted Tree	79.89%
		1	Random Forest	80.29%
	until 2-2.5HRs	2	Boosted Tree	82.25%
		2	Random Forest	81.95%
	until 3-3.5HRs	3	Boosted Tree	83.98%
		3	Random Forest	83.77%
	until 4-4.5HRs	4	Boosted Tree	84.77%
		4	Random Forest	84.98%
	until 5-5.5HRs	5	Boosted Tree	85.91%
		5	Random Forest	85.94%
	until 6-6.5HRs	6	Boosted Tree	87.08%
		6	Random Forest	86.97%
	until 7-7.5HRs	7	Boosted Tree	88.05%
		7	Random Forest	87.99%
	until 8-8.5HRs	8	Boosted Tree	88.74%
		8	Random Forest	88.68%
	until 9-9.5HRs	9	Boosted Tree	89.67%
		9	Random Forest	89.54%
	until 10-10.5HRs	10	Boosted Tree	90.87%
		10	Random Forest	90.43%
	until 11-11.5HRs	11	Boosted Tree	91.46%
		11	Random Forest	91.40%
	until 12-12.5HRs	12	Boosted Tree	92.46%
		12	Random Forest	92.05%
	until 13-13.5HRs	13	Boosted Tree	93.38%
		13	Random Forest	92.83%
	until 14-14.5HRs	14	Boosted Tree	94.10%
		14	Random Forest	93.68%
	until 15-15.5HRs	15	Boosted Tree	95.23%
		15	Random Forest	94.40%
	until 16-16.5HRs	16	Boosted Tree	95.80%
		16	Random Forest	95.05%
	until 17-17.5HRs	17	Boosted Tree	96.33%
		17	Random Forest	95.71%
	until 18-18.5HRs	18	Boosted Tree	96.92%
		18	Random Forest	96.28%
	until 19-19.5HRs	19	Boosted Tree	97.19%
		19	Random Forest	96.65%
	until 20-20.5HRs	20	Boosted Tree	97.38%
		20	Random Forest	96.90%
	until 21-21.5HRs	21	Boosted Tree	97.30%
		21	Random Forest	97.26%
	until 22-22.5HRs	22	Boosted Tree	97.84%
		22	Random Forest	97.73%
	until 23-23.5HRs	23	Boosted Tree	98.23%
		23	Random Forest	97.99%

__IMG0__

TABLE 12.22 The Cumulative Urine Output for each hourly bucket is used to predict the Cumulative Urine at hour 6, given that we only know the Cumulative Urine Output between the first 30 minutes to 3 hours.

Model Quality Summary												
1 Model name	2 Data n	3 Sour	4 Data u	5 Mean erro	6 Absolute me	7 Mean squa	8 Root Mean sq	9 R squ	10 Adjusted R	11 Sum of squa	12 Sum Squared	13 Vali
1 CUM6-6.5HRS - IndepUO3HRs - Boos		Data		3.347	164.4(	5247(	229.0	0.8160	0.8160	6021(	32731(	114
2 CUM6-6.5HRS - IndepUO3HRs - C&R		Data		-0.0000000000	84.99	13244	115.0	0.9998	0.9998	42383	32731(	3
3 CUM6-6.5HRS - IndepUO3HRs - Rand		Data		2.152	164.3	56763	238.2	0.800	0.800	6513(	32731(	114

The best performing model is the Misclassification with an $R^2 = 81.60\%$ accuracy, meaning, that at hour 3, we can predict the Urine Output at hour 6 with around 82% accuracy.

Table 12.23 below, shows the predictor importance, which means, in descending order of importance, how much each predictor contributes Urine Output:

TABLE 12.23 This table of data analysis output shows how each predictor variable contributes to Urine Output for Boosted Trees and Random Forest algorithm, at the 3-6 hour interval.

Hr 3 - 6HOUR

BOOSTED TREE

	Predictor importance (De Response: 0-6 Total Cum	
	Variable Rank	Importance
0-3 Total Cumm	100	1.000000
0-2 Total Cumm	65	0.652084
0-1 Total Cumm	39	0.394945
drugtaken	36	0.362915
0-0.5 Total Cumm	25	0.250242
dosage_amt_normalized	19	0.186704
input_weight	15	0.145874
age	12	0.117339
Potassium (serum)	10	0.096055
Creatinine (serum)	8	0.077767
Glucose finger stick (range 70-100)	3	0.026672

RANDOM FOREST

	Predictor importance (De Response: 0-6 Total Cum	
	Variable Rank	Importance
0-3 Total Cumm	100	1.000000
0-2 Total Cumm	60	0.598955
0-1 Total Cumm	21	0.209328
drugtaken	15	0.149146
0-0.5 Total Cumm	12	0.120159
dosage_amt_normalized	6	0.055003
age	5	0.047810
input_weight	3	0.030092
Potassium (serum)	2	0.019764
Creatinine (serum)	2	0.017137
Glucose finger stick (range 70-100)	1	0.012283

For example, as seen in Table 12.23 for the Boosted Tree Regression, the 0−3 hours Cumulative Urine Output has an importance value of 100, relative to Drug/Diuretic Administered (drugtaken) at 36 and Input Weight at 15.

Hour 6 results or predicting hour 12 The Cumulative Urine Output for each hourly bucket is used to predict the Cumulative Urine at hour 12, given that we only know the Cumulative Urine Output from the first 30 minutes−12 hours, Patient Weight, Diuretic (Drug Taken), and Dosage Amount, as well as all the added variables outlined above.

Table 12.24 below shows a model/algorithm comparison at hour 6 for predicting hour 12 of urine output:

TABLE 12.24 Data analysis output that show that how the Cumulative Urine Output for each hourly bucket is used to predict the Cumulative Urine at hour 12, given that we only know the Cumulative Urine Output from the first 30 minutes 12 hours.

Model Quality Summary												
1 Model name	2 Data r	3 Sour	4 Data u	5 Mean e	6 Absolute m	7 Mean squa	8 Root Mean s	9 R squ	10 Adjusted R	11 Sum of squ	12 Sum Squared	13 Vali
1 CUM12-12.5HRS - IndepUO6HRs - B		Data		9.716	321.3	20424	451.9	0.783(	0.783(	5573	25763	27
2 CUM12-12.5HRS - IndepUO6HRs - C		Data		6.77283	374.3	3086(	555.5	0.893(	0.893(	2756(	25763	89
3 CUM12-12.5HRS - IndepUO6HRs - R		Data		19.41	293.8	1949(	441.5	0.793	0.793	5320	25763	27

The best performing model is the Random Forest Regression with an $R^2 = 79.34\%$ accuracy, meaning, that at hour 6, we can predict the Urine Output at hour 12 with around 79% accuracy.

Hour 8 results or predicting hour 12 The Cumulative Urine Output for each hourly bucket is used to predict the Cumulative Urine at hour 12, given that we only know the Cumulative Urine Output from the first 30 minutes—12 hours, Patient Weight, Diuretic (Drug Taken), and Dosage Amount, as well as all the added variables outlined above.

Table 12.25 below shows a model/algorithm comparison at hour 6 for predicting hour 12 of Urine Output:

TABLE 12.25 The model/algorithm comparison at hour 6 for predicting hour 12 of Urine Output, based on the Cumulative Urine Output from 30 minutes through 12 hours Boosted Trees, Classification & Regression Trees, and Random Forests are the 3 algorithms compared.

| Model Quality Summary | | | | | | | | | | | | |
1 Model name	2 Data n	3 Sourc	4 Data u:	5 Mean error	6 Absolute me	7 Mean squar	8 Root Mean squ	9 R squa	10 Adjusted R	11 Sum of squa	12 Sum Squared	13 Valid
1CUM12-12.5HRS - IndepUO8HRs - Boo		Data		11.757	255.22	13490	367.29	0.8571	0.8571	36815	25763·	27
2CUM12-12.5HRS - IndepUO8HRs - C&F		Data		-0.0000000000	341.35	27981	528.97	0.9030	0.9030	24987	25763·	89
3CUM12-12.5HRS - IndepUO8HRs - Ran		Data		1.049?	238.04	13584	368.56	0.8561	0.8561	37071	25763·	27

The best performing model is the Boosted Regression Tree with an $R^2 = 85.71\%$ accuracy, meaning, that at hour 6, we can predict the Urine Output at hour 12 with around 86% accuracy.

Hour 12 results or predicting hour 24 The Cumulative Urine Output for each hourly bucket is used to predict the Cumulative Urine at hour 24, given that we only know the Cumulative Urine Output from the first 30 minutes—23 hours, Patient Weight, Diuretic (Drug Taken), and Dosage Amount, as well as all the added variables outlined above.

Table 12.26 below shows a model/algorithm comparison at hour 12 for predicting hour 24 of Urine Output rst 30 minutes through 23 hours:

TABLE 12.26 This software data analytic output shows a model/algorithm comparison at hour 12 for predicting hour 24 of Urine Output, based on the known Cumulative Urine Output from the from the first 30 minutes through hour 23; Boosted Trees, Classification & Regression Trees, and Random Forests algorithms are compared.

| Model Quality Summary | | | | | | | | | | | | |
1 Model name	2 Data n	3 Sourc	4 Data us	5 Mean error	6 Absolute me	7 Mean squar	8 Root Mean squ	9 R squa	10 Adjusted R	11 Sum of squa	12 Sum Squared I	13 Valid
1CUMM 24HRS - IndepUO12HRs - Boos		Data		15.249	556.48	56023	748.49	0.8028	0.8028	152888	775567	27
2CUMM 24HRS - IndepUO12HRs - C&R		Data		-0.0000000000	688.36	10420	1020.8	0.8800	0.8800	93058	775567	89
3CUMM 24HRS - IndepUO12HRs - Ran(		Data		31.643	547.18	56644	752.62	0.8006	0.8006	154584	775567	27

The best performing model is the Boosted Regression Tree with an $R^2 = 80.28\%$ accuracy, meaning, that at hour 12, we can predict the Urine Output at hour 24 with around 80% accuracy.

Hour 18 results or predicting hour 24 The Cumulative Urine Output for each Hourly Bucket is used to predict the Cumulative Urine at hour 24, given that we only know the Cumulative Urine Output from the first 30 minutes—23 hours, Patient Weight, Diuretic (Drug Taken), and Dosage Amount, as well as all the Added Variables outlined above.

Table 12.27 below shows a model/algorithm comparison at hour 18 for predicting hour 24 of Urine Output:

TABLE 12.27 STATISTICA Predictive Analytic Data Analysis software output that shows a model/algorithm comparison at hour 18 for predicting hour 24 of Urine Output.

| Model Quality Summary | | | | | | | | | | | | |
1 Model name	2 Data na	3 Sourc	4 Data us	5 Mean error	6 Absolute mea	7 Mean square	8 Root Mean squa	9 R squa	10 Adjusted R s	11 Sum of squar	12 Sum Squared R	13 Valid
1CUMM 24HRS - IndepUO18HRs - Boosted		Data		4.0831	362.74	238274	488.13	0.91615	0.91615	65025	775567	27?
2CUMM 24HRS - IndepUO18HRs - C&RT S(		Data		-0.0000000000	572.15	805057	897.24	0.90730	0.90730	718910	775567	89
3CUMM 24HRS - IndepUO18HRs - Random		Data		4.3689	355.70	267204	516.91	0.90597	0.90597	729200	775567	27?

The best performing model is the Boosted Regression Tree with an $R^2 = 91.61\%$ accuracy, meaning, that at hour 18, we can predict the Urine Output at hour 24 with around 92% accuracy.

Predictor importance for the above Table 12.28 shows the predictor importance, which means, in descending order of Importance, how each predictor contributes to Urine Output:

TABLE 12.28 Predictor importance variables.

HR6 - 24 HRS

BOOSTED TREE			RANDOM FOREST		
Predictor importance (Dep Response: 0-24 Total Cumm			Predictor importance (Dep Response: 0-24 Total Cum		
	Variable Rank	Importance		Variable Rank	Importance
0-6 Total Cumm	100	1.000000	0-5 Total Cumm	100	1.000000
0-5 Total Cumm	87	0.869524	0-6 Total Cumm	94	0.943660
0-4 Total Cumm	83	0.825178	0-4 Total Cumm	73	0.731108
0-3 Total Cumm	73	0.730778	0-2 Total Cumm	66	0.656528
0-2 Total Cumm	61	0.607925	0-3 Total Cumm	61	0.612991
0-0.5 Total Cumm	56	0.563408	0-0.5 Total Cumm	43	0.426719
0-1 Total Cumm	56	0.563051	0-1 Total Cumm	36	0.363029
input_weight	46	0.459737	drugtaken	24	0.242114
age	39	0.390248	age	20	0.195631
dosage_amt_normalized	36	0.358288	input_weight	18	0.184183
Potassium (serum)	33	0.332430	dosage_amt_normalized	13	0.129251
drugtaken	32	0.318831	Potassium (serum)	10	0.103471
Creatinine (serum)	17	0.165346	Creatinine (serum)	9	0.090629
Glucose finger stick (range 70-100)	16	0.164083	Glucose finger stick (range 70-100)	8	0.081399

H12 - 24 HRS

BOOSTED TREE			RANDOM FOREST		
Predictor importance (De Response: 0-24 Total Cun			Predictor importance (De Response: 0-24 Total Cun		
	Variable Rank	Importance		Variable Rank	Importance
0-12 Total Cumm	100	1.000000	0-10 Total Cumm	100	1.000000
0-11 Total Cumm	89	0.888054	0-12 Total Cumm	91	0.909618
0-10 Total Cumm	82	0.821879	0-7 Total Cumm	77	0.774440
0-9 Total Cumm	78	0.777482	0-9 Total Cumm	77	0.770934
0-8 Total Cumm	74	0.738860	0-11 Total Cumm	76	0.759023
0-7 Total Cumm	69	0.694912	0-8 Total Cumm	70	0.695983
0-6 Total Cumm	66	0.659795	0-6 Total Cumm	63	0.628038
0-5 Total Cumm	63	0.626614	0-4 Total Cumm	60	0.598712
0-4 Total Cumm	62	0.624307	0-5 Total Cumm	59	0.590425
0-3 Total Cumm	58	0.581038	0-3 Total Cumm	39	0.385731
0-2 Total Cumm	52	0.517658	0-0.5 Total Cumm	32	0.320494
0-0.5 Total Cumm	39	0.393952	0-2 Total Cumm	31	0.313986
0-1 Total Cumm	39	0.390021	0-1 Total Cumm	31	0.308614
input_weight	27	0.274652	drugtaken	13	0.125581
age	24	0.242355	age	11	0.109540
Potassium (serum)	22	0.217430	input_weight	9	0.090463
drugtaken	20	0.201018	dosage_amt_normalized	6	0.059878
dosage_amt_normalized	12	0.119213	Potassium (serum)	5	0.053867
Glucose finger stick (range 70-100)	11	0.113730	Creatinine (serum)	4	0.043083
Creatinine (serum)	11	0.109429	Glucose finger stick (range 70-100)	4	0.037183

HR18 -24HRS

BOOSTED TREE			RANDOM FOREST		
Predictor importance (De Response: 0-24 Total Cun			Predictor importance (De Response: 0-24 Total Cun		
	Variable Rank	Importance		Variable Rank	Importance
0-18 Total Cumm	100	1.000000	0-16 Total Cumm	100	1.000000
0-17 Total Cumm	95	0.948909	0-18 Total Cumm	86	0.862209
0-16 Total Cumm	90	0.899338	0-15 Total Cumm	84	0.839522
0-15 Total Cumm	86	0.856107	0-17 Total Cumm	77	0.774794
0-14 Total Cumm	82	0.819330	0-14 Total Cumm	59	0.591511
0-13 Total Cumm	77	0.770138	0-11 Total Cumm	57	0.573587
0-12 Total Cumm	72	0.723617	0-8 Total Cumm	56	0.564574
0-11 Total Cumm	70	0.697386	0-12 Total Cumm	55	0.553216
0-10 Total Cumm	66	0.659395	0-4 Total Cumm	51	0.512893
0-9 Total Cumm	63	0.628710	0-13 Total Cumm	51	0.512399
0-8 Total Cumm	61	0.606444	0-9 Total Cumm	50	0.498971
0-7 Total Cumm	57	0.570147	0-10 Total Cumm	49	0.486281
0-6 Total Cumm	54	0.535142	0-7 Total Cumm	48	0.476386
0-5 Total Cumm	51	0.514473	0-5 Total Cumm	47	0.472188
0-4 Total Cumm	50	0.498170	0-3 Total Cumm	42	0.419576
0-3 Total Cumm	47	0.473829	0-6 Total Cumm	41	0.409776
0-2 Total Cumm	43	0.426063	0-2 Total Cumm	31	0.308136
0-0.5 Total Cumm	33	0.331216	0-0.5 Total Cumm	23	0.232739
0-1 Total Cumm	33	0.330769	0-1 Total Cumm	20	0.200290
input_weight	13	0.126633	drugtaken	9	0.094736
age	11	0.110022	input_weight	8	0.080525
drugtaken	9	0.093413	age	6	0.055412
Potassium (serum)	8	0.076925	dosage_amt_normalized	4	0.036868
dosage_amt_normalized	7	0.070622	Potassium (serum)	4	0.036602
Glucose finger stick (range 70-100)	5	0.047681	Creatinine (serum)	2	0.020519
Creatinine (serum)	4	0.043588	Glucose finger stick (range 70-100)	2	0.019921

Algorithm step C: calculation of targets As explained above for Step C for the algorithms above, we were looking to calculate a target (to simulate the "Target" a clinician might set) for us to predict whether a particular patient is "Yes"—likely to reach this target, or "No"—not likely to reach this target at the respective Urine Output time buckets.

To calculate these "Targets," we have looked at the population of patients available, based on the selection criteria we defined above. Given these criteria and population, we had calculated for each Urine Output time bucket, a target based on the 75th and 95th Percentiles of Urine Output that our population of patients were likely to reach. We thus had created simulated "Targets" to be able to illustrate how our algorithms may perform in predicting a Yes or No as per above.

As an example: To calculate the 75th percentile of Urine Output in mL at hour 3, one would find the Urine Output Value which 75% of patients had reached in the research population at hour 3. This would then be used as the simulated "Target" (which would normally be entered by the clinician). We did the same for the 95th Percentiles at each hourly time bucket. This gave us the option of using either the 75th or 95th percentiles as targets, depending on how sensitive we wanted the model/algorithm to be.

This allowed us to code our data with a Yes or No—to use as an output/dependent variable to train our algorithm for Step D—given the same input variables used in Step A and B.

Algorithm step D: predicting if patients would reach targets (yes/no)—target at 75th percentile As described above, we are looking to predict a Yes or No—for the Patient reaching the simulated target calculated from a mix of continuous (Numerical) and categorical Independent/Input variables.

With the same criteria described above we decided on the CART, Boosted Trees, and Random Forest models/algorithms, which, given the data structures and sample sizes available in buckets, were performing well (Table 12.29).

TABLE 12.29 Given the data structures and sample sizes available in buckets, CART, Boosted Trees, and Random Forest models/algorithms were performing well, as indicated by the 'Descriptive Statistics' provided in this table.

Variable	Descriptive statistics									
	Valid N	Mean	Median	Minimum	Maximum	Lower quartile	Upper quartile	Percentile 5%	Percentile 95%	Std. Dev.
0–1 Total Cum	28970	27.698	0.000	0.000000	2050.00	0.000	0.000	0.0000	185.000	87.399
0–2 Total Cum	28970	85.281	0.000	0.000000	2500.00	0.000	100.000	0.0000	450.000	178.456
0–3 Total Cum	28970	160.654	25.000	0.000000	3825.00	0.000	230.000	0.0000	680.000	261.258
0–4 Total Cum	28970	247.873	140.000	0.000000	4300.00	0.000	350.000	0.0000	890.000	322.166
0–5 Total Cum	28970	344.039	225.000	0.000000	5100.00	82.000	485.000	0.0000	1085.000	374.531
0–6 Total Cum	28970	446.778	320.000	2.000000	5980.00	150.000	610.000	49.0000	1270.000	416.919
0–7 Total Cum	28970	544.603	405.000	2.000000	7780.00	210.000	740.000	75.0000	1465.000	469.996
0–8 Total Cum	28970	636.260	495.000	2.000000	8030.00	270.000	855.000	105.0000	1630.000	512.738

(Continued)

TABLE 12.29 (Continued)

Variable	Descriptive statistics									
	Valid N	Mean	Median	Minimum	Maximum	Lower quartile	Upper quartile	Percentile 5%	Percentile 95%	Std. Dev.
0–9 Total Cum	28970	732.933	580.000	2.000000	8330.00	325.000	990.000	130.0000	1850.000	566.852
0–10 Total Cum	28970	832.161	675.000	2.000000	9230.00	380.000	1115.000	160.0000	2050.000	621.879
0–11 Total Cum	28970	930.954	760.000	4.000000	9680.00	435.000	1250.000	180.0000	2250.000	679.509
0–12 Total Cum	28970	1034.473	860.000	4.000000	10130.00	500.000	1385.000	210.0000	2460.000	733.582
0–13 Total Cum	28970	1143.945	965.000	4.000000	11450.00	570.000	1530.000	230.0000	2650.000	788.008
0–14 Total Cum	28970	1257.772	1082.500	4.000000	13750.00	645.000	1680.000	260.0000	2860.000	839.829
0–15 Total Cum	28970	1367.942	1190.000	4.000000	15000.00	720.000	1815.000	285.0000	3055.000	893.732
0–16 Total Cum	28970	1472.148	1290.000	4.000000	15000.00	785.000	1950.000	310.0000	3250.000	943.798
0–17 Total Cum	28970	1573.870	1390.000	4.000000	15250.00	850.000	2085.000	330.0000	3450.000	994.878
0–18 Total Cum	28970	1672.839	1480.000	4.000000	15250.00	915.000	2220.000	355.0000	3650.000	1045.873
0–19 Total Cum	28970	1767.429	1565.000	4.000000	15470.00	970.000	2345.000	375.0000	3840.000	1096.555
0–20 Total Cum	28970	1855.344	1650.000	4.000000	15520.00	1023.000	2460.000	395.0000	4025.000	1143.998
0–21 Total Cum	28970	1944.528	1727.500	4.000000	15535.00	1075.000	2580.000	415.0000	4200.000	1194.099
0–22 Total Cum	28970	2032.301	1807.000	4.000000	15575.00	1125.000	2705.000	426.0000	4385.000	1246.947
0–23 Total Cum	28970	2117.930	1882.000	4.000000	15635.00	1172.000	2830.000	445.0000	4575.000	1300.041
0–24 Total Cum	28970	2204.988	1960.000	4.000000	15635.00	1218.000	2948.000	458.0000	4765.000	1353.311

Predicting a yes/no for hour 24 at hour 6, 12, and 24 *The Detailed Misclassification Matrix for each of the above is shown below* (Table 12.30):

TABLE 12.30 The Detailed Misclassification Matrix for each of the algorithms is shown in these tables.

(A)

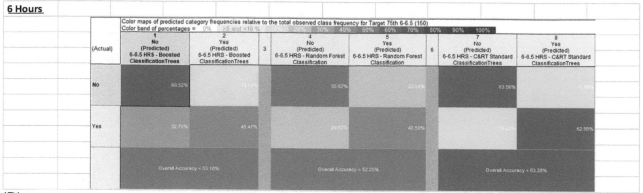

6 Hours

(B)

12 Hours

(C)

24 Hours

Algorithm step D: predicting if patients would reach targets (yes/no)—target at 95th percentile For classification type models, to evaluate Model Accuracy, one looks at the misclassification error rate, as well as the confusion matrix, which shows the number and percentage of patients/cases which are correctly and incorrectly classified in each output class. In our case these are either Yes (a patient will reach their Urine Output Target), or No (a patient will not reach their Urine Output Target).

Predicting a yes/no for hour 24 at hour 6, 12, and 24 *The Detailed Misclassification Matrix for each of the above is shown below* (Table 12.31):

TABLE 12.31 The Detailed Misclassification Matrix for predicting Yes/No at hour 24 for each of the 6, 12, and 24 hour cumulative buckets, is shown in these tables.

(A)

(B)

(C)

Algorithm accuracy, sensitivity, and specificity for the classification trees above

As described above, to evaluate model accuracy for classification type models, one looks at the misclassification rate, as well as the confusion matrix, which shows the number and percentage of patients/cases that are correctly and incorrectly classified in each output class. In our case these are either Yes (a patient will reach their Urine Output Target), or No (a patient will not reach their Urine Output Target).

A further evaluation criteria for classification type models, and specifically to evaluate sensitivity and specificity, an ROC curve can be used to evaluate the goodness of fit for a binary classifier. An ROC curve is a plot of the true positive rate (rate of events that are correctly predicted as events) against the false positive rate (rate of nonevents predicted to be events) for the different possible cut-points.

An ROC curve demonstrates the following:

1. The trade-off between sensitivity and specificity (any increase in sensitivity will be accompanied by a decrease in specificity).
2. The closer the curve follows the left border and then the top border of the ROC space, the more accurate the test.
3. The closer the curve comes to the 45-degree diagonal of the ROC space, the less accurate the test.

ROC curve for algorithm step D: predicting if patients would reach targets (yes/no) — target at 75th percentile

The ROC Curves illustrated in Fig. 12.17A−C, show the following Sensitivity/Specificity Area Under the Curve for each of the three model−algorithms at *hour 6*:

1. Boosted Classification Tree—0.725
2. Random Forest Classification—0.722
3. CART—0.905

The ROC Curves in Fig. 12.18A−C below, show the following Sensitivity/Specificity Area Under the Curve for these three model−algorithms, respectively, at *hour 12*:

1. Boosted Classification Tree—0.703
2. Random Forest Classification—0.449
3. CART—0.968

The ROC Curves illustrated in Fig. 12.19A−C below, show the following Sensitivity/Specificity Area Under the Curve for each of these three model−algorithms at *hour 24*:

1. Boosted Classification Tree—0.685
2. Random Forest Classification—0.699
3. CART—0.884

ROC curve for algorithm step D: predicting if patients would reach targets (yes/no) — target at 95th percentile

The ROC Curves illustrated I Fig. 12.20A−C below, show the following Sensitivity/Specificity Area Under the Curve for each of the three model−algorithms at *hour 6*:

1. Boosted Classification Tree—0.762
2. Random Forest Classification—0.763
3. CART—0.987

The ROC Curves illustrated in Fig. 12.21A−C below, show the following Sensitivity/Specificity Area Under the Curve for each of these three model−algorithms at *hour 12*:

1. Boosted Classification Tree—0.739
2. Random Forest Classification—0.764
3. CART—0.991

The ROC Curves illustrated in Fig. 12.22A−C below, show the following Sensitivity/Specificity Area Under the Curve for each of these three model−algorithms at *hour 24*:

1. Boosted Classification Tree—0.731
2. Random Forest Classification—0.746
3. CART—0.997

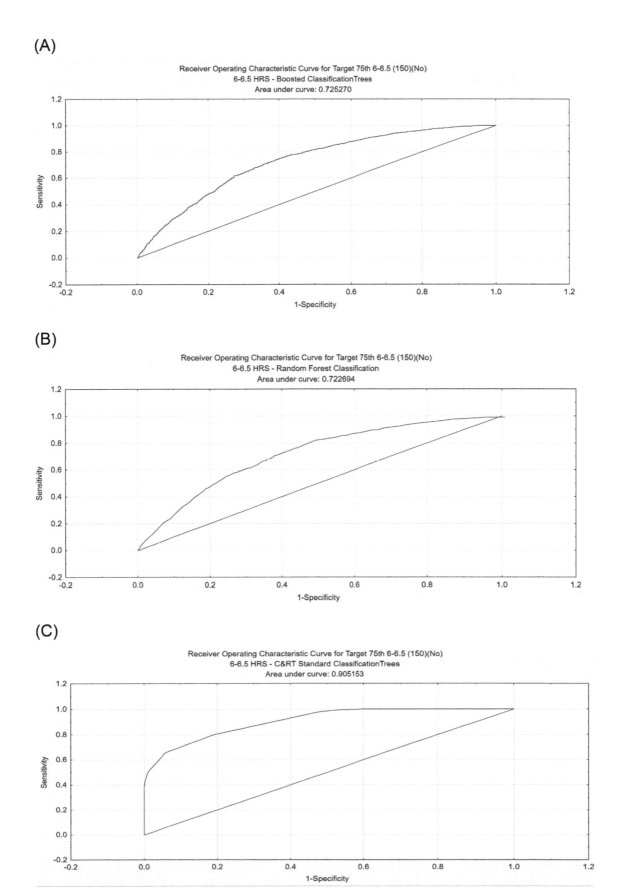

FIGURE 12.17 ROC curve for algorithm step D: predicting if patients would reach targets (yes/no)—target at 75th percentile at hour 6.

(A)

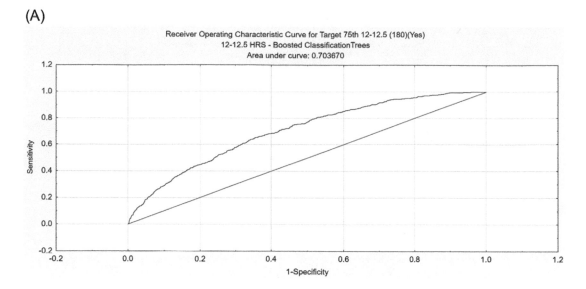

(B)

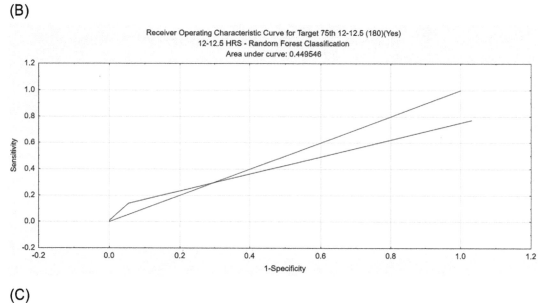

(C)

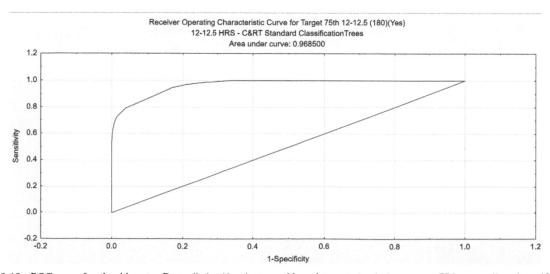

FIGURE 12.18 | ROC curve for algorithm step D: predicting if patients would reach targets (yes/no)—target at 75th percentile at hour 12.

(A)

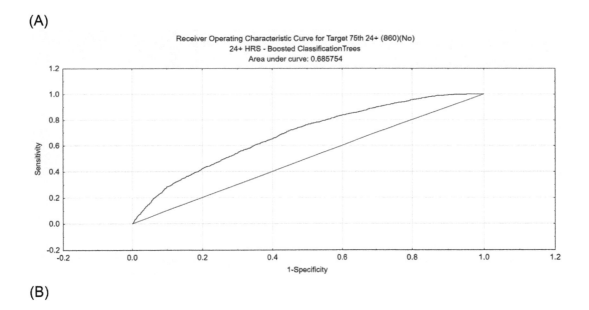

(B)

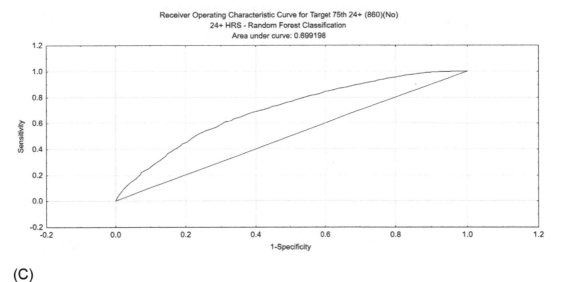

(C)

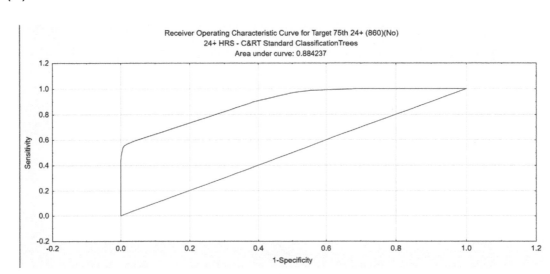

FIGURE 12.19 ROC curve for algorithm step D: predicting if patients would reach targets (yes/no)—target at 75th percentile at hour 24.

(A)

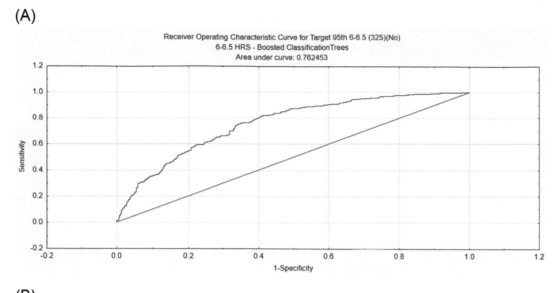

(B)

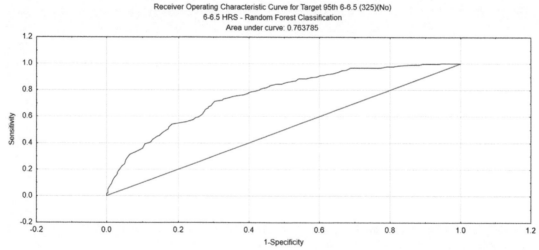

(C)

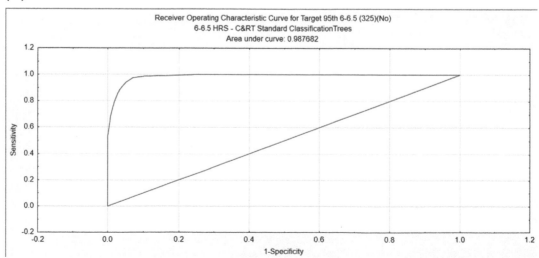

FIGURE 12.20 ROC curve for algorithm step D: predicting if patients would reach targets (yes/no)—target at 95th percentile at hour 6.

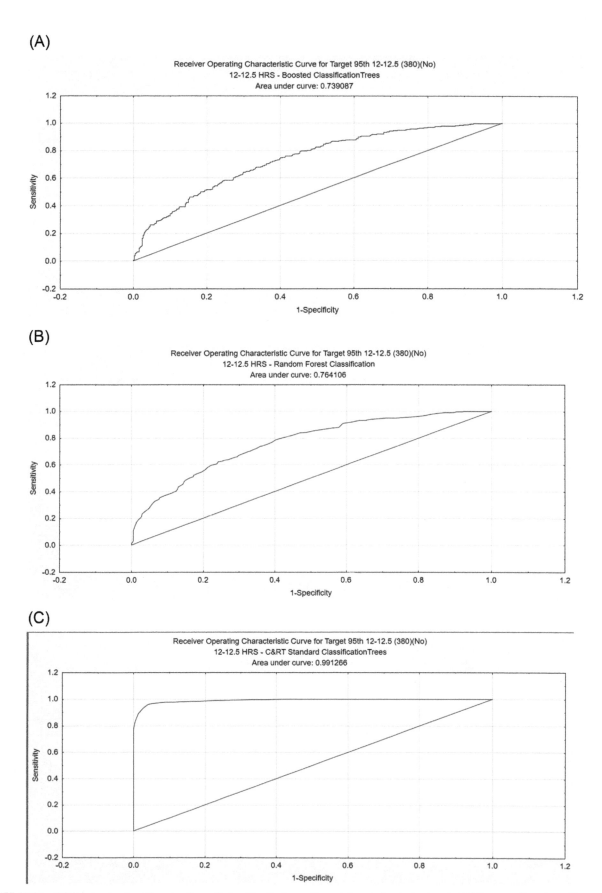

FIGURE 12.21 ROC curve for algorithm step D: predicting if patients would reach targets (yes/no)—target at 95th percentile at hour 12.

(A)

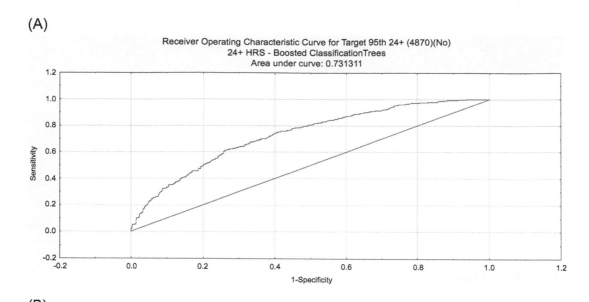

(B)

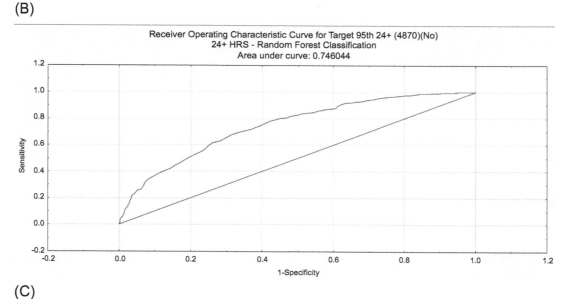

(C)

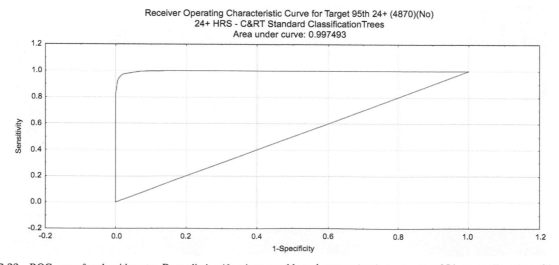

FIGURE 12.22 ROC curve for algorithm step D: predicting if patients would reach targets (yes/no)—target at 95th percentile at hour 24.

Algorithm version 3

Introduction

The image below describes the logic of the algorithm built as Version 3. As opposed to Versions 1 and 2, in Version 3, we have removed all variables, except the Urine Output variable, to see how the final algorithm would perform when only having the bucketed urine (per hour) output (Fig. 12.23).

1. *Step 1: Data preparation for modeling*

    Using the criteria as described in the Data Demographics section above, we brought data into our datasheet from the MIMIC-IV database using Google Big Query, and obtained the following variables:

a. Urine Output Records per patient
b. Date/Time of Urine Output

    We then used the Date-Time of Urine Output and pivoted this to create Urine Output values bucketed in hourly increments up to 36 hours, for understanding demographics and creating visualizations of the data. For modeling of Urine Output however, we are using cumulative figures per hour for up to 24 hours.

2. *Step 2: Algorithm Step A—effects on urine output in first 3 hours*

    We used regression trees (Boosted, Random Forests and Neural Networks) to analyze UO buckets (first 3 hours in hourly buckets—Cumulative UO) and the effect of only the Urine Output variable. Results of this are shown below.

3. *Step 3: Algorithm Step B—effects on urine output after 6, 12, 24, 36 hours*

    We use regression trees (CART, Boosted, Random Forests) to analyze UO buckets (6, 12, 24, 36 hours in hourly buckets—Cumulative UO) and effect of first 3, 6, 12 hours cumulative UO, as well as only the Urine Output variable. Results of this are shown below.

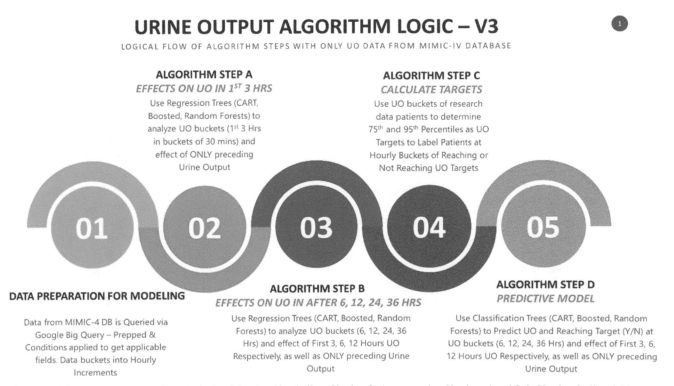

FIGURE 12.23 The diagram describes the logic of the algorithm built as Version 3. As opposed to Versions 1 and 2, in Version 3 all variables were removed except the Urine Output variable. The purposes of this was to see how the final algorithm would perform when only having the bucketed urine (per hour) output.

4. *Step 4: Alorithm Step C—calculate urine output targets*

We used hourly Urine Output buckets of patient research data to determine 75th and 95th percentiles as Urine Output targets to label patients at hourly buckets of reaching or not reaching Urine Output targets. Results of this are shown below.

5. *Step 5: Algorithm STEP D—predictive model*

We used classification trees (Boosted, Random Forests and Neural Networks) to predict cumulative UO and reaching target (Y/N) at cumulative UO buckets (6, 12, 24 hours) and the effect of first 3, 6, 12 hours cumulative UO. Results of this are shown below.

Algorithm selection

Algorithms are selected depending on the business/research questions one wants answered, as well as the structure of the data available, to answer these questions.

Selection Criteria are the same as in 4.1.2.

Short description of each algorithm

Algorithm descriptions are done in detail in section "Short description of each algorithm".

Algorithm results

Summary of R-squared values projected to predict UO at hour 6, 12, and 24

The summaries below were created to evaluate the performance of the respective models to predict unknown UO from hour 1 onwards, and the value of R-squared, as it increases with more learning.

How is R-squared used and interpreted

The *R-squared* value usage and interpretation is described in detail in section "Algorithm results."

For this model—and as an example below—if we were looking at hour 3, we are at an R-squared of .80 or 80%. This means that at this point, we can explain and predict 80% of the variability in Urine Output at what it is likely to be by hour 6. Of course, the closer we get to hour 6 in reality and knowing the Urine Output of preceding hours, this passes 90%.

To evaluate and investigate the example used above, for any other hours or R-squared values, full details are presented below (Figs. 12.24—12.26).

Algorithm step C: calculation of targets As explained above for Step C for the algorithms, we were looking to calculate a Target (to simulate the "Target" a clinician might set) for us to predict whether a particular patient is "Yes"—likely to reach this target, or "No"—not likely to reach this target at the respective Urine Output time buckets.

To calculate these "Targets," what we have done is to look at the population of patients available, based on the selection criteria we described above. Given these criteria and population, we had calculated for each Urine Output time bucket, a target based on the 75th and 95th percentiles of Urine Output our population of patients were likely to reach. We thus had created simulated "Targets" to be able to illustrate how our algorithms may perform in predicting a Yes or No as per above.

As an example: To calculate the 75th Percentile of Urine Output in mL at hour 3, one would find the Urine Output value which 75% of patients had reached in the research population at hour 3. This would be used as the simulated "Target" (which would normally be entered by the clinician). We did the same for the 95th percentiles at each hourly time bucket. This gave us the option of using either the 75th or 95th percentiles as targets, given how sensitive we wanted the model—algorithm to be.

This allowed us to code our data with a Yes or No—to use as an Output/Dependent variable to train our algorithm for Step D—given the same input variables used in Step A and B (Table 12.32).

TABLE 12.32 The Descriptive Statistics tabulated here, allowed codeing the data with a Yes or No—to use as an Output/Dependent variable to train our algorithm for Step D—given the same input variables used in Step A and B.

Variable	Descriptive statistics									
	Valid N	Mean	Median	Minimum	Maximum	Lower quartile	Upper quartile	Percentile 5%	Percentile 95%	Std. Dev.
0−1 Total Cum	28970	27.698	0.000	0.000000	2050.00	0.000	0.000	0.0000	185.000	87.399
0−2 Total Cum	28970	85.281	0.000	0.000000	2500.00	0.000	100.000	0.0000	450.000	178.456
0−3 Total Cum	28970	160.654	25.000	0.000000	3825.00	0.000	230.000	0.0000	680.000	261.258
0−4 Total Cum	28970	247.873	140.000	0.000000	4300.00	0.000	350.000	0.0000	890.000	322.166
0−5 Total Cum	28970	344.039	225.000	0.000000	5100.00	82.000	485.000	0.0000	1085.000	374.531
0−6 Total Cum	28970	446.778	320.000	2.000000	5980.00	150.000	610.000	49.0000	1270.000	416.919
0−7 Total Cum	28970	544.603	405.000	2.000000	7780.00	210.000	740.000	75.0000	1465.000	469.996
0−8 Total Cum	28970	636.260	495.000	2.000000	8030.00	270.000	855.000	105.0000	1630.000	512.738
0−9 Total Cum	28970	732.933	580.000	2.000000	8330.00	325.000	990.000	130.0000	1850.000	566.852
0−10 Total Cum	28970	832.161	675.000	2.000000	9230.00	380.000	1115.000	160.0000	2050.000	621.879
0−11 Total Cum	28970	930.954	760.000	4.000000	9680.00	435.000	1250.000	180.0000	2250.000	679.509
0−12 Total Cum	28970	1034.473	860.000	4.000000	10130.00	500.000	1385.000	210.0000	2460.000	733.582
0−13 Total Cum	28970	1143.945	965.000	4.000000	11450.00	570.000	1530.000	230.0000	2650.000	788.008
0−14 Total Cum	28970	1257.772	1082.500	4.000000	13750.00	645.000	1680.000	260.0000	2860.000	839.829
0−15 Total Cum	28970	1367.942	1190.000	4.000000	15000.00	720.000	1815.000	285.0000	3055.000	893.732
0−16 Total Cum	28970	1472.148	1290.000	4.000000	15000.00	785.000	1950.000	310.0000	3250.000	943.798
0−17 Total Cum	28970	1573.870	1390.000	4.000000	15250.00	850.000	2085.000	330.0000	3450.000	994.878
0−18 Total Cum	28970	1672.839	1480.000	4.000000	15250.00	915.000	2220.000	355.0000	3650.000	1045.873
0−19 Total Cum	28970	1767.429	1565.000	4.000000	15470.00	970.000	2345.000	375.0000	3840.000	1096.555
0−20 Total Cum	28970	1855.344	1650.000	4.000000	15520.00	1023.000	2460.000	395.0000	4025.000	1143.998
0−21 Total Cum	28970	1944.528	1727.500	4.000000	15535.00	1075.000	2580.000	415.0000	4200.000	1194.099
0−22 Total Cum	28970	2032.301	1807.000	4.000000	15575.00	1125.000	2705.000	426.0000	4385.000	1246.947
0−23 Total Cum	28970	2117.930	1882.000	4.000000	15635.00	1172.000	2830.000	445.0000	4575.000	1300.041
0−24 Total Cum	28970	2204.988	1960.000	4.000000	15635.00	1218.000	2948.000	458.0000	4765.000	1353.311

(A)

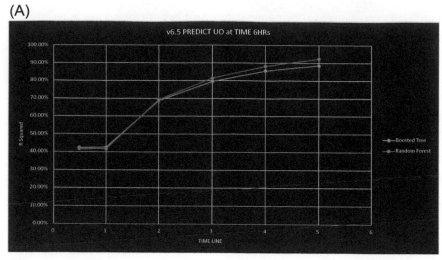

FIGURE 12.24 A and B. Predicting hour 6, hourly after 1−6 hours of data collected.

(B)

<u>DEPENDENT</u> **CUM 6HRs UO Summary**

Basic V plus Independent Variables below			v6.5
TIME	max time interval	**Method**	**R^2**
0-0.5HRs	0.5	Boosted Tree	42.32%
	0.5	Random Forest	41.69%
until 1-1.5HRs	1	Boosted Tree	42.61%
	1	Random Forest	41.76%
until 2-2.5HRs	2	Boosted Tree	68.83%
	2	Random Forest	69.33%
until 3-3.5HRs	3	Boosted Tree	79.57%
	3	Random Forest	81.38%
until 4-4.5HRs	4	Boosted Tree	85.64%
	4	Random Forest	88.23%
until 5-5.5HRs	5	Boosted Tree	88.68%
	5	Random Forest	92.39%

Algorithm step D: predicting if patients would reach targets (yes/no)—target at 75th percentile As described above, we are looking to predict a Yes or No—for the patient reaching the simulated target calculated above, from a mix of continuous (Numerical) and categorical independent/input variables.

With criteria described above we decided on the CART, Boosted Trees, and Random Forest models/algorithms, which, given the data structures and sample sizes available in buckets, were performing well.

Predicting a yes/no for hour 24 at hour 6, 12, and 24

Hour 6. Table 12.33 shows a model/algorithm comparison at hour 6 for predicting hour 24 of Urine Output:

TABLE 12.33 Boosted Trees, Random Forest, and Classification Trees model/algorithm comparison at hour 6 for predicting hour 24 of Urine Output.

	Model Quality Summary							
	1	2	3	4	5	6	7	8
	Model name	Data name	Source	Data usage	Misclassification error rate	Chi-square statistic	G-square statistic	Valid N
1	6-6.5 HRS - Boosted ClassificationTrees		Data		0.378571429	2013.20094	5643.42814	5460
2	6-6.5 HRS - Random Forest Classification		Data		0.378205128	2029.45846	5648.32199	5460
3	6-6.5 HRS - C&RT Standard ClassificationTrees		Data		0.384249084	6086.20078	7635.49467	5460

(A)

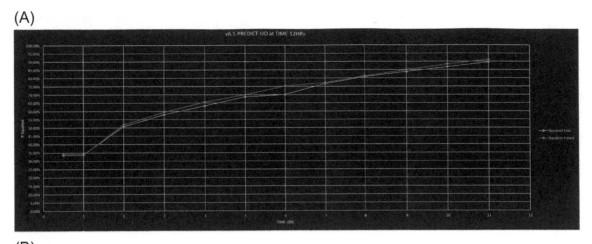

(B)

<u>DEPENDENT</u> **CUM 12HRs UO Summary**

Basic V plus Independent Variables below			v6.5
TIME	max time interval	Method	R²
0-0.5HRs	0.5	Boosted Tree	33.88%
	0.5	Random Forest	33.30%
until 1-1.5HRs	1	Boosted Tree	34.00%
	1	Random Forest	33.72%
until 2-2.5HRs	2	Boosted Tree	50.75%
	2	Random Forest	51.89%
until 3-3.5HRs	3	Boosted Tree	57.82%
	3	Random Forest	59.45%
until 4-4.5HRs	4	Boosted Tree	63.21%
	4	Random Forest	65.53%
until 5-5.5HRs	5	Boosted Tree	68.64%
	5	Random Forest	70.01%
until 6-6.5HRs	6	Boosted Tree	69.98%
	6	Random Forest	75.15%
until 7-7.5HRs	7	Boosted Tree	76.67%
	7	Random Forest	77.31%
until 8-8.5HRs	8	Boosted Tree	80.71%
	8	Random Forest	81.26%
until 9-9.5HRs	9	Boosted Tree	83.81%
	9	Random Forest	84.99%
until 10-10.5HRs	10	Boosted Tree	86.70%
	10	Random Forest	88.43%
until 11-11.5HRs	11	Boosted Tree	89.60%
	11	Random Forest	91.22%

FIGURE 12.25   A and B. Predicting hour 12, hourly after 1–12 hours of data collected.

(A)

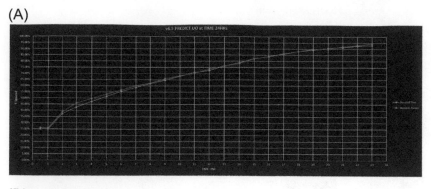

FIGURE 12.26 A and B. Predicting hour 24, hourly after 1−24 hours of data collected.

(B)

CUM 24HRs UO Summary

Basic V plus Independent Variables below			v6.5
TIME	max time interval	Method	R^2
0-0.5HRs	0.5	Boosted Tree	25.08%
	0.5	Random Forest	26.02%
until 1-1.5HRs	1	Boosted Tree	25.13%
	1	Random Forest	25.88%
until 2-2.5HRs	2	Boosted Tree	37.26%
	2	Random Forest	38.88%
until 3-3.5HRs	3	Boosted Tree	42.62%
	3	Random Forest	46.00%
until 4-4.5HRs	4	Boosted Tree	47.22%
	4	Random Forest	49.75%
until 5-5.5HRs	5	Boosted Tree	51.37%
	5	Random Forest	53.27%
until 6-6.5HRs	6	Boosted Tree	55.32%
	6	Random Forest	56.62%
until 7-7.5HRs	7	Boosted Tree	58.58%
	7	Random Forest	60.08%
until 8-8.5HRs	8	Boosted Tree	61.89%
	8	Random Forest	62.22%
until 9-9.5HRs	9	Boosted Tree	64.56%
	9	Random Forest	65.84%
until 10-10.5HRs	10	Boosted Tree	67.91%
	10	Random Forest	68.27%
until 11-11.5HRs	11	Boosted Tree	70.96%
	11	Random Forest	70.23%
until 12-12.5HRs	12	Boosted Tree	72.41%
	12	Random Forest	73.75%
until 13-13.5HRs	13	Boosted Tree	76.13%
	13	Random Forest	76.09%
until 14-14.5HRs	14	Boosted Tree	78.77%
	14	Random Forest	78.18%
until 15-15.5HRs	15	Boosted Tree	82.27%
	15	Random Forest	81.97%
until 16-16.5HRs	16	Boosted Tree	83.81%
	16	Random Forest	83.97%
until 17-17.5HRs	17	Boosted Tree	85.93%
	17	Random Forest	86.24%
until 18-18.5HRs	18	Boosted Tree	88.10%
	18	Random Forest	88.02%
until 19-19.5HRs	19	Boosted Tree	89.27%
	19	Random Forest	89.97%
until 20-20.5HRs	20	Boosted Tree	90.77%
	20	Random Forest	89.90%
until 21-21.5HRs	21	Boosted Tree	91.32%
	21	Random Forest	92.06%
until 22-22.5HRs	22	Boosted Tree	92.49%
	22	Random Forest	92.93%
until 23-23.5HRs	23	Boosted Tree	93.07%
	23	Random Forest	94.53%

The best performing model is the Random Forest Classification with a misclassification error rate of 37.82%, which means that around 62% of the time it predicts a Yes or No accurately versus around 38% predicting it inaccurately.

Hour 12. Table 12.34 shows a model/algorithm comparison at hour 12 for predicting hour 24 of Urine Output:

TABLE 12.34 Boosted Trees, Random Forest, and Classification Trees model/algorithm comparison at hour 12 for predicting hour 24 of Urine Output.

	Model Quality Summary							
	1 Model name	2 Data name	3 Source	4 Data usage	5 Misclassification error rate	6 Chi-square statistic	7 G-square statistic	8 Valid N
1	12-12.5 HRS - Boosted ClassificationTrees		Data		0.366071429	1098.98053	4528.6101	4928
2	12-12.5 HRS - Random Forest Classification		Data		0.364448052	1094.57238	4508.17321	4928
3	12-12.5 HRS - C&RT Standard ClassificationTrees		Data		0.458198052	42408.3847	14142.0819	4928

The best performing model is the Random Forest Classification with a misclassification error rate of 36.44% which means that around 64% of the time it predicts a Yes or No accurately versus around 36% predicting it inaccurately.

Hour 24 +. Table 12.35 shows a model/algorithm comparison at hour 24 for predicting hour 24 of Urine Output:

TABLE 12.35 Boosted Trees, Random Forest, and Standard Classification Trees model/algorithm comparison at hour 24 for predicting hour 24 of Urine Output.

	Model Quality Summary							
	1 Model name	2 Data name	3 Source	4 Data usage	5 Misclassification error rate	6 Chi-square statistic	7 G-square statistic	8 Valid N
1	24+ HRS - Boosted ClassificationTrees		Data		0.403040205	8043.38282	10118.0909	7039
2	24+ HRS - Random Forest Classification		Data		0.389543969	6196.33326	9180.74397	7039
3	24+ HRS - C&RT Standard ClassificationTrees		Data		0.425770706	33691.1309	16332.4872	7039

The best performing model is the Random Forest Classification with a misclassification error rate of 38.95%, which means that around 61% of the time it predicts a Yes or No accurately versus around 39% predicting it inaccurately (Table 12.36).

TABLE 12.36 A, B, and C. detailed misclassification matrix.

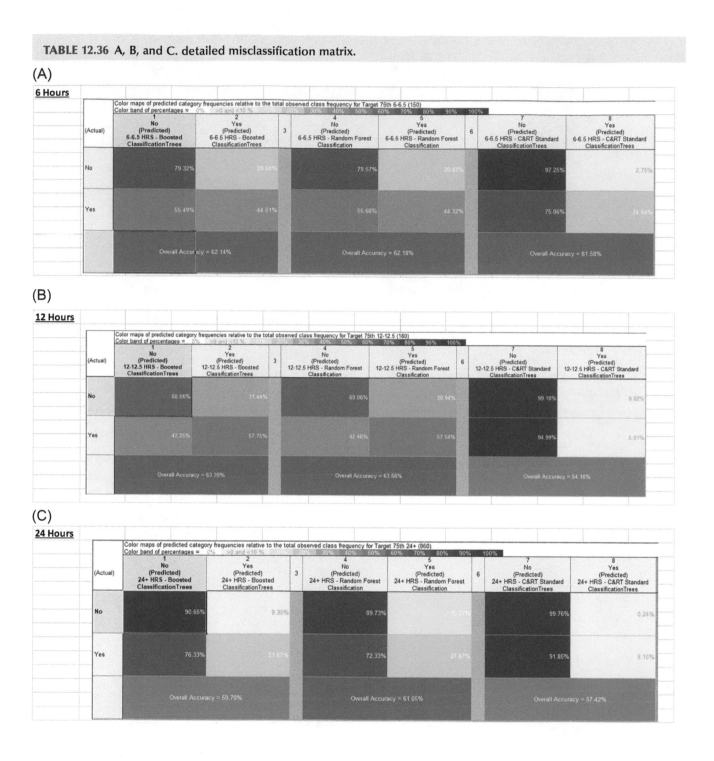

(A)

6 Hours

(B)

12 Hours

(C)

24 Hours

The Detailed Misclassification Matrix for each of the above is shown below:

Algorithm step D: predicting if patients would reach targets (yes/no)—target at 95th percentile For classification type models, to evaluate model accuracy, one looks at the misclassification error rate, as well as the confusion matrix, which shows the number and percentage of patients/cases which are correctly and incorrectly classified in each output class. In our case these are either Yes (a patient will reach their Urine Output Target), or No (a patient will not reach their Urine Output Target).

Hour 6, 12, and 24 +

Hour 6. Table 12.37 shows a model/algorithm comparison at hour 6 for predicting hour 24 of Urine Output:

TABLE 12.37 Predicting hour 24 UO from hour 6.

	Model Quality Summary							
	1 **Model name**	**2** Data name	**3** Source	**4** Data usage	**5** Misclassification error rate	**6** Chi-square statistic	**7** G-square statistic	**8** Valid N
1	6-6.5 HRS - Boosted ClassificationTrees		Data		0.37992278	566.661098	1392.63458	1295
2	6-6.5 HRS - Random Forest Classification		Data		0.383011583	661.887243	1452.58995	1295
3	6-6.5 HRS - C&RT Standard ClassificationTrees		Data		0.388416988	403.321936	1323.82135	1295

The best performing model is the Boosted Classification Tree with a misclassification error rate of 37.99% which means that around 62% of the time it predicts a Yes or No accurately versus around 38% predicting it inaccurately.

Hour 12. Table 12.38 shows a model/algorithm comparison at hour 12 for predicting hour 24 of Urine Output:

The best performing model is the Standard CART Classification Tree with a misclassification error rate of 24.68% which means that around 75% of the time it predicts a Yes or No accurately versus around 25% predicting it inaccurately.

TABLE 12.38 Predicting UO at 24 hours from the values of hour 12.

	Model Quality Summary							
	1 **Model name**	**2** Data name	**3** Source	**4** Data usage	**5** Misclassification error rate	**6** Chi-square statistic	**7** G-square statistic	**8** Valid N
1	12-12.5 HRS - Boosted ClassificationTrees		Data		0.383711167	297.411821	1160.88892	1191
2	12-12.5 HRS - Random Forest Classification		Data		0.350965575	226.442873	1029.93159	1191
3	12-12.5 HRS - C&RT Standard ClassificationTrees		Data		0.246851385	180.839493	737.208101	1191

Hour 24 +. Table 12.39 shows a model/algorithm comparison at hour 24 for predicting hour 24 + of Urine Output:

TABLE 12.39 Hour 24 UO predicting greater than 24 hours UO.

	Model Quality Summary							
	1 **Model name**	**2** Data name	**3** Source	**4** Data usage	**5** Misclassification error rate	**6** Chi-square statistic	**7** G-square statistic	**8** Valid N
1	24+ HRS - Boosted ClassificationTrees		Data		0.451631702	5100.82257	3583.79086	1716
2	24+ HRS - Random Forest Classification		Data		0.398601399	817.220275	1951.17915	1716
3	24+ HRS - C&RT Standard ClassificationTrees		Data		0.264568765	309.701163	1159.03571	1716

The best performing model is the Standard CART Classification Tree with a Misclassification error rate of 26.45% which means that around 74% of the time it predicts a Yes or No accurately versus around 26% predicting it inaccurately (Table 12.40).

TABLE 12.40 A, B, and C. detailed misclassification matrix for hours 6, 12, and 24 + .

(A)

6 Hours

Color maps of predicted category frequencies relative to the total observed class frequency for Target 95th 6-6.5 (325)
Color band of percentages = 0% >0 and <10 % 20% 30% 40% 50% 60% 70% 80% 90% 100%

(Actual)	1 No (Predicted) 6-6.5 HRS - Boosted ClassificationTrees	2 Yes (Predicted) 6-6.5 HRS - Boosted ClassificationTrees	3	4 No (Predicted) 6-6.5 HRS - Random Forest Classification	5 Yes (Predicted) 6-6.5 HRS - Random Forest Classification	6	7 No (Predicted) 6-6.5 HRS - C&RT Standard ClassificationTrees	8 Yes (Predicted) 6-6.5 HRS - C&RT Standard ClassificationTrees
No	82.84%	17.16%		85.16%	14.84%		73.11%	26.89%
Yes	58.80%	41.20%		61.73%	38.27%		50.77%	49.23%
	Overall Accuracy = 62.01%			Overall Accuracy = 61.70%			Overall Accuracy = 61.16%	

(B)

12 Hours

Color maps of predicted category frequencies relative to the total observed class frequency for Target 95th 12-12.5 (380)
Color band of percentages = 0% >0 and <10 % 20% 30% 40% 50% 60% 70% 80% 90% 100%

(Actual)	1 No (Predicted) 12-12.5 HRS - Boosted ClassificationTrees	2 Yes (Predicted) 12-12.5 HRS - Boosted ClassificationTrees	3	4 No (Predicted) 12-12.5 HRS - Random Forest Classification	5 Yes (Predicted) 12-12.5 HRS - Random Forest Classification	6	7 No (Predicted) 12-12.5 HRS - C&RT Standard ClassificationTrees	8 Yes (Predicted) 12-12.5 HRS - C&RT Standard ClassificationTrees
No	56.59%	43.41%		65.88%	34.12%		58.28%	41.72%
Yes	33.39%	66.61%		36.06%	63.94%		7.85%	92.15%
	Overall Accuracy = 61.63%			Overall Accuracy = 64.90%			Overall Accuracy = 75.31%	

(C)

24 Hours

Color maps of predicted category frequencies relative to the total observed class frequency for Target 95th 24+ (4870)
Color band of percentages = 0% >0 and <10 % 20% 30% 40% 50% 60% 70% 80% 90% 100%

(Actual)	1 No (Predicted) 24+ HRS - Boosted ClassificationTrees	2 Yes (Predicted) 24+ HRS - Boosted ClassificationTrees	3	4 No (Predicted) 24+ HRS - Random Forest Classification	5 Yes (Predicted) 24+ HRS - Random Forest Classification	6	7 No (Predicted) 24+ HRS - C&RT Standard ClassificationTrees	8 Yes (Predicted) 24+ HRS - C&RT Standard ClassificationTrees
No	96.97%	3.03%		80.40%	19.60%		55.66%	44.34%
Yes	87.19%	12.81%		60.07%	39.53%		8.61%	91.39%
	Overall Accuracy = 54.84%			Overall Accuracy = 60.14%			Overall Accuracy = 73.54%	

The Detailed Misclassification Matrix for each of the above is shown below:

Algorithm accuracy, sensitivity and specificity for the classification trees above

As described above, for classification type models, to evaluate model accuracy, one looks at the misclassification rate, as well as the confusion matrix, which shows the number and percentage of patients/cases which are correctly and incorrectly classified in each output class. In our case these are either Yes (a patient will reach their Urine Output Target), or No (a patient will not reach their Urine Output Target).

Further evaluation criteria for classification type models, and specifically to evaluate sensitivity and specificity, an ROC curve can be used to evaluate the goodness of fit for a binary classifier. It is a plot of the true positive rate (rate of events that are correctly predicted as events) against the false positive rate (rate of nonevents predicted to be events) for the different possible cut-points.

An ROC curve demonstrates the following:

1. The trade-off between sensitivity and specificity (any increase in sensitivity will be accompanied by a decrease in specificity).
2. The closer the curve follows the left border and then the top border of the ROC space, the more accurate the test.
3. The closer the curve comes to the 45-degree diagonal of the ROC space, the less accurate the test.

ROC curve for algorithm step D: predicting if patients would reach targets (yes/no)—target at 75th percentile
The ROC Curves illustrated in Fig. 12.27A−C show the following sensitivity/specificity area under the curve for each of the three model−algorithms, at *hour 6*:

1. Boosted Classification Tree—0.67
2. Random Forest Classification—0.65
3. CART—0.64

The ROC Curves illustrated in Fig. 12.28A−C show the following sensitivity/specificity area under the curve for each of the three model−algorithms, at *hour 12*:

1. Boosted Classification Tree—0.69
2. Random Forest Classification—0.70
3. CART—0.60

The ROC Curves in Fig. 12.29A−C show the following sensitivity/specificity area under the curve for each of the three model−algorithms, at *hour 24*:

1. Boosted Classification Tree—0.63
2. Random Forest Classification—0.67
3. CART—0.65

ROC curve for algorithm step D: predicting if patients would reach targets (yes/no)—target at 95th percentile
The ROC Curves illustrated in Fig. 12.30A−C show the following sensitivity/specificity area under the curve for each of the three model−algorithms, at *hour 6*:

1. Boosted Classification Tree—0.690
2. Random Forest Classification—0.687
3. CART—0.668

The ROC Curves illustrated in Fig. 12.31A−C show the following sensitivity/specificity and area under the curve for each of the three model−algorithms, at *hour 12*:

1. Boosted Classification Tree—0.672
2. Random Forest Classification—0.721
3. CART—0.771

The ROC Curves illustrated in Fig. 12.32A−C show the following sensitivity/specificity area under the curve for each of the three model−algorithms, at *hour 24*:

1. Boosted Classification Tree—0.611
2. Random Forest Classification—0.684
3. CART—0.761

The champion algorithms

Given the results above, there are two approaches one could take, depending on the clinical needs of the customer. Typically, there is always a sliding scale between:

(A)

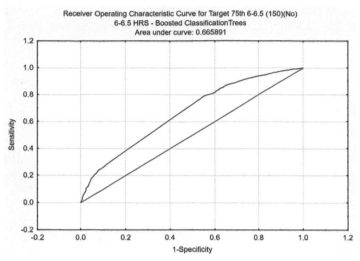

(B)

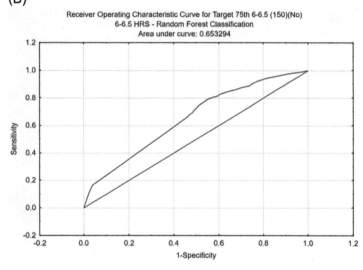

(C)

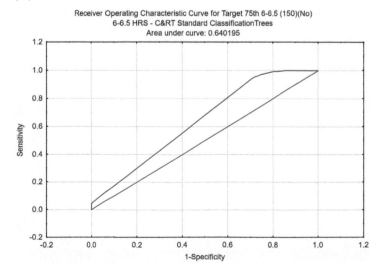

FIGURE 12.27 ROC curve for algorithm step D: predicting if patients would reach targets (yes/no)—target at 75th percentile at hour 6.

(A)

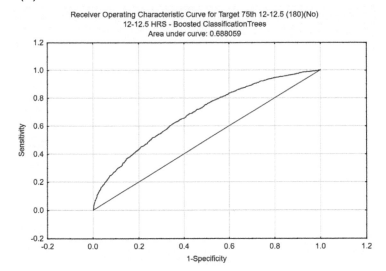

(B)

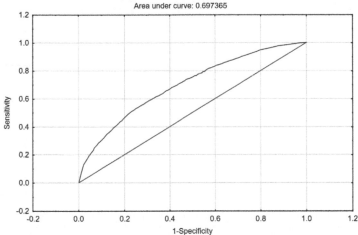

(C)

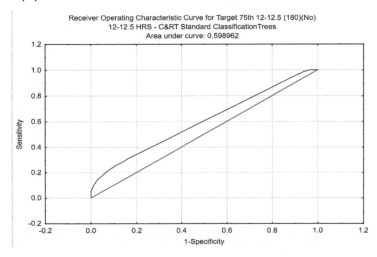

FIGURE 12.28 ROC curve for algorithm step D: predicting if patients would reach targets (yes/no)—target at 75th percentile at hour 12.

(A)

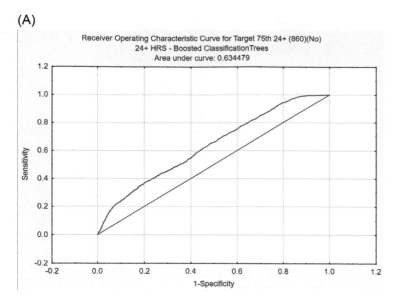

(B)

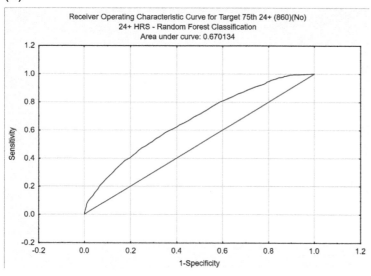

(C)

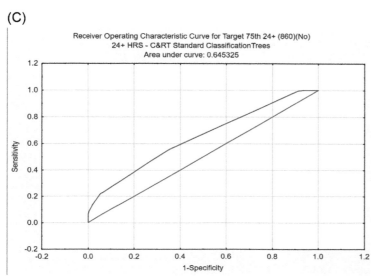

FIGURE 12.29 ROC curve for algorithm step D: predicting if patients would reach targets (yes/no)—target at 75th percentile at hour 24.

(A)

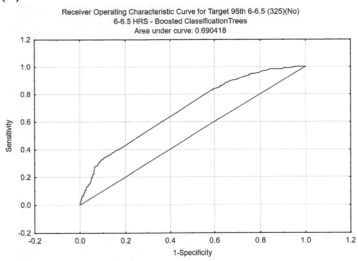

(B)

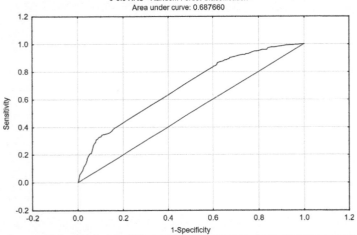

(C)

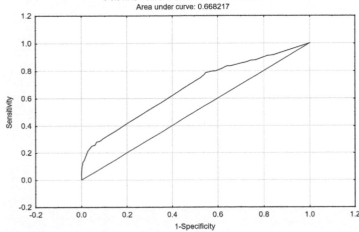

FIGURE 12.30 ROC curve for algorithm step D: predicting if patients would reach targets (yes/no)—target at 95th percentile at hour 6.

(A)

Receiver Operating Characteristic Curve for Target 95th 12-12.5 (380)(No)
12-12.5 HRS - Boosted ClassificationTrees
Area under curve: 0.672241

FIGURE 12.31 ROC curve for algorithm step D: predicting if patients would reach targets (yes/no)—target at 95th percentile at hour 12.

(B)

Receiver Operating Characteristic Curve for Target 95th 12-12.5 (380)(No)
12-12.5 HRS - Random Forest Classification
Area under curve: 0.721899

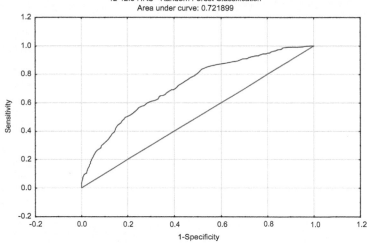

(C)

Receiver Operating Characteristic Curve for Target 95th 12-12.5 (380)(No)
12-12.5 HRS - C&RT Standard ClassificationTrees
Area under curve: 0.771066

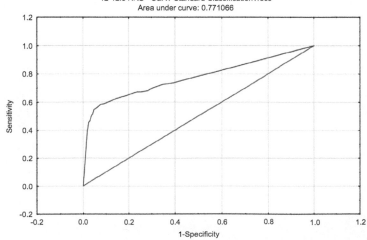

(A)

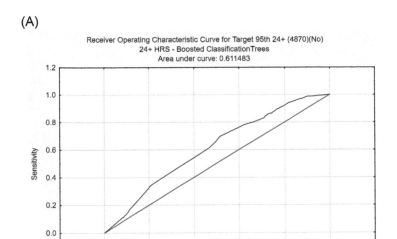

(B)

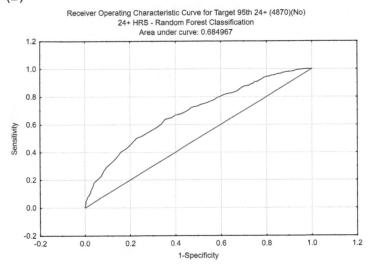

(C)

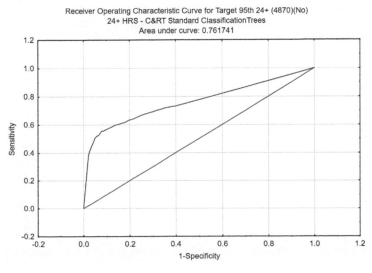

FIGURE 12.32 ROC curve for algorithm step D: predicting if patients would reach targets (yes/no)—target at 95th percentile at hour 24.

1. How *Accurate* a model is at predicting exact values on an Output or Dependent Variable.
2. How well it *Generalizes* across multiple uses and applications.
3. For Classification Models, one also looks at the ROC Curve to determine False Positives and Negatives.

In general, and as per the approach above, if one fits an algorithm to a particular set of data and it becomes extremely accurate, one might run into the problem that it overfits. Overfitting happens when we have an algorithm which fits a particular data set so well, that when one feeds it new data, it does not do a very good job at making predictions on it. This is partially prevented by working with Training/Test data sets to see how well the data predicts on a "Test" set of data which it hasn't seen, as opposed to the "Training" set of data which it used to learn from. This can also be prevented by using techniques like V-Fold Cross-validation. In this case, one builds a model by using the entire data set but iterating through it by slicing pieces of the data out at each iteration (V-folding), with V being the number of pieces we slice the data into for iteration. This means that for example, I could use a set of data with 1000 rows/cases and pick a V number of 5. It would then split the data into five pieces and iterate through building the model by each time leaving out one of the pieces/slices. This helps to make algorithms "Generalize" well, across populations.

One also has a few choices when picking Champion Algorithms to enhance accuracy and generalizability:

1. For the algorithm as a whole, picking the technique (i.e., Boosted Trees or Random Forest) with the highest accuracy overall, across all possible Urine Output buckets.
2. For each algorithm using the respective technique (i.e., Boosted Trees or Random Forest) with the highest accuracy for each respective technique, across the respective Urine Output buckets.
3. Use what is called "Ensemble" modeling, in which case one uses, for example, all three techniques used in this study, and averaging their predictive output.
4. Using "Voting," which means one could use more than one model and then take a "vote among them", in classification models when one is, for example, using a Yes or No; in this case we let the techniques run against each other and if 2 of them say Yes and 1 No, then the prediction will be Yes. In essence one votes for the response with the highest frequency between models.

Each of the above options has benefits and drawbacks. These will need to be discussed and tested in a pilot phase before production to assess the optimal option.

In general, for our algorithm Step 2 (Regression Type), where we are predicting Urine Output, the Boosted Trees and Random Forest Algorithms consistently outperform the simple CART modeling technique, and they intermittently score highest in accuracy. This "intermittent scoring" happens depending on the actual data available in each of the respective Urine Output buckets over time. For algorithm Step 5 (Classification Type), the CART technique turns out to perform very well, especially when tree rules found are simple and noncomplex.

Thus, depending on project needs, one should pick the best fit for purpose among the above techniques, as well as how these impact on the experience of clinicians.

Further research not published here—a champion emerges

Further research has allowed us to relook at Targets initially used and add in responsiveness variables, to further enhance the work done in the initially. In essence, we used the initial research data, to create variables labeling patients' actual responsiveness to diuretics (i.e., do they respond quickly and well, or not at all) and included these back into the above three algorithms.

The results when reapplied to a larger dataset from MIMIC-IV, as well as using the additional inputs, show a great improvement in model quality from the initial research. It has furthermore conclusively shown a Champion Model emerging when the final results could be evaluated side by side.

The conclusions on our champion algorithm

Comparison of algorithm 1 and 3, with and without responsiveness

1. A comparison of the first 12 hours UO after diuretic administration, shows that if the responsiveness at hour 3 is included in the models, that we reach 80% model quality/accuracy quicker (between hour 6 and 7 respectively) versus when responsiveness is not included, in which case we reach 80% accuracy between hour 7 and 8.
2. Furthermore, Version 1 of the algorithms, where the patient weight, diuretic received, dose of diuretic received (normalized for IV and med bolus, as well as Bumetanide and Furosemide) and UO are used, reach 80% accuracy quicker than algorithm Version 3 where only the Patient UO is used as input.

3. The same is true when one looks at hour 13–23 for reaching 95% accuracy far earlier (at hour 17 when responsiveness is included) versus hour 21 without responsiveness; and similar differences are observed between Version 1 and 3 of the algorithm.
4. The results also show that the boosted tree models perform best across algorithm versions.
5. Thus a recommended *Champion Model* has emerged using a boosted tree model for algorithm 1, where the responsiveness is included.

Examples to illustrate model performance for actual patients

In the section below, we are illustrating in a dashboard some actual results for a few patients, given algorithm Version 1 above.

As an example: Fig. 12.33 illustrates, for patient "A," the journey of their first 24 hours of 1 × ICU stay. During this time, they were started on an IV with Metropolol at hour 1 at 5 mg/hour, until hour 13. They also received a med bolus of Furosemide at hour 9. Given this, and their weight, the chart below shows:

1. In PINK—The prediction for hour 6, hourly from hour 1–5. The actual Urine Output at hour 6 was 265 mL. The Predicted hour 6 Urine Output at hour 3 was 389 mL (31% over predicted), and at hour 5 was 235 mL (11% under predicted).
2. In GREEN—The prediction for hour 12, hourly from hour 1–11. The actual Urine Output at hour 12 was 930 mL. The Predicted hour 12 Urine Output at hour 6 was 1123 mL (17% over predicted), and at hour 8 was 932 mL (0.02% over predicted) and at hour 11 was 935 mL (0.05% over predicted). Close from hour 7 onwards!
3. In PURPLE—The prediction for hour 24, hourly from hour 1–24. The actual Urine Output at hour 24 was 1780 mL. The predicted hour 24 urine output at hour 6 was 1667 mL (6% under predicted), and at hour 12 was 1822 mL (5% over predicted), and at hour 23 was 1620 mL (9% under predicted).

Looking at the results below and given an actual patient, the models, especially over 12 and 24 hours, seem to perform quite well at predicting urine output and as expected from the R-squared values explained in previous sections (Figs. 12.33 and 12.34).

These can be explored further on the actual dashboards themselves, but there are more examples visualized below as follows:

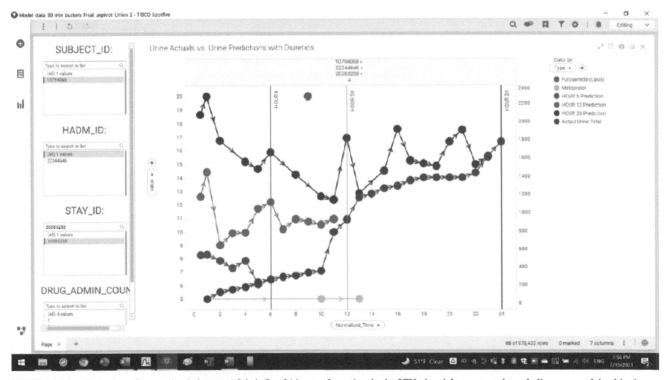

FIGURE 12.33   Illustration of one patient's journey of their first 24 hours of one time in the ICU; the pink, green, and purple lines are explained in the text.

Urine Actuals vs. Urine Predictions with Diuretics

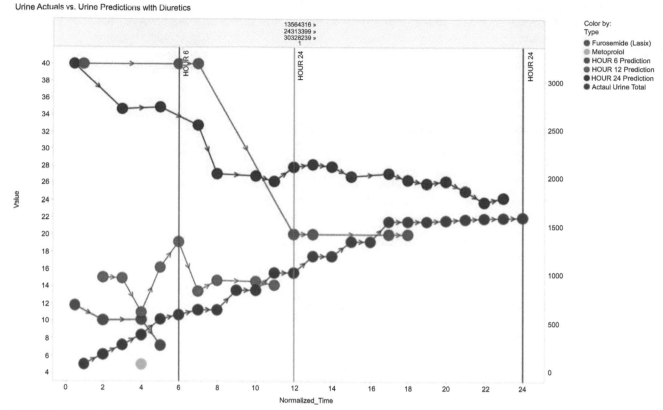

FIGURE 12.34   Actual urine output of a patient, compared to those predicted by the 6 hour, 12 hour, and the 14 hour models.

For patient "B," the journey of their first 24 hours of $1 \times$ ICU stay is shown below. During this time, they were started with a med bolus of Furosemide at hour 1 at 40 mg, with another 40 mg at hour 6 and 7, and then another 20 mg dosage at hour 12, 13, 17, and 18. A small dose of Metropolol of 5 mg is also recorded at hour 4. Given this, the chart below shows some interesting patterns:

1. The 6 hour Prediction seems to be quite close at hour 4, but then drops off significantly at hour 5.
2. The 12 hour Prediction seems to get close at hour 5, but then spikes at hour 6 and drops at hour 7. It seems to be influenced by the Urine output flattening out. It then gets close at the 10- and 11-hour marks.
3. For the 24-hour Prediction, it seems to start high at over 3000 mL of UO (most likely because of the diuretic dosages), and stays relatively high, until hour 8 where it starts dropping closer to 2000 mL (with diuretic dosages lowering but continuous). It starts flattening out and getting closer to the actual UO until getting close at hour 22. This seems to ring true when we look at variable importance as we progress in hours in our model results. The more urine output values we have, they start outperforming the other variables (like diuretic dosage) in Predicting UO over time.

Conclusions and further recommendations

The study allowed us to find a useful research data set, discover appropriate variables, and confirmed our approach and choice of algorithms to be used for building the models for our research goals; more importantly for patient-centered health care, it also allowed us to assess the feasibility of implementing such algorithms in medical devices.

Conclusions

From the predictive models discussed in this chapter, it becomes clear that we should be able to build feasible models for the prediction of urine output for clinicians' real-time use, given data available from both medical devices and electronic medical records. Given even considerably basic input data, models can be built to predict urine output. These models become exceedingly accurate as they receives actual urine output over the first 24 hours of a diuretic drug administration in the ICU.

Recommendations

Given the conclusions above, it appears that there are various other input and output variables that could be studied around this topic and would be highly informative and supportive of future research.

Some future research endeavors could include but are not limited to:

1. Developing algorithms around acute kidney injury, effects on length of hospital and ICU stay (and associated costs).
2. Effects on various inputs around volume responsiveness and other variables on mortality.
3. Studying the above from the perspective of integration of other data and technologies. Namely data from other sources (electronic medical records, CCE data, infusion data) could be made available to the models/algorithms.
4. Most efficient strategies around getting the predictions and decisions from algorithms to devices (the edge), as well as communicating results, data, and dashboards to other systems and to providers, payors, patients, and other stakeholders efficiently.

These are immediate by-products of this study which come to mind; the above recommendations are well worth further study with an aim of developing a better set of Best Practices.

Postscript

This case study was based on a research program directed by a vendor external to the Beth Israel Deaconess Medical Center. After completion of the project, the study of treatment program design was continued at the medical center for patients of acute kidney failure. The overriding importance of this chapter and also chapter 11 on glaucoma IOP patient monitoring is that they show how such studies could lead to the development of effective, real-time devices that the patient can use at home in monitoring and predicting "next steps" in the treatment of a disease; this is all part of the wearable devices use in patient care that is expected to explode in use during 2022 and subsequent years. Use of such "wearable devices" is one type of innovation in medicine that providers can develop and put into use easily in today's world, thus these will be invaluable in the years to come in both the diagnosis and treatment of many medical conditions, plus they provide tons of new data to use in perfecting the "individualized patient analytic models." In conclusion, both the doctors and the patients have new modern effective "tools" to enhance the care of the individual person.

Further reading

Claure-Del Granado, Mehta, 2016. BMC Nephrol. 17, 109. Available from: https://doi.org/10.1186/s12882-016-0323-6.

McCoy, et al., 2019. Crit. Care Expl. 1, e0021. Available from: https://doi.org/10.1097/CCE.0000000000000021.

Pattharanitima, et al. Blood Purif; Available from: https://doi.org/10.1159/000513700.

Seitz, et al., 2020. J. Intensive Care 8, 78. Available from: https://doi.org/10.1186/s40560-020-00496-7.

Shen, et al., 2019. Crit. Care 23, 9. Available from: https://doi.org/10.1186/s13054-019-2309-9.

Zhang, et al., 2019. Crit. Care 23, 112. Available from: https://doi.org/10.1186/s13054-019-2411-z.

Chapter 13

Prediction tool development: creation and adoption of robust predictive model metrics at the bedside for greatly benefiting the patient, like preterm infants at risk of bronchopulmonary dysplasia, using Shiny-R

John B.C. Tan, PhD, Rebekah M. Leigh and Fu-Sheng Chou, MD, PhD

Chapter outline

Prelude

This book is focused on promoting personalized medical diagnoses and treatments. But one segment of this population of patients are those with very special problems—preterm infants. This case study describes briefly the pressing need to assess the risk of infants to develop Bronchopulmonary Dysplasia (BPD). Sufficient data is not available yet for build such a predictive model. The case study describes the challenges of data collection and data wrangling that must be met with data cleaning and visualization operations that must be accomplished before a predictive model can be built. Analysts may be tempted to start building a model as soon as appropriate data are available. This is almost always a big mistake. This case study describes an approach to exploratory data analysis (EDA) with analytical tools based on the R programming language.

Author's note

This chapter assumes that the reader is familiar with R syntax and has R and RStudio installed. Download links will be provided in the appendix.

Practical Data Analytics for Innovation in Medicine. DOI: https://doi.org/10.1016/B978-0-323-95274-3.00023-3

Rationale

Exploratory data analysis for health data

Imagine you've just been given access to a large healthcare dataset that contains various demographics and common morbidities of preterm infants across the entire country. What's your next move to turn this gigantic collection of data into meaningful conclusions and relevant data? Rightly, you decide to first dig into the files and determine what demographics and diseases are included in the dataset before you begin the data analysis and statistical tests needed. This first step or initial assessment of this—or any—dataset has been termed "exploratory data analysis". EDA is essential in any data-driven project to mentally and visually organize the information given to gain valuable insight and better understanding of the data.

Differences between traditional analysis and exploratory data analysis driven analysis

The approach to any dataset wasn't always this way. An important distinction between traditional analysis and EDA is open-mindedly exploring the data and making hypotheses based on what the data reveals. In traditional analysis, data collection is followed by implementation of a model which you would use to analyze and test the data. In comparison, EDA immediately analyzes the data collected to decide what model would best fit. In other words, EDA lets the data speak up for what statistical model (like regression model or analysis of variance) would be appropriate. Choosing a statistical model can be difficult because fitting a real-world phenomenon into the constraints of a mathematical formula is easier said than done, yet can be greatly improved by understanding your data through EDA.

Established in 1977 by John W. Tukey in his book *Exploratory Data Analysis* (David and Tukey, 1977), this field of data exploration has continued and evolved likely due to the development of new technology, access to more and bigger data, and increased availability of computing power and data analysis software (Bruce and Bruce, 2017). EDA could be described more as an art than a protocol, in that there is no predetermined formula or set of steps to reach the desired result. An important component of EDA is visualizing your dataset through graphs or tables to give you a clear picture of potential relationships between variables and any outliers that would affect your analysis. The location, variability, distribution, density of numerical data can be plotted using histograms, scatter plots, stem-and-leaf diagrams, etc., while categorical variables can be noted in bar charts or pie charts.

Prediction tool development

Using EDA gives statisticians and clinicians the power to make important inferences and notice patterns or trends based on real-time clinical data. This power of datasets has been harnessed to develop prediction models that aim to use pre-existing healthcare data to predict an outcome that has not yet been observed (Leisman et al., 2020). Clinical variables are selected for the prediction model if they are presumed to be prognostic of the outcome of concern. The ultimate goal of prediction models is to guide healthcare teams in customized clinical decision-making by stratifying the risk, likelihood, or prognosis of a disease with the intent of improving patient outcome (Shipe et al., 2019). For example, a prediction model developed by the Eunice Kennedy Shriver National Institute of Child Health and Human Development (NICHD) predicts the absence or presence and severity risk of BPD, a common pulmonary morbidity of preterm infants. Clinicians can administer important treatments to these infants guided by the level of disease severity predicted by the online estimator (Cuna et al., 2018).

Furthermore, using prediction tools in healthcare can positively impact disease outcomes by advising healthcare professions to administer influential treatment interventions before it's "too late." That is, physicians can predict the onset and severity of a disease before the usual symptoms manifest to provide treatment and individualized care well before the patient would have received it with a traditional diagnosis. Utilizing prediction tools in this manner is especially useful in preterm infants, noting the concept of antenatal and postnatal factors significantly affecting long-term health outcomes and diseases manifesting in adulthood. This hypothesis is termed "Developmental Origins of Health and Disease (DOHaD)" which originated in a 1990s study that discovered a strong linkage between reduced growth in early life and noninsulin dependent diabetes (Hales et al., 1991). Overall, the idea postulates that any factors (e.g., tobacco, maternal stress, infection, etc.) affecting the period prior to birth or early development shows an strong association and critical contribution to increased risk and development of adult disease (Mandy and Nyirenda, 2018). Overall, the most important part of prediction tools is whether it is clinically useful and can guide physicians in treatment decisions. One hope with this chapter is to show you how we upgraded an easy-to-use online prediction tool for BPD into a more clinically informative guide for physicians and patients.

The need for a prediction tool for premature infants at risk of bronchopulmonary dysplasia

To give you some background, BPD continues to be one of the most common diseases of extreme preterm birth, complicated and potentially caused by the intense ventilation typically given to support a preterm infant's immature lungs. Although many other neonatal morbidities have shown a decrease in incidence, BPD incidence rates have not improved, perhaps due to its multifactorial nature (Thébaud et al., 2019). These infants struggle with inspiration as evidenced by common symptoms like rapid, labored breathing, nasal flaring, grunting, and recruiting abdominal muscles to aid in the work of breathing. Conclusive diagnostic criteria for BPD remain elusive, yet many classify the disease by the amount and type of (i.e., more or less intense) oxygen therapy required after 28 days of life. BPD is associated with neurodevelopmental delay (Trittmann et al., 2013), higher in-hospital mortality rate, increased length of hospital stay, greater healthcare costs, and many additional health care encounters (Mowitz et al., 2019). Furthermore, infants who survive after severe BPD have been shown to have significant pulmonary complications in adulthood (Carraro et al., 2013; Eber and Zach, 2001). In short, BPD will affect an infant's cognition, health, and life long after they leave the neonatal intensive care unit.

A multitude of pre- and postnatal variables are hypothesized to contribute to the characteristic disruption in lung development including infection, oxygen toxicity and mechanical ventilation, patent ductus arteriosus, low birth weight, and preterm birth (Pasha et al., 2018; Thébaud et al., 2019). The link between all these factors is the final common pathway of BPD—inflammation—which provides a very narrow therapeutic window for antiinflammatory treatments like postnatal corticosteroids (Thébaud et al., 2019). Importantly, despite still being debatable, some studies reported that the incidence of BPD is reduced if prompt treatment is given in the first few weeks of life, necessitating the accurate and prompt diagnosis of BPD using prediction tools. Yet, here's the tricky part. Corticosteroid use has been associated with impaired neurodevelopment, cerebral palsy, increased risk of intestinal perforation with indomethacin exposure, and late-onset sepsis (Barrington, 2001; Shaffer et al., 2019). Given the potential for both benefit and harm, corticosteroid use is recommended if the severity risk of BPD is considerable (>50%) (Thébaud et al., 2019; Doyle et al., 2017). Thus, individually tailored therapies for BPD have been guided by severity risk estimators like NICHD's BPD outcome estimator. Premature infants suffering from BPD would greatly benefit from the development of a robust prediction model to help clinicians make informed decisions, especially when it comes to the timing of corticosteroid administration.

Modeling output from the estimator to improve prediction

NICHD's BPD online estimator provides the probabilities for five possible outcomes (no BPD, mild, moderate, severe, or death) based on demographic variables and respiratory support required in a single postnatal day. An assessment of the estimator was conducted using the probabilities generated for data collected from 469 infants admitted to the neonatal intensive care unit between 2015 and 2020 at Loma Linda University Children's Hospital (Leigh et al., 2021). Surprisingly, the results showed a poor correlation between predicted clinical outcome and the actual clinical outcome with an overall accuracy rate of 29%. As part of EDA, the probabilities were modeled using generalized additive mixed modeling (GAMM). The curves generated for each clinical outcome group (i.e., mild, moderate, etc.) continuously overlapped as the predicted clinical outcome severity increased, except for the no BPD group. Through GAMM modeling and data visualization, we realized that the estimator would not be able to provide accurate outcome probabilities, negating any possibility to change the course or trajectory of these infants developing BPD. That is, the infants with severe BPD or with a high possibility of death would not be able to receive the life-changing treatments in time. Noting that the estimator's most useful clinical function relies on its prediction of BPD severity or death in preterm infants, we thought of a way to improve its accuracy (Fig. 13.1).

Here's where the power of modeling gets really cool and can be unabashedly used to improve lives and aid clinicians! Given that infants predicted to have severe BPD or death need the most intervention, we dichotomized this target population from infants in the other severity groups. Using a receiver operator characteristic curve to identify the most likely cutoff score to maximize sensitivity (lower false negatives) and specificity (lower false positives) with a grid search of the nearby values, we noted an optimal cutoff point of 21 which provided the balanced performance for sensitivity, specificity, positive predictive value, negative predictive value, and overall accuracy in 80% of the data—the training dataset. We then tested this cutoff value on the remaining 20% of the data (the testing dataset) to demonstrate that combining the probability scores generated from the severe and death categories accurately predicts the clinical outcome in 75%–80% of neonates (if score is more than 21). Remembering that multiple factors play a role in the pathogenesis of BPD, our future goal is to include these variables in a model to improve severity prediction and guide lifesaving treatments and ameliorative care for countless preterm infants. Finally, our goal in this chapter is to take you

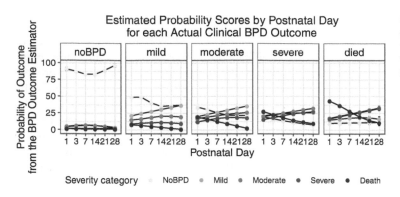

FIGURE 13.1 The trajectory of probability scores for each actual clinical BPD outcome. Besides the "No BPD" group, we found that the probability scores generated by the estimator had poor predictive capabilities, regardless of postnatal day.

through some steps in the process of EDA and prediction tool development based on real-time clinical data collected by the NICHD BPD outcome estimator.

Methods

Obtaining and processing data

Obtaining data for a predictive modeling project is no easy task. While novice data scientists or data scientists without access to their own data may benefit from perusing Kaggle, Google's Dataset Search, or PhysioNet to explore freely available datasets, researchers who want to answer a specific research question without a dataset would need to obtain that data themselves.

First, upon embarking on a new and innovative predictive modeling project, one must decide *which* data to collect. Furthermore, the researcher must decide the *form* that their variables are going to take, since this may be an important factor in their predictive model. For example, a variable called *age* can be represented in multiple ways: in years, months, or even days. These a priori decisions all play a large role in all stages of the modeling process.

Once the specifics of which data to collect is decided, the researcher must find where the data is located. In our example, data was in one main location: the *electronic medical record* (EMR). The EMR contains a lot of private patient information—institutions are subject to a host of regulatory and legal rules to ensure patient confidentiality. Thus, the researcher must do their due diligence and go through the proper channels to obtain this highly sensitive information. Access to the data must be approved by the institutional review board.

After access is granted, the data can be extracted manually by going into each patient's chart and individually recording data. However, this method tends to be inefficient and wastes time and resources. A more appropriate solution would be to ask the institution's IT team or data warehouse engineer (or another entity with similar responsibilities) to batch extract the data that is necessary for your predictive model. Admittedly, this more efficient approach is a luxury; not all researchers will be lucky enough to batch export their patient data. However, if the institution plans to support predictive modeling and other big data type projects, it would be prudent to deploy a data extraction pipeline from the EMR.

Data tidying

After the required data is extracted from the EMR, postprocessing of the data is necessary. Because the EMR contains data that is (1) input by humans and (2) formatted to prioritize human readability rather than machine readability, the data will most likely be hard to utilize for modeling until the researcher tidies that data. The "messy" data will need to be "tidied" in such a way that statistical analysis and predictive modeling tools are able to understand the data to provide the expected output. The data tidying strategy will largely depend on the research question and hypothesis, though there are general guidelines that one can follow to ensure ease of analysis and predictive modeling.

Sometimes, data tidying may seem as intuitive as making sure that there are no typographical errors, or that all "Yes" and "No" answers in a column are consistently spelled, as most coding languages and software are unable to differentiate between "Yes," "yes," "y," and "Y." Other times, the messy data may need more structured tidying. Dr. Hadley Wickham, the prominent developer for many indispensable R packages, wrote an article discussing the principles of tidying data. In his article, he states that the three properties of tidy data are that (1) each variable forms a column, (2) each observation forms a row, and (3) each type of observational unit forms a table (Wickham, 2014). These properties contribute to the ease of analyzing data and statistical modeling.

Our data collected from the EMR was originally organized in a one encounter (any interaction between a healthcare provider and the patient) per row basis. We had to tidy this data so that it would be organized from a one encounter per row basis to a one observation per row basis. The data was tidied and filtered according to the variables necessary to use NICHD's Neonatal BPD Outcome Estimator. This data includes gestational age in weeks, birth weight in grams, sex, race/ethnicity, postnatal day, ventilator type, and FiO_2. Because the Neonatal BPD Outcome Estimator only accounts for postnatal days 1, 3, 7, 14, 21, and 28, we filtered our dataset to only include those days. This is how the tidy preliminary data set was created.

Using R Shiny for efficient data input and visualization

Our next step was to obtain the BPD probability scores. To expedite this process, we created an interactive web app using R's Shiny package. Shiny is a very flexible package that can be used for multiple purposes although its primary purpose is to display data in a dashboard format. It can also be made to be interactive. Our Shiny web app's main purpose was to maximize the efficiency by which we could input the Neonatal BPD Outcome Estimator probability score for each subject and each day. To achieve this, our Shiny web app had to: (1) display the clean preliminary dataset, (2) embed and display the Neonatal BPD Outcome Estimator as an iframe to increase efficiency, and (3) add the resulting probability score to a new dataset and be able to save the results.

The Shiny package allows us to create a web app to process information in a *reactive* manner. The basic structure of a Shiny app is made up of two main sections: *UI* and *server*. The details of creating a Shiny web app are subject to change in the future and as such are outside the scope of this chapter. An in-depth guide to Shiny can be found here (Sievert et al., 2021), and some examples of code can be found later in the chapter.

After obtaining the finalized clean data

Training and testing dataset

After the final clean data was obtained, the dataset was split into training and testing datasets to estimate the performance of the predictive model. The most common type is a random split where the data is split into an arbitrary ratio, which is most commonly an 80:20 split. That means that 80% of the data is randomly set aside as the training dataset and 20% will be used to evaluate the model created by the training dataset. This is known as a *train-test split*.

Depending on the research question and the data, the random split may not be an optimal split to evaluate your predictive model. In some cases where there is an imbalanced dataset or a dataset that spans across multiple years, it is important to have representation for the values of variables, or years in both the training and the testing set. One way to ensure this is to perform a *stratified split*, where the dataset is split in such a way that the training dataset and the testing dataset have a similar percentage of samples for each value of a variable.

TRIPOD guidelines highlights

Prediction models in medicine are becoming increasingly popular. However, a standardized way to report prediction models was not established until 2015, when the Transparent reporting of a multivariable prediction model for individual prognosis or diagnosis (TRIPOD) Guidelines were developed. These standardized guidelines are important because their utility strongly contributes to an increased quality in prediction models and increases their chances to be used for real-life applications. For example, many prediction models have been recently developed to accurately assess risk for contracting the infamous severe acute respiratory syndrome coronavirus 2 (SARS-CoV 2). Most of these prediction models have been strongly criticized because of the lack of transparency in their prediction model development and validation. Therefore, it is important to be clear and transparent when reporting prediction models.

The TRIPOD Guidelines provide a checklist on what aspects of the prediction model development and validation need to be included to fully understand the context of the prediction model. These guidelines include important sections such as sources of data, a description of the participants, outcomes, development, validation, etc. To increase the transparency of the reporting of a prediction model study, it is vital to follow the TRIPOD recommendations. These recommendations may also serve as a guide for novice researchers starting out in predictive modeling on how to communicate their methodology and results. The low barrier of entry for creating predictive models and high rise in popularity have led to an increased number of prediction models being published, consequently highlighting the need to follow standardized recommendations, such as the TRIPOD Guidelines, to ensure that only the highest quality prediction models are utilized for real-life applications (Fig. 13.2).

TRIPOD Checklist: Prediction Model Development and Validation

Section/Topic	Item		Checklist Item	Page
Title and abstract				
Title	1	D;V	Identify the study as developing and/or validating a multivariable prediction model, the target population, and the outcome to be predicted.	
Abstract	2	D;V	Provide a summary of objectives, study design, setting, participants, sample size, predictors, outcome, statistical analysis, results, and conclusions.	
Introduction				
Background and objectives	3a	D;V	Explain the medical context (including whether diagnostic or prognostic) and rationale for developing or validating the multivariable prediction model, including references to existing models.	
	3b	D;V	Specify the objectives, including whether the study describes the development or validation of the model or both.	
Methods				
Source of data	4a	D;V	Describe the study design or source of data (e.g., randomized trial, cohort, or registry data), separately for the development and validation data sets, if applicable.	
	4b	D;V	Specify the key study dates, including start of accrual; end of accrual; and, if applicable, end of follow-up.	
Participants	5a	D;V	Specify key elements of the study setting (e.g., primary care, secondary care, general population) including number and location of centres.	
	5b	D;V	Describe eligibility criteria for participants.	
	5c	D;V	Give details of treatments received, if relevant.	
Outcome	6a	D;V	Clearly define the outcome that is predicted by the prediction model, including how and when assessed.	
	6b	D;V	Report any actions to blind assessment of the outcome to be predicted.	
Predictors	7a	D;V	Clearly define all predictors used in developing or validating the multivariable prediction model, including how and when they were measured.	
	7b	D;V	Report any actions to blind assessment of predictors for the outcome and other predictors.	
Sample size	8	D;V	Explain how the study size was arrived at.	
Missing data	9	D;V	Describe how missing data were handled (e.g., complete-case analysis, single imputation, multiple imputation) with details of any imputation method.	
Statistical analysis methods	10a	D	Describe how predictors were handled in the analyses.	
	10b	D	Specify type of model, all model-building procedures (including any predictor selection), and method for internal validation.	
	10c	V	For validation, describe how the predictions were calculated.	
	10d	D;V	Specify all measures used to assess model performance and, if relevant, to compare multiple models.	
	10e	V	Describe any model updating (e.g., recalibration) arising from the validation, if done.	
Risk groups	11	D;V	Provide details on how risk groups were created, if done.	
Development vs. validation	12	V	For validation, identify any differences from the development data in setting, eligibility criteria, outcome, and predictors.	
Results				
Participants	13a	D;V	Describe the flow of participants through the study, including the number of participants with and without the outcome and, if applicable, a summary of the follow-up time. A diagram may be helpful.	
	13b	D;V	Describe the characteristics of the participants (basic demographics, clinical features, available predictors), including the number of participants with missing data for predictors and outcome.	
	13c	V	For validation, show a comparison with the development data of the distribution of important variables (demographics, predictors and outcome).	
Model development	14a	D	Specify the number of participants and outcome events in each analysis.	
	14b	D	If done, report the unadjusted association between each candidate predictor and outcome.	
Model specification	15a	D	Present the full prediction model to allow predictions for individuals (i.e., all regression coefficients, and model intercept or baseline survival at a given time point).	
	15b	D	Explain how to the use the prediction model.	
Model performance	16	D;V	Report performance measures (with CIs) for the prediction model.	
Model-updating	17	V	If done, report the results from any model updating (i.e., model specification, model performance).	
Discussion				
Limitations	18	D;V	Discuss any limitations of the study (such as nonrepresentative sample, few events per predictor, missing data).	
Interpretation	19a	V	For validation, discuss the results with reference to performance in the development data, and any other validation data.	
	19b	D;V	Give an overall interpretation of the results, considering objectives, limitations, results from similar studies, and other relevant evidence.	
Implications	20	D;V	Discuss the potential clinical use of the model and implications for future research.	
Other information				
Supplementary information	21	D;V	Provide information about the availability of supplementary resources, such as study protocol, Web calculator, and data sets.	
Funding	22	D;V	Give the source of funding and the role of the funders for the present study.	

*Items relevant only to the development of a prediction model are denoted by D, items relating solely to a validation of a prediction model are denoted by V, and items relating to both are denoted D;V. We recommend using the TRIPOD Checklist in conjunction with the TRIPOD Explanation and Elaboration document.

FIGURE 13.2 TRIPOD checklist form (https://www.equator-network.org/reporting-guidelines/tripod-statement/).

Code examples and tutorial

Data cleaning and TidyR examples

The data tidying strategy is largely dependent on the data source and the initial impression of the data when viewing it manually. Furthermore, only information and variables related to your research questions are important to the researcher. The raw data must be processed into tidy data. The package *tidyr*, a component of the collection of packages known as *tidyverse*, incorporates such functions to structure the data into tidy data (Wickham et al., 2019). One of these functions is to *pivot* the dataset.

Pivoting is the concept of transforming the dataset between *wide* and *long* formats. A common example of a wide format may include one subject per row and includes longitudinal information across the columns (Table 13.1).

Note: You may notice that the column titles contain no spaces—this is an important caveat when naming headers. Accounting for spaces in headers can cause problems down the line.

TABLE 13.1 Example of a wide dataset.

Subject	LabValue_Day1	LabValue_Day2	LabValue_Day3
1	44.5	67.2	23.6
2	20.5	43.2	62.3
3	24.6	26.2	42.2

While this *wide* format may be more readable to humans, most analysis and modeling functions would prefer a *long* format to make it more machine readable. We want to make the dataset *longer* by increasing the number of rows while decreasing the number of columns. We will use the function *pivot_longer()* from the tidyr package to achieve our goal.

```
1.  # assuming that the dataset is saved under the variable "data"
2.  library(tidyr)
3.  data <- data %>% pivot_longer(!ID,
4.                  names_to = c(".value", "Day"), # .value says that the first item after
    the split will be used as the column name for values
5.                  names_sep = "_") # separate by underscore
6.  data
```

This will produce a table like so (Table 13.2).

TABLE 13.2 Example of a long dataset.

Subject	Day	LabValue
1	Day1	44.5
1	Day2	67.2
1	Day3	23.6
2	Day1	20.5
2	Day2	43.2
2	Day3	62.3
3	Day1	24.6
3	Day2	26.2
3	Day3	42.2

As you can see, with just a few lines of code, we are able to transform our dataset to make our data more analyzable. For example, we can easily call the column *LabValue* and perform mathematical operations such as calculating the arithmetic mean.

```
1.  mean(data$LabValue) # $ selects only that column
```

Initializing an R Shiny web app

An R Shiny web app can be initialized directly from RStudio. Within RStudio, click File > New File > Shiny web app. We recommend that beginners start with a single file application type. At the time of this writing, the default template for your *app.R* file may look like this:

```
1.  #
2.  # This is a Shiny web application. You can run the application by  clicking
3.  # the 'Run App' button above.
4.  #
5.  # Find out more about building applications with Shiny here:
6.  #
7.  #    http://shiny.rstudio.com/
8.  #
9.
10. library(shiny)
11.
12. # Define UI for application that draws a histogram
13. ui <- fluidPage(
14.
15.     # Application title
16.     titlePanel("Old Faithful Geyser Data"),
17.
18.     # Sidebar with a slider input for number of bins
19.     sidebarLayout(
20.         sidebarPanel(
21.             sliderInput("bins",
22.                         "Number of bins:",
23.                         min = 1,
24.                         max = 50,
25.                         value = 30)
26.         ),
27.
28.         # Show a plot of the generated distribution
29.         mainPanel(
30.             plotOutput("distPlot")
31.         )
32.     )
33. )
34.
35. # Define server logic required to draw a histogram
36. server <- function(input, output) {
37.
38.     output$distPlot <- renderPlot({
39.         # generate bins based on input$bins from ui.R
40.         x    <- faithful[, 2]
41.         bins <- seq(min(x), max(x), length.out = input$bins + 1)
42.
43.         # draw the histogram with the specified number of bins
44.         hist(x, breaks = bins, col = 'darkgray', border = 'white')
45.     })
46. }
47.
48. # Run the application
49. shinyApp(ui = ui, server = server)
```

Following the instructions in the comment blocks, hitting Run App at the top will open a local browser that will show a slider and a histogram. Notice how the slider bar changes the bin size of the histogram.

The code structure for a Shiny web app has two parts:

1. The User Interface

```
ui <- fluidPage()
```

2. The Server

```
server <- function(input, output) {}
```

and lastly, we need to run the application by calling these two objects.

```
shinyApp(ui = ui, server = server)
```

The user interface

The user interface of the Shiny web app contains elements that will be the face of your web app. For example, in the template above, the title, slider, and the histogram output all have corresponding panels. The title has its own title panel (line 5). The sidebar panel (lines 9–15) and the main panel (lines 18–20) are objects that exist within *sidebarLayout* (lines 8–20). Within the sidebar panel argument there is a *sliderInput* (line 10) whose options determine the slider's options. feel free to change some of these numbers around to see how it affects the slider. In the default template, the main panel is rather simple. It contains a function to plot the histogram. This function is defined in the server portion.

```
1.  # Define UI for application that draws a histogram
2.  ui <- fluidPage(
3.
4.      # Application title
5.      titlePanel("Old Faithful Geyser Data"),
6.
7.      # Sidebar with a slider input for number of bins
8.      sidebarLayout(
9.          sidebarPanel(
10.             sliderInput("bins",
11.                     "Number of bins:",
12.                     min = 1,
13.                     max = 50,
14.                     value = 30)
15.         ),
16.
17.         # Show a plot of the generated distribution
18.         mainPanel(
19.             plotOutput("distPlot")
20.         )
21.     )
```

The server

The server portion of the Shiny web app is responsible for the code that is not shown to the user. This code could be calculations, algorithms, saving algorithms, connecting or writing to databases, or anything that interfaces with the data. In this instance, the code to produce the histogram is included in the *renderPlot* function and saved within *distPlot*. This allows our UI to call *distPlot* and make a new histogram every time the slider changes.

```
1.  # Define server logic required to draw a histogram
2.  server <- function(input, output) {
3.
4.      output$distPlot <- renderPlot({
5.          # generate bins based on input$bins from ui.R
6.          x    <- faithful[, 2]
7.          bins <- seq(min(x), max(x), length.out = input$bins + 1)
8.
9.          # draw the histogram with the specified number of bins
10.         hist(x, breaks = bins, col = 'darkgray', border = 'white')
11.     })
12. }
```

Loading and saving onto a SQL database

After tidying our data, we moved the data to a data warehouse system known as PostgreSQL. This data warehouse system uses Structured Query Language (SQL), which is a useful tool for handling structured data, as one might encounter when extracting data from the EMR. For projects that involve writing new data, SQL is the preferred tool. This is because for Excel or CSV files, insertion or removal of data is impossible without rewriting the whole file. Using a SQL database allows multiple users to read and write to only one database, which maintains the integrity of the data and reduces the chances of error and duplicating work. Thus, we used SQL to insert more data to our preliminary dataset. Specifically, we used SQL to (1) load the preliminary dataset, and (2) insert the probability scores after using the Estimator. While staying within our Shiny web app, we can connect to the SQL database. This is because SQL has been integrated into R via packages and even some built-in functionality. Some of these packages include DBI, dplyr, and odbc. Using a SQL database along with your Shiny web app means that users can store additional data into your database while maintaining the integrity of the rest of the data. Connecting to the SQL database belongs in the *server* portion of your web app.

Showing and interacting with data

Our goal was to facilitate the process of creating another dataset comprising the probability score obtained from the BPD Outcome Estimator. For this strategy, we employed the use of the DT (DataTables) package and inline frames, or *iframe*.

The DT package allow us to view the preliminary dataset as a table. By default, DataTables allows the user to interact with the rendered table which has functionality for sorting, filtering, and searching.

The iframe allows you to load a different website or document while you're looking at your original website. We directly loaded the NICHD Neonatal BPD Outcome Estimator in our Shiny app. When loading a site through an iframe, please be sure to check the Terms and Conditions of that website to stay legally compliant. In the example below, we created an iframe in the server portion and chose to render the iframe in the side panel. Meanwhile, an example DataTable is shown in the main panel (Fig. 13.3).

```
1.  library(shiny)
2.
3.  ui <- fluidPage(titlePanel("iFrame Example"),
4.
5.              sidebarLayout(
6.                  # Side panel with the iFrame
7.                  sidebarPanel(
8.                      fluidRow(
9.                          htmlOutput("bpd_calculator")
10.                     )
11.                 ),
12.
13.                 # Main panel with example dataset
14.                 mainPanel(
15.                     fluidPage(DTOutput('tbl')),
16.                 )
17.             )
18. )
19.
20. server <- function(input, output) {
21.
22.     output$bpd_calculator <- renderUI({
23.
    tags$iframe(src='https://neonatal.rti.org/index.cfm?fuseaction=BPDcalculator.survival',
24.                 height = 700, width = 310)
25.     })
26.
27.     # render the iris dataset, a popular beginner's dataset
28.     output$tbl <- renderDT(
29.         iris, options = list(lengthChange = FALSE))
30.
31. }
32.
33. shinyApp(ui, server)
```

iFrame Example

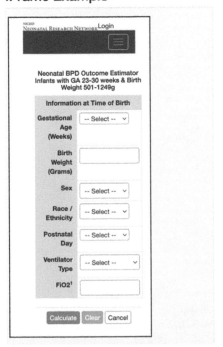

	Sepal.Length	Sepal.Width	Petal.Length	Petal.Width	Species
1	5.1	3.5	1.4	0.2	setosa
2	4.9	3	1.4	0.2	setosa
3	4.7	3.2	1.3	0.2	setosa
4	4.6	3.1	1.5	0.2	setosa
5	5	3.6	1.4	0.2	setosa
6	5.4	3.9	1.7	0.4	setosa
7	4.6	3.4	1.4	0.3	setosa
8	5	3.4	1.5	0.2	setosa
9	4.4	2.9	1.4	0.2	setosa
10	4.9	3.1	1.5	0.1	setosa

Showing 1 to 10 of 150 entries   Previous 1 2 3 4 5 ... 15 Next

FIGURE 13.3 Example of how to incorporate an iframe while simultaneously viewing a data table on the same page. This greatly increases data pipeline efficiency.

Reactive expressions to account for computationally or temporally expensive algorithms

One of the benefits of using a Shiny web app is its fast responsiveness. However, at one point your web app may be so complex that the processing time may start slowing down your app, resulting in a poor user experience. One way to combat this is to use *reactive* expressions in the server portion of your Shiny web app.

```
1.  server <- function(input, output) {
2.
3.    dataInput <- reactive({
4.      # computationally expensive or time consuming algorithms go here
5.    })
```

Wrapping R expressions around *reactive* tells R to cache the values when it is first run and check if it changes so that that the calculation won't be needlessly repeated. Reactive expressions check if their values need to be updated by checking if widgets that depend on that reactive expression have changed. If the values are outdated compared to its corresponding widget(s), then the computation is repeated, but it will not perform the calculation if no changes have been detected. Reactive expressions are a powerful tool when your web app becomes more complex.

Conclusion

It was not until relatively recently that predictive modeling has become more accessible. Predictive modeling can easily be misinterpreted and can have severe consequences if the information gleamed from a model is prematurely clinically implemented. The pipeline described in this chapter is only a fragment of the overall picture. EDA can be used to assess the dataset in a manner that is conducive to choosing the appropriate statistical model for the research question. Transparent reporting of a multivariable prediction model for individual prognosis or diagnosis (TRIPOD) guidelines should be followed to ensure total transparency and allow for evaluation of reported prediction models. Another aspect of the data pipeline is the utilization of the Shiny web app. Shiny is a fantastic tool to use for efficient communication, visualization, and data entry. *Currently, there is a great need for patient bedside metrics that employ predictive model methods to help the clinician make informed decisions, especially when the risk for mortality and morbidity is high.* The creation and adoption of robust predictive model metrics at the bedside would greatly benefit the patient population, like preterm infants at risk of BPD. Therefore, we hope that this chapter has piqued your interest to learn more about predictive modeling and its potential benefits for mankind.

Appendix

Download links

- R can be installed here: https://www.r-project.org/.
- RStudio can be installed here: https://www.rstudio.com/

Versions of software and packages

- R version 4.0.4
- RStudio version 1.4.1106
- Tidyr version 1.1.3
- Shiny version 1.60
- DT version 1.14.0

Postscript

Construction of a predictive model to predict the risk of BPD would alert the neonatal physicians and nurses to practice special care in treatment of the preterm infant. In Chapter 14, the rationale for building a model to predict the risk for the presence of precancerous colon polyps is presented, along with a case study of the model that was built to predict this risk.

References

Barrington, K.J., 2001. The adverse neuro-developmental effects of postnatal steroids in the preterm infant: a systematic review of RCTs. BMC Pediatr. 1, 1.

Bruce, P., Bruce, A., 2017. Practical Statistics for Data Scientists. O'Reilly Media, Sebastopol, CA.

Carraro, S., Filippone, M., Dalt, L.D., Ferraro, V., Maretti, M., Bressan, S., et al., 2013. Bronchopulmonary dysplasia: the earliest and perhaps the longest lasting obstructive lung disease in humans. Early Hum. Dev. 89 (Suppl. 3), S3–S5.

Cuna, A., Liu, C., Govindarajan, S., Queen, M., Dai, H., Truog, W.E., 2018. Usefulness of an online risk estimator for bronchopulmonary dysplasia in predicting corticosteroid treatment in infants born preterm. J. Pediatr. 197 (June), 23–28.e2.

David, F.N., Tukey, J.W., 1977. Exploratory data analysis. Biometrics 33 (4), 768.

Doyle, L.W., Cheong, J.L., Ehrenkranz, R.A., Halliday, H.L., 2017. Late (> 7 days) systemic postnatal corticosteroids for prevention of bronchopulmonary dysplasia in preterm infants. Cochrane Database Syst. Rev. 10, CD001145.

Eber, E., Zach, M.S., 2001. Long term sequelae of bronchopulmonary dysplasia (chronic lung disease of infancy). Thorax 56 (4), 317–323.

Hales, C.N., Barker, D.J., Clark, P.M., Cox, L.J., Fall, C., Osmond, C., et al., 1991. Fetal and infant growth and impaired glucose tolerance at age 64. BMJ (Clin. Res. Ed.) 303 (6809), 1019–1022.

Leigh, R., Tan, J.B., DeGiorgio, S., Cha, M., Kent, C., Yeh, H.-W., et al., 2021. Combining probability scores to optimize clinical use of the NICHD neonatal BPD outcome estimator. Neonatol. Today 16 (9), 3–13.

Leisman, D.E., Harhay, M.O., Lederer, D.J., Abramson, M., Adjei, A.A., Bakker, J., et al., 2020. Development and reporting of prediction models: guidance for authors from editors of respiratory, sleep, and critical care journals. Crit. Care Med. 48 (5), 623–633.

Mandy, M., Nyirenda, M., 2018. Developmental origins of health and disease: the relevance to developing nations. Int. Health 10 (2), 66–70.

Mowitz, M.E., Ayyagari, R., Gao, W., Zhao, J., Mangili, A., Sarda, S.P., 2019. Health care burden of bronchopulmonary dysplasia among extremely preterm infants. Front. Pediatr. 7, 510.

Pasha, A.B., Chen, X.-Q., Zhou, G.-P., 2018. Bronchopulmonary dysplasia: pathogenesis and treatment (review). Exp. Therap. Med. 16 (6), 4315–4321.

Shaffer, M.L., Baud, O., Lacaze-Masmonteil, T., Peltoniemi, O.M., Bonsante, F., Watterberg, K.L., 2019. Effect of prophylaxis for early adrenal insufficiency using low-dose hydrocortisone in very preterm infants: an individual patient data *meta*-analysis. J. Pediatr. 207 (April), 136–142.e5.

Shipe, M.E., Deppen, S.A., Farjah, F., Grogan, E.L., 2019. Developing prediction models for clinical use using logistic regression: an overview. J. Thorac. Dis. 11 (Suppl. 4), S574–S584.

Sievert, C., Chang, W., and Schloerke, B., February 1, 2021. Shiny 1.6. Studio-Blog: https://www.rstudio.com/blog/shiny-1-6-0/.

Thébaud, B., Goss, K.N., Laughon, M., Whitsett, J.A., Abman, S.H., Steinhorn, R.H., et al., 2019. Bronchopulmonary dysplasia. Nat. Reviews. Dis. Prim. 5 (1), 78.

Trittmann, J.K., Nelin, L.D., Klebanoff, M.A., 2013. Bronchopulmonary dysplasia and neurodevelopmental outcome in extremely preterm neonates. Eur. J. Pediatr. 172 (9), 1173–1180.

Wickham, H., Averick, M., Bryan, J., Chang, W., McGowan, L., François, R., et al., 2019. Welcome to the tidyverse. J. Open. Source Softw. 4 (43), 1686.

Wickham, H., 2014. Tidy data. J. Stat. Softw. 59 (1), 1–23.

Further reading

1.1.2. How Does Exploratory Data Analysis Differ from Classical Data Analysis? n.d. <https://www.itl.nist.gov/div898/handbook/eda/section1/eda12.htm>.

NICHD Neonatal Research Network. n.d. <https://neonatal.rti.org/index.cfm?fuseaction = BPDCalculator.start>.

Chapter 14

Modeling precancerous colon polyps with OMOP data

Robert A. Nisbet

Chapter outline

Prelude

The best way to determine whether a patient has cancer in a body part is to do a biopsy of it. For colon cancer, suspected colon polyps can be excised during a colonoscopy procedure, and then biopsied to determine if the polyp is precancerous or cancerous. Removal of precancerous polyps prevents colorectal cancer. The problem is that most people

Practical Data Analytics for Innovation in Medicine. DOI: https://doi.org/10.1016/B978-0-323-95274-3.00018-X

try to avoid having a colonoscopy at any age. A favorite retort about some obnoxious activity is, "That is about as much fun as having a colonoscopy!"

This chapter describes a screening method driven by an analytical model to predict the presence of precancerous polyps, which can be used as a basis for recommending a colonoscopy for a patient regardless of age. It is true that the use of screening methods like this will move some patients to have a colonoscopy that would not otherwise happen. It is also quite likely that this screening method may provide also some evidence for delaying a colonoscopy, which will be welcomed by most patients.

Chapter purpose

The purpose of this chapter is to demonstrate the set of data preparation and modeling operations necessary to predict the presence of precancerous colon polyps using only common demographic data and clinical data (no images; with no data from special tests) (e.g., FIT or FOBT). Age was not used as a predictive factor for two reasons:

1. because all of the patients in the data set were preselected for age, in that they were older than the guideline age (50) for having a regular colonoscopy, and;
2. because this study was performed to develop a screening tool for use for patients of any age who have not had a prior colonoscopy

This chapter shows how to build a predictive model directly on data available in normal patient files (e.g., birth date) or in the EPIC database (in OMOP format) at the University of California at Irvine Medical Center. The OMOP Common Data Model (CDM) (introduced in Chapter 4) was used to transform data elements from input devices and install them into this data structure. Problems related to disparities at the OMOP data level (e.g., different levels of aggregation and historical records for many data elements) still exist, however. These problems must be solved and many candidate predictor variables must be derived from base OMOP data elements at appropriate levels of aggregation to produce a patient-centric data structure consisting of one row per patient with all appropriate candidate predictor variables in it.

Introduction

The age cutoff for screening colonoscopy was age 50 for the past four decades. Mortality due to colon or rectal cancer (CRC) in adults over the age of 50 has declined by 1% each year from 2008 to 2017 (American Cancer Society, 2021). This decline in CRC mortality is due presumably to the effect of an increasing incidence of colonoscopy exams, and the consequent excision of precancerous polyps. The CRC mortality rate of those people below age 50, however, has been increasing by 2% per year during the period from 2014 to 2018. Therefore, American Cancer Society (ACA) guidelines recently lowered the screening age to 45. Colorectal cancer (CRC) is estimated to be the fourth most diagnosed type of cancer among people aged 30–39 in the United States. There is no indication in the medical literature to explain this increasing rate of CRC among young people. Therefore, physicians need some source of rationale to support the recommendation that certain people under the current minimum age for colonoscopy examination should have a colonoscopy. But what is the appropriate rationale for the appeal and to whom should it be applied? Some diagnostic tests can be performed like Fecal Immunoglobulin Test (FIT), the Fecal Occult Blood Test, or the Cologuard test. But results from these tests provide only preliminary indication that a colonoscopy might be necessary. This study will look for patterns in the EPIC data stored in the UCI Epic data warehouse that can be related to the presence of precancerous colon polyps. Analyses with data analytics tools can provide this rationale and its target population by predicting which patients without a prior colonoscopy have a sufficiently high risk of colon cancer, based solely on available demographic data and data in clinical medical records. Some clinical data must be excluded from the analysis, including fecal tests (FIT and FOBT), because these are not required tests, and many patients lack data for them in their medical records. The model is built with data from patients who have had a prior colonoscopy, but it is to be applied to patients with no prior colonoscopy, therefore, any clinical data gathered from or after the colonoscopy was excluded from the analysis. GI physicians already know that patient age is the most important single factor in the risk of developing cancerous colon polyps. They need, however, another estimate of cancer risk to use as the basis for recommending that a patient has a colonoscopy procedure regardless of age.

The University of California, Irvine Colonoscopy Quality Database

The UCICQD was developed at the University of California Medical Center in 2012 to provide a database to store measurements and documentations related to colonoscopy quality at UC-Irvine. It contains limited patient demographic information (e.g., Age), and all quality metrics of colonoscopies including indications, extent of exam, prep quality and detailed findings including all polyps and cancers, and their pathology diagnoses.

The UCI Colon Polyp Project

Dr. William Karnes MD, AGAF Clinical Professor of Medicine served as Principal investigator of the UCI Colon Polyp Project, from which this case study was abstracted. Dr. Erica Duh (Chief Resident in the Department of Medicine) and Dr. Christopher Rombaoa (Chief Fellow, Division of Gastroenterology and Hepatology, UC Irvine Health) shared the primary responsibility for the development of protocols to integrate the Epic Data Warehouse with the UCICQD system, from which the data set used in this case study was generated.

Previous colon cancer risk screening and predictive modeling programs

Many studies of colon cancer risk have been performed using questionnaires to provide input data for analysis. These studies will not be discussed here, because questionnaires were not used for this case study, and because of the intentional limitation of the study to use only patient demographic and clinical data sources.

A relatively few studies have built predictive models based on clinical data exclusively. One such model build by Labianca et al. (2013) was based on a heuristic that included:

- Personal history of adenoma, colon cancer, inflammatory bowel disease (Crohn's disease or ulcerative colitis);
- Significant family history of colon polyps or CRC;
- Inherited syndrome related to familial adenomatous polyposis.

This study provided no predictive analysis, but rather used this heuristic as evidence of the likelihood of the presence of precancerous polyps in an individual. This model could not be used to score patients with no prior colonoscopy, because some important predictor variables (e.g., Crohn's disease and a prior history of adenoma) would be populated only with data from a colonoscopy.

Three other recent studies did develop analytical models of precancerous colon polyps using only clinical and patient demographic data. A final list of predictor variables in these studies is shown in Table 14.1.

Murchie et al. (2019) did not include BMI or alcohol use for diabetes data, but did include the FIT test (Fecal Immunochemical Test). Park et al. (2019) included data inputs for BMI and diabetes, and from a FIT test. The predictive model was built with a logistic regression model of clinical factors. These inputs were used to calculate risk scores to prompt the recommendation of a colonoscopy. An initial univariate logistic analysis identified the factors shown in Table 14.1. Subsequently, a multivariate logistic regression analysis of these univariate factors produced a model with an Area-Under-the-Curve (AUC)=0.72, but this analysis did not consider any interaction effects.

A more recent study included all of the variables used by Park et al. (2019), and added alcohol use to the list of important predictors of precancerous polyps (Soonklang et al., 2021), generating an AUC=0.77.

The studies of Park et al. and Murchie et al. were based on a very few variables identified from univariate analyses. The UCI case study of colon cancer prediction included over 2000 variables identified in a variable frequency analysis of OMOP-CDM patient data, instead of a univariate statistical analysis of a select few. A feature selection process was performed to generate a short-list of 69 predictor variables.

All three studies listed above used Age as a predictor variable. Also, all three studies used FIT (Fecal Immunochemical Test) as a primary predictor. The UCI study was designed for use without Age or FIT data, but included as candidate predictors all the other variables listed in Table 14.1.

TABLE 14.1 Important predictor variables and the plotted area under the ROC curve (AUC) for three studies that used clinical data as inputs.

Primary authors	Age	Sex	Smoking	BMI	Alcohol	Diabetes	FIT test	AUC
Murchie et al. (2019)	X	X	X				X	0.63
Park et al. (2019)	X	X	X	X		X	X	0.72
Soonklang et al. (2021)	X	X	X	X	X	X	X	0.77

OMOP data

The OMOP CDM is designed to facilitate the systematic analysis of disparate observational medical data in a common storage database format to permit consistent reports and analysis across different organizations (Overhage et al., 2012). See Chapter 4 for more information on the OMOP CDM.

This case study will show how to use OMOP-CDM data in the UCI Medical Center to build a model to predict the presence of precancerous colon polyps. Results from this model can be used as the rationale for physicians to recommend that specific patients have a colonoscopy, regardless of their age.

Caveat

The following data engineering processes are presented here to help the reader understand how to process OMOP-CDM data to build a predictive analytics model. The input OMOP-CDM data set cannot be just submitted to the modeling algorithm directly, it must be processed through a long series of operations to facilitate the building of the optimal model by the modeling algorithm. As useful as the OMOP-CDM data is for reporting and descriptive analyses, it does not support a plug-n-play modeling operation. Extensive data preprocessing operations are described below. Readers can adapt these data preprocessing operations for use with OMOP data in their patient data systems. Most of the work must be done manually, because of the many operations required for using EPIC data are similar, but different from those used in this study. Consequently, these data preprocessing operations are difficult to automate. See the section on Automation of Data Preparation for Medical Informatics for a discussion of those operations that can be safely automated and those that cannot.

Modeling objective

The objective of this case study was to predict the presence of precancerous colon polyps with average risk of being cancerous in patients using only patient demographic, UCICQD data, and clinical data stored in the EPIC data base at the UCI Medical Center. Outcomes of first-time colonoscopies were used to determine whether or not precancerous polyps of average risk of being cancerous were present. The clinical data profile for patients with average risk precancerous polyps prior to a colonoscopy was simulated by excluding all EPIC data on or after the date of colonoscopy, and excluding other relevant factors in the patient record (e.g., FHx of CRC or inflammatory bowel disease).

Methods

Major tasks of data preparation of OMOP data for modeling

Data preparation of OMOP data is a very complex process. The major tasks that must be accomplished before modeling can begin include:

1. integration of data elements from potentially many sources;
2. derivation of target variable classes;
3. deletion of inappropriate records and fields;
4. generation of high frequency OMOP code lists;
5. aggregation to the patient level of data in multiple records for each variable for each patient;
6. derivation of one-hot binary variables from the lists of high frequency codes;
7. feature elimination of "unnecessary" codes to produce the variable "short-list";
8. balancing the data set to produce an equal numbers of rows for each target class; and
9. positive unlabeling of unrepresentative Target=0 rows.

Data access

The analysis of this project was conducted remotely in an off-campus location. This off-campus analysis could constitute a violation of HIPAA regulations. UCI data regulations in compliance with the 1996 Health Insurance Portability and Accountability Act (HIPAA) require that UCI personal health information not be disclosed to information systems outside official campus Health Services facilities. Therefore, all items of personal information, including the UCI patient ID were removed from the data set. Only general information (e.g., Gender, and Age), OMOP-CDM codes, and a common key (unrelated to Patient ID) remained in the data set accessed in the remote location.

A data set consisting of 10,228 records of deidentified data elements from the UCI gastroenterology database and EPIC electronic health records (EHRs) for matching patient IDs were extracted at the UCI Medical Center, including a special record key person-ID, which could be translated to the official UCI patient ID only on a secure UCI Health Services system.

Five data sets were extracted from the EPIC database of EHRs for each person-ID, along with explanation text for each code in each concept category for:

- *Condition codes*, referring to the specific OMOP-CDMmedical conditions (e.g., Kidney disease) for each service instance for a given patient;
- *Procedure codes*, referring to the specific OMOP-CDMprocedure (e.g., appendectomy) performed for each service instance for a given patient;
- *Drug codes*, referring to the specific OMOP-CDMdrug (e.g., aspirin) administered for each service instance for a given patient;
- *Observation codes*, referring to the specific OMOP-CDMobservation (e.g., heavy smoker) for each service instance for a given patient;
- *Measurement codes*, referring to the specific OMOP-CDMlaboratory measurement (e.g., blood pressure) determined during each service instance for a given patient.

The modeling tool

The KNIME Analytics Platform was chosen for the data engineering and modeling operations of this project. KNIME was developed as an open source application at the University of Konstanz, Germany in 2006. It has over 5000 users in over 500 organizations in the world. It is supported by a volunteer collaboration of developers similar to the volunteer support structure for Unix in the 1980s. This support structure is provided primarily in the form of a very active KNIME Forum blog to help individual users solve problems in KNIME workflows (strings of processing nodes in a graphical programming application). Many workflows were built to process separate processing activities in this project. The Variable engineering workflow was built and processed on a UCI server, using raw data provided by the deidentification process. Subsequent processing was performed on a remote server on the integrated and aggregated data set of unique customers assigned a person_id unrelated to the original patient ID number. The selected modeling approach was an ensemble consisting of the fusion of predictions of the Tree Ensemble algorithm (a form of the Random Forest algorithm), a Stochastic Gradient Boosted Trees algorithm, and an XGBoosted Trees algorithm.

Data integration

Many medical operations store data in several different formats in several different places; few have data warehouses. This situation requires identification of the storage format and location of every data source required for analytics operations. Once discovered, information stored in different formats must be reformatted and/or transformed into a common format to serve data requirements for analytics.

UCI data sources consisted of a Gastroenterology Department patient database (extracted to the UCICQD file) and five separate files for the OMOP-CDM concept categories listed above, which were extracted from the EPIC data warehouse. These data sources were integrated in the KNIME Variable Engineering workflow. A portion of this workflow is shown in Fig. 14.1.

Each joining operation requires the specification of selected columns from one data source with a list of selected columns from the other data source according to a selected join type (e.g., left-outer join). The join type used is the left-join, which joins the selected list of columns for all rows of the first data source (the top Joiner process) with all selected columns of rows of the second data source that have matching join keys.

Five OMOP-CDM concept categories (described above) represented by five source data files were integrated to form a single row for each patient. The problem is that there are many records (rows) for each patient with different values in each of their fields. For example, the conditions data source may contain many dozens of records for a given patient, each one of which has different values for blood pressure. Which blood pressure value should be installed in the combined record? Also, there is much information available in the many blood pressure values,

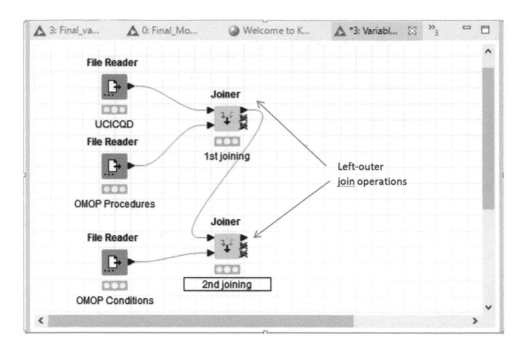

FIGURE 14.1 An example series of join operations, using the left-outer join type to join data from three input data files. The join key is the anonymous person_id field in each input file.

such that a mean blood pressure might be predictive of the outcome, or the range of blood pressure values might be predictive. Both of those summary variables should be derived before the multiple records for a given patient are aggregated.

After appropriate summary variables are derived for conditions, procedures, and drugs, the codes for each concept category must be concatenated to form a comma-separated string of unique codes for further processing. KNIME provides an aggregation option to form a comma-separated text string of unique OMOP-CDMcodes associated with a given patient (list fields). Later, codes comma-separated strings for each concept category are searched for specific strings in the code text fields for each record. This is one way of combining text mining operations to derive many candidate predictor variables for each OMOP concept category (see the Unique Code Determination section below). Another way to use these code lists is in variable frequency analysis (see the Text Mining Frequency Analysis section below).

This string of unique OMOP-CDM codes can be installed in a column of the record for each patient. In the following data preparation processing, this code string is queried to the presence of a specific code in the concatenated text field of a given concept category for each patient. These selected codes can be used to derive the one-hot variables (1=present; 0=absent) for each code selected. This operation is one of the automated operations available in KNIME. The KNIME One-to-Many node will derive a new binary variable (1 or 0) for each class in each categorical variable. This feature saves an enormous amount of time compared to deriving about 2000 binary variables manually.

Target variable definition

One of the early operations in data preparation for analytical models is the derivation of the outcome to be predicted (the *Target* variable). In this case study, a binary target variable was defined as 1 if the pathology from an endoscopic procedure in the UCICQD data for a given patient record contained any Type-1 adenoma:

1. "Tubular adenoma"
2. "Serrated adenoma"
3. "Tubulovillous"
4. "CA" (indicating cancer of some kind)
5. "Villous"
6. "Flat adenoma"

Otherwise, Target was set to 0.

Data type changes

Changing of data type is often necessary to serve analytical purposes. For example, the Target variable was changed to a string data type to satisfy the input requirements of the classification modeling algorithms. Initially, the data type of the 2025 variables was set automatically to String during the one-hot variable derivation process. These variables were reset to the Integer data type for use in modeling.

Data quality assessment and resolution

Sometimes, the assigned dates were invalid (e.g., a future date such as 2022–9–12). These records were deleted, because they provided no valid date information necessary for the data exclusion operations.

Data exclusions

Some data records were judged as not appropriate for the prediction of the Target variable.

1. If the colon benign neoplasm (or colon malignant neoplasm) codes reflect the findings on the current colonoscopy they should not be included (e.g., if entered on day of or after the colonoscopy). If they predate the colonoscopy, they are clear risk factors for the presence of polyps in the current colonoscopy and cannot be ignored.
2. OMOP-CDM codes pertaining to the colonoscopy procedure or the results of it must be excluded (e.g., procedure code 45381).
3. Several codes were found that could only be learned from results of a colonoscopy (e.g., presence of malignant neoplasm in the colon). Therefore, a screening model for patients with precancerous polyps before a colonoscopy could not use colonoscopy results data as a source of predictor variables, therefore, these variables were excluded from further analysis.

The data exclusion process is very important to assure that the patterns for the Target=1 and 0 classes represent those necessary to support the deployment of the model. In this project, the deployment population will be patients under the age of 45 who have never had a colonoscopy. The data exclusions remove any predictors that are related to results of a colonoscopy. For example, the drug Mesalamine is used to treat ulcerative colitis, the treatment of which presumes a colonoscopy. Drugs and procedures that are related to the colonoscopy procedure or treatment resulting from the procedure must be removed. If one drug or procedure related to a colonoscopy sneaks through to the list of predictors, model accuracy will be artificially high, but deployment of the model will cause misclassifications.

Aggregation to the patient level

After all appropriate data have been excluded from the analysis, the data set can be aggregated to the patient level. The problem is that some patients have hundreds of records (rows) in the detailed source data. Each row has a different procedure, condition, or drug code associated with it. These multiple rows per patient were grouped, and each feature (column) in the record was aggregated according to the appropriate aggregation heuristic. For example, all body mass index (BMI) values were grouped together to calculate a mean BMI value for a given patient. All other features were handled similarly.

Unique code determination

Multiple records exist often for a given patient, each one of which contains OMOP code values for a specific condition, drug, or procedure. Which code value should be assigned to the aggregated record for a patient? Fortunately, KNIME provides an aggregation function to generate a comma-separated string of unique values for each patient. This function is used to generate a new variable in the aggregation operation for condition code, procedure code, and drug code that contains unique values among all of the records for a given patient. These string variables were named procedure_code_list, condition_code_list, and drug_code_list, respectively. The most important features in the patient aggregated record are the patient age at the time of the procedure (AgeProc) and these code list variables, because it is from these strings of unique codes that thousands of candidate predictor variables are derived in the automated procedure to generate binary variables for each unique code.

Text mining frequency analysis

KNIME provides some useful text mining operations that can be used to generate new predictor variables for modeling. The unique code lists strings for each patient were used to build a KNIME text mining workflow shown in Fig. 14.2.

The comma-separated codes string (e.g., condition_code_list) was changed to a document data type in the strings to document node. The unique codes in the code lists were used to generate a bag-of-words in the Bag of Words Creator node, and the terms in the bag of words were changed back to a string format in the Term to string node. Extraneous rows in the output data flow were removed in the Rule-based Row Filter. The Statistics node was used to perform a frequency analysis on the term strings (codes) across all records of data. The code frequencies were sorted in descending order in the Sorter node, and the Row Filter node was used to limit the codes passed to the 675 codes with the highest frequency of occurrence. This operation was performed for the Condition codes, the Procedure codes, and the Drug codes to produce a combined list of new 2025 variables.

Manual variable derivation

Initially, many one-hot variables were derived individually before an automated process was designed. There were a number of instances where the same code was used to derive variables in both the manual operation and the automated operation. Later, any duplicate variables were removed.

The initial derivation of new candidate predictor variables occurred in four phases:

1. Based on variables in the general gastroenterology database;
   a. Gender;
   b. Ethnicity;
   c. Zip-region (the first number of a Zipcode refers to a region of the US).
   d. Age was derived, but it was not used as a variable in this analysis, it appeared in preliminary models to mask predictive effects of other variables. Colon cancer physicians already know that age is an important factor. This study was designed to provide a screening tool for factors other than age.
2. New variables derived from Conditions codes, based on a literature search of correlates of colon cancer:
   a. Acute neoplasm of pancreas
   b. Antiphospholipid
   c. Vitiligo
   d. Abnormal weight
   e. Acute gas
   f. Acute pancreatitis
   g. Alcohol
   h. Drug-induced
   i. Mycosis fungoides
   j. Herpes
   k. Non-Hodgkin disease
   l. Obesity
   m. Liver disease

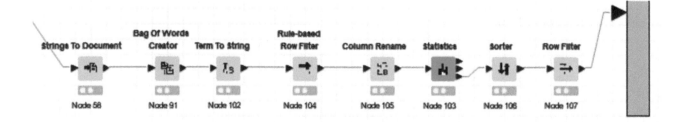

FIGURE 14.2   The text mining component of the KNIME data engineering workflow.

3. New variables derived from the Measurements dimension data elements:
   a. Systolic blood pressure range;
   b. Mean BMI;
   c. High erythrocyte sedimentation, and low erythrocyte sedimentation;
   d. High urea nitrogen;
   e. Sum of the high measurements and low measurements;
   f. High and low leukocytes;
   g. High lymphocytes;
   h. Low hemoglobin;
   i. Low platelets;
   j. Low erythrocytes;
   k. Low phosphate;
   l. Low hematocrit.
4. New variables derived from the Observations data elements.
   a. BMI change;
   b. Systolic blood pressure change;
   c. Diastolic blood pressure change;
   d. Heart rate change;
   e. Glucose change;
   f. LDL cholesterol change;
   g. Erythrocyte sedimentation change;
   h. Alcohol use change.

Derivation of one-hot (binary) variables

The three unique list columns were searched in an automated variable derivation loop in KNIME to produce 2025 binary variables (instead of the potential 43,552 for all OMOP-CDM codes), using a series of column expression nodes in KNIME. This list of binary variables was integrated with the 123 variables derived previously to generate a working data set of 2148 variables.

Feature selection process

Occam's Razor. A 13th century clergyman, William of Occam, proposed a principle expressed in the original Latin as:

"Numquam ponenda est pluralitas sine necessitate,"

Literal English translation: "Plurality must never be posited without necessity."

This principle is just a restatement of the older Law of Parsimony, which in turn is based loosely on writings of Aristotle and Ptolemy, proclaiming that the simplest explanation is usually the best (Franklin, 2001). In Data Science, the application of this principle posits that the model of any entity with the fewest predictor variables necessary to express the signal of the target variable is usually the best model. Additional variables would be "posited without necessity," and they just get in the way of the efficient functioning of the algorithm. This principle works out in analytical modeling such that the model with the fewest number of variables that are necessary to define the predicted entity is the best model.

The mathematical reason for this principle is that each variable represents a dimension in the multidimensional decision space (called a Hilbert space). Increasing the dimensionality of the decision space increases the complexity of the problem to find the optimum location on the decision space to maximize the prediction accuracy. A moderate complexity of the model might be necessary to express the "necessary" effects of many nonlinear variables on the target variable outcome. The challenge before the modeler is to find which predictor variables are necessary in the optimum model to accomplish that task. The best model might not be the one with the fewest variables, but it almost always is not the model with the most possible variables.

The "short-list"

The list of "necessary" variables is called the "short-list" by some data science practitioners. This short-list of important predictors is very valuable as a starting point for developing predictive models. A case in the 4th Circuit Court of

Appeals in Hagerstown, MD found in 2003 that the short-list of predictor variables developed by DTM Research for AT&T for phone call modeling was ruled as a trade secret violated by the defendant (DTM Research LLC vs. AT&T Corp, 2001). Arguments in the case continued until July, 2003. Casteneda (2003) reported in the Washington Post that the judgment for the plaintiff was $12.1 million.

Given the high importance of the short-list in mathematical, legal, and monetary terms, how can we specify this short-list of predictor variables?

Methods of feature selection

The activity of feature selection in modeling eliminates "unnecessary" variables by one of three groups of methods: (1) filtering methods; (2) wrapper methods (using an algorithm); and (3) embedded methods (filtering or wrapper methods embedded in the operation of an algorithm). Filtering methods and wrapper methods were used in this study to reduce the number of predictor variables. No embedded method was used.

Variable filtering

Three filtering methods for feature selection were used in the project:

1. Constant value filtering (3 variables deleted)
2. Low variance filtering (0 variables deleted)
3. High correlation filtering (208 variables deleted with simple correlation coefficients > 0.8)

Wrapper methods

A wrapper method was used (with a Random Forest algorithm) to train a series of models in a recursive loop, removing predictor variables with an importance value (IV)=0, until none were left. The Accuracy table report of the algorithm was used to calculate variable IV. This feature selection operation output a file with 139 variables (out of a total of 2217 potential variables) to use as predictors and the results were recorded.

Data conditioning

After all features (or variables) have been selected, it is often necessary to modify the complement of rows and the values in the columns (features).

Balancing the data set

Machine learning algorithms look for patterns in the Target variable classes related to values in the set of predictor variables. If the number of instances (rows) with a one Target class is greater than the number of instances of another Target class, the prediction will be biased toward the class with the largest number of instances.

The percentage of records labeled as Target=1 records in the modeling data sets was 66%, and that of Target=0 records was 34%. To create the desired balance in target class instances, the Synthetic Minority Oversampling Technique (SMOTE) method was applied to the data set to generate a 50:50 balance. SMOTE uses a nearest-neighbor analysis to estimate the value for each variable in enough "synthetic" rows for the minority target class to equal the number of records in the majority class. This addition of synthetic rows might appear to inject information into the modeling data set, but the information used is already in the data set; it is just recombined to form the extra rows. The SMOTE technique is a popular balancing technique in biomedical studies (Blagus and Lusa, 2013; Mohammed et al., 2020).

Unrepresentative negatives

Kim et al. (2017) has shown that among the over 15,000 colonoscopies in China, that 17%–28% missed precancerous polyps. Anecdotal information from the UCI Medical Center staff estimates that this colon polyp miss rate is about 20%. Each of the patients with missed colon polyps would be labeled as Target=0, then the label should be 1. We can expect that any of these patients included in the modeling data set will be classified as Target=0 (a "negative"), even though they were labeled as Target=0. These patients are not good representatives of the Epic data pattern associated with normal Target – 0 patients. Positive Unlabeled Learning is one way to identify then unrepresentative negatives and remove them from the data set.

Positive unlabeled learning

This methodology (described in greater detail in Chapter 9) removes from the data set those Target=0 rows predicted with a low probability of confidence (Liu et al., 2003; Hernández-Fusilier et al., 2015). The effect of the adaption of this method was to remove Target=0 rows predicted as Target=1 (false-positives) if the probability of classification as target=0 was <0.3. The strategy is to remove uncharacteristic Target=0 records before training the modeling algorithms. The hypothesis is formulated by assuming that the ensuing model will be a better predictor of Target=0 rows, because the discriminating power of the model trained without the unrepresentative Target=0 rows would be increased. This hypothesis was tested by applying the new model to predict those Target=0 rows that were removed. Results of the test showed that the model trained without the unrepresentative Target=0 rows (Target=0 rows predicted as 1) was 25% accurate to predict the unrepresentative Target=0 rows correctly. This corresponds to the range of the rate of missed polyps during a colonoscopy (Kim et al., 2017). This correspondence doesn't mean that all unrepresentative target=0 rows were misclassified initially as a result of polyp misses, but we will make the assumption that those of lower than 0.3 probability of actually be Target=0 represent those patients with missed colon polyps.

The result of the positive unlabeling process was to reduce the number of Target=0 rows to 492, to be used with the 2,286 rows with Target=1 as the unbalanced input data set. This data set was balanced with the SMOTE technique such that an equal number of positive (Target=1) and negative (Target=0) rows were submitted to the modeling algorithm.

Modeling

Modeling algorithms

Three classification algorithms available in the KNIME Analytics Platform were selected for inclusion into an ensemble: Stochastic gradient boosted trees, Tree Ensemble (a form of Random Forest); and XGBoosted trees (using the XGBoost method).

Cross-validation

Each model with 139 predictor variables was trained within a 10-fold cross-validation loop, used to remove sampling bias in random selection of records in the training and validation data sets. The KNIME cross-validation loop is shown in Fig. 14.3.

In the cross-validation process shown in Fig. 14.3, the X-Partitioner node divides the data set randomly into a number of parts (folds). In our case, 10 folds (subsets) are generated. The process trains a model on nine folds and validates it on the 10th fold. Then, another fold is held out of the training process and used to validate predictions in a second model. This process continues until 10 models are trained on different 9/10 portions of the input data set and 10 different validation data subsets.

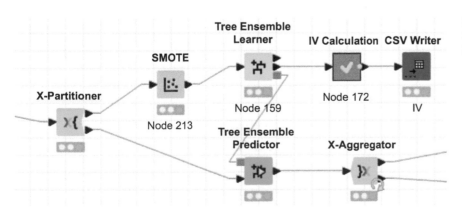

FIGURE 14.3 The cross-validation modeling loop as implemented in the KNIME Analytics Platform.

The cross-validation loop starts with the X-Partitioner node in Fig. 14.5, and ends with the X-Aggregator node. The X-Partitioner node divides the input data into 10 folds randomly, and then it divides each fold into two parts, with 70% of the fold data is balanced by the SMOTE node, and used to train the model. The other 30% of the data is directed to the Tree Ensemble Predictor node, where predictions are made for that data fold. The X-Aggregator node collects predictions for all 10 folds, and passes them through the top output port of the X-Aggregator node. The entire data set with predictions is directed to the KNIME Prediction Fusion node (see Fig. 14.4 below). The bottom port carries prediction error information for further analysis.

The cross-validation process removes any sample bias associated with the random selection processes in the X-Partitioner node for generating the 10 data folds.

Ensemble modeling

The final modeling approach (Fig. 14.4) was an ensemble of algorithms (Stochastic Gradient Boosted Trees algorithm and a Tree Ensemble algorithm (a form of a Random Forest)). Each algorithm was trained separately with cross-validation, and their predictions and associated probabilities were fused with the KNIME Prediction Fusion node, using the prediction with the maximum confidence value to produce a final prediction for each row. An output report of the Tree Ensemble algorithm was used to calculate IV for variables in the model.

Results and discussion

Model evaluation

The Model Evaluation node for the fused model shown in Fig. 14.3 produced information from which prediction accuracy values are calculated and shown in Table 14.2. A confusion matrix is produced by the KNIME Scorer node, which

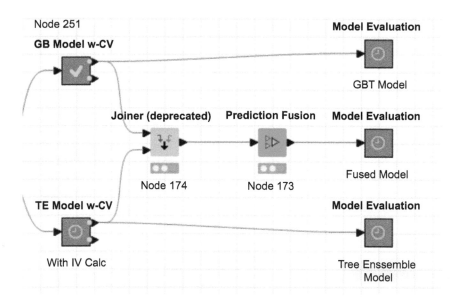

FIGURE 14.4 The final ensemble modeling process to generate the fused model.

TABLE 14.2 The confusion (or classification) matrix of the fused model.

		Actuals	
Predicted		0	1
	0	504	308
	1	392	1973

included the numbers of true-positive predictions (TP), false-positive predictions (FP), true-negative predictions (TN), and false-negative predictions (FN) as shown in Table 14.2.

Multiple accuracy metrics are output in another table, based on the confusion matrix, to evaluate the accuracy of the following accuracy metrics:

1. Overall accuracy=(TP + TN)/(TP + TN + FP + FN)
2. Sensitivity accuracy=TP/(TP + FN)
3. Specificity accuracy=TN/(TN + FP)

Prediction accuracies

The accuracy metrics of the fused model with different numbers of predictive variables are shown in Table 14.3.

Receiver operator characteristic curve

Another accuracy metric was calculated, the area under the Receiver Operator Characteristic (ROC) curve. The ROC curve is plotted as Sensitivity values (or the TP rate) on the Y-axis for each quantile of the 1–Specificity values (or the False-positive rate) along the X-axis. The ROC curve is shown in Fig. 14.5.

The primary property of the ROC curve is the AUC. AUC values of 0.7 to 0.8 are considered by many statisticians as acceptable, 0.8–0.9 values are considered excellent, and values greater than 0.9 are considered as outstanding (Hosmer and Lemeshow, 2000). Therefore, the AUC value of 0.83 shown in Fig. 14.5 can be considered as excellent, and it exceeds values in similar studies cited in Table 14.1.

Other important aspects of the trained model

Important predictor variables

The list of predictor variables output from the modeling process is shown in Table 14.4, in which only those variables with IVs of > 0 were included in the list. A KNIME workflow was built to assign a text explanation to each important predictor variable automatically. Table 14.3 shows the list of the top 10 predictor variables in the fused model according to their IVs and simple correlation coefficients with the Target variable.

Note: variables annotated with an asterisk (*) have correlation coefficients that are >=0.1 which show relatively strong individual effects on the outcome for the fused model, while variables with a hash mark (#) have coefficients < 0.1 and are examples of emergent effects.

Variable IVs in Table 14.4 were calculated by summing the number of times a given variable participated in the first three splits in the building of the KNIME decision tree. This list of important variables does not mean that every variable is associated with every classification as Target=0 or 1, rather it means that the most important variables associated with the classification for a patient are contained in this list.

Emergent properties

Studies of complex systems have revealed that most systems exhibit properties that become obvious only when the system is complete and functioning normally (Anderson, 1994). These properties are called *emergent properties*. These emergent properties may cause effects that are quite obvious—*emergent effects*. Aristotle focused his attention on phenomena in the world that could be sensed directly by sight, sound, or touch. He explained the nature of the whole as the sum of its parts. Plato (Aristotle's teacher for 20 years) focused, however, on phenomena like anger and love, which

TABLE 14.3 Prediction accuracies of the fused model with 139 predictor variables.

Overall accuracy	78.0%
Sensitivity	83.4%
Specificity	62.1%
AUC	0.83

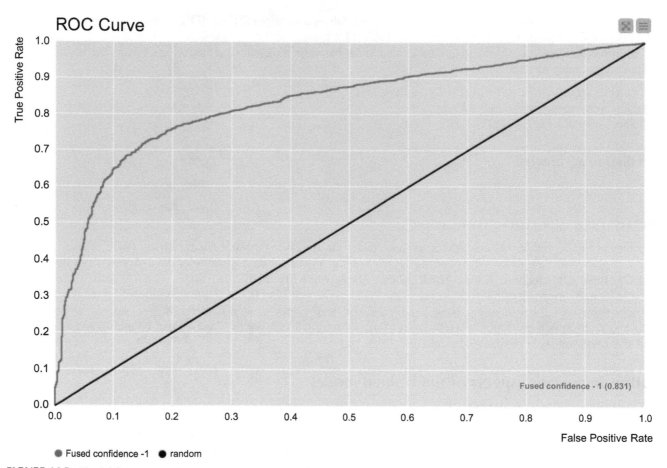

FIGURE 14.5 The ROC curve, showing the curve plot for the fused ensemble model, and the proportion of the area under the curve (AUC) above the X-axis as 0.921 (see red circled value).

TABLE 14.4 Top 10 predictor variables with importance values and simple correlation coefficients.

Variable	Importance value	Correlation cf with target
Live Zoster vaccine	26.3	0.14*
Periodic comprehensive medical exam and evaluation	21.6	0.09#
Actinic keratosis	15.9	0.09#
Unhealthy alcohol use	10.4	0.07#
Impotence of organic origin	10.3	0.09#
PSA test	8.4	0.12*
Low sirolimus in blood	8.4	0.04#
BMI	18.0	0.13*
Pain assessment	7.6	0.07#
Dehydroepiandrosterone (DHEA)	7.5	0.06#

transcended the senses, in which the whole of a phenomenon (e.g., retribution) was greater than the sum of its parts. Thus Aristotle focused on direct sensory effects of sight and sound, while Plato focused on emergent effects (e.g., love and hate). Modern analytical modelers must combine relatively strong direct effects (e.g., Zoster vaccine) with the more indirect emergent effects (e.g., actinic keraktosis) when they build a model of a phenomenon that may exhibit both kinds of effects. This situation may pertain to the modeling of almost all phenomena in the natural world.

The concept of emergent properties can be explained best in biological terms. The healthy heart pumps blood throughout the biological system of the organism. The heart is composed of many heart cells, but an individual heart cell cannot pump blood by itself. Many cells must compose the whole heart and its parts must function in the proper sequence before it can pump blood in the right direction through arteries (outflow) and veins (inflow). Many complex systems in nature are like this. When we define the presence of the target condition of precancerous polyps using OMOP-CDM data, we can expect that combinations of variables (like combinations of heart cells) are likely to produce some emergent effects. Three of the top 10 important variables in the fused model are annotated with a hash mark (#) in Table 14.3. These variables have relatively strong individual correlations with the Target variable. The other seven variables (with an asterisk, *) have a relatively high IVs, but have relatively small simple correlation coefficients with the target variable (correlation coefficients values below 0.1). The combination of these variables with other variables, however, appears to produce relationships that have significant power to predict the target variable. From a statistical analysis perspective, emergent effects are represented by statistical interactions. In this way, machine learning algorithms (like the forms of decision trees used in this study) can express effects of variable interactions in terms of variable IVs.

The presence of emergent effects in this OMOP-CDM data set means that we should not place much significance on the effects of single variables. Rather, we should concentrate on the combined effect of all important predictor variables. There is, however, some justification to focus on those variables with high IVs that also have relatively high simple correlation coefficients (variables in Table 14.4 with a hash mark, #). That focus is reflected in the current recommendation that people 45 years old and older should have regular colonoscopies. The combination of all predictor variables (both direct and indirect in operation) in the model can be used as a screening method to recommend a colonoscopy to patients regardless of age.

Automation of data preparation for medical informatics?

Some analytical modeling tools claim that much or all of data preparation preprocessing operations can be automated. This notion appeals to some decision-makers who have only a rudimentary understanding of medical analytics. The significant data problems associated with the modeling of information of electronic medical records are described in this chapter. These challenges require at least a general knowledge of the medical data domain for the models. Some operations in data preprocessing can be automated (e.g., filling of missing values and derivation of "dummy" variables), but most cannot—currently. The integration of many data preprocessing operations into the EMR data structures may permit an increased amount of automation in the future.

Conclusions

1. Ordinary clinical data can be used to build predictive models of the presence of precancerous polyps.
2. Processing of OMOP-CDM data to build analytical models facilitates data integration, but it requires a number of data preprocessing operations to yield a data set suitable for prediction with machine learning algorithms.
3. Currently, most of the data preprocessing operations must be done manually with the help of a data domain expert, rendering them difficult to automate with the current state-of-the-art in Artificial Intelligence.
4. Text mining analysis can help to derive predictive variables from free-form text data in OMOP-CDM data sets.
5. Predictive analytics models of the presence of precancerous polyps present several examples of emergent properties related to specific predictive variables.
6. This study demonstrates the need to consider the combined effect of all important predictor variables, not just the independent effects of a few variables.

How this chapter facilitates patient-centric medical health care

- This case study provides a processing framework to facilitate the consistent integration of disparate medical data sets

- The sequence of data preparation processed presented in this chapter provides guidance for transforming OMOP-CDM data sources into a patient-centric file for analytical modeling.
- The analytical modeling operations documented in this chapter can provide physicians and medical researchers with a powerful system for diagnosis and treatment of patients with strategies tuned to their individual clinical medical conditions and treatment history.

Postscript

This chapter demonstrates the critical importance of data engineering operations in the successful modeling of medical clinical data. In many ways, medical data elements in electronic medical records are not stored in the proper format nor with the proper relationships with other elements to permit the immediate use of them in building analytical models of complex medical conditions. For the immediate future, the analyst with data domain knowledge must perform the necessary data engineering operations prior to building a model, rendering them difficult to automate. Some data preprocessing tasks can be automated at present, but eventually, many more of these operations can and will be automated for specific medical analysis platforms.

References

American Cancer Society, 2021. Cancer Facts & Figures 2021. American Cancer Society, Atlanta, GA.

Anderson, P.W., 1994. More is different: broken symmetry and the nature of the hierarchical structure of science. Science 177 (4047), 393−396.

Blagus, R., Lusa, L., 2013. SMOTE for high-dimensional class-imbalanced data. BMC Bioinforma. 14, 106. Available from: https://doi.org/10.1186/1471-2105-14-106.

Casteneda, R., 2003, July 31. Laurel firms gets $12.1 million. Washington Post Local Headlines Newsletter. <https://www.washingtonpost.com/archive/local/2003/07/31/laurel-firm-gets-121-million/09a622f2-ca45-43c7-8dfc-913195941f1a/>

DTM Research LLC vs. AT&T Corp. 2001. 245 F. 3d 327,332 58 U.S. P.Q. 2d (BNA) 1236 (4th Cir. Md 2001). Civil docket for Case #: 8:96-cv-01852-PJM (judgment, 7/2003).

Franklin, J., 2001. The Science of Conjecture: Evidence and Probability Before Pascal. The Johns Hopkins University Press, p. 241, Chapter 9.

Hernández-Fusilier, D., Montes Gómez, M., Rosso, P., Guzmán Cabrera, R., 2015. Detecting positive and negative deceptive opinions using PU-learning. Inf. Process. Manage. 51 (4), 433−443. Available from: https://doi.org/10.1016/j.ipm.2014.11.001. Available from: https://doi.org/10.1016/j.ipm.2014.11.001.

Hosmer, D.W., Lemeshow, S., 2000. Chapter 5 Applied Logistic Regression, second ed. John Wiley and Sons, New York, pp. 160−164.

Kim, N., Jung, Y., Jeong, W., Yang, H., Park, S., Choi, K., et al., 2017. Miss rate of colorectal neoplastic polyps and risk factors for missed polyps in consecutive colonoscopies. Intest. Res. 15 (3), 411−418.

Labianca, R., Nordlinger, B., Beretta, G., Mosconi, S., Mandala, M., Cervantes, A., et al., 2013. Early colon cancer: ESMO clinical practice guidelines for diagnosis, treatment and follow-up. Ann. Oncol. 24, 64−72.

Liu, B., Dai, Y., Li, X.L., Lee, W.S., Philip, Y., 2003. Building text classifiers using positive and unlabeled examples. In: ICDM 2003, Third IEEE International Conference on Data Mining, November 2003.

Mohammed, A., Hassan, M., Kadir, D., 2020. \Improving classification performance for a novel imbalanced medical dataset using SMOTE method. J. Adv. Trends Comp. Sci. 9 (3), 3161−3172. Available from: http://www.warse.org/IJATCSE/static/pdf/file/ijatcse104932020.pdf.

Murchie, B., Tandon, K., Hakim, S., Shah, K., O'Rourke, C., Castro, F., 2019. A new scoring system to predict the risk for high-risk adenoma and comparison of existing risk calculators. Clin. Gastroenterol. 51, 345−351.

Overhage, J., Ryan, P.B., Reich, C.G., Hartzema, A.G., Stang, P.E., 2012. Validation of a common data model for active safety surveillance research. J. Am. Med. Inform. Assoc. 19, 54−60. Available from: https://doi.org/10.1136/amiajnl-2011-000376 [PMC free article] [PubMed] [Google Scholar].

Park, C., Suk, Y., Kim, N., Park, J., Park, D., Sohn, C., 2019. Usefulness of risk stratification models for colorectal cancer based on fecal hemoglobin concentration and clinical risk factors. Gast. Endoscopy 89, 1204−1211.

Soonklang, K., Siribumrungwong, B., Siripongpreeda, B., Auewarakul, C., 2021. Comparison of multiple statistical models for the development of clinical prediction scores to detect advanced colorectal neoplasms in asymptomatic Thai patients. Medicine 100 (20), e26065.

Further reading

Hernández-Fusilier, D., Guzmán-Cabrera, R., Montes-y-Gómez, M., Rosso, P., 2013. Using PU-learning to detect deceptive opinion spam. In: Proceedings of the 4th Workshop on Computational Approaches to Subjectivity, Sentiment and Social Media Analysis, June 2013, pp. 38−45.

Lee, Y., Bang, H., Kim, D., 2016. How to establish clinical prediction models. Endocrinol. Metab. 31 (1), 38−44. Published online: 16 March 2016.

Chapter 15

Prediction of pancreatic and lung cancer from metabolomics data

Robert A. Nisbet

Chapter outline

Prelude

The role of metabolites in cancer diagnosis was explored at a high level in Chapter 9. Chapter 15 presents an example of how metabolomics data from blood serum samples can be used to build highly accurate models of cancer presence. This technology is poised to transform the cancer and disease diagnosis landscape in medicine, and the blood analysis industry in the process. Cancer cells and cells infected with disease (e.g., kidney disease) excrete metabolite products at levels completely beyond the ranges of those metabolites in normal healthy patients. These metabolites form a pattern of peak signatures in gas chromatograph–mass spectrometer (GCMS) outputs that may be extremely predictive of the presence of the disease or cancer. The case discussed in this chapter describes the many data preparation operations necessary to build a multicategorical model to predict the presence of either lung cancer or pancreatic cancer based on output data from GCMS analyses of blood serum.

Purpose of this chapter

This chapter is designed to present an overview of the modeling process in KNIME for modeling the presence of lung cancer or pancreatic cancer in patients from GCMS analysis of their blood serum. The purpose of each node in the modeling workflow will be discussed, but not documented in the form of a tutorial; this chapter represents a summary of the study only, with emphasis on the data preparation operations necessary to generate the modeling data set. The workflow built in this study is not available on the book web page for experienced KNIME modelers to use, because it uses data that include confidential information.

Practical Data Analytics for Innovation in Medicine. DOI: https://doi.org/10.1016/B978-0-323-95274-3.00009-9

Introduction

Cancer deaths in the United States

A recent CDC update on cancer deaths in the United States reported the number of deaths from cancer in 2019, as shown in Table 15.1.

Cancer metabolites

Metabolites have been defined as molecules less than 1500 daltons molecular weight, which are produced endogenously by cellular metabolism (Fiehn, 2002). Cells altered by cancer or disease produce additional metabolites or significantly higher than normal levels of existing metabolites. The quantitative analysis of these metabolites can be related to the presence/absence of the cancer or disease in a patient.

After cancer and disease metabolites are emitted, they dissolve into organic fluids in the cell and bloodstream. Many of these volatile organic compounds (VOCs) are subject to ionization prior to being chromatographed in GCMS. The number of ions in these VOCs can be counted in representative peaks and related to normal and abnormal (diseased or cancerous) conditions. This set of highest ion count peaks can serve as markers to indicate the presence of lung cancer, breast cancer, liver cancer, pancreatic cancer, and colon cancer.

VOCs have been used to predict cancer presence with high accuracy in hundreds of studies during the last 10 years. This chapter will present results from a study of blood plasma from a group of lung cancer, pancreatic cancer, and normal patients. These results were given in a presentation at Predictive Analytics World (Nisbet et al., 2009). This chapter will present additional information.

Methods

Metabolites from blood plasma were analyzed in a GCMS system to provide counts of the number of ions of various molecular weights flowing through the system. Output data were preprocessed with a fast-Fourier transform to convert data from the time domain to the frequency domain. A Python program was used to further preprocess the GCMS output data to generate the data sets for building the analytical models in the KNIME analytical platform. The first 10 rows in the data set were omitted, to remove effects of false readings during the sensor equilibration process. Remaining data rows were randomized and data values were normalized with a z-score function. Missing values were imputed with the value = 0.

The modeling process

A multicategory classification model was built to predict the target class for lung cancer, pancreatic cancer, or normal. An overview workflow was built in KNIME for combined model using the KNIME Tree Ensemble algorithm (a form of Random Forest) as shown in Fig. 15.1.

The high-level view of the workflow was built with a data input node, two KNIME *metanodes* (colored green), and an output node. A KNIME metanode is a collection of KNIME nodes collapsed into a single node icon to generalize

TABLE 15.1 The number of deaths by cancer type in the United States during 2019.

Cancer type	Number of deaths
Lung	139,603
Colorectal	51,896
Pancreatic	45,886
Breast	42,281
Prostate	31,638
Liver	27,959

Source: https://www.cdc.gov/cancer/dcpc/research/update-on-cancer-deaths/index.htm.
Lung cancer and pancreatic cancer are the first and third most common causes of deaths in the United States (Table 15.1). An analysis of the metabolic by-products of cancer cells in blood serum (a type of metabolomics data) provides an opportunity to design an early warning system to predict the presence of lung cancer or pancreatic cancer using multicategorical machine learning modeling.

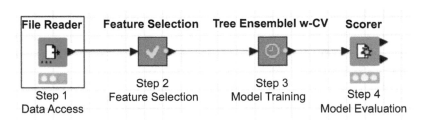

FIGURE 15.1 The KNIME modeling workflow operational steps.

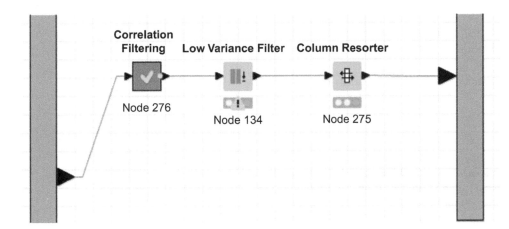

FIGURE 15.2 Two feature selection operations: correlation coefficient and low variance filtering.

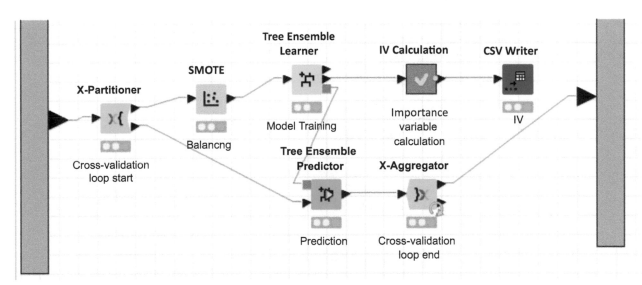

FIGURE 15.3 The model training operation in a cross-validation loop.

the detailed workflow architecture to form a high-level overview of the flow of data through a series of processes. Double-clicking on a metanode expands the display to show the component nodes contained in it.

Step 1. This step consists of a file loading operation with a File Reader node. Configuration of this node specifies the file name and location of the input data set.

The Step 2 process is shown as the expanded metanode in Fig. 15.2.

Correlation filtering calculates the simple linear correlation coefficient for each pair of predictor variables, and eliminates one member of each pair of variables correlated above 0.8 (the variable with the lowest number of correlated rows of data). The Low Variance Filter node filters out columns (variables) with below level of variance specified in the configuration of the node.

The expansion of the Step 3 metanode is shown in Fig. 15.3.

Nodes in the Modeling metanode shown in Fig. 15.3 include:

1. the cross-validation loop is initiated by the X-Partitioner node, in which the node configuration includes the number of iterations for the loop. The number of iterations is set often to 10 (for 10-fold cross-validation), but for this relatively small data set, the number of iterations was set to 3;
2. the SMOTE (Symbolic Minority Oversampling Technique) node balances the imbalanced data by synthesizing new records of the minority data set with a nearest-neighbor method of estimating values for each column. SMOTE is the balancing technique most commonly used in medical informatics;
3. the Tree Ensemble Learner node trains the model with a form of a Random Forest algorithm;
4. the IV calculation metanode calculates the variable importance values from data outputs of the tree building process;
5. the X-Aggregator node collects and calculates the average prediction probabilities for each target class for data from each iteration of the cross-validation process and assigns the target class with the highest mean prediction probability; and
6. the Scorer node calculates a confusion matrix and number of accuracy metrics.

Results

Model accuracy

The RowID in Table 15.2 contains the code for 0=normal; 1=lung cancer, and 2=pancreatic cancer. The overall accuracy is 0.963 or 96.3%. The Sensitivity accuracy for lung cancer (1) is 0.889 (88.9%) and for pancreatic cancer (2) is 1.0 (100%).

Table 15.3 shows the output of the IV Calculation (Importance Value calculation) metanode.

These important variables are not listed according to their value in predicting a specific cancer type (lung or pancreatic). They are listed to describe the combined model only.

TABLE 15.2 The accuracy statistics report of the Tree Ensemble algorithm.

Row ID	TruePo...	FalsePo...	TrueNe...	FalseN...	Recall	Precision	Sensitivity	Specificity	F-meas...	Accuracy	Cohen'...
0	10	0	17	0	1	1	1	1	1	?	?
1	8	0	18	1	0.889	1	0.889	1	0.941	?	?
2	8	1	18	0	1	0.889	1	0.947	0.941	?	?
Overall	?	?	?	?	?	?	?	?	?	0.963	0.944

TABLE 15.3 The list of important variables and their Mean_IV values in the multiclass model.

Variable (peak)	Importance value
Ion_MW_120_Peak_2	7.0
Ion_MW_276_Peak_2	6.0
Ion_MW_298_Peak_3	6.0
Ion_MW_328_Peak_3	6.0
Ion_MW_308_Peak_3	5.7
Ion_MW_326_Peak_1	5.3
Ion_MW_297_Peak_3	5.0
Ion_MW_310_Peak_1	5.0
Ion_MW_299_Peak_3	4.7
Ion_MW_120_Peak_3	4.0

For example, the peak is the second highest peak of an ion with a molecular weight (MW) of 120 daltons.

TABLE 15.4 The top three modeling variables according to their importance value in the model.

Lung cancer	Pancreatic cancer
Ion peak 325_1	Ion peak 349_3
Ion peak 342_1	Ion peak 348_1
Ion peak 248_3	Ion peak 331_2

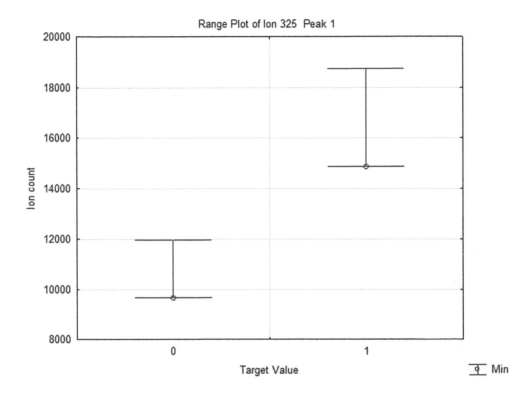

FIGURE 15.4 The range plots for Ion 325 peak #1. Notice the disjunction distribution of data points between normal (Target=0) and cancer patients (Target=1).

Specific models for lung cancer and pancreatic cancer

Models were trained to predict lung cancer and pancreatic cancer separately. Both models were 100% accurate. Table 15.4 shows the top three ion peaks according to their importance values for each model.

Fig. 15.4 shows the completely disjunct distributions of ion counts of the highest peak of the ion with a molecular weight of 325 (Ion peak 325_1 in Table 15.4) for normal patients and lung cancer patients.

Discussion

These results may appear too high to be valid. Normally, perfect models point to errors in methodology (e.g., including a predictor related to the target variable by definition). Bu, this is not the case here, or in many other recent studies of cancer prediction from analysis of VOC data. Wong et al. (2019) found similar results in models of lung cancer and pancreatic cancer, with results for some other cancers ranging to 99% accuracy of prediction. Models for lung cancer based on analysis of VOCs were 100% accurate in studies by Gordon et al. (1985), Westhoff et al. (2009), and Qian et al. (2018). Therefore, we are forced to conclude that results of this case study are valid.

How can these models be that accurate? The reason is that the abnormally high levels of metabolites specific to each cancer type are very consistent among all patients with a given cancer type. In addition, Fig. 15.4 shows that the ranges of ion counts for ion 325 are very disjunct (separated by over 2000 ion counts), therefore, it is easy for the modeling algorithm to make a prediction for lung cancer that is right for each patient. This pattern of disjunction is seen in many of the ions with high importance values for the lung cancer model (5 ions) and for the pancreatic cancer model (31 ions).

Implications of this case study for future medical diagnosis

The power of metabolite molecules for predicting the presence of cancer types (some perfectly) may move many physicians to order GCMS analysis of patient blood samples in the future. In this event, blood analysis laboratory organizations (of which there are one to many in every city) must include GCMS analysis in their offerings to satisfy physician demands. An analytical model was used to identify the pattern of disjunctive ranges in ion counts between normal and cancer patients, but the model will not be needed for deployment of a screening system based on model results. All that a blood analysis technician will have to do is to compare GCMS ion counts for specific ions between patients and a normal control sample. Obtaining a good normal sample to use may be a challenge for the laboratory.

Conclusions

1. Analysis of metabolites in blood serum samples paired with machine learning modeling is a powerful methodology for very early and minimally invasive detection of cancer in humans.
2. Analysis of VOCs can be used to build combined models, and may be expanded to include many other types of cancer.
3. Individual models for some cancer types will be 100% accurate.
4. These combined models can become the basis for a panel of new blood tests in medicine.

How this chapter facilitates patient-centric healthcare

Based on the power of machine learning analysis of blood samples for very early minimally invasive detection of cancer, patients in the future may be able to take control of their own diagnostic screening programs by requesting blood analyses that output GCMS counts of specific ions known to be linked closely to cancer presence. Someday, patients will be able to go to a testing laboratory like those that do routine blood analyses today, and order a blood test to predict the presence of a large panel of cancer types. Such early diagnosis will facilitate the success of surgery and treatment programs in preventing death from dreaded cancers and diseases.

Postscript

The models in applications described in Chapters 14 and 15 illustrate how important machine learning algorithms are for finding patterns of relationships for guiding medical diagnosis. Chapter 16 will show how traditional display graphics can be used to gain insights which help understand the nature of the COVID-19 pandemic.

References

Fiehn, O., 2002. Metabolomics—the link between genotypes and phenotypes. Plant. Mol. Biol. 48 (1–2), 155–171.

Gordon, S., Szldon, J., Krotoszynski, B., Gibbons, R., O'Neill, H., 1985. Volatile organic compounds in exhaled air from patients with lung cancer. Clin. Chem. 31 (8), 1278–1282.

Nisbet, R., Elder, J., Miner, G., 2009. first ed. Handbook of Statistical Analysis and Data Mining Applications, 864. Academic Press, London.

Qian, K., Wang, Y., Hua, L., Chen, A., Zhang, Y., 2018. New method of lung cancer detection by saliva test using surface-enhanced Raman spectroscopy. Thorac. Cancer 9, 1556–1561.

Westhoff, M., Litterst, P., Freitag, L., Urfer, W., Bader, S., Baumbach, J., 2009. Ion mobility spectrometry for the detection of volatile organic compounds in exhaled breath of patients with lung cancer: results of a pilot study. Thorax 64, 744–748.

Wong, K., Chen, J., Zhang, J., Lin, J., Yan, S., Zhang, S., et al., 2019. Early cancer detection from multianalyte blood test results. iScience 15, 332–341. Available from: https://doi.org/10.1016/j.isci.2019.04.035. published online.

Chapter 16

Covid-19 descriptive analytics visualization of pandemic and hospitalization data*

Robert (Bob) Nisbet

Chapter outline

Preamble

This chapter is an example of *descriptive analytics*. Descriptive analytics uses averages, trend, and the, show the dynamic pattern of the time-course of change in a system, rather than predictive algorithms to provide estimates of the future state of the system. The distinction between descriptive analytics and predictive analytics is given briefly in Chapter 22, and in greater detail in Chapter 25.

Introduction

This tutorial is designed to show how to develop some insights from data visualizations (line graphs) of COVID-19 data by country. It includes an introduction to the KNIME analytics platform, and it shows how to train a predictive model, based on the insights gained in the visualization operations. In Part I, you will use KNIME to select a country to visualize: (1) new cases, new deaths; (2) hospitalizations; and (3) the scaled number of full vaccinations vs. the % Death rate. You can visualize data from several countries sequentially by changing the Country variable in the String Configuration node, and compare elements of the curves (e.g., peaks). In Appendix-A, you can train a predictive model of new deaths due to COVID-19, using insights gained in Chapter 16.

3 KNIME workflow data streams

This chapter will discuss three of the four the workflow data streams in the workflow World_Country_Hospitalizations_Model.knwf provided on the book's web page.

---

*READERS – Please NOTE: There is an extension of this tutorial how to use KNIME with this COVID dataset located in the APPENDIX of this book. Also note that on this book's COMPANION WEB PAGE are instructions on HOW TO DOWNLOAD KNIME to your computer, in addition to the instructions provided below. KNIME is free to use, and a very good predictive analytic software platform.

Practical Data Analytics for Innovation in Medicine. DOI: https://doi.org/10.1016/B978-0-323-95274-3.00010-5

1. New cases and deaths
   a. This data stream is preconfigured;
   b. All you have to do is to execute the stream, and view the line chart output according to instructions;
2. Hospitalization and ICU beds
   a. This data stream is preconfigured;
   b. All you have to do is to execute the stream and view the line chart output according to instructions;
3. Scaled number of full vaccinations vs. % Death rate
   a. This data stream is preconfigured;
   b. All you have to do is to execute this data stream and view the line chart output according to instructions.

The data set used in this tutorial is updated daily by the data provider (https://ourworldindata.org/) from external sources:

1. Johns Hopkins University (for COVID-19 cases and deaths); and
2. the European CDC and the Covid Tracking Project (for hospitalizations).

The latest data update can be downloaded automatically from the Internet by the File Reader node. Instructions are provide in the tutorial for updating the data from the Internet. You can continue to use this workflow on a daily basis, by refreshing the data set and executing the entire data stream.

Preparatory steps for using this tutorial

1. Download and install KNIME on your computer.
   a. Go to KNIME.com and select the Software tab at the top of the screen
   b. Scroll down to the KNIME Analytics Platform section, and click on the Download KNIME button.
   c. Follow the prompts to install KNIME on your computer.
      i. You will be prompted for user information.
      ii. Follow the prompts to download and install the KNIME Analytics Platform.
2. Practice with your own workflow.
   a. Click on the File tab on the top-left corner of the workflow, and select the New option.
   b. Highlight the New KNIME Workflow option as shown in Fig. 16.1;
   c. Click on Next at the bottom of the screen (Fig. 16.2);
   d. Fill in the name of the workflow and the destination folder to store it, then click Finish.
   e. A blank workflow canvas will display in the Workflow Editor pane.

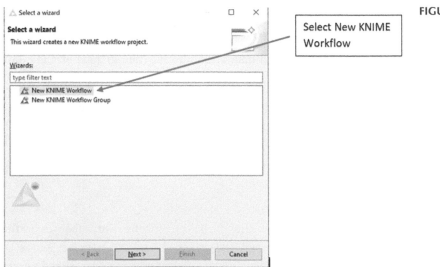

FIGURE 16.1 Selecting a new KNIME workflow.

FIGURE 16.2   Naming the workflow and destination selection.

3. Getting to know the KNIME interface shown in Fig. 16.3
   a. Each information pane can be minimized.
   b. Make sure that at least the minimum information panes are visible:
      i. KNIME Explorer.
      ii. Node Repository.
      iii. Workflow Editor.
      iv. Description
   Before starting the introduction to KNIME section, import and load the World_Country_Hospitalizations_Model.knwf from the book webpage
4. Import the World_Country_Hospitalizations_Model.knwf workflow.
   a. Download the World_Country_Hospitalizations_Model.knwf workflow from the book's web page to some location on your computer where you can find it.
   b. Load the KNIME application showing the set of panes in the graphical interface shown in Fig. 16.1. Initially, your workflow editor pane will be blank (Fig. 16.1). The panes available in the KNIME graphical interface.
   c. In the KNIME graphical interface, click on the File tab at the top left of the screen.
   d. Select the <Import KNIME Workflow> option as shown in Fig. 16.2.
      In the "Select File" box, click on the Browse button to navigate to where you saved the downloaded the workflow, and select the workflow to be imported and the destination (where you want to store it in the KNIME workflow list) in the KNIME Explorer pane.
   e. Enter in the Select File box, the path to the downloaded workflow, and its name, *World_Country_Hospitalizations_Model*.
   f. If desired, change the location in the KNIME Explorer listing where you want to save the imported workflow in the KNIME Explorer window.
5. In the KNIME Explorer window, double-click on the World_Country_Hospitalizations_Model workflow to load it into the modeling workspace.

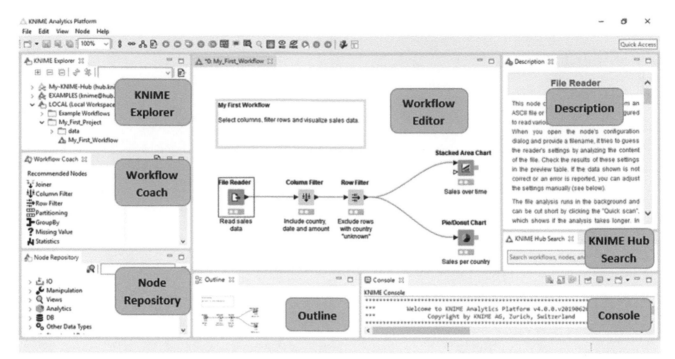

FIGURE 16.3 The KNIME workflow display showing 8 information panes (in orange).

The workflow is shown in Fig. 16.4. Workflow canvas showing visualizations for Cases & Deaths, Hospital & ICU beds, and Tests & Positive Test Rate.

General introduction to KNIME

1. KNIME is a graphical programming tool. Fig. 16.3 shows the architecture of a set of processing operations represented by icons called *nodes*. These nodes are arranged on the programming screen (called a *workflow)* and connected with arrows to represent the processing data flow through the sequence of operations.
2. If you work with each node separately, there are four steps you should follow to operate a node properly:
   a. Step 1. Click on any node to highlight it, and then read the description of the operations of the node and the elements of it (in the Description pane on the left of the Expression Editor).
   b. Step 2. Double-click on the node to configure the node properly, then click OK to return to the workflow canvas.
   c. Step 3. Right-click on the node to bring up the Options menu, and select the <Execute> option to Execute the node.
   d. Step 4. Right-click on the executed node, and click on the last option in the list to look at the output table or chart to see what happened. Click the X in the upper-right corner of the screen to return to the workflow.
3. When you want to add a node to the workspace:
   a. You can select it from the menu in the Node Repository pane of the screen; or
   b. You can enter part of the node name in the Node Repository box next to the magnifying glass icon. You can even miss-spell the name, and KNIME does a fuzzy matching operation to display a list of candidate nodes.
   c. Connect nodes by clicking on the output black triangle on the right side of one node and holding the left mouse button, drag an arrow to the black triangle on the left side of another node, then release the mouse button.
4. Importing a workflow:
   a. Click on the File option on the top menu, and select the *Import KNIME* workflow oprion.
   b. The selection screen shown in Fig. 16.4 is displayed. (Note: display Fig. 16.4 here).
   c. Click on the Browse button, and navigate to where you saved the (World_Country_Hospitalization_Model.knwf) workflow.
   d. Select the folder in the KNIME Repository where you want to save the workflow.
   e. Click *Finish* to complete the operation.

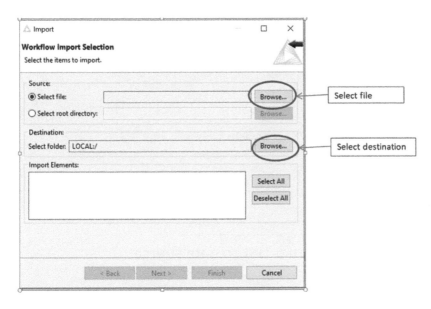

FIGURE 16.4   Workflow import selection screen.

Data access—the file reader node

1. Double-click on the File Reader node to see that it is configured to load data from the Internet to use in this tutorial.
2. You don't have to change anything; click OK to return to the workspace.
3. Right-click on the File Reader node to display the options screen, and select the <Execute > option.
4. Right-click on the File Reader node again, and select the <File Table> option.
5. Review all of the variables (Fig. 16.5).

Data understanding

1. Right-click on the Statistics node and select the <Execute> option. Note the progress bar below the node. When the execution finishes, the progress bar will show a "green light" symbol on the "traffic light" icon in the progress bar.

2. Right-click on the Statistics node again, and select the <Statistics Tables> option to see common descriptive statistics for each variable in the input data file.

Country selection

The country to be analyzed is specified in the String Configuration node.

A. This is the operation where you can select the country to analyze according to your choice.
B. Double-click on the String Configuration node to open the configuration menu screen shown in Fig. 16.6.
C. Notice that the Parameter/Variable name is specified as "Country"; don't change it. This parameter is called a "flow variable," and it is used later in the workflow to build labels for the graphics.
D. The Default Value box contains the name of the selected country (currently specified as United States). You can change this country specification to any country in the world. See the file table listing of the File Reader node to check the country spelling.
E. If you change the Default Value in this configuration screen to specify some other country, you must re-execute all nodes in the workflow.
   1. Change the Default Value from "United States" to "South Africa" to see the data for that country;
   2. Click OK to exit the configuration menu;
F. Back in the workflow editor, click on the red button on the top menu with two white triangles in it, as shown in Fig. 16.7.

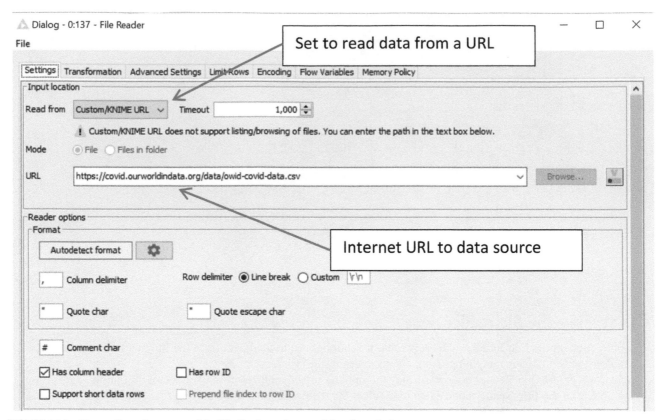

FIGURE 16.5   The configuration screen of the File Reader node to read data from the Internet.

FIGURE 16.6 The configuration screen for the string configuration node.

Flow variables function as global variables for downstream nodes. Values in these variables can be accessed via the Flow variable tab in the node configuration menus of downstream nodes. In order to input a new flow variable into the downstream data stream, the node option <Show Flow Variable Ports> must be activated to show the red dots to be used as flow variable connection ports for any node. We will practice doing so on the Rule-based Filter node. Two other flow variables will be derived later to hold text strings to be used for two line chart titles.

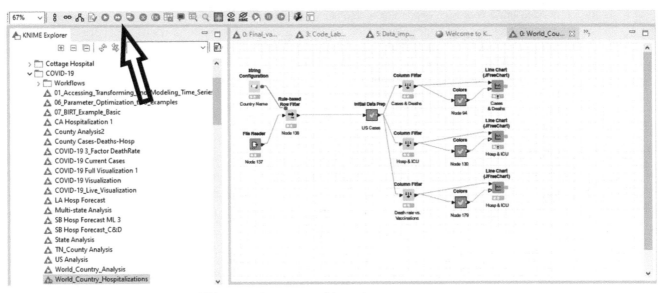

FIGURE 16.7 The location of the Execute-All button on the top menu of the workspace.

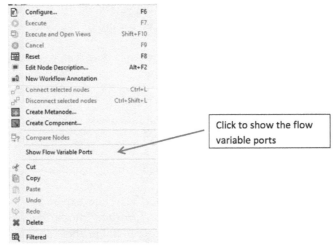

FIGURE 16.8 The options menu for the rule-based filter node.

G. Rule-based Filter node—the selected country is used to limit the downstream rows by the Rule-based Row Filter node.

1. Double-click on the node to display the node configuration screen.
2. Notice the definition of the logic statement limiting the rows to only those for the selected country.
3. The expression used to specify the row filter is: $location$=$${SCountry}$$=>TRUE (Note: the single $ symbols enclose column names and the $${...}$$ symbols enclose flow variable names) The logic of this expression is: If the column named location is equal to the selected country (flow variable named Country), the pass the row (=>TRUE), otherwise do not pass the row. This logic will pass all rows (one row per date) for the specified country.
4. Right-click on the node to bring up the node options screen (Fig. 16.8).
5. Click on the option for "Show Flow Variable Ports."
6. See that a second red dot (port) appears to the right of the one on the left which is already connected.
7. Right-click again on the node, and click again on the "Hide Flow Variable Ports" option. See that the right-hand red dot disappears (the left-hand red dot port remains, because it is connected to the String Configuration node).
8. Double-click on the Rule-based Filter node again to display the configuration screen.

a. If you want to build your own expression in another workflow, this is where you would write your expression.
   i. To build a new expression, double-click on the <location> variable in the Column list, enter a space, entering an=sign, entering another space, and then double-click on the flow variable <Country> in the Flow Variable list.
   ii. This expression builder is active for this node and for the Column Expressions node used later in the workflow. Notice that column names are enclosed in single dollar signs, and flow variables are enclosed in squiggly brackets and double dollar signs, which are added automatically.

Visualization data stream

The visualization data stream begins with the Rule-based Filter node (already described). The node connected to the Rule-based Filter node is the green Initial Data Prep node. This node is actually a collection of nodes symbolized by the green node color, and it is called a "metanode." To see the collection of nodes in the metanode, double-click on the metanode to see a group of five included nodes shown in Fig. 16.9.

A. Data Prep metanode
   1. Included node #1: String to DateTime node
      a. Double-click on the node to see that the string variable Date (with an S symbol next to it) is included for processing, and all other string variables are excluded.
      b. This date in the form of yyyy-MM-dd in this string type variable will be reformatted in the form of a date variable with a date data type.
         i. This operation changes only the data type, not the data in it;
         ii. The date data type is necessary to permit proper sorting on the date variable, if the user desires to perform some side analysis based on date.
         iii. The reason for the change from a string data type to a date data type is because string sorts are based on the ASCII number of the string characters, which do not sort numerically using the number designated by the digits in the data string.
      c. Double-click on the node, and execute it.
      d. Look at the Output table of the node (right-click on it and select the Output table option). See the new date data type icon displayed next to the Date variable.
   1. Included node #2: The Variable Expressions node
      a. This node permits you to define a new flow variable, using expression statements similar to those used in the Column Expressions node.
      b. Double-click on the Variable Expressions node to view the expression in the Expression Editor for the highlighted flow variable Hosp_Title: *"Total & ICU Hospitalizations/million in", variable("Country")*
         i. This expression builds a text string from a string literal element and the Country flow variable, defined in the String.
         ii. Click on the Cases_Title flow variable and see its definition in the Expression Editor.
   2. Included node #3: The RowID node renumbers the row number
      a. Executing this node will issue a warning that there was not column specified to use as the new RowID.
      b. Ignore this error; without specifying a column to use as the new RowID, the node just renumbers existing RowID, and that is exactly what is required here.

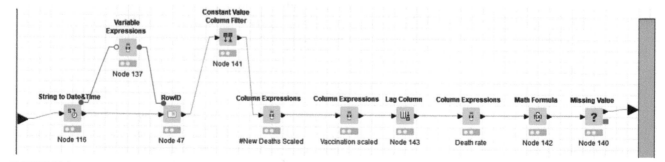

FIGURE 16.9 The data preparation metanode showing five included nodes.

3. Included node #4: The Constant Value Column Filter
   a. This node removes any column that has a common value in all rows.
   b. The node does not need any configuration.
4. Included node #5: Column Expressions node—#New Deaths scaled
   a. This node uses an expression to define a new column value, similarly to how the Variable Expressions node defines a new flow variable.
   b. Double-click on the node to see the Expression Editor as shown in Fig. 16.10.
      i. Fig. 16.10 shows the new variable name (new-deaths/MM*60), which defines the new variable as 60 times the actual new deaths number (using the * as a multiplication sign).
      ii. The new column expression is defined as: column("new_deaths_smoothed_per_million")*60
5. Included node #6: Column Expressions node—Vaccinations Scaled.
   a. This node uses an expression to define the #Vaccinations to make it easier to compare to the Death rate variable (Fig. 16.11).
   b. Notice the definition of the new Vaccinations_scaled variable as: *column("total_vaccinations_per_hundred")*25*.
   c. Click OK to exit the configuration screen.
6. Included node #7: Lag Column node
   a. This node generates a temporal abstraction variable called a *lag variable*.
   b. The logic shown in Fig. 16.12 shows that the lag interval is set to 14, which means it copies the value of the variable 14 time periods (days) before the current day.
   c. Double-click on the Lag Column node to see the configuration screen.
   d. The direction for derivation of this lag variable is provided by the insight gathered by inspecting Fig. 16.13, namely that it appears that the number of new deaths lags behind the number of new cases by about 10 days to 14 days.
   e. The 14-day lag was chosen to be conservative in representing the lag of new deaths behind new cases.
   f. Click OK to exit the configuration screen.

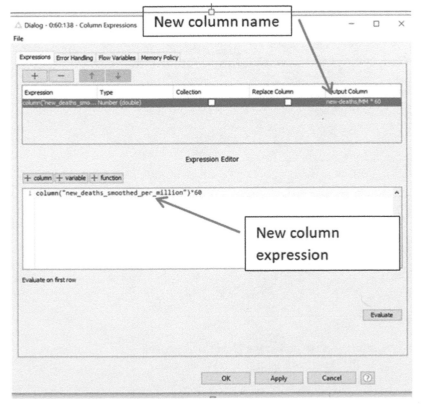

FIGURE 16.10 The expression editor screen in the column expressions node.

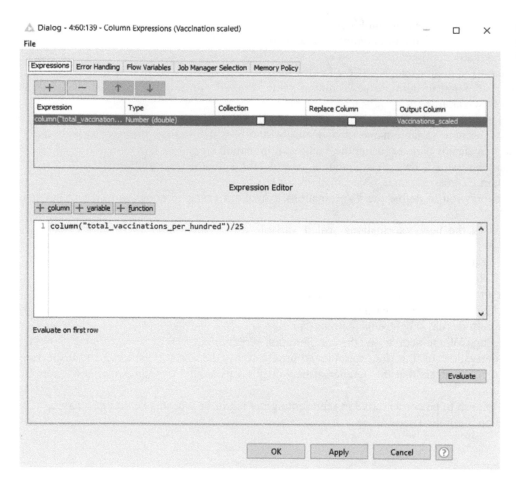

FIGURE   16.11 Configuration screen do derive the new vaccinations_scaled column.

FIGURE 16.12 The configuration screen of the lag column node, in which the lag interval is set to 14 days before the current day.

7. Included node #8: Using the Column Expressions node to derive the Death rate variable.
   a. This node is configured in a similar way to the previous Colum Expressions node.
   b. Double-click on the node to see the configuration.
   c. Click OK to exit the configuration screen.

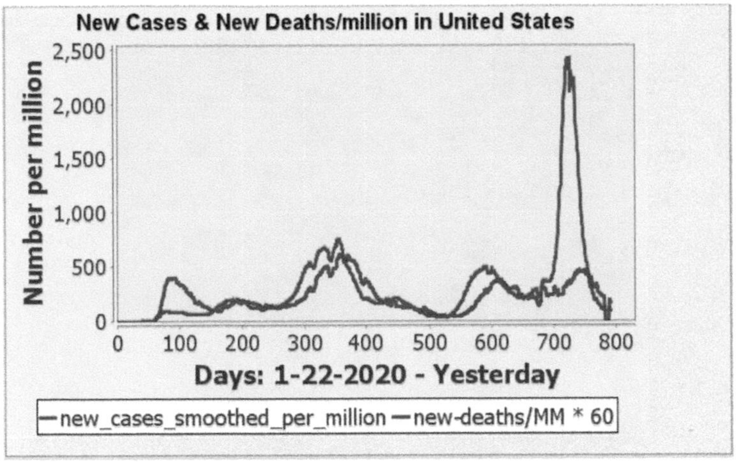

FIGURE 16.13 The daily new smoothed new cases and new deaths (X 60) between 7/15/2020 and 3/22/2022 ("Yesterday").

8. Included node #9: Math Formula node.
   a. This node is used to remove the NaN data symbol in the Death Rate variable caused by division by zero.
   b. Double-click on the node to see that the Death Rate variable is selected next to the Replace variable radio button (checked).
   c. This setting will replace the Death Rate variable values all rows with a decimal number, which by default removes the NaN data symbol, and replaced it with 0.0.
9. Included node #10: Missing Values node.
   a. This node replaces missing values in string and decimal number variables with 0.
   b. Double-click on the node to see the configuration screen.
   c. Click OK to return to the metanode w.
10. Click the X on the yellow highlighted metanode name (Initial Data Prep) to return to the main workflow.
B. Execute all executable nodes in the workflow. To execute all other properly configured (ergo executable) nodes in the workflow, click on the red circle with two white triangles in it located on the top menu, as shown in Fig. 16.4. This action will execute all executable nodes in the workflow. If you work with each node separately, there are three operations you should perform to operate a node:
   1. Configure the node properly Click on any node to highlight it, and then read the description of the operations of the node and the elements of it (in the Description pane next to the Expression Editor).
   2. Execute the node.
   3. Right-click on the executed node to display the options menu and look at the output table or chart to see what happened.
      a. The last option in the list is usually the table view option (variously named).
      b. Chart nodes are viewed by selecting the "View: Line Chart" option.
         i. Back in the workflow diagram, notice that the top Column Filter node includes only the new-cases_smoothed_per_million and the new-deaths/MM *60 variables.
         ii. The Colors metanode contains formatting and color specification data used in the Line Chart node of the JFree Chart node. The user can view the configuration of the nodes to see what is performed in them.
C. The output of the top line chart JFree Chart node is shown in Fig. 16.13.
   1. Right-click on the node.
   2. Select the option for: *View: Line Chart*.
D. The line plot for the total and ICU hospitalizations in the US is shown in Fig. 16.14.
   1. Right-click on the node.
   2. Select the option for: *View: Line Chart*.
   3. This plot for US hospitalizations serves as an introduction to the modeling of hospitalizations in 27 countries of the world for which hospitalization data is available.
   4. Notice that the time-course of the ICU hospitalizations follows that for the total hospitalizations, but at a lower level.
E. Line plot for the Death rate vs. vaccinations plot.
   1. Right-click on the JFree Chart node for Death Rate vs. Vaccinations to see Fig. 16.15.
F. Insights gained in the descriptive analytics visualizations:

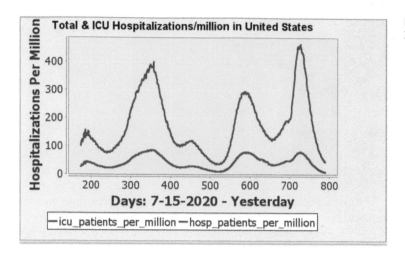

FIGURE 16.14   The time-course of all hospitalizations and ICU cases from 7/15/2020 to 3/22/2022 ("Yesterday").

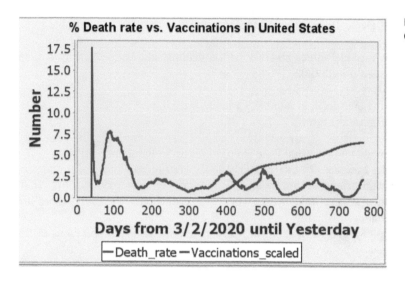

FIGURE 16.15   Line plot of the death rate vs # vaccinations (scaled) 7/15/2020 to 3/22/2022 ("Yesterday").

1. Fig. 16.13 shows evidence of 4 previous waves of the pandemic.
2. The new deaths curve in Fig. 16.11 appears to lag behind the new cases curve by about 10–14 days through the previous waves. This insight will be used to derive a new variable in Appendix-A expressing the number of new deaths 14 days previously. This new variable will be used as a predictor in building a predictive model for new deaths.
3. The plot shows that the number of new cases in the current wave (presumably due to the Omicron variant) exceeds the highest previous peak by about three times, and is now at an historical low.
4. Consequently, this visualization workflow provides a basis for building a predictive model.

Using the workflow for another country

1. Change the selected country to United Kingdom (or any other country of choice).
2. Right-click on the File Reader node, and select the Reset option to prepare it for loading in current data.
3. Click on the Execute All red button the top of the workflow screen.
4. Compare the data for the new country with those for the United States for:
   a. new cases and new deaths;
   b. hospitalization and ICU beds used; and
   c. death rate and vaccinations.

How this chapter facilitates patient-centric healthcare

This case study shows how to track the time-course of occurrence and hospitalization of cases of COVID-19 in each country. The full tutorial is available in this book's Appendix to permit readers to change the country specification to any country of the world. Trends and insights gained in case studies like this can help to guide preparations for the disease treatment appropriate for each patient, according to their needs.[1]

Postscript

The data visualization examples shown in this chapter are part of the larger phase of the CRISP-DM analytics process model for data understanding introduced in Chapter 4. Much can be learned in this phase of predictive modeling before new variables are derived for predicting an outcome. In this example of visualization, we learned that the time-course of new deaths from the COVID-19 virus lags behind the time-course of new cases by about 2 weeks. Armed with this information, we might decide to derive a new variable, New_cases-2, to be used to predict the number of expected ICU hospital beds for the next week. This type of derived variable is called a *lag variable*. Such lag variables have been shown in many data domains (e.g., telecommunications) to be useful predictors of important outcomes (e.g., customer attrition, or "churn").

Further reading

Wright, S., 1921. Correlation and causation. J. Agric. Res. 20, 557–585.

---

[1]NOTE: This Tutorial–Case Study is continued in this book's Appendix-A, for those readers who would like to learn more about this topic and also use the free software program KNIME for Predictive Analytics.

Chapter 17

Disseminated intravascular coagulation predictive analytics with pediatric ICU admissions

Linda A. Miner, Harsha Chandnani, MD, MBA, MPH[1], Mitchell Goldstein[2], Mahmood H. Khichi, MD, FAAP, MBA[2] and Cynthia H. Tinsley, MD, FAAP, FSCCM[3]

[1]Pediatric Intensivist, Department of Pediatrics] Loma Linda University, School of Medicine, Children's Hospital, Loma Linda, CA, United States, [2]Critical Care Provider,Texas Children's Hospital, TX, United States, [3]LOMA LINDA UNIVERSITY Children's Hospital, Pediatric Critical Care Medicine Division Chief, CA, United States

Chapter outline

Prelude

Chapters 14 and 15 presented case studies of technology to predict the presence of cancer by analyzing clinical data in the pediatric patient record upon admission to a hospital. This chapter presents a case study of the development of a predictive model to predict the risk of survival or death from pediatric DIC (disseminated intravascular coagulation). Predictive models are based on the wealth of patient clinical data that can be leveraged to provide early warning signs of presence of a condition, disease, or cancer before any formal diagnosis begins. Such predictive models do not take the place of formal diagnostic procedures, but can be used to support the recommendation of whether specific diagnostic procedures (e.g., a colonoscopy) should be performed even though other elements of the medical specialty guidelines are not satisfied.

Introduction

In the first edition of this book, the original idea came from Tutorials J1, Predicting survival or mortality with disseminated intravascular coagulation (DIC) and/or critical illnesses, p. 558 and J2, Decisioning for DIC, p. 603, both by Mahmood Khichi, Wanda Parsons, and Linda Miner. In that tutorial, the reader may see specific instructions. This chapter is a Case Study, and as such, does not go into the specifics of how to use the program. The program used was Statistica, version 13.2, in 2017, and in particular, much of the below used the data miner recipes from Statistica. Statistica, at the time, was owned by Dell, but now is owned by Tibco. The companion web pages show tutorials on the use of Statistica.

Practical Data Analytics for Innovation in Medicine. DOI: https://doi.org/10.1016/B978-0-323-95274-3.00024-5

Background (from first edition)

DIC is a pathologic condition of system-wide hemorrhaging and clotting. Many underlying conditions, such as having surgery, cancer, and sepsis, can trigger inflammatory chemicals that cause microclots throughout the body (Dressler, 2004).

It seems a bit counterintuitive that clotting could cause bleeding. However, as the clotting increases, the clotting factor, fibrin, can become so otherwise engaged that it is not available where it might be needed in wounds, such as surgery sites, pick lines, IV lines, and so on. Bleeding can therefore ensue as secondary to coagulopathy. The clots might also keep blood from vital organs, which can then begin to die. Extreme clotting, hemorrhaging, and death are real possibilities with DIC.

The coagulopathy cascade is initiated in a couple of ways: intrinsically and extrinsically (McCance, Huether, Brashers, and Rote, 2010, p. 977). The intrinsic method, or contact activation pathway, generally starts with injury to the inside (endothelium) of the blood vessel, which then sends out negatively charged subendothelial substances. The extrinsic method, or the tissue factor (TF) pathway, starts with TF. The protein CD142 (a protein that is TF) can initiate the formation of thrombin. CD142 protein is found in leukocytes and the subendothelium. Once the cascade begins, it is difficult to stop, and soon the process is happening all over the body.

Prothrombin or thrombin helps make clots, and antithrombin helps break down clots. Both are made in the liver. Thrombin converts fibrinogen into fibrin. There are also chemicals in the blood that break down clots: Protein C, plasminogen, and antithrombin break down clots. If there is too much thrombin, then too much clotting occurs. Too much internal antithrombin can encourage bleeding. Causes of DIC may be from trauma or illness or even from using prescription blood thinners in over-the-counter medications like aspirin or ibuprofen. The possibility of DIC is why patients are often instructed not to take these before an operation. Fig. 17.1 shows the cascading coagulation process.

There are several ways in which the body reduces blood flow. Blood vessels constrict, clots form or plugs form from platelets (Dressler, 2012). If the endothelium of a blood vessel is damaged and exposed, then platelets start their action by changing their shape from disks to spheres. They then become sticky, clump, and plug the hole. If the hole is large, platelets become more of a foundation for fibrin strands to knit more of a platform to stop the bleeding.

Furthermore, the tissue that has been damaged produces TFs to stop the bleeding. Inflammation, caused by such trauma, or even viruses or bacteria, can induce cytokines that serve to enhance coagulation. This can rarely happen, thank goodness, even with covid vaccines increasing the formation of blood clots (Goodman, 2021). The problem is that the inflammatory response brings on white blood cells, vasodilation, and spreading of the inflammatory response down the line. As seen by the schematic in Fig. 17.1, the processes of either stopping upflow or increasing flowing are complicated, and if something goes awry in any of the links, clotting abnormalities and/or excessive bleeding can occur. In general, when the sequence gets out of control, things do not go well.

Quick action is required to regulate the processes. Predicting which patients are most likely to either die or be discharged may help speed up the regulation. The original study in Tulsa attempted to find such an algorithm, but the number of patients was small, so it was important to try again with more patients. See Tutorials J1 (p. 558) and J2 (p. 603).

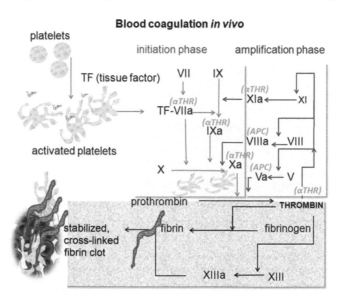

FIGURE 17.1   Blood coagulation in vivo. *Originally downloaded from Wikipedia. The proper contributions by authors may be found at this link http://en.wikipedia.org/wiki/File:Coagulation_in_vivo.png.*

The example

This example outlines the procedures we used to analyze the data for 11,569 pediatric patients at the Loma Linda Hospitals. These children were admitted between January 1, 2009, and December 31, 2015. with a diagnosis of DIC.

We attempted to develop a prediction model for survival or death, given the measurements of various tests *upon admission*. The hospitals were using two known prediction instruments, the Pediatric Risk of Mortality (PRISM3) (Khajeh et al., 2013; Pollack, Patel, and Ruttimann, 1996) and the Pediatric Index of Mortality (PIM2) (Choi et al., 2005; Shann, Pearson, Slater, and Wilkinson, 1997). We wanted to see if we could determine a smaller number of lab tests (variables) to predict survival or death.

This example goes through the processes used to analyze the data. Linda Miner analyzed the data, Dr. Chandnani wrote up the findings for her residency program research requirements, And Dr. Chandnani was supervised by Drs. Tinsley and Goldstein for the final papers of Dr. Chandnani (Chandnani et al., 2018a; Chandnani et al., 2018b). Dr. Khichi was instrumental in the first project and provided valuable insights as the team wrote back and forth.

Data files

After obtaining the IRB approval, it was over a year before we heard anything and then I, Linda, received four files from the team which needed to be analyzed within a week for a draft of a paper that was to be sent to meet deadlines for upcoming meetings, allowing Dr. Chandnani to show her research.

There were four data files sent:

1. A demographics file of 11,569 cases. This one had 11,569 cases, with 15 variables for each case. The demographics file also had the *outcome variable of survived or died*.
2. A diagnosis file with 167,384 rows with 24 variables for each case. These rows repeated cases from one entry for some children and up to 15 or 16 entries for other children. Fig. 17.2 shows a sample of what they looked like with groupings indicated by the red circle around them. I decided to put the demographics file aside until there was time to work on them using text mining instead of the files with continuous variables.
   a. A "PIM" data file with the 11,569 cases, each with 11 variables, presumably some of the variables were needed to arrive at a PIM score.
   b. A "PRISM3" file with the 11,569 cases, each with 94 variables, some of which I assumed were necessary to derive a PRISM score.

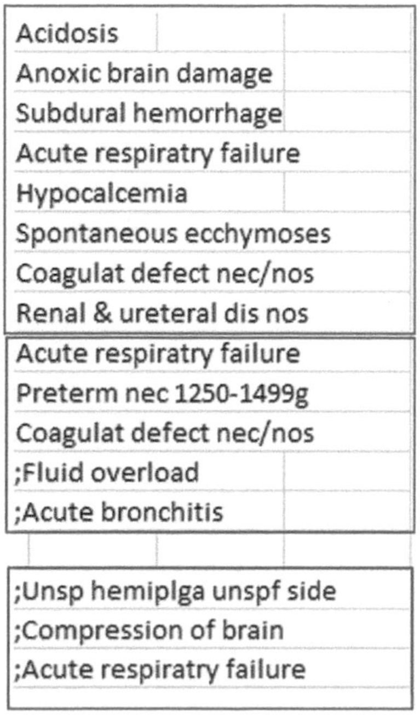

Acidosis		
Anoxic brain damage		
Subdural hemorrhage		
Acute respiratry failure		
Hypocalcemia		
Spontaneous ecchymoses		
Coagulat defect nec/nos		
Renal & ureteral dis nos		
Acute respiratry failure		
Preterm nec 1250-1499g		
Coagulat defect nec/nos		
;Fluid overload		
;Acute bronchitis		
;Unsp hemiplga unspf side		
;Compression of brain		
;Acute respiratry failure		

FIGURE 17.2 Diagnoses grouped by case.

It may be interesting to the reader to know up-front that we determined three variables to be the final best predictive variables after all the analyses. Also, the reader should know that we were not terribly trusting in the final model, and the reasons will be evident through the case study. The variables ended up being those variables Dr. Chandnani presented in the papers and were high Ptt, low bicarbonate, and base excess.

Ptt stands for partial thromboplastin time, which is the time it takes for a clot to form. So high Ptt means it could take longer for bleeding to stop. Bicarbonate is a base. Low bicarbonate could indicate metabolic acidosis or that the blood is too acidic. Low bicarbonate can affect fibrinogen (making it degrade faster), and fibrinogen is needed to form clots. Metabolic acidosis can, therefore, cause bleeding. Base excess is another value of how much acid is in the blood and indicates the acid to the base ratio (Higgins, 2018). If blood is normal, the pH is normal, po2 is normal, and the temperature is normal, the base excess would be zero.

All three of these values, Ptt, bicarbonate, and base excess, could interact to cause either bleeding or clotting. Other values also enter the DIC cascade, and if any are out of whack, it is incumbent upon the clinicians to correct the situation. Sometimes that correction becomes impossible. These values may be good indicators of the tipping point and logically could make good predictors of death or survival.

First week of analysis

The first thing done to the data was that the files, PIM, PRISM, and demographics files were merged by case. Having only a few days to get something to the team, I knew that the data were complicated, missing, and perhaps redundant. We wanted to see if we could predict well using only a few lab tests that could be collected from the child upon admission.

One other thing that was apparent just by eyeballing the data set was there were many missing values.

To get something to the team right away, I randomly broke the 11,569 data set into training testing and validation sets to do a data mining recipe (DMR, in Statistica). Fig. 17.3 shows the feature selection on the training data. What

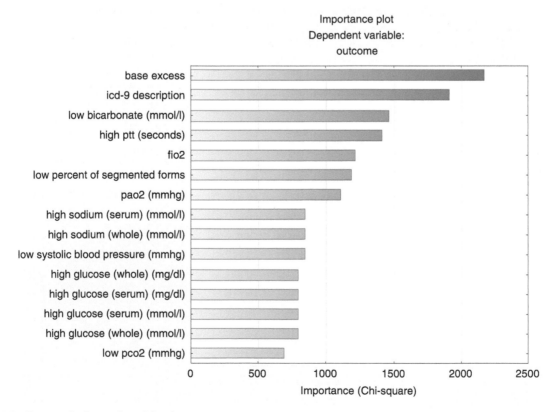

FIGURE 17.3   Feature selection on the training data.

one does in feature selection, is enter the outcome variable as death or survival and then use as many other variables as seem that they might be predictive in as the independent/predictor variables. I used all the variables except for the PIM and PRISM scores and percentages. What feature selection does is that it iteratively determines the variables that seem to predict the outcome variable the most. These would then be good to try using in a predictive model. The longer the bar in Fig. 17.3, the more the variable seems to predict. (Again, to see a tutorial demonstrating the use of Statistica see the original DIC tutorial on the companion web pages.)

Data mining recipes using statistica

All the crucial variables from the feature selection were put into the DMR, again using only the training data. The DMR tests many predictive models (predictive algorithms) and then tells which model seemed to predict the best. It is a semiautomated approach. One can find a number of tutorials from the first edition of the book in the companion web pages that show step by step how to use the DMR. It tests many common learning algorithms at the same time and then tells the tester which predictive algorithm seems to predict best, by showing frequency tables for each algorithm. Then the tester is able to decide which of them should be used interactively. Interactive means that the tester is able to use more parameters, including v-fold cross-validation, which helps to guard against overfitting models.

I particularly liked C&RT because the tree graphs were usually easy to understand, and the C&RT had the least error in the predictive models (besides Neural Networks, which was nearly impossible to interpret using a graph). Besides, the C&RT had great predictions, it seemed. Fig. 17.4 shows the predictions from the training group of the model. (The model further split the training data into training and testing groups—randomly, of course.)

Another way of looking at Fig. 17.4 table is by considering *sensitivity* and *specificity*, both conditional probablilities.

Sensitivity is the proportion of observed survivals that were predicted to survive (670, no error), given the proportion of predicted survivals that actually died (20, error).

Specificity is the proportion of observed deaths that were predicted to be deaths (253, no error), given the proportion of predicted deaths that actually survived (55, error).

For Fig. 17.4, we can make another box containing the elements needed for sensitivity and specificity.

$$\text{a } 670 \quad \text{b } 55 \quad \text{Sensitivity} = a/(a+c) = 670/690 = .971$$
$$\text{c } 20 \quad \text{d } 253 \quad \text{Specificity} = d/(b+d) = 253/308 = .821$$

Generally, however, when one runs a predictive algorithm, one looks at the two figures pointed to with the arrows in Fig. 17.4 (92.41 and 92.67) to get a good estimate of the value of the prediction, if the data are good. And that is a big if. In this case, the predictions seemed too good to be true. However, if one only has data that are not the best, one works at the best prediction possible, given the probable errors and then continues in the future to refine and redo the prediction model. In our case the data were sparce and so in the future, to test and refine our final predictive algorithm, we should collect all the needed data for the model from each patient.

	Summary Frequency Table (Prediction) Table: outcome(2) x Model-1-Prediction(2) Training C&RT			
	outcome	Model-1-Prediction Survived	Model-1-Prediction Died	Row Totals
Count	Survived	670	55	725
Column Percent		97.10%	17.86%	
Row Percent		92.41%	7.59%	
Total Percent		67.13%	5.51%	72.65%
Count	Died	20	253	273
Column Percent		2.90%	82.14%	
Row Percent		7.33%	92.67%	
Total Percent		2.00%	25.35%	27.35%
Count	All Grps	690	308	998
Total Percent		69.14%	30.86%	

FIGURE 17.4 Training C&RT classification matrix shows that 92.41% (true positives or 670/725) of the children who did survive were correctly predicted, and 92.67% (true negatives, or 253/273) of the children who died were correctly predicted. This seemed too good to be true.

	Summary Frequency Table (Prediction) Table: outcome(2) x Model-1-Prediction(2) Testing C&RT			
	outcome	Model-1-Prediction Survived	Model-1-Prediction Died	Row Totals
Count	Survived	220	43	263
Column Percent		91.29%	45.74%	
Row Percent		83.65%	16.35%	
Total Percent		65.67%	12.84%	78.51%
Count	Died	21	51	72
Column Percent		8.71%	54.26%	
Row Percent		29.17%	70.83%	
Total Percent		6.27%	15.22%	21.49%
Count	All Grps	241	94	335
Total Percent		71.94%	28.06%	

FIGURE 17.5 Testing C&RT classification matrix shows that 83.65% of the children who survived were correctly predicted, and 70.83% of the children who died were predicted correctly. These were lesser predictions than the training, but still very good—and likely too good to be true.

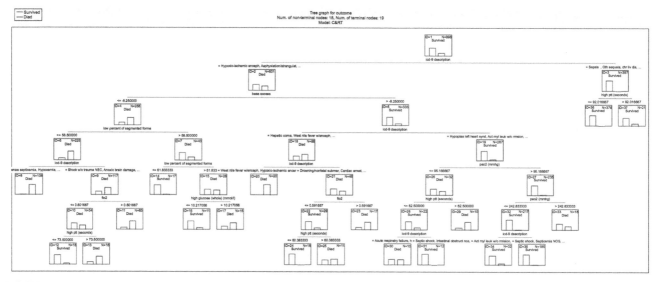

FIGURE 17.6 Complicated rules—many branches and sparse data.

Next, Fig. 17.5 shows proportions using the testing data in the DMR.

The tree graph was very complicated, and it seemed there were just too many variables in the model to be useful for what we wanted to do—predict with fewer tests (Fig. 17.6). Moreover, there were several missing entries.

Looking back 5 years, I believe these analyses ended what was done in the first week and sent to the team to look at. I was still worried about the missing values; more analyses would be performed.

Data imputation

Next, the Statistica Data Health Check was run to look closely at the data. As suspected, the data looked redundant and sparse (See Fig. 17.7).

I wondered if imputation by k nearest neighbors (KNN) could work when there were so many missing values. Two studies that used computer simulation trials said that KNN likely could work accurately with many missing entries (Beretta and Santaniello, 2015; Jönsson and Wohlin, 2004). How many missing was the question? One went up to 30% of missing values and still worked. How many neighbors also seemed to be an issue. A good number of neighbors seemed to be three to best preserve the data structure and reduce imputation error (Beretta and Santaniello,

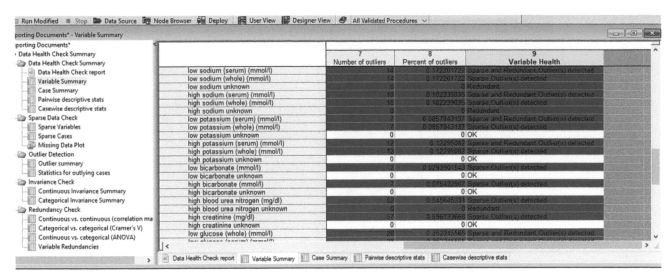

FIGURE 17.7   Data health check summary.

2015). I decided to use three and see what would happen. Statistica imputation was completed, and I used the 11,569 imputed data set.

Using the 11,459 imputed file—training data

I randomly divided the data into training, testing, and validation. The training-only file had about 50% of the cases, the testing group, 25%, and the validation group 25%. First, I used feature selection on the training data, without PIM and PRISM scores, and this time, I found that high ptt, low bicarb, and high base excess may be three major predictors of survived and died. I figured those PIM and PRISM should be predictive, but they required additional tests, and we were trying to get as good a prediction as those two with fewer lab tests. The following feature selection resulted in Fig. 17.8 for the outcome variable of died vs survived.

We first got the idea that these three might be good predictors for the algorithm. However, these were from the imputation of many missing values.

Imputation of large amounts of missing data could introduce bias into the system (Seijo-Pardo et al., 2019). Just about anything we would try would *not* be ideal, as the analyses would be if we had all the values or near all the values in the data set. Besides imputation, there were few options—we could drop cases or try a method such as filling in the missing values with means from all the other cases. However, the KNN procedure results are much more accurate than the mean values. These were the data we had, and we proceeded the best we could, understanding that any resulting predictive algorithm may or may not be accurate for new cases outside this data set, and any predictive algorithm we found should be continuously tested in the future for revision, as should be done for all prediction models.

Figs. 17.9–17.11 show mean plots of those variables in the feature selection. At the time of the analysis, I did not know what it would mean, for example, for there to be a difference between survived and died in base excess. I trusted that the team would know. I did not receive any information from the team when I sent analyses at the end of the week, so I figured everything was okay at this point. Now, when I have had more time to study the interactions, as mentioned above, aided by better searches given 5 years of Google development, it seems obvious that base excess should be closer to zero for the group that survived. It makes sense that those who died would have a value further away from zero. Concerning, for example, high ptt, higher values would indicate more bleeding, and those who died likely had dangerous levels of bleeding. Low bicarbonate could indicate acidosis, which would mean too much acid in the blood, an imbalance in the acid to base ratio, perhaps even meaning problems with the kidneys. So now, the data make more sense, for which I am grateful because I was working in the dark at the time. Philosophically, maybe it is better not to know—I could not have any biases.

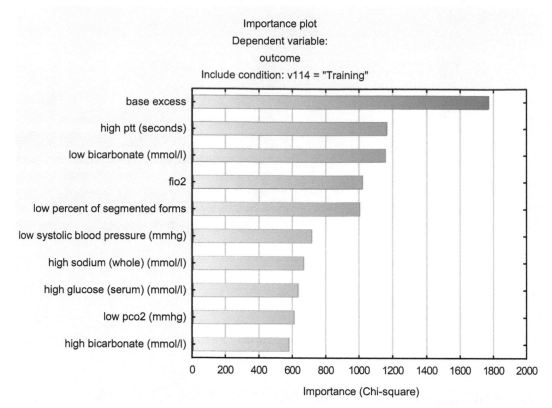

FIGURE 17.8 Feature selection of the most critical continuous variables for the outcome of died or survived, which appeared to be base excess, high ptt, and low bicarbonate. Note that they had the longest bars, although one could have gone for more variables.

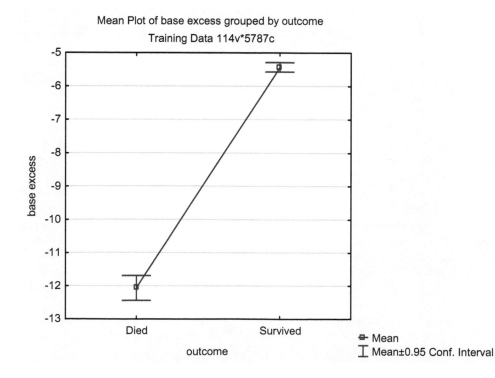

FIGURE 17.9 Base excess mean plot, imputed data, training group.

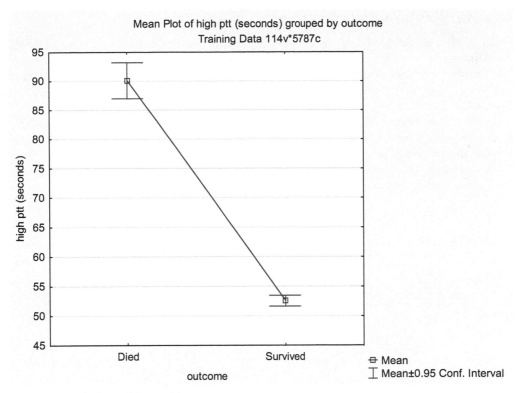

FIGURE 17.10   High ptt mean plot imputed data, training group.

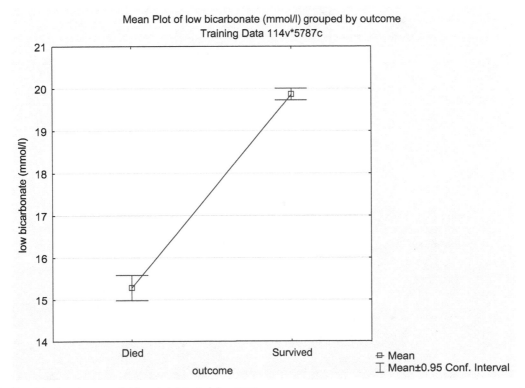

FIGURE 17.11   Low bicarbonate mean plot imputed data, training group.

Training data (11,569 imputed) continued

Again, a Statistica DMR was conducted on the training data, which was 50% of the original 11,569 files with KNN data imputation. I used the outcome as the dependent variable and the three variables, base excess, high ptt, and low bicarbonate, as the independent variables. Of the methods tried, Boosted Trees had the lowest error rate (Fig. 17.12).

Fig. 17.13 shows the classification chart for boosted trees.

Fig. 17.14 shows the classification chart of the parceled-out testing data. (Again, this was the testing group produced by the boosted trees program *within* the training data of the entire 11,589 data set; I used only the training data to leave the testing group and the validation group untouched until the end.)

There is a high agreement between training and testing looking at these two matrices. That agreement is considered an indication of a stable prediction model.

	Summary of Deployment (Error rates) (Training Data_Validation)				
	2-Random forest	3-Boosted trees	5-Neural	1-C&RT	4-SVM
Error rate	0.160653	0.142612	0.175258	0.175258	0.169244

FIGURE 17.12   Error rates of DMR for training data. For training group (subgroup within the training group data). Boosted trees had the least amount of error which was 14.26%.

	Summary Frequency Table (Prediction) Table: outcome(2) x model-3-Prediction(2) Training			
	outcome	Model-3-Prediction Died	Model-3-Prediction Survived	Row Totals
Count	Died	742	378	1120
Column Percent		77.45%	10.31%	
Row Percent		66.25%	33.75%	
Total Percent		16.05%	8.18%	24.23%
Count	Survived	216	3287	3503
Column Percent		22.55%	89.69%	
Row Percent		6.17%	93.83%	
Total Percent		4.67%	71.10%	75.77%
Count	All Grps	958	3665	4623
Total Percent		20.72%	79.28%	

FIGURE 17.13 The classification matrix for data mining recipe (DMR) uses only the training data and boosted trees. Looking at the chart, 66.25% of those who died were predicted to die by the model. 93.83% of those who survived were predicted to survive by the model.

	Summary Frequency Table (Prediction) Table: outcome(2) x Model-3-Prediction(2) Testing			
	outcome	Model-3-Prediction Died	Model-3-Prediction Survived	Row Totals
Count	Died	199	101	300
Column Percent		75.38%	11.22%	
Row Percent		66.33%	33.67%	
Total Percent		17.10%	8.68%	25.77%
Count	Survived	65	799	864
Column Percent		24.62%	88.78%	
Row Percent		7.52%	92.48%	
Total Percent		5.58%	68.64%	74.23%
Count	All Grps	264	900	1164
Total Percent		22.68%	77.32%	

FIGURE 17.14 Classification of the testing data for the DMR. Here we see that 66.33% of those who died, were predicted to die; 92.48% of those who survived were predicted to survive.

A problem

After the papers were presented, Dr. Chandnani told me someone had asked her about possible collinearity. I wondered what the person was referring to. Now I do understand, as I have a better understanding of the three variables.

However, just because two variables are correlated does not mean that they both cannot be predictive, only that the more they are related, the less one of them is needed. This is especially true if the correlation is quite high. I should have deleted one (low bicarb) and used the next variable, fo2.

At the data analysis, I was still very much bothered by the missing data and wanted to do something about the problem, as I thought missing data or imputation could somehow be causing overfitting or bias. I went after the missing data problem.

According to Tibco using boosted trees: "When, during model building, missing data are encountered for a particular observation (case), then the prediction for that case is made based on the last preceding (nonterminal) node in the respective tree. So, for example, if at a particular point in the sequence of trees, a predictor variable is selected at the root (or other nonterminal) node for which some cases have no valid data, then the prediction for those cases is simply based on the overall mean at the root (or other nonterminal) node. Hence, there is no need to eliminate cases from the analysis if they have missing data for some of the predictors, nor is it necessary to compute surrogate split statistics (e.g., see also the documentation for the number of surrogates option on the Interactive Trees Specifications dialog box—Advanced tab for C&RT)." Boosted Trees for Regression and Classification Overview (Stochastic Gradient Boosting)—Basic Ideas (tibco.com)

I had already imputed values, so there should not have been many, if any, missing data for the program to have to deal with, but I was still quite worried. My gut said that the data might not really represent the whole group of 11,569 because of the large imputation. I suggested that we should drop cases that were more than 40% missing. Forty percent was also stretching it, but I felt it would be better than doing nothing. One other thing that Dr. Goldstein suggested was to try deriving missing values using other variables that might exist for a case to increase usable cases. I found that this would be possible for base excess and bicarbonate using the formulas. From the Cornell site that Dr. Goldstein sent me (http://www-users.med.cornell.edu/~spon/picu/calc/basecalc.htm), I was able to put these into Statistica.

These formulas were:

$$\text{Base Excess:} // \text{B.E.} = 0.9287 * \text{HCO3} + 13.77 * \text{pH}124.58$$
$$\text{Bicarbonate:} // \text{HCO3} = 0.03 * \text{pCO2} * 10 * *(\text{pH}6.1)$$

I should have thought right then that HCO_3 is bicarbonate, and therefore would be related to base excess if it was in the derivation formula. Of course, looking now at the project, if one measure is highly predictive, then the other likely would be as well. The other thing I am thinking about now is the quote from Tibco above that the program has its way of imputing missing data. Would the method already in the DMR program be more valid than using its imputation program? We shall finish the story of what was done in 2017–18.

We decided to derive base excess and low bicarb to increase cases and, after that, to eliminate all the cases that had more than 40% missing data. Statistica did that elimination for us. The final data set had 6184 of the original cases.

And for interest, now, in 2022, when I checked the correlation between our derived base excess and our derived bicarbonate, it was $r=.91$. The correlation for the original base excess and original bicarbonate was $r=.69$, which was lower, and below the common rule of thumb of .80 for a cutoff of usability. Probably using the original values would have been better than the derived values. Regardless, it is likely that the prediction model did not need one of the two variables.

Randomly separating the data and new data mining recipe

Back to the analysis: For the *new data* set of 6184 cases, from the original file of 11,569—not imputed, I randomly separated the 6184 cases into two sets, training, testing. DMR was performed using the training data. A feature selection was made using the training set, and the top five variables were used in the DMR prediction model (Fig. 17.15).

Fortunately, the derived variables were not used in the DMR, *accidentally avoiding the mistake of using the most highly correlated variables*. Whew!

Using the DMR and the first five variables, the learning model, Boosted Trees, had not the lowest but close to the lowest error rate. They were close, and I wanted to compare these results with the previous analysis (Fig. 17.16).

In addition, the predictions of death using Support Vector Machines were not as accurate as with Boosted Trees, even though both did not predict the deaths well as did the previous imputed large data file. Fig. 17.17 and 17.18 show training and testing frequencies using Boosted Trees.

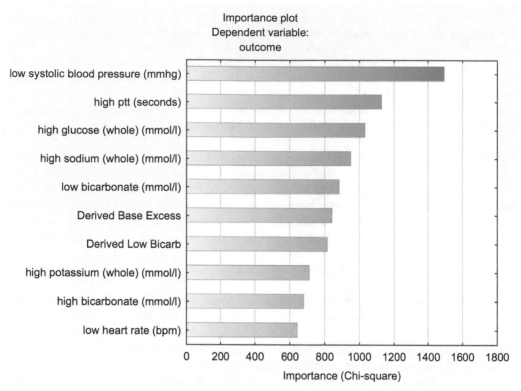

FIGURE 17.15 Feature selection using the smaller data set removes cases with >40% missing. And including the derived low bicarb and derived base excess resulting in 6184 cases.

	Summary of Deployment (Error rates) (All - demogr - Prism - PRIM Diagoses Derived scores 40 minus duplicates_Validation)						
	2-Random forest	**3-Boosted trees**	5-Neural	1-C&RT	4-SVM		
Error rate	0.269737	0.203947	0.217105	0.276316	0.194079		

FIGURE 17.16 Note the 1% difference between Boosted Trees and Support Vector Machines. Boosted Trees was the method used.

	Summary Frequency Table (Prediction) Table: outcome(2) x Model-3-Prediction(2) Training			
	outcome	Model-3-Prediction Died	Model-3-Prediction Survived	Row Totals
Count	Died	112	139	251
Column Percent		76.71%	14.62%	
Row Percent		44.62%	55.38%	
Total Percent		10.21%	12.67%	22.88%
Count	Survived	34	812	846
Column Percent		23.29%	85.38%	
Row Percent		4.02%	95.98%	
Total Percent		3.10%	74.02%	77.12%
Count	All Grps	146	951	1097
Total Percent		13.31%	86.69%	

FIGURE 17.17 Training DMR Boosted trees, using low systolic blood pressure, high ptt, high glucose, high sodium, and low bicarb (not derived). While the survival predictions were high at 95.98%, the deaths were not well predicted, with only 44.62% correctly predicted.

Our previously high mood from the original predictions lowered. However, one good thing, as stated, was that neither the derived base excess nor the derived bicarbonate was used in the DMR.

The training and testing matrices are shown in Figs. 17.17 and 17.18.

Summary Frequency Table (Prediction) Table: outcome(2) x Model-3-Prediction(2) Testing				
	outcome	Model-3-Prediction Died	Model-3-Prediction Survived	Row Totals
Count	Died	28	41	69
Column Percent		57.14%	16.08%	
Row Percent		40.58%	59.42%	
Total Percent		9.21%	13.49%	22.70%
Count	Survived	21	214	235
Column Percent		42.86%	83.92%	
Row Percent		8.94%	91.06%	
Total Percent		6.91%	70.39%	77.30%
Count	All Grps	49	255	304
Total Percent		16.12%	83.88%	

FIGURE 17.18 Testing DMR Boosted trees, using low systolic blood pressure, high ptt, high glucose, high sodium, and low bicarb (not derived). Note that while the survival predictions were still quite high at 91.06%, the deaths were predicted even worse, with only 40.58% correctly predicted.

Classification matrix (Training and Testing Data from original imputed data 10-28) - Analyis Sample Response: outcome Analysis sample;Number of trees: 200					
	Observed	Predicted Survived	Predicted Died	Row Total	
Number	Survived	1881	310	2191	
Column Percentage		92.52%	34.83%		
Row Percentage		85.85%	14.15%		
Total Percentage		64.35%	10.61%	74.96%	
Number	Died	152	580	732	
Column Percentage		7.48%	65.17%		
Row Percentage		20.77%	79.23%		
Total Percentage		5.20%	19.84%	25.04%	
Count	All Groups	2033	890	2923	
Total Percent		69.55%	30.45%		

FIGURE 17.19 Training classification matrix. This looked pretty good (likely too good)! 86% accuracy for those that survived; 79% accuracy for those that died.

Final analysis—a return to the past

I pondered which data to use ultimately, imputed or cases removed? The real answer is probably not either of them because of the missing data likely causing the data sets used not to truly represent the children in the study. We needed to try in any future study to devise some method of ensuring that most of the data points were recorded.

In the end, I decided that the imputed data would at least have all the children included and perhaps represent them better than taking away nearly half. An interactive boosted trees with the three original variables was completed using the training and testing data from the large, imputed file of the 11,569 cases. As luck may have it, we were therefore using the base excess and low bicarbonate from the original numbers with $r=.69$ (a smaller correlation than the derived variables of $r=.91$). We likely would have an error in the final algorithm due to all the missing data imputed, but, again, the resulting model might better represent all the children from those years and allow us to work on the algorithm over time. We would have errors no matter what we did using these data. As suggested earlier, we could then revise the model to result in a better prediction model.

Again, I formed training, testing, and holdout data. The training-only file had about 50% of the cases, the testing group, 25%, and the validation group, 25%, randomly divided. This time I put in all the sensible admission variables. The feature selection revealed that high ptt, low bicarb, and high base excess were the three major predictors of survival or death.

Next, I used a DMR to see which form of predictive models would be the most accurate and found that Boosted Trees, again, was deemed most accurate. Choosing Boosted Trees was good because we would be able to get a predictive percentage score on the died or survived outcome for each case.

Finally, for the training/testing data, *interactive Boosted Trees* was conducted—interactive Boosted Trees allowed for cross-validation to be used, helping to decrease overfitting. Figs. 17.19 and 17.20 show the resulting classification matrices.

	Classification matrix (Training and Testing Data from original imputed data 10-28) (Test sample) Response: outcome Test set sample;Number of trees: 200						
	Observed	**Predicted Survived**	**Predicted Died**	**Row Total**			
Number	Survived	3722	645	4367			
Column Percentage		92.40%	36.67%				
Row Percentage		85.23%	14.77%				
Total Percentage		64.32%	11.15%	75.46%			
Number	Died	306	1114	1420			
Column Percentage		7.60%	63.33%				
Row Percentage		21.55%	78.45%				
Total Percentage		5.29%	19.25%	24.54%			
Count	All Groups	4028	1759	5787			
Total Percent		69.60%	30.40%				

FIGURE 17.20 Testing sample classification matrix. Again, the algorithm used on the testing sample showed 85% correct prediction for survival and 78% for death. If accurate, this model would be outstanding.

The model was consistent but suspect due to the amount of imputation. With this prediction matrix (for testing) matching the training data matrix and seeing how the predictions were better than using the first Classification and Regression Trees, I saved the PMML for future deployment.

Conclusion—personal ending thoughts

It was difficult looking at all the cases of children that the data file said had died (about one out of four). Truly, this is a horrible condition to acquire, and I felt so sorry for all the children who had suffered—and their parents—for those children who died, and for those children who survived but also for those who may have been left with disabilities after discharge. I felt sorry for all the medical teams that had worked with these children. It had to be hard.

At the same time, I was so grateful for the dedicated doctors, nurses, and other health professionals who worked with these children and were able to help save three out of four—dedicated doctors like Dr. Chandnani, Dr. Goldstein, Dr. Tinsley, and Dr. Khichi. It was truly a privilege to work with them.

My thought now is this: would it make a difference if a medical team believed (because of an algorithm) that a child was predicted to die or to live? Would they try harder if they thought the child would die or would live? Would they try just as hard regardless? (This is what I believe is the case, given all the fine doctors I have known and worked with.)

But the next question I am not certain: How about insurance companies? Would they approve or disapprove of expensive treatments if they thought the child would die or live? Perhaps this research effort is best not continued after all.

Postscript

This case study leaves one with the exciting prospect of predicting outcomes of patients with DIC, but it raises several ethical questions related to how different functionaries are likely to respond to the results of the predictive models. The case study presents an exciting prospect of using predictive analytics to aid in the diagnosis and treatment of DIC. On the other hand, it raises questions that could decrease the level of health care. What if a patient is in the error bounds of the models? All physicians have taken the oath to "do no harm" in their practice of medicine. This technology begs the question, "will medical informatics models like this one lead to harm of some patients"? This may be an aspect of the downside risk of personalized medicine. There is much to be worked out in the guidelines associated with the implementation of technology like this. The downside risk of implemtation of this technology is not, however, cause to reject it. We must head into the new "world" of personalized medicine with our eyes open.

References

Beretta, L., Santaniello, A. 2015, Nov. 7–9. Nearest neighbor imputation algorithms: a critical evaluation. In: Proceedings of the Fifth Translational Bioinformatics Conference. Tokyo.

Chandnani, H., Goldstein, M., Khichi, M., Miner, L., Tinsley, C. 2018a. Prediction model for mortality in patients with disseminated intravascular coagulopathy based on pediatric ICU admission: three factors identified. In: Proceedings of the Ninth World Congress on Pediatric Intensive and Critical Care, Singapore. PICC8-O690.

Chandnani, H., Goldstein, M., Tinsley, C., Khichi, M., Miner, L. 2018b. Developing a prediction model for mortality in patients with disseminated intravascular coagulopathy based on pediatric ICU admission. In: Proceedings of the Pediatric Academic Societies Annual Meeting, Toronto, Canada. PAS2018: Poster Presentation.

Choi, K.M.S., Ng, D.K.K., Wong, S.F., Kwok, K.L., Chow, P.Y., Chan, C.H., et al., 2005. Assessment of the Pediatric Index of Mortality (PIM) and the Pediatric Risk of Mortality (PRISM) III score for prediction of mortality in a paediatric intensive care unit in Hong Kong. Hong. Kong Med. J. 11, 97−103. Available from: hkmj.org.

Dressler, D.K., 2004. Coping with a coagulation crisis. Nursing 34 (5), 58−62.

Dressler, D.K., 2012. Coagulopathy in the intensive care unit. Crit. Care Nurse 32 (5), 48−60. Available from: https://doi.org/10.4037/ccn2012164.

Goodman, B., April 22, 2021. Scientists Find How AstraZeneca Vaccine Causes Clots. WebMD. <webmd.com>.

Higgins, C., December 2018. Understanding Base Excess. A Review Article. <acutecaretesting.org>.

Jönsson, P., Wohlin, C., September 2004. An evaluation of k-nearest neighbor imputation using likert data. In: Proceedings of the Tenth International Symposium Software Metrics, pp. 108–118, Chicago, USA.

Khajeh, A., Noori, N.M., Reisi, M., Fayyazi, A., Mohammadi PhD, Miri-Aliabad, G., 2013. Mortality risk prediction by application of pediatric risk of mortality scoring system in pediatric intensive care unit. Iran. J. Pediatr. 23.

McCance, K.L., Huether, S.E., Brashers, V.L., Rote, N.S., 2010. Pathophysiology: The Biologic Basis for Disease in Adults and Children, sixth ed. Mosby, Elsevier, Maryland Heights, MS.

Pollack, M.M., Patel, K.M., Ruttimann, U.E., 1996. PRISM III an updated pediatric risk of mortality score. Crit. Care Med. 24 (5), 743−752. Available from: https://doi.org/10.1097/00003246-199605000-00004.

Seijo-Pardo, B., Alonso-Betanzos, A., Bennett, K.P., Bolón-Canedo, V., Josse, J., Saeed, M., et al., 2019. Biases in feature selection with missing data. Neurocomputing 342, 97−112ISSN 0925–2312. Available from: https://doi.org/10.1016/j.neucom.2018.10.085.

Shann, F., Pearson, G., Slater, A., Wilkinson, K., 1997. Paediatric index of mortality (PIM): a mortality prediction model for children in intensive care. Intensive Care Med. 23 (2), 201−207. Available from: https://doi.org/10.1007/s001340050317.

Prologue to Part IV

Part IV covers advanced topics in administration and delivery of healthcare including practical predictive analytics for medicine in the future. For this 2nd Edition of this book, we have added new chapters that include Chapters 20 and 21 written by coauthor Thomas Hill, Chapter 22 cowritten by all of the authors, Chapter 23 written by Mitchell Goldstein and Gary D. Miner with assistance from Thomas Hill, Chapter 24 written by Scott Burk and Chapter 25 written by Scott Burk.

Chapter 18 (Chapter 19 in 1st Edition), Chapter 19 (Chapter 20 in 1st Edition), and also Chapter 26 (also 1st Edition Chapter 26) remain in this 2nd edition with updated comments.

The following chapters of the 1st Edition are omitted in the paper pages of this 2nd Edition, but can be found in their entirety on the Elsevier companion web page (https://www.elsevier.com/books-and-journals/book-companion/9780323952743) for this 2nd Edition book, for those readers that have an interest; these 1st Edition chapters include:

- 1st Edition Chapter 16—*Predictive Analytics in Nursing Informatics*
- 1st Edition Chapter 17—*The Predictive Potential of Connected Digital Health*
- 1st Edition Chapter 18—*Healthcare Fraud*
- 1st Edition Chapter 21—*Introduction to the Cornerstone Chapters of this Book, The "Three Processes": Quality Control, Predictive Analytics, and Decisioning*
- 1st Edition Chapter 22—*The Nature of Insight from Data and Implications for Automated Decisioning*
- 1st Edition Chapter 23—*Platform for Data Integration, Analysis and Publishing Medical Knowledge as Done in a Large Hospital in Poland*
- 1st Edition Chapter 24—*Decisioning Systems & Predictive Analytics in a Real Hospital Setting at the University of Iowa*
- 1st Edition Chapter 25—*IBM Watson for Clinical Decision Support*

Part IV

Advanced topics in administration and delivery of health care including practical predictive analytics for medicine in the future

Chapter 18

Challenges for healthcare administration and delivery: integrating predictive and prescriptive modeling into personalized—precision healthcare

Nephi Walton, Gary D. Miner and Mitchell Goldstein

Chapter outline

Prelude

You, the reader, have probably just arrived here after reading Chapter 17, which was the last of seven real case studies either involving application of good data analytics or showing the need for good predictive analytics (AI, ML, and Deep Learning methods among others). If you are interested in getting a taste for and seeing an outline of "Challenges in Healthcare Delivery" (Chapter 18) and "Challenges in Research in Healthcare Analytics" (Chapter 19), read or skim over these two chapters. But if your interests are primarily on understanding how predictive analytics apply to medicine, you may wish to skip these two chapters, and continue on with Chapters 20—26.

Introduction to challenges in healthcare delivery

We face many challenges as we integrate predictive modeling into personalized—precision healthcare. Some of these challenges raise questions for which we "do not have answers" at present. Nevertheless, we need to consider them and strategize to develop methods to find answers. There are likely many more challenges that we discuss below, but this list provides the concerns most on the minds of the authors of this book.

Challenge #1

A new infrastructure must be created for healthcare

As personalized medicine based on individual characteristics becomes the norm, healthcare providers and pharmacists must be educated about these measures and how to prescribe and manage patients in a less standard manner. This transition will also require additional education in the use of complicated tools to help manage this information.

Practical Data Analytics for Innovation in Medicine. DOI: https://doi.org/10.1016/B978-0-323-95274-3.00001-4

The structure of electronic medical record (EMR) systems in their present configuration cannot incorporate or manage even genetic information, let alone integrate all the other variables such as metabolomics, epigenetics, and transcriptomics (as we wrote in 2013). However, today (2022) there are a few healthcare systems that have incorporated a limited subset of genomic information into the EMR as structured (computable) data. This consists primarily of pharmacogenomic data and a small number of the ACMG reportable incidental findings. This is still not widespread, and is more the exception than the rule.

Challenge #2

Who will pay for gene sequencing of every individual for healthcare purposes?

The cost of whole genome sequencing as a service is now less than $500. However, the interpretation is far more expensive, ranging from $5000 to more than $20,000. Many of the variants that are well established and useful in a patient's health can be computed purely bioinformatically, forgoing the need for human interpretation, given that we had a regulatory framework that would allow for this. Despite the low cost and the value this information holds throughout the life of the patient, payers are not on board. It is hard to imagine convincing insurance companies to pay for a relatively expensive test on an otherwise healthy individual, considering the difficulties in gaining approval for testing of sick people. There are several reasons payers have been slow to adopt preventative genetic testing, the primary reason being that people switch insurance companies frequently and most of the value in genomic testing is realized over time. While finding pathogenic variants in cancer predisposition genes, and genes that cause premature heart failure have been shown to decrease mortality and decrease the overall cost to the healthcare system, genetic testing may actually increase short-term costs due to the increase in preventative or prohylatic measures. More studies need to be done on the economics of preventative genetic testing as we already know there are significant positive effects on patient health. If such studies are done and fail to convince payers of the value, this may be a case where regulatory measures need to be implemented to encourage payers to pay for a one-time test for the benefit of the system as a whole.

Challenge #3

Who will regulate predictive models that that directly affect patient care? How will these models be tested for clinical use?

The possibility of FDA regulation for these models is a consideration, but it raises some serious concerns. If the FDA becomes involved with predictive modeling, will the costs of testing and regulation drive up the prices? Probably, they will. In addition, can we expect the FDA to interpret and integrate predictive modeling properly in their deliberations? Probably, we cannot, at least in the near future.

Serious questions are raised already about the regulation (Butler et al., 2020) of genetic testing in the wake of new offerings by many companies to provide genetic testing services directly to consumers. These and many other questions regarding genetic testing remain open as we enter this era of predictive medicine.

Challenge #4

What effects will predictions of health outcomes have on an individual's mental health and general overall daily anxiety levels?

Knowing that there is a certain probability of developing a debilitating disease could have a significant impact on an individual's mental state (Watson et al., 2004). Patient response to genetic risk assessments is highly variable, depending on the disease and many other issues. Strategies have been developed to present this information effectively, but more research is needed in this area as it develops (Lautenbach et al., 2013; Ciucă et al., 2022). These questions are even more complex and confounded when they enter the pediatric domain, where studies are in their early stages (Savatt et al., 2020).

Challenge #5

Does the ability to predict someone's health outcome change their behavior?

For years, it has been known that the effects of smoking have caused cancer, cardiovascular disease, and it contributes to a myriad of other diseases. Yet people continue to smoke. Diet and exercise can also reduce your health burden significantly on a number of different fronts, yet the rate of obesity continues to rise in the United States. Millions of people smoke, millions are overweight, and millions of people have high cholesterol. Regardless of the many warnings in schools and the media that these practices are associated with high risks of serious diseases, many people continue the practices. Even though we are constantly bombarded about the benefits of a healthy diet and exercise, we continue to get fatter and maintain unhealthy diets. Most early evidence from studies in this area suggests that genetic risk alone has a minimal to no effect on behavior or in motivating changes in behavior (Henrikson et al., 2009; Li et al., 2016). Despite these gloomy findings, there are studies that demonstrate effectiveness in changing behavior when coupling genetic information with the delivery of specific planned behavioral interventions, particularly with regard to nutrition (Horne et al., 2018). Based on this information, significant effort must be made to create plans for effective interventions that have solid grounding in behavioral theory for personalized medicine to be effective in its truest form. Despite the possibility that people may not change their behavior, predicting risk does give the physician an advantage for prescribing the proper medications, ordering the appropriate surveillance studies, or recommending prophylactic procedures for people on the edge of disorders, such as diabetes, hypertension, or cancer. Strong predictions by these models might induce the physician to prescribe drugs earlier than otherwise or take other more drastic measures such as mastectomy in the case of breast cancer risk with *BRCA1*. Currently, we need more information to determine the most effective ways that genetic risk information can be delivered in conjunction with specific lifestyle plans to motivate healthy behavior. Identifying the settings in which genetic risk can motivate healthy behavior, and identifying which people are likely to respond to this information are important goals for predictive analytics (Henrikson et al., 2009; Li et al., 2016).

Challenge #6

With the possibility of building clinical risk prediction models also comes the possibility of legal liability

There has been little litigation that addresses the use of risk prediction models. However, as they become more widely used, the prospect of legal liability and lawsuits will only increase. There is an existing body of litigation on family history and genetics that would have some relevance to the use of risk prediction models, but even this domain leaves us with many unanswered questions (McGrath et al., 2021). The possibility of a "bad prediction" is a legal concept that presents some difficulty, and it is very difficult to judge its significance at this early stage of the implementation of predictive analytic risk models. Outside of blatant misuse or absolute failure to use a required risk prediction model, there is very little legal guidance to avoid the risk for medical liability (Black et al., 2012; McGrath et al., 2021).

Challenge #7

Many of the biological (body) specimens that are required for full "omics analysis" are not easily obtained

More effective and less invasive methods of gathering this information will be required to get a good assessment of the "omics" of the body. For example, requiring a brain biopsy and undergoing general anesthesia at every checkup is not a likely possibility considering the risk, time, and cost involved, yet some markers of neurological disease using today's technology would require this information. While this is certainly the case now, we are getting better at detecting biological signatures in the blood from different organs or processes in the body. This is well demonstrated with our ability to detect and utilize fetal DNA in the mothers blood for prenatal testing and our ability to detect cancer through circulating tumor DNA. As we learn more about how to utilize different "omics" in clinical care, the method of capturing this information from a patient in a minimally invasive way must be considered (McGrath et al., 2021).

Challenge #8

The technology required for personalized medicine is certainly not perfect, and many refinements and further developments are needed

For example, despite the hype about the value of whole genome sequencing, the current clinical form of this test misses a significant portion of the actual genome as we now know through the use of long-read sequencing. In fact, it was not until recently that we had a complete map of the X chromosome (Miga et al., 2020). The integration of genomic data into the patient health record leaves even more to be desired. Only recently has there even been the ability to store genetic data in the health record in a structured form, and early implementations have many challenges that have yet to be addressed (Walton et al., 2020).

Challenge #9

It is likely that the effective implementation of predictive—precision medicine models in medicine will require more support staff and new clinical workflows

Such an implementation is likely to require more genetic counselors and other medical staff to deliver preventative care. Consideration must be given for creating a model for these interactions and determining how the care processes will fit into our healthcare delivery and reimbursement systems.

Challenge #10

The scientific disciplines covered by personalized—predictive medicine are numerous

And the tremendous amount of data and computing power required to employ them is unprecedented. New devices and techniques must be developed to collect and analyze this data. The complete implementation of personalized—predictive medicine will require the concerted work of teams of scientists and engineers from many fields, along with guidance from ethical and legal professionals to develop guidelines for implementing this technology. The task is daunting but also very exciting.

Challenge # 11

Accurately capturing data from the medical record for use in predictive models

Much of the phenotypic data that exists in the medical record is unstructured and currently requires manual processes to convert it into a structured computable form. The EMR was designed around billing and structured diagnoses are often added primarily for billing purposes and will often not be assigned to the patient if not needed for billing. This leaves the majority of phenotypic data residing in unstructured notes. Given the time constraints on physicians we should not rely on them to input all the details in a structured way rather what is needed is the development of automated processes to capture important elements from unstructured data and convert it into a structured computable form.

There are other challenges also, but those discussed above are the ones of most concern in 2022 in the minds of the authors of this book. You, the reader, can follow these concerns and new ones that develop by keeping a close watch on what blog writers and others are posting on the internet.

Postscript

If you, the reader, were interested enough in the "Challenges in Healthcare Delivery" (Chapter 18), you may also want to read or skim over the few challenges for researchers doing work utilizing modern predictive analytics in medical research. If so continue on with Chapter 19. However, if your interests are primarily on understanding how predictive analytics apply to medicine, you may wish to skip this next Chapter 19, and continue on with Chapters 20—26, which discuss advanced topics in data analytics for medicine and healthcare.

References

Black, L., Knoppers, B.M., Avard, D., Simard, J., 2012. Legal liability and the uncertain nature of risk prediction: the case of breast cancer risk prediction models. Public Health Genomics 15 (6), 335—340. Available from: https://doi.org/10.1159/000342138. Epub 2012 Sep 12.

Butler, E., Collier, S., Boland, M., Hanhauser, Y., Connolly, E., Hevey, D., 2020. Self-concept and health anxiety relate to psychological outcomes for BRCA1/2 carriers. Psycho-Oncol. 29 (10), 1638—1645.

Ciucă, A., Moldovan, R., Băban, A., 2022. Mapping psychosocial interventions in familial colorectal cancer: a rapid systematic review. BMC Cancer 22 (1), 8. Available from: https://doi.org/10.1186/s12885-021-09086-8. PMID. 34980016.

Henrikson, N.B., Bowen, D., Burke, W., 2009. Does genomic risk information motivate people to change their behavior? Genome Med. 1 (4), 37.

Horne, J., Madill, J., O'Connor, C., Shelley, J., Gilliland, J., 2018. A systematic review of genetic testing and lifestyle behaviour change: are we using high-quality genetic interventions and considering behaviour change theory? Lifestyle Genomics 11 (1), 49—63.

Lautenbach, D.M., Christensen, K.D., Sparks, J.A., Green, R.C., 2013. Communicating genetic risk information for common disorders in the era of genomic medicine. Annu. Rev. Genomics Hum. Genet. 14, 491—513.

Li, S.X., Ye, Z., Whelan, K., Truby, H., 2016. The effect of communicating the genetic risk of cardiometabolic disorders on motivation and actual engagement in preventative lifestyle modification and clinical outcome: a systematic review and *meta*-analysis of randomised controlled trials. Br. J. Nutr. 116 (5), 924–934.

McGrath, S.P., Peabody Jr, A.E., Walton, D., Walton, N., 2021. Legal challenges in precision medicine: what duties arising from genetic and genomic testing does a physician owe to patients? Front. Med. 8.

Miga, K.H., Koren, S., Rhie, A., Vollger, M.R., Gershman, A., Bzikadze, A., et al., 2020. Telomere-to-telomere assembly of a complete human X chromosome. Nature 585 (7823), 79–84.

Savatt, J.M., Wagner, J.K., Joffe, S., Rahm, A.K., Williams, M.S., Bradbury, A.R., et al., 2020. Pediatric reporting of genomic results study (PROGRESS): a mixed-methods, longitudinal, observational cohort study protocol to explore disclosure of actionable adult-and pediatric-onset genomic variants to minors and their parents. BMC Pediatr. 20 (1), 1–13.

Watson, M., Foster, C., Eeles, R., Eccles, D., Ashley, S., Davidson, R., et al., 2004. Psychosocial impact of breast/ovarian (BRCA 1/2) cancer-predictive genetic testing in a UK multi-centre clinical cohort. Br. J. Cancer 91 (10), 1787–1794.

Walton, N.A., Johnson, D.K., Person, T.N., Reynolds, J.C., Williams, M.S., 2020. Pilot implementation of clinical genomic data into the native electronic health record: challenges of scalability. ACI Open. 4 (02), e162–e166.

Further reading

Chen, R., Snyder, M., 2013. Promise of personalized omics to precision medicine. Wiley Interdiscip. Rev. Syst. Biol. Med. 5 (1), 73–82.

Chen, R., Mias, G.I., Li-Pook-Than, J., Jiang, L., Lam, H.Y., Chen, R., et al., 2012. Personal omics profiling reveals dynamic molecular and medical phenotypes. Cell 148 (6), 1293–1307.

Chapter 19

Challenges of medical research in incorporating modern data analytics in studies

Nephi Walton, Gary D. Miner and Linda A. Miner

Chapter outline

Prelude

You, the reader, have probably just arrived here after reading the previous chapter, number 18. Chapter 18 was as list of challenges to healthcare delivery in general. This Chapter 19 is a short listing of some of the major challenges to researchers in medicine and healthcare delivery. Discovering how to make accurate diagnoses for the individual patient and prescribing the "right medicine or treatment plan" the "first time" for the "right patient" is critical if we want to have precision medicine practiced in our society. We are only giving a small sampling of challenges to medical researchers in this Chapter 19, but if you are interested in getting a taste for this, read on. However, if your interests are primarily on understanding how predictive analytics apply to medicine, you may wish to skip this chapter and continue on with Chapters 20–26, which describe advanced topics in data analytics for both large healthcare organizations and also the individual person ("top-down" and "bottom-up" frameworks).

Introduction—challenges to medical researchers

We face many challenges as we design research to answer the medical questions facing us both now and in the future. Some of these challenges raise questions for which we don't have answers at present. But we cannot forget the main goals: discovering and developing pathways of knowledge to make accurate diagnoses for the individual patient the first time presented and prescribing the "right medicine or treatment plan" the "first time." Giving this "right patient" the "right diagnosis" and the "right treatment" the "first time presented" is critical if we want to have precision medicine practiced in our society in the future.

Challenge #1. Getting the "best datasets." PCORI, the new National "Patient Centered Outcomes Research Institute," which has been very active during the years 2010—present day (2022), has made "National networks of medical data" one of its primary goals; the success of this will depend on the cooperation of both academic, nonprofit, and for-profit organizations sharing this data so that scientific medical researchers can use it to produce new knowledge. However, as we write this 2nd edition in 2021–22, not much of significance has happened along these lines; yes, there are efforts "here-and-there," but no really good comprehensive national networks of medical data that are readily available to all researchers.

Practical Data Analytics for Innovation in Medicine. DOI: https://doi.org/10.1016/B978-0-323-95274-3.00007-5

Challenge # 2. Getting good outpatient data. The majority of health outcomes analysis of Medicare data is based on in-hospital data. The reason is that such data is much easier to conjoin into individual patient records as well as to episodes of disease or disability and treatment. Outpatient health data has been notoriously difficult to track since it is collected at various sites which typically do not share records. To better assess the efficacy of outpatient disease episodes and treatment, ways must be developed to create and store patient-level inpatient and outpatient episode data.

Challenge # 3 Accurate frontline data capture (i.e., at the clinic, and also hospital bedside) is particularly problematic. The patient's condition is" described" by the ICD codes chosen by the clinician. Higher level, nonspecific codes are frequently chosen for the sake of time. EMRs have not made this necessarily any easier even though there may be built-in coding assistance. It still requires more steps than using a more generic code which may be memorized or chosen from the patient's problem list. For example, 250.00 is the code for type II diabetes under good control. This code is commonly used even though we know that the vast majority of diabetics are not controlled and have one or more complications. The code 250.02 is for type II diabetes not controlled but with no complications. For each body system affected by the diabetes (kidneys, peripheral nerves, cardiovascular, eyes, etc.) there is a specific fourth and fifth digit with an accompanying "Buddy code" from the particular system involved.

And this is just for diabetes!

Also, for the majority of patients seen, the diagnostic code does not affect reimbursement [in Medicare Advantage Plans (HMOs)]. Instead, insurance companies ares capitated (paid on a per member per month basis) based on the severity of illness of the member (higher SI factor = higher payment) so there is a financial incentive for more accurate coding. So as long as the doctors function as data entry clerks it is likely that inaccuracies in coding will continue to be the norm. Perhaps new technologies like "natural language processing" will help solve this dilemma (as one of this book's authors commented on this "challenge": *"Excellent! Yes, like we all want doctors to be "data entry clerks," in the wee hours of the night! I wonder if the EMR creators include all types of physicians? PCPs (Primary Care Physicians) would have to learn the most. How many codes are there for ICD-10? Is it 140,000? And the old one was 20,000? Watson will have to be the clerk!"*)

Challenge #4. Development of new antibiotics, especially for Gram-negative infections and methods for reducing the approval time (in keeping with continued safety concerns). Development of new treatment for bacterial infections (antibiotics) is a continuing health delivery concern and will be for the foreseeable future.

Challenge #5. Development of innovative methods for the treatment of infections and other diagnoses with specific focus on genetic considerations and biomarkers (for example) to target treatments to individual patients. This is where precision medicine will really come into play.

Challenge #6. Research into diagnostic procedures/tests more specific to medical conditions

Challenge #7. Good data. A major problem in doing accurate research in medical diseases is the problem of obtaining good data for analysis, especially considering noncommunication between EMRs and HIPPA as it currently stands. One chapter in Part I of this book suggests that informed consent could be obtained as patients enter clinics and hospitals, but what if they don't consent? How will researchers be able to get data that are valid and reliable for making the predictions, especially moving prospectively. Large quantities of good data will be needed for making accurate predictive models, and thus accurate diagnoses and treatment plans; this data will be needed ideally from all people. (An alternative that will serve some areas of medicine is the "N-of-1" type of studies that are discussed at length in Chapter 26, and also discussed in Chapter 10 briefly.)

Challenge #8. Paying for "all of it." As coauthor Nephi Walton pointed out (in Chapters 3 and 6), paying for genomic research will be a challenge, but also medical payments in general will be a problem. The United States governmental "reimbursement system" pays about $8200 per person per year and our nearest neighbor in terms of amounts paid is Norway with $5388. (These are 2014-dollar figures; today in 2022 it is higher, of course.) How we morph this with other countries that have different healthcare plans will be a big issue for healthcare, and therefore predictive research in the years ahead.

Trends that we might want to know about

An article in News Everyday (2020), wrote about 10 cutting-edge trends in data science. They were, "Automation and Machine Learning (AutoML), Blockchain, Conversational AI, Data Fabric, Data Security, Digital Twins, Edge Computing, Graph Analytics, Persistent Memory, Real-Time Intelligence." The article explained each. Below are some of those trends, explained with applications to medical research.

Automation and machine learning (AutoML)

The News Everyday example of AutoML was very much as we likely are familiar with—the automatic data mining recipes used in Chapter 17 as an example, using data health and cleaning, feature selection, and machine learning algorithms. Sidey-Gibbons and Sidey-Gibbons (2019) also provided examples of machine learning for researchers and clinicians. Their article explained that machine learning comprises all three disciplines: statistics, computer science, and mathematics, to help medicine in both diagnostics and prediction of outcome (such as in Chapter 17, predicting survival or death for DIC). *Data fabric* was one way in which the AutoML could be performed. Data fabric was described as a platform integrating various modules for analysis, and platforms such as KNIME (Chapter 14), Shiny R (Chapter 13), and Statistica (Chapter 17) achieve such integration.

Blockchain

Technology used in cryptocurrencies, blockchain enables users to be anonymous. This would be a big boost for medical research if the technology could be used for patients so their data could be used without possibility of being identified. Such technology could provide the data security needed for compliance with standards and regulations.

Conversational artificial intelligence

Text analysis is becoming useful with the development of machine learning and natural language processing (NLP). (Note to the reader, the first DIC tutorial, J1, found on the companion web page, used text mining to enhance the predictive model.) The News Everyday (2020) article moved into other areas to gather such text, including AI interfaces drawing from chat boxes and extending beyond text to voice inflections and gestures (as in cameras on smartphones, perhaps?). All our data are being collected and stored. Perhaps those data can be used to diagnose diseases and predict outcomes.

   Deo (2015) said that with the "unprecedented wealth of data," the time is ripe that such data could be turned to medical problems. Deo mentioned the Framingham Risk Score for heart disease being used in the 1970s as a precedent. In this case, the mechanisms were more known, and the task was one of classification. According to Deo, a more complicated task occurs when there are not known outcomes or mechanisms, allowing researchers to "identify novel disease mechanisms" (p.3). What comes to mind is the left-brain, right-brain thoughts of Chapter 26. Regardless, accurate models in medicine are hard to attain, with overfitting on training data leading to poorer fit for testing data (as could be seen in Chapter 17). Deo did emphasize the value of extracting features in a data set that could be likely predictors in an algorithm (such as feature selection in Statistica, Chapter 17). Getting more interest in searching for predictive algorithms is key.

Digital twins

This is basically using simulations to analyze data. Simulations have been used for testing precision for quite a while (as in the imputation simulations mentioned in Chapter 17). Simulations could be combined with medical competitions to encourage novel ideas in the field of medicine and for making better predictions.

Medical competitions

One excellent way of achieving good predictions in medicine is by holding competitions, such as the Sage Bionetworks/DREAM Breast Cancer Prognosis Challenge (2012). Competitions such as this allow teams to predict medical outcomes using the same data source. The winner is the team that predicts the best.

Conclusion

There are very significant challenges and new interesting trends to conducting accurate data analytics on medical research data. The primary cause of this is the scattered nature of health data, from difficult to read hand scribblings in clinic and doctor's offices paper records (mostly from years in the past, but also some individual doctor's practices currently) which are badly needed for long-term data analysis, to electronic health records (produced by different companies for hospitals and clinics) that do not talk to one another, and other ancillary but important health data coming from pharmacy and other allied health records (which again, usually are not able to communicate with each other, as the companies developing the software programs that run these systems purposely make proprietarily function in that manner). Until these challenges are overcome little will change in the accuracy of data analysis from these sources. In the meantime, studies on just one person, on "you as an individual patient," may be the best way to go; these are called "N-of-1" studies. We have presented several of these types of studies in this book with the most thorough example being the glaucoma N-of-1 data analysis presented in the Chapter 11 case study.

Postscript

Discovering how to make accurate diagnoses for the individual patient and prescribing the "right medicine or treatment plan" the "first time" for the "right patient" is critical if we want to have precision medicine practiced in our society. This chapter has considered some of the challenges to researchers in making this goal become a reality. Now you can continue reading Chapters 20–26 where we provide in-depth discussions of some of the things that are needed and happening now to make personalized medicine a reality for all of us.

References

Deo, R.C., 2015. Machine learning in medicine. Circulation 132 (20), 1920–1930. Available from: https://doi.org/10.1161/CIRCULATIONAHA.115.001593.

News Everyday. March 26, 2020. Top ten cutting-edge data science trends you will need in 2020. <https://www.newseveryday.com/articles/64442/20200326/top-10-cutting-edge-data-science-trends-you-will-need-in-2020.htm>.

Sidey-Gibbons, J., Sidey-Gibbons, C., 2019. Machine learning in medicine: a practical introduction. BMC Med. Res. Methodol. 19, 64. Available from: https://doi.org/10.1186/s12874-019-0681-4.

Further reading

IBM. June, 2012. Seeking the wisdom of crowds through challenge-based competitions in biomedical research. <deainfo.nci.nih.gov/advisory/bsa/archive/bsa0313/SingerStolovitzky.pdf>.

Chapter 20

The nature of insight from data and implications for automated decisioning: predictive and prescriptive models, decisions, and actions

Thomas Hill

Chapter outline

Prelude

The usual focus of predictive modeling is focused on the accuracy of the predicted outcome. The problem is that this outcome is worthless (even if it is highly accurate) unless it can be put to use in some operational system in the organization. The case studies presented in Chapters 13—17 focus on the direct results of the outcomes modeled. But even the potential benefits associated with those outcomes might be dwarfed by insights gained by the analysis of the model outcome in the context of business operations. These insights can be crucial to the success of the company. For example, the initial models built in support of the presidential campaign of Barak Obama in 2012 identified a group of swing voters susceptible to change. Rather than target them directly, the modelers gained an insight from analyzing the social media of this group, which suggested that they should identify the influencers of that list of swing voters, contact them, and ask them to suggest to their friends to vote for Obama. Using this strategy in 2012, he gained about 10 million more voters in the 20—30-year age range than he received in 2008. This gain

Practical Data Analytics for Innovation in Medicine. DOI: https://doi.org/10.1016/B978-0-323-95274-3.00020-8

approximately equaled the number of votes by which Obama won the 2012 election. Such insights can be extremely useful in medical informatics also.

Overview

Predictive modeling methods based on artificial intelligence (AI) and machine learning (ML) have transformed industries and altered day-to-day life in remarkable ways. As you search for specific products such as laptops on the internet, your clicking experience at various websites can be customized automatically to this specific product type. As you peruse various online sellers of laptops, the particular specifications and interests of your searches can be broadcasted to all online merchants, who have customized their offerings to present to you exactly what you are looking for.

Using AI models, these systems can:

1. retrieve automatically relevant information to predict future behavior or outcomes;
2. create customized predictions about the propensity for specific decisions, behaviors, or outcomes; and
3. provide suggestions to guide best next actions.

Such capabilities can deliver tremendous value in healthcare domains, and offer promise to improve the efficacy of information and guidance provided to patients after medical procedures or hospitalization (see, e.g., Palmer, 2021; Ngiam and Khor, 2019).

Intelligent systems can monitor all processes and actions during hospitalization and at the point of discharge. They can also provide postdischarge data captured during follow-up visits, to generate relevant guidance about likely next outcomes, best next actions and treatments, accurate predictions about readmission to expensive emergency care, and offer ways to prevent it. Today, these methods and technologies exist in various industries, along with the tooling to validate, govern, maintain, and update AI-driven support for critical decisions (see also Chapter 21, Model management and ModelOps: managing an AI driven enterprise).

The purpose of this chapter

The purpose of this chapter is to explore in some detail *how* modern AI and ML technologies can facilitate new insights and inform better decisions in healthcare. More specifically, the discussion will focus on the nature of insights, causal inference, and explainability.

It is widely understood that *correlation does not imply causation*. AI and ML methods are (almost always) applied to historical data, in order to detect "relationships," that is, correlations. Does that mean that AI/ML models do *not* lend themselves to be explained, drive insight, and support better informed decisions? Actually, they *do*.

Another chapter(see Chapter 25 of this book) discusses recent work on causal inferences (causal "precedence") from correlational data (see Pearl, 2009, 2010). The current chapter first discusses how much of human expertise to "do-things" (e.g., that of an expert surgeon, radiologist, etc.) is acquired through interaction with complex historical and correlational data. Next it will provide an overview of methods and algorithms now commonly in use to *explain* results and predictions from AI/ML models.

The nature of insight and expertise

In many ways, experts do not (consciously) "know" what they are doing. At first, this may seem to be an odd statement, but in fact day-to-day observation is consistent with this statement. For example, an expert violinist cannot write a manual or convey through verbal instruction to others about how to play the violin. Expert fishermen cannot easily verbalize the rules that allow those unfamiliar with a specific lake or part of the ocean to be equally successful.

In practically all highly skilled professions—including and in particular medical professions—a critical part of the training is an extended internship (or practicum) that provides trainees with a rich and diverse environment and various structured and unstructured learning experiences.

Procedural and declarative knowledge

In cognitive psychology, the distinction between *declarative* and *procedural* knowledge describes the difference between the things we know and can verbalize explicitly as rules (*declarative* knowledge), and the things we know how to *do* or *judge intuitively* (*procedural* knowledge; see, e.g., Lewicki, 1986a,b; Lewicki et al., 1992). The knowledge of the alphabet is declarative, and we can recite the alphabet. However, knowledge of how we judge distances intuitively, determine if we find another person interesting or attractive, or decide quickly that "something is not right" with the medical profile of a patient is not easily verbalized, and is procedural in nature: we can *do* it, but cannot describe how *it is done*.

Lewicki et al. (1986a), Lewicki et al. (1992) have accumulated a large amount of evidence demonstrating not only that much of the knowledge that experts use to make effective and efficient decisions is procedural in nature and not accessible to conscious scrutiny, but also never *was* accessible to conscious awareness in the first place.

Nonconscious acquisition of knowledge

In short, Lewicki and others have demonstrated over a wide range of human experiences and expertise, that exposure to complex and rich stimuli, consisting of large numbers of sensory inputs and high-order interactions between the presence or absence of specific features, will stimulate the acquisition of complex procedural knowledge without the learners' conscious awareness. Hence the acquisition of such knowledge is best characterized as *nonconscious* information acquisition and processing. For example, when humans look at sequences of abstract pictures, faces, or tracking targets over seemingly random locations on the screen, carefully calibrated measures of procedural knowledge (e.g., based on response times) will reflect the acquisition of knowledge about complex covariations and rules inferred from the rich and complex stimuli.

The nature of the "nonconscious." It is important to note at this point that the notion of "nonconscious" here is quite different from the commonly used "unconscious" in the psychoanalytic tradition. This research is about information processing, and the mechanism that enables humans to extract common cooccurrences of features and interactions between features from extremely complex experiences (stimuli). These low-level cognitive (learning) mechanisms are the building blocks from which complex knowledge and expertise is built. The notion of "nonconsciousness" here pertains to the fact that the respective expertise is almost entirely *procedural* in nature, rather than *declarative*. Hence, experts cannot verbalize or consciously scrutinize how rapid decisions, choices, and predictions are made in the presence of very complex stimuli—but they can make those decisions and predictions with remarkable accuracy. Conscious reasoning, it appears, rather is reserved for explicit problem-solving in the presence of novel situations or detailed evaluations, for example, when solving math problems, thinking about a strategy for success, etc.

Conclusion: expertise and the application of pattern recognition methods

The conclusions from this research are highly relevant for understanding how large amounts of high-dimensional information, consisting of complex interactions between numerous parameters, can be derived efficiently through systematic exposure to relevant stimuli and exemplars. Specifically:

1. It appears that knowledge about complex interactions and relationships in rich stimuli are the result of the repeated application of simple covariation-learning algorithms, which detect patterns and cooccurrences between certain stimuli and combine them into complex interactions and knowledge.
2. In human experts, most of this knowledge is procedural in nature, not declarative; in short, experienced experts can be effective and efficient decision-makers, but are poor at verbalizing *how* those decisions were made.
3. When the covariations and repeated patterns in the rich stimulus field change, so that previously acquired procedural knowledge is no longer applicable, experts are slow to recognize this, and often are confused and reluctant to let go of "old habits."

Human expertise and effective decision-making can be remarkable in many ways:

1. It is capable of leveraging "big data," that is, is remarkably capable with respect to the amount of information and stored knowledge that is used.
2. It is capable of coping with high-velocity data, that is, it is very fast, with respect to the speed with which information is synthesized into effective, accurate decisions.
3. It is very efficient, with respect to how little energy our brain requires to process vast amount of information, and make near-instant decisions.

From the perspective of the analytic approach, these capabilities are accomplished through the repeated application of simple learning algorithms to rich and complex stimuli to identify repeated patterns that allow for accurate expectations and predictions regarding future events and outcomes. This approach is quite different from statistical hypothesis and significance testing. These differences will be highlighted in the next section.

Statistical analysis versus pattern recognition

There is a fundamental difference in approaches to data analysis between statistical hypothesis testing and pragmatic predictive modeling and AI, or "learning-from-the-data." This difference has been pointed out by numerous authors, such as the classic paper by Breiman (2001) on *Statistical modeling: The two cultures* (see also Nisbet et al., 2009; Miner et al., 2012).

Fitting a priori models

In traditional clinical research, statistical data analysis methods have been applied widely to build predictive models. A common example of such a method is a multiple linear regression model that is fitted to the data, following the form

$$y = b_0 + b_1 \times x_1 + \ldots + b_n \times x_n + \varepsilon,$$

where y is some variable or outcome that is to be predicted (e.g., propensity for hospital readmission after discharge), x_1 through x_n are predictor variables, b_0 through b_n are coefficients of the linear prediction model, and ε is the error variability that cannot be accounted for by the prediction model.

In general (and without going into details about the theory of statistical inference), the approach for building such models is to estimate the parameters of the model from a subsample of cases, and then to perform statistical significance tests to decide if the model parameters and predictions from the model are more accurate than some baseline expectation (typically, that there are no relationships between the x's and y).

Statistical probabilities. The general approach to classic statistical data analysis and modeling is to test a priori expectations regarding possible functional relationships between predictors and outcomes. In that context, statistical significance (probability) is usually reported, to reflect how likely it is that a specific result predicted by the statistical model (hypothesis) occurred by chance alone, when in reality the model is not true. If that probability is found to be very small (less than some criterion of *statistical significance*; e.g., $P < .05$), then it is concluded that the respective model *is* supported, and *true*—at least until it is falsified by future research. So fundamentally, classic statistical modeling is about hypothesis testing, and the rejection or acceptance of a priori hypotheses about the data, and how important outcomes can be predicted from available inputs (predictor variables). (Note that Bayesian statistics approaches the problem of inference and explanation differently; see the chapter on predictive models vs. prescriptive models; causal inference and Bayesian networks).

Pattern recognition: data are the model

Modern predictive modeling based on AI/ML is based on a different approach: *No* a priori hypotheses are tested, but instead the goal of the analysis is to extract from the data set repeated patterns and relationships that are useful in the prediction of future outcomes. Thus this approach closely resembles how expertise and procedural knowledge is acquired by human experts when interacting with complex and rich (sensory) inputs.

The data are the model

Suppose that some important outcome—such as hospital readmission—was 100% predictable from a set of input variables, measured accurately and reliably (without error). In that case, given a data set of historical experience with other patients, a prediction about hospital readmission would be easy to make. Simply find for each patient to be discharged the one or more most similar patients with respect to the available inputs. The prediction then would be the outcome observed in those similar cases.

This example is illustrated in the simplified graph shown in Fig. 20.1.

It is assumed in Fig. 20.1 that there are three relevant predictors. Furthermore, in the historical data existing patients based on observed outcomes were assigned to either *Low*, *Moderate*, or *High* risk groups. How should a new patient, as shown in the Fig. 20.1, be predicted? It would make sense to assign this patient to the *High Risk* group, as shown in Fig. 20.2.

Assigning/Predicting a New Patient to a Risk Group

FIGURE 20.1 Predicting a new patient from similar patients.

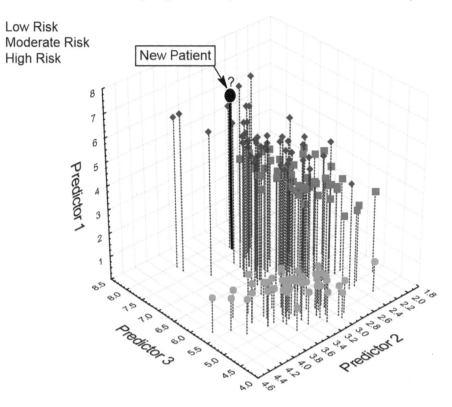

Assigning/Predicting a New Patient to a Risk Group

FIGURE 20.2 Predicting a new patient using k-nearest neighbors.

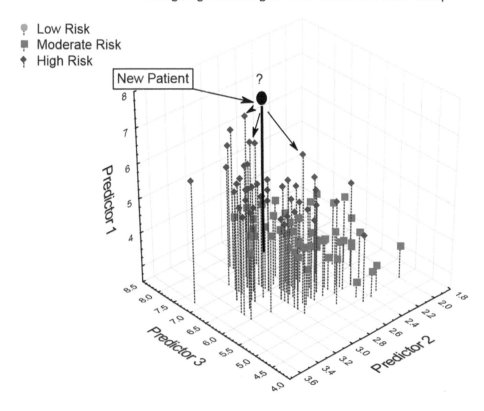

The *New Patient* in Fig. 20.2 is closest to other patients in the historical data that were classified as *High Risk*, with respect to three important *Predictors* of risk. Therefore, the *New Patient* will be assigned to the *High Risk* group.

The method of assigning new observations to the most likely outcome among the nearest "neighbors" in the space defined by the predictors is a well-known predictive modeling algorithm: k-Nearest Neighbors.

During the "learning" process of this algorithm, the task is to find a good subset of representative and informative exemplars with respect to important predictor variables. The prediction simply is to assign a new observation to the similar "neighbors."

Pattern recognition in artificial intelligence/machine learning: general approximators

In predictive modeling, unlike in statistical analyses, no a priori hypotheses or functional relationships are tested with the data. Instead, learning algorithms capable of approximating any kind of relationship in the data are applied to extract the repeated and consistent patterns across all of the data, which are relevant and useful for the prediction of important outcomes. Therefore, such algorithms are also called *general approximators*.

To reiterate, this general approach is most similar to the mechanism by which the human cognitive system learns and acquires knowledge of patterns from diverse and rich sensory inputs with remarkable efficiency, even when those repeated patterns are very complex.

All effective AI/ML algorithms such as Deep Learning and Boosted Tree algorithms will apply pattern recognition methods to approximate the repeated patterns in data (see, e.g., Hastie et al., 2017). These algorithms have revolutionized risk modeling and prediction, fraud detection, demand forecasting, to name only a few of the domains where data mining and predictive modeling are common. These algorithms are also transforming the way patient outcomes are modeled and predicted, and how best-next-action plans with respect to treatment regimens can be determined (see, e.g., Ngiam and Khor, 2019).

Pattern recognition and declarative knowledge: interpretability of results

To summarize, in statistical modeling, a priori model-based expectations about relationships in the data are tested using statistical methods; in AI/ML-based predictive modeling, pattern recognition algorithms are applied to extract repeated relationships and patterns from the data, to enable accurate prediction.

Explainability of artificial intelligence/machine learning models

One of the common criticisms and often barriers to the adoption of AI/ML techniques is that they are "black-boxes." For example, deep learning or the k-nearest neighbor method described earlier may yield very accurate predictions regarding hospital readmission, but provide little information about *why* a patient is likely to be readmitted. In order to provide actionable information regarding effective treatments and the best-next-action, it is important that prediction models also provide insights into what to do or change to affect an outcome.

Global and local explainability

The term "explainability" (of a model) is somewhat ambiguous. There are really two important questions: first, what are the "important variables" in my prediction model, *overall*? Second, given a specific prediction, what specific input or predictor values are responsible for the specific prediction (e.g., why is the prediction for a specific individual different from the average prediction)? The first question addresses the *global* importance of predictors in the respective prediction model, over all observations and predictions; the second question addresses the *local* importance of predictors in the respective model, at the specific "location" in the multivariate input space, for example, for the specific input or predictor values for a person.

To clarify, consider a simple hypothetical example of a classification tree. Suppose the prediction model—a binary classification tree—consists of these decision points and leafs (see Fig. 20.3):

The goal is to predict (classify) observations into one of two categories for a person's *Life Satisfaction*: "High," or "Low," based on predictors: *Age, Career Satisfaction,* and *Family Satisfaction*. In this hypothetical example model, which predictor is most important?

In short, this is really the "wrong question," because "it depends." Neither *Career Satisfaction* nor *Family Satisfaction* by itself allow for a perfect prediction, but for older respondents (*Age > 50), Family Satisfaction* is

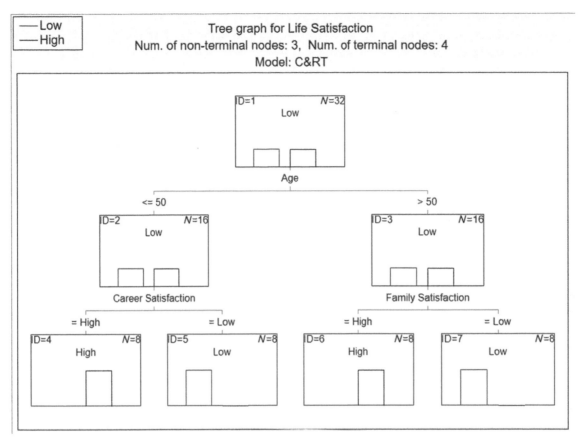

FIGURE 20.3 A classification tree algorithm analysis output.

important, while for younger respondents (*Age < = 50*), *Career Satisfaction* is important and allows one to accurately predict *Life Satisfaction*. *Globally*, and over the entire data set, all predictors are important; for specific predictions—that is, individuals of a certain age—either *Career Satisfaction* or *Family Satisfaction* are important.

Global predictor importance and explainability refers to the importance of a predictor overall, with respect to the average prediction over all observations in the respective sample. *Local predictor importance* refers to the importance of a predictor with respect to a specific prediction, with specific input values. For example, for a person older than 50 years of age, in the model above *Family Satisfaction* "explains" *Life Satisfaction*; but for a person younger than 50 years of age, *Career Satisfaction* explains *Life Satisfaction*.

Statistical models, and reason scores for linear models

Using linear (or logistic, generalized linear) prediction models simplifies the task of identifying the important predictors driving outcomes, because local and global predictor importance are the same: those predictors associated with the largest parameters (regression weights) in the respective linear prediction model are the ones that are most important in driving the outcomes. Using statistical and mathematical reasoning alone, it is possible to determine why a specific outcome is predicted to occur, and hence, help to determine what could be done about it.

For example, suppose you had a linear equation in a simple linear model that predicted the likelihood of hospital readmission based on numerous predictor inputs. Given the specific values for the predictor variables, for each patient it will be obvious which predictor variable and specific value contributes most to the expectation of elevated readmission risk. If blood pressure is an important predictor variable in the linear prediction equation and a patient shows elevated blood pressure, then it is clear and apparent that steps should be taken to lower blood pressure in order to lower the risk of hospital readmission.

Thus statistical models are usually more easily interpretable, providing useful *reason scores* for why a particular predicted value was computed, and what to do to change it.

What-if, and reason scores as derivatives

"Reason scores" are a way to explain the prediction of analytic models and to identify the root causes driving a specific prediction. The term emerged in the risk scoring domain, where a final score is computed as the sum of reason scores from multiple predictors (see, e.g., Siddiqi, 2006). In general and regardless of the prediction model, reason scores are computed as first order partial derivatives of the parameter under consideration (e.g., partial derivative for hospital readmission probability with respect to parameter *Patient Age*; see, e.g., Hill et al., 2013). Thus these scores address the question of by how much the outcome variable of interest will change if the values of a specific predictor variable changes or is changed.

Parameter estimates of linear regression models such as logistic regression models describe a *slope* and thus a derivative: how much will the respective outcome variable of interest change when the values in a specific predictor variable change?

Explainability of nonlinear models, artificial intelligence/machine learning models

Increasingly, people-facing applications of AI and ML models are governed by regulation (see e.g., Benjamin et al., 2021). Prediction models that allocate to people credit, promotions, or medical treatments must be fair, and that means accessible and explainable. As a consumer and target of AI/ML prediction, it is not acceptable to be told that some benefit, credit limit, or medical treatment cannot be provided because of the prediction of an opaque AI/ML model. Therefore, specific approaches have been proposed and are now widely in use in order to assess global and local explainability and predictor importance for any prediction model, including models that are highly nonlinear.

Local interpretable model-agnostic explanations

One way to approach the problem of generic *explainability*—regardless of the nature of the prediction model—is to fit linear surrogate models at and near ("around") the location of the predictor values for a specific prediction. LIME, which stands for *Local Interpretable Model-Agnostic Explanations*, is one such approach (see Ribeiro et al., 2016; see also Nayak, 2019; Molnar, 2022). Note that implementations of LIME are readily available in Python.

A detailed definition of the approach and algorithm is beyond the scope of this chapter, but the general approach is simple: For each observation (instance) that is to be predicted, a new sample of points is generated with values close to the observation that is to be predicted. A weighted linear model is then fitted to that sample, weighted by the distance of the sample points to the respective point (observation) under consideration. As a result, linear regression coefficients can be computed for each predictor based on the model predictions at the sample points close to the actual observation and prediction to be explained. Those coefficients are again effectively derivatives at the respective point in the input space, and provide insight and explainability of the specific prediction for the respective observation at that point, that is, for the input values for the respective observation.

Global LIME explainability values can then be computed by adding the absolute values of individual predictor weights for each observation (the local predictor importance values) over all observations in the sample.

Shapley additive exPlanations

SHAP stands for *Shapley Additive exPlanations*, and is an approach to general AI/ML model explainability introduced by Lundberg and Lee (2017). This approach does not rely on a surrogate model, but is based on game theory and the notion of "contribution" to an outcome. Suppose you have three players A, B, and C who are together building a tower from a fixed number of blocks as quickly as possible. One could set up multiple trials where either all three players work together, only A and B work together (C does not participate), only B and C work together (A does not participate), and so on. In other words, one can try all possible combinations or "coalitions" of players A, B, and C—including individual players (A, B, or C) building towers each by themselves. One can then compute for each player the difference in the *average time* it took to build the tower over all trials vs. the average time it took for trials that excluded a specific player. For example, one could compare the average time it took to build the tower over all trials with the average time it took for all trials that excluded player A.

In general, Shapley explainability identifies (estimates) the specific difference between the average predictions and predictions that include only specific predictor values for the respective observation. In practice, the algorithm uses Monte Carlo simulations to randomly replace specific predictor variable values for an observation or instance under consideration, with randomly chosen other values found in the dataset or sample. To reiterate, the idea is to determine how different a specific prediction is when a specific predictor value is "toggled out," or replaced by other randomly chosen values also found in the data set. Put another way, the Shapley explainability value is *not* the difference between the prediction with or without a predictor, but the *contribution* of a variable to the difference between the actual prediction and the mean prediction.

Global Shapley explainability values can then be computed by adding the individual absolute Shapley explainability importance values for each predictor, over all observations in the sample. Note that implementations of SHAP are readily available in Python.

Comparing local interpretable model-agnostic explanations and Shapley additive explanations

To reiterate, an explanation of the computational details (and "tricks") involved in the implementations LIME and SHAP is beyond the scope of this chapter. If *explainability* is important for a specific project—either for understanding and insight, or in order to comply with regulatory requirements—it is recommended to study the references provided earlier or other sources to fully appreciate the specific computations involved.

To summarize, LIME applies (and assumes) a surrogate linear model at the specific point in the predictor input space, estimated from sample points similar to ("near") the respective observation being predicted. The weights of that linear model are effectively first-order partial derivatives that indicate how the respective output function will move if one moves the input values for a predictor in a particular direction.

SHAP does not rely on a surrogate model, but effectively evaluates the contribution of a variable to the difference between the actual prediction and the mean prediction from all possible combinations of values. This comparison is based on systematically (or via Monte Carlo simulation) evaluating the predictions based on all possible input values, in/excluding (or changing to randomly chosen values) the specific input values in the observation under consideration. SHAP is computationally more expensive (takes more time to compute), but does not depend on the validity of a surrogate linear model ("approximation") at the point of a prediction as does LIME.

Regardless of what method of "explainability" is used, another aspect of it is whether or not it allows predictions of what would actually happen if a particular input value were to be changed. That will be discussed in the next paragraphs.

Caution: inverse predictions can be very risky

So far, the discussion has almost exclusively focused on predicting future outcomes from observed inputs. For example, a typical application might be to predict the number of days that a specific patient is expected to remain in a hospital, based on various measurements obtained for that patient during an emergency visit.

However, there is another use case that is common: How should or could inputs be *changed* in order to arrive at a desired outcome? This is the inverse-prediction problem.

Inverse prediction

Consider a general prediction model $y = f(X)$, that is, some outcome y is a function of predictor values (vector) X. If one wanted to achieve a certain outcome y, how should one set the predictor values X? For example, if we know from modeling that *smoking* (a variable *smoking = yes*) is an important predictor in terms of *explainability* for heart disease, does that mean that *as soon as* a patient stops smoking, that risk reverts to that of a nonsmoker? Obviously not.

There are at least two issues to consider: correlation does not mean causation, and the evidence at the proposed or desired point in the space of input values.

Correlation is not necessarily causation

This commonly known fact can sometimes be difficult to recognize. For example, it may be true that people with longer hair are likely to be shorter (not as tall) overall. Such a correlation would be the result if one were to take a large sample from the US population, as long as fashion favors longer hair styles among self-identified *females* compared to *males* (assuming that self-identified *males* tend to be taller than self-identified *females*). Naturally, cutting one's hair will not make a person grow. What is missing here is that there are several nonobserved mediating variables, many of which a person has no control over.

In practice and real-world applications it is often very difficult to detect such problems, because humans can be very creative to find seemingly obvious causal explanations for such correlations. It is important to keep in mind that just because one identified through AI/ML a variable X that clearly and beneficially affects an outcome Y in a sample *does not mean* that *changing* the values of variable X will result in those favorable outcomes. Perhaps some unobserved and unknown variables and dimensions are involved that affect both X and Y, but that are not causally dependent on X.

Lack of evidence at the specific point in the input space

Another common mistake is to suggest or make changes to input predictor values that result in combinations of values rarely or never encountered in any of the training data on which the models are built.

Extrapolation. It is usually well known and taught that *extrapolation* of models is not justified, and "risky" in the sense that, effectively, predictions based on no actual evidence in the training data are really pure speculation. For example, suppose one built a model relating a patient's age to blood pressure, based on data of 20−30-year-old patients. Using that model, one then predicts blood pressure for 80−90-year-old patients. Obviously, the predictions will be wrong, and doing this is not justified.

The role of a strong theory. To be clear, any predictions based on extrapolation of models from predictor values outside of what was observed and used during model building can *only* be justified if there is a "strong theory," that is, a theory supported by much empirical evidence. Without going into the details of philosophy of science, common sense suggests that if there are hundreds of studies that identify a specific functional relationship between some variable X and an outcome Y, confirmed in samples of different ages, ethnicities, genders, cultures, and so on, then extrapolating a model derived from a specific sample seems more justified: there already exists a lot of evidence that the relationship between X and Y is as expected across various populations and subpopulations.

Interpolation. What is less often recognized is that *interpolation* can be as "dangerous" as extrapolation. Suppose you built a model for predicting blood pressure from age, based on a sample of 20−30-year-olds *and* 70−90-year-olds, and you want to use that model to predict blood pressure of a 50-year-old. Making that prediction for the 50-year-old patient requires interpolation of input values for *Age*. The prediction *may* be correct, if indeed the functional relationship between the input variable (*Age*) and *Blood Pressure* holds in the specific "place" or "neighborhood" at/around the specific age of the patient that is to be predicted. Thus such predictions require interpolation and "faith" in a strong theory or just "faith" that "this should work."

Interpolation with many predictors. In practice, the real problem is that a model may be based on many predictor variables. Suppose a model was built to predict *Risk* from a sample consisting of patients of various ages, ethnicities, and genders. A new patient for whom *Risk* is to be predicted might be within the age range observed in the training sample (from which the model was built); the new patient may also be of an ethnicity and gender observed in the training sample. But the *combination of specific age, ethnicity, and gender* may not have been encountered during model building in the training sample. Making a prediction for that new patient based on said model requires interpolation of input values in the multivariate input space to the location of the vector of input values for the respective new patient. Again, the prediction *may* be correct, if indeed the relationships between the input variables (the predictor variables) to *Risk* (the predicted outcome variable) holds in the specific place and neighborhood at/around the respective new patient in the multivariate input space. In short, predictions require interpolation and "faith" in a strong theory or just "faith" that "this should work."

So what to do. In practice, when applying models to score new observations (patients), careful model monitoring is a known best practice. Careful comparisons of the characteristics of the sample from which a model was built against the population where it is deployed—or monitoring for *population stability*—is one important aspect of using prediction models carefully and responsibly.

Optimization of inputs to achieve a desired output

The problem described above—the interpolation of models with high-dimensional predictor inputs—is exacerbated when models are used to make inverse predictions. Models are often built from observed data describing the complex, high-dimensional, continuous processes of living organisms. Those models are then used explicitly or implicitly (by identifying "important variables") to recommend or try to *change* inputs in the hopes of achieving specific outputs, as predicted by the model. This *can* work in many situations, but is likely to fail when there is little actual observed supporting evidence in the training data (from which the model was built) at the specific combination of input values that is recommended. In effect, at that point combinations of settings are tried that have never been seen before, and it takes a "strong theory" or "faith" to make such a prediction with confidence.

Naive explanations

To summarize, the issue is that interpolation sometimes is required in order to make predictions for specific "locations" in a high-dimensional predictor variable input space where no empirical evidence had previously been observed during model building. It should be noted that for many if not most people-facing applications of AI/ML prediction models, the targets (persons) of the prediction will be unaware of the complexities discussed here. In short, if a model predicts that a person is not eligible for some medical procedure, because some coefficient of explainability indicates that person's *Weight* was too high, then that person would reasonably conclude that losing weight would increase the chances for eligibility. However, first, it is not clear if changing one's *Weight* would not also change other predictor variable values, leading the model to again predict person as *in*eligible; second, by changing *Weight* alone it is possible that a person "sets" a configuration of input variable values that is so unique that the prediction of the model has a great chance of being wrong.

In real-world applications, people seek *explainability* because it confers control over what a model is accurately predicting. That however may not be the case—neither with respect to what the model will predict nor the accuracy of the prediction.

Summary

This chapter consists of two major parts.

1. The first part discussed how human complex decision-making and expertise is the result of mostly nonconscious learning from diverse exemplars, which is organized to derive repeated patterns that allow for better-than-random prediction of future outcomes, and the identification of factors responsible for those outcomes. Modern predictive modeling methods and systems are functionally very similar to how human expertise is acquired, and the implementation of such systems to improve the effectiveness of health care delivery holds enormous promise.
2. The second part examines the nature of "insight" and "explainability." A number of techniques have been proposed and are now available through open-source Python libraries, for example, to explain the predictions of Machine Learning and Artificial Intelligence (ML/AI) models. The approach to *explainability* of models distinguishes between *local* explainability and *global* explainability: The former is an explanation of why a specific prediction was made, given the specific values of predictors into the model. The latter is an explanation of what drives the average model prediction over all predictions. Finally, *explainability* does not necessarily mean *causation*, and caution must be exercised when models and their explanations are used to *set*, *recommend*, or *affect* changes to the input values of predictors, with the goal of achieving some desired outcomes expected by the model.

Postscript

This chapter describes how humans make decisions, and how they can learn to explain those decisions in terms others can understand. A machine learning algorithm like a decision tree is not human, even though its architecture was inspired by human learning patterns. Results of a ML algorithm like a decision tree may be very difficult to interpret. The reader can add the two explainability methods described (LIME and SHAP) to other variable importance measures like the use of regression coefficients to show individual variable importance. Keep in mind, however, that most ML models have significant variable interaction effects. Therefore, it is difficult to assess importance in the presence of significant interaction effects.

References

Benjamin, M., Buehler, K., Dooley, R., Zipparo, P., 2021. What the Draft European Union AI regulations mean for business; McKinsey; extracted July 30, 2022 from: https://www.mckinsey.com/business-functions/quantumblack/our-insights/what-the-draft-european-union-ai-regulations-mean-for-business.

Breiman, L., 2001. Statistical modeling: the two cultures. Stat. Sci. 16 (3), 199–231.

Hastie, T., Tibshirani, R., Friedman, J., 2017. The Elements of Statistical Learning: Data Mining, Inference, and Prediction, second ed. Springer (Springer Series in Statistics).

Hill, T., Rastunkov, V., Cromwell, J.W. 2013. Predictive and prescriptive analytics for optimal decisioning: hospital readmission risk mitigation. In: Proceedings of the IEEE International Conference on Healthcare Informatics (ICHI), Philadelphia, PA.

Lewicki, P., 1986a. Nonconscious Social Information Processing. Academic Press, New York.

Lewicki, P., 1986b. Processing information about covariations that cannot be articulated. J. Exp. Psychol. Learn. Memory Cogn. 12, 135–146.

Lewicki, P., Hill, T., Czyzewska, M., 1992. Nonconscious acquisition of information. Am. Psychol. 47, 796–801.

Lundberg, S.M., Lee, S.-I., 2017. A unified approach to interpreting model predictions. Available from: https://doi.org/10.48550/arXiv.1705.07874.

Molnar, C., 2022. Interpretable machine learning: a guide for making black box models explainable. <https://christophm.github.io/interpretable-mL-book/>.

Nayak, A., 2019. Idea behind LIME and SHAP: the intuition behind ML interpretation models. Towards Data Sci. Dec 22. Available from: https://towardsdatascience.com/idea-behind-lime-and-shap-b603d35d34eb#: ~ :text = Shapley%20value%20guarantees%20a%20fair,not%20have%20any%20such%20assumptions.

Ngiam, K.Y., Khor, I.W., 2019. Big data and machine learning algorithms for health-care delivery. Lancet Oncol. 20, 262–273.

Nisbet, R., Elder, J., Miner, G., 2009. Handbook of Statistical Analysis & Data Mining Applications. Academic Press, Burlington, MA.

Miner, G., Delen, D., Elder, J., Fast, A., Hill, T., Nisbet, B., 2012. Practical Text Mining and Statistical Analysis for Non-Structured Text Data Applications. Elsevier/Academic Press, New York, NY.

Palmer, M., 2021. Singapore's AI-driven healthcare: the Singapore National University Healthcare System shows how ModelOps data helps answer "Medical Phone Calls" every day. Techno Sapien Oct 11. Available from: https://techno-sapien.com/blog/nuhs.

Pearl, J., 2009. Causal inference in statistics: an overview. Stat. Surv. 3, 96–146.

Pearl, J., 2010. In introduction to causal inference. Int. J. Biostat. 6. Available from: https://doi.org/10.2202/1557-4679.1203.

Ribeiro, M.T., Singh, S., Guestrin, C., 2016. Why should I trust you? Explaining the predictions of any classifier. Available from: https://doi.org/10.48550/arXiv.1602.04938.

Siddiqi, N., 2006. Credit Risk Scorecards: Developing and Implementing Intelligent Credit Scoring. Wiley & Sons, New York.

Chapter 21

Model management and ModelOps: managing an artificial intelligence-driven enterprise ☆

Thomas Hill

Chapter outline

Prelude

The case studies presented in Chapters 11 through 17 show how predictive analytics technology can be applied to answer crucial medical questions about current patient health conditions. After the model has been trained and validated, it can be deployed in the organization's systems. But the deployment of the model brings a number of questions and problems into focus. This chapter explores these issues of model deployment and interface with organizational systems.

Introduction

Increasingly, it is clear that in order to be effective and successful any organization must embrace Machine Learning (ML), Artificial Intelligence (AI), advanced analytics, AI-infused applications, and automation. In practically all applications and domains, deep insight and accurate foresight are necessary to be effective and competitive, and also to adapt rapidly to changing and increasingly competitive environments. This is also true for healthcare providers, insurers, and practitioners.

Operationally, this means that effective management of analytics, AI, and ML life cycle processes and operations is now of critical importance. What are the best practices, technology considerations and tools that will ensure repeatability of effective analytics and AI/ML, management of teams and responsibilities, and compliance

---

☆ Note: TIBCO and Omni-HealthData are trademarks or registered trademarks of TIBCO Software Inc. and/or its subsidiaries in the United States and/or other countries.

Practical Data Analytics for Innovation in Medicine. DOI: https://doi.org/10.1016/B978-0-323-95274-3.00026-9

with evolving regulatory oversight frameworks? Also, with large numbers of models in use (in "production"), how to assess and manage the significant and growing risks associated with analytics and AI/ML, with respect to applications in healthcare but also and more generally when applied to predict the behavior or propensities of customers, clients, applicants, and so on?

This chapter explores these issues from the perspective of Model Management and Operations (ModelOps). ModelOps is required tooling to achieve efficient and agile model deployment, monitoring, and recalibration. Any organization deploying advanced analytics and AI to optimize their services, processes, or products *must* implement demonstrated best practices with respect to the model-deployment life cycle, governance, monitoring, and risk management. In short, organizations must fully appreciate the scope of end-to-end analytic processes.

The model building/authoring life cycle

The requirements for successful model management and ModelOps go far beyond the simple—deploy—monitor—recalibrate predictive model life cycle, as it is depicted in the still popular *Cross-Industry Standard Process for Data Mining* (*CRISP-DM*) model proposed over 20 years ago (see Wikipedia, "Cross-industry standard process for data mining," 2021). This model is shown in Fig. 21.1.

Effective and future-proof ModelOps also is much more than just efficiently moving a "model" into production, and using it in production at scale for many customers and clients based on large amounts of available data. Careful consideration of the many and diverse ways how models and model pipelines are deployed (as Apps, to infuse intelligent BI, through smart automation, as advisory tools for critical decisions), and the UI/UX requirements for different personas and roles, stakeholders, and end-users/consumers/decision-makers are important.

Key definition

Advanced Analytics. Any statistical or rules-based analytics analytic process, converting numeric, textual, image, or other data into information to inform a process, decision, or understanding. For the purposes of this chapter, Advanced Analytics encompasses ML including Deep Learning Neural Networks as well as all statistical analyses.

Artificial Intelligence (AI) Model. Usually a broader concept that includes ML, but also encompasses the notion of adaptive continuous learning similar to or simulating human learning; like ML Models, AI models can be used to predict or forecast with some probability the values of other variables, discrete events, or identify and predict anomalies.

Machine Learning (ML) Model. An algorithmic methodology that can identify patterns in numeric or other data, including text, speech, or images; those patterns can serve as ML Models to predict or forecast with some probability the values of other variables, discrete events, or identify and predict anomalies.

Model. Throughout this chapter, the term *Model* refers to a repeatable systematic, algorithm or rules-based step or steps that transform data. The goal usually is to reduce data to information and actionable insight. Therefore, this definition encompasses AI and ML models, Advanced Analytics, or any set of rules that transforms data into predictions, recommendations, warnings, alarms, and so on.

Model scoring flows. A set of connected steps that can include data transformation, preparation, validation steps, feature engineering (creation of derived variables from actual measured variables in order to maximize the diagnostic accuracy of the prediction model), and postmodel-prediction steps in order to convert predictions into useful information or decisions (e.g., convert prediction probabilities into actual predictions); sometimes model pipelines are composed of reusable steps, data preparation models (e.g., for replacing missing data), rules, validation steps that can prevent subsequent steps from executing, and so on. Model flows are themselves abstractions or *Models* because they can be connected to different inputs and outputs.

Model scoring pipelines. *Model Scoring Flows* connected to specific input data or views into input data, and connected to specific outputs (where outputs, decisions, information, etc. will be rendered, written, or delivered).

ModelOps. Gartner (2021a) defines ModelOps as follows:

ModelOps (or AI model operationalization) is focused primarily on the governance and life cycle management of a wide range of operationalized artificial intelligence (AI) and decision models, including machine learning, knowledge graphs, rules, optimization, linguistic and agent-based models. Core capabilities include continuous integration/continuous delivery (CI/CD) integration, model development environments, champion-challenger testing, model versioning, model store and rollback.

Throughout this chapter, this definition is expanded to include not only AI Models, but also ML and Advanced Analytics models.

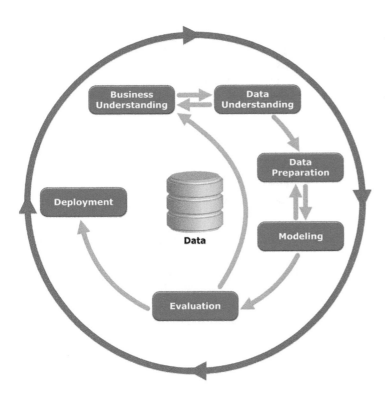

FIGURE 21.1 Process diagram showing the relationship between the different phases of CRISP-DM* (Cross-industry standard process for data mining," 2021). *Extracted from: Wikipedia, Cross-industry standard process for data mining, 2021). Licensed under the Creative Commons Attribution-Share Alike 3.0 Unported license. Author Keneth Jensen.*

Overview: managing the life cycles for thousands of models

Effectively, modern organizations today cannot run without analytics. The definition of "model" in this context (see above) is much wider than what is usually considered as AI, ML, or Predictive Analytics models.

Types of analytic models

Any analysis of data collected by an organization, when summarized, plotted, correlated, etc. represents a model of, for example, what an expected average (mean) is, what a distribution of data looks like, or how measurements are related to other measurements:

1. *AI and ML applications.* Of course, the use of modern AI and ML techniques has proliferated in recent years, and prediction models and scoring pipelines are used to anticipate consumer preferences and behaviors, predict maintenance requirements or impending process issues before they materialize, and so on. In fact, modern e-commerce, automated manufacturing, banking, weather forecasting, and practically all conveniences and features of contemporary life are not possible without such models to predict, guide, automate, and optimize processes. This trend of AI-everywhere is unlikely to abate, as such model-scoring pipelines are increasingly built from reusable processing steps authored in diverse tools and formats—including R, Python, Tensorflow, or as self-contained scoring services—and deployed or "placed-into" applications, smart home-appliances and devices, medical devices, etc. (edge computing).
2. *Traditional BI dashboards.* However, an analytic model driving critical decisions can also be encapsulated into traditional BI (Business Intelligence) dashboards, for example summarizing cost-per-patient, by quarter and region over the last two fiscal years. Such a dashboard really is a model and expectation of what "normal" and "no-change" looks like, against which cost and operational performance can be measured.
3. *Process monitoring, statistical analyses, SPC.* In manufacturing, statistical techniques are common in order to characterize and monitor the measurable outcomes of processes. These could be measured indicators of quality, counts of defects; these could also be counts of patients per day or week or any other measured property of a process. Statistical process control methods and control charts (SPC line charts) are the standard tools and models of what a stable, in-control, high-quality (manufacturing) process looks like, against which current quality can be monitored. In general, Statistical process/quality control charts provide statistical models to detect anomalies, trends, changes in the variability in measurements, percentages, etc.

4. *Compliance reporting.* In many regulated industries, such as pharmaceutical manufacturing, financial services, and of course health-care related industries, periodic analytic summary reports must be created for regulatory agencies to verify that requirements are met with respect to quality standards, equal treatment of customers, clients, patients, applicants, and so on. Again, those analytic summary reports are really models of what is desired and—in many applications—*required* by applicable laws or regulations. Compliance with regulatory standards or organizational policies are monitored and evaluated against those models.

Given this broader view of "models," based on AI, ML, or statistical analyses and reporting, it is not surprising that organizations today are challenged to manage hundreds or thousands of models that are guiding or informing critical decisions and automating processes regarding hiring, business forecasting, the delivery of services, and so on. Those models need to be monitored continuously; also, when needed, models must be updated (recalibrated), improved, or rebuilt. Management of models, and the responsibilities and tasks associated with their deployment (continued use) to improve real-world outcomes requires ModelOps. For example:

1. models (scoring pipelines, dashboards, reports, etc.) must be vetted through a careful life cycle of testing, validation, and reviews for compliance, risks, and expected impact;
2. models must then be deployed efficiently and quickly based on the right data and data-streams;
3. results must be delivered when and where they are most critical in a way that augments the respective organizational processes and tasks; and
4. models must be monitored for performance, value, continuous compliance, risks, and costs.

Managing the risks of analytics, artificial intelligence

The management of analytics, AI, or AI-infused analytic BI is an activity that can carry risks if "done wrong." Because of the risks, those activities are usually governed by policies describing best practices informed by long experience, previous successes, and of course occasional failures.

1. *Healthcare-related applications.* In medicine, pharmaceutical, or medical device manufacturing, analytic models that demonstrate the consistent quality of outcomes and products are obviously critical. When the quality of people-facing products or procedures is critical to ensure safety, the need for managed and governed consistency of those analytics is obvious, and governmental regulatory agencies and frameworks usually dictate specific analytic models, approaches, and reporting to ensure compliance. Failure to comply can invalidate processes, and threaten the very existence of a business.
2. *Personnel decisions.* If prediction models and analytic approaches to hiring the best qualified candidate for a position are wrong, the wrong candidates get hired. A capable, well-trained, skilled, motivated, and happy workforce or team is a necessary prerequisite for achieving exceptional outcomes. A lack of policies, consistency, and controlled quality of the process and analytics used to screen job applicants can cause "bad hires," and easily result in arbitrary and discriminatory hiring practices, ineffective teams, reputational damage, legal challenges, and regulatory attention.
3. *Complex processes.* In modern hi-tech manufacturing, prediction models are critical to control very complex processes sometimes consisting of thousands of processing steps and tools. In fact, those models themselves represent part of the critical Intellectual Property (IP) that allow manufacturers to remain competitive. Those models and the IP they represent must be managed and nurtured through systems that accumulate and retain experience, knowledge, and expertise, even if critical resources and human experts retire or leave. Else, when knowledge about the details and workings of critical analytic models disappears, the consequences can threaten the very continuity of an organization. Service delivery processes in general can also be extremely complex, and models predicting resource requirements and timing are often used to optimize efficient service delivery. Those models are part of critical IP for enabling organizational viability and success.
4. *People facing analytics.* In general, there is risk whenever predictive analytic models or AI is applied to allocate or guide resources to people. For example, consider a lower insurance rate that an analytic model offers to some people but not others. While it is true that probably no business would intentionally disadvantage, offend, or "inflict harm" to its customers, doing so inadvertently can easily happen: if evidence of bias or discriminatory impact surfaces, the damage to the organization can be substantial or even existential, regardless if the (e.g., financial) harm that was done was not anticipated. Again, governance, policies, and consistent procedures with respect to models, how they are evaluated, tested, and documented is a critical aspect of AI, when put into practice at-scale and enterprise wide.

Do-no-harm

Solutions that manage models must provide visibility into their performance, lineage, operations, and consequences. ModelOps should be thought of as a required "operating system" to manage the risks that are always associated with analytics, in particular with people-facing predictive analytics. Regulatory frameworks are rapidly evolving around the world to require specific best practices when building and deploying models, and also how they must be monitored in order to guarantee adherence to the dictum: first, *do no harm.*

All regulatory frameworks such as FDA 21 CFR Part 11, privacy laws and regulation such as the emerging strict European AI regulations, or the laws and regulations to ensure nondiscriminatory practices and impact of analytic models aim to ensure that no harm is done (see, for example, Burt, 2021, for an overview of regulatory frameworks proposed or in effect for Ai/ML around the world).

Industry certifications

Obviously, it is important that the ModelOps tools used to manage AI/ML models are themselves of high quality and built consistent with best practices. Industry certifications such as ISO9001 (Quality), ISO27001 (Security), SOC 2 (Customer Data), HITURST (HIPAA compliance), and others were established and are often required for ModelOps software. Software vendor audits are already common and required for such software before it can be used in pharmaceutical or medical device manufacturing, or in financial services. The deployment of AI/ML models will require the same diligence, and quality certifications of ModelOps software are increasingly important.

ModelOps scope

ModelOps manages the models and deployment, monitoring, and model rebuilding organization-wide, for a few critical, or hundreds or thousands of models. ModelOps enables collaboration between teams from different backgrounds contributing different expertise (sales, manufacturing, insurance underwriting, data science, statistics, risk-management), serving in different roles (IT, Data Engineering, Security, Compliance), and with different responsibilities (executive decision-making, operations, quality control and engineering).

Shown in Fig. 21.2 is a diagram organizing the typical life cycle of AI/ML models, from model building through deployment, monitoring, and rebuilding (recalibration).

The two-sided arrow on the right side of Fig. 21.2 shows an "approximate" scope of ModelOps tools. In later sections of this chapter the question of *scope* will be discussed in greater detail. In general, ModelOps definitely entails the deployment (application) of models and their continuous evaluation. However, data access, data preparation, feature engineering (creation of useful numeric indicators), and postmodel decision steps (rules) can and almost always *are* part

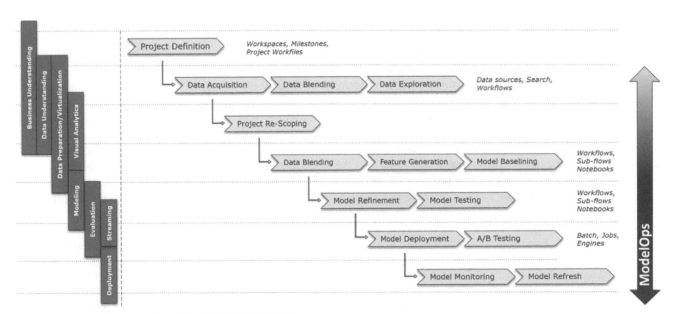

FIGURE 21.2   Complete ModelOps life cycle (Modified CRISP).

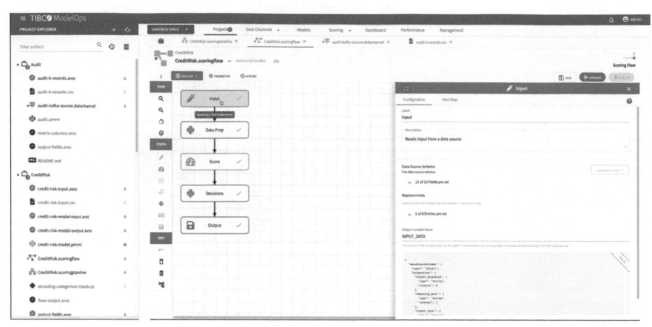

FIGURE 21.3   A simple scoring pipeline in TIBCO ModelOps.

of the complete deployed model *pipeline* that transform data into insights, decisions, or automated actions. Fig. 21.3 shows such a complete Model Scoring Pipeline in TIBCO ModelOps.

Throughout this chapter those pipelines will be referred to as *Model Flows* or *Model Pipelines*. The former is defined as an abstract flow of steps that transform data, apply an analytic or AI/ML model, and apply postmodel decision steps and transformations; the latter is defined as model flows connected to specific inputs and outputs (see also the *Key Definitions* box).

ModelOps details: managing model pipelines and reusable steps

ModelOps can be thought of as the "operating system" to deliver information derived from models in a reliable manner to the right location when it is needed. Fig. 21.4 shows an example dashboard in TIBCO Omni-HealthData, for delivering predictive readmission information summaries to decision-makers and stakeholders.

In practice, there are a number of challenges that must be addressed.

The tools and languages of artificial intelligence/machine learning

Data scientists use various tools—commercial, open source, or models purchased from third parties. Also, Scoring Pipelines are built from multiple reusable steps, for data preparation, feature engineering, scoring, and postscoring decisioning. So support for various model formats—and pipelines built from steps authored in different formats—is *required*.

To reiterate, a model is a transformation of data to indicators, values, or other outputs that contain the useful information derived from data. Those transformations can be formatted and "encapsulated" in many different ways. Constraining a ModelOps solution and processes to only a subset of them will limit what models can or cannot be used.

Currently, Python and different Python Integrated Development Environments are the major programming language and authoring tools that data scientists (and programmers) use to build models. Support for the deployment of (reusable) Python data preparation steps, transformations, or features (result of feature engineering) is important. Statisticians also often require specific functionality only found in R (the popular programming language for statistical modeling).

However, data scientists, programmers, and statisticians may *not* be the primary authors of models. Increasingly, models are built by business or process experts and users themselves (sometimes called the "Citizen Data Scientist"; see, for example, Gartner, 2021c). Their models may be made available to ModelOps in some proprietary formats (for example, as TIBCO Data Science workflow/project), or as an application programming interface and via function calls supported by a third-party web service. There are also several formats for prediction models that are portable across

FIGURE 21.4 Example dashboard delivering predictive readmit info via TIBCO Omni-HealthData Solution.

applications, such as predictive model markup language and portable format for analytics (see Data Mining Group, 2021), or ONNX (Open Neural Network Exchange; see ONNX, 2021).

In general, effective ModelOps solutions must support an extensible model format framework, so that any and all such current and future formats can be supported.

Reusable steps, building intellectual property

The value of AI/ML and analytic models is a function of how much better they are than current practice. Also, models should align with the unique capabilities, positioning, KPIs, values, and goals of an organization or business. Building out repositories of proven and reusable steps for scoring pipelines that work across several domains and applications effectively means building unique IP critical for the organization's success and perhaps survival.

In this chapter "models" are defined to include steps to prepare the data, transform them to useful indicators ("features"), to apply equations or rules to make predictions, to transform predictions into insights and actions, and to deliver the insights and actions to the right place at the right time. This way of looking at Model Scoring Pipelines as previously defined will unlock the real value for AI/ML. Put another way, if every insurer would predict insurance risk exactly in the same way there would be little if any room for insurers to specialize, and differentiate on premiums to insure those risks. The same is true for healthcare providers, and virtually any other organization.

To reiterate, ModelOps can be thought of as the operating system that enables advanced analytics and AI/ML as key organizational activities, and encourages the cumulative development of an organization's IP. As discussed, models can come in various forms and formats, and include steps for preparing data, rules for detecting outliers, and other "special ways" to process data. Reusable steps in different formats can be efficiently assembled to support large numbers of Model Scoring Pipelines. Effective ModelOps allows steps in any format to be assembled into Model Scoring Pipelines. In short, when done correctly, ModelOps will be the organizing deployment system for an organization's all-things-analytics-IP, driving most if not all critical organizational processes.

Avoiding analytic silos, big code

To be clear, experience has shown that without proper organization, data science, AI/ML, and similar advanced analytics projects can and likely will fail. A frequently cited statistic originally reported by Gartner is that 85% of all big-data "data science" projects fail (see, for example, O'Neil, 2019). Venturebeat (2019) estimated that 87% of Data Science projects never make it to full deployment to a "production" environment, to solve actual problems. And this picture may not improve very much in the immediate future: Through 2022, only 20% of analytic insights may actually deliver real business outcomes (according to White, 2019).

There may be many possible reasons for this, but for sure one of them is that models are too often delivered by data scientist specialists, who develop voluminous sets of functions and subroutines authored in a big Python notebook, for example. The Python code may intertwine data preparation, validation, model scoring, and decision rules—all in one big document that is then handed to IT for implementation. It is still not uncommon for some organizations who follow this approach to require over a year to implement, for example, an insurance fraud model. Obviously, the creativity of fraudsters is less encumbered by "process," and the resulting fraud models in productions are not very effective.

Siloed data science

When models are created in silos of data science specialists and programmers, and deployed as large "chunks" rather than made up of small reusable steps authored or at least informed by experts in specific aspects of organizational processes, agile management of such models becomes near-impossible.

Managing input data and output sinks

Another aspect of effective model management and ModelOps tools is that they should manage not only the individual steps of scoring or analytic pipelines, but also the data inputs and outputs, as well as the modes of delivery of the results.

Validating data

For example, so-called *Data Drift* describes unexpected and undocumented changes to input data formats, scaling, etc. Such changes to upstream data—how they are collected, encoded, scaled, etc.—will break the compatibility with the model and how it was built, and cause wrong predictions and results. As a simple example, suppose the measurement of *Weight* that is a critical input to a model changes from ounces (oz) to grams (g). One oz equals 28.35 g—obviously, predictions from a model built against a variable *Weight* measured in ounces but scored against data measured in grams would be very wrong, and potentially dangerously wrong. Modern IT architectures and integration technologies make data drift a common problem that needs to be addressed for reliable model deployment. Effective ModelOps must therefore not only manage the data processing steps required for scoring, but also validate and manage the data inputs or virtual views (e.g., approve and lock down for all phases of the model life cycle, including deployment).

Validating outputs

The same is true for the outputs from models, and how they are consumed downstream. A complete Model Scoring Pipeline consists of validated input data, as well as output data, including reports, dashboards, and so on. A correct result and accurate prediction can lose its usefulness if incorrectly formatted or displayed in a report, or rendered in an ambiguous manner in a dashboard. The actual outputs of model predictions themselves must be managed and governed.

Managing model life cycles

A typical deployment of a Model Scoring Pipeline is not a single-step process; instead, it is a process that typically involves multiple roles and approvals. The final deployment of models to real data in order to drive actual decisions or automation is the final step of multiple stages. Note that the entire process is actually similar to the best-practice methods used in software development, to support so-called continuous integration–continuous deployment software development and delivery (see Sacolick, 2020, for an overview).

Proposing a new model, updating a model

As mentioned before, models, data preparation steps, or complete Model Scoring Pipelines are authored via one or multiple commercial or open-source tools. Often, the authoring process has a few constraints such as what types of variables *can* or *cannot* be used for modeling. For example, variables such as gender, race, ethnicity, and so on cannot be used in some applications, although when predicting medical outcomes they may be important variables to include. Sometimes these variables are not available in the input data at all, in order to prevent any bias to enter into the models. Although, when evaluating models for discriminatory impact and bias, they must be carried along and available for other monitoring and reporting analytic Scoring Pipelines.

But in general, when a new model, operating step, or Scoring Pipeline has been authored, it is committed as an "artifact" to the ModelOps platform.

Deployment environments

Typically, the respective model, operating step, or Scoring Pipeline is initially tested in some "sandbox" or testing environment. That means that the respective Scoring Pipeline is run against test data, historical data, or most recent data in order to observe and verify performance and issues; however, results are *not* used in order to affect any business decisions or operations. Also, different computing environments are usually provisioned for each environment so as not to affect the performance of critical systems and processes, and the models that support them.

In many ways this is simply a continuation of the typical diligence and vetting of models that normally happens as part of the model building process. However, in ModelOps the respective model artifacts can be connected to the parts of Scoring Pipelines that have not changed (input data, output data, dashboards, preprocessing steps, postprocessing steps, etc.). Also, in many real-world ModelOps scenarios the Scoring Pipelines themselves are scheduled to run in a particular sequence, allowing successive Scoring Pipelines to incorporate results from those that precede it.

For example, a model Scoring Pipeline may estimate a patient's risk for developing complications in response to a particular drug or treatment; a downstream Scoring Pipeline may use that risk estimate as one variable to predict overall risks. Also, a number of testing and validation reports may need to be executed to verify the robustness of the updated model with respect to missing data, out-of-bounds inputs, and so on.

One or multiple scoring environments must be set up to perform the vetting and validation of models or improvements to models, without impacting any decisions or automated steps affecting real customers, patients, or processes. Models move through multiple environments, and eventually into a preproduction environment where they are applied against actual data, and where their performance and value against actual data can be monitored—again, without actually affecting real decisions, patients, or customers. Actually, so-called *champion-challenger* comparisons of new models with a current model used in production (to make actual decisions from real data) can also be managed through ModelOps and/or automated in such an environment.

Roles, personas, approvals

In larger organizations and when managing hundreds or thousands of models, different roles and personas will perform their respective tasks in the environment to which they have access. For example, an individual or role responsible for compliance reporting may run standard compliance reports (i.e., Scoring Pipelines) incorporating the new model in a preproduction environment. There is also usually a formal role of "approver," and often that role is specifically limited and not permitted to contribute new models or other artifacts to an environment. Instead, the approver can only approve models (steps) for a specific environment. This is typical best-practice in order to ensure the independence of approvers.

Granular permissions

In general, ModelOps solutions permit granular permissions regarding who (what role, group of individuals) can contribute models, operating steps, etc. to what specific project(s), who can use certain environments (Development, Testing, Preproduction, Debug, Production, etc.) and resources (input data or "data channels," output data or "sinks," reusable processing steps, etc.). This allows an organization to create and implement policies and procedures regarding how specific models, data sets, reports, dashboards, and so on are managed, who can/not review them, approve them, modify them, etc.

In short, ModelOps facilitates the implementation and enforcement of best practices and compliance with respect to the deployment of Analytic/Model-Scoring pipelines, and enables the robust and validated *life cycle* (steps, approvals) required to ensure consistency, efficiency, quality, and compliance of models and deployments.

Model monitoring

When organizations embrace analytics and AI/ML at scale, that is, using hundreds or thousands of models to improve or automate decisions and organizational processes, models, processing steps, and Scoring Pipelines are not managed as single entities. Instead, they consist of multiple steps, and are "surrounded" by multiple other reports, test and validation scripts or dedicated Scoring Pipelines, and various dedicated Scoring Pipelines that perform computations that monitor model performance.

Beyond model accuracy

Model accuracy is an obvious and important aspect of prediction models that should be monitored. Beyond simple accuracy measures (percentage correct classifications, residual sums of squared deviations of predictions from observed values, etc., see Dhinakaran, 2021) and others used also during the model building process (such as lift, gain, etc.), in practice it is equally and often more important to identify value and bias.

Differential accuracy

The overall accuracy of a model may not be the same across all possible stratifications of observations. For example, genetic males in a specific age group may be consistently predicted with a lower accuracy when compared with other customers, clients, or patients.

Linking model predictions to value

The *value* of a model is a function of the goals of a project and model, the specific value of specific outcomes when the model makes correct predictions, the cost when it makes incorrect predictions, and other dimensions related to the organization's values and mission. Also, different roles and users of the predictions or results from models may assess the value of models differently. A healthcare worker may want to see a dashboard of the accuracy of recent diagnoses derived via a model from medical images. A financial officer in the same organization will want to see the value of models in terms of the cost and benefits to the organization. A compliance officer will need to see standard reports that attest to the lack of discriminatory impact or differential accuracy of models (harming certain groups of patients). A ModelOps solution and tooling needs to provide the flexibility to deliver this information reliably to the respective stakeholders in the most useful format, when it is required.

Monitoring for model drift, shift

Model accuracy in general often deteriorates over time, for a number of reasons.

Concept drift

For example, the relationships between the input variables may change over time. Suppose that a specific infectious disease over time becomes more dangerous and thus predictive of future health outcomes in a specific age group. Also, as lifestyles, habits, and social conditions change, the health risks and relationships between risks will likely change over time. For example, geographic distance from the immediate family may have been attributable to a person's personal preference, choice, and personality, and have no or only moderately impact on mental health. However, during times of pandemics that restrict social interactions during lockdowns the lack of close friends or family members nearby may change that. In general, the relationships between variables and underlying "concepts" may change and drift over time, impacting the accuracy and value of models.

Population stability and drift

Also, when the demographic characteristics of the population to which models are applied deviates from the population characteristics of the sample from which the models were built, model accuracy and value can be impacted. Over time, an organization may deal with customers, clients, or patients that are different than a month or a year ago. This could be attributable to migration of new groups of individuals to a neighborhood or community, the relocation of a large employer, and so on.

Data drift

As mentioned earlier, data drift is a general and in practice particularly important problem to monitor for and catch. Complex organizations manage complex data integration and problems, and it is not uncommon that measurements and fields in databases change, invalidating models that use those data. For example, suppose income of a customer, client, or patient had been measured in terms of the *annual* income during model building, but while a model was deployed in production that field was changed to reflect *monthly* income. Obviously, the model predictions will be impacted, and likely inaccurate or nonsensical.

Effective monitoring of models, and the ModelOps platform and tools that support it, must be able to accommodate flexible analytic Scoring Pipelines, reports, and automation as necessary to implement periodic scoring of recent data to detect all of these problems. Data drift is a particularly serious and "instant" problem, so processing steps should be inserted into (or precede) all Scoring Pipelines to validate data ranges, and stop the scoring process to prevent grossly erroneous predictions and results. For that purpose, validation *rules* can be created and approved as reusable data preprocessing steps for the relevant production models.

Monitoring risks

So far, the discussion of capabilities has still focused on the monitoring of model accuracy, common causes for accuracy changes (data drift, concept drift, population shift), and the impact of reduced accuracy (value, cost). But applying AI/ML models or any analytics in people-facing applications, in particular in the context of medical applications, implies risk.

A previous section discussed many obvious risks in terms of "first, do-no-harm." Wrong prediction can have costly consequences, noncompliance with regulatory requirements can be costly, and discriminatory practices and biases will have costly consequences once they come to light.

However, a model can be perfectly accurate, nondiscriminatory with respect to its predictions, and consistent with regulatory oversight—yet be risky and cause damage the organization that deploys it.

Unexpected consequences

A famous example was the case of a national retailer who successfully predicted who among their customers was likely pregnant, and provided those customers (customer *households*) with relevant offers (coupons for diapers, baby food). The unanticipated problem was that those customers in some cases may not have been aware of their state themselves, nor did their loved ones, guardians, or parents (in the case of minors; see Hill, 2012). The publicity that ensued was definitely not anticipated, unwanted, and likely damaging.

The consequences of applying AI/ML prediction models can be embarrassing, offensive. Also, any model can be inadvertently discriminatory in a way not seen before, and unfairly withhold some benefit, prevent a capable job applicant from finding a job, inflict pain on a patient, and so on.

Anticipating risk

To reiterate, the tools and methods for deploying and managing models and Scoring Pipelines should be able to surface the risks that may be associated with their deployment, *before* they are put into production. That aspect of predictive models and the consequences of their deployment have received relatively little attention so far, but will become important as thousands of models are managed that are driving all critical processes of an organization.

For starters, a flexible "smart" BI tool should be integrated with or at least connected to models, their metadata, and their predictions in historical and most recent data. Predictive model accuracy is only one measure that *can* be important; in reality it is rarely the most important metric. Predictive models—even when they are accurate—can cause grave damage to an organization and its reputation. Note that TIBCO is actively researching this specific area of ModelOps*.

Efficiency, agility, elasticity, and technology

AI/ML models can be deployed in various ways, and increasingly are finding their way into smart applications (e.g., cell phone apps, smart devices and sensors) and automated (e.g., IoT edge) processes. The tooling that allows for efficient management of models is based on modern cloud technologies and architectures.

A discussion of cloud technologies in general is beyond the scope of this chapter. However, a few general key points should be considered.

Cloud architecture

Cloud architecture is the way technology components combine to build resources that are componentized and orchestrated with shared services. Cloud architecture relies on shared services that can be sized as needed, providing elasticity with respect to resource needs.

---

*TIBCO recently filed a patent on *Systems and methods to screen a predictive model for risks of the predictive model*.

Private and public cloud

Cloud applications and the services that support them can be shared across multiple organizations, and accessed as services in a *public cloud* and SaaS (Software as a Service) computing environment. Salesforce or Google apps are examples of such an environment. A *private* cloud is a cloud computing environment that is dedicated solely to a specific organization; the resources in a private cloud can be hosted and managed in various ways, and often are supported by resources and infrastructure already present in an organization (e.g., on-premise, or provided by a third-party organization).

On-prem cloud solutions

On-prem or on-premise cloud computing environments are implemented on hardware that is located on-site at the respective organization's location(s).

Cloud architecture and containerization

Increasingly, modern cloud-architected solutions leverage containerization, that is, virtualized computing environments that can be efficiently and flexibly deployed and taken down (for an overview, see Haley, 2019). Thus containerization makes it possible to run distributed computing/processing applications efficiently.

With respect to ModelOps, containerization enables parallel execution of multiple analytic and AI/ML Scoring Pipelines efficiently, with flexible resourcing based on demand and performance requirements. Fig. 21.5 shows an example of such an architecture for the TIBCO ModelOps solution (see TIBCO, 2021). In this environment, multiple scoring runs can be efficiently executed in parallel via containers, orchestrated in a Kubernetes (also called K8s) architecture. Kubernetes is an open-source system for automating deployment, scaling, and management of containerized applications (see Kubernetes.io).

In this illustration, the individual boxes denote processes running in separate containers. For example, *Data Sources*, which are managed artifacts in TIBCO ModelOps that can be approved for use in certain environments by authorized groups of users, can be defined as views into underlying multiple data sources integrated via the TIBCO Data Virtualization solution, or can be a real-time streaming data services implemented through Apache Kafka (an open-source distributed event streaming platform).

Managing models for data-at-rest and data-in-motion

The majority of the discussion of AI/ML and ModelOps in general is still mostly focused on use cases for batch data. The traditional (e.g., CRISP) model for AI/ML described earlier defines a cycle of understanding and learning from historical data, and then applying those insights to score new batches of data. For example, an insurer may score all claims

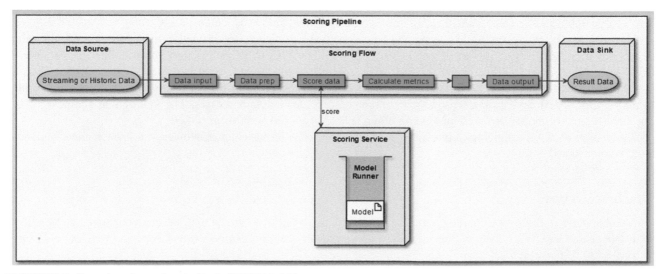

FIGURE 21.5 Execution of a scoring pipeline in TIBCO ModelOps.

nightly to update the predicted cost (required reserve) and complexity of claims (to plan resources); or a hospital may score all current patients nightly to predict availability of hospital beds and resource needs in the future.

Addressing real-time scoring requirements

Of course, AI/ML models derived from historical data sets (batches) are often applied to real-time scoring use cases. For example, an insurance reserving (cost) risk model can be called from an application for accepting insurance claims online and in real time. The critical metric in that case is the maximum acceptable latency between the request for scoring (to a scoring service) and the receipt of results, including risk scores, expected cost, immediate acceptance/denial of a claim, and auxiliary information such as reasons (so-called rason-scores) if more information is needed. Obviously, low-latency and high-throughput "fast" scoring is important in this case, ideally in the lower 100s of milliseconds per request; other applications require scoring latencies in the low 10s of milliseconds, or single milliseconds. The latter is required to support intelligent advanced automation solutions.

Many organizations will build separate services to support low-latency scoring applications. However, this adds complexity that may not be necessary. Given the right ModelOps solution and architecture, low-latency scoring services can be supported directly by the ModelOps servers and services, without the need to build out a separate application or service.

Responding to a dynamically changing world

Modeling and model management gets more complicated when conditions and relationships change quickly between inputs and outputs that are to be predicted or anticipated. Previous sections discussed model drift, and the need to monitor the accuracy of models, which will inevitably decay over time. In many applications, model drift can happen very quickly. Technically, many processes in many domains are (increasingly) dynamically unstable, that is, the relationships between variables, concepts, and process inputs and outputs can change very quickly and in novel ways.

For example, the recent changes due to the COVID-19 pandemic in 2020 and 2021 with respect to work, consumer behavior, travel, and of course all aspects of healthcare happened quickly, were unprecedented, and mostly unexpected. But changes to health risks, consumer preferences, fashions, and fads happen all the time. In automated scheduling processes, for example, a large number of automated steps and measurements interact to create complex and dynamically unstable processes where new quality problems and inefficiencies can be impossible to anticipate from historical data that are more than a few weeks old (or less).

Effectively, platforms for building and deploying models must be able to adapt quickly, and sometimes learn directly and in real time from the data streams generated by some process. Industry analysts at Forrester[†] acknowledged the value and requirements for real-time analytics in the recent Streaming Analytics Wave report (Gualtieri, 2021). While this is an area of ongoing development and research (see, for example, the brief application overview at Palmer, 2021), the need to adjust quickly to new conditions and realities across various domains and industries is making support for dynamic learning from real-time data an important requirement for ModelOps tools.

Fig. 21.6 shows an illustration of a Dynamic Learning model implemented through TIBCO Streaming, with model predictions shown in TIBCO Spotfire Streaming, in real time. The TIBCO Streaming flow editor defines the data preparation steps and includes continuous dynamic learning against streaming data recording customers' actions in response to best-next-action offers (e.g., to purchase insurance on some expensive electronics). The Spotfire Streaming display on the right illustrates how *concept drift* and model decay affect the accuracy (depicted in the line graph at the top) and value (depicted via the bar plot in the lower left corner of the dashboard; the cumulative value is depicted to the right of that graph). About two-thirds through the display, the accuracy decreases for both a static (traditional) prediction model (see the yellow line/bars) and the dynamic learning prediction model (see the cyan/green line/bars). However, the dynamic learning model (cyan/green line/bars) quickly relearns and returns to previous accuracy and value. More importantly, the value of the dynamic model remains high, leading to significantly great cumulative value.

In general, efficient support for flexible scoring and analytics against low-latency real-time streaming data is important now, will become more important in the future, and will be a must-have for organizations aiming to differentiate their AI/ML-infused capabilities and processes from competitors.

[†] Forrester is a registered trademark of Forrester Research, Inc.

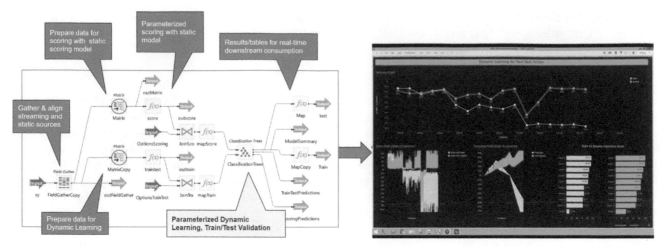

FIGURE 21.6 Dynamic/adaptive learning model and deployment: next-best-action.

Conclusion

AI/ML and the application of advanced analytics has transformed many industries, it will continue to do so, and it is hard to see how a modern organization or business can survive without it. The focus with respect to AI/M and analytics has definitely shifted away from computational algorithms and methods, to considerations of best practices, value, risk, and ROI. Modern cloud-based architectures have also made it possible to scale up (or down) quickly and efficiently applications of Scoring Pipelines, automation solutions, scoring services, analytic reports or intelligent BI solutions, or AI/ML-infused applications.

ModelOps and its various facets must be part even of small first steps, because otherwise analytic silos, "gurus" (critical individual resources who "have the knowledge"), competing interests, disconnected data science teams, and any number of issues will quickly fester.

Postscript

One of the most distressing facts in the early history of development of predictive analytics applications in organizations is that many models developed were never put into practice, they just "sat on the shelf." The primary reasons for this failure were either production data were not available to feed the model, or model outcomes could not be integrated easily into operational systems. The only way to prevent this logical disconnect is to develop processes to generate the necessary data required by the models in operational business systems, and design the business processes required to use them.

References

Burt, A., 2021. New AI regulations are coming. Is your organization ready? Harvard Business Review, April 30. <https://hbr.org/2021/04/new-ai-regulations-are-coming-is-your-organization-ready>.

Dhinakaran, A., 2021. The playbook to monitor your model's performance in production: performance monitoring of ML models. towardsdatascience.com, March 4. <https://towardsdatascience.com/the-playbook-to-monitor-your-models-performance-in-production-ec06c1cc3245>.

Gualtieri, M., 2021. The Forrester Wave™: streaming analytics, Q2 2021. <https://www.forrester.com/report/the-forrester-wave-streaming-analytics-q2-2021/RES161547>.

Haley, J., 2019. Demystifying containers, Docker, and Kubernetes. Microsoft Open Source Blog, July 15. <https://cloudblogs.microsoft.com/opensource/2019/07/15/how-to-get-started-containers-docker-kubernetes/>.

Hill, K., 2012. How target figured out a teen girl was pregnant before her father did. Forbes. <https://www.forbes.com/sites/kashmirhill/2012/02/16/how-target-figured-out-a-teen-girl-was-pregnant-before-her-father-did>.

O'Neil, B., 2019. Failure rates for analytics, AI, and big data projects = 85%—yikes! <https://designingforanalytics.com/resources/failure-rates-for-analytics-bi-iot-and-big-data-projects-85-yikes/>.

Palmer, M., 2021. Dynamic Learning 101: adjust machine learning models based on real-time inputs like a child learning to walk. <https://technosapien.com/blog/dynamic-learning>; Sept. 09.

Sacolick, I., 2020. What is CI/CD? Continuous integration and continuous delivery explained. InforWorld, Jan 17. <https://www.infoworld.com/article/3271126/what-is-cicd-continuous-integration-and-continuous-delivery-explained.html>.

TIBCO®. 2021. TIBCO® ModelOps documentation. <https://docs.tibco.com/pub/modelops/1.0.0/doc/html/index.html>.

Venturebeat. 2019. Why do 87% of data science projects never make it into production? <https://venturebeat.com/2019/07/19/why-do-87-of-data-science-projects-never-make-it-into-production/>.

White, A., 2019. Our top data and analytics predicts for 2019. Gartner, 2019. <https://blogs.gartner.com/andrew_white/2019/01/03/our-top-data-and-analytics-predicts-for-2019/>.

Cross-industry standard process for data mining. 2021, December 21. Wikipedia. <https://en.wikipedia.org/wiki/Cross-industry_standard_process_for_data_mining>.

Data Mining Group. 2021, December 21. <http://dmg.org/>.

ONNX. 2021, December 21. Open neural network exchange—the open standard for machine learning interoperability. <https://onnx.ai/>.

Gartner. 2021a, December 21. Gartner glossary—ModelOps. <https://www.gartner.com/en/information-technology/glossary/modelops>.

Gartner. 2021c, December 21. Citizen data scientists can boost an organization's business value and analytics maturity; however, in most cases, their capabilities remain underutilized. <https://www.gartner.com/smarterwithgartner/how-to-use-citizen-data-scientists-to-maximize-your-da-strategy>.

Further reading

FDA, 2011. Guidance for industry—process validation: general principles and practices. <https://www.fda.gov/files/drugs/published/Process-Validation-General-Principles-and-Practices.pdf>.

FDA, 2021. Food and drug administration—21 CFR Part 11 Electronic Records; Electronic Signatures. <https://www.accessdata.fda.gov/scripts/cdrh/cfdocs/cfcfr/CFRSearch.cfm?CFRPart = 11>.

Firn, M. 2020. Why every marketer should consider uplift modeling: predicting the future is helpful. predicting what strategies you should take to influence the future is better. Towards Data Science <https://towardsdatascience.com/why-every-marketer-should-consider-uplift-modeling-1090235572ec>.

O'Connell, M., Sweenor, D., Hillion, S., Rope, D., Kannabiran, D., Hill, T. 2020. ML Ops: Operationalizing Data Science. O'Reilly.

Gartner. 2021b, December 21. Gartner analytics and business intelligence platforms reviews and ratings—what are analytics and business intelligence platforms? <https://www.gartner.com/reviews/market/analytics-business-intelligence-platforms>.

Chapter 22

The forecasts for advances in predictive and prescriptive analytics and related technologies for the year 2022 and beyond

Gary D. Miner, Linda A. Miner and Scott Burk

Chapter outline

Prelude

Descriptive analytics presents data about what happened in the past, either in textual or graphical form. **Predictive analytics** builds mathematical models based on past data to predict what might happen in the future and associated probabilities of occurrence (risk). **Prescriptive analytics** goes beyond predicting what will happen with associated risks of occurrence, it provides some guidance for deciding what can be done about it. More information on the various kinds of analytics is presented in Chapter 25. This chapter focuses on prescriptive analytics in healthcare. This activity is analogous to the specification of medical treatments following diagnosis.

Section I: specific technological trends predicted for 2022–2023+

Before we get into specific predictions for 2022 and the years beyond, let's look at what are the primary predictive analytic tools and prescriptive analytic methods:

What is predictive analytics, and what are the most frequently used methods (or algorithms) in predictive analytics?

Predictive analytics is the use of data, statistical algorithms, and machine learning techniques to **identify the likelihood of future outcomes based on historical data**. The goal is to go beyond knowing what has happened to be providing an understanding of what is currently happening, and also providing a best assessment of what will happen in the future.

Practical Data Analytics for Innovation in Medicine. DOI: https://doi.org/10.1016/B978-0-323-95274-3.00003-8

Three of the most widely used predictive modeling techniques are **decision trees, regression, and neural networks**. Regression (linear and logistic) is one of the most popular methods in statistics. Regression analysis estimates relationships among variables (SAS, A,B, retrieved, March, 2022).

What is prescriptive analytics, and what is an example of prescriptive analytics?

Prescriptive analytics goes beyond simply predicting options in the predictive model to suggesting a range of prescribed actions and the potential outcomes of each action. **Google's self-driving car, Waymo**, is an example of prescriptive analytics in action (Ohio University, 2020).

Overall, **Predictive analytics** is often defined as predicting at a more detailed level of granularity than for an entire population (predicting to a group is what traditional statistical modeling does); in other words, predictive analytics can generate predictive scores (probabilities) for each individual. This feature distinguishes predictive analytics from forecasting.

Focusing on healthcare:

Prescriptive analytics can assist providers to become more effective in their clinical care diagnosis and treatment. This means providers can achieve better patient satisfaction giving patients more accurate healthcare with the potential of tailoring treatment to the individual person's unique genetic and environmental situation.

Prescriptive analytics can help pharmaceutical companies expedite their drug development by identifying patient cohorts that are subgroups of the total population, thus a grouping by shared "genetic and environmental" factors. Again, such subgrouping can achieve better patient outcomes.

Part I—healthcare: what trends can we expect in the year 2022 and beyond?

The healthcare "industry" has been slow to act, slow to change—at least in the United States—but today with the advances of abundant data storage systems and the advancement of computer systems that can handle large amounts of data, even data in real time, the healthcare industry is challenged to act and make good use of data and data analytics information.

Below we will succinctly summarize what many of the forecasters predicted in late 2021, as the year 2022 was approaching. We will not discuss these in detail, but just present the "trends or predictions" that each forecaster made, then provide a few sentences of explanation here-and-there, and finally at the end of this chapter provide a paragraph to summarize. For the reader that has greater interest in these forecasts, the references are provided, mostly web blogs, which you, the reader, can easily access for more details, if you would like to know more.

Martin (2021) discusses what he saw as 10 emerging trends in healthcare during 2021, which he named "The New Normal":

1. More strategic and agile supply chains
2. Competition as a viable strategy
3. Patient consumerization
4. Personalization of care
5. Workforce diversity and safety
6. Virtual care
7. Artificial Intelligence (AI) and automation
8. Revenue diversification
9. Mergers and integration
10. Payor shifts.

Another forecaster (Healthcare-Management-Degree-Guide, 2022) wrote about what was seen as "Five Trends in US Health Care System" near the end of 2021:

1. The drive to repeal the Patient Protection & Affordable Care act
2. The shift toward value-based care
3. TECHNOLOGY=technology—and more technology
4. The rise of collaboration (doctors working in hospitals rather than in single-office or 2–3-person group-offices)
5. The move toward single-payor.

There have been great efforts in the past number of years to undo the "Affordable Care Act," but along with that "value-based care" has become more of the norm. Prior to the ACA, healthcare was reimbursed more for "quantity" of services performed; but with value-care, the quality of what the patient received in services became more important and payment for services were geared toward this quality (Healthcare-Management Degree Guide, 2022).

Another forecaster (Taranto, 2021) put forth six trends to watch in 2022, including that healthcare will finally get fully "Patient-Centric," or in the terms we are using in this book, it will become personalized medicine, patient-centered medicine, and patient-directed medicine. Here are the six trends Taranto listed in his December 20, 2021 blog post:

1. Telemedicine will change how chronic conditions are treated.
2. Digital therapeutics will rewrite the future of healthcare.
3. Social determinants of health will result in greater health equity.
4. Remote health monitoring will improve outcomes and lower costs.
5. Real-world data will deliver real-world results.
6. Healthcare will get truly patient-centric.

Telemedicine, data, and patient-centric are the terms that stand out to the authors of this book, as these are all critical to use of modern data analytics to provide doctors with the information they need to make "person-centered" diagnoses and treatment decisions. Digital therapeutics and remote health monitoring stand out to the authors as being critical to patient-directed healthcare. Patients will be collecting more data than before and helping to share the information needed for accurate diagnoses and treatments.

Another forecaster (Burroughs, 2021) described three things she thought would be the trends to watch in 2022. These involved communication, cybersecurity, and data analytics, as follows:

1. Streamlining clinical communications and collaborations
2. Bolstering visibility and automation in cybersecurity amid continued threats
3. Improving data analytics and visualization—to address population health.

These are general process improvement types of considerations, although the underlying intent may be out of date if it is directed at population health, as population health concepts are a thing that had their peak of interest maybe about 5–7 years ago. Today the concern is the individual, e.g., patient-centric healthcare. There is a difference as patient-centric is based on the unique genetic constitution of an individual person, whereas population health refers to groups of people, like the group of diabetics, or the group having cardiovascular health issues. Not everyone in such a group will have the same genetic constitutions, thus the population health concept is not truly patient-centric, even when it forms groupings within the population.

Siwicki (December 6, 2021) believed that the following three things will be required in 2022 by healthcare:

1. Augmented intelligence
2. IoT and wearables data
3. Information coming from outside EHRs (electronic health records).

Definition of Augmented Intelligence: "Augmented intelligence is **a design pattern for a human-centered partnership model of people and artificial intelligence** (AI) working together to enhance cognitive performance, including learning, decision making and new experiences" (Gartner, 2022).

Examples of Augmented Intelligence: "*Augmented intelligence and deep learning have applications in any industry that mines big data for patterns and predictive indicators. Examples include the following:* **Online stores using data analytics to predict customer preferences. Political think tanks using big data analytics to identify undecided voters**" *(IEEE, 2022).*

What do these three things mean?

Using Augmented Intelligence is like gaining X-ray technology, as suddenly the doctor can see what is "visible and explainable." Here is an analogy:

Think, for example, of the 1881 shooting of US President James Garfield. In this pre-X-ray era, neither the best physicians nor Alexander Graham Bell could find the bullet lodged in the president's body. As a result, Garfield suffered through weeks of futile examinations by unwashed hands before succumbing to sepsis. In 1895, X-ray was invented and would easily have saved President Garfield. It took World War I to begin to make the technology accessible. Today, we expect this technology to be available to save or ease the suffering, not only of presidents, but of our pets.

Siwicki (2021)

Now during 2022, healthcare has a similar opportunity as in the analogy above; augmented intelligence can be used to see previously obscured information.

Using information coming from the IoT (Internet of Things), wearable data sources, and patient portals adds data from outside the hospitals and clinics and doctor's offices. Bringing this data from outside the medical facilities means we are moving more toward an "outpatient-centric model," or using the terminology of this book, a patient-centered framework. COVID-19, since 2020, has accelerated this trend by greater adoption of telehealth and remote patient monitoring.

2022 will see "data governance" and "data orchestration" become important. What does this mean? What do these two terms mean?

1. Data governance: the application of standard meanings to different types and different qualities of data coming from patient portal, wearables, and other sources.
2. Data orchestration: the management and infrastructure needed to handle data coming in from sources external to the clinic and hospital. This may mean moving data to the cloud (from where it was previously stored in "warehouses" or "lakes") to now, what may be called a "data lake house," e.g., a blend of these new formats with the traditional formats.

In addition to a "data lake house," some of these new capabilities include the ***data mesh***," which comprises loosely coupled services architecture to allow organizations to integrate and share data at scale (Siwicki, 2021).

And thirdly, data beyond the EHR must be used in 2022 and future years to get a full understanding of a patient's health. Things like a patient's physical environment, genetics, biology, level of medical care, social circumstances, and the patient's behavior among other things like any mobile apps, smart health wearables, and other sources of external data need to be taken into account. The traditional EHR now only contains about 11% of the health data needed for any one individual (Siwicki, 2021). **The need for more information makes patient–healthcare professional collaboration increasingly vital. Patients and physicians need to work together for true patient-centered healthcare.**

And finally, another healthcare forecaster, predicts her three trends for 2022 (Gupta, 2021):

1. ESG* Strategy focused on innovation
2. Data Analytics to accelerate biotechnology innovation
3. Consumer-facing telemedicine and digital care solutions.

*ESG represents Environment, Sustainability, and Governance. The life science industry is moving toward customized therapies for specialized patient populations. With this change, pharma and biotech are having trouble financing such efforts, so they are looking for partners in manufacturing and the supply chain to meet these patient-centric goals of the future. Data analytics are becoming more and more important as genetics–genomics understanding of the individual increases.

What does this all mean?

Telemedicine and other virtual care methods that allow doctors to monitor patients remotely are becoming more important, and this will most likely accelerate in 2022 and beyond. Telemedicine and related processes will become more mainstream in 2022, as more ritual options are likely to be also made available (Gupta, 2021).

Part 2—In general: PA and business intelligence trends for 2022

Companies, hospitals, and users such as medical personnel and even patients need to be able to use already developed methods for analyzing their data so they don't have to hire in people to code or to analyze their data for them. Companies that go into organizations with their black boxes are not what people will want in the future. They will want modules, and visual modules, that are "no code/low code" so they can construct the analytics they need and maintain the integrity of their data to stay current and relevant to their own organization. This is particularly true for medical organizations that cannot make public their patient data. Self-service analytics will therefore occupy the future, according to Hiter (2021).

In addition to organizations building their own sequences, individual patients will want to put their own N-of-1 data into a program developed just for their body app. A program that is simple to understand (visual) and that gives them the answers they need about their own individualized treatments.

According to Hiter (2021), future demand will be for analytics that are possible to do in situ by themselves, to have automatic analytics (embedded) that are comprehensive (IOPs), that provide integrity and compliance for their data (particularly for medical data), and that provide security such as using blockchain for performing analytics that therefore could not be hacked.

TOP 10 analytics and business intelligence trends for 2022

Calzon's (2021) list of BI trends looks a lot like what is already being done. However, she would like them to be done better and more reliably. In addition, she agreed with others that data analysis will increasingly belong to everyone and not just analysts, using the term data democratization, which will increase customization. She stressed the vital nature of quality data and data compliance.

Key artificial intelligence and data analytics trends for 2022 and beyond

Oh my, after reading even a little of Pool's (2021) blog I couldn't help but think about the cyborgs, the future police, and Keanu Reeves in The Matrix. Are we on the road to managing risk, low costs, and data integrity (accurate and digestible) or something else? Who will be the future model makers? People or machines? Who can own and use 1.6 trillion parameters in their deep learning models, but people like Bezos and Musk? The authors of this book want to help medicine get better, not control the world; control pathogens, not a competitive edge. We might fear that the learning machines we are building will soon build themselves, taking over the minds of even the superwealthy.

But then, Pool comes back to Earth by stating something that we all feel intuitively. Why use a hugely complex model when a simple one will do? Why should we use a complex and inscrutable neural network when a simple model (Pool said Bayesian) model will do? Perhaps the cyborgs will fold in on themselves. Until then, the robots may start working side by side with humans. This seems a bit scary, frankly. As one who created an in-class game, Land, Castles, and Blood, A Feudalism Simulation, while teaching sixth graders in 1967 (Miner, 1972), it is a bit frightening to think that Dungeons and Dragons is now based on a 175-billion parameter model. When will the clouds reign?

Returning to earth, it was good to see Pool talking about the usability of Python for neural networks in deep learning. Humans may not be as involved in the deep recesses of neural networks as they are with easily graphed models, but at least the use of neural networks can help businesses progress to the top.

The data are many, and increasing in the future. Therefore when using trillions of parameters, "multipronged approaches" are useful for efficiency and automation, as Pool said. However, Pool's last paragraph said it all, "With accessibility, democratization, and automation becoming key priorities for the data industry moving forward, creators should aim to keep their models easy to understand and, where possible, future-proof" (Pool, 2021).

Cindy Howson made the following predictions for what she thought would trend for 2022:

1. People analytics rises to the top of analytics priorities.
2. The analytics engineer displaces the data scientist as the world's sexiest job.
3. Data mesh, data fabric, and data lakehouse dethrone the data warehouse.
4. Insight to action becomes a reality with cloud ecosystems.
5. Data sharing goes from competitive advantage to essential.
6. Environmental, social, governance (ESG) data, and data for good go mainstream.
7. Power returns to the people in the battle for personal data … maybe (https://go.thoughtspot.com/ebook-seven-data-analytics-ai-trends-predictions-and-resolutions.html).

Cindy Howson goes on to state that:

Oh, what a glorious year 2022 is destined to be! After 30 years of us dabbling with data, this industry's moment has arrived. The worst of the pandemic will be behind us, but the effects of accelerated digital transformation remain. Businesses are moving faster than ever before, and the only fuel that can power this speed is data. But it's not all roses. The hype for certain technologies far outweighs their real-world value. New roles are needed to take advantage of these capabilities, lest companies drown in the growing data deluge. All the while, our own demands for data, whether for more transparency into data for good and diversity initiatives or more privacy for our personal data, continue to evolve rapidly.

ARSR Technologies, in an October 2021 LinkedIn posting, made the following predictions for big data and analytics for the year 2022:

1. Increasing rise in use of predictive analytics
2. AI will continue to evolve but humans will remain crucial
3. Increased adoption of BI (Business Intelligence) in industries
4. The need for cloud-based analytics solutions
5. Increased BI budget in various organizations in 2022 (ARSR Technologies, 2021).

ARSR Technologies is a software development firm in the San Francisco area with specific expertise in big data services, including cloud and big data strategies.

Brian Wallace in a December 23, 2022 post wrote about what he saw as the top data and analytic trends he expected to see develop during 2022:

1. In-memory computing will become popular
2. Predictive Analytics will increase in usage
3. AI will keep evolving
4. Automation will be at the forefront of the data analytic process
5. Augmented Analytics will spur major developments
    a. including Augmented Data Preparation
6. Data-as-a-Service will gain in popularity (Brian Wallace, 2021).

Brian concludes his predictions by stating:

The digital economy relies more on data today. Businesses need reliable data for decision-making. Data is like an engine that propels businesses and industries. Thus, businesses should invest in data and analytics. You should adjust their strategies if you want to increase your returns. Whether it is IoT, AR, AI, or VR, knowing how these new technologies impact your businesses is fundamental to achieving growth and success.

Brian Wallace, B. (2021). Top data and analytics trends for 2022, 23 December. Available at: https://www.webpronews.com/analytics-trends/.

Kumar Goswami in his December 22, 2021 article gave what he saw as the top six trends in data and analytics that he expected would be evident in the year 2022:

1. Data lakes get more organized, but the unstructured data gap will continue.
2. Citizen science will be an influential related 2022 trend.
3. "Right Data" analytics will surpass Big Data analytics as a key 2022 trend.
4. Data analytics "In Place" will dominate during 2022.
5. Storage-agnostic data management will become a critical component of the modern data fabric.
6. Multicloud will evolve with different data strategies.

(Goswami, 2021).
As Kumar Goswami pointed out:

For decades, managing data essentially meant collecting, storing, and occasionally accessing it. That has all changed in recent years, as businesses look for the critical information that can be pulled from the massive amounts of data being generated, accessed, and stored in myriad locations, from corporate datacenters to the cloud and the edge. Given that, data analytics—helped by such modern technologies as artificial intelligence (AI) and machine learning—has become a must-have capability and in 2022, the importance will be amplified. Enterprises need to rapidly parse through data—much of it unstructured—to find the information that will drive business decisions. They also need to create a modern data environment in which to make that happen.

Goswami, K. (2021). Top 6 trends in data and analytics for 2022, 22 December. Available at: https://venturebeat.com/2021/12/22/top-6-trends-in-data-and-analytics-for-2022/.

Thus, these six trends are among those that must/will happen to make this all possible.
Eric Avion on December 22, 2021, gave his five BI trends to watch during 2022:

1. The automation trend continues during 2022

2. Natural Language Processing, which we can term "Augmented Intelligence capability," will pick up in 2022, but it is slowly blossoming, so its adoption will continue on in 2023 etc.
3. AutoML (automated machine learning) will advance in 2022
4. NEW AREAS OF EMPHASIS: ESG; scenario planning (e.g., "what-if" moves...)
5. Money movement (e.g., consolidation of industry vendors into larger entities by acquisitions, as vendors, even smaller vendors, raise significant capital) (Avion, 2021).

As Eric Avion pointed out:

Automation is expected to be one of the biggest business intelligence trends of 2022. But not merely process automation, which is becoming common in data management to reduce certain tasks that need to be done repeatedly. Instead, industry insiders predict 2022 will be the year analytics vendors add more automated insight delivery capabilities to their platforms, enabling users to take action in real time. In addition to automation, BI users can expect to see further development of natural language processing and AutoML capabilities, scenario planning tools and more money moves in the business analytics market.. There's a lot of money being poured into data and analytics companies, and it will be interesting to see where some of that gets spent.

Avion (2021)

Erin Cunningham on December 14, 2021 made predictions about AI and predictive analytics, especially related to federal governmental agencies, that included:

1. AI, analytics can help agencies like health, agriculture, and housing with forecasting.
2. Predictive Analytics can aid in climate and economic and national security modeling.

For example, predictive analytics can help identify risk factors for housing and food insecurity, addiction and mental health needs, allowing federal agencies and legislators to adopt a data-driven approach to not only providing relief and access to resources but also implementing policies aimed at prevention and intervention.

From: Cunningham, E. (2021). 2022 tech trends: predictive analytics and AI will make agencies more efficient, Fed-Tech Magazine, 14 December. Available at: https://fedtechmagazine.com/article/2021/12/2022-tech-trends-predictive-analytics-and-ai-will-make-agencies-more-efficient.

Aanchal Iyer in a November 2021 statement discussed "What can we expect from Big Data Analytics in 2022?" (Iyer, 2021). The following points were emphasized:

- Human jobs will remain unaffected despite the development of AI.
- Predictive Analytics will grow.
  - "The market share for predictive analytics was valued at USD 7.2 billion in the year 2020 and it is projected to grow up to USD 21.5 billion by the year 2025. Currently, the predictive analytics market is growing at a Compound Annual Growth Rate of 24.5%."
- The AI market will reach a record breaking high with regards to implementation and usage.
- Self-Service Analytics will become even more critical to BI.
  - Meaning: "...more organizations will adopt truly self-service tools that enable nontechnical business users to securely collect insights from data," and
- Wholesale/retail, technology organizations, and financial services will increase their BI budgets over 50% during the year 2022.

The conclusion from these predictions is that organizations need to get on board with these data analytics trends of 2022 to stay relevant (Iyer, 2021).

If you, the reader, will take another look at the predictions from the various forecasters, above, you will see that there are similar threads in all of them. Thus we should take these seriously and watch during the remainder of 2022, and also into 2023 and 2024 to see how many of these projections materialize.

Section II: overriding philosophies which will guide trends over the next 10 years

Regardless of all the predictions of forecasters, there are some foundational structures and philosophies that will endure, either through a resurrection in their everyday use or in an embellishment of their current status in the analytic world. These include:

1. Science will again reign in discovering and understanding truth. Science has been increasingly misunderstood over the past 40 years. This has become readily apparent in the most recent years, as media, politicians, and others in

power have corrupted our educational and research systems by declaring belief or poorly done science as "truth." Fortunately, the public has lost faith in these "proclamations," and has been quick to again look for and demand scientific truth. A new era in science is coming.

2. We predict a "Statistics Renaissance"! Statistics is (most likely) always the tip of the spear in data analytics yet can be misquoted as parametric modeling. Statistics will hit a new wave, from basic statistics to nonparametric statistics, including Bayesian statistics and experimental design. Computing power will never bridge the gap of data quality and lack of data design and analytics design. Statistics will always be required, unless in those rare instances where one can look at the "entire universe of data" and thus "see" what is there (e.g., if one can see an "entire forest" then there is no need for statistics, as statistics are only needed when we are sampling a portion of the forest).

3. **To be very specific about medicine and healthcare:** we predict that Directed Acyclic Graphs (DAG) (for a good definition and understanding of DAG, please see Hazelcast Glossary, 2022; Barrett, 2021) will dominate medical and related research. Computing performance is overtaking many of the challenges to make this possible.

4. **Also specific to medicine**: nontraditional data will become mainstream in analytics in medicine. On-premises clinical data are largely obsolete in understanding, predicting prescribing medicinal interventions that truly take into account the individual person. Data that are social, environmental, geospatial, and genetic will dominate the new signals.

What this means, among other things, is that person (patient)-wearable data sensors will explode in 2022 and subsequent years, allowing what is probably the most important variables (data) important to the individual person to be coming from outside of a doctor's office, outside of a hospital, and outside of any clinical physical setting. Instead, this data will be coming from each of us, the individual person, the individual patient. Hospitals, clinics, and individual doctors will be relying on this data to make the best diagnostic and treatment decisions.

Postscript

Chapter 2 presented the history of methods and systems for storing and analyzing data during the last 5000 years. The rise of Data Science and associated mathematical tools (statistical analysis, machine learning, and deep learning neural networks) in the 21st century have provided us with the means to make some sense out of a growing body of digital data. This presently large body of data is about to become a tsunami of data output from mobile and wearable devices. Our healthcare systems must be redesigned to capture, store, and analyze, and apply this hoard of information.

References

ARSR Technologies. (2021). Top 5 big data analytics trends and predictions 2022, 8 October. ARSR Technologies. Available at: https://www.linkedin.com/pulse/top-5-big-data-analytics-trends-predictions-2022-arsr-technologies/?trk=organization-update-content_share-article; https://arsr.tech/top-5-big-data-analytics-trends-and-predictions-for-2022/.

Avion, E. (2021). 5 Business intelligence trends to watch in TechTarget Network 2022, 22 December. Available at: https://searchbusinessanalytics.techtarget.com/news/252511319/5-business-intelligence-trends-to-watch-in-2022.

Barrett, M. (2021). An introduction to directed acyclic graphs, 10 October. Available at: https://cran.r-project.org/web/packages/ggdag/vignettes/intro-to-dags.html.

Brian Wallace, B. (2021). Top data and analytics trends for 2022, WebProNews, 23 December. Available at: https://www.webpronews.com/analytics-trends/.

Burroughs, A. (2021). Three healthcare trends to watch in 2022. *Health-Tech*, 8 December. Available at: https://healthtechmagazine.net/article/2021/12/3-health-tech-trends-watch-2022.

Gartner (2022). Gartner Glossary. <https://www.gartner.com/en/information-technology/glossary/augmented-intelligence> (accessed 1.02.22.).

Goswami, K. (2021). Top 6 trends in data and analaytics for 2022, 22 December, Venture Beat: DataDecisionMakers. Available at: https://venturebeat.com/2021/12/22/top-6-trends-in-data-and-analytics-for-2022/.

Gupta, A. (2021). The future of health: three healthcare trends for 2022. *Forbes Business Council*, 14 December. Available at: https://www.forbes.com/sites/forbesbusinesscouncil/2021/12/14/the-future-of-health-three-healthcare-trends-for-2022/?sh=4d64ff64541c.

Hazelcast Glossary. Directed acyclic graphs (DAG). <https://hazelcast.com/glossary/directed-acyclic-graph/> (accessed 2.03.22.).

Healthcare-Management-Degree-Guide. (2022). Five trends in the U. S. health care system. 5 Trends in the U.S. Health Care System—Healthcare Management Degree Guide, 2 March. Available at: healthcare-management-degree.net.

Hiter, S. (2021). Data analytics market trends 2022, July, Datamation. Available at: https://www.datamation.com/big-data/data-analytics-trends/.

IEEE Digital Reality. What is augmented intelligence? <https://digitalreality.ieee.org/publications/what-is-augmented-intelligence> (accessed 1.02.22).

Iyer, A. (2021). What can we expect from big data analytics in 2022?, ARETOVE. 7 November. Available at: https://www.aretove.com/what-can-we-expect-from-big-data-analytics-in-2022.

Martin, G. (2021). Top 10 Emerging Trends in Health Care for 2021: The New Normal. American Hospital Association. Available at: https://trustees.aha.org/top.

Miner, L. (1972). Land, Castles, and Blood, Stem Publications. (no longer available). Available at: <https://www.amazon.co.uk/castles-Student-education-material-publication/dp/B0006WQE66>.

Ohio University, (2022). Predictive vs prescriptive analytics: What's the difference? <https://onlinemasters.ohio.edu/blog/predictive-vs-prescriptive-analytics-whats-the-difference/>.

Pool, D. (2021). 10 key AI and data analytics trends for 2022 and beyond. <https://www.kdnuggets.com/2021/12/10-key-ai-trends-for-2022.html > .

SASa (2022). Predictive analytics; What it is and why it matters. SAS Insights AE. <https://www.sas.com/en_ae/insights/analytics/predictive-analytics.html > (accessed 7.31.2022).

SAS—B <https://www.sas.com/en_us/insights/analytics/predictive-analytics.html> (accessed 7.31.2022).

Siwicki, B. (2021). 2022 will require augmented-intelligence, IoT, and wearables data, and info outside EHRs. Healthcare IT News. <https://www.healthcareitnews.com/news/2022-will-require-augmented-intelligence-iot-and-wearables-data-info-outside-ehrs > .

Taranto, B. (2021). The growing power of digital health: 6 trends to watch in 2022, 20 December. <https://techcrunch.com/2021/12/20/the-growing-power-of-digital-healthcare-6-trends-to-watch-in-2022>.

Chapter 23

Sampling and data analysis: variability in data may be a better predictor than exact data points with many kinds of Medical situations

Mitchell Goldstein and Gary D. Miner

Chapter outline

Prelude

Personalized medicine is enabled by personal measurement data, but it matters how this data is gathered. Individual measurements of a dynamic system (e.g., blood pressure) are very likely to express the system norm. Even the system norm might differ at night, compared to the daytime (e.g., COVID-19 vaccine response). This chapter presents some sampling and analysis issues that can invalidate the use of the data gathered for predictive and prescriptive analysis. These issues must be acknowledged and accounted for in medical analytics and diagnoses.

Sampling and data analysis issues

Purpose summary of this chapter

There are serious real issues of measurements and modeling in healthcare. What is needed is a "decisioning system" that clearly says "Do This OR Do That," especially for situations where the variability in data may be much greater than the isolated selected data sampling points (e.g., pulse oximetry saturation; blood pressure; and eye IOP). Use of measures other than "exact specific date points in time," are needed. Such measures are averages over a period of time,

Practical Data Analytics for Innovation in Medicine. DOI: https://doi.org/10.1016/B978-0-323-95274-3.00013-0

also kurtosis, and the standard deviation. Using these three statistics, for example, the sampled averaged value, the standard deviation, and kurtosis, in concert, may better indicate the central tendency within the range and thus lead to more accurate diagnoses and treatments. These methods of central tendency should become the "Gold Standards" of 2022 and future years.

One issue—electronic health record and specific measures taken on patients

The electronic health record (EHR) provides a consistent means of accessing patient care data. It is the repository of extensive information about our patients, trends in disease processes, and a reliable source of high-fidelity data. The companies involved in this endeavor work tirelessly to ensure that the data is backed up with a high degree of integrity to both cloud and local sources. However, this integrity may be subject to interpretation. Although a patient may have a reported heart rate of 52, and this rate may be correct for the instant in time when sampled, what about the next hour, minute, or second? When we sample is as critically important as what we sample. "Even a broken clock is right twice a day" (Gravenstein et al., 1989).

Sampling can be defined as how often we measure the same value. In the case of the clock, if we only sample at noon and the hands are stuck at noon, we would never know that the clock is broken. Sampling every 12 hours masks the inadequacy of the device and does not allow enough resolution to tell time through the day accurately. The same can be said for the heart rate. What happens if the patient has a heart attack, and the EHR is programmed to sample only every 3 hours? It is an extreme example, but if the ECG leads only sample to the EHR at that frequency, the patient might be dead for over 2 hours before EHR registers the "0" heart rate.

ECG leads display a continuous readout in an ICU or critical care environment. The premise is that there will be an individual capable of responding to a crisis in moments, but the EHR may be oblivious to these changes if not set up to receive data from the monitor frequently. Someone remotely monitoring the EHR could not tell that the patient was receiving emergency treatment until orders were entered or the EHR polled the ECG. Indeed, manual entry is possible, but anyone remote from the patient location looking at the standard readout of the EHR would be oblivious to the crisis.

Remote monitoring solutions improve upon the periodic monitoring offered by most EHR solutions but not every patient is remotely monitored, and the resolution and bandwidth to make this data instantly available to any willing provider are not universally available. Further, these technologies are subject to disruption because they are not necessarily transmitted to the EHR in real time, and they are only as reliable as the individual monitoring remotely. Although various elevation tools help ensure that increasingly responsible individuals have eyes on the remote activation, there is no assurance that these events make it to the EHR (Bhatia et al., 2010; Gubbi and Amrutur, 2015; Park et al., 2018; Zapata et al., 2014). Although these solutions claim compatibility with myriad EHR products, the implementation of compatibility is as varied as the EHRs themselves.

Looking further down the chain, each of these remote monitoring solutions transmits continuous patient variables. Because of differences in sampling between the EHR and the monitoring technologies, the same patient data may result in different information transmitted to the user. For example, real-time data may show a heart rate that varies from 0 to 70 beats per minute over a defined epoch, and these values may be transmitted in real time by the remote monitoring solution. The EHR may only indicate a heart rate of 70. Each is correct because the sampling mechanisms have been defined as appropriate for the platform, but does it inform optimal patient care? Although it may be argued that stable patients may not require the degree of granularity provided by real-time monitoring, what happens when the stable patient changes and requires more acute monitoring? Can the EHR keep pace? That transition represents the vulnerability in the system (Sjostrand et al., 2013; Tautan et al., 2015).

However, what of the technologies that require real-time monitoring? Most solutions look at pulse oximetry, respiratory rate, heart rate, temperature, and blood pressure. Other parameters include perfusion index, pleth variability index, noninvasive hemoglobin, EEG monitoring, transcutaneous oxygen and CO_2, near-infrared spectroscopy, and glucose. Each has its optimal sampling interval. Where sampling every five minutes may be satisfactory for real-time glucose monitoring, it is unsatisfactory for heart rate monitoring (Gravenstein et al., 1989; Dantas et al., 2010; De Felice et al., 2006; Ichimaru and Kuwaki, 1998; Lacher, 1989; Menssen et al., 2009; Schmidt et al., 2014). Pulse oximetry may be monitored beat to beat or using different averaging intervals. These generated values are then sampled by the monitoring system before being transmitted. A beat-to-beat pulse oximeter solution sampled at one cycle per second may still only transmit a third of the available data if the patient is a tiny baby with a heart rate of 180 beats per minute.

In contrast, the EHR that samples every 60 minutes represents only 0.01% of the data (i.e., 1/(60 minutes180 beats/ minute)). Different manufacturers employ different sampling intervals, which are not necessarily user definable. The sampling interval is uncertain with these solutions. With the addition of error management technologies that may not be validated in vivo, it becomes even more challenging to understand what data have been lost.

It is commonplace and more "palatable" clinically to look at data that have been postprocessed, providing that data have been processed in a way that clinical information is not lost. Data are not presented in their raw form. Oxygen saturation is computed from a ratio of ratios. The pulsating transmission of red light, as reported by the transducer measuring the light transmitted through a finger or toe, is divided by the same portion of the waveform that is not pulsating. This value is then divided by the pulsating transmission of infrared light divided by the same portion of its waveform that is not pulsating. The result is then placed into an industry-standard lookup table (Aoyagi and Miyasaka, 2002). Additional "engines" process the data through simple bandpass filtration as well as adaptive filtration methods that employ recognition of signature patterns in the signal. Additional proprietary filtration is often necessary to arbitrate values that otherwise report spurious nonsensical data. First in, first-out processing (commonly referred to as FIFO) allows a data segment to be treated as a single value. In situations where the data varies enough to be distracting but not enough to warrant clinical action, it is helpful to refer to a composite value to establish patient stability. However, composite valuation can make significant events "disappear" through averaging if the value varies wildly. In small patients with high heart rates, small blood values, and occasional apneic (breathing pauses) events, a 32-second averaging time can smooth out a desaturation episode that would otherwise provoke clinical concern (Bhatia et al., 2010; Aoyagi and Miyasaka, 2002; Bolanos et al., 2006). As many manufacturers produce pulse oximeters of different capabilities, knowing the particulars of these devices is vital in evaluating the sampling interval and the veracity of the data received. Technologies from the early 1990s that are still in use may report values that are incorrect, freeze, or report the last "good" value recorded. More recent technologies may reject these issues and appropriately not report data when there is none (i.e., when the patient does not have a pulse)—having both newer and older technologies in a unit or evaluating different locations on the same patient may produce results that are unreliable at best and potentially life-threatening.

Pulse oximetry data measurements, as an example

Introduction

Pulse oximetry has been broadly used in the neonatal intensive care unit. Its appropriate use has been credited with reducing chronic lung disease, improvement in retinopathy of prematurity, and decreased mortality (De Felice et al., 2006; Butterwegge, 1997; Chow et al., 2003; Luttkus et al., 1995). Consistency in range and range selection has improved outcomes. Despite landmark improvements, pulse oximetry averaging can represent different vignettes. Although much attention has been focused on defining the appropriate saturation or saturation range, additional central-based saturation parameters are not adequately described by using the mean alone (Gubbi and Amrutur, 2015; Zapata et al., 2014; Schmidt et al., 2014). Using common averaging strategies to improve alarm management may cause a clinical paradigm where no saturation values are actually in range, despite high compliance to an average specified range. The use of the median vs. mean saturation may have little bearing on this phenomenon. A single central saturation will suffice as a median for compliance to a range with literally no other values within that range.

Objective

Is there a better way of qualifying saturation values that describe saturation's centricity in a way not imparted by measuring average saturation alone?

Methods

To define variable saturation contribution of different sampled saturation values, we modeled a pulse oximeter set for 16 second averaging and a tachycardic neonate with a pulse rate of 180 beats per minute. Forty-eight points were averaged. Sixteen scenarios were modeled with varying interval saturations that would produce a 90% saturation when averaged to illustrate diversity in saturation values. In addition to average saturation, standard deviation, kurtosis, and skew were measured (Dell, Inc., 2016, STATISTICA data analysis software system v13, http://www.statsoft.com).

Results

Negative kurtosis occurred with less variability around the mean. A negative kurtosis may correlate with a more "pointed" center fit curve, as shown, but with decreased outliers (Tables 23.1, 23.2, and Fig. 23.1). Increased skewness indicated a large difference between mean and individual values (Tables 23.1, 23.2, and Fig. 23.1). A relatively low standard deviation associated with a mean within the desired range and kurtosis that approached -1.53 (see Table 23.2) was predictive of a saturation consistently within range.

TABLE 23.1 Example excel sample spreadsheet of raw pulse oximetry data.

	A	B	C	D	E	F	G	H	I	J	K	L	M	N	O	P
1	80	70	90	60	40	90	0	89	88	87	86	85	84	83	82	10
2	80	70	90	60	40	90	0	90	90	90	90	90	90	90	90	100
3	80	70	80	60	40	90	0	91	92	93	94	95	96	97	98	100
4	80	70	80	60	40	90	0	89	88	87	86	85	84	83	82	100
5	80	70	80	60	40	90	98.18	90	90	90	90	90	90	90	90	100
6	80	70	80	60	40	90	98.18	91	92	93	94	95	96	97	98	100
7	80	70	80	100	100	90	98.18	89	88	87	86	85	84	83	82	100
8	80	70	80	100	100	90	98.18	90	90	90	90	90	90	90	90	100
9	80	70	80	100	100	90	98.18	91	92	93	94	95	96	97	98	90
10	80	70	80	100	100	90	98.18	89	88	87	86	85	84	83	82	10
11	80	70	80	100	100	90	98.18	90	90	90	90	90	90	90	90	100
12	80	70	80	100	100	90	98.18	91	92	93	94	95	96	97	98	100
13	80	80	80	100	100	90	98.18	89	88	87	86	85	84	83	82	100
14	80	80	80	100	100	90	98.18	90	90	90	90	90	90	90	90	100
15	80	80	80	100	100	90	98.18	91	92	93	94	95	96	97	98	100
16	80	80	80	100	100	90	98.18	89	88	87	86	85	84	83	82	100
17	80	80	80	100	100	90	98.18	90	90	90	90	90	90	90	90	100
18	80	80	80	100	100	90	98.18	91	92	93	94	95	96	97	98	100
19	80	80	80	100	100	90	98.18	89	88	87	86	85	84	83	82	100
20	80	80	80	100	100	90	98.18	90	90	90	90	90	90	90	90	90
21	80	80	80	100	100	90	98.18	91	92	93	94	95	96	97	98	10
22	80	80	80	100	100	90	98.18	89	88	87	86	85	84	83	82	100
23	80	80	80	100	100	90	98.18	90	90	90	90	90	90	90	90	100
24	80	80	80	100	100	90	98.18	91	92	93	94	95	96	97	98	100
25	100	90	80	90	95	90	98.18	89	88	87	86	85	84	83	82	100
26	100	90	80	90	95	90	98.18	90	90	90	90	90	90	90	90	100
27	100	90	100	90	95	90	98.18	91	92	93	94	95	96	97	98	100
28	100	90	100	90	95	90	98.18	89	88	87	86	85	84	83	82	100

29	100	90	90	90	90	90	90	90	90	98.18	90	95	90	100	90	100
30	90	98	97	96	95	94	93	92	91	98.18	90	95	90	100	90	100
31	10	82	83	84	85	86	87	88	89	98.18	90	95	90	100	90	100
32	100	90	90	90	90	90	90	90	90	98.18	90	95	90	100	90	100
33	100	98	97	96	95	94	93	92	91	98.18	90	95	90	100	90	100
34	100	82	83	84	85	86	87	88	89	98.18	90	95	90	100	90	100
35	100	90	90	90	90	90	90	90	90	98.18	90	95	90	100	90	100
36	100	98	97	96	95	94	93	92	91	98.18	90	95	90	100	90	100
37	100	82	83	84	85	86	87	88	89	98.18	90	95	90	100	100	100
38	100	90	90	90	90	90	90	90	90	98.18	90	95	90	100	100	100
39	100	98	97	96	95	94	93	92	91	98.18	90	95	90	100	100	100
40	90	82	83	84	85	86	87	88	90	98.18	90	95	90	100	100	100
41	90	90	90	90	90	90	90	90	91	98.18	90	95	90	100	100	100
42	90	98	97	96	95	94	93	92	89	98.18	90	95	90	100	100	100
43	90	82	83	84	85	86	87	88	90	98.18	90	95	90	100	100	100
44	90	90	90	90	90	90	90	90	90	98.18	90	95	90	100	100	100
45	90	98	97	96	95	94	93	92	91	98.18	90	95	90	100	100	100
46	90	82	83	84	85	86	87	88	89	98.18	90	95	90	100	100	100
47	90	90	90	90	90	90	90	90	90	98.18	90	95	90	100	100	100
48	90	98	97	96	95	94	93	92	91	98.18	90	95	90	100	100	100

TABLE 23.2 Descriptive statistics of pulse oximetry data showing minimum and maximum values, as well as skewness and kurtosis, which may be better indicators to use in making diagnoses and treatment plans for an individual patient.

Variable	Descriptive statistics (Sheet1 in APoxOnU)									
	Valid N	Mean	Median	Minimum	Maximum	Range	Variance	Std. Dev.	Skewness	Kurtosis
A	48	90.00000	90.0000	80.00000	100.0000	20.00000	102.1277	10.10582	0.00000	−2.08889
B	48	85.00000	85.0000	70.00000	100.0000	30.00000	127.6596	11.29865	−0.00000	−1.37685
C	48	89.58333	85.0000	80.00000	100.0000	20.00000	97.6950	9.88408	0.08606	−2.03262
D	48	90.00000	90.0000	60.00000	100.0000	40.00000	153.1915	12.37705	−1.68615	1.99050
E	48	90.00000	95.0000	40.00000	100.0000	60.00000	370.2128	19.24091	−2.27214	3.44746
F	48	90.00000	90.0000	90.00000	90.0000	0.00000	0.0000	0.00000		
G	48	90.00000	98.1800	0.00000	98.1800	98.18000	752.0031	27.42267	−3.11326	8.02530
H	48	90.00000	90.0000	89.00000	91.0000	2.00000	0.6809	0.82514	−0.00000	−1.53261
I	48	90.00000	90.0000	88.00000	92.0000	4.00000	2.7234	1.65207	−0.00000	−1.53261
J	48	90.00000	90.0000	87.00000	93.0000	6.00000	6.1277	2.47541	0.00000	−1.5321
K	48	90.00000	90.0000	86.00000	94.0000	8.00000	10.8936	3.30055	−0.00000	−1.5321
L	48	90.00000	90.0000	85.00000	95.0000	10.00000	17.0213	4.12568	0.00000	−1.5321
M	48	90.00000	90.0000	84.00000	96.0000	12.00000	24.5106	4.95082	0.00000	−1.5321
N	48	90.00000	90.0000	83.00000	97.0000	14.00000	33.3617	5.77596	−0.00000	−1.5321
O	48	90.00000	90.0000	82.00000	98.0000	16.00000	43.5745	6.60110	−0.00000	−1.5321
P	48	90.00000	100.0000	10.00000	100.0000	90.00000	612.7660	24.75411	−2.95076	7.36787

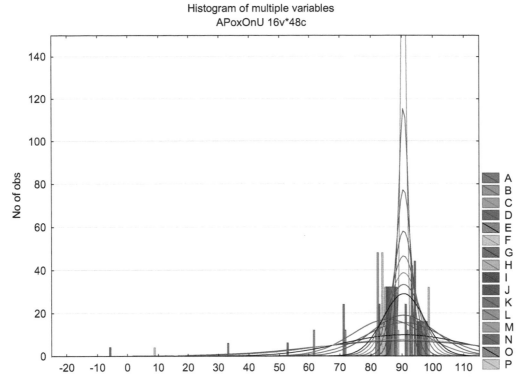

FIGURE 23.1 Histograms of pulse oximetry data on multiple patients showing variations in means, skewness, and kurtosis.

Discussion

Pulse oximetry is based on the analysis of a pulsed waveform that is usually obtained from the natural pulsation of blood in the arteries of the distal extremities. As it is a "pulsed" technology, the heart rate is an essential determinant of the sampling rate. Pulse oximeters may sample three times as much pulsatile data in the neonate as in the adult, where the average heart rate is in the 60s. There is intrinsic variation in the neonate as well but in a higher and narrower range. Although a sampling interval may be a convenient way of looking at oximetry and minimizing the effect of small transient desaturations, the "n" of sampled pulses varies from patient to patient as well as from minute to minute in the same patient (Johnston et al., 2011; King et al., 2019; Pena-Juarez et al., 2019; Sahni et al., 2003).

Saturation ranges are similarly confusing as they may be based on a 4-, 8-, or 16-second average which uses a FIFO buffer to measure saturation at any point during time; or an analysis of the beat-to-beat variation recorded to a data recorder and analyzed post hoc (King et al., 2019; Sahni et al., 2003; Hay et al., 2002; Nizami et al., 2015; Phattraprayoon et al., 2012; Rosychuk et al., 2012). Because the FIFO buffer will prevent the extreme high values and extreme low values from being counted, "clinical range" sampling, as obtained from the EHR, reflects less deviation and a tighter range than the beat-to-beat analysis would otherwise indicate.

Various alarm management techniques further complicate analysis by building in an alarm delay when saturations fall outside the range for a defined length of time that the clinician has determined to be in keeping with the clinical goals of monitoring. SatSeconds (Nellcor) will only generate an alarm when a certain period of time in seconds times the depth of the desaturation exceeds a certain duration/depth threshold throughout a 60-second epoch. Range considerations (Hay et al., 2002) are difficult to implement when SatSeconds impede normal notification parameters (Ahlborn et al., 2000; Grieve et al., 1997).

Finally, using the median saturation can be misleading as pulse oximetry data may be significantly skewed across both extremes, and the median saturation may only occur in a single instance. Range implies that saturation values fall within a defined interval, but because of FIFO buffering and the potential for rapid variation in neonatal saturation, a neonate whose saturations are right in the middle of the desired set point clinically may not have had a single saturation value fall within the range (Tables 23.1 and 23.2).

Conclusion on Pulse Oximetry Example

Although specifying a saturation range may be a convenient way of specifying compliance to a metric, there is no guarantee that the patient spends any time in the specified saturation range when an averaging interval is selected. Using the sampled averaged value, the standard deviation and kurtosis in concert may better indicate the central tendency within the range (Fig. 23.1).

Eye-intraocular pressure measurements: a personal example by one of the authors to illustrate the problem of when and how data is collected

This book's coauthor, Gary Miner, has been having treatment for glaucoma, particularly in one eye, the left, for the past 10 years. The prime variable that the eye doctors need at each visit is the IOP (intraocular pressure) of each eye. The Goldman Apparatus machine is used to get this IOP, and the clinical eye profession considers this "Goldman" reading the Gold Standard.

However, at the first Dx of high pressure (higher than 21 in any eye.) most **ophthalmologists (MD or DO with graduate work in medical care of the eye)** initially only want the patient to visit for an IOP check every 6 months, or every 3 months at the most. This goes on until, at an appointment, the IOP level is "way out of bounds," and then they may check every 1 month for a while "to be sure" before making a definitive diagnosis of the progression of the glaucoma, and what should be the next treatment plan.

Gary had already lost about one third of the "visual field—3D high definition" in his left eye before this glaucoma (high L-eye pressure, for example, consistently greater than 21; normal levels should be between 15−20) was discovered, at which time he was referred to an ophthalmologist in the Midwest (where the patient was living). Various eye drops used to lower IOP were used for several years, keeping the pressure below 20, but eventually, most of these stopped working, and the IOP of L-eye started "bouncing around" between high and lower pressures (like between 20, to 30, even up to 40 at one point). Thus it was time for surgical intervention.

After minimally invasive surgery (Khaimi, 2015) in 2015, which kept the IOP at a 14−17 level (our goal was 13−15) for about 3 years, in 2019 the L-eye IOP popped up to 28 or 31 a couple of times. The eye doctor in the Midwest had followed Gary and his glaucoma since about 2010. But Gary was spending 40% of his time living in the eastern United States, so he had to work between ophthalmologists. In the fall of 2019, the Tulsa doctor, sensing that the L-eye was acting up again, discussed three different invasive surgical interventions and wanted to have the L-eye checked at least once a month if not more frequently on the East Coast, so we could get a "good reading" on what "really was going on," and schedule surgery "as soon as possible" if warranted. But my East Coast doctor refused, saying every 3−4 months was good enough (not sure why; his large medical clinic decision-makers may have put a mandate in place for such, as this clinic was not even letting their technicians take the IOP prior to the doctor taking it; such practice to have both the technician take it initially and then the Dr. to take it minutes later as a confirmation is the standard in the industry, and is still practiced in most clinics across the United States).

So, during the year 2020 (also the "COVID year") Gary was stuck in the eastern United States most of the time. The readings every 3−4 months were generally a bit high, but still under 20, but then in the September 3-month check, the L-eye IOP was about 31. So then finally, this doctor wanted to check it every 3−4 weeks. By January 2021, we decided that it was "too high/too often," and we needed a local glaucoma specialist to assist.

BUT by this time, the visual field test clearly showed that Gary had lost about another one third or more of his visual field. Thus the L-eye was "focusing on a haze," although the peripheral vision was still pretty good. (One arrogant old-school ophthalmologist seen as a third opinion even went so far as to exclaim, "You only have about 100,000 rods left of the millions you had … you need immediate trebuculectomy—the only thing we do here—and my young colleague is the one to do it." It turned out this was the only type of glaucoma invasive surgery this colleague did. What does this say about "good Dx and good treatment"?)

By getting a fourth and finally a fifth opinion, Gary finally found an excellent young eye surgeon in April of 2021. Thus further surgery in 2021 (invasive this time; Ahmed valve shunt implantation to pull fluid out of the eye to reduce IOP) brought down the IOP, but this type of surgery requires several months to 6 months or more to get the IOP stabilized. At this point, the doctors wanted the IOP checked every week for the first few weeks following surgery. (Interesting, why was this weekly IOP check not needed in the months and years before).

But what is the "big point" here from the above story? Gary had lost about 2/3 visual acuity in the L-eye over a several-year period because there were not sufficient sampling data to indicate early on what was needed to keep the

IOP level in the 15−18 range, and thus prevent further death of vision cells. Why was the profession not sampling IOP adequately to tell the full story of each patient?

Solution—and this is most important for this story/this chapter: Gary wanted to take his own IOP readings and had been working with his Midwest doctor over the previous 2 years to get some sort of apparatus that would do this. Several were offered onto the market, but some got taken off-market, and others were extremely expensive. Realistically, the only one available was the expensive one—about $3000 if purchased directly from the company (but lots of "red tape" to enable an individual patient to do this), but one third off at $2000 if purchased through a doctor. My Tulsa doctor made an arrangement with the i-CARE HOME company to do this for me, me being his "first such patient" and an experiment for my Tulsa doctor to see if this would be a good thing for other patients of his. He was so concerned that he set an "alarm" that would send him an immediate email every time my IOP was higher than 20.

To make a long story short: Gary has i-CARE HOME readings for the past 6 months. The readings taken by Gary on the days that he had ophthalmologists' visits were almost always identical or at the most 1 point off either higher or lower in several cases (this included all three of Gary's doctors' readings—East Coast surgeon; East Coast follow-up ophthalmologists, and Midwest ophthalmologists).

Yet, one of the East Coast doctors insisted that the Goldman Apparatus he used was the "gold standard" and that the i-CARE HOME that I was using was way off in its readings, clearly showing his ignorance (e.g., his view, and the view of many eye doctors in the glaucoma field, that "Goldman is the standard" and only what it reads in the clinic setting "is correct" is in reality garbage, a "falsehood"):

Example of comparison of Goldman with i-CARE HOME intraocular pressure readings

On December 16 follow-up visit with one of the East Coasts doctors, this doctor got L-eye 18, but R-eye 22 (he had always gotten about 15 or 16 for both eyes; except in the few weeks following surgery when the patient was on high-level/high-dose eye steroid drops, which elevates IOP. This doctor insisted that I take another IOP-lowering eye drop, wanting Vyzulta, which I am allergic to (makes the eye inflamed and only worsens things, as my East Coast surgeon doctor found out months previously when she had to take me off it after a few days; and I knew even a year before that, having tried this eye drop with previous doctors), OR he said I'd need surgery on the R-eye. (***Surgery on the R-eye for just ONE SAMPLING of 22? When even 22 is not that bad?***). But here's the real story: this R-eye 22 was a "fluke" reading; Gary had readings over the previous 6 months with an average of R-eye 18 with "very tight error bars" (e.g., Confidence Limits); and if the data was deleted for those periods where a high-level/high-dose steroid (like Duroal every 2 hours), the R-eye was close to an average of 15−16 with a range between 13 and 20 (13 being the minimum value and 20 being the high value). **"Poor decisioning process leading to BAD DIAGNOSIS !"**

Steroid drops were needed for the L-eye because of invasive surgery and the need to keep a fluid outflow port open and not healed over; during part of the past 6 months Gary was on very high-level steroids for L-eye, but it "seeps over to the R-eye" and ups the IOP in that eye also.

Worthy of note is that steroid use in one eye appears to "leak over" and affect the IOP pressure in the other eye; see Fig. 23.2 below which illustrates this.

The above December 16 incident is ***only one of several*** among the doctors over the past 8 months that showed how "off" these doctors were in their Dx and treatment decisions. Luckily my Midwest doctor really knows better than any of them what is going on, listens to the patient, and he has been my "Referee." The surgeon doctor is young; the young glaucoma surgeon doctors are required to learn all the types of glaucoma surgeries, and there are many these days, and apparently, because of that, they do not get sufficient training in the use of eye drop medications. The glaucoma surgeon I had was top-level—she did all the glaucoma surgeries and did them with excellence. (My East Coast follow-up doctor, who lives closer to me and is easier to get to, had done the "old-fashioned" Glaucoma surgery for 20 years but recently got out of it, stating that he feels best to leave this type of surgery to the younger generation, as they are "not afraid to do anything" and are having success.)

In conclusion

In the glaucoma eye case, we need "averages and minimum and maximum values of readings taken at different times of day over a specified length of time" to use in making Dx and treatment decisions and we should probably also look at the kurtosis in this data in making diagnostic and treatment decisions. But to get enough data to have these "sampling

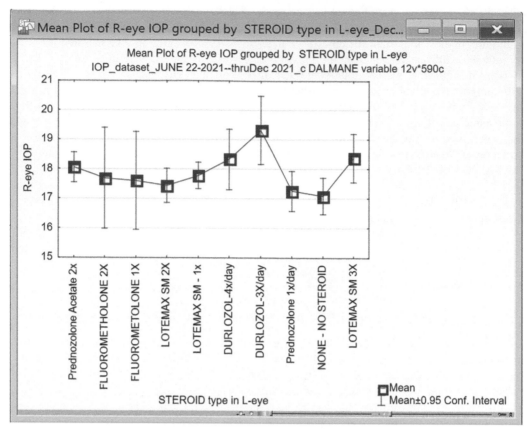

FIGURE 23.2 (same as Graph 11.2 in Chapter 11). Right eye IOP from June 22, 2021, through December 5, 2021: showing it was affected by the L-eye steroid drop, which apparently "leaked" over to the right eye. R-IOP however was consistently below 20, with means (average) of 17.5 with LOTEMAX SM either 1x/day or 2x/day, each having a "tight" error bar, thus showing that the variance was small, compared to high variances (wide error bars) in the Durozol and Fluorometholone steroids (please see Chapter 11 for additional tables and graphs of IOP eye data, if interested).

averages and kurtosis stats", data needs to be obtained frequently, even multiple times a day in some cases. This data is not going to be taken by visits to the eye clinic, it seems; and even if taken by the eye clinic, it would require many trips by the patient, maybe daily for a few weeks. So, in comes "digital health"—which is invading our society today. We need home devices, at reasonable cost, like the $79 "smartphone looking" device that reads heart EKG, and other similar home devices.

The i-CARE HOME is about the only "usable" IOP eye device available today. The eye patient has no alternative. But most eye doctors do not mention it to their patients—the patient has to be proactive and the CEO of their care, and bring it up, and insist upon this device. However, this device should not be costing the patient $2000−3000, forcing most patients today to have to go to the doctor just to get IOP readings (*What does this say about current "health care delivery"?*). CMS provides a "code" for having a doctor's technician take the IOP on a "walk-in basis," which my East Coast primary care eye doctor refused to comply with; yet my Midwest doctor had suggested it to me, prior to my getting the i-CARE HOME.

This glaucoma personal example illustrates minimally two things:

1. the need to have averaged IOP readings over time, and maybe also kurtosis stats, to make accurate Dx and treatment decisions; and
2. the need to have personalized individual healthcare, with home digital devices as "par for the norm" of healthcare.

Types of data analysis that may be helpful in solving the types of issues presented in this chapter

The following provides ideas on types of data analysis that may be useful for working with healthcare variables to get accurate information to make good decisions.

The methods below replace the "old-fashioned" way of getting just individual point values, for example, where "random—at-doctor's office/Emergency Room" sampling may not give the "real story" (i.e., the "real value") of what is going on with a patient.

In the end, one needs a "decisioning system" that clearly says: "Do this, because of that pattern which has a 99% probability to lead to that undesirable outcome." At the most bureaucratic end, such a system might need to be practically automated, and MUST be validated, and acceptable to medical practitioners. But from the patient-centered and "patient as CEO of their own healthcare" viewpoint, many of the data variables could be measured with home devices by the patient—person. So we need to look at both sides of this.

Statisticians experienced with the problems of sampling rates, reliability, variability, nonnormality, and decisioning systems suggest some of the following ways of looking at the data:

Reliability of inputs determines the validity of models

First, statistically speaking, the validity of any prediction model is bounded by the reliability of the measured inputs. In other words, if the average of a variable is 100% accurate for predicting some outcome, then if there is a lot of error variability in the data (variability NOT related to the outcome), then a single measurement will lead to very bad predictions. That is because the single measurement is not a reliable indicator of the true average.

The answer then simply is to take **many measurements** and *average* them.

However, it gets more complicated

What if the *variability* of the respective measurement is what truly predicts the outcome? This is the case in practically all manufacturing with respect to quality. It was Genichi Taguchi, the Japanese Statistician, who revolutionized manufacturing with Robust Design Methods, and who demonstrated that quality is not a function of how close you come to the specifications (the desired "mean") but how much variability you accept/observe around the spec.

The cost of bad quality is an exponential function of variability around the spec.

Translated to these health problems: It makes sense that *it is the variability* in "something" (blood pressure, pulse oximetry saturation, eye IOP, etc.) that is related to health outcomes and overall health.

Put another way, if your blood pressure is way too high today, and way too low tomorrow then on average it is great, but you are in bad health (I assume). And this is clearly true for eye IOP—if the IOP if bouncing around, high today, low tomorrow, this bouncing consistently will be killing retinal cells, thus bringing about further loss of vision.

You, the reader, if more interested in this, may also want to look into Taguchi's signal-to-noise ratios that he proposed and are still widely used to characterize variability-around-spec/means; see for example here.

But then, it gets even more complicated

In automated process manufacturing, like semiconductor manufacturing, it is the *patterns* and their *interactions* that matter, that is, these two things, and not specific data values themselves, are strongly related to process yield and quality.

In other words, one can have hundreds (or many more) sensors that measure deposition rates and conditions in chambers as layers of circuitry are created on computer chip wafers. This leads to multiple time series with distinct highly nonlinear patterns, and the nature of those patterns is significantly related to root causes of quality and yield.

If you, the reader, are interested in this research and methods, look up "SAX encoding" (see here for a comprehensive list of papers)—it is a technology and set of algorithms to detect, search for, and process time series patterns at scale.

The **bottom line** is that we would definitely advocate looking at the *moments-of-measurements* (variance, skewness, kurtosis, and maybe signal-to-noise ratios) and identify if it is the variability and shape of measurement distributions that predicts important outcomes. We suspect it will be!

We also think that the pattern detection methods (e.g., like SAX, and lately also LSTM deep learning models which are now widely used in automated batch/process manufacturing) will likely be very useful to lead to new insights and innovations in healthcare.

After all, life (and health) is a dynamic and constantly evolving systems, and not static "rocks" that can be characterized by simple means.

Clinical Dx and treatment needed changes for true patient-centered care

The "mode of operation" of the clinical practitioner (MD, OD., nurses, technicians, etc.) will have to change to a frame of reference where the "patient is at the center" of the diagnostic and treatment process, meaning among other things that (1) the patient has to be fully listened to and their complaints and interpretations of how drugs and other treatments affect them, as a "genetically unique person," and (2) the medical establishment must comprehend and act upon the fact that "people are different"—different because of their unique genetic makeup and also their unique environmental inputs, including things like exercise and nutrition among many other variables.

In Chapter 22 of this book, we described some of the predictions that various experts were proclaiming in their blogs during December of 2021, as the new year 2022 approached; among those was the set of six trends that Bill Taranto (December 20, 2021) proclaimed in late December 2021. Three of these predictions speak directly to what is beginning to happen and what has to happen in the years ahead to get the type of diagnostic and treatment changes needed to fully enable a patient-centered care modality. These include:

1. **Real-world data will be essential to deliver real-world results**: "The number of mobile devices, wearables and sensors that collect health-related data is growing rapidly. This real-world data enables us to answer questions previously thought unknowable, and we can turn that data into insights that lead to better clinical care and outcomes. The companies that amalgamate, aggregate and integrate health data will continue to prosper, the ones to drive value will be the firms with fully annotated, longitudinal multimodal date. The coming year will see a line of difference between the companies providing access to data and those that provide insights" (Taranto, 2021).
2. **Telemedicine will change how chronic conditions are treated**. "The pandemic showed how telemedicine could change how we think about care interactions, with virtual visits increasing almost 40 times, according to data from McKinsey. Most of these interactions were centered around acute care. But for telemedicine to achieve its full potential, it will need to *engage patients more frequently*, especially for certain chronic conditions. The companies that succeed will be the ones that change the way patients interact with the healthcare system by *building their entire operation around the patient experience*" (Taranto, 2021).
3. **Healthcare will get truly patient-centric**. "The consumerization of healthcare will accelerate next year. *We'll see the market recognize the importance of patients and put them at the center of care*. The companies that succeed will be the ones that change the way patients interact with the healthcare system by building their entire operation around the patient experience and ensuring patients get the best care at the best price. This includes offering value-add services like online self-scheduling tools that make it easier for patients to book appointments, as well as *interactive solutions that guide patients to the right appointment with the right provider based on their needs*. It also includes giving people greater control over their healthcare decisions and allowing them to share in the financial benefits of those decisions. Next year will also reward companies that are building out *personalized medicine engines to get patients quickly to the right treatment*. We can expect to see personalized medicine startups creating *effective drug therapies based on users' individual specifications*, which should be safer, cheaper, and more effective because they're *fine-tuned to each person's unique genetic makeup*" (Taranto, 2021).

In summary, healthcare is being forced to reinvent itself in order to bring better healthcare to patients. The use of "accurate predictor variables," such as the variables that make proper use of "sampling needs," as we described above in this chapter, will be one of these things that healthcare has to "reinvent," to make "real Gold Standards" based on objective data.

Postscript

This chapter suggests that the combination of three measurements be constituted as the "Gold Standard" of medical data operations:

1. *Mean value* (a measure of central tendency of the data set);
2. *Standard deviation* (a measure of the distribution of the data set); and
3. *Kurtosis* (a measure of the "clumpedness" of the data set).

Provision of this Gold Standard of measurement is necessary to characterize sufficiently the dynamic system which generated them. Only then can personalized medical diagnosis and treatment be properly designed, which is truly representative of the complexity of human physiological function related to the measurements, to guide personalized medical diagnosis and treatment.

References

Ahlborn, V., Bohnhorst, B., Peter, C.S., Poets, C.F., 2000. False alarms in very low birthweight infants: comparison between three intensive care monitoring systems. Acta Paediatr. 89 (5), 571–576. Available from: https://doi.org/10.1080/080352500750027880. Epub 2000/06/14 PubMed PMID: 10852195.

Aoyagi, T., Miyasaka, K., 2002. Pulse oximetry: its invention, contribution to medicine, and future tasks. Anesth. Analg. 94 (1 Suppl.), S1–S3. Epub 2002/03/20 PubMed PMID: 11900029.

Bhatia, V., Rarick, K.R., Stauss, H.M., 2010. Effect of the data sampling rate on accuracy of indices for heart rate and blood pressure variability and baroreflex function in resting rats and mice. Physiol. Meas. 31 (9), 1185–1201. Available from: https://doi.org/10.1088/0967-3334/31/9/009. Epub 2010/07/29 PubMed PMID: 20664161.

Bolanos, M., Nazeran, H., Haltiwanger, E., 2006. Comparison of heart rate variability signal features derived from electrocardiography and photoplethysmography in healthy individuals. Conf. Proc. IEEE Eng. Med. Biol. Soc. 2006, 4289–4294. Available from: https://doi.org/10.1109/IEMBS.2006.260607. Epub 2007/10/20 PubMed PMID: 17946618.

Butterwegge, M., 1997. Fetal pulse oximetry and non-reassuring heart rate. Eur. J. Obstet. Gynecol. Reprod. Biol. 72 (Suppl. S63–6). Available from: https://doi.org/10.1016/s0301-2115(97)02720-6. Epub 1997/03/01 PubMed PMID: 9134415.

Chow, L.C., Wright, K.W., Sola, A., Group, C.O.A.S., 2003. Can changes in clinical practice decrease the incidence of severe retinopathy of prematurity in very low birth weight infants? Pediatrics 111 (2), 339–345. Available from: https://doi.org/10.1542/peds.111.2.339. Epub 2003/02/04 PubMed PMID: 12563061.

Dantas, E.M., Goncalves, C.P., Silva, A.B., Rodrigues, S.L., Ramos, M.S., Andreao, R.V., et al., 2010. Reproducibility of heart rate variability parameters measured in healthy subjects at rest and after a postural change maneuver. Braz. J. Med. Biol. Res. 43 (10), 982–988. Available from: https://doi.org/10.1590/s0100-879x2010007500101. Epub 2010/10/15 PubMed PMID: 20945039.

De Felice, C., Goldstein, M.R., Parrini, S., Verrotti, A., Criscuolo, M., Latini, G., 2006. Early dynamic changes in pulse oximetry signals in preterm newborns with histologic chorioamnionitis. Pediatr. Crit. Care Med. 7 (2), 138–142. Available from: https://doi.org/10.1097/01.PCC.0000201002.50708.62. Epub 2006/02/14 PubMed PMID: 16474255.

Dell, Inc., 2016, STATISTICA data analysis software system v13. Available from: http://www.statsoft.com.

Gravenstein, J.S., de Vries, A., Beneken, J.E., 1989. Sampling intervals for clinical monitoring of variables during anesthesia. J. Clin. Monit. 5 (1), 17–21. Available from: https://doi.org/10.1007/BF01618365. Epub 1989/01/01 PubMed PMID: 2926463.

Grieve, S.H., McIntosh, N., Laing, I.A., 1997. Comparison of two different pulse oximeters in monitoring preterm infants. Crit. Care Med. 25 (12), 2051–2054. Available from: https://doi.org/10.1097/00003246-199712000-00025. Epub 1997/12/24 PubMed PMID: 9403758.

Gubbi, S.V., Amrutur, B., 2015. Adaptive pulse width control and sampling for low power pulse oximetry. IEEE Trans. Biomed. Circuits Syst. 9 (2), 272–283. Available from: https://doi.org/10.1109/TBCAS.2014.2326712. Epub 2014/07/12 PubMed PMID: 25014964.

Hay, Jr.W.W., Rodden, D.J., Collins, S.M., Melara, D.L., Hale, K.A., Fashaw, L.M., 2002. Reliability of conventional and new pulse oximetry in neonatal patients. J. Perinatol. 22 (5), 360–366. Available from: https://doi.org/10.1038/sj.jp.7210740. Epub 2002/06/26 PubMed PMID: 12082469.

Ichimaru, Y., Kuwaki, T., 1998. Development of an analysis system for 24-hour blood pressure and heart rate variability in the rat. Psychiatry Clin. Neurosci. 52 (2), 169–172. Available from: https://doi.org/10.1111/j.1440-1819.1998.tb01011.x. Epub 1998/06/17 PubMed PMID: 9628132.

Johnston, E.D., Boyle, B., Juszczak, E., King, A., Brocklehurst, P., Stenson, B.J., 2011. Oxygen targeting in preterm infants using the Masimo SET radical pulse oximeter. Arch. Dis. Child. Fetal Neonatal (Ed.) 96 (6), F429–F433. Available from: https://doi.org/10.1136/adc.2010.206011. Epub 2011/03/08 PubMed PMID: 21378398.

Khaimi, M.A. (2015). Ab interno canoloplasty: a minimally invasive and maximally effective glaucoma treatment, *Glaucoma Today in Surgical Pearls*, November–December <https://glaucomatoday.com/pdfs/gt1115_surgpearls.pdf>.

King, C., Mvalo, T., Sessions, K., Wilson, I., Walker, I., Zadutsa, B., et al., 2019. Performance of a novel reusable pediatric pulse oximeter probe. Pediatr. Pulmonol. 54 (7), 1052–1059. Available from: https://doi.org/10.1002/ppul.24295. Epub 2019/03/27 PubMed PMID: 30912314.

Lacher, D.A., 1989. Sampling distribution of skewness and kurtosis. Clin. Chem. 35 (2), 330–331. Epub 1989/02/01 PubMed PMID: 2914396.

Luttkus, A., Fengler, T.W., Friedmann, W., Dudenhausen, J.W., 1995. Continuous monitoring of fetal oxygen saturation by pulse oximetry. Obstet. Gynecol. 85 (2), 183–186. Available from: https://doi.org/10.1016/0029-7844(94)00353-F. Epub 1995/02/01 PubMed PMID: 7529914.

Menssen, J., Colier, W., Hopman, J., Liem, D., de Korte, C., 2009. A method to calculate arterial and venous saturation from near infrared spectroscopy (NIRS). Adv. Exp. Med. Biol. 645, 135–140. Available from: https://doi.org/10.1007/978-0-387-85998-9_21. Epub 2009/02/21 PubMed PMID: 19227462.

Nizami, S., Greenwood, K., Barrowman, N., Harrold, J., 2015. Performance evaluation of new-generation pulse oximeters in the NICU: observational study. Cardiovasc. Eng. Technol. 6 (3), 383–391. Available from: https://doi.org/10.1007/s13239-015-0229-7. Epub 2015/11/19 PubMed PMID: 26577369.

Park, Y.H., Lim, S., Kang, H., Shin, H.Y., Baek, C.W., Woo, Y.C., 2018. Comparison of the accuracy of noninvasive hemoglobin monitoring for preoperative evaluation between adult and pediatric patients: a retrospective study. J. Clin. Monit. Comput. 32 (5), 863–869. Available from: https://doi.org/10.1007/s10877-017-0098-8. Epub 2018/01/06 PubMed PMID: 29302896.

Pena-Juarez, R.A., Chavez-Saenz, J.A., Garcia-Canales, A., Medina-Andrade, M.A., Martinez-Gonzalez, M.T., Gutierrez-Cobian, L., et al., 2019. Comparison of oximeters for the detection of critical congenital heart diseases. Arch. Cardiol. Mex. 89 (1), 159–166. Available from: https://doi.org/10.24875/ACME.M19000039. Epub 2019/11/09 PubMed PMID: 31702739.

Phattraprayoon, N., Sardesai, S., Durand, M., Ramanathan, R., 2012. Accuracy of pulse oximeter readings from probe placement on newborn wrist and ankle. J. Perinatol. 32 (4), 276–280. Available from: https://doi.org/10.1038/jp.2011.90. Epub 2011/07/09 PubMed PMID: 21738120.

Rosychuk, R.J., Hudson-Mason, A., Eklund, D., Lacaze-Masmonteil, T., 2012. Discrepancies between arterial oxygen saturation and functional oxygen saturation measured with pulse oximetry in very preterm infants. Neonatology 101 (1), 14–19. Available from: https://doi.org/10.1159/000326797. Epub 2011/07/28 PubMed PMID: 21791935.

Sahni, R., Gupta, A., Ohira-Kist, K., Rosen, T.S., 2003. Motion resistant pulse oximetry in neonates. Arch. Dis. Child. Fetal Neonatal (Ed.) 88 (6), F505–F508. Available from: https://doi.org/10.1136/fn.88.6.f505. Epub 2003/11/07 PubMed PMID: 14602699.

Schmidt, B., Roberts, R.S., Whyte, R.K., Asztalos, E.V., Poets, C., Rabi, Y., et al., 2014. Impact of study oximeter masking algorithm on titration of oxygen therapy in the Canadian oxygen trial. J. Pediatr. 165 (4), 666–671. Available from: https://doi.org/10.1016/j.jpeds.2014.05.028. e2. Epub 2014/06/29 PubMed PMID: 24973289.

Sjostrand, F., Rodhe, P., Berglund, E., Lundstrom, N., Svensen, C., 2013. The use of a noninvasive hemoglobin monitor for volume kinetic analysis in an emergency room setting. Anesth. Analg. 116 (2), 337–342. Available from: https://doi.org/10.1213/ANE.0b013e318277dee3. Epub 2013/01/11 PubMed PMID: 23302975.

Tautan, A.M., Young, A., Wentink, E., Wieringa, F., 2015. Characterization and reduction of motion artifacts in photoplethysmographic signals from a wrist-worn device. Annu. Int. Conf. IEEE Eng. Med. Biol. Soc. 2015, 6146–6149. Available from: https://doi.org/10.1109/EMBC.2015.7319795. Epub 2016/01/07 PubMed PMID: 26737695.

Taranto, B., 2021. The growing power of digital health: 6 trends to watch in 2022 <https://techcrunch.com/2021/12/20/the-growing-power-of-digital-healthcare-6-trends-to-watch-in-2022/>.

Zapata, J., Gomez, J.J., Araque Campo, R., Matiz Rubio, A., Sola, A., 2014. A randomised controlled trial of an automated oxygen delivery algorithm for preterm neonates receiving supplemental oxygen without mechanical ventilation. Acta Paediatr. 103 (9), 928–933. Available from: https://doi.org/10.1111/apa.12684. Epub 2014/05/13 PubMed PMID: 24813808.

Further reading

STEM LSTM=Long Short-Term Memory Long short-term memory (LSTM) is an artificial recurrent neural network (RNN) architecture used in the field of deep learning. Long short-term memory. In Wikipedia: <https://en.wikipedia.org/wiki/Long_short-term_memory>.

SAX ENCODING. Welcome to the SAX (Symbolic Aggregate approXimation); Keogh, E., Lin, J., and Fuhttps. A. 2005. Available from: https://www.cs.ucr.edu/~eamonn/SAX.htm; and: https://www.cs.ucr.edu/~eamonn/discords/, <https://www.cs.ucr.edu/~eamonn/SAX.htm>.

What if the VARIABILITY truly predicts the OUTCOME – See: Taguchi methods. Wikipedia: https://en.wikipedia.org/wiki/Taguchi_methods; and also the reference list in this publication.<https://en.wikipedia.org/wiki/Taguchi_methods>.

Chapter 24

Analytics architectures for the 21st century

Scott Burk

Chapter outline

Prelude

A major axiom in project management is to plan the work, then work the plan. This axiom can and should be expanded to include all elements of the organization system that contribute data for analysis ("the needs") and those elements that use analytical results ("the feeds"). The needs and feeds of any system must be coordinated to provide the right data elements at the right time. This chapter discusses three elements of the analytic architecture that must be coordinated in any medical organization to permit the collection of personalized patient data and produce personalized diagnoses and treatments.

Introduction

Purpose/summary

Artificial Intelligence (AI) and analytics efforts are too often just extensions or isolated additions to medical practice and research. It could be an extension of research efforts, part of a medical degree curriculum or residency program to extend scholarly pursuits. It could be hospital systems wishing to gain operation efficiency and control. It could be physician groups-of-practice trying to mitigate risk or improve financial performance. While isolated efforts may provide some benefit, leaders understand that exceptional results require a paradigm shift in thinking and the infrastructure to support the new technologies and innovation.

In this chapter, we present three pillars that lay the foundation for success in medical research and results in the 21st century. Successful participants will have designed and implemented the following architectures:

Practical Data Analytics for Innovation in Medicine. DOI: https://doi.org/10.1016/B978-0-323-95274-3.00017-8

1. Social and organizational architectures
2. Data architectures
3. Analytics architectures

The reader may find it beneficial to envision a large integrated academic medical center for illustration, regardless of their current position. At an academic medical center, education, research, and clinical care are combined to provide the best possible clinical care that uses cutting-edge technologies, resources, and therapies. They are typically composed of a medical school, a residency program and at least one allied health professional school, and either own or are affiliated with a teaching hospital or healthcare system. It is this type of center that will gain the most from these recommendations, yet any medical enterprise wishing to improve their clinical quality and operational efficiency will gain from this information.

Successful predictive and prescriptive analytics programs consist of these three cornerstones:

1. Social and Organizational Design for Success

Every organization has an organizational structure and an accompanying culture. Funding often determines how organizations are structured. Unfortunately, these often create silos and "pockets" of innovation. There needs to be a base foundation of data literacy and collaboration across the entire organization to reach maximum potential in data driven technologies. This is enabled by opening up the organization for cross-functional collaboration and education. An example of organizational structure that enables this is the creation of an Analytics Center of Excellence. This center supports an innovation-oriented culture.

Data Design for Success No organizational plan, educational prowess, or analytics platform can be successful without meaningful data. Data has become complicated and there is much hype about platforms and technologies. We present some basic data architecture considerations in this chapter and provide references for future review. We survey data movement, storage, and consumption. Determining the data infrastructure that will be necessary to support all enterprise initiatives for AI, ML, and analytics is essential.

Analytics Technology Design for Success Determining the AI platforms, functionality and infrastructure that will be necessary to support all the enterprise initiatives for AI and analytics. Architectural considerations for the data scientist, understanding the differences of academic vs. professional practice, and the analytic model maturity framework are all important aspects to success.

We briefly describe these three pillars and provide references and additional reading suggestions.

Organizational design for success

Some say it starts with data, it doesn't

Analytics success requires more than technology. In fact, in many ways the determination of types of technology to deploy and use are the easy part. Technology is much more predictable than the social operating mechanisms that are often part of organizations.

Organizational alignment

Organizational structure is dynamic. It normally occurs by agglomeration of departments and staff over time that have specific key roles. IT departments did not exist at one time. They were added to support specific needs. Informatics and statistical research departments were added as well, as these needs arose. As we mentioned earlier, the way an organization is structured often follows the funding. Academic medical centers are often funded by a combination of "university funds," "hospital funds," "clinic funds," and third party funding of grants and research dollars. Thus, at times they look like Lego blocks stuck together with no real design in mind—they are joined, but there is no esthetic or useful structure in the end. Each block is doing their own thing.

However, organizations that provide exceptional results in medical research and practice realize that data and analytics must become part of the organizational DNA if they really want to gain extraordinary results. A great analogy is to see what has transpired in digital technology over the years. At one time, computers and information technology were closely held assets by a select few with specialized knowledge to support very specific niche needs and requirements. Leading centers of medical innovation and practice realized very early that this technology could set them apart from other centers and therefore made digital technology part of the entire organization. Moreover, they promoted the IT department to the highest level in the organizational chart, created new roles like the CIO and CTO and did not consider any strategy or tactic without considering IT.

Most importantly, structures and cultures were created based on technology. The way the organizations thought and did business changed. People changed how they accomplished virtually every task. A new "computer" literacy developed and it changed the way people thought and behaved, that is, it changed the culture of the organization.

The same is happening in leading organizations as AI, ML, and analytics become important. These were once niche tools used by a select few that speak a special language. But things are changing. To turn the corner and make the most of these revolutionary technologies, organizations will need to create the organizational structure and literacy to make them part of the everyday practice and the way people work, think, and act, that is, fully integrated as part of the culture.

Framework for trustworthy and ethical AI and analytics

Medicine has always been a leader in the area of ethics, beginning with Hippocrates. It is widespread. First, clinical workers are every present in thinking about ethical treatment of patients. Second, internal review boards have overseen the ethical treatment of test subjects in research.

However, now with new AI and machine learning methods for predictive and prescriptive medicine, it is important that the underlying models are transparent in their nature. For example, some rules-based methods such as recursive partitioning and tree-based methods can yield transparent and traceable results so that a proprietor can see why the algorithm built the tree (and subsequent rules) it did and corroborate the results with documented medical science (Kuhn et al., 2016).

However, there are methods that yield models and recommendations that are not transparent. One can understand the mathematics and intent of the original algorithm, but why the clinical model was produced is not directly traceable by a human and therefore very diffident to validate with medical science. You can only determine the importance of factors in the final model post hoc and by external model diagnostics. Neural networks and deep learning methods fit in this category. They may yield highly accurate predictive models that are actionable. However, the models are "black box" in nature. A proprietor cannot trace and make predictions outside a computer, unlike many tree methods where predictions can be made with pen and paper. They cannot interpret why the model does what it does. They would only know what the key factors were and if they could measure these input factors for a new patient.

Data design for success

Data design is paramount to any successful clinical improvement. Without complete data, our understanding is incomplete. Using the wrong data, improper design of data collection, gaps in actual collection, or poor data quality—any of these issues may lead to nonoptimal understanding and results or potentially critical failures.

The COVID pandemic offers a compendium of examples of the tales of poor data—either incomplete or misleading. An entire book should be written about the differences that timely and meaningful data could have made. For example, the right samples and statistics could have been more extremely worthwhile if that data had been made available to all scientists to make inferences and decisions. Often, the media blamed epidemiologists and their models; however, it was not the epidemiologists or their modeling that was at fault, it was the data or lack thereof.

Why is data so important?

A process is a series of actions or steps taken in order to achieve a particular end goal. These can be biological and chemical processes that occur millions of times each day inside the human body. It could be operational or business processes that impact the success or failure of our institutions. We have government agencies that have processes to provide oversight and administrate. Anything that you desire to understand or improve involves a process. Data are just the artifacts of a process. Therefore, data are key to any consciously based improvement. Data is a requirement in all statistical methods including machine learning.

The potential of data is insight and action

Collecting data is simply a cost and a liability. First, it takes money to input data. In addition, it takes money to create the connectors to move data. This could be creating pathways or pipes into databases or data lakes through extract, transform, and load processes. More common now is the creation of application programming interfaces (APIs) that bring data from the web (services), cloud, software as a service (SaaS), and other data connections.

In addition to the cost of creating and supporting the connections, there is the cost of the pipes themselves, fiber, networks, and wireless devices. There is the cost of storage. These costs are not just buying or renting the infrastructure, but the labor associated with keeping everything running—fail-over redundancy and backup recovery plans. This runs into billions of dollars in the United States. You could say that this an organizational requirement, one that is expected of the enterprise in running the business.

But, not really, as most of this data sits idle. We collect far more than we ever analyze.

More importantly, all of this excess data is a *liability*. Enterprises are responsible for *all* the data that sits in legacy databases, not just protected health information. Any information that can be hacked and used is a liability. Moreover, there is legal liability from (former) employees on email systems and related databases that are never purged, and thus discoverable in lawsuits and other legal proceedings.

According to Forbes (Meehan, 2016) it is estimated that as much as 90% of big data is never analyzed. In fact, the amount of data never analyzed is so large that Gartner has coined a term called "Dark Data." Dark Data describes the information assets organizations collect, process, and store during regular operational activities, but generally fail to use for other purposes like, for example, analytics, operational relationships, and direct monetizing. Most organizations fail to use this data; however, many times it can and should be used. Data is a strategic asset.

We have a problem:

We have all this data being collected at a great expense, but we are not gaining value from it. The objective should be taking data that took money to collect, provision, and store and turning it into actionable insights that drive value by decreasing costs, driving revenue, and improving quality. In addition, the value of data decreases with time, so data sitting in silos decrease in value every day.

Data value is the return on investment (ROI) for sourcing, transforming, securing, and protecting data. Collecting and storing data is a huge cost. Therefore, the only way to get a positive return on data is to offset this cost by applying and operationalizing analytics, predictive analytics, and prescriptive analytics.

Data and analytics literacy are requirements to successful programs

You cannot create value from data and analytics without knowledge, specialized knowledge. Burk and Miner (2020) covered the foundations of AI, analytics, and data science for professionals. They also described the fundamentals of BI, machine learning, data mining, and statistics. It covers data uses and types of data for developing analytics. Analytics literacy and data literacy are cornerstones of a data-driven organization. You need to speak the same language in order to communicate.

Kasey Panetta (2019) stated this clearly: "…[Our Need is to]… Champion data literacy and teach data as a second language to enable data-driven business." Different specialties will have certain specialized knowledge and there will be accompanying specialized vocabularies, but everyone needs to speak "data and analytics" at a basic level.

Panetta also points out that in 2020 half of all organizations will lack data literacy skills that are needed to achieve business value. In addition, Valerie Logan, Senior Director Analyst, Gartner, points out another important fact:

Prevalence of data and analytics capabilities, including artificial intelligence, requires creators and consumers to "speak data" as a common language… Data and analytics leaders must champion workforce data literacy as an enabler of digital business and treat information as a second language.

It is common for various practice groups to speak their own language when it comes to data and analytics. Premier medical organizations will have two things that separate them:

1. education in basic data literacy across the entire organization; and
2. translators to navigate across specialized data dialects—physicians, nurses, statisticians, informaticists, data scientists, finance, BI developers, and IT.

Everyone in the organization needs basic education in the language of data.

Brief considerations in data architecture

Data architecture is a framework of rules, policies, procedures, standards, and systems that govern data in the organization. Today, data and people are the two most valuable and strategic assets in the enterprise. Data architecture links collected data to the business rules that translate it into meaningful contexts. Data architecture specifies what data is accessible, whether it is stored and how it is used. It provides data governance which covers access, security, change management, scale, reliability, and sustainability.

ProcessesDataAnalyticsImprovementsProcessesData

At the heart of data is a process, at the heart of improvement is analytics. Here a short story of improvement.

1. Processes and systems are everywhere
2. Data are generated by a process
3. We collect some of this data
4. We analyze data and create models
5. Analytics and models provide insight
6. Insights inform action
7. Action improves life

Processes, systems, and data

Analytic models are trained and made operational with data via algorithms. For algorithms to generate models, or execute models, we must feed them data. We collect data at a dizzying rate these days. We collect data in a variety of ways—manual entry into smartphones, computers, and notepads. Operational systems that run organizations are generating data. Devices collect most of the data without us even knowing—cameras, sensors, voice, electronic controls, machines, devices, and telemetry. At the heart of this collection are again systems and processes. We do not think about it most of the time, but when we do, we realize life consists of processes. Therefore, we have a great opportunity to improve life by understanding these processes via analytics.

There are millions of things that could be collected as data. However, we only collect a small fraction. The amount is increasing at an exponential rate. Most of the current increase is as images, video, sound, and various forms of nonstructured information. X-rays, CT, CAT, fluoroscopy, MRI, MRA, PET, and ultrasound are among the sources of this data. Additionally, all the data collected 24 hours a day by patient sensors and monitors becomes part of the data warehouse. Also included are data from Smart Watches, continuous glucose monitors, EKGs, and ECGs.

As said by a famous statistician (Deming, 2018) a long time ago, *"In God we trust, all others bring data."* We can do that more than ever today. It is well-known that much of medical practice today still lacks scientific data approaches. Data has the ability, if used correctly, to separate anecdote and hyperbole from facts.

The five Vs of data:

Today's biggest data challenges are due to one or more of the "Data Vs."

Initially there were three:

1. Data Volume
2. Data Variety
3. Data Velocity

Then two more qualities were later added to make it the five Vs of Data:

4. Data Value
5. Data Veracity

Data volume

Data volume is the sheer amount of data. Burk, Sweenor, and Miner estimate that more data was generated per day in 2020 than was produced in the entire year of 2005. The volume is daunting, overwhelming, and most of it goes unused. It should be of great value. Why collect if you are not going to use it? Yet, according to *Harvard Business Review* (Davenport, 2017)

• Less than half of structured data is actively used in decision-making.
• Less than 1% of unstructured data is analyzed or used at all.

Data variety

Data Variety describes the disparate data sources, data types, different structures, and formats of data. We source data for analytics via traditional database connections, web data, SaaS, and data lakes. We source data in a wide variety of formats and data structures. Enterprises must have access to both on-premises and cloud data. Data Virtualization (DV) is a technology that is useful to overcome these challenges.

Data velocity

Data Velocity describes how fast data is being added to systems and how often it is refreshed. Many forms of data today are streaming or event-based information. We provide valuable information on these technologies in the following sections, "Microservices" and "Streaming Data."

Data value

Data value is the ROI for sourcing, transforming, securing, and protecting data. Collecting and storing data is a huge cost. Therefore, the only way to get a positive return on data is to offset this cost by applying and operationalizing analytics.

Data veracity

Data Veracity is the quality, reliability, and trustworthiness of the data. Data quality is one of the biggest limiting factors in medicine today. Enterprises need technology to help remediate the data in virtually every link in the data chain and the accompanying analytics tools. Associated paradigms are Data Governance, Data Management, and Metadata Management.

We present some basic considerations for data architecture and provide some references for further review.

Connecting and moving data—data in motion

Data that is stored in a large repository, like an enterprise data warehouse, a data lake, or some Data as a Service made possible with DV, enable most AI and Analytics. Before we look at these data stores, let us briefly cover some of the ways data are connected and moved.

Understanding the piping, the way that data flows (moves), is important to a data architecture plan. We will not present this information at a detailed or technical level. However, everyone, as part of their analytics education, should have a basic understanding of some of the technologies that support the operational and decision-making capabilities of the organization.

Much of this *Data Design for Success* involves where you "park" or store the information; these "parking" or "storage" areas are the focus of the next section, Data Stores and Virtualized Data.

This section is about the highways: when and how you move the vehicles along and where they are allowed to go. Technologies are developed all the time for new platforms like smartphone and tablets, quicker application development, innovations in process tracking, and speed. All of this becomes part of the "pipeline" and "storage" of data and has to be modified as new technologies come into use.

We now briefly cover three technologies that are gaining attention in predictive and prescriptive medicine. Why are these so important? Because historically many of the predictive and prescriptive models have been built from on-premise data sourced from the EHR and clinical systems. Yet, many predictive model determinants are outside these systems. Take asthma for example, predicting a critical asthma event (one that might send a patient to the ER) accurately requires many variables external to the clinic systems. Air quality, weather data, social determinants of health, and personal events can lead to an asthma attack and this data is not available with on-premise data. Therefore, we need connectors into this data.

Three technologies that need to be understood and incorporated into prestigious healthcare enterprises should be:

Application programming interfaces and management

Healthcare enterprises have been using APIs to move data internally within their systems for several years. External APIs are gaining traction with examples like surgical registry information, CMS, and CDC data. To unlock the full power of these external APIs, IT must allow for API management systems that allow individual personnel to create and share applicable APIs for institutional use. Use cases such as asthma critical event prediction are examples where participants of particular projects need specific data interfaces. These users should be allowed to create these interfaces and share them with the enterprise for other projects and applications.

Microservices

An API is a contract that provides guidance for a consumer in using the underlying service. A microservice architecture is an architectural development style that allows building an application as a collection of small autonomous services developed for an enterprise domain. A microservice is an architectural design that separates portions of an (usually monolithic) application into small, self-containing services. Each service runs a unique process and usually manages its own database.

This architecture is normally developed and maintained by IT, however it is made available to the larger organization. It is especially important in the deployment, production, and maintenance of predictive and prescriptive models. It speeds up the development and application time. It allows for a secure, flexible, and scalable design.

Streaming data

Streaming data are data that are continuously generated by different sources. Unlike most data sourcing that requires a "fetch," a query or request to "pull" data for collection and use, streaming data are pushed into a database as they are created. Examples include sensors in hospitals, clinics, and home health devices.

Medical care is far behind in the application and use of streaming data. Many industries have been applying the power of this data for the improvement and control of processes for many years. Analytics based upon streaming analytics enables healthcare providers to apply predictive analytics to data in motion for continuous decisions, allowing them to capture and analyze data—all the time, just in time. The end goal is to save lives, shorten hospital stays, and build healthier communities through preventive care.

There are a few notable exceptions/examples where this data is used:

1. Elon Musk, the founder of Tesla and SpaceX is investing $100 million into a somewhat secretive venture called Neuralink that will employ some of the leading scientists to create a brain-to-device-network. This news was announced in summer of 2019. Musk said the Neuralink system would allow for a chip to be implanted into the brains of willing subjects and would allow humans to achieve "symbiosis with artificial intelligence." This is an Internet of Things (IoT) application of streaming data.
2. Stanford Healthcare–New Stanford IoT Hospital. More than 10 years in the making at a cost of $2.1 billion, Stanford opened the doors of a new 824,000-square-foot medical facility October 23, 2019. The hospital includes a variety of new technologies from patient-centric controls in rooms to modular architecture, to a robotic pharmacy, and IoT-based technology. According to the Wall Street Journal (see Rosenbush, Nov 2019):

   Sensors will track the location of staff and equipment in real time, improving efficiency and inventory control. The infrastructure can support 120,000 connected devices streaming 4K high-definition video. The infrastructure will be able to accommodate upgrades such as 5G wireless. Magnetic resonance imaging equipment and other systems will be integrated with one another in new ways.

   ...Doctors and nurses will be able to monitor multiple patients from a single remote location. Alerts and alarms will go directly to secure mobile devices carried by nurses and doctors, instead of sounding at nursing stations—reducing noise levels. ...[What's Next] .... Two patient rooms will test a bedside computer-vision system that uses depth and thermal sensors to improve patient safety.

3. AI and IoT are also helping gain freedom from disabilities. Whether it is assisting veterans walk again or helping the blind see, these technologies are enhancing humans. One example is a new app that Microsoft launched in 2018 called "Seeing AI." It helps people who are blind or with other visual disabilities interact with the world by using the smartphone's camera; the phone will audibly describe what it is seeing, read written documents, scan product barcodes, and even recognize faces (Microsoft, 2018; Bureau of Internet Accessibility, 2018). Even things like the Apple enabled "Siri" allow people with spinal cord injuries and the inability to use their hands to dial a phone number, communicate with others by speaking to their smartphone and verbally telling it who to call—just with a voice command. Microsoft in Australia appears to be a leader in advancing AI and IoT. Globally there are 1.3 billion people dealing with disability needing some kind of AI/IoT support. Of this 1.3 billion worldwide number, about 4 million are Australians, and thus Deloitte in Australia has estimated there is a market worth about $1.4 billion by 2020 just for Australia (Microsoft, April 2019).

Data stores and limitations of the enterprise data warehouses

We just covered connecting and moving data, that is, data in motion. Data stores are where we park this data when it is not moving. Traditional data stores used for analytics are Operational Data Stores (ODS), Data Marts, and the

Enterprise Data Warehouses (EDW). This data is relational and structured. Big data architectures used semistructured data to store data in data lakes. Now lake houses are beginning to become popular. They combine the flexibility, cost-efficiency, and scale of data lakes with the data management of data warehouses.

EDWs still serve as the backbone of most predictive analytics in medicine. They present very stable sources of data that are easily accessible. However, there are challenges, and we note a few:

Timeliness of data

Some data are aggregated by the week or even month for some variables. This makes some analysis difficult.

Data additions are slow and difficult

Request for additions to the EDW have to go through IT. These requests are often difficult to complete and may take 6 months to make it into the EDW.

Some data will never be available in the enterprise data warehouses

SaaS is becoming more and more popular. Users of AI and analytics often query these system for the data they need on an ad hoc basis. There is no way to pull the data from an entire SaaS system into an EDW.

Ancillary and cloud data

As we illustrated in the predictive model for the asthma critical event, web APIs are required to source some of the data requirements. This data is typically merged with EDW prior to model development. This merging and validation requires a great amount of labor and system resources to perform the necessary data management responsibilities each time new models are developed. This data management approach does not scale.

Data virtualization

DV allows you to overcome these shortcomings of the EDW. DV was created not to supplant the EDW, but to extend its reach and capabilities. With virtualization, there is no physical recreation or copying of data, as needed with data warehousing.

Rick F. van der Lans (van der Lans, 2012) offers this definition:

> Data virtualization *is the technology that offers data consumers a unified, abstracted, and encapsulated view for querying and manipulating data stored in a heterogeneous set of data stores.*

He covers a detailed treatise with definitions of abstraction, encapsulation, and federation. These details are beyond the scope of this chapter. However, we recommend Burk et al. (2021), as the authors go through DV in greater detail and offer additional references.

A graphic illustration of DV is provided in Fig. 24.1. At the bottom of the figure are the various data sources. These are rich sources, beyond traditional data sources. In the middle of the figure is the virtualized layer. This is not a persistent copy of data, but instead it is data brought into fast, virtual memory, thus making it immediately available for the user. When a user or application (we label both here as a data consumer) requests data, this data is queried from the source and presented into this virtual layer.

To be clear, an important feature of DV is that a separate "write to disk" (persistent) copy of this data is never made, thus the data is not duplicated. While there are a myriad of data sources being queried and brought into this virtualized layer, **the consumer perceives that they are accessing one very large integrated data source**.

DV is a platform that connects to "virtually" any data source. However, that is not the reason for its name. The process is termed "Data Virtualization" because there is no persistent copy of the data. The sources of data you can acquire seem nearly limitless. These can be data warehouses, data marts, ODSs, cloud-based platforms, SaaS, Streaming sources, web and Restful APIs, local databases, network drives, and much, much more. Here are a few:

A platform with searchable data and rich metadata

Users can easily search all available enterprise data that they have permission to access. If there are multiple instances of the same data elements (for example, patient or payer ID) they can easily see where the different elements are sourced and which they should use for each analytic application.

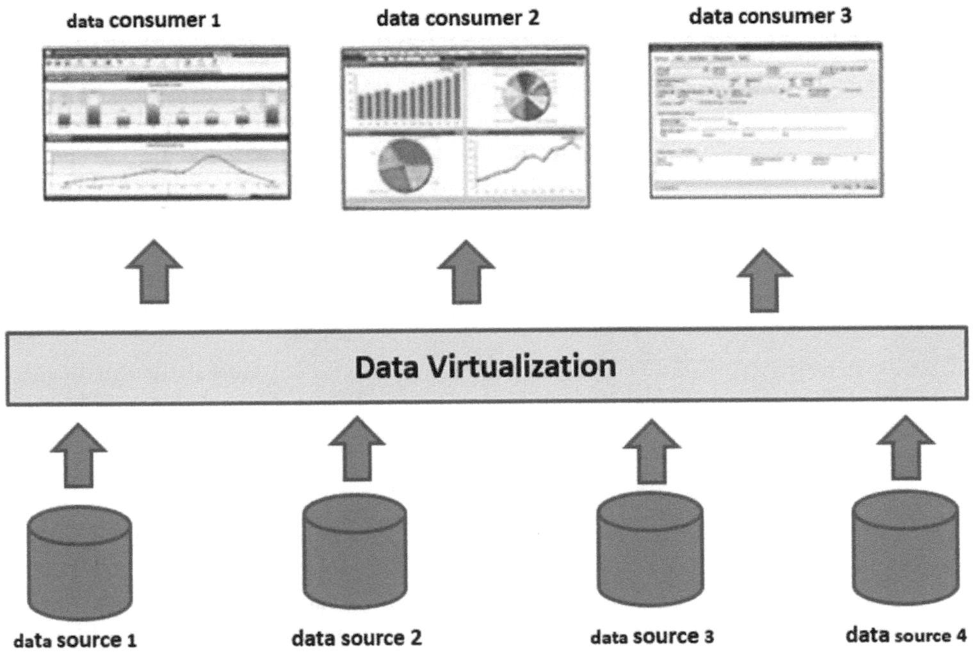

FIGURE 24.1 What is data virtualization?

A collaboration tool for functional areas and users

The system catalog also allows clinical users, IT users and even third parties to communicate and collaborate in regimented/defined ways. You can make forums open to everyone or allow only group restricted conversations. This offers many benefits, but one of the great benefits is a boost to data quality and IT productivity.

A pathway for new systems and system migration

One of the major hurdles that organizations face is moving from old, existing (legacy) systems to new systems. Often this migration is delayed for years because it requires so much time and money. There is a huge amount of effort mapping old data to new data and quality checks. DV can greatly reduce the time and effort and therefore money to bring on new systems.

An IT tool for rapid prototyping

Another way that DV saves IT time and money is by its quick prototyping of data management processes. As stated, DV is not meant to replace the EDW, but to augment it and extend its reach. So, new elements are always being sourced into the EDW and this takes time and effort of which DV can greatly assist.

A system for enhanced security of data

DV allows for additional security between the data consumer and the data source system. When the data is made available in DV, the administrators can give access permission to individuals, or to groups of users. These security measures can be defined in very granular ways.

Data governance and data management

Data governance involves people, processes, and information technology; all these are required to create a consistent and proper handling of an organization's data across the enterprise. Data management and data governance work hand-in-hand. Data is a vital asset to every organization and therefore it is critical to manage and secure it effectively.

We briefly consider two major dimensions of data governance:

- the policies, procedures, and process side of data governance; and then
- the technology that supports data governance.

If we have one takeaway we would say that a consistent, strong message from leadership on how vital data and information are to the organization is essential. The most successful enterprises in the last 20 years are ones that have capitalized and seized opportunity from their investment of data. Moreover, accessible, usable data is everyone's responsibility.

Mid-to-large size organizations often have a data governance program that includes a governance team. This governance team consists of a steering committee that acts as the governing body, and additionally data stewards that continually examine their incoming and stored data to make sure it complies with the governance policies. Both the steering committee and the data stewards work together to create standards and policies for governing data. It is very important to have participation from the leadership team and every functional operation represented. IT should not be considered solely responsible for data governance. Everyone in the organization has a responsibility for ethics. Data quality is everyone's responsibility as well.

If the organization has a chief data officer (CDO) they will be the senior executive who oversees a data governance program. The CDO is responsible for the design and budget of the program and will routinely report to the board of directors on the program's status. If the organization does not have CDO, the program is typically led by another c-level executive, possibly the CIO. If the CIO is the leader, it is imperative that the program not be considered an IT program; instead it must be a corporate program.

The **data governance team** (sometimes called the data governance office) may be led by the CDO or by another executive that has some operational authority with data (director of data management, director of data quality, etc.). This data governance team is dedicated full-time to data governance activities. It coordinates the process, leads meetings and training sessions, tracks metrics, manages internal communications, and carries out other management tasks.

The **data governance committee** is a separate entity that is responsible for setting policy and procedural standards. This committee is primarily made up of executives and other data owners. The committee approves the foundational data governance policy and associated policies and rules on things like data access and usage, plus the procedures for implementing them. It also resolves disputes, such as disagreements between different departments over data definitions and formats. The data governance team often attends meetings and makes recommendations to the data governance committee, but does not have a formal vote in the process.

Data stewards have the responsibility of overseeing data sources and data stores of the enterprise. These data sources include operational systems on premise, in-the-cloud, and data purchased from third party providers. They also ensure that the policies and rules approved by the data governance committee are understood by the enterprise and followed.

Professionals with knowledge of particular data assets and domains are generally appointed to handle the data stewardship role. It might be a full-time or part-time job depending on the size of the organization. Stewardship works best when there are a mix of IT data stewards and other department data stewards.

It should be noted that everyone in the organization should be a data steward at a fundamental, informal level. Each person is part of the organization's success and must know that useful, available data is core to the way an enterprise operates, competes, and improves.

Goals of data governance

The goals of data governance are multifaceted. They must be tailored specifically to the organization's mission and strategy. Following are some goals of data governance that we commonly see in large organizations:

- Data integrity
- Data security
- Data consistency
- Data confidence
- Compliance of regulations and data privacy laws
- Adherence to organizational ethics and standards
- Risk management and data leakage
- Data distribution
- Value of good data
- Moving data quality upstream reduces costs
- Technology and structure that minimizes the cost of the DG program
- Data literacy education

Technology to support data management and governance

There are companies that market dedicated data governance platforms. These platforms may cover some of the needs of certain organizations. However, in our experience, most organizations require several platforms across vendors to support all their functions and activities. We believe it is most beneficial to the reader to outline process and technology requirements that will bring the organization forward in becoming a "data-driven" organization. Data management goes hand-in-hand with data governance. We outline technology that supports both these activities; specifically, we cover technology considerations for data management, data quality, and security.

Data management

Ask 100 people what data management means and you are likely to get 100 answers. According to DATAVERSITY (dataversity.net)

> *Data Management is a comprehensive collection of practices, concepts, procedures, processes, and a wide range of accompanying systems that allow for an organization to gain control of its data resources. Data Management as an overall practice is involved with the entire life cycle of a given data asset from its original creation point to its final retirement, how it progresses and changes throughout its lifetime through the internal (and external) data streams of an enterprise.*

We have already covered some of what is encapsulated by this definition. What we have not covered is metadata (*data about data*) management or specifically reference data management and master data management. These are extremely useful technologies that are available and should be considered by any data governance program. Master data, along with reference data, and metadata, are key organizational data assets.

Master data

A simple example of master data in a clinical system would be the data that represents an order or transaction. When a patient gets a laboratory test in a hospital, there are multiple tables of master data that "marry" the data across functional and business entities. For example, if a patient has a laboratory test there is a hospital laboratory code master table that identifies the test for the laboratory. There is an accompanying charge code master in the financial system that "cross walks" or matches the internal code to external billing entities like CMS and other insurance payers. There may other master table codes that match the internal lab code to external lab codes for tests that are performed at labs outside the hospital.

What complicates things further is that codes change over time and it is necessary to keep up with these changes and be able to refer to what code was used at a particular point in time.

Imagine doing an analytic study and the code changed three times based on different date ranges. As a statistician or data scientist, you do not want to have to keep track of all these changes, master data management makes these challenges moot as the systems tracks all the changes and makes data acquisition a snap.

From an analytical standpoint, how do you view patient interactions across locations, departments, and providers? You cannot create reports, dashboards, and predictive models without relevant and consistent master data. Moreover, from a regulatory standpoint or governance perspective, master data allows you to implement and enforce controls.

Reference data

Reference data is another form of metadata that is sometimes equated with master data. However, it is a newer and a separate data form. Reference data management is about managing classifications and hierarchies across systems and business lines. For effective reference data management, organizations must set policies, frameworks, and standards to govern and manage both internal and external reference data. Platforms are available to manage all these structures.

Some examples of reference data are ICD-10-CM codes, hierarchical condition categories, geo codes, units of measure, SNOMED CT, legal entities, charts of accounts, and conversion rates. Most often, there are dependencies within these classification systems. As organizations grow and time evolves reference data are extremely important to analytics supporting the business. These include things like acquisitions, mergers, sales, and organizational changes. Reference data are constantly evolving; good reference data management allows for consistent analytics across time.

Metadata

IoT has been a major contributor to the data growth explosion by capturing, sensor, audio, and vision data. Another major contributor to this growth is the creation of metadata. Metadata is simply data about data. It means it is a description and context of the data acquired. It helps to organize, find, and understand data. For example, Digital Imaging and Communications in Medicine (medical imaging information) data files contain metadata that provide information about the image data, such as the size, dimensions, bit depth, modality used to create the data, and equipment settings used to capture the image.

Data quality

The authors have been involved with analytics for decades. Data quality was a problem when we started and it remains a problem today. It can be introduced anywhere along the chain of clinical processes and once it is introduced it normally persists. For commercial enterprises, it is difficult to remediate some of these issues. Local, state, and federal government remediation can be much more difficult. Adding the interexchange that many organizations have with other organizations increases the complexity, and it just gets extremely difficult to ensure quality in data.

One thing that organizations can do to alleviate data quality issues is planning and process. As Benjamin Franklin said, "An ounce of prevention is worth a pound of cure." Additionally, a paradigm endorsed by Edward R Deming, the father of quality management, is to adopt the practice of moving data quality issues upstream, that is, identifying them early before they are perpetuated throughout all the systems. This work may be assisted with technology. Data needs to be easily tagged in any system and annotated. The technology should allow collaboration between users identifying problems and system administrators with oversight responsibilities. Convenient messaging with annotated data should be communicated between these parties as well as the data stewards.

Security

No one needs to be told how big an issue data security is for modern enterprises. From small to the largest enterprises, from public to private organizations, securing your data is paramount. Good policy and procedure are vital to data security.

Controls to safeguard your data should be created at every level of the firm. It is not best practice for users to have to have individual credentials/access to every system. Most employees have access to dozens of systems. Technologies that employ a domain controller that authenticates and authorizes all users and computers in a domain type network allow for seamless access and control in assigning and enforcing security policies for all computers, and installing or updating software. Multifactor authentication is also becoming a common security protocol; this is an electronic authentication method in which a user is granted access to an application only after successfully presenting two or more pieces of identification.

Data governance and data management summary

No organization of any size can compete or be run effectively today without a solid data strategy. Data governance is the glue that binds the second most important asset of the enterprise; the first being the people involved.

Data governance is not just a technology platform. It is planning, strategy, and controls around data and the people that interact with that data. It is about ensuring that not only all the investment that has been made to source, transform, and store that information and make it available is recouped, but also the organization actually makes a return on that investment.

Analytics design for success

In the next chapter, we will cover analytics maturity and some key fundamental requirements for using various analytic models. In this section we overview some important technical capabilities and tools that premier organizations have in place to create analytics, predictive, and prescriptive models.

Technology to create analytics

In the first book of their series, *It's All Analytics*, Burk and Miner (2020) reviewed the CRISP-DM (cross-industry standard process for data mining) model that was popularized in the mid-1990s. Although the methodology is still widely used, they presented several refinements to the model in their second book (Burk et al., 2021) based on their experience. Fig. 24.2 outlines the common steps associated with the data science and ML process.

Data Science and ML Process

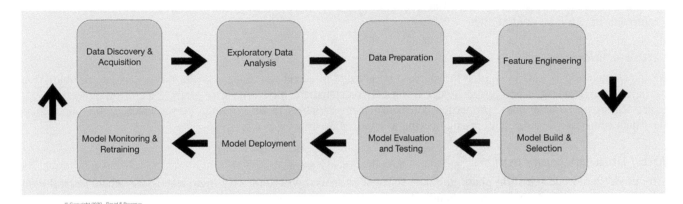

FIGURE 24.2 Steps for the data science and machine learning process.

Since the basic elements of the data science and ML process are well documented in other works, we will not provide a review of what happens for each of these process steps. However, as it relates to architecture, we will provide some considerations and things to think about for each of the steps.

Data discovery and acquisition

With a solid understanding of the clinical decision to be made and the ramifications to current processes, the project team needs to understand the various data sets available to make that decision while respecting privacy rules, internal policies, and procedures, as well as external regulations. Before building models, the team should understand any limitations or restrictions that may impede the integration of the final model into the clinical process. For example,

- Will the data be available in the runtime environment?
- Is the data sourced internally, externally, or both?
- Are you legally and ethically entitled to use the data?

Exploratory data analysis

During this phase, the data scientist will use statistics and visualizations to gain an overall understanding of the structure and nature of the data used for analysis. Many statistical techniques have assumptions on the underlying structure of the data. The data scientist will need to examine the data and make sure that the data are of the right shape and format for methods that will be employed. A few considerations are:

- How similar are the training and test data to the data that we will see in real-life when the model is in production?
- Will the production model be able to access similar data structures to those that occur in the training and test environment?
- Are there redundant variables?
- Is there target or data leakage that may not be available at the time of prediction? In other words, are their correlated variables to the predictor variable that are in the training and test data set that may not be available in the run-time environment at the time of prediction?

Data preparation

When preparing the data for training and testing, the data scientist may impute missing values, remove values (or select outliers of interest), apply transformations to the variables, and group or bin variables. Additionally, there may be specialized processing required for nontraditional data such as text, image, and audio data. Architectural considerations include:

- How performant does the data prep sequence need to be in the production environment?
- Where will the data be processed? For example, if you are doing image analysis and recognition at the edge, is there enough processing power?
- Is there sensitive or personally identifiable information present in the data? Does it need to be masked, obscured, removed, or can it be used at all?

Feature engineering

Feature engineering is a process of using domain knowledge to select and transform the most relevant variables from raw data when creating a predictive model using ML or statistical modeling. The goal of feature engineering is to extract features (characteristics, properties, attributes) from variables to improve the performance of ML algorithms. Many data scientists consider feature engineering as much of an art as it is a science.

Model build and selection

In the model build and selection phase, the data scientist or analyst will iteratively train, test, and evaluate several models in their relation to the problem at hand. They may select one or more models or may even combine the models which is known as ensemble modeling (also known as voted prediction). There are several things we want to know about these models in this process:

- How sensitive are the models to changes in the incoming data streams?
- Are the models scalable to large data sets? Do they need to be?
- How complex are the models, are they understandable and interpretable?

Model evaluation and testing

In the model evaluation and testing phase, the analytic professional will move from the sandbox environment to one that is more consistent with the production run-time environment in performance and technical characteristics. Architectural considerations include:

- What are the clinical, accuracy, and performance requirements for the model inference or prediction?
- How big or complex are the models? Can they be processed on an edge device if needed?
- Are the models free from bias?

For example, with regard to performance requirements, if the target is predicting a cancer in a patient, we want to place limits on the false positive or false negatives rates before moving models to production. If it is a prescriptive model for cancer treatment, the project team, especially clinical staff, will have to balance the false positive and false negatives rates and may need to revisit the model selection process or even the feature engineering process.

Model deployment

Model deployment generally consists of taking the necessary code and exporting it in a language that is appropriate to the target run-time environment. For example, if the model is expected to run in a database the optimal score code may be Structured Query Language or Predictive Model Markup Language. Other formats may include Java, C#, C++, PFA (Portable Format for Analytics), and ONYX (Open Neural Network Exchange).

Aspects of concern in model deployment include:

- How many computational resources will we require to run the models?
- Will the data pipelines be included in the score code or are they separate?
- What dependencies does the model require to run?
- Where will the results go? A dashboard, another control system, a database, an automated process?

Model monitoring

After the model is deployed, it is important to have a system and framework in place to monitor the model. Monitoring may include tests for model accuracy, performance metrics, and business metrics.

Architectural considerations include:

- What metrics will we be using to monitor model performance?

- Do our results need to be reproducible? Do we need to version the models and the data?
- Where will we store the metadata associated with the model runs?
- What are the criteria for which we will update models, either retraining or complete remodeling? Is the population changing or is it a pipeline data failure, that is, the data has changed?
- Who responds to model failures? Do we have data scientists or engineers on call?

Legality and ethical use of data

In addition to the considerations of each step in the data science and ML process, there are specific legal and ethical considerations that one must think about throughout the process. For example,

- Are we allowed to use certain variables given geography and jurisdiction requirements?
- Are we compliant with the data protection regulations?
- Are we compliant with federal, state, local, and internal regulations?

Technology to communicate and act upon analytics

We have discussed the AI and ML architectural components that are vital to the creation of predictive and prescriptive models. We now focus on technology that will be needed to communicate and act upon analytics. To communicate and act upon analytics, there are a few essential factors to consider. Organizations need to be able to deploy and monitor analytic workflows, they need to be able to monitor the performance of models using business, technical, and performance metrics, and they need to establish and adhere to an ethical and governance framework for analytics.

A new paradigm for AI and analytics is the notion of "composable apps" (Panetta, 2020) which are a grouping of data and analytics capabilities (e.g., DV + machine learning + data visualization) assembled in a coherent manner to make a business decision or solve a specific clinical problem. To assemble these applications, one can think of putting the various capabilities together like Lego bricks.

Through the confluence of these capabilities, organizations will assemble them in unique ways to build analytic applications to support digital decisions. These applications may be deployed as reports and dashboards, they may be embedded in custom apps, the output could feed other business systems and data sources, or the predictions could deliver results to other analytic systems. However, simply creating and deploying the technical underpinnings of the analytics applications is not enough. Even though they may technically produce output that can be consumed and acted upon by other systems or people, more is needed. This is where an understanding of data storytelling can come in handy; we recommend the book "Resonate" by Nancy Duarte (Duarte, 2010) and (Barnes, 2013) as additional references.

Model Ops

A final consideration of your analytics design for success is "Model Ops." Model Ops "is a cross-functional, collaborative, continuous process that focuses on operationalizing data science by managing statistical, data science, and machine learning models as reusable, highly available software artifacts, via a repeatable deployment process" (Sweenor et al., 2020).

Why does an organization need Model Ops? Quite simply put, without it, your organization's digital decision making will be suboptimal. Many organizations are increasing their spending on data, analytics, and AI technology. In order to maximize the investment, organizations must correspondingly invest in model ops. They need to have visibility on what models are deployed, they need to understand how the models perform, they need to understand if the models are having the desired impact on their business. "A good rule of thumb is that you should estimate that for every $1 you spend developing an algorithm, you must spend $100 to deploy and support it" (Redman and Davenport, 2020).

In addition to understanding what models are being used and how they are performing, Model Ops provides a process and framework to allow professionals to deploy models in a repeatable manner using best practices. With the Model Ops process, framework, and technology, organizations can deploy models faster than ever before. Many surveys suggest that about half of organizations take over 6 months to deploy models and there are about a quarter of organizations that take 9 months or longer to deploy.

Conclusion

The purpose of this chapter has been to provide some high-level considerations for

- Social and cultural design considerations for analytics success

- Data architecture design considerations for analytics success
- Analytics architecture design considerations for analytics success

These considerations are vital to gain optimal results in your organization's predictive and prescriptive model using AI, ML, and related technologies.

Postscript

In some organizations, predictive analytics solutions just die on the vine, because the operational systems necessary to implement them do not exist. Before analytics solutions are designed, the corporate infrastructure must exist to use the solutions. By analogy, the reciprocating action of the piston in an automotive engine is just a lot of wasted motion if the piston is not attached to the crankshaft, which in turn is attached to the transmission and drive line to make the wheels go around. Many analytics solutions are finely tuned engines that just sit on the bench and do nothing, because they are not installed in a vehicle designed to use them. Organizations must reinvent themselves to be able to run on predictive analytics outcomes. This chapter provides the high-level blueprint for building effective analytics architectures in an organization to put results into action.

References

Barnes, H., 2013. Messaging in Technology Marketing Stinks, But Improvement Is Possible. Gartner, Inc. <https://blogs.gartner.com/hank-barnes/2013/08/28/messaging-in-technology-marketing-stinks-but-improvement-is-possible/>.

Burk, S., Miner, G.D., 2020. It's all Analytics!: The Foundations of AI, Big Data, and Data Science Landscape for Professionals in Healthcare, Business, and government. CRC Press.

Burk, S., Sweenor, D.E., Miner, G.D., 2021. It's All Analytics—Part II: Designing an Integrated AI, Analytics, and Data Science Architecture for Your Organization. *CRC Press.*

Davenport, T.H., April 2017. What's your data strategy?, *Harvard Business Review.* <https://hbr.org/webinar/2017/04/whats-your-data-strategy>.

Deming, W.E., 2018. The New Economics for Industry, Government, Education, third ed MIT Press, Cambridge.

Duarte, N., 2010. Resonate: Present Visual Stories that Transform Audiences. John Wiley & Sons, Cop.

Kuhn, M., Johnson, K., Springer Science+Business Media, 2016. Applied Predictive Modeling. Springer, New York.

Meehan, M. (2016). Where data goes to die, *Forbes*, 8 December. <https://www.forbes.com/sites/marymeehan/2016/12/08/where-data-goes-to-die-big-data-still-holds-answers-but-theyre-not-where-your-looking-for-them/#5624533a5896>.

Panetta, K., 2019. A Data and Analytics Leader's Guide to Data Literacy. Gartner. <https://www.gartner.com/smarterwithgartner/a-data-and-analytics-leaders-guide-to-data-literacy/>.

Panetta, K., 2020. Gartner Keynote: The Future Of Business Is Composable. <https://www.gartner.com/smarterwithgartner/gartner-keynote-the-future-of-business-is-composable>.

Redman, T.C., Davenport, T.H. (2020). Getting serious about data and data science, *MIT Sloan Management Review*, 8 September. <https://sloanreview.mit.edu/article/getting-serious-about-data-and-data-science/>.

Rosenbush, S. 2019. New stanford hospital takes holistic approach to technology, The Wall Street Journal CIO Journal, <https://www.wsj.com/articles/new-stanford-hospital-takes-holistic-approach-to-technology-11573905600>.

Sweenor, D., Hillion, S., Rope, D., Kannabiran, D., Hill, T., O'Connell, M., 2020. ML Ops: Operationalizing Data Science. O'Reilly Media, Inc. Available from: https://www.oreilly.com/library/view/mL-ops-operationalizing/9781492074663/.

van der Lans, R.F., August 2012. Data Virtualization for Business Intelligence Systems: Revolutionizing Data Integration for Data Warehouses. Morgan Kaufmann.

Chapter 25

Predictive models versus prescriptive models; causal inference and Bayesian networks

Scott Burk

Chapter outline

Prelude

The primary types of analytics operations were introduced in Chapter 22. This chapter will expand on them and add several more types. On this platform is cast one of the most vexing of issues in analytics—correlation of two variables does not necessarily imply causation. What is causation? And how can we specify it and use it to guide prescriptive medicine? Those are the central questions of this chapter.

Introduction

Predictive models and prescriptive models in medicine are compelling and are changing the way healthcare is delivered around the globe. Examples range from clinical to operational improvements. These include the power to predict the likelihood of a disease, readmission to the hospital, and the correct staffing and resource levels in a clinical setting. Additionally, these include power to prescribe the optimal clinical therapy, and more importantly for person-centered medicine, the power to prescribe the best therapy for this patient rather than what is best for a generalized population. These advanced analytics are indeed changing the way medicine is practiced.

Medicine is becoming less a "practice" and more of a "science."

While this is great news for medicine, misapplication of these models can lead to undesirable results. The formulation and application of predictive and prescriptive analytics require different philosophical and mathematical assumptions. This chapter is about revealing different model assumptions—especially the difference between predictive and prescriptive models. We also highlight the Design of Experiments (DoE) and Bayesian networks to bridge the gaps of model assumptions.

Practical Data Analytics for Innovation in Medicine. DOI: https://doi.org/10.1016/B978-0-323-95274-3.00022-1

Classification of AI and ML models in medicine

There are a myriad of analytics available—visual analytics, text analytics, big data analytics, advanced analytics, and of course, predictive and prescriptive analytics. In fact, there are many more. Burk and Miner offer an "Analytics Mega List" of 50 commonly use types of analytics and write that there are many more (see Burk and Miner, 2020). However, there is a commonly referenced classification of analytics often attributed to Gartner, a major research firm for tech and innovation. This framework describes five categories that most types of analytics fall within:

1. descriptive analytics,
2. diagnostic analytics,
3. predictive analytics,
4. prescriptive analytics, and
5. process optimization.

Fig. 25.1 represents an increasing level of complexity for analytics classification by value received by that classification. For example, descriptive statistics require less machine processing time to generate results while more human mental effort is required to make a decision or take action. As we move to the right to predictive and prescriptive analytics, much more computer processing is required to generate the associated analytics; yet the human mental effort is reduced and offset by machine processing.

Note: It is imperative to note that the model requirements of each type of these analytics are not the same. We will dive deeper in this chapter, but here are some important differences in modeling assumptions as an introduction. We start with fundamental definitions and the accompanying assumptions.

Descriptive analytics

Descriptive analytics has a very long historical foundation. These analytics are the easiest to implement but often do not offer the value of other methods, which are generally more complex and sophisticated. Descriptive analytics typically fall into one of the following categories:

1. Descriptive statistics and numerical summaries
2. Visual representation

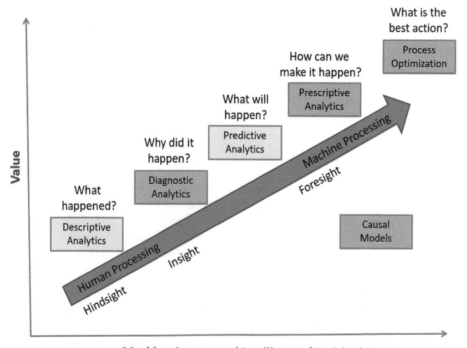

FIGURE 25.1 A visual reference of a generally accepted classification scheme for analytics.

According to Fig. 25.1, descriptive analytics describes analytics as answering the question, "**What Happened?**" However, descriptive analytics go well beyond this question. They form the foundation for all the other analytics categories, meaning that diagnostic, predictive analytics etc., all start with descriptive analytics.

Model Types and Assumptions—Retrospective, reactive, no mathematical requirements for variable relationships.

Diagnostics analytics

Diagnostic analytics is a form of advanced analytics that examines data or content to answer the question, "**Why did it happen?**" It is causal and retrospective. It has causal meaning in that a change in an attribute (a cause) will generally or specifically result in an observable change in a second attribute (the effect). Moreover, the cause must precede the effect. It is essential to note the difference in association or correlation with causation (see gray box below on "Causation vs Correlation").

Model Types and Assumptions—Retrospective, reactive, mathematical requirement of inputs and outcomes—causative.

Predictive analytics

Predictive analytics answers the question, "**What is the probability this event will happen?**" Predictive analytics is about assigning a probability or likelihood to an outcome based on some type of analysis or model. Many predictive models today are generated with AI/ML algorithms. When given a set of inputs, these models will generate the probability of an outcome.

For example, the University of Iowa Hospitals and Clinics used patient and operating room data to generate predictive models to determine the likelihood of surgical site infections (SSIs). The most interesting and innovative thing about this model was that it was executed while the patient was still under anesthesia in the operating room. The resulting probability in "real time" gave the care team information to make a unique decision based on this unique patient and the corresponding factors of this surgery. Alternative therapies based on this model could be normal closure, delayed surgical wound closure, or negative pressure therapy (Siwicki, 2018). The relationship between targets and predictors in the predictive model may be correlated or causal.

Model Types and Assumptions—Proactive/forward-looking, active, the mathematical requirement of inputs and outcomes—correlative.

Prescriptive analytics

Gartner notes prescriptive analytics is a form of advanced analytics, which examines data or content to answer the question "What should be done?" or "**What can we do to make XYZ happen?**" While it is important that the clinical staff validate the medical decision, the decision is made by the algorithm.

As an example, let us return to the University of Iowa Hospitals and Clinics SSI use case. With enough quality data —a long history that included competing for alternative therapies, relevant outcomes, and the statistical validation that the inputs cause the clinical outcome, i.e., not merely correlated, a prescriptive model could be generated. This model would return a qualitative result—a prescribed therapy for this patient given their specific circumstances. In this example, computational power is actually making the decision based on the data. The clinical team is just validating and acting upon this decision.

Model Types and Assumptions—Proactive/forward-looking, active, the mathematical requirement of inputs and outcomes—causative.

Process optimization

Optimization is a word overused a lot these days, "I optimized my schedule." We use the word optimization in a very strict, mathematically rigorous way in analytics. A mathematical construct is created, including an objective function, goals, and constraints. The objective function states whether we want to maximize or minimize this target (decision) variable. This is optimization. Constraints cannot be violated, and goals are penalties to the objective function. We then search for solutions that maximize or minimize our function and do not violate our constraints. The solution that has the highest or lowest value of the objective function is the optimal solution.

Model Types and Assumptions—Proactive/forward looking, active, mathematical requirement of inputs and outputs for interventions—causative.

This chapter focuses on the difference between predictive analytics and prescriptive analytics. Of all the analytics methods just covered, these models are most often misunderstood. **Prescriptive analytics requires causality**. It is essential to understand the differences. In the next section, we cover an example in an upcoming gray box, "Predictive Models versus Prescriptive Models—They are not the same!" but first offer a brief insight into causal inference.

What is Causal Inference?

In his own words, in an interview with Ron Wasserstein (Amstat News, 2012) Judea Pearl (Turing Prize winner, 2011) defines causal inference as:

"Causal inference is a methodology for answering causal research questions from a combination of data and theoretical assumptions about how the data are generated. Typical causal questions are the following: What is the expected effect of a given treatment (e.g., drug) on a given outcome (e.g., recovery)? Can data prove an employer guilty of hiring discrimination? Would a given patient be alive if he had not taken the drug, knowing that he, in fact, did take the drug and died? The distinct feature of these sorts of questions is that they cannot be answered from (nonexperimental) frequency data alone, regardless of how many samples are taken; nor can they be expressed in the standard language of statistics, for they cannot be defined in terms of joint densities of observed variables. (Skeptics are invited to write down a mathematical expression for the sentence, 'The rooster's crow does not cause the sun to rise.')"

The variables that "prescribe" a course of treatment must be **open to manipulation** prior to the expected results (see the gray box, Predictive Models versus Prescriptive Models—They are not the same!). Additionally, the manipulated variables must be **causal**, not correlative.

Causation—the most misunderstood concept in data science today

In every Statistics 101 class, every student learns that "correlation is not causation." With good reason as one of the author's favorite quotes is *"if you torture the data long enough they will admit to anything."* This is consistent with what most students learn, what statisticians call happenstance or spurious correlation. Tyler Vigen has constructed dozens of these examples where he shows extremely high correlation statistics between two variables, and yet common sense tells us that the variables cannot be causal in nature. Examples are "US spending on science versus suicides," "# of people drowned after falling out of a fishing boat versus marriage rate in Kentucky," and many more. See Tyler's website, "Curious Correlations" at http://tylervigen.com/spurious-correlations.

This is only one type of misunderstanding we make in determining correlation/association with causation. Another example is *temporal precedence*. For one variable, say x, to cause another variable, say y, then x must occur prior to y. A highly predictive model for determining the likelihood of rain is the presence of mud! The reader may laugh, but many predictive models have been placed in practice that suffer from this misunderstanding.

A final example that may cause issues in understanding correlation/association versus causation (others exist) is the presence of a **confounding variable**. It is true that sleeping with shoes on is highly correlated with waking up with a headache. This is not a spurious correlation. This is not an issue with temporal precedence. This is an example where a third variable, a common cause, accounts for the relationship—drunkenness. Drunkenness is a confounding variable. It causes an increased likelihood of wearing shoes to bed and an increased likelihood of waking up with a headache. (Example attributed to Brady Neal, 2020.)

These examples are elementary and straightforward. In practice, this is much more difficult. This author, a trained statistician and Bayesian for more than 30 years, still makes mistakes and has to constantly fact check his beliefs—science is founded in constant doubt.

Practically, what is the big difference in correlation/association versus causation? It depends on what you are trying to do! If you want to make a prediction, all you need is correlation/association. If you want to understand *why* then you need causation. Precisely diagnosing a condition requires knowing what factors are present to cause that condition. Accurately prescribing a therapy requires you to know the causal factors. An optional example of predictive versus prescriptive models is presented in the following gray box.

Predictive Models versus **Prescriptive Models—They are not the same!**

It is amazing how many articles written today describe prescriptive analytics as just a different business case of predictive analytics with no additional requirements from the data side or the modeling side. It is absolutely one of the least understood paradigms out here. **You cannot simply use observational data and a machine learning algorithm and declare it to be a prescriptive model. No. We are sorry to be so contentious here, but there are hundreds of articles and references that imply you can without potentially serious consequences**. We have even had conversations with people that know that causation is not correlation but still think there are no requirement differences for answering the question, "what is the probability this event will happen?" versus "why did it happen?" or "how can we make it happen?" The following is a very simple example of why (there are many reasons) you cannot take a predictive model and then declare it as a prescriptive model. Consider a simple five-variable model to predict obesity by measuring BMI (Body Mass Index). BMI is a weight-to-height ratio, calculated by dividing one's weight in kilograms by the square of one's height in meters and used as an indicator of obesity and underweight. Suppose we use the following five variables to predict BMI:

1. Gender (male/female)
2. Number of hours per week in the gym or fitness center
3. Amount of diet soda intake per week
4. Age
5. Geographic location (place of current residence)

Those five factors would do a decent job **predicting BMI** via a machine-learning model given that we can train the model correctly and with enough data. This model will likely generalize, meaning that it will predict BMI of individuals with future data (given ordinary, common assumptions). **Good predictive model—PASS!** Now, if you believe the many posts, articles, and misdirection out on the internet and in some books, you can use this model to **PRESCRIBE** actions to lower your BMI. **No**, you cannot relocate someone to Colorado or Hawaii and expect it to have a practical effect. Colorado and Hawaii states have lower BMIs on average due to culture and genetics—not geography. **Culture and genetics are confounding variables (they relate to location and BMI) and they are causal the causal mechanisms NOT geography. Good prescriptive model—FAIL!** This is just one reason a predictive model may not be prescriptive.

This example is simple and straightforward; moving someone from Mississippi to Colorado will not likely change his or her BMI, at least in the near term. It is much more difficult to understand the underlying associative mechanism in medical models, therefore, practitioners must be careful and understand the difference between association and causation.

The following section will discuss options for transforming predictive models into prescriptive models. First, let us briefly state some requirements for each of these advanced analytics.

Some basic assumptions for predictive modeling

1. Correlation or association between what is being predicted (target, dependent variable, predictions) and the predictors (independent variables, factors, features).
2. There is sufficient quality data available to train and assess the accuracy and relative metrics of the model. The algorithm used to create the model must be able to discriminate the signal (goal) from the noise (additional meaningless information) in the data.
3. The data used to train the model are representative of the population under study.
4. The model generated is extensible to new/future data. It is robust to small changes in population stability/variation.
5. Temporal precedence of data inputs (independent variables, factors, features) for the predictive model exist and are available for model execution, i.e., making predictions.

Some basic assumptions for prescriptive modeling

1. Causation is established or assumed between what is being predicted (target, dependent variable, predictions) and the *manipulated* predictors (independent variables, factors, features). Note: You can have additional variables in the model that are merely correlated. However, the variables you *manipulate* must be causal to the outcome.
2. There is sufficient quality data available to train and assess the accuracy and relative metrics of the model. The algorithm used to create the model must be able to discriminate the signal (goal) from the noise (additional meaningless information) in the data.
3. The data used to train the model are representative of the population under study.
4. The model generated is extensible to new/future data. It is robust to small changes in population stability/variation.

5. Temporal precedence of data inputs (independent variables, factors, features) for the prescriptive model exist and are available for model execution, i.e., making predictions.

Using a predictive model for prescription purposes

The end goal of all analytics is to make better decisions and take better actions. As shown in Fig. 25.1, prescriptive models afford clinicians more time to serve patients by taking away the thought process of evaluating different treatment options or operational decisions. Instead of these mental calculations based on the output of the predictive model, a prescriptive model provides the next best action to serve patients. The main task of medical staff is the validation, given their medical knowledge and related resources, that this action is consistent with their medical acumen.

Therefore, we would like to use predictive models in more prescriptive ways. We will not go into model development efforts like this chapter's classification of competing therapies. Instead, we provide three essential considerations to transform a predictive model into prescriptive methods.

Three considerations for transforming a predictive model into a prescriptive model

1. Assume causation based upon on well-documented scientific studies
2. DoE and randomized clinical trials (RCTs)
3. Causal inference methods that may include Bayesian networks

Assume causation based on well-documented scientific study

If there is evidence to conclude the **manipulated** predictors (independent variables, factors, features) cause the prediction (target, dependent variable) then a proprietor of the model may cautiously move forward.

We cannot express enough that extreme caution should be used with this method. A practitioner should know the assumptions and limitations of the study. They should also keep in mind the multivariate and interactive nature of ML and consider the differences of isolated variables versus extracting features and the impact of interactions between variables. Moreover, better statistical methods validate a prescriptive action, as we will see.

Design of experiments and randomized clinical trials—the traditional gold standard

The golden standard for years in determining whether a clinical intervention was causal in nature upon a medical outcome has been the RCT. These trials require good data and a rigorous statistical design process. They are expensive, are proactive, and interventional, so they require a great deal of time to conduct. Furthermore, while the gold standard, these studies have potential problems.

For example, RCT requires treatment and control groups to be comparable. If there is a confounding relationship of the treatment itself with the treatment outcome, we have a problem. Going back to our previous example, if we want to know if wearing shoes to bed causes a headache in the morning, we need comparable test and control groups. However, the shoe sleepers were not comparable to the nonshoe sleepers because most of the shoe sleepers were drinking the night before. Conversely, most nonshoe sleepers were not drinking the night before. Therefore, you cannot simply randomize the control and groups based on shoe sleepers. You would have to manipulate the sleepers (experimental units), take some shoes off sleepers who had been drinking, put on shoes for nondrinkers and then randomize.

Any time you manipulate experimental units, you can cause extraneous factors that can ruin the experiment's validity. Fortunately, as we will see, there are better ways to determine causation when a cofounding variable exists. RCTs require comparable groups. Randomization works well if you have large enough samples sizes and the ability to assign experimental subjects into test and control groups. Randomization creates an even playing field that averages the impact of confounding factors or unseen (lurking) variables and allows for testing the treatment difference between the test and control. See de Finetti (2017) exchangeability and the i.i.d. statistical model for more information.

Causal inference methods that may include Bayesian networks

We will soon discover there are techniques that overcome confounding variables with the outcome, such as causal inference methods that may include Bayesian networks. We will cover these methods and dedicate a section to these methods, and we will provide an example at the end of the chapter.

Some important notes on observational studies

Before we move to causal interference, we should make some important points about observational studies. Most studies are observational, saying that there is no random assignment of test subjects into test and control groups. Furthermore, most often, there is no manipulation of the intervention or treatment.

According to a BMC Medical Research Methodology (Mueller et al., 2018), around 80%–90% of published clinical research is observational in design. The authors state, "many research questions cannot be investigated in RCTs for ethical or methodological reasons." We expand upon this comment and add reasons why observational studies are often preferred to RCTs:

1. **Ethical reasons**. It is unethical to randomize people into treatment groups. For example, suppose we want to determine the effect of smoking on cancer type or rates. It would be unethical to assign people to smoke randomly.
2. **Infeasibility**. There are certain factors you cannot control or manipulate. Suppose you want to measure the effect of capitalism versus communism on GDP. You would have to be a dictator of the world as you'd have just to assign whole countries to economic systems.
3. **Impossibility**. Suppose you want to determine the likelihood of cancer given a genetic sequence of DNA. It would be impossible (at this time, and potentially unethical at any time) to alter a baby's DNA sequence.
4. **Practicality**. RCTs are complex, expensive, and very time-consuming. Observation is an everyday practice we all engage in. If we can collect data observationally and then model it to account for conditions we cannot control, we have a principal advantage.

It makes sense why observational studies are dominant. Yet, with new analytical techniques, scientists and researchers can provide more credible results versus the way most of these studies have historically been analyzed. Observational studies are often plagued with confounding conditions. Historically, these conditions are just assumed away or neglected. Yet, with modern causal inference methods, these factors in observational studies can often be isolated, and a causal mechanism can be specified.

Causal inference and why it is important

For some applications, prediction is all that a clinician or analyst requires to make a decision or to take action. Returning to Fig. 25.1 They want to know:

What happened?
What is the likelihood that this event will happen?

In other circumstances, the scientist wants to know to understand the underlying mechanisms that produce an outcome. They want to know:

Why did it happen?
How can we make it happen?
What is the best action?

These are very different questions and require a different set of mathematical rigor.

Burk and Miner (2020) explain that all predictive modeling is based upon statistical modeling. Statistical models are simply trying to understand the characteristics of a population of interest. That population often contains unforeseen, future outcomes. We collect samples and generate mathematical characterizations to model or understand these unseen observations, called statistics. *Any* function of a sample is a *statistic*.

We now offer a very high-level overview of some bridges from statistics to causal models. These models are useful in understanding the second set of presented questions—the why, the how, and what is the best action? We previously presented an example attributed to Brady Neal where we want to know if going to bed with shoes on causes a headache in the morning. We spoke of some of the difficulties one would face trying to perform a RCT to determine whether this effect was causal due to the confounding variable drunkenness that influences both whether someone goes to bed with their shoes on and whether they wake up with a headache. To view this relationship, we may want to construct a causal diagram, Fig. 25.2.

This diagram may also be called a directed acyclic graph meaning there is a directed flow, and the process is not cyclical. These graphs are useful in diagraming and understanding the relationships between mechanisms/variables.

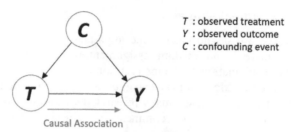

T : observed treatment
Y : observed outcome
C : confounding event

FIGURE 25.2 A causal diagram of treatment, outcome, and confounding mechanisms.

We can see that while the proposed causal mechanism is *T* (going to bed with shoes on) describes the outcome *Y* (waking up with a headache), there is a third confounding mechanism C (drunkenness) that influences both.

> **Directed Acyclic Graphs are a powerful tool for causal inference**
> Andrew Heiss states
>
> *DAGs (Directed Acyclic Graphs) are a powerful tool for causal inference because they let you map out all your assumptions of the data generating process for some treatment and some outcome. Importantly, these causal graphs help you determine what statistical approaches you need to use to isolate or identify the causal arrow between treatment and outcome. One of the more common (and intuitive) methods for idenfifying [sic] causal effects with DAGs is to close back doors, or adjust for nodes in a DAG that open up unwanted causal associations [sic] between treatment and control. By properly closing backdoors, you can estimate a causal quantity using observational data.*
>
> (https://www.andrewheiss.com/blog/2021/09/07/do-calculus-backdoors/)

Causal diagrams look much like Bayesian networks (not a coincidence since they were both pioneered by Judea Pearl). They do not compete with ML and AI. They complement AI and ML.

Why is causal inference so important? It is essential to science. It answers why or how things happen, that is, causal likelihoods. Rather than the mere likelihood, something will occur, that is the statistical likelihood. They inform actions we can take to extend life and improve the quality of health. Causal inference is essential for optimal decision-making.

We do not merely want to understand the PROBABILITY this event will happen—we want to know WHY it might happen and WHICH action we can take to make something better happen!

Bridging the causal models to statistical models—causal inference

Our end goal is to determine causation and create prescriptive models. We can take the model presented in Fig. 25.2 and present it statistically. Furthermore, and more importantly causal inference allows us to determine causation in observational studies, which are by far and wide the majority of scientific studies and virtually all business studies. At an extremely high level, we take a "Causal Estimand" and through a process of "Identification" we formulate a "Statistical Estimand" which through a process of Estimation we express as an Estimate. This is illustrated in Fig. 25.3.

While this is not a book on Bayesian networks, causal inference, or directed acyclic graphs, we offer a brief overview of these techniques as they offer superior methodologies to overcome some of the weaknesses of RCTs and observational studies we have mentioned. We also provide recommendations, references, and computing techniques for further review.

Bayesian networks

There is a common misconception that Bayesian networks by themselves establish causation. This is misplaced in the literature, blogs, and posts. Judea Pearl, the inventor of Bayesian networks, says in *The Book of Why* (Pearl & Mackenzie, 2020):

FIGURE 25.3 The identification–estimation flowchart.

"Though I am delighted with the ongoing success of Bayesian networks, they failed to bridge the gap between artificial intelligence and human intelligence. I'm sure you can figure out the missing ingredient: causality."

Pearl points out that Rev. Thomas Bayes, the creator of Bayesian Statistics, did not understand that when he mathematically showed his rule for inverse probability, it represented the simplest form of a Bayesian network. This type of network takes the form:

Rain Mud

Fire Alarm

Hypothesis Evidence

Unlike causal diagrams like we saw in Fig. 25.2, these simple Bayesian networks carry no assumption that the arrow has any casual meaning. The arrow simply signifies we have a forward probability. P(Alarm | Fire), "the probability of alarm given there is a fire." Belief propagation works the exact same way whether the direction is causal or noncausal.

Bayesian networks are extremely powerful; however, they can be tremendously computationally intensive. They can be used to calculate simple probabilities or test causal mechanisms. We end this section with a summary of some of the benefits and drawbacks of these networks.

Benefits of Bayesian networks

1. They provide a mechanism to forgo the complicated and time-consuming requirements of RCTs.
2. They are useful in retrospective, observational studies.
3. Given the graphic nature of these networks, it is easy to understand the relationships and dependencies of the variables of the model.
4. Models are transparent, and all parameters are interpretable.
5. They are a great learning device for new practitioners given the model transparency. Simple and intuitive examples are available for this purpose.
6. Predictions can be made about any variable in the model. They are not limited to dependent (Y) variables.
7. They are robust to missing or sparse data.
8. They can be used with a retrospective, observational studies.

Drawbacks of Bayesian networks

1. As stated, these models can be very computationally intensive.
2. They perform poorly on high-dimensional data. These networks will not support dozens, hundreds, or thousands of variables like many other ML methods.
3. These networks require much more analyst time and effort to set up. Therefore, they can be expensive.
4. They require expertise. While understanding their behavior is intuitive, it requires experience and knowledge to know how to set them up properly. They take much more effort to set up versus many other ML methods. Therefore, other ML methods are preferred if the goal is supervised predictive models.
5. They are acyclic and therefore cannot support a recursive or reinforcement loop.

Causal inference and the do-calculus

The do-calculus is an axiomatic system for replacing ordinary conditional probabilities with probability formulas containing the do operator. It was first developed by Judea Pearl (see Pearl, 2009, 2012).

In simple terms, this calculus is a way to express a causal diagram mathematically using conditional probabilities. Note that we express what we know or believe in a causal diagram and then symbolically express it a mathematical formulation of conditional probabilities to express *what we want to know*.

In RCTs, we often test the efficacy of a clinical intervention or treatment. If we are interested in the effect of drug (D) on life span (L), then our query could be expressed symbolically as $P(L \,|do(D))$. What is the probability (P) that a typical patient would survive L years if made to take the drug? In an RCT, this would be the test group and we can express the control group as $P(L \,|do(not\text{-}D))$. This do-operator signifies we have an active intervention and not merely an observation. There are no classical statistics operators that function in this manner. In classic statistics, we answer a question like, if the patient takes a drug, what is the likelihood they will survive L years, $P(L \mid D)$?

There is no do-operator in classical statistics. The power of classical statistics in a RCT is that by randomizing and taking sufficiently large samples, we are washing out extraneous factors and the test and control groups are similar except for the intervention/treatments. This allows us to compare the test versus control specifically for the intervention.

Causal inference and the do-calculus allow for a completely new language and new questions to be asked. As Pearl states "$P(L \mid D)$ may be totally different from $P(L \mid do(D))$. This difference between seeing and doing is fundamental and explains why we do not regard the falling barometer as the storm's cause. Seeing the barometer fall increases the probability of the storm, while forcing it to fall does not affect this probability." This confusion of seeing and doing can form a set of paradoxes if one does not consider causes. A more significant proportion of people die when they go to the hospital than if they do not go to the hospital. Without considering causal factors, this type of logic is misleading at best.

Causal models can improve study designs by providing new language and clear rules for deciding which variables need to be included, which variables need to be controlled for, and what variables may be confounding factors. This is a far different practice from what some naïve ML practitioners take, which is to throw a bunch of variables in and let the computer figure it out! This process for designing studies and the DoE has recently been discounted or replaced by sheer volumes of data and computational power. Good studies and good advanced analytics require good design.

Causal models allow certain questions to be answered from existing observational data without needing an interventional study like the RCT. As previously noted, there are several reasons that interventional studies are inappropriate - ethical reasons, infeasibility, impossibility and practical reasons. This means that some hypotheses cannot be tested without a causal model.

Observational Calculus versus the Do-Calculus

Causal calculus differentiates between two types of conditional distributions one might want to estimate. Given the traditional statistical setup, we are sampling from some joint distribution $p(x,y,z,...)$. We have large data sets, and we have the tools and machinery to estimate the joint distribution fully and thus any property therein (marginal, conditional), i.e., p is known and tractable. We are interested in the conditional of y given x. There are two calculus methods, and which one we use determines whether we can claim association or causation.

In traditional AI or ML methods, we use the **observational** conditional distribution for supervised learning, i.e., we use the traditional statistical method for the conditional, $p(y|x)$. That is the ratio of the joint $p(y, x)$ divided by the marginal $p(x)$.

For causal interference we use the **interventional** conditional distribution $p(y|do(x))$ that answers the question, "what is the distribution of Y if I were to set the value of X to x?" This describes the distribution of **Y** we would observe if we intervened in the data generating process by artificially forcing the variable **X** to take value x, but otherwise simulating the rest of the variables according to the original process that generated the data. The data generating procedure is **not** the same as the joint distribution $p(x,y,z,...)$, which is an important detail.

These conditionals are not the same. Imagine an instrument that produces a "real" measurement, **Y**. Suppose that instrument has a gauge that displays that the estimated measurement, **X**. While **X** is very useful for estimating the real measurement **Y**. **Y** is not caused by **X**.

NOTE: Lattimore and Rohde propose to circumvent the Do-Calculus with Bayes Rule (see Lattimore and Rohde, 2019).

A summary example of causal modeling

We now offer an example that ties together several of the concepts we have covered (observational studies, causal diagrams, confounding, do-calculus), as well as some new ideas like Simpson's Paradox. This discussion is an extended example, based on one offered by Brady Neal in his Causal Inference course (see Neal, 2020). In this example, Neal supposes a new fictitious disease called COVID-27 and there are two treatments (T) that are being investigated, A(0) and B(1). There are two conditions (C) of the disease, mild (0) or severe (1) and we are interested in a binary outcome Y, alive(0) or dead(1). It is important to note that treatment B is more expensive than treatment A, such that treatment A is administered at a rate of about three times as often as treatment B.

If we look a simple comparison of Treatment A versus B and the proportion of patients that die, we have 240 of 1500 patients or 16% dying with treatment A and 105 of 550 patients or 19% dying with treatment B. This is shown in Table 25.1 with the outcomes being simple proportions equivalent to the Expectation of the outcome Y given the treatment t, $E[Y|t]$.

Treatment A in this table seems to be preferred, with only 16% of patients dying versus 19% for treatment B.

Now if we subgroup the mortality table by condition, mild and severe (Table 25.2), something very interesting happens. While the overall total is the same, Treatment B now appears to provide a lower death rate for both mild and severe conditions. The mild condition group is 10% for Treatment B versus 15% for Treatment A. In addition, for the

TABLE 25.1 Simple comparison of Treatment A versus Treatment B, $E[Y|t]$.

	Total
Treatment A	16% (240/1500)
Treatment B	19% (105/550)
	$E[Y/t]$

TABLE 25.2 Comparison of Treatment A versus Treatment B by condition subgrouping.

	Mild	Severe	Total
Treatment A	15% (210/1400)	30% (30/100)	16% (240/1500)
Treatment B	10% (5/50)	20% (100/500)	19% (105/550)
	$E[Y/t, C=0]$	$E[Y/t, C=1]$	$E[Y/t]$

severe condition group, it is 20% for Treatment B versus 30% for Treatment A. Thus, each subgroup for Treatment B is lower, yet the total is higher. This is not a mathematical trick. It is Simpson's Paradox and it obviously has to do with the nature of the subgroups. You can see the associated conditional expectation formulas at the bottom of the columns.

As stated previously, treatment B is more expensive and therefore is used one third as often as treatment A. Additionally, 1400 of the 1500 people who received treatment A had a mild condition, whereas 500 of the 550 people who received treatment B had a severe condition. Because people with a mild condition are less likely to die, this means that the total mortality rate for those with treatment A is lower than what it would have been if mild and severe conditions were equally split among them. The opposite bias is true for treatment B.

This is the paradox. It appears that if we know the condition we should provide treatment B and if we do not know the condition we should provide treatment A. If you have to make the decision, what should you do? Causal inference can help. By marrying the medical knowledge with statistical modeling of expectations, you can improve your decision-making. To do this, you can model your medical knowledge into a casual diagram.

You know that we are tracking three variables, condition, treatment, and outcome. From your medical knowledge, what is the suspected relationship of these variables? By collecting some additional information, you determine there is a temporal effect, time to receive treatment. What is happening is that preliminary studies suggested that treatment B might be more effective in treating more severe cases. That is why it is more expensive. Moreover, since it is more expensive, clinicians have to receive prior authorization from insurance companies before it can be administered. Thus patients often have to wait 1 or 2 days before receiving it. It might be that the time lag is contributing to the death rate. For sure, the condition appears to be causal to the selection of the treatment. Therefore, a plausible causal relationship is presented in Fig. 25.4. The condition of the patient (mild, severe) is hypothesized to be a causal factor for both treatment selection (A, B) and the patient outcome (alive, dead).

Before we calculate these new, causal probabilities, we can express the two proportions in the "Total" column of Table 25.2 for each treatment as:

For Treatment A, (1400/1500)(0.15)+(100/1500)(0.30)=0.16

For Treatment B, (50/550)(0.10)+(500/550)(0.20)=0.19

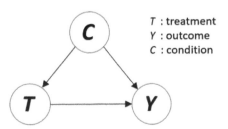

T : treatment
Y : outcome
C : condition

FIGURE 25.4 A causal diagram of treatment, outcome, and condition where condition is a confounding variable.

TABLE 25.3 Comparison of Treatment A versus Treatment B with causal modeling and the do-calculus.

	Mild	Severe	Total	Causal
Treatment A	15% (210/1400)	30% (30/100)	16% (240/1500)	19.4%
Treatment B	10% (5/50)	20% (100/500)	19% (105/550)	12.9%
	$E[Y/t, C=0]$	$E[Y/t, C=1]$	$E[Y/t]$	$E[Y|do(t)]$

For treatment A, the row total is 1500 and the "weight" or number of patients in each cell is 1400 for mild condition and 100 for severe condition. The mortality rate is 0.15 for mild conditions and 0.30 for severe conditions. For treatment B, the row total is 550 and the "weight" or number of patients in each cell is 50 for mild condition and 500 for severe condition. The mortality rate is 0.10 for mild conditions and 0.20 for severe conditions.

These agree with the original (naïve) totals of 16% for Treatment A and 19% for Treatment B.

If we want to calculate the mortality rates based on our causal model we simply need to manipulate the weights as follows:

For Treatment A, (1450/2050)(0.15)+(600/2050)(0.30)=0.194

For Treatment B, (1450/2050)(0.10)+(600/2050)(0.20)=0.129

The weights now are simply the number of patients in the column (1450 for mild and 600 for severe conditions) divided by the total of patients in the study, 2050.

Putting it all together, we get Table 25.3, which allows us to compare the naïve "Total" mortality rates versus the "Causal" mortality rates for the respective treatments.

If the condition of the patients influences both the treatment prescribed as well as the outcome then treatment B is preferred with an expected mortality rate of 12.9%. You can see the associated symbolic formulas for the conditional expectations at the bottom of the columns. We also see the calculation we just made for the causal mortality rate is essentially just applying the conditional expectation and the do-operator.

Therefore, we see that the determination of the most effective treatment is dependent on the formulation of the problem. Without constructing the causal diagrams and related formulas we cannot resolve Simpson's paradox. By constructing the causal diagrams and related formulas we can resolve Simpson's paradox.

Conclusion

As I said in my opening note, a key to success in life is to have many tools in your bag. One thing I emphasize more than anything else to my data science students is "The No-Free-Lunch Theorem" (no machine learner is universally optimal). There is no statistical, AI, or ML methodology that is universally optimal. We need options that fit the problem at hand. We need a large tool bag.

Yet, these days we are always looking for the killer app, the killer classifier, the killer method. Good science and analytics take time and a deep understanding of the problem. We should not rush in. Although I am not sure if it is true, one of my favorite stories is of someone asking Einstein if he had an hour to solve a problem how he would approach it. He replied that he would spend the first 55 minutes deeply understanding the problem and all its intricacies, conditions, and dependencies and the last 5 minutes devising a solution. I agree. I think we should spend more time understanding the problems we face before we try to solve them. We should start with questions—what are we attempting to answer? Do we want to know:

What happened?
Why did it happen?
What is the likelihood this event will happen?
How can we make it happen?
What is the optimal action we can take?
We can then reach into our bag of tools—those tools developed over decades and apply them appropriately and responsibly.

Postscript

The statement was made earlier that medicine is becoming less a "practice" and more of a "science." The practice of medicine for over 5000 years was based on what has happened to other people in the past, and the "practice" of assuming that previous medical outcomes will apply to new patients. Rather, the application of science (the art of gaining knowledge in the present) to individual patients turns this practice around in the 21st century, and can point to individual data from which the best treatments can be *prescribed*, and not *practiced*. The implications of this paradigm shift are explored and discussed in Chapter 26.

References

Burk, S., Miner, G.D., 2020. It's All Analytics!: The Foundations of AI, Big Data, and Data Science Landscape for Professionals in Healthcare, Business, and Government. CRC Press.

de Finetti, B., 2017. In: Machí, A., Smith, A. (Eds.), Theory of Probability: A Critical Introductory Treatment. John Wiley & Sons.

Lattimore, F., Rohde, D., 2019. Replacing the Do-Calculus With Bayes Rule. In Cornell University-axie Blog: https://arxiv.org/abs/1906.07125.

Mueller, M., D'Addario, M., Egger, M., et al., 2018. Methods to systematically review and *meta*-analyse observational studies: a systematic scoping review of recommendations. BMC Med. Res. Methodol. 18, 44. Available from: https://doi.org/10.1186/s12874-018-0495-9.

Neal, B., 2020. Introduction to Causal Inference From a Machine Learning Perspective. https://www.bradyneal.com/causal-inference-course.

Pearl, J. as interviewed by Wasserstein, R. (2012) Turing award winner, longtime ASA member talks causal inference, *Amstat News*, November. Available at: https://magazine.amstat.org/blog/2012/11/01/pearl/.

Pearl, J., 2009. Causality: Models, Reasoning, and Inference, second ed. Cambridge University Press, Cambridge.

Pearl, J., Mackenzie, D., 2020. The Book of Why: The New Science of Cause and Effect. Basic Books, New York.

Siwicki, B., 2018. Machine learning helps University of Iowa reduce surgical site infections, 10 September. Available at: https://www.healthcareit-news.com/news/machine-learning-helps-ui-health-care-reduce-surgical-site-infection-74-save-12-million.

Further reading

Pearl, J., 2012. The Do-Calculus revisited. In: Proceedings of the Twenty-Eighth Conference on Uncertainty in Artificial Intelligence. UAI'12. AUAI Press, Arlington, VA, pp. 3–11. https://dl.acm.org/doi/10.5555/3020652.3020654.

Chapter 26

The future: 21st century healthcare and wellness in the digital age [☆]

Gary D. Miner and Linda A. Miner

Chapter outline

Prelude

The central value of this book is to provide you with a framework for conducting and evaluating medical research, which will enable you to read research and to question methods and assumptions in the process of making your own decisions about reality. You might conclude that research is difficult to conduct, appropriate data to serve it are often not available (perhaps missing or inaccurate), or the data may not be representative of the case at hand. Researchers certainly do strive for excellence, but flaws can often be found in most studies. In spite of these problems, progress is made through repetition and refinement of models. The overall goal is to strive for perfection and do the best we can to eliminate diseases.

Evidence of this progression is clear when we look back over the history of medicine. For example, we no longer employ bloodletting for headaches, we no longer sequester people as "inmates" of "lunatic asylums," nor believe that the brain is the repository of mucus. No longer do we search for a bullet with dirty fingers in a wound, as was done for President Garfield, causing him to die of sepsis. We have made progress in the eradication of many contagious diseases, progress in the treatment of many cancers and other diseases. We are making progress in fitting treatments to individuals as we use predictive research and learning algorithms. Now, you can use new techniques in your own research, even if your research concerns you as a patient conducting your own individualized study.

[☆]. With contributions by Thomas Hill, Robert Nisbet and Nephi Walton, and Scott Burk and Mitch Goldstein (e.g., "Your Authors").

Practical Data Analytics for Innovation in Medicine. DOI: https://doi.org/10.1016/B978-0-323-95274-3.00012-9

Above all, your reading should prompt questions as well as stimulate your desire to become a part of new kinds of research. You may now be more interested in doing your own exploratory and predictive research. We hope so!

Overview

This chapter attempts to summarize much of the proceeding 1st edition of the book, plus focuses the reader on the future. In the first edition, we predicted dramatic changes and improvements in the field of healthcare and wellness within the next 6 years: a new 21st century high technology healthcare and wellness movement would shake the existing healthcare industry and institutions.

Here we are 6 years later (and more). Listed below are the expectations and the reality that happened:

1. That the healthcare industry would be reorganized into patient-centered wellness and medical teams supported by digital and mobile health technologies (by 2020). **But here is the reality of early 2022**: This has become partly true. Many medical practices now have patient portals, some better than others, but most are attempting to reach out to patients electronically and individually. Patient-centeredness seems to be a bit spotty, but the idea is out there and becoming part of the vocabulary. There may be some fear in the patient-directed part, perhaps due to fears that patients will avoid doctors, and perhaps in part because it could be threatening to the old status of doctors as the source of all medical information, and "what if patients find out we really do not know everything?"

2. By 2020, the first comprehensive health and wellness sensors and predictive analytics hardware tools will be common: embedded in eyeglasses, clothing, architecture, facilities, and nanotechnologies in the human body. **BUT here is the reality of early 2022**: There are now common wearable devices such as i-phone watches that track heartrates, blood pressure, and sleep; medical alerts wearables for the elderly that send information to monitors even when a person is about to fall, as well as if the person needs help and can send an alert to the monitoring agency; GPS tracking devices for Alzheimer's patients, home defibrillation machines, smartwatch ECGs, glucose monitors, at home EKG machines, and author Gary's I-Care IOP machine. Computer screens have been fitted into eyeglasses, but that might not be a health benefit. Home devices and wearable devices will continue to proliferate, if not explode during 2022 and 2023, we predict. In addition, we predict that smart data scientists and IT folks will be designing apps for patients to conduct their own N-of-1 studies, and making a lot of money doing so, as suggested in Chapter 7.

3. In 2015 (1st Edition) we stated: Our 21st century wellness movement will be focused on real-time analytics and feedback to achieve peak wellness through *optimal nutrition, exercise, stress-reduction, and measurements of continuous human health improvements* and medical treatments, when needed. By 2020, science based personalized medicine and wellness will become the standard: most (if not all) wellness and medical treatments will be based upon individual genome studies combined with a full array of digital health information with real time analytics guiding individuals and medical support teams. **BUT here is the reality of early 2022**: Oh my, we don't know what it will take to help people achieve this goal. Obesity alone continues to rise, in the United States, as well as many countries. Other obesity-related diseases are on the rise as well, often due to inflammation. The use of genome information in "real time" in the EHR for assisting clinicians in making rapid and more accurate diagnoses and treatment decisions has NOT happened yet. This is due partly to the fact that the computing power needed is not yet readily available (see last past of Chapter 10). For individual people, many are changing their eating habits and increasing exercise, at least they are thinking of it. It is hoped that the concept of N-of-1 studies will help people figure out what works for them as individuals. Perhaps if doctors catch on to the idea, they can prescribe "studies" for their patients and help them learn to analyze data. This was the goal of Chapter 7—to have the means on patient portals for patients to input their data and produce graphs or other helpful basic statistics for doctors and patients.

Background and need for change

We started this book in its initial chapters (see particularly Chapter 1) with the need for changes in healthcare delivery and especially the need for predictive analytics in medicine. We reiterated that need for change and point to the fact that change is happening already. But this change has not been without some dead-ended failures.

One of these failures, which was much heralded since the early 2010s and especially from 2015 to 2019, was the IBM Watson Health solution. But there have been lawsuits from major hospitals over the past 5 years, like from MD Anderson which adopted IBM Watson Health at a cost of millions of dollars. They brought these lawsuits because of the exorbitant amount of money they spent with promises from IBM that Watson would do what it promised it would do in terms of monetary returns (Pringle, 2022). However, the Watson Health solution did not generate the revenue predicted. After years of frustration with Watson Health, IBM put it up for sale, but it did not sell. Finally, IBM Watson was put up for sale again in early 2022 (Pringle, 2022), and evidently it sold within 2 weeks at a very low value compared to what IBM had invested into Watson Health.

Faults of IBM healthcare were flaws rooted in data collection and flaws in interoperability (Sweeney, 2017a). In addition, at that time, the predictions to individuals may not have been accurate, which could also affect revenues. Jaklevic (2017) reported that even if IBM could view all the medical literature available, not all articles were of the same quality, and implied that the machine viewed all of them as equally valid. The assumption that the machine could not differentiate between qualities of articles would not be true, if the authors of the system successfully taught the machine to evaluate the quality of research articles. One other way to do this, might have been if the builders preselected only those articles that were of top caliber to input into the system. Answering correctly in Jeopardy is not the same as evaluating evidence to use in making an accurate prediction.

As Jonathan Chen MD and Steven Asch MD from Stanford University's Department of Medicine pointed out in the New England Journal of Medicine:

> *AI's true value will be in decision support, Chen and Asch added. Current AI applications often reach conclusions that are already well known to patients. For AI to make an impact on clinical care, the technology needs to provide predictions that can influence care decisions enough to improve current practice... But first, the industry will have to tackle a significant hurdle in making patient data more accessible and available in real time. As Chen and Asch point out, clinical data for predictive purposes has a shelf life of about four months.*

(Sweeney, 2017b)

The authors of this book are not surprised that IBM got rid of IBM Healthcare but are surprised that they found a buyer for it at $1B. One billion dollars is still a gigantic loss for IBM, but we question that value to the buyer, as our medical book describes a landscape of needed changes in the healthcare industry (see Chapters 5 and 6) before the use of analytics will hit the mainstream. Watson was too much too soon, before machines could think for themselves, and applied in the wrong way.

Comparative effectiveness research and heterogeneous treatment effect research

Many of the developed world's national healthcare systems use a structured process to synthesize scientific studies; comparative effectiveness research (CER), and heterogeneous treatment effect research (HTE) to help determine optimal courses of care and improve science- based medical treatments that have been proven to be effective. CER compares the benefits and harms of alternative methods to prevent, diagnose, and treat a clinical condition and monitor care; and HTE looks at different "groupings" of individuals based on genetics and other parameters that separate them into distinct groups; the focus is on methods to improve the delivery of care including the effectiveness of drug treatments and dosage, for the individual patient.

In 2010, the Patient Protection and Affordable Care Act established the Patient-Centered Outcomes Research Institute (PCORI, 2014), a nonprofit organization, to conduct research to provide information about the best available evidence to help patients and their healthcare providers make more informed decisions. PCORI has focused on CER and HTEs, and is building a national infrastructure to conduct CER and HTE research (PCORI, 2022a,b).

PCORI's targeted grants were helping to build this CER and HTE national infrastructure and contribute to the overall PCORI Mission and Vision as of their 2014 "Mission and Vision" statement (which is no longer available on their website), but today PCORI has simplified their Mission and Vision statement:

> *Our vision*
>
> *Patients and the public have information they can use to make decisions that reflect their desired health outcomes.*
>
> *Our mission*
>
> *PCORI helps people make informed healthcare decisions, and improves healthcare delivery and outcomes, by producing and promoting high-integrity, evidence-based information that comes from research guided by patients, caregivers, and the broader healthcare community.*

(PCORI, 2022c)

In the opinions of the authors of this book, PCORI's Mission and Vision need to be enlarged to fully include "patient-centered healthcare," which means that individual patient's needs (e.g., individual genetics emphasized) are held as the utmost importance, which goes beyond "comparative effectiveness population groupings" and "heterogeneous treatment population groupings."

As the authors of this book wrote the final chapter of this book for its 1st edition back in late May of 2014, we saw an interesting phenomenon reminiscent of the "Silicon Valley Bubble" of the 1990s—many MDs were leaving their practice or CMO (Chief Medical Officer) positions in major companies to start their own, small "startup" companies.

We wrote back in 2015:

> *These start-ups are pursuing various medical technology advanced products and processes, including mobile, genetic and other avenues. Apparently venture capital is flowing into these start-ups, meaning that investors see a potential market. Only the best of these products and processes will likely prevail; the best of these will become part of our routine healthcare procedures, making healthcare more cost-effective and individualized. (p. 1045, first edition).*

Well, for the most part this appeared to be happening in 2021 and we expect that the patient "at-home use" mobile aps and devices will only grow exponentially in 2022 and subsequent years.

New technology and 21st century healthcare: health startup firms

Building the "Star Trek" tricorder

What is a tricorder? A multifunction hand-held device that **performs sensor environment scans, data recording, and data analysis**—hence the word "tricorder" to refer to the three functions of

1. sensing (variables in its environment);
2. recording data; and
3. computing (data analytics).

We wrote this back in 2014 for the first edition of this book

Remember the TV and movie series "Star Trek" and how Dr. Bones McCoy used a tricorder device to help diagnose all medical problems?

In 2012, the X Prize Foundation announced a long-term, worldwide contest in which $10 million will be awarded to the three teams that produce the best "tricorder type" devices for detecting and monitoring a variety of diseases and conditions. Teams registered in 2013 and had to show their devices and test results in early 2014. By the summer of 2013, the Qualcomm Tricorder X Prize had drawn hundreds of competitor teams. In 2014, the original 300 teams had been "whittled" down to 10 (Dvorsky, 2014) (Fig. 26.1).

According to the contest website, the Qualcomm Tricorder XPRIZE would provide $7 million to the winning team and prize monies to the second and third placed teams:

— The Qualcomm Tricorder XPRIZE is a $10 million global competition to stimulate innovation and integration of precision diagnostic technologies, making reliable health diagnoses available directly to "health consumers" in their homes....

— "Advances in fields such as artificial intelligence, wireless sensing, imaging diagnostics, lab-on-a-chip, and molecular biology will enable better choices in when, where, and how individuals receive care, thus making healthcare

FIGURE 26.1   Tricorde. *Reproduced from Dvorsky (2014).*

more convenient, affordable, and accessible. The winner will be the team whose technology most accurately diagnoses a set of diseases independent of a healthcare professional or facility, and that provides the best consumer user experience with their device" (Qualcom, 2014, pars 2 and 3).

The winning Tricorder device was to be a tool capable of capturing key health metrics that could lead to diagnosing a set of 15 diseases; among the core set of diseases that had to be diagnosed, were the following:

1. Anemia
2. Urinary tract infection, lower
3. Diabetes, type 2
4. Atrial fibrillation (AFIB)
5. Stroke
6. Food borne illness
7. Shingles
8. Melanoma
9. Strep throat
10. Cholesterol screen
11. HIV Screen
12. Osteoporosis
13. Absence of condition

The Vital Signs Set Tricorder requirements included the following:

1. Blood pressure
2. Electrocardiography (heart rate/variability)
3. Body temperature
4. Respiratory rate
5. Oxygen saturation

The winning Tricorder device was to collect large volumes of data: from various sources; from ongoing measurements of health data; using wireless sensors, and imaging technologies. Further the Tricorder was to use portable, noninvasive laboratory test replacements. Of the 300 teams, several leading medical sensor technology companies registered to join the contest to win the Tricorder prize monies. For example, 10 teams of firms registered from the San Diego area.

In this competition, teams were to leverage technology innovation in areas such as artificial intelligence and wireless sensing. Much like the medical Tricorder of Star Trek(R) fame, doctors, nurses, and individuals/patients can use the Tricorder devices and a smartphone to make preliminary medical diagnoses independent of a physician or healthcare provider. The goal of the competition is to drive the development of devices that will give consumers access to their state of health using the new devices connected to smartphones to share data in the cloud.

The 10 teams that were left were each working to improve medical care by taking the devices to the homes and to the patients themselves. Diagnoses can be made regardless of the location of the individual (Dvorsky, 2014). Our future doctors could be robots! And predictive analytics should be at the center of those processers.

Well did this all happen as predicted? Not quite

But here is what did happen, as we write this final chapter in March 2022 for this second edition of this book:

*The original call for submissions attracted more than 300 pre-registrations. By late 2013, when the X Prize announced the teams for the first time, **there were 34**. The diversity of the teams was impressive—from Silicon Valley startups to large public companies to garage operations, there really was no common thread. They came from all over the world. There were college teams (one of which made it to the semifinals) and even a high school team.*

*By the end of 2014, the 34 teams were winnowed down **to 10 semifinalists**, which eventually became seven (https://www. xprize.org/prizes/tricorder/teams) after **two teams merged and two dropped out** (one by choice and one because its prototype was tragically held up in customs). These 10 built prototypes and had to demonstrate their accuracy in clinical tests.*

At the end of year 2016 only DBG and Final Frontier remained to move on to the final phase, where the devices were left alone in a room with consumer testers, to see if they were sufficiently user friendly.

(Comstock, 2017)

According to a 2017 report (Comstock, 2017) a winner of the Tricorder competition was found, but work on tricorders will continue.

The winner was Final Frontier Medical Devices group (Final Frontier) (https://www.xprize.org/prizes/tricorder/teams/final_frontier_medical_devices).

The runner up was the Dynamical Biomarkers group (DBG) (https://www.xprize.org/prizes/tricorder/teams/dynamical_biomarkers_group).

Final Frontier won the top prize of $2.6 million, and DBG won $1 million. They were somewhat alike, with modular devices containing various sensors. DBG used a smartphone and Final Frontier used a touchscreen called DxtER. One thing that the Final Frontier did that was different from all the other teams was to build their diagnostic AI up front and then develop the devices.

The Qualcomm company decided at the end of the contest, to continue funding innovative ideas and put up $3.8 million towards the "promotion of the digital health ecosystem, including the continuation of consumer testing, guidance for the Qualcomm Tricorder XPRIZE teams, and support for further development of tricorder devices" (Comstock, 2017, par. 25).

Can DstER actually work? (Bacha, 2018; A Noninvasive device that can diagnose several medical conditions: Dxter is a new tricorder that can detect various medical conditions https://strammer.com/en/dxter/). The Final Medical devices team developed a Tricorder sensor that can detect 34 different medical conditions and monitor 5 vital signs. It can detect illnesses such as diabetes, AFIB, pneumonia, tuberculosis, and many more with the use of Bluetooth sensors connected to an Apple iPad-based system. In the future, DxtER could ultimately be time saving by eliminating the overcrowding issue in hospitals. Patients will be able to self-monitor their chronic conditions, therefore, avoiding hospital bills. It will also stop people from having to rush to the hospital for something that is nonlife-threatening or immediately urgent. In parts of the world where there are few healthcare workers, this innovation will provide lifesaving insight (see YouTube video of it at: https://youtu.be/9Yj37PUN_VU).

Scientific American in a June, 2017 article (Moschou, 2017; How close are we to a real star-trek style medical tricorder?; https://www.scientificamerican.com/article/how-close-are-we-to-a-real-star-trek-style-medical-tricorder/#) described the two winners of the Tricorder Prize:

> The main aim of the two prizewinners is to integrate several technologies in one device. They haven't created an all-in-one handheld machine but they do both represent significant steps forward. The main winner, known as DxtER and created by US firm Basil Leaf Technologies, is actually an iPad app with artificial intelligence. It uses a number of non-invasive sensors that can be attached to the body to collect data about vital signs, body chemistry and biological functions. The runner-up technology from Taiwan's Dynamical Biomarkers Group similarly connects a smartphone to several wireless handheld test modules that can analyze vital signs, blood and urine, and skin appearance. The judges said both devices nearly met the benchmarks for accurately diagnosing 13 diseases including anaemia, lung disease, diabetes, pneumonia and urinary tract infection… Part of the success is due to the development of a variety of technologies that make up such all-in-one systems, although they still have some way to go. Probably the most advanced are mobile vital signs monitoring devices. …that feed all the signals wirelessly to desktop or mobile device, with the same accuracy as conventional intensive care equipment.
>
> But while there has been significant progress in the developing bits and pieces of a tricorder, there is still work to do putting them altogether in a genuinely handheld package. Various equipment needs to be miniaturised and we need more progress in portable computers so they can handle all the information and data required for a complete picture of a patient's health condition… We may not have a tricorder in our hands yet, but we are definitely getting closer.

From: Moschou, 2017; How close are we to a real star-trek style medical tricorder?; Scientific American: https://www.scientificamerican.com/article/how-close-are-we-to-a-real-star-trek-style-medical-tricorder/#.

The following video also points out clearly why a tricorder type of device that the patient can control and use themselves is critical to the future of healthcare; Daniel Kraft, MD (founder of Singularity University) is one of the speakers featured in this video: https://youtu.be/rdpdWJdx5CE (2018).

But according to the Basil Leaf Technologies website, as of March 2022, the DifTER device is still under construction to reach its ultimate goals in self-monitoring of health signals (Basil Leaf Technologies; March, 2022; http://www.basilleaftech.com/dxter/).

And the runner-up in the Tricorder XPrize, the Dynamical Biomarkers Group of Taiwan, seems to have fallen "off the map" since 2016, with even their Facebook page having its last posts in December of 2016 (https://www.facebook.com/dynamicalbiomakers/?ref = page_internal), and searches on more recent activity bring nothing.

Why have we spent so many pages on the "tricorder"? Because if it was working, as planned, today, we'd have an excellent device that the individual person (patient) could use to keep track of their health, at home, without numerous trips to doctors and healthcare clinics.

No matter what the final status of the tricorder becomes later in 2022 or beyond, the "X-Prize.Org" page posted the following in 2021, which rings true for healthcare delivery, as it is also the most pertinent message of this 2nd edition of our book:

EMPOWERING PERSONAL HEALTHCARE.
   THE GRAND CHALLENGE
   Very few methods exist for consumers to receive direct medical care without seeing a healthcare professional at a clinic or hospital, creating an access bottleneck … In virtually every industry, end consumer needs drive advances and improvements. EXCEPT IN HEALTHCARE!!! Very few methods exist for consumers to receive direct medical care without seeing a healthcare professional at a clinic or hospital creating an access bottleneck. Despite substantial investment to improve the status quo, even average levels of service, efficiency, affordability, accessibility and satisfaction remain out of reach for many whom the system was intended to help.

---

from: 2021: https://www.xprize.org/prizes/tricorder—Final approvals expected in 2021.

Listing of other e-items in this "outside of healthcare facilities" category but within at least the partial control of patients

- *Mobile Health and Social Media*—patients can (and are) forming groups (support groups) based on either risk profiles or chronic diseases.
- *Digital Health Sensors and Tracking*—health sensors can be put into buildings, homes, bathrooms, and other places to keep track of patients symptoms.
- *Wearable Computing Machines*—such as Google Glasses.
- *In-Body Sensors*—these are either ingested or implantable computerized devices.

Of the above categories, we will give primary focus to "wearable devices" and "N-of-1" studies for the remainder of Chapter 26.

Examples of wearable devices that are working for people today

Atrial fibrillation wearable watch sensors

As mentioned in Chapter 7, devices, such as watches, are now providing valuable medical information to wearers, such as information about stress, sleep problems, breathing problems, buzzing the wearer whenever the person sits too long, sleep apnea, AFIB, even providing ECGs to the wearers. There are many smartwatches from which to pick (Saneesh, 2021). Naturally, one would work with one's physician with these, as with any device intended to give one medical information.

Dr. Roger Seheult (2022) talked about watches currently available that patients can use to see if they are experiencing AFIB. The video is one of Seheult's excellent MedCram lectures. (See Chapter 7 for examples of other topics.) Seheult explained the 12 electrical pathways of a normal EKG that one might have in a doctor's office. The smartwatches, he explained, provide only one pathway, however, that pathway happens to be the one that is needed to report AFIB. Those who wear such a watch and check when they sense that their heartbeat is abnormal, might be able to avoid problems such as blood clots, strokes, or even cardiac arrest. They could also use their phones to conduct N-of-1 studies to learn what personal conditions, such as stress or poor sleep, associate with their AFIB.

I thought to myself, perhaps some company can develop an app for smartphones with sensitive screens that could detect AFIB. In looking it up I was surprised to find that there evidently *is* a smartphone app for Androids and iPhones (Lifestyle, 2014). The app uses one finger on the screen of the phone and another finger on the camera. Of course, one would have to determine if the apps were approved by the FDA—and one's doctor, who also might be very interested in helping establish one's N-of-1 study. It would also be good if insurance companies provided such watches, or apps for a smartphone to people with a diagnosis of AFIB.

Augmented reality app

A very interesting application (Ortega, 2022) was called an augmented reality app. It was not related to medicine, but augmented reality as a concept certainly could be developed to be used by patients and medical personnel. As the subtitle of the article states, the purpose of the app was to "add scientific accuracy to paleoart."

Scientists wanted to help bring the La Brea Tar Pits and Museum animals to life for visitors to the museum. The researchers worked with all the scientific information, and bones, they could find on each animal of interest, such as mammoths, sloths, mastodons, and other extinct animals from the Pliocene era when there were tar pits, some 40,000 years ago. (Reminiscent of IBM Watson?) The researchers hired and worked with a development company that created video games, which then created the virtual animals and animated them. The final app was designed so that most smartphones could display the animations. The visitors look on their phones of various animated animals superimposed on the actual tar pits, creating a more realistic virtual reality.

Such apps for medicine could help explain diseases and processes in the body for patients (and training medical personnel), helping them to understand what was happening where they could not see. Again, as suggested in Chapter 7, teams of doctors, data scientists, IT people, and animators could create such apps, get patents, and sell the apps. Combined with predictive analytics, such applications might be individually generated. One that comes to mind is predicting what one's health will be like if one does not modify their lifestyle, or it they do. Such views might convince someone to change life habits.

EKG—home monitoring digital system

EKG and heart rate can easily be monitored at home by a person using either Kardia or one of the advanced series of the Apple Watch. This can be very helpful for a cardiologist doing an evaluation via telemedicine. It is generally more accurate than a Fitbit or similar heart rate (pulse) monitors, and allows examining the rhythm of the heart. These devices have built-in AI (Artificial Intelligence) that tries to interpret the tracings mainly to try to see if one has "atrial fibrillation." The recordings have to be emailed to the doctor's office (KARDIA: Caren, and Urman, 2020; AliveCor, 2022; APPLE WATCH; January 18, 2022).

Eye pressure (IOP) home measurement devices

Measurement of IOP (Intraocular Pressure of the eye) can be measured at home by the individual person. There are several devices available, but one of the easiest to use is the i-CARE HOME device. Your senior author of this book has had IOP issues for over 10 years; in Fig. 26.2 you see this author being trained in how to use this device; the training took about 20 minutes and your author was using it flawlessly within 2 days of practicing. When used properly we found its accuracy to be almost identical to the eye doctor's office "Goldman IOP Measuring Machine" which has been considered the "Gold Standard"; we found the i-CARE HOME to be identical or within 1 IOP point to the Goldman when measured seconds apart. Please see Chapter 11 in this book for an in-depth look at IOP measuring devices.

(See Chapter 11 for complete story on use of i-CARE HOME intraocular eye pressure monitoring.)

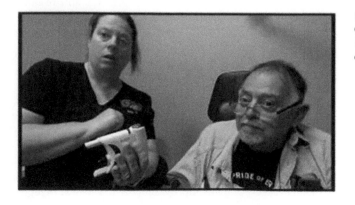

FIGURE 26.2 Gary Miner and his trainer, contributing author, Billie Corkerin with the iCare machine. *Thumbnail by Linda Miner for YouTube video, Gary Trains on the iCare Machine: https://youtu.be/6nakad5Y3HY.*

Nonautomatic vital health signal measuring devices

In addition to the digital vital health signal devices that either automatically (or with a click on the Smartphone app) send the recorded data to a doctor or health clinic, there are other home devices that can easily be used. These do not automatically digitally send data to one's doctor, but the individual person can record these on paper or an electronic spreadsheet to be provided to one's doctor.

Blood pressure devices

Green, Anderson, and Cook (2022) ran a randomized, controlled study of 510 adults from Kaiser Permanente Care Centers in Western Washington. The patients were randomly assigned to clinic readings, at-home readings, or kiosk readings for blood pressures. The at-home group used an Omron blood pressure machine (which I call, Big Bertha). After the 3-week intervention, all participants used 24-hour ambulatory monitoring (ABPM). When the researchers compared the three methods to the ABPM they found no significant difference only with the at-home method. They concluded that blood pressure readings were most accurate when taken over time at home and that home was the best place to have continuing monitoring for hypertension diagnoses. See Fig. 26.5 for photo of home blood pressure monitor device.

Oxygen level home monitors

A home oxygen monitor is handy, especially during pandemics. As long as the Covid virus and variants are around, it would be good to determine oxygen levels at home. It is also handy for monitoring any lung condition, such as chronic obstructive pulmonary disease, to measure oxygen saturation in the blood (Pathak, 2020). Pathak recommended the clip-on style as they tend to be more reliable. Levels below 90% could be a sign for concern and for the Covid times in 2020, was an indication that one should call one's doctor for advice. See Fig. 26.8 for photo of a home oxygen level monitor.

Trends and expectations for the future of health IT and analytics

A. Top-down "mega-sized" institutionalized AI-ML platforms, and
B. Bottom-up "small sized" but working individually controlled data gathering & instant analytics output systems.

This book has provided a background to predictive analytics, provided methodologies for predictive analytics, and provided examples of predictive analytics. It also went beyond that to the next step—prescriptive analytics and decisioning systems that lead to action. *In today's world data comprise the new gold.* Acquiring appropriate data has "become the supreme challenge for efforts at prediction. Anyone who can obtain permission to use the data has the best chance of producing individualized models and generally those individuals likely are those closest to the data, such as physicians, medical groups, and teaching hospitals. It does no good if those data are unavailable, however, even if they are structured in Electronic Medical Record (EMRs) that allow communication between departments and institutions but not among medical researchers. Patients are the closest to the data, but HIPPA generally stands in the way between the data and the research. Providing voluntary permission for one's data to be used in research could be one of the many forms that new patients fill out. National groups such as PCORI can aid in joining patient data to researchers that need them. New IRB practices could include generalized informed consent to cover research both within and between institutions. New predictive techniques of analysis are imperative to the goal of individualizing practice and reliance upon the older methodologies must be combined with newer techniques to produce better outcomes, which constitute the prize.

In addition to the "futuristic ideas" discussed, in this book we have presented *three phases* in the development of "accurate, nonerror, cost-effective healthcare":

1. **PHASE I**: Quality Control/Six-Sigma applied to medicine
2. **PHASE II**: Predictive Analytics applied to medicine
3. **PHASE III**: Decisioning systems (also now called by many "Prescriptive Analytics") applied to medicine
   Will there be a fourth phase?
   The "fourth phase" of innovation for workable Healthcare Delivery:
4. **Phase IV of medical innovation**: putting "Phase I, II, and III" all together along with all aspects of healthcare delivery into a deep-learning-ML-AI all-inclusive package.

A. Top-down "mega-sized" institutionalized AI-ML platforms:
   a. Singapore NUHS AI-ML platform
   b. Mayo Clinic AI-ML platform

Since the writing of the 1st edition of this book (e.g., 2014), various organizations have attempted to develop what we might call "mega-sized institutionalized AI-ML platforms" to carry out all data analytics on the Electronic Medical Records data (the digital information on patients that all clinics and hospitals use today) and other associated patient data that doctors and clinicians document as they work up a diagnosis and treatment plan for a patient. We are calling this PHASE IV in the development of digital-age healthcare.

a. Singapore NUHS AI-ML platform

One of the largest and most comprehensive of these "mega" top-down platforms has been (and is currently under continued development) the "Singapore NUHS AI-L platform" (Ngiam and Khor, 2019). Back in 2019 they named this the Discovery AI platform; the essence of this platform (as envisioned in 2019) is illustrated in Figs. 26.2 and 26.3.

The overall processes, and some of the specifics of the Discovery platform, are illustrated in Figs. 26.4 and 26.5. One can see that almost (if not everything) of importance in healthcare delivery is included, including diagnosis, readmissions, complications, disease progression, and even finance concerns (we assume of both the patient and the healthcare organization?). Clinical data, research data, and everything of the EMR goes into a "data cloud" from which predictive analytics (apparently "Deep Neural Networks," also called "Deep Learning" is primarily used) is applied in order to generate new information that shows how the diagnosis and treatment plan are proceeding and even give warning signals if the analysis indicates something wrong may be happening.

This Discovery AI platform was developed at the National University Health System (Singapore) as a production system that houses modular machine learning tools (Fig. 26.3). The platform uses daily data from the electronic health record system to make predictions about patients. Simultaneously, it also acts as a research sandbox by linking and aggregating multiple clinical and research databases to facilitate the joint development of machine learning tools. The training aspects of developing a machine learning healthcare tool, such as automated data preprocessing, clinical data curation, and deep learning best practices, are integral to the development process of machine learning tools within the platform. Putting various machine learning modules all into one system allows fewer alerts to clinicians, thus reducing alert fatigue for doctors. The system can also gather valuable feedback from doctors, simultaneously educating in the use of these new tools.

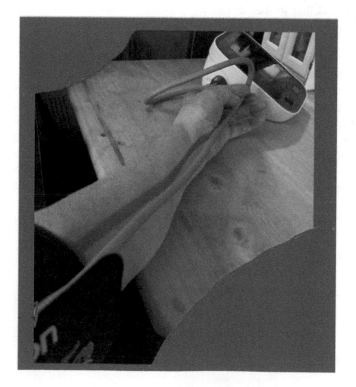

FIGURE 26.3 Home blood pressure monitor. *Photo by Linda Miner.*

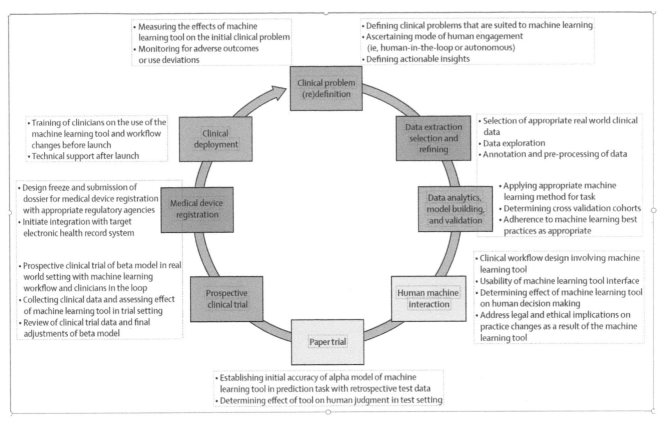

FIGURE 26.4 Discovery AI platform processes overview. *Discovery AI houses various modular AI tools that process clinical and research data to make predictions, which can then be sent as clinical alerts to the electronic health record system (https://www.sciencedirect.com/science/article/pii/ S1470204519301494-*©*2019 Elsevier Ltd. All rights reserved).*

One example of a machine learning application that has come out of the Singapore project is a cancer drug dosage delivery system tailored to the individual patient. This application is called CURATE.AI. It was developed jointly between the National University of Singapore and the University of California—Los Angeles. Its purpose is to optimize drug dosages, especially when more than one drug is used in what is termed "combination therapy." It does this by using data from the patient's own drug doses and responses to continually predict optimal drug dosing for that patient, e.g., this is "personalized or precision medicine" (Blasiak et al., 2019). Fig. 26.6 shows nicely how these interactions are processed via the ML-AI application to make a person-centered treatment.

So where is this Singapore NUHS Discover AI platform today, in early 2022? Is it performing as one would have expected after three more years of efforts? Well, we are not sure: for real precision medicine, genomics data is critically important—so where is it on this front?

By the end of 2021 it was not clear if the Singapore NUHS was finally ready for real-time integration of genomics and AI (Thomas, 2021; BioSpectrum-Asia Edition, 2021); however it was clear that they had added a new "hardware/ software" tool called Nvidia DGX A100 system that was given the name **Endeavour platform** to work with the Discovery AI platform at incredibly increased computing speeds, such that real-time data-streaming allowed physicians instant access to important information needed to making real-time patient diagnoses and treatment decisions. Even though the new NVIDIA DGX A100 has just gone live, NUHS is already looking forward to the next generation of the system to tackle the expected growth in datasets and speed needed to process those data. The Singapore group has invested in programs that look at genomics not just as gigabytes, but terabytes of genomics data per day. More computing power is needed to handle this genomic data in real time, however, which may require the addition of more NVIDIA DGX A100 units to get the GPUs needed.

b. Mayo Clinic AI-ML platform

On April 14, 2021 the Mayo Clinic (Rochester, Minnesota; also Scottsdale, Arizona and Jacksonville, Florida) announced their launch of a digital technology platform that would revolutionize healthcare delivery. At that time they

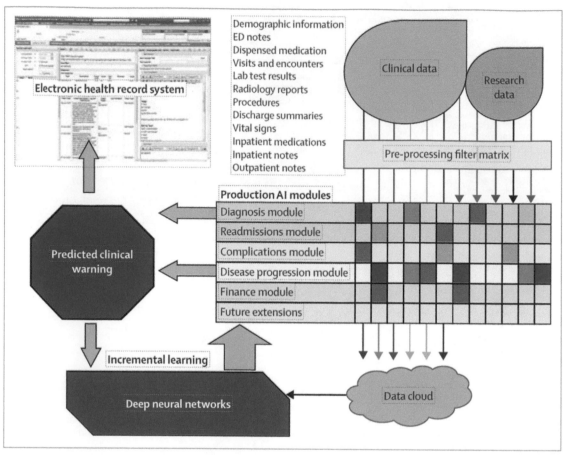

FIGURE 26.5 Discovery AI platform processes specifics. *https://www.sciencedirect.com/science/article/pii/S1470204519301494*-ⓒ*2019 Elsevier Ltd. All rights reserved.*

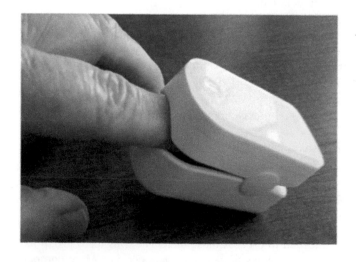

FIGURE 26.6 Pulse/oximeter machine for home. *Photo by Linda Miner.*

called it "The Remote Diagnostics and Management Platform" (RDMP) (Anastasijevic, 2021; Pearson, 2021). The purpose of this new platform is to provide the next generation of clinical decision tools to help clinicians make faster and more accurate diagnoses and provide truly continuous care to patients. This RDMP provides what we can call "event-driven medicine," i.e., insights in the right context, at the right time.

The Mayo Clinic launched two new companies, with partners, to support this newly created digital platform (RDMP). One of these new companies is called Anumana, Inc.; its purpose is to develop and commercialize AI-enabled algorithms. Anumana will initially focus on developing "state-of-the-art" Neural Network analytics applied to the Mayo Clinic's billions of pieces of heart health data. There are many heart conditions, including silent arrhythmias, where evidence-based treatments already exist that can prevent heart failure, stroke, or death; but the key is to detect these before symptoms develop. The new RDMP digital analytics should be able to provide analytic modeling that can make these predictions prior to symptoms.

The second of these new companies associated with Mayo is Lucem Health, Inc. Lucem Health's purpose is to collect, orchestrate, and curate data from any device—yes any device! This is to provide a platform for connecting remote patient telemetry devices with the AI-enabled algorithms in the Mayo digital platform, in order to generate the necessary information to make real-time diagnoses and treatment decisions (Anastasijevic, 2021; Pearson, 2021) (Fig. 26.7).

The Mayo Clinic Platform strategy has several components, including (1) a remote diagnosis and management platform, (2) a virtual care platform (advanced care at home), and (3) clinical data analytics platform (Halamka, 2021).

A subpart of the Mayo Clinic Digital Platform is the "Mayo Platform_Deliver" (Mayo Clinic Platform-Deliver, Accessed March, 2022b). The purpose of this is to make use of remote information from wearable devices and other sources. Mayo Clinic Platform states the "Deliver" part of the platform as follows:

- **Mayo Clinic Platform_Deliver** uses data curated from devices, wearables, and multiple data sources and sends insights and recommendations seamlessly to the point of care or the origin of the data.
- **Mayo Clinic Platform_Deliver** makes data and AI insights actionable. It simplifies clinical and operational processes, which will directly improve patient care and reduce cost.

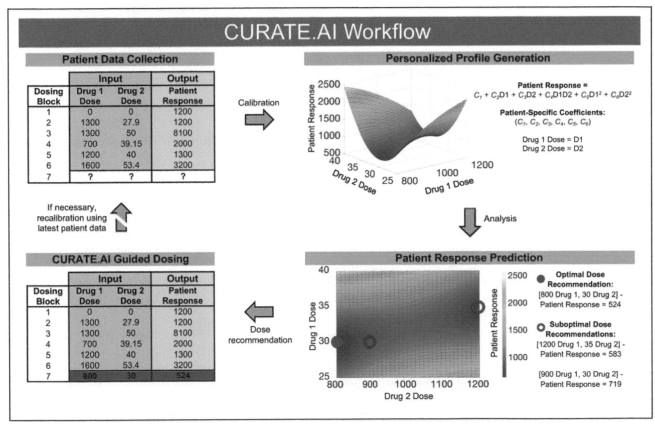

FIGURE 26.7 CURATE.AI-guided dosing workflow for a two-drug optimization. Through calibration of patient data from an initial physician-guided six-dosing block period, CURATE.AI generates a personalized profile using a second-order quadratic correlation between drug doses (input 1, input 2), and a phenotypic patient response (output). The generated second-order equation contains patient-specific coefficients (*C1, C2, C3, C4, C5, C6*), which CURATE.AI then analyzes to predict dose combinations that guide patient responses toward a preferable outcome. In this particular case, CURATE.AI aims to minimize the phenotypic output. CURATE.AI recommends several effective dose combination options for physician review (*open circles*) while highlighting an optimal dose (*closed circle*). In the case of external interventions such as regimen changes, CURATE.AI recalibrates the generated personalized profile using the latest available patient data to accommodate changes in patient state. Though drug doses serve as inputs in this case, inputs may be other interventions with varying intensities. *From Blasiak, A., Khong, J., Kee, T., 2019. CURATE.AI: optimizing personalized medicine with artificial intelligence. SLAS Technol.: Transl. Life Sci. Innov. 25. 247263031989031. https://doi.org/10.1177/2472630319890316.*

- **Deliver** uses the latest interoperability standards which means data and diagnostic signals can be transported from *any source* directly into a specified clinical workflow, thus creating customized care plans for improved care of the individual patient.
- **Mayo Clinic Platform_Deliver** provides critical infrastructure and services to the rapidly growing space of remote diagnostics (Halamka, 2021; Cusano, 2021).

What is happening at Mayo in 2022?

According to the Mayo Clinic Platform homepage (Accessed March 10, 20202: https://www.mayoclinicplatform.org/), the Mayo Clinic Platform will launch a new spoke of their project, to be known as the Mayo Clinic Platform_Accelerate in March 2022 as a 20-week program for health tech startups to go from concept to market with guidance from leaders in the business. The program is different than the usual accelerator program, in that it will drive a startup company with in-kind investments, from datasets to validation tools to clinical workflow (Mayo Clinic Platform_Accelerate, 2022a; Halamka and Cerrato, 2022).

Additionally, Mayo Clinic at Rochester, Minnesota has been developing a "hospital-at-home" program which launched during the past year, led by Michael Maniaci, MD (Gonzalez, 2022a,b). This adds another avenue of providers and payors putting the "patient at the center" of attention. It will be interesting to see how this works out during 2022 and the following years.

There are other examples of machine learning platforms systems that have been/are being developed in a "top-down large organization" fashion; we list two of them below, but will not go into detail about them—for those readers interested, beginning references are provided next:

- **Philips (Amsterdam, Netherlands) HealthSuite Insights platform**, which brings together machine learning tools and analytic capabilities in diagnostic imaging, patient monitoring, genomics, and oncology to facilitate their deployment for clinical and research applications. Healthcare professionals can access different types of patient data, curate the data, and apply it to personalized diagnosis and treatment (Philips, 2018, 2022).
- **GE Healthcare's (Chicago, IL) Edison**, connects data from millions of imaging devices such as MRI and CT machines, facilitating faster and more accurate diagnoses based on imaging data. One of the major uses of Edison is in updating existing equipment in hospitals by plugging them into applications and algorithms with data analytics capabilities (GE, 2016, 2021).

Are there other avenues to change healthcare delivery using a top-down approach? Have any been attempted?

1. **One that failed**
2. **One that is suggested for the future**
3. **One currently working at three places**

1. **The one that failed:** Berkshire-Hathaway (Warren Buffet), Amazon, and JP Morgan Chase partnered in about 2017 to develop a modern cost-effective and patient-centered efficient approach for delivering healthcare. This was called the Haven venture. We had high expectations for this project, and wrote about it glowingly in our 2019 book (Miner et al., 2019).

   However this Haven project "bit the dust" in 2021; the reasons for this are given below in the next section.

2. **The one suggested:** Suggested possibility for demonstration project of a "Person-Centered Healthcare System":

   Big tech buying a large established healthcare system and reorganizing it for the 21st century:

   Such a demonstration project could start a snowball effect for a comprehensive real modern change in healthcare delivery.

   According to Jain (2022), big tech companies such as Apple, Amazon, and Google, have wanted to improve medicine, but none of them has done it. In fact, he pointed out that the Haven venture partnered by Buffet's Berkshire Hathaway, Amazon, and JP Morgan Chase, bit the dust last year.

   Jain offered reasons why such giants have not been successful thus far. The reasons he gave, involved all the variety of things done in our healthcare system—insurance, drugs, providers, as in clinics, hospitals, physical therapy, and so on. This diversity of medical services is complicated and not easily summarized under one roof.

   Further, the current medical system pays when people are sick, and so wellness is not reimbursed. That would have to change if organizations want to increase health—they would be working against themselves. Perhaps they need to start reimbursing for health and the maintenance of health.

   In addition, profit margins are small (not something that big tech companies are used to) and profits in the present medical system are risk managed by insurance companies, involving another complicated element. Jain thought big tech companies could figure out all these drawbacks and problems but were unlikely to do so soon.

One idea might be for the big tech companies to adopt some form of the subscription model, the direct pay model, highlighted in Chapter 7. Or seriously figure out a way to pay for maintaining health.

Jain stated:

> *It's time for the tech companies to get serious about changing American healthcare. They need to show us* **the art of the possible by changing healthcare from within** *not as customers, but as owners. And the way they can do that is buy* **acquiring a large health system and integrating advanced, tech-driven health solutions with solid risk management operations.** *Under this model, they could demonstrate how operations could* **get leaner, how whole-patient care could supplant fee-for-service care, how payers and providers can be integrated**, *and how administrative burdens can be lowered.*

(Jain, 2022, par. 14).

3. **One now working at least three places**: **"Hospital-at-Home" movement**. Hospital-at-home programs are currently developing across the country (USA). Some are being helped by companies like "Medically Home" and others are developing internally out of hospital systems. Waivers to regular hospital operating procedures have been made during the COVID pandemic which have made these in-home programs workable via hospitals. We will take a brief look at three here:

   a. Mayo Clinic Advanced Care at Home: Michael Maniaci, MD, the physician leader at Mayo Clinic Advanced Care at Home (Rochester, Minnesota), says that such programs across the country have demonstrated success in providing safe care, reducing hospital readmissions, improving patient mobility, and achieving high levels of patient satisfaction. The Mayo hospital-at-home system really got going during the COVID-19 pandemic, with a side-benefit of releasing needed hospital beds for COVID patients (Gonzalez, 2022a,b).

   b. Hospital-at-Home Program at Presbyterian Healthcare Services (Albuquerque, NM) was one of the first hospitals to adopt the hospital-at-home method, starting it in the year 2008. Elizabeth De Pirro, MD, who directs this program, says "We saw this as an innovative and successful program for acutely ill patients who could be treated and recover in the comfort of their own homes, with pets and family nearby for support. Our results show higher patient satisfaction, lower rate of falls, reduced mortality and 42% lower costs than what we would expect for hospitalized patients with similar conditions" (Gonzalez, 2022a,b).

   c. Post-Acute Care Services at Michigan Medicine (Ann Arbor). Grace Jenq, MD, who directs the clinical care at this "hospital-at-home" program, says "We started planning our hospital care at home in August of 2018 as a partnership with Blue Cross Blue Shield of Michigan; and then we started our pilot project for this July of 2020. So it took two years to get it off the ground and develop all the different pieces to be able to support enrolling patients. Our goal is to continue to grow the program … What we recognize is that there are patients who are in our hospital beds, who can receive very basic IV medications, IV fluids, laboratory tests, and a combination of in-person and virtual visits in the home that are equal to what they get in the hospital. We've had actually zero readmissions, zero emergency center visits, and patient satisfaction that has been outstanding… We've had no adverse events. So … our pilot has been very, very promising in terms of outcomes" (Gonzalez, 2022a,b). **When Dr. Jenq was asked "What advice would you give to hospitals that wish to establish such a program?"** She replied: "I think one of our key pieces is having this be the strategic initiative of the year, so the entire health system has to be completely committed to driving this hospital care home program forward, or else you will just end up with pieces, here and there, where people are willing to develop, but then you're going to be held up in another area. We need everyone, not just the doctors, not just the nurses, but you need everybody in HIT to build platforms in a timely manner, you need people in regulatory compliance, to move forward to make sure that we're doing things correctly. You can't underestimate how many hundreds and hundreds of hours that are put in at the system level, just to be able to enroll even just a few patients right off the bat. **It takes everybody so you have to have the commitment of the entire health system in order to get something like this off the ground.**" We think Dr. Jenq's advice is most appropriate, and fits 100% with what we describe for successful endeavors in our "Executive Guide… to Successful Execution" in today's world (Burk and Miner, 2022a), i.e., that the entire organization must be committed to new programs in order to make them successful.

Bottom-Up "small-sized" but working individually controlled data gathering and instant analytics output systems

We predict that patients will become more informed consumers of medical care and will become more actively involved in their own healthcare, including using personalized predictive analytics. Much of this will happen through wearable devices and similar types of devices that measure medical body signals that the individual person can own, use, and control; as a result of this the data will be available for N-of-1 studies. These N-of-1 studies may become the most important analytic revolution, producing accurate diagnostic and treatment plan information in the future. This will mostly (if not all) happen from "bottom-up" individuals or small-sized groups working to create these new patient-controlled medical devices.

HISTORICAL PERSPECTIVE: (The following was written in the first edition of this book: writing was done in the year 2014): One of the authors of this book has had an eye condition that could eventually result in blindness. This author spent 18 months going to five different eye clinics, getting different diagnoses but more concerning were the five different treatment plans (the treatment plans were all different at each facility). At the fifth, the "renowned real expert," the author was told that "this is a no-brainer," there is only one thing to do first! Really? Thus, the author finally had to "take things into his own hands," do his own literature/scientific research, and discover for himself what was the "best treatment plan." (It turned out that this fifth expert was "right"; the "no-brainer" treatment was the first of five possible treatments, the least invasive and simplest, also least costly, and it in itself would also be "diagnostic"—if it worked we knew the cause of the condition; if it did not work, then we knew that the cause was different and we had to go to step number two treatment). *Five different and conflicting medical opinions! before a good, "right" answer!* If this author had taken some of the early "treatment plans" presented it would have been the absolutely "wrong thing to do." Additionally, eye pressure checks were needed frequently (every 2 weeks) during this period for "safely monitoring" this condition; this took considerable time and expense to be always running to the eye clinic. The author kept wondering to himself, "why is there not a 'smart contact lens' that can do this monitoring, sending continuous readings to the clinic?" If this author had had a smartphone device or ap and/or a self-monitoring IOP device, he would have probably saved 17 of those 18 months of searching among different clinics, arrived at the "right answer" much faster, and gotten treatment earlier, and all at considerably less medical cost! (remember, this was written back in 2014.)

Now in this second edition, you, the reader, can go to Chapter 11 where this author's "eye story" has been continued (written in 2021) and see that the availability of a self-monitoring eye pressure device was not made easy for the patient, over the intervening 7 years since 2014.

This example from this book's author's experience illustrates first-hand how "fragmented" our medical knowledge and application is in the US healthcare delivery system, and what needs to be done to rectify problems like this.

Where will the next innovations in medicine come from?

Innovations in healthcare in 2022 and the following few years will probably not come from the top-down large, institutionalized organizations but instead from the bottom up via individuals and small groups that are unrestricted by regulations, and thus can easily make devices and systems that the individual person can use (without the need of visiting a health clinic). The individual person is very important—the experience and knowledge gained through experience obtained by the individual is where most new innovation originates—see the following "offset box" that complements this bottom-up resource:

> The fact is that great ideas most often come from the bottom of the organizations, not the top. People who are closest to the customer know better what the customer wants than the executive leadership team. This is a critical idea that must be incorporated in your AI and analytics program. The people in service or manufacturing operations know what the bottlenecks are and have ideas on how to improve efficiency.
>
> They are the idea generators. It is the executive team that needs to listen to these ideas, work with staff to crystalize them, bolster them, and evaluate the organization's capacity and budget required to implement them. Leaders need to get the right people and resources engaged, provide the structure to make sure there is a constant tracking of trajectory to target, correct where needed—to drive results—simply the ability to execute (Burke and Miner, 2022b, p. 27).

N-of-1 studies—the future for person-centered healthcare

You may ask:

What is a N-of-1 clinical trial?

Also:

What is an N-of-1 randomized trial?

An N-of-1 trial is a clinical trial in which a single patient is the entire trial, a single case study, thus the sample size is 1 (e.g., N = 1).

An N-of-1 randomized controlled trial is one in which random allocation can be used to determine the order in which an experimental and a control intervention are given to a single patient, thus the sample size is also 1 (e.g., N = 1).

Research studies using a sample size of 1 (e.g., one person, the individual, the patient) were discussed in the first edition of this book, written back in 2014. We can call these "N-of-1" studies. "N" represents the sample size in statistics, thus an N-of-1 means sample size of one. Statisticians may initially reactive negatively to such a concept. However, because it is extremely difficult to get good population-based studies in medical research (HIPPA and other privacy concerns plus other factors inhibit development of good datasets), N-of-1 studies offer an alternative to get needed information on a patient. Others have begun to suggest N-of-1 studies in recent years, and the momentum to use this format seems to be increasing (Lillie et al., 2011; Nikles and Mitchell, 2015; Huang and Hood, 2019; Cerrato and Halamka, 2017; Percha et al., 2019; Cerrato and Halamka, 2022; Halamka, 2022).

We think that a percentage of the predictive analytic medical innovations in 2022 and future years will be nonregulated apps and devices. Examples of these are detecting skin cancer with your smartphone (Louie et al., 2018) and speech recognition to detect cognitive problems (Haulcy and Glass, 2021).

So advisory systems that will enable individuals to use consumer electronics (cell phones?) is where the next round of innovation in medicine and healthcare will be found.

These wearable and other devices and apps are all things the individual person can use at home to take data on themselves for their use and/or use by their doctors in making the best diagnostic and treatment decisions—again data from N-of-1 methods.

The notion of N = 1 studies is intriguing! If you think about it, **modern automated manufacturing processes are effectively N = 1 situations:** you try to control, understand, predict, and optimize for quality and most efficient functioning and "health," a specific and one-of-kind N = 1 specific production line. In many situations we have worked, even though two factories or production lines may nominally be doing the same thing, in practice, because of different serial numbers and vintages of furnaces or reactors **each production process is one-of-a-kind**. And you optimize/monitor/understand one process at a time by instrumenting and analyzing it comprehensively and continuously.

This can be done by individual patients taking data using wearables and other apps.

Some of us authors of this book would like to instrument ourselves this way!

Styles of thinking—how brain laterality affects innovation in healthcare

Human "thinking styles" may be part of the solution to what is needed to help bring about innovative problem solving and a true reinvention of healthcare:

Alameddine et al. (2021) conducted a resilience study in Lebanon of 511 registered nurses working with patients. (One wonders immediately if those who were still working in 2021 might have self-selected for higher resilience; those with lower resilience might have already left the service.) They found nurses with lower resilience scores were saying that they wanted to leave their jobs. They also found some associations with resilience, for example, having a master's degree associated with higher resilience. Personal burnout, work-related burnout, and client-related burnout associated with lower resilience. Is it possible that the nurses with higher degrees might have been those who were giving the orders and the burned-out nurses might have been following the orders? The study found that those who wanted to quit had the lower resilience scores, which was not surprising.

People were leaving medicine due to the pandemic lockdowns, overcrowded hospitals, mandates, burnout and so on. People in charge of others can be difficult to work with, as reported to this author by nurse educators. In fact, Edmonson and Zelonka (2019) explored "horizontal hostility" of nurses in their article. And bullying can go in both directions, top to bottom and bottom to top. Resentments, texts, under the breath comments, and so on can make the workplace unbearable. Nurses with advanced degrees often are placed over nurses without advanced degrees, and harassment of those beneath is evidently common. Things happen in medicine that are beyond the confines of prediction models. Should they be? What impact does such behavior have on medical outcomes?

Brain functioning and problem solving: In the book, The Master and His Emissary, McGilchrist (2018) related brain functioning to the kinds of attention and problem solving a person might do. In the fable that McGilchrist related purportedly from Nietzsche, an intuitive master of a large domain was a visionary, and an excellent leader of many peoples and concerns. The master was a wholistic thinker and visionary. The master needed helpers to carry out the plans. These were his emissaries. One emissary highlighted in the fable was an efficient convergent thinker, clever and ambitious, someone who could categorize quickly and tidily conduct business. In his ambition the emissary used his position to further his own position, finally taking over from the Master (that he believed weak). He at last usurped the Master and happily became the Master himself. The consequences were bad. Eventually, because he did not understand the people, could not anticipate problems, and did not lead well, the kingdom collapsed. McGilchrist then related that story to the functioning of the right and left hemispheres of the brain.

The "right brain": In describing the structures and functions of right and left hemispheres in the brain, McGilchrist spoke of greater dendritic overlapping in the cortical columns in the right brain (p. 33). This allows the right hemisphere to make more connections and quickly. Such faster transfers allow for greater interconnectivity and more wholistic, divergent thinking. The right was also able, because of its unique structure, to be more alert to peripheral vision and that direction from which new experiences were likely to come. The right brain is more alert to whatever comes from the edges of awareness and to noticing what else exists "out there" rather than what it *expects* to observe. The increased white matter of the right brain allows for easy transfers across regions, for cross thinking. The right brain thinks broadly, flexibly, and attentively, and often messily.

The "left brain": The left brain is more convergent, gathering bits of information and putting them in preset categories. The bits and pieces thinker enjoys being in charge because their ideas are tidy and fit the slots that have been formed through study or bureaucratic rules. The messy right-brained thinker causes consternation to the bits and pieces thinker.

The right and left brain working together: Obviously both hemispheres work together, and both are needed. The left has more focused and divided attention, narrow rather than broad, and wants to finish things quickly by categorizing. On the other hand, the left is more likely to perseverate mistakes along with right categorizations, continuing a single solution, that fits what it already knows. Holistic thinkers are more likely to bring forth new ideas about how things are by seeing around the edges and not by the converging thinking of the tidy thinker; however, wholistic, divergent thinkers never seem to know when to stop and slow down corporate decision-making, regardless of how good their ideas are.

Duality of function in organizations: If one looks, one can see this duality of function in most aspects of organizations, whether they be medical, educational, commercial, or governmental. Tidy thinkers might be better accepted by bureaucratic organizations and often make it up the power chain easier. Educational systems can truly value tidy thinkers, particularly around accreditation time. The system becomes smoother and expedient with a gaggle of tidy thinkers. That kind of institution likely hires people like themselves, tidy. Thinking thoughts that go along with the flow are readily accepted and rewarded. The problem is, as McGilchrist stated, "because the left hemisphere is drawn by its expectations, the right hemisphere outperforms the left *whenever predictions are difficult*" (p. 40).

Perhaps McGilchrist would say that if a nurse or a doctor who was a typical "left-brainer," working on a procedural level, was overseeing a "right-brainer," intuitive underling tensions could arise. For example, a Master's-degreed nurse director might not appreciate a lower level but intuitive nurse. There could be tension and ultimately, undue stress resulting in burnout for either party.

Further, **tidy thinkers might miss difficult diagnoses and make wrong predictions**, operating after their own preset constructs. On the other hand, **intuitive thinkers could take a team astray and cause a miss in diagnosis as well**. Tidy thinkers might miss nuances that could lead to uncommon or difficult diagnoses, particularly in a fast-paced practice in which payments are based on how quickly they can move patients through.

Now let us consider the field of predictive analytics. Could some statistical procedures be "tidier" than others? Would those be preferred in research? Why is it that often one sees lots of contrasts in the conclusions of medical research papers? With a list of 25 pairs how often do we see "probability" levels compared. Generally, as mentioned in Chapters 1 and 7, if this is done, and the researchers pick out the pairs that have a probability of less than .05 to highlight and conclude, they could be committing the family-wise error.

Do we prefer statistical methods that match our preferred hemisphere? Would someone who worked in regression all one's career, trying to achieve smaller and smaller groups to mimic prediction to the individual, be reluctant to move over to using machine learning algorithms? And this author, who admitted in Chapter 17 to enjoying Classification and Regression Trees models rather than Neural Networks because it was easier to "see" the relationships, chose an algorithm other than Neural Networks for the final prediction? Hmm.

So, if we are predicting what should happen, it seems to this author that we need both the right and the left-brain functions in our predictive endeavors. Predicting to the individual is an incredible ideal. Patient-centered, patient-directed are worthy goals as we move more to decision-making for the individual. How much should we listen to algorithms? Which kinds should we use? Convergent? Divergent? In our business/medical organizations, how much should we seek out intuitive workers' ideas? How can we get along and how can we work without becoming petty or political?

Will we want to have machines make the decisions?

Final concluding statements

How much should we listen to algorithms?—Should machines make the decisions?

Predicting to the individual is an incredible ideal. Patient-centered, patient-directed are worthy goals as we move more to decision making for the individual. How much should we listen to algorithms? Which kinds should we use? Convergent? Divergent? In our business/medical organizations, how much should we seek out intuitive workers' ideas? How can we get along and how can we work without becoming petty or political? Will we want to have machines make the decisions? Will those machines be converging or diverging? What will happen when machines learn to remake themselves and become so much smarter than we are that we don't know what they are doing? And, if we don't know what they are doing should we accept their predictions, even if they can predict accurately? We are moving into territory in which there are many more questions than answers … as we follow the science. Or should the doctors and clinicians and the patients together make the final decisions based on both the AI and Machine Learning algorithm predictions—prescriptions AND the "Gold-Standard" best practices available?

Genomics and AI will start exploding in 2022 and subsequent years, and thus we need to be prepared

As we look to the future of healthcare and predictive analytics, the importance of genomics cannot be understated. Today we are just starting to integrate structured genomic data into the electronic health record and this integration is limited currently to a few Mendelian diseases and pharmacogenomics. There is a substantial amount of work to do even in these two fields alone, however.

What lies just beneath the surface is a rapidly growing body of work around polygenic risk scores and their potential impact on health and treatment. As whole genome sequencing becomes more ubiquitous with a decrease in price, our knowledge of genomics increases substantially, in particular our knowledge of biological pathways and gene functions. The amount of data that is being generated in this domain is rapidly outpacing our ability to utilize it. As this increase in knowledge grows and we better understand in detail the biological underpinnings of human health, we will reach a critical mass where there will be so much useful information that our minds cannot comprehend it all in the context of the patient sitting in front of us. **It is critical that analytical models and technical infrastructures are available to make use of this information and deliver it in a ways that maximize its impact on patient care**.

Patient-centered (precision) health for the future

It is interesting to note that as we are finishing writing this book in February and early March of 2022, PCORI came out in early February 2022 with their goal of making healthcare delivery a "patient-centered affair." See Fig. 26.8 for the PCORI illustration that kicked off their 2022 healthcare research year.

This PCORI "patient centered" model (Fig. 26.8) comes at the right time for the release of this book you are reading. The goals are similar: "patient-centered," "patient-directed", "personalized medicine," and a reengineering of healthcare delivery. Along with this is the prediction by many that "wearables" and similar devices and apps generating N-of-1 data acquisition, which are controlled by the person/patient outside of healthcare facilities, will become some of the best sources of individualized medical data and the basis of medical diagnosis and treatment decisions in the future. This bottom-up approach is "workable" as the individual person has the ability to "move and get it done" accurately and efficiently. But interestingly this will also meet the goals of the top-down platforms, like the Mayo Digital Platform which also has personalized/precision medicine as their top goal—top need. In future years as fewer doctors are available for an enlarging patient population, medicine will be to a large extent practiced outside the "walls of clinics"—telemedicine, hospital-at-home programs—and individual wearable health vital signals data sources will be essential.

Postscript

The analysis of any complex system requires a merger of approaches from the bottom-up and the top-down. The reason is that human brains tend to be dominated by concerns focused on one approach to truth or the other. The underlying

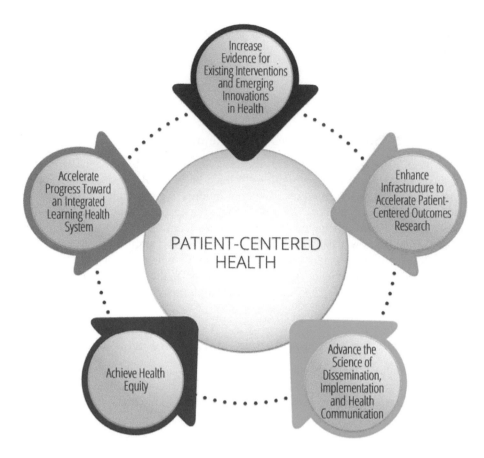

FIGURE 26.8 Even PCORI in early 2022 is putting "patient-centered health" at the center of their goals, as a healthcare delivery research funding organization. *(Adapted from and credited to Cook, 2022).*

problem is that there is too much to learn between the very top-level concerns and the very bottom-level concerns. The need to combine the bottom-up approach of PCORI with the top-down approach of the Mayo Digital Platform will require the formulation of what engineers call "transfer functions." A transfer function links two sets of functions together in terms of cause and effect, when exact formulations of the true cause and effect relationship are not known. Instead, assumptions are made to express the relationships. The transfer functions will be represented by work-arounds necessary to move information between the top-down processes and the bottom-up processes to support personalized (precision) medicine.

References

Alameddine, M., Clinton, M., Bou-Karroum, K., Richa, N., Doumit, M., 2021. Factors associated with the resilience of nurses during the COVID-19 pandemic. Worldviews Evid. Based Nurs. 18 (6), 320–331. Available from: https://doi.org/10.1111/wvn.12544.

AliveCor, 2022. <https://store.kardia.com/>.

Anastasijevic, D. November 9, 2021. Mayo Clinic co-leads a new coalition to improve patient care through community-level clinical trials. https://newsnetwork.mayoclinic.org/discussion/mayo-clinic-co-leads-a-new-coalition-to-improve-patient-care-through-community-level-clinical-trials/.

APPLE WATCH, January 18, 2022. Apple-Support <https://support.apple.com/en-us/HT208955>.

Bacha, 2018; A Noninvasive device that can diagnose several medical conditions: Dxter is a new tricorder that can detect various medical conditions. https://strammer.com/en/dxter/.

Basil Leaf Technologies" - here is the URL link to a You-Tube video about them: https://youtu.be/rdpdWJdx5CE.

BioSpectrum-Asia Edition, Dec 1, 2021. NUHS builds AI production platform using NVIDIA DGX A100 for better healthcare predictions, <https://www.biospectrumasia.com/news/45/19420/nuhs-builds-ai-production-platform-using-nvidia-dgx-a100-for-better-healthcare-predictions-.html>.

Blasiak, A., Khong, J., Kee, T., 2019. CURATE.AI: optimizing personalized medicine with artificial intelligence. SLAS Technol. Transl. Life Sci. Innov. 25. Available from: https://doi.org/10.1177/2472630319890316.

Burk, S., Miner, G., 2022a. The Executive's Guide to AI and Analytics: The Foundations of Execution & Success in the New World. Routledge Press/CRC Press (release date expected Jue 2022).

Burke, S., Miner, G., 2022b. The Executive's Guide to AI and Analytics: The Foundations of Execution & Success in the New World. Rutledge Press—CRC PRESS—Taylor & Francis—Expected release date: June 2022, p. 27.

Caren, J., Urman, M., April 5, 2020. Taking an EKG on yourself. CorMedical GroupBlog. <https://www.cormedicalgroup.com/blog/taking-an-ekg-on-yourself/>.

Cerrato, P., Halamka, J., 2017. Realizing the Promise of Precision Medicine. Elsevier-Academic Press. <https://www.elsevier.com/books/realizing-the-promise-of-precision-medicine/cerrato/978-0-12-811635-7>.

Cerrato, P., Halamka, J., 2022. The Digital Reconstruction of Healthcare: Trasitioning from Brick and Motar to Virtual Care. Rutledge Press. <https://www.routledge.com/The-Digital-Reconstruction-of-Healthcare-Transitioning-from-Brick-and-Mortar/Cerrato-Halamka/p/book/9780367555979#>.

Comstock, J., April 13, 2017. The Qualcomm Tricorder X Prise has its winner but work on tricorders will continue. MobileHealthNews. <https://www.mobihealthnews.com/content/qualcomm-tricorder-x-prize-has-its-winner-work-tricorders-will-continue>.

Cook, N.L., Feb 2, 2022. 'Opeing remarks': PCORI-2022-and-beyond-webinar-presentation-slides-020222. <https://www.pcori.org/sites/default/files/PCORI-2022-and-Beyond-Webinar-Presentation-Slides-020222.pdf>.

Cusano, D., April 23, 2021. Mayo clinic creates AI-powered clinical decision—diagnostics support platform, two digital health portfolio companies. <https://telecareaware.com/tag/remote-diagnostics-and-management-platform-rdmp/>.

Dvorsky, G., Mar. 14, 2014. Meet the teams who are building the world's first tricorder. <http://io9.com/meet-the-teams-who-are-building-the-worlds-first-medic-1543000639>.

Edmonson, C., Zelonka, C., 2019. Our own worst enemies: the nurse bullying epidemic. Nurs. Adm. Q. 43 (3), 274–279. Available from: https://doi.org/10.1097/NAQ.0000000000000353.

GE, 2016. Big data, analytics, and artificial intelligence. <https://www.gehealthcare.com/static/pulse/uploads/2016/12/GE-Healthcare-White-Paper_FINAL.pdf>.

GE, November 28, 2021. GE healthcare unviels new AI & digital technologies and solutions to help sove healthcare's most pressing problems. <https://www.ge.com/news/press-releases/ge-healthcare-unveils-new-ai-and-digital-technologies-and-solutions-to-help-solve>.

Gonzalez, G., March 9 2022a. Hospital at home okatbook lessons from mayo clinic, Michigan-medicine and presbyterian-health. Becker's Hospital Review. <https://www.beckershospitalreview.com/strategy/hospital-at-home-playbook-lessons-from-mayo-clinic-michigan-medicine-and-presbyterian-health.html?linkId = 155946074&s = 03>.

Gonzalez, G., March 9 2022b. Hospital at home playbook lessons from mayo clinic, Michigan-medicine and presbyterian-health. Becker's Hospital Review. <https://www.beckershospitalreview.com/strategy/hospital-at-home-playbook-lessons-from-mayo-clinic-michigan-medicine-and-presbyterian-health.html?linkId = 155946074&s = 03>.

Halamka, J., 2021. Mayo clinic platform portfolio strategy. <https://www.mahealthdata.org/resources/Documents/CIO%20Forum/CIO%20Forum%202021/MCP%20core%20deck%20Jan%202021.pdf>.

Halamka, J., 2022. N-ofof-1 studies can make patient care more personalized. <https://www.mayoclinicplatform.org/2022/02/08/n-of-1-studies-can-make-patient-care-more-personalized/>.

Haulcy, R., Glass, J., 2021. Classifying Alzheimer's disease using audio and text-based representations of speech. Front. Psychol. Vol 11, . Available from: https://www.frontiersin.org/article/10.3389/fpsyg.2020.624137, http://doi.org/10.3389/fpsyg.2020.624137.

Huang, S., Hood, L., 2019. Personalized, precision, and n-of-one medicine: a clarification of terminology and concepts. Perspect. Biol. Med. 2019 62 (4), 617–639. Available from: https://doi.org/10.1353/pbm.2019.0036, https://pubmed.ncbi.nlm.nih.gov/31761797/.

Jain, S.H. Feb. 15, 2022. What big tech should actually do in healthcare. Forbes. <https://www.forbes.com/sites/sachinjain/2022/02/15/what-big-tech-should-actually-do-in-healthcare/?sh = 3e570c8251f6>.

Jaklevic, M.C. Feb. 23, 2017. MD Anderson Cancer Center's IBM Watson project fails, and so did the journalism related to it. Health News Revies. <healthnewsreview.org>.

Lifestyle, 2014. Smartphone apps for AFIB. newLifeoutlook AFIBJuly 23 AFib Apps Helpful Smartphone Apps Monit. Track. Symptoms.

Lillie, E.O., Patay, B., Diamant, J., Issell, B., Topol, E.J., Schork, N.J., 2011. The n-of-1 clinical trial: the ultimate strategy for individualizing medicine? Personal. Med. 8 (2), 161–173. Available from: https://doi.org/10.2217/pme.11.7, https://www.ncbi.nlm.nih.gov/pmc/articles/PMC3118090/.

Louie, D.C., Phillips, J., Tchvialeva, L., Kalia, S., Lui, H., Wang, W., et al., 2018. <https://pubmed.ncbi.nlm.nih.gov/30554501/>.

Mayo Clinic Platform_Accelerate, 2022a. <https://www.mayoclinicplatform.org/accelerate/>.

Mayo Clinic Platform-Deliver <https://www.mayoclinicplatform.org/deliver/> (accessed March 2022b).

McGilchrist, I., 2018. 2009 The Master and His Emissary: The Divided Brain and the Making of the Western World. Yale University Press, New Haven and London.

Miner, G.D., Linda, M.A., Dean, D., 2019. Healthcare's Out Sick—Predicting a Cure. Productivity Press, Rutledge.

Moschou, 2017. How close are we to a real star-trek style medical tri-corder?. < https://www.scientificamerican.com/article/how-close-are-we-to-a-real-star-trek-style-medical-tricorder/# >.

Ngiam, K.Y., Khor, I.W., 2019. Big data and machine learning algorithms for health-care delivery. Lancet Oncol. 20 (5), e262–e273. Available from: https://doi.org/10.1016/S1470-2045(19)30149-4, https://www.sciencedirect.com/science/article/pii/S1470204519301494—© 2019 Elsevier Ltd. All rights reserved.

Nikles, J., Mitchell, G., 2015. The Essential Guide to N-of-1 Trials in Health. Springer Publishers. <https://link.springer.com/book/10.1007/978-94-017-7200-6>.

Ortega, R.P., Mar 4, 2022. Augmented reality brings back extinct ice age animals: the new models add scientific accuracy to "paleoart.," American Association for the Advancement of Science (AAAS), Science Shots—Plants and Animals.

Pathak, N., April 28, 2020. What is a pulse oximeter and can it help against COVID-19? WebMD Blogs. <webmd.com>.

PCORI, About us. <http://pcori.org/about-us/mission-and-vision/> (accessed 17.05.14).

PCORI, 2022a. Research we support. <https://www.pcori.org/research/about-our-research/research-we-support#content-3697> (accessed 4.03.22.).

PCORI, 2022b. Standards for heterogeneity of treatment effects. 5: standards for heterogeneity of treatment effects (accessed 2.03.22.).

PCORI, 2022c. Our Vision and Mission. Our Vision & Mission. PCORI.

Pearson. D. | April 15, 2021. 1 new platform, 2 new companies sprout in Rochester, Minn. In HEALTH-EXEC. <https://healthexec.com/topics/healthcare-management/healthcare-economics/1-new-platform-2-new-companies-sprout-rochester>.

Percha, B., Baskerville, E.B., Johnson, M., Dudley, J.T., Zimmerman, N., 2019. Designing robust N-of-1 studies for precision medicine: simulation study and design recommendations. Journal of Medical Internet Research 21 (4), e12641. Available from: https://doi.org/10.2196/12641. Available from: https://www.ncbi.nlm.nih.gov/pmc/articles/PMC6462889/.

Philips, 2018. Philips launches AI platform for healthcare. <https://www.philips.com/a-w/about/news/archive/standard/news/press/2018/20180301-philips-launches-ai-platform-for-healthcare.html>.

Philips, 2022. Philips recognized in 2021 gartner guide for digital health platforms. GlobeNewswire, March 10, 2022 <https://www.globenewswire.com/news-release/2022/03/10/2400555/0/en/Philips-recognized-in-2021-Gartner-Market-Guide-for-Digital-Health-Platforms.html>.

Pringle, S., Jan. 5, 2022. Scoop: IBM tries to sell Watson Health again. Axios.

Qualcomm, 2014. Qualcomm tricorder prize. <http://www.qualcommtricorderxprize.org/competition-details/overview>.

Saneesh, V.S. Dec. 10, 2021. Best smartwatch with ECG approved by FDA. Reviews Break <reviewsbreak.com>.

Seheult, R., Jan. 2022. ECG Watch: How it Works (Apple, Samsung A Fib Watches/EKG). MedCram—Medical Lectures Explained CLEARLY. (30) ECG Watch: How it Works (Apple, Samsung A fib Watches/EKG)—YouTube.

Sweeney, E., 2017a. Experts say IBM Watson's flaws are rooted in data collection and interoperabilitySept. 6Health Tech . Available from: https://www.fiercehealthcare.com/analytics/ibm-watson-s-flaws-trace-back-to-data-collection-interoperability.

Sweeney, E. June 29, 2017b. Bringing expectations for AI back down to earth. <https://www.fiercehealthcare.com/analytics/stanford-researchers-bring-expectations-for-ai-back-down-to-earth>.

Thomas, W., Nov 24, 2021. NUSH builds new architecture for AI-based patient care. CDO-TRENDS. <https://www.cdotrends.com/story/16041/nuhs-builds-new-data-architecture-ai-based-patient-care?refresh = auto>.

XPRIZE Foundation, 2022 < https://www.xprize.org/prizes/tricorder/teams/final_frontier_medical_devices >.

Further reading

Agus, D.B., 2011. The End of Illness. Free Press (A division of Simon & Schuster), New York.

AMA, 2013. ACOs and Other Options: A "how-to" Manual for Physicians Navigating a Post-Health Reform World, fourth ed. Practice Management Center. http://www.acponline.org/running_practice/delivery_and_payment_models/aco/physician_howto_manual.pdf.

Bastawrous, A., 2014. TED <http://www.ted.com/talks/andrew_bastawrous_get_your_next_eye_exam_on_a_smartphone?utm_campaign = &awesm = on.ted.com_quBe&utm_source = facebook.com&utm_medium = on.ted.com-facebook-share&utm_content = awesm-publisher>.

Berry, L.L., Seltman, K.D., 2008. Management Lessons from Mayo Clinic. McGraw Hill, New York.

Boulos, M.N.K., Wheeler, S., Tavares, C., Jones, R., 2011. How smartphones are changing the face of mobile and participatory healthcare: an overview, with example from eCAALYX. Biomed. Eng. 10, 24. Available from: http://www.ncbi.nlm.nih.gov/pmc/articles/PMC3080339/.

Boult, C., Giddens, J., Frey, K., Reider, L., Novak, T., 2009. Guided Care: A New Nurse-Physician Partnership in Chronic Care. Springer Publishing Company, New York.

Brawley, O.W., Goldberg, P., 2011. How We Do Harm: A Doctor Breaks Ranks about Being Sick in America. St. Martin's Press, New York.

Carey, R.G., Lloyd, R.C., 2001. Measuring Quality Improvement in Healthcare: A Gjide to Statistical Process Control Applications. ASQ—American Society for Quality—Quality Press, Milwaukee, WI.

Centers for Disease Control and Prevention, Costs of falls among older adults. <http://www.cdc.gov/homeandrecreationalsafety/falls/fallcost.html> (accessed 6.5.14.).

Chernova, Y., 2014. Oh, Baby. Wearables track infants' vital signsMay 13 Wall Str. J. 263 (111), B1.

Cleveland Biomedical Job Fair, Explorys. <http://www.biomedicaljobfair.com/Members_Only/Explorys-2> (accessed 17.05.14.).

Clifton, G.L., 2009. Flatlined: Resuscitating American Medicine. Rutgers University Press, New Brunswick, NJ.

Demiris, G., Oliver, D., Dickey, G., Skubic, M., Rantz, M., 2008. Findings from a participatory evaluation of a smart home application for older adults. Technol. Health Care 16 (2), 111–118.

Drucker, P.F., 2006. What executives should remember. Harv. Bus. Rev. 84 (2), 144–152.

Duncan, I., 2011. Healthcare Risk Adjustment and Predictive Modeling. ACTEX Publications, Winsted, CT.

Explorys, 2014. About us: Who we are. <https://www.explorys.com/about-us/who-we-are>.

Gawande, A., 2010. The Checklist Manifesto: How to Get Things Right. Picador (Henry Holt and Company), New York.

Goldhill, D., 2013. Catastrophic Care: How American Health Care Killed my Father And How We Can Fix It. Alfred A. Knopf, New York.

Green, B.B., Anderson, M.L., Cook, A.J., et al., 2022. Clinic, home, and kiosk blood pressure measurements for diagnosing hypertension: a randomized diagnostic study. J. Gen. Intern. Med. Available from: https://doi.org/10.1007/s11606-022-07400-z.

Halamka, J., Cerrato, J., January 17, 2022. Precision medicine, omics, and the black box <https://www.mayoclinicplatform.org/2022/01/17/precision-medicine-omics-and-the-black-box/>.

Hay, T., August 12, 2013. Google glass could become a fixture in the operating room. <http://blogs.wsj.com/venturecapital/2013/08/12/google-glass-could-become-a-fixture-in-the-operating-room/>.

HIMSS Analytics, 2014. <http://www.healthcareitnews.com/directory/himss-analytics>.

HIMSS Healthcare Global, 2014. <http://www.healthcareglobal.com/healthcare_technology/himss-2014-uncovers-5-new-mhealth-trends>.

HIMSS transforming health through IT, 2014. <http://www.himss.org/ResourceLibrary/NewsDetail.aspx?ItemNumber = 29708&navItemNumber = 18600>.

HIMSS14, January 3, 2014. 7 trends healthcare experts anticicpate in 2014 <http://www.himssconference.org/GenInfo/NewsDetail.aspx?ItemNumber = 26634>.

Hotz, R.L., 2014. A future where bionics track your healthApril 22 Wall Str. J. 263 (93), D2.

IBM, 2012. Solutions for healthcare: IBM content and predictive analytics for healthcare <http://public.dhe.ibm.com/common/ssi/ecm/en/zzb03009u-sen/ZZB03009USEN.PDF>.

IBM, Retrieved May 17, 2014. Analytics gives care providers new insights to improve individual outcomes. <http://www.upmc.com/media/newsreleases/2012/pages/upmc-personalized-medicine-investment.aspx>.

Kaiser Family, 2011. Snapshots: health care spending in the united states & selected OECD countries, Exhibit 4A. <http://kff.org/health-costs/issue-brief/snapshots-health-care-spending-in-the-united-states-selected-oecd-countries/> (Accessed 16.5.14.).

Kudyba, S.P., 2010. (Edirtor) Healthcare Informatics: Improving Efficiency and Productivity. CRC Press (Taylor and Francis Group, Boca Raton, FL.

Mace, S., Oct. 30, 2013. Readmissions drop like a rock with predictive modeling. <http://www.predictiveanalyticsworld.com/patimes/readmissions-drop-like-a-rock-with-predictive-modeling/>.

Makary, M., 2012. Unaccountable: What Hospitals Won't Tell You and How Transparency Can Revolutionize Health Care. Bloomsbury Press, New York.

Marcus, A.D., 2014. A hidden data treasure trove in routine checkups May 13 Wall Str. J. 263 (111), D1.

Memorial Sloan-Kettering Cancer Center, 2013. IBM Watson helps fight cancer with evidence-based diagnosis and treatment suggestions. <http://www-03.ibm.com/innovation/us/watson/pdf/MSK_Case_Study_IMC14794.pdf>.

Nield, D., Oct. 18, 2020. Six health conditions a smartwatch can detect before you can: It's almost like wearing your doctor around your wrist. Popular Science. Six health conditions a smartwatch can detect before you can (popsci.com).

PCORI, Dec. 3, 2013. Steven Clauser to join PCORI as improving healthcare systems program director. <http://www.pcori.org/2013/steven-clauser-to-join-pcori-as-improving-healthcare-systems-program-director/>.

Pronovost, P., Vohr, E., 2010. Safe Patients, Smart Hospitals: How One Doctor's Checklist Can Help Us Change Health Care from the Inside Out. Plume (published by the Penguin Group), New York.

Reid, T.R., 2010. The Healing of America: A Global Quest for Better, Cheaper, and Fairer Health Care. A Penguin Book, New York.

Rock Health, 2013. Startup list. <http://rockhealth.com/resources/digital-health-startup-list/>.

Scanadu. November 11, 2013. Press releases. <https://www.scanadu.com/pr/>.

Scanadu, 2014. Scanadu Scout: Progress and Scanathon. <https://www.scanadu.com/blog/>.

Sheehan, B., 2013. Doctored. Sandy, UT: Jeff Hays Films. <http://www.doctoredthemovie.com/>.

Skubic, M., Alexander, G., Popescu, M., Rantz, M., Keller, J., 2009. A smart home application to eldercare: current status and lessons learned. Technol. Health Care 17 (3), 183−201. Available from: https://doi.org/10.3233/THC-2009-0551.

Stenovec, T., June 20, 2013. Scandu Scout, medical tricorder, passes $1 million on Indiegogo, but doesn't need the money. <http://www.huffingtonpost.com/2013/06/20/scanadu-scout_n_3468894.html>.

Topol, E., 2012. The Creative Destruction of Medicine: How the Digital Revolution Will Create Better Health Care. Basic Books, New York.

Truffer, C.J., Keehan, S., Smith, S., Cylus, J., Sisko, A., Poisal, J.A., et al., 2010. Health spending projections through 2019: the recession's impact continues. Health Affairs (Millwood) 29 (3), 522−529.

UMPC, 2012. UMPC Fosters 'Personalized Medicine' with $100 million investment in sophisticated data warehouse and analytics. <http://www.upmc.com/media/newsreleases/2012/pages/upmc-personalized-medicine-investment.aspx>.

UMPC, 2014. About UMPC. <http://www.upmc.com/about/partners/icsd/locations/Pages/default.aspx> (accessed 16.5.14.).

U.S. health care costs <http://www.kaiseredu.org/Issue-Modules/US-Health-Care-Costs/Background-Brief.aspx>.

Wang, S.S., 2014. New technologies to help seniors age in placeJune 3 Wall Street J. D 2.

Weinberg, J.E., 2010. Traching Medicine: A Researcher's Quest to Understand Health Care. Oxford University Press, New York.

Wicklund, E., (Eds.), March 12, 2014. 7 mHealth trends at HIMSS 14. mHealthNews. <http://www.mhealthnews.com/news/7-mhealth-trends-himss14>.

Appendix A

Modeling new COVID-19 deaths*

Introduction

In Chapter 16, KNIME was used in the *World_Country_Hospitalization_Model* workflow to visualize the relationships in three of four displayed data streams:

1. new cases and new deaths;
2. COVID-19 hospital beds and ICU beds; and
3. death rate versus total number of vaccinations.

In this exercise to build a predictive model, we will use a fourth data stream as shown in Fig. A1.

The fourth data stream (at the bottom of the workflow) builds a model to predict the number of new deaths from a set of predictor variables, including a new variable based on the Chapter 16 documentation of this workflow. This exercise will explain the workflow to you. This document is not designed as a step-by-step training exercise to show you how to build the processing operations in KNIME from scratch. Rather, it shows you what to do and explains why it is done.

Processing steps in this workflow

1. If you have not done so already, download the *World_Country_Hospitalization_Model.knwf* workflow from the book's companion web page (https://www.elsevier.com/books-and-journals/book-companion/9780323952743) to a folder on your computer.
2. Import the workflow according to the instructions in Chapter 16.
3. Load the workflow into the KNIME modeling screen.
4. Familiarize yourself with the visualization operations to display the three line plots.
   a. Right-click on the top JFree line plot of the numbers of new cases and new deaths. Notice that the number of new deaths lags behind the number of new cases from 10 to 14 days (Fig. A2).
   b. Based on that insight, a new derived variable (a "lag variable") was generated by the Lag Variable node in the Initial Data Prep metanode (described in Chapter 16). This variable expresses the number of new cases 14 days (to be conservative) before the number of new deaths on a given day.
   c. We will use the lag variable (*new_cases_smoother_per_million(-14)*) as a predictor variable name in a model to predict the number of new cases together with other candidate predictor variables.
   d. The nodes in the workflow are not executed yet; we will do that later.
5. The Final Data Prep metanode. The Initial Data Prep node is connected to the Final Data Prep metanode to carry the output data stream to be prepared further for modeling. A metanode in KNIME is an icon that symbolizes a string of connected nodes that performs a specific task and is collapsed to form the metanode colored green.
   a. Double-click on the Final Data Prep metanode to expand it to see five included nodes
   b. The Column Filter node
      i. The operation performed by this node is to restrict the data flow to only those variables that are viable candidates as predictors.
      ii. All variables related to deaths are removed, because they are related by definition to the target variable *new-deaths/MM*60*.

---

*This continues the "Case Study-like-tutorial" presented in Chapter 16 in this book.

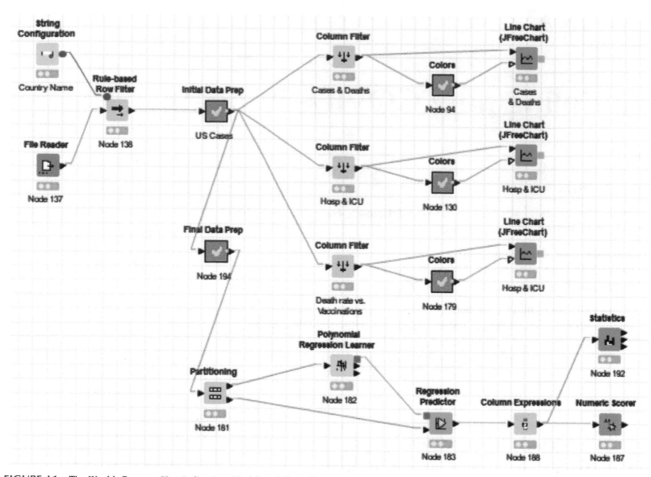

FIGURE A1   The *World_Country_Hospitalization_Model* workflow, showing the four data streams.

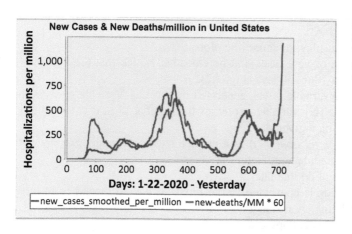

FIGURE A2   Line plot of the numbers of new cases and new deaths since 1/22/2020 and "Yesterday" (=3/26/2022). Note: the term "Yesterday" was included in the X-axis label because when the data accessed by the workflow is refreshed on a given date, the data series will extend to the previous day, or the "yesterday" of that date.

iii. Only those forms of other variables that are expressed per hundreds, thousands or millions population are passed to permit comparisons between countries with significantly different populations.

iv. This operation reduces the number of candidate variables from 51 to 21. Right-click on the Column Filter node, and select the *Filtered table* option to see the output table shown in Table A1.

c. The Normalizer node

i. *Weekly_hosp_admissions_per_million* variable is downloaded as a string data type, but it is really a number. We must convert it to numeric format in order to use it as a predictor variable in the Polynomial Regression algorithm.

TABLE A1 Output table of the column filter node.

Row ID	D total_c...	D new_ca...	D new_ca...	D reprod...	D icu_pati...	D hosp_p...	S w
Row0	0.003	0	0	0	0	0	0
Row1	0.003	0	0	0	0	0	0
Row2	0.006	0.003	0	0	0	0	0
Row3	0.006	0	0	0	0	0	0
Row4	0.015	0.009	0	0	0	0	0
Row5	0.015	0	0	0	0	0	0
Row6	0.015	0	0	0	0	0	0
Row7	0.018	0.003	0.002	0	0	0	0
Row8	0.018	0	0.002	0	0	0	0
Row9	0.024	0.006	0.003	0	0	0	0
Row10	0.024	0	0.003	0	0	0	0
Row11	0.024	0	0.001	0	0	0	0
Row12	0.033	0.009	0.003	0	0	0	0
Row13	0.033	0	0.003	0	0	0	0
Row14	0.033	0	0.002	0	0	0	0

File Edit Hilite Navigation View

Table "default" - Rows: 791 Spec - Columns: 21   Flow Variables

    ii. Double-click on the node to see the configuration screen, showing:
      1. exclusion of new-deaths/MM*60 (the target) variable; and
      2. inclusion of all other numerical variables.
      3. The Z-Score normalization technique is selected at the bottom of the screen. This setting is most compatible with the regression mathematics.
      4. Click the Cancel button at the bottom (changes were made) to return to the expanded metanode data flow screen.
   iii. Normalization of predictor variables removes any possible bias in the numerical processing of variables with significantly larger ranges of values compared to other variables.
   iv. The target variable (the Y-variable) is not normalized because the mathematics does not process it in the same way it does the predictors (the X-variables), and potential bias is not a problem, therefore, normalization is not necessary. The other reason not to normalize the target value is to permit the relationship of predicted values with the target variable in terms of its native numbers for easy comparison with other population statistics.
    v. The reason for this issue lies in one of the assumptions of the theoretical Parametric Model (to which Polynomial Regression subscribes), in which the variances of all predictor variables are *homogeneous (or homoscedastic)*, meaning that they are roughly comparable. The variance of a variable with a significantly different range is most likely to have a variance that is distinctly higher than that of other predictor variables. That situation will cause the variable with the larger range to have a greater effect on the total sum-of-squares operation in the regression processing. This effect constitutes a bias toward that variable in its effect on predicted value. Normalization of all predictor variables removes this potential bias.
  d. Double-click on the Linear Correlation node to see the configuration screen.
    i. Notice that all of the numerical variables (except the target variable new-deaths/MM*60) are selected in the right panel.
    ii. Click the Cancel button on the bottom to return to the data stream.
   iii. Right-click on the node again and select the *Execute* option as shown in Table A2.
   iv. This execution operation also executes all previous executable nodes up to that point in the data flow of the workflow.
  e. Double-click on the Correlation Filter node to see the configuration screen.
    i. This node is configured to remove one of a pair of variables that are correlated at or above 0.8.

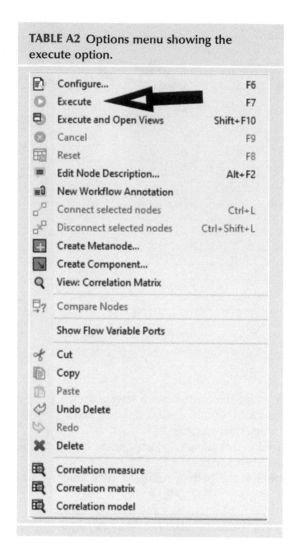

TABLE A2 Options menu showing the execute option.

ii. Why? Because highly correlated predictors prevent the regression matrix from being inverted (an operation in the regression mathematics), causing the algorithm to issue an error, and quit.

iii. The node will delete the member of a pair of highly correlated variables with the lowest confidence value among the pair.

iv. Click OK to exit the configuration screen.

v. Right-click on the node, and select the **Filtered data from input** option to see that the number of variables has been reduced to 12 (the target variable plus 11 viable candidate predictor variables).

vi. Click on the X on the upper-right of the report table to exit the table.

f. Click on the X in the yellow heighted box on the top of the metanode data stream to return to the main workflow.

g. Execution of all nodes in the workflow by clicking on the red button on the top menu with two white triangles in it ( ).

6. Modeling operation

a. The Partitioning node selects a hold-out data set for use in model accuracy evaluation.

i. This hold-out data set will not be used for training the model, it will be used to evaluate the accuracy of the model.

ii. Model accuracy evaluation should not be performed on data that was used in any way to train the model.

  iii. Double-click on the node to see that the ***Relative[%]*** option is set to 70, which means that 70% of the data will be randomly selected to be output at the top output port of the node, and 30% will be output at the bottom output port of the node.

  iv. Click on Cancel to exit the configuration screen

 b. The Polynomial Regression Learner node builds a curvilinear regression using all input variables and their squares.

  i. The square of each input variable has an exponent of 2.

  ii. The regression will be built to estimate coefficients with exponents = 1 and 2.

   1. Output coefficients labeled with an exponent = 1 are raw input data values.

   2. Output coefficients labeled with an exponent = 2 are squares of raw input data values.

   3. Double-click on the node to see the configuration screen shown in Fig. A3.

   4. Notice that the target column is selected as *new-deaths/MM*60*.

   5. Notice also that all candidate predictor variables are included in the input list of candidate predictor variables.

   6. Click Cancel to return to the workflow

 c. The Regression Predictor node accepts the trained model as input (via the blue square data port) and the holdout data set to be scored by the model for assessment of the prediction accuracy. There is no need to configure this node.

7. The Accuracy Estimate operation

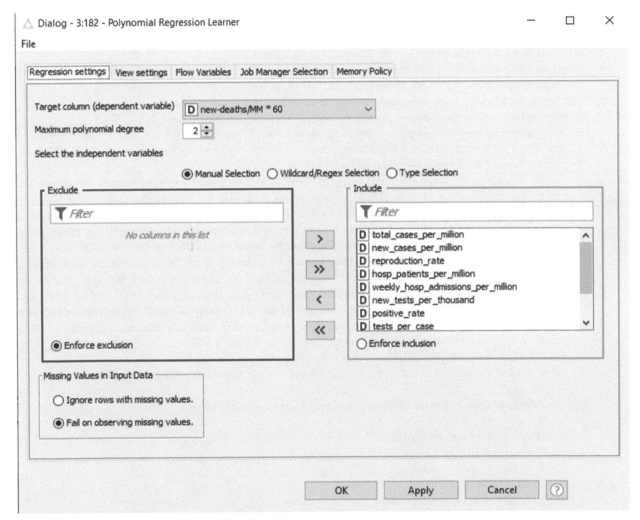

FIGURE A3 The polynomial regression learner node configuration screen.

TABLE A3 The output table of the numerical scorer node.

Row ID	D Predicti...
R^2	0.867
mean absolute error	41.628
mean squared error	2,725.714
root mean squared error	52.208
mean signed difference	5.545
mean absolute percentage error	147.532
adjusted R^2	0.867

a. Column Expressions node.
   i. This node replaces target = 0 values with target = 0.01, to prevent a division-by-zero error and permit division by the target value in the calculation of the mean absolute percent error in the Numerical Scorer node. This expression is:

   if(column("new-deaths/MM * 60") == = 0) 0.01
   else column("new-deaths/MM * 60")

   ii. The logic in this if-then-else expression means "if the target = 0, replace it with 0.01, otherwise, uses the existing target value."
b. The Statistics node calculates summary statistics values for the input data stream.
   i. Right-click on the node and select the **Statistics table** option.
   ii. Scroll down to the target variable on the left (new-deaths/MM*60) and over to the Mean column to see that the mean target value = 210.86. Remember this value, we will use it later.
   iii. The Numerical Scorer node calculates various estimates of prediction accuracy.
      1. Right-click on the node, and select the **Statistics** option to see the output table in Table A3.
      2. Your numbers may vary slightly from those shown above, because of the random process involved in the row selection operation of the Partitioning node.
      3. Notice that the R^2 value (R^2) is 0.867.

      An old adage in statistical analysis is that correlation does not necessarily imply causation. The R^2-value was invented by Wright (1921), a population ecologist, to express the estimate of the causative component of the correlation coefficient (R). He proposed that the square of the R-value was a good estimate of the degree to which the R-value represented causation. This assumption has never been validated, but rather, it has been accepted initially by agronomists (being published in the Journal of Agronomy), and then it was adopted rather blindly by scientists in all fields as a metric to express causation in regression analyses. Even if the R^2-value was valid for linear regression, it is not valid for nonlinear regression methods such as polynomial regression. The reason is that one of the assumptions of regression mathematics is that the total variance of the relationship is equal to the sum of the predictor variance and the error variance. This assumption is reflected in the mathematical operations of the calculation of the correlation coefficient (the generation of the total sums-of-squares term). But, it is true in the calculation operations only for *linear regression*, not for nonlinear regression.
      4. With this information in mind, what metric should we use to evaluate the accuracy of a nonlinear regression model?
      5. The Numerical Scorer node outputs a number of accuracy estimates:
         a. R^2
         b. Mean absolute error (MAE)
         c. Mean squared error
         d. Root mean squared error
         e. Mean signed difference
         f. Mean absolute percentage error
         g. Adjusted R^2

TABLE A4 The coefficients and statistics table output by the polynomial regression learner node.

ıble "Coefficients and Statistics" - Rows: 23   Spec - Columns: 6   Properties   Flow Variables

Row ID	S Variable	I Exponent	D Coeff.	D Std. Err.	D t-value	D ▲ P>|t|
Row7	positive_rate	1	83.693	9.407	8.896	0
Row10	stringency_index	1	88.345	8.161	10.826	0
Row22	new_cases_smoothed_per_million(-14)	2	-25.455	2.807	-9.067	0
Row23	Intercept	0	249.302	7.673	32.493	0
Row11	new_cases_smoothed_per_million(-14)	1	141.21	17.277	8.173	0
Row3	reproduction_rate	1	-66.268	8.232	-8.05	0
Row16	weekly_hosp_admissions_per_million	2	-71.649	11.811	-6.066	0
Row15	hosp_patients_per_million	2	68.761	11.831	5.812	0
Row14	reproduction_rate	2	12.019	2.199	5.465	0
Row20	new_vaccinations_smoothed_per_million	2	-19.062	3.659	-5.21	0
Row21	stringency_index	2	19.105	3.812	5.012	0
Row9	new_vaccinations_smoothed_per_million	1	44.871	9.715	4.619	0
Row18	positive_rate	2	-9.549	3.337	-2.862	0.004
Row19	tests_per_case	2	-5.984	2.708	-2.21	0.028
Row2	new_cases_per_million	1	-24.544	11.579	-2.12	0.034
Row17	new_tests_per_thousand	2	5.915	3.438	1.72	0.086
Row4	hosp_patients_per_million	1	40.376	24.124	1.674	0.095
Row5	weekly_hosp_admissions_per_million	1	-34.93	23.117	-1.511	0.131
Row8	tests_per_case	1	13.896	9.68	1.435	0.152
Row1	total_cases_per_million	1	9.892	11.324	0.874	0.383
Row13	new_cases_per_million	2	1.179	1.377	0.856	0.392
Row12	total_cases_per_million	2	-2.615	7.057	-0.371	0.711
Row6	new_tests_per_thousand	1	2.119	6.175	0.343	0.732

6. The accuracy metrics that are easiest to relate to prediction accuracy in real-world terms are:
   a. R^2
   b. MAE
   c. Adjusted R^2
7. Because of the problems associated with using R^2 with regression algorithms, we will use the MAE of prediction to estimate the prediction accuracy by the expression:

$$\text{Accuracy} = 1 - (\text{MAE/Mean target value}) = 1 - (41.628/210.86) = 0.803 \qquad (1)$$

   a. This expression relates the MAE to the mean target value as an estimate of the overall error of prediction.
   b. 1−(error of prediction) = accuracy of prediction.
   c. Because we build the estimated accuracy value on mean values, we can expect that the actual accuracy values of rows will be higher or lower, but the average accuracy can be estimated by Eq. (1).
   d. The prediction accuracy estimated from the MAE (0.803) is less than the accuracy estimated by $R^2 = 0.867$ (or the Adjusted R^2), but it is probably closer to the correct value.
 c. Important predictor variables.
    i. Right-click on the Polynomial Regression Learner node, and select the *Coefficients and Statistics* table shown in Table A4. Expand the Variable column to read the entire variable names.
   ii. Click on the header of the P > [t] column, and select the *Sort Ascending* option.
  iii. All variables with P-values < = 0.05 are important predictors.
   iv. Notice that the square of the derived lag variable (new-cases-smoothed-per-million(-14) with an exponent = 1 or an exponent = 2) is among the most important predictors of new deaths.
    v. Only the *positive_rate* and the *stringency_index* exceed the lag variable in importance. Note: top 12 predictor variables show as 0 in the P > [t] column, but the actual values range between 0.0 and 0.009.

Summary

1. This exercise has explained the operations required to prepare COVID-19 time-series data for building a Polynomial Regression model of the number of new deaths to be expected on any date in the future.
2. The importance of the derived lag variable in the model illustrates the value of visualization steps to provide insights useful in data preparation operations prior to modeling.
3. This approach to building a predictive model can be adapted to model other dependent variables to build predictive models.

Reference

Wright, S., 1921. Correlation and causation. J. Agric. Res. 20, 557–585.

Index